AF863041

Reef Fish Spawning Aggregations: Biology, Research and Management

FISH & FISHERIES SERIES

VOLUME 35

Series Editor: David L.G. Noakes, Fisheries & Wildlife Department, Oregon State University, Corvallis, USA

For further volumes:
http://www.springer.com/series/5973

Yvonne Sadovy de Mitcheson • Patrick L. Colin
Editors

Reef Fish Spawning Aggregations: Biology, Research and Management

Springer

Editors
Yvonne Sadovy de Mitcheson
Division of Ecology and Biodiversity
School of Biological Sciences
University of Hong Kong
Pok Fu Lam Road
Hong Kong SAR
yjsadovy@hku.hk

Patrick L. Colin
Coral Reef Research Foundation
PO Box 1765
96940 Koror
Palau
crrf@palaunet.com

ISBN 978-94-007-1979-8 e-ISBN 978-94-007-1980-4
DOI 10.1007/978-94-007-1980-4
Springer Dordrecht Heidelberg London New York

Library of Congress Control Number: 2011939476

© Springer Science+Business Media B.V. 2012
No part of this work may be reproduced, stored in a retrieval system, or transmitted in any form or by any means, electronic, mechanical, photocopying, microfilming, recording or otherwise, without written permission from the Publisher, with the exception of any material supplied specifically for the purpose of being entered and executed on a computer system, for exclusive use by the purchaser of the work.

Printed on acid-free paper

Springer is part of Springer Science+Business Media (www.springer.com)

A small group of spawning Nassau grouper, the dark female leading a cluster of males. (Copyright Doug Perrine/SeaPics.com)

To George, love and laughter
To Lori, partner in love, work and life

Foreword

In the world of fish and fisheries there are few things more spectacular than spawning aggregations, and the most dramatic of those are in marine reef fishes. This volume is a landmark in studies of reef spawning aggregations. This is the first comprehensive consideration of this phenomenon. It presents a wealth of opportunities for those interested in the evolution of the behaviour and life histories of fishes. The Editors were among the first to study reef fish spawning aggregations and the first to recognize the combination of fundamental biology and practical management in this phenomenon. They have clearly defined the phenomenon of reef spawning aggregations and resolved a great deal of confusion from earlier reports of this behaviour. For the first time we have an operational framework for both practical and theoretical studies. From their comprehensive review of earlier published descriptions and accounts they have compiled a definitive list of reef fishes with aggregative spawning. They show how studies of reef fish spawning aggregations are a particularly clear example of the progressive development of science, from initial descriptive studies to correlational analyses to experimental studies designed to test hypotheses. In this volume the authors consider reef fish spawning aggregations from physiological, ecological and evolutionary perspectives. They also include the practical implications and applications of traditional ecological knowledge and management to reef fish spawning aggregations. There are extensive case histories of many of the best – known species characterized by spawning aggregations.

Some of the most fundamental questions about spawning aggregations remain to be addressed and are highlighted in this volume. In particular, the adaptive significance and evolution of aggregative spawning have yet to be resolved. The life history of these species often includes a planktonic larval interval with the possibility of far ranging dispersal. At the same time the juveniles and adults are typically associated with reefs. Many have recognized the complexity of those life histories, but we do not yet have a clear resolution of those basic questions.

The timing of this volume could not be more critical for all concerned with the designation and establishment of marine protected areas and marine fish conservation. The conservation implications of reef fish spawning aggregations are now widely recognized. Resource managers have long appreciated the importance of spawning aggregations for artisanal and commercial harvest. Spawning aggregations of reef fishes are some of the clearest, and most pressing examples of threatened fishes, with urgent need for conservation measures. It is surprising to note that many declines of major marine fisheries involve aggregating species. It is sobering to realize that little has been done to protect those species from further declines. Marine Protected Areas are now much in vogue in a number of jurisdictions. However, as this volume shows, those protected areas rarely include fish spawning aggregations. In fact the situation is often to the contrary. Notable examples of excessive exploitation, whether by artisanal, recreational or industrial fishing, threaten numerous aggregating reef fish species.

Certainly a major contribution of this volume will be the increased awareness of the conservation status of aggregative spawning reef fishes. The species in question are recognized and documented, and the biology and life history of many species are described in detail. The threats to these fishes are admirably explained, and the recommendations for conservation and management are clearly outlined. The research needs for understanding the important biological aspects of spawning aggregations are clear. The need for sustainable management decisions is urgent.

Dr. David L.G. Noakes
Editor, Springer Fish and Fisheries Series
Professor of Fisheries and Wildlife
Oregon State University
Senior Scientist, Oregon Hatchery Research Center
Corvallis, Oregon 97331–3803
USA

Preface

Why a Book on Reef Fish Spawning Aggregations, Why Now?

Spawning aggregations are extravagant biological events known to occur in many reef fish species. They are a key factor in population regeneration yet at the same time they appeal as extremely attractive fishing opportunities. Such aggregations are spectacles of nature, in the same class as mass gatherings of animals as diverse as wildebeest, flamingos and monarch butterflies. As we come to better understand these important reproductive events, and see them increasingly exploited globally, we are discovering that many have diminished following human disturbance, in particular uncontrolled fishing. Some aggregations no longer appear to form at all where once they annually contained tens of thousands of fish. Since many species with this habit are particularly desirable as food, heavily sought and otherwise intrinsically vulnerable to fishing due to their life history and general absence of management, it is clear that they merit considerably more research and management attention.

It was with a mix of fascination and concern over what we, independently, were observing in our respective studies and parts of the world that first inspired us to embark upon this book. We hope to share what is being discovered and caution over what could be lost if current trends continue. In this introduction we briefly outline what we believe to be key issues in the characterization, biology, ecology, phylogeny, history, and fishery management of reef fishes that have the habit of aggregation-spawning. These issues are addressed by the chapters in this book and revisited in the final, Discussion, Chap. 13.

It is not possible to explore the many issues around aggregation-spawning species without first defining what 'spawning aggregations' are. With this foundation we can explore possible patterns and processes that distinguish them from non-aggregating species to consider the possible adaptive significance of this reproductive habit, and seek means to best preserve their formation and functionality. As a starting point we have compiled a list of species for which there is unquestioned to strongly suggestive evidence of aggregative spawning (Appendix). This Appendix of information was used as the working basis for many chapters in this book, but is neither complete nor final. It will change as information increases, and it highlights those species for which further work is needed. Nonetheless, at this writing it is believed to be reasonably thorough. Clear definitions of other terminology commonly used in association with aggregations would greatly clarify discussions and we suggest ways in which commonly used, but often poorly defined, terms such as 'catchment area', or 'group-spawning' can be applied more systematically (Glossary).

The chapters of this book build from the specifics of defining aggregations to the many different angles of their biology, ecology, use and preservation. We have also included a chapter on species-specific accounts for the better known aggregating species. Chapter 1 addresses the definition and classification of aggregations across a wide range of fish taxa. While the focus of the book is reef fishes, we occasionally use examples from non-reef, even non-fish, species where these are particularly illustrative or interesting (e.g. Chap. 3). Chapters 2–7 explore questions that ask why, when and where aggregation spawning occurs, the ecological and evolutionary processes associated with the habit, the oceanography and early life-history associated with these species, and ecosystem links. Chapters 8–11 look at the human angle of the exploitation, impacts and study of aggregating species, including their commercial and traditional use, study methodologies, economics, perceptions and attitudes, culminating in conservation and management. And we complete the book by summarizing major findings, gaps and research areas that need to be addressed, ending by suggesting next steps (Chap. 13). Chapter 12 covers the better known or newly studied aggregating fishes plus some of their relatives and includes much in the way of novel information and perspectives.

In the process of preparing and compiling this volume, we were struck by two, related and disturbing, issues that reaffirmed our initial concerns and highlighted some worrying perspectives. The first is how little effective conservation and management there is in place for spawning aggregations, and how infrequently they are considered in either fishery management planning or in marine protected area (MPA) designation in conservation efforts. The second issue is a general absence of management attention on tropical reef-associated fisheries in general, on effective MPA management, and on aggregations in particular; there appears to persist in many places a deep-seated, often unspoken, belief that commercially exploited fishes will somehow continue to supply coastal communities without significant intervention; fishery management is not a high enough priority in most countries where they continue to be important sources of food and livelihoods. Whereas, for example, there is little question that the nesting colonies of certain seabirds or the beaches where turtles congregate to lay eggs need protection (even if such goals are difficult to achieve in practice), fish spawning aggregations are commonly viewed as *opportunities* for fishing rather than a life history phase to be *preserved* and management is rarely enforced. Legislation to protect berried (egg-bearing) lobsters has long been in force in many parts of the world, but we know of no examples where female fish, visibly full of eggs for the brief annual spawning season, are conferred protection. We hope, through the chapters of this book, to demonstrate the need for a fundamental change in attitude regarding this life history phenomenon from being a *focus for fishing* to one needful of *stewardship and management* for long-term biologically, and hence economically, sustainable use.

Introductory Primer on Aggregating Species and Their Management

Despite the wide taxonomic diversity of reef fishes that aggregate to spawn, most share several core biological attributes. All have a two-phase, or bipartite, life cycle in which eggs, usually pelagic, hatch into planktonic larvae. This means that while the juvenile and adult phases are associated with reef ecosystems, the egg and/or larval phase is planktonic, potentially able to disperse widely and likely a major determinant of connectivity. However, adult movements between home reefs and transient aggregation sites can also be considerable and may be another important aspect of population structure. In this respect, some reef fishes that aggregate differ from the 'sedentary' form that tends to characterize most reef fishes and, indeed, is a basis for the applicability of MPAs to reef ecosystem conservation. Evolutionary pressures acting in these two phases of the life-cycle are likely to be very different, and both must be considered when examining hypotheses about where, when and why aggregations form (Chap. 5).

Given the broad taxonomic diversity, ecology and geographic distribution of aggregating species, it is not surprising that there are substantial differences in many characteristics of their biology, including longevity, age of sexual maturation, maximum size, diet, spawning mode, etc. This diversity partly accounts for the range of different aggregation types we observe (Chaps. 2 and 4) and is also a primary reason that some species are intrinsically more vulnerable to overfishing than others (Chaps. 8 and 11).

Looking at the bigger picture, what we understand relatively little about is the role of aggregations and aggregating species in the reef ecosystem generally. The range and number of species, and the large biomasses involved, their mass seasonal movements and the use of aggregations by egg and adult predators are just some of the considerations explored in Chap. 2.

Work is still in the discovery phase for aggregations and regarding the way we should best be using and managing aggregating species. In many areas, such as the Pacific and Indian Oceans, they are poorly documented and, not surprisingly, little managed but there is still time to manage and study relatively intact gatherings. In some parts of the world, on the other hand, such as much of Southeast Asia and parts of the Caribbean that are very heavily fished, aggregations have probably been lost, with unknown prospects of future recovery. We have only recently begun to appreciate the overall economic value of species with this habit, how prevalent such species are in coastal fisheries, and what would be lost without appropriate management (Chap. 10). We have come to realize that not only commercial and recreational fisheries need moderation but that even artisanal use can lead to overexploitation of aggregating species (Chap. 8). We have learned that most MPAs do not encompass aggregation sites, or are typically too small to accommodate reef fishes that migrate away from their home reefs each year to reproduce. And, we are aware that while conventional fishery management addresses species level management, it rarely focuses on controls in aggregation-fishing (Chap. 11).

Good Science and Management Practice

A thorough understanding of aggregating species can only develop on a foundation of good science, sound scholarship and effective management practices. This calls for a greater rigour and attention to detail in studying, monitoring and reporting on this phenomenon. For example, detailed and important information from the early literature is sometimes ignored or misreported, so caution is needed to refer back to primary literature, rather than rely solely on reviews or secondary sources. Speculative comments should not be treated as facts, and literature that has not yet passed through the peer review process should be treated with care.

Critically important for monitoring and adaptive management is the development and application of sound sampling and surveying methodologies (Chap. 9). Aggregations are highly dynamic phenomena that often occur in areas that are difficult to work in. As many workers have come to discover, however, counting large numbers of fish over short periods of time when their numbers change hourly or daily, working at diving sites that are deep or have strong currents, or maintaining consistent dive schedules during specific lunar cycles, represent unique challenges. Tailored sampling methods are often needed for individual aggregation sites and for different species. Logistics include reaching dive sites and safely with enough fit and able divers to consistently survey numbers of a few critical days when numbers are peaking on a regular basis. Experience shows that comprehensive and consistent sampling is a major challenge and could better be served by putting effort into major surveys that include expert input every few years, rather than hobbling together less than ideal yearly initiatives.

Yet, we are also aware that much can be learned about a fishery using a range of simple and inexpensive techniques. Sampling fish from markets or fish landings sites can easily and cheaply provide valuable information on fish sizes and species diversity in catches, catch rates, reproductive biology and size of sexual maturation, especially if it is part of a long-term monitoring programme (Chapter 9). Interviews with fishermen and knowledgeable biologists or fishery department personnel can yield invaluable current and historic information and perspectives. A word of caution, however, is that despite the underlying simplicity of such approaches, much money, time and effort can be wasted due to poorly conceived interview surveys or sampling protocols and there is no substitute for training, experience and careful planning to ensure scientific rigour (Chap. 10). Moreover the handling and use of information from interviews may call for due diligence to ensure confidentiality of fishing site locations and respect the knowledge collected from interviewees. With ready public access to the internet, site-based information of active aggregation sites, for example, is just as readily available to those who may wish to exploit aggregations as it is to those who wish to manage them.

Unfortunately, many of the species that aggregate to spawn are difficult to study in the field and scientists must be creative in addressing key questions. Large body size and wide-ranging movements can make such species challenging to track, while the brief duration and challenging field conditions of annual spawning events seriously limit many field-based reproductive studies, especially of the experimental or manipulation types. In the case of commercially important species, natural populations may already be much reduced or fish may be particularly wary. For some research questions, smaller and/or non-commercial species, therefore, are useful for testing hypotheses or conducting experiments. Many such examples are provided in the chapters of this book, ranging from the small bluehead wrasse to various species of smaller parrotfishes and surgeonfishes. There are also opportunities to use technological methods to describe spawning sites or to count fish. Nonetheless, considerable care is needed in the interpretation of such data unless it is thoroughly ground-truthed. As one example, hydroacoustic surveys show much promise for assessing fish numbers in aggregations without the need for diving. However, fish numbers thereby obtained are valid only following demonstration (ground-truthing) that the methodology is unquestionably reliable for the target species.

Information needed to manage aggregating species ranges from basic biology to reproductive seasonality, migration distances, physical characteristics and number of aggregation sites in a population or fishery, to regular monitoring and the food and economic value of aggregations to the local and national economy. Management of aggregating species may be necessary on both aggregating and non-aggregating (i.e. non-reproductive) phases. Spatial data are important for MPA designation and should include catchment area and migration routes, if applicable. The period between spawning and recruitment, the early life history, is virtually unknown yet important for questions that explore the possible significance importance of spawning times and locations (Chaps. 6 and 7). For fishery management, both fishery-dependent and –independent data of aggregation and non-aggregation catches are needed to provide the most comprehensive information. Understanding the value of catches, selectivity and size/age-related mortality and reproductive schedules is also important. Given the realities of data collection, however, much of this information is unlikely to be forthcoming, and making best use of available data and using precedents and similar species are good starting points.

Finally, a better general appreciation by a wider public of the significance of aggregations to fisheries and their vulnerabilities could lead to support or pressure for the necessary protective policies and address the undeniable need for management. One option open to the conservation community could be to use aggregations as 'indicators' of fishery condition by developing an index that integrates intrinsic (biology of the species, type of aggregation) and extrinsic (fishery impacts) factors; this possibility is explored in Chap. 8. Such an index would help to highlight their overall importance in marine ecosystems and coastal fisheries. To manage these species may call for a major shift in thinking, from viewing aggregations as ideal fishing opportunities to preserving them for their fundamental role in population replenishment and persistence.

We hope that this book is useful to all those with interest in fish spawning aggregations, from managers and conservation practitioners, to biologists, teachers, artists, writers, students and to others who wish to understand more about these incredible natural events, and how to appreciate and preserve them far into the future.

Society for the Conservation of Reef Fish Aggregations

In 2000, following a mini-symposium on aggregating species, the Society for the Conservation of Reef Fish Aggregations (SCRFA) was formed by a small group of biologists concerned about what they were witnessing in their respective parts of the world in relation to aggregating fish species. Since 2000 the SCRFA has developed, and continues to work towards the effective management and conservation of reef fish aggregations (www.SCRFA.org). Our participation in the SCRFA, the funding received, particularly from the David and Lucile Packard Foundation that has enabled us to develop our work and bring to it wide international attention as well as local action, has been a major motivating and focusing factor in bringing this book project to completion and in producing a range of data reported in many of the chapters and species accounts.

Acknowledgements

We are very grateful to the many people and institutions involved in variously supporting the development and realisation of this book. Without their assistance, funding, contributions encouragement, inspiration, sacrifice and patience, this book would never have been completed.

Over the last decade the Society for the Conservation of Reef Fish Aggregations (SCRFA), with the support of the David and Lucile Packard Foundation, has been invaluable and instrumental in helping us to focus our efforts and conduct much new work that has greatly increased our understanding and appreciation for coral reef fish spawning aggregations. SCRFA Board members, present and past who have facilitated these efforts include Chairmen Martin Russell and Michael Domeier, and members Enric Sala, Brian Luckhurst, Kenyon Lindeman, Janet Gibson, Richard Hamilton, Richard Nemeth, Terry Donaldson, Brad Erisman and Jos Pet. Lori Bell Colin has greatly assisted with financial matters. Tim Daw, Richard Hamilton, Kevin Rhodes, Liu Min and Andy Cornish conducted many of the fisher interview surveys, while Joy Lam and Leath Muller have supported operations in many ways. Various other support has come from International Union for Conservation of Nature (IUCN), especially through the Groupers & Wrasses Specialist Group, the Ocean Foundation, the Kingfisher Foundation and the University of Hong Kong. We are also grateful to the many fishers, biologists and divers who have shared their knowledge on aggregations with us.

For images and other visual materials we thank Mandy Etpison, Min Chandiramani, Michael Minigele, Michael Berumen, Octavio Oburto, Charlie Arneson, Rachel Graham, Paul McKenzie, and Doug Perrine of SeaPics.com.

Yvonne would like to acknowledge support from many institutions and people. Research over the years on reef fishes, their reproductive biology and aggregations, has variously been supported by the National Geographic Society, IUCN, National Oceanographic and Atmospheric and Administration (NOAA in the USA), and. Research Grants Council (Hong Kong). Early work was conducted at the University of Puerto Rico, the Laboratorio de Investigaciones Pesqueras (Fishery Research Laboratory) of the government of Puerto Rico, and the Caribbean Fishery

Management Council: special thanks go to Graciela Garcia-Moliner, Miguel Rolon and Douglas Shapiro. The Gulf and Caribbean Fisheries Institute has been an excellent forum for presenting aggregation work over many years with its dedicated sessions. Work in Palau over many years was greatly facilitated by the Coral Reef Research Foundation, and I am especially grateful to Lori Bell Colin and Pat Colin for years of friendship. In Palau, Bob Johannes was involved in early aggregation work, while more recently Asap Bukurrou and Scotte Kiefer of the Palau Conservation Society provided field companionship and much other support. In Fiji, Aisake Batibasaga, and colleagues in the Fishery Research section of government, Waisalima resort, Wildlife Conservation Society, and Loraini Sivo have variously assisted supported and participated in field-work. It has been a pleasure partnering with WWF on many occasions. A very special mention goes to Mariella Rasotto for years of friendship, field-work and shared passions.

In Hong Kong, Yvonne would like to thank David Dudgeon and other colleagues in the former Department of Ecology & Biodiversity, with special recognition to Rachel Wong for assistance and support over many years, Lily Ng for her computer expertize, and her many wonderful post-graduate students, particularly Andy Cornish, Liu Min, William Cheung Allen To and Stanley Shea for variously assisting in field and research work. A special thanks to Mariella Rasotto for years of friendship and shared passions. From the local salsa scene, she particularly thanks JE, DT, CW, SC, FW, XP and Sugar for music and dance that inspired.

George Mitcheson has encouraged, nurtured, and entertained for over 30 years, making this and so much more possible. To Pamela, Jane, John, George, Eleanor and Brian for being in my life and the strength that brings me.

Pat would like to acknowledge support from a variety of institutions and granting agencies over many years. His first aggregation work on Atlantic surgeonfishes was supported by grants from the National Geographic Society (NGS) in 1977–1979 with support equipment provided by a grant from the National Science Foundation, Biological Oceanography. The NGS also provided support for later work on the goliath grouper (1991) in Florida. Additional funds to study specific species were provided by the National Fish and Wildlife Foundation (NFWF) and the International Union for Conservation of Nature (IUCN) for humphead wrasse, and the Coral Reef Research Foundation (CRRF) for work on numerous species is Palau.

Work has been facilitated by various institutions where Pat has worked including the Department of Marine Sciences of the University of Puerto Rico Mayaguez, the Mid-Pacific Research Laboratory (Enewetak), the Motupore Island Research Centre (University of Papua New Guinea) and more recently by the Coral Reef Research Foundation. Work on Atlantic groupers and other fishes was supported in 1987–1991 by the NOAA National Undersea Research Program through the Caribbean Marine Research Center. The Nature Conservancy East Asia Pacific programme provided funds for the first implementation of GPS density surveys of grouper aggregations in Palau. The Caribbean Coral Reef Institute provided funds and support to revisit Puerto Rico sites in 2011.

Finally, Pat would finally like to thank colleagues who have been integral in field and laboratory work: Yvonne Sadovy, Michael Domeier, Terry Donaldson, Don Demaria, Ileana Clavijo, Doug Shapiro, Debbie Weiler, Charlie Arneson, Paul Collins, Mandy Etpison, and Bert Yates. Lori Jane Bell Colin has provided major support and assistance over 30 years towards work on aggregations, as well as many other subjects.

Contents

1 **Revisiting Spawning Aggregations: Definitions and Challenges** 1
Michael L. Domeier

2 **Ecosystem Aspects of Species That Aggregate to Spawn** 21
Richard S. Nemeth

3 **Why Spawn in Aggregations?** 57
Philip Patrick Molloy, Isabelle M. Côté, and John D. Reynolds

4 **Spawning Aggregations in Reef Fishes; Ecological and Evolutionary Processes** 85
John Howard Choat

5 **Timing and Location of Aggregation and Spawning in Reef Fishes** 117
Patrick L. Colin

6 **Oceanography of the Planktonic Stages of Aggregation Spawning Reef Fishes** 159
William Marion Hamner and John Louis Largier

7 **Aggregation Spawning: Biological Aspects of the Early Life History** 191
Patrick L. Colin

8 **Fishery and Biological Implications of Fishing Spawning Aggregations, and the Social and Economic Importance of Aggregating Fishes** 225
Yvonne Sadovy de Mitcheson and Brad Erisman

9 **Studying and Monitoring Aggregating Species** 285
Patrick L. Colin

10 The Role of Local Ecological Knowledge in the Conservation and Management of Reef Fish Spawning Aggregations 331
Richard Hamilton, Yvonne Sadovy de Mitcheson, and Alfonso Aguilar-Perera

11 Management of Spawning Aggregations 371
Martin W. Russell, Brian E. Luckhurst, and Kenyon C. Lindeman

12 Species Case Studies 405
Rachel Pears, Richard S. Nemeth, Beatrice P. Ferreira, Mauricio Hostim-Silva, Athila A. Bertoncini, Felicia C. Coleman, Christopher C. Koenig, Kevin L. Rhodes, Yvonne Sadovy de Mitcheson, Scott A. Heppell, Patrick L. Colin, Melita A. Samoilys, Jiro Sakaue, Hiroshi Akino, Hitoshi Ida, Gary Jackson, Robert R. Warner, R.J. Hamilton, J.H. Choat, Mandy T. Etpison, Paul Collins, Ileana J. Clavijo, and Ann Hillmann Kitalong

13 Conclusion 567

Abbreviations and Acronyms 585

Glossary 587

Appendix: Species That Form Spawning Aggregations 595

Index 605

Chapter 1
Revisiting Spawning Aggregations: Definitions and Challenges

Michael L. Domeier

Abstract The term spawning aggregation was first formally defined in 1997. Since that time, the original definition has been cited over 200 times and modified definitions proposed. Spawning aggregations are both unique from a behavioural ecology perspective as well as important in terms of fisheries management discussions. A single definition that recognizes both of these factors is important to researchers and resource managers. Here the original definition of the spawning aggregation phenomenon is improved to correct misinterpretation while also using language to recognize spawning aggregations of non-fish species: *A Spawning Aggregation is a repeated concentration of conspecific marine animals, gathered for the purpose of spawning, that is predictable in time and space. The density/number of individuals participating in a spawning aggregation is at least four times that found outside the aggregation. The spawning aggregation results in a mass point source of offspring*. Different types of spawning aggregations are also recognized, for example, some species travel relatively large distances to gather at the spawning site while others make more frequent, short migrations. Also, some species spawn demersal eggs that then may/may not be guarded by one or both of the parents, while other (most) species spawn pelagic eggs that are given no care. Many intriguing theoretical questions remain unanswered with respect to spawning aggregations, and it is very difficult to test the many differing hypotheses proposed to explain observations. The author's favoured hypotheses are discussed and hypothetical evidence proposed.

M.L. Domeier (✉)
Marine Conservation Science Institute, 2809 South Mission Road,
Fallbrook, CA 92028, USA
e-mail: ml.domeier@gmail.com

Y. Sadovy de Mitcheson and P.L. Colin (eds.), *Reef Fish Spawning Aggregations: Biology, Research and Management*, Fish & Fisheries Series 35,
DOI 10.1007/978-94-007-1980-4_1, © Springer Science+Business Media B.V. 2012

1.1 Introduction

Animals are known to gather *en masse* for shelter or physiological benefits (i.e. conservation of heat or water), to avoid predators, to migrate, to feed, and to reproduce. One such reproductive aggregation known to occur in marine reef fishes, as well as other fishes and invertebrates, is the spawning aggregation. Although the term spawning aggregation has been used in the literature for decades, it was not formerly defined until 1997 (Domeier and Colin 1997). While the 1997 paper reviewed just 82 publications spanning 73 years, in contrast the Domeier and Colin (1997) review has been cited over 200 times in just 14 years, indicating a recent and dramatic focus on spawning aggregations.

Spawning aggregations are predictable in time and space and therefore often the target of intense directed fisheries. Much of the recent focus on spawning aggregations can be attributed to the deleterious effect that fishing them can have on the target species (Sala et al. 2001; Sadovy and Domeier 2005). The challenge of regulating reef fisheries has brought the science of spawning aggregations to the forefront of management discussions, often in relation to the designation of marine protected areas (MPA) in tropical regions. The large increase in published spawning aggregation related data, and a proposed alternative definition of the phenomenon (Claydon 2004), warrants a re-examination of the general definition of a spawning aggregation, as well as considering the different types of aggregations. In addition, there is a need for establishing very clear criteria to properly document a spawning aggregation, particularly when a new example is being added to the worldwide list of species known to form them (Appendix). The study of the spawning biology and behaviour of fishes should be encouraged, but discipline and restraint are necessary before labelling a new spawning aggregation. Definitions are not intended to constrain discussion, instead they are meant to streamline communication, and in the current case, implicate special fisheries circumstances in relation to management and conservation considerations.

1.2 General Spawning Aggregation Definitions

Domeier and Colin (1997) defined a spawning aggregation "as a group of conspecific fish gathered for the purpose of spawning, with fish densities or numbers significantly higher than those found in the area of aggregation during the non-reproductive periods (Fig. 1.1). For fishes, such as jacks (Carangidae), mullet (Mugilidae), rabbitfishes (Siganidae) and surgeonfishes (Acanthuridae) which normally occur in dense schools, when in a spawning aggregation they must occur in significantly greater number and take up significantly more space than non-reproductive fish." The authors discuss the need for a more quantitative definition but acknowledge a lack of data to do so. They suggest requiring a greater than three-fold increase in fish density to exclude streak spawning events (a single spawning pair joined by a solitary male during the spawning rush).

Fig. 1.1 Spawning aggregation site that accommodates several species that gather to spawn on a predictable basis in Polynesia. Surgeonfish, *Acanthurus* sp., and camouflage grouper, *Epinephelus polyphekadion*, and other species spawn here closely watched by sharks (Photo: Paul and Paveena McKenzie/wildencounters.com)

Claydon (2004) argued that the Domeier and Colin (1997) definition was too restrictive. He proposed a definition that does not require spawning individuals to occur in greater numbers or higher density than normal: "spawning aggregations are temporary aggregations formed by fishes that have migrated for the specific purpose of spawning." The Claydon (2004) definition, therefore, simply requires the participants to migrate to a specific spawning site while the Domeier and Colin (1997) definition requires both a migration and an increase in density or numbers. Domeier and Colin (1997) recognized this type of spawning but they termed it Simple Migratory Spawning: "migration and spawning of pairs or small groups of fishes from a non-spawning area to a spawning area." Furthermore, the Claydon definition is circular, in that it uses the term "aggregation" to define a spawning aggregation. This makes interpretation of the definition quite subjective other than to effectively consider groups of three or more fish that come together temporarily to spawn as spawning aggregations, a definition that includes the majority of reef fish species and would include streakers with pair-spawners.

The intent of the Domeier and Colin (1997) definition was to differentiate a unique phenomenon of behavioural ecology where an entire sub-population of individuals halt their normal routine, migrate, gather and spawn. Not only is this a biologically significant event, but it is also an economically important event with unique management implications. For example, the increased numbers of fish predictably available increases catchability and makes aggregations a specific target for fishing; this calls for specific action. I would argue that Claydon's (2004) unrestrictive definition of a spawning aggregation does not adequately differentiate the very unique act of spawning in large numbers from Simple Migratory Spawning. In fact, the growing use of the term 'spawning aggregation' in management plans and MPA designations puts greater importance on constructing relatively unambiguous language to describe and define the event. Spawning aggregations are particularly vulnerable to over exploitation simply due to the fact that they constitute particularly large concentrations of fish that are repeatedly predictable in time and space. The language of the original Domeier and Colin (1997) definition does not directly limit spawning aggregations to events that repeatedly occur at specific times and locations. Although this criterion was implied when we distinguished between types of aggregations (see below), I propose the following modified general definition for the sake of clarity:

> A Spawning Aggregation is a repeated concentration of conspecific marine animals, gathered for the purpose of spawning, that is predictable in time and space. The density/number of individuals participating in a spawning aggregation is at least four times that found outside the aggregation. The spawning aggregation results in a mass point source of offspring.

The term 'spawning aggregation' has most widely been applied to coral reef fish examples despite the fact it has never explicitly excluded non reef fishes or invertebrates; the above modified definition substitutes the word 'animal' for 'fish' to acknowledge that spawning aggregations can occur across a wide spectrum of marine organisms and habitats. In fact, a recent review of spawning aggregations included decapods, elasmobranchs and an anadromous catfish (Nemeth 2009, Chapter 3). Another subtle change proposed in this definition is language relative to the observed increase in number/density of animals: from "greater than a three-fold increase" to "at least a four-fold increase." Domeier and Colin (1997) recognized that selecting a density/number criterion was somewhat arbitrary, but the intent was to be inclusive while excluding non-aggregating mating strategies like streak spawning, which could involve just three fish. This criterion has often been cited in error with authors omitting the words "greater than;" changing this to "at least four times greater" will eliminate the confusion.

Claydon (2004) listed far fewer species 158 species as forming spawning aggregations, while a more recent paper (Sadovy de Mitcheson et al. 2008) listed only 67 species (see also Appendix). How can there be such a discrepancy? Sadovy de Mitcheson et al. (2008) used the Domeier and Colin (1997) definition of spawning aggregation while Claydon (2004) used his new definition. However, upon closer examination of the two papers, the choice of definition was not the major factor that created the large discrepancy; instead, it was Claydon's use of an unpublished list of spawning fishes which first appeared as an appendix to a report from the Great Barrier Reef Marine Park Authority (GBRMPA) (Russell 2001). The report appendix lists species purportedly known to form spawning aggregations along the Great Barrier Reef, but many of the species listed are from an unpublished list of fishes cited as "Squire and Samoilys unpublished." This unpublished list is actually a list of species that were believed or observed to spawn at the same aggregation site as the aggregating leopard coralgrouper, *Plectropomus leopardus*. However, many of these species had never been documented to form a spawning aggregation (Melita Samoilys personal communication). The unfortunate use of this informal list from the GBRMPA report has erroneously perpetuated a large number of species as examples of aggregate spawners. If these species, and those from another anecdotal source (Johannes 1981), are removed from Claydon's review, approximately two thirds of the species now drop off his list of species that aggregate to spawn. This is an important example of why great care need be taken when discussing and identifying species that aggregate to spawn and why the primary literature needs to be referred to and evaluated in such cases.

The definition of a spawning aggregation proposed in this chapter is more restrictive than Claydon's (2004) definition and clarifies the original intent of Domeier and Colin (1997); however, the substitution of the word 'animal' for 'fish' is much less restrictive than previous definitions and allows the consideration of additional phyla. When considering new phyla, the word 'spawn' may appear too restrictive since it is generally used to describe the release of eggs and sperm or a large number of offspring; this would exclude mating/breeding aggregations that may only involve copulation. From strictly a management perspective, a periodic, predictable mass gathering of economically valuable marine organisms presents similar challenges, regardless of the function of the aggregation, but it is the mass gathering of adults, release and subsequent dispersal of large numbers of offspring that make this phenomenon biologically unique and distinct from other types of aggregations. The selection of a unique spawning site, or time, that ensures a relatively high recruitment success for the offspring may be the single most important driving force behind the evolution of the phenomenon (although there are many competing hypotheses, Chaps. 4, 6, and 7). Aggregations that occur solely for the purpose of copulating in the absence of releasing offspring (e.g. elasmobranchs) are therefore not considered spawning aggregations in this definition, but single-sex gatherings for the purpose of releasing offspring (e.g. female decapods) are considered spawning aggregations since they meet all criteria (Fig. 1.2a). Nonetheless, from a management perspective, some non-spawning gatherings may also require action because the concentration of adults may attract excessive fishing pressure, as in the case of nurse shark, *Ginglymostoma cirratum* (Pratt and Carrier 2001).

Fisheries management plans are beginning to focus on spawning aggregations and MPA planning is beginning to incorporate them. This fact underscores the importance of creating a definition that both adequately describes a unique biological phenomenon, while also strengthening the phrase "spawning aggregation" from a fisheries perspective. A less restrictive definition fails to distinguish a phenomenon where a species is particularly vulnerable to intense fishing pressure from other general gathering behaviours, thereby losing its value in fisheries discussions while relinquishing its power to examine the proximal and ultimate factors that might be involved in its evolution.

Fig. 1.2 (**a**) Rays, *Taeniura melanospilus*, aggregating to mate, Cocos Island, Costa Rica, Pacific (Photo: © Mark Conlin/SeaPics.com). (**b**) Temporary gatherings of fish can only be confirmed as formed for spawning using clear indicators. An aggregation of king angelfish, *Holocanthus passer*, was observed in the Galapagos Islands but no spawning was observed and gonads could not be collected for inspection. This is important since no pomacanthid (angelfish) has been reported as an aggregation spawner. See text for detail (Photo: © Michael L. Domeier)

1.3 Types of Spawning Aggregations

Domeier and Colin (1997) subdivided spawning aggregations into two distinct types, *Resident* and *Transient*, based upon (1) the frequency with which the spawning aggregation occurs, (2) the length of time the aggregation persists, (3) the site specificity of the aggregation, and (4) the distance that individual fish travel to the aggregation site. The original definitions are as follows:

Resident spawning aggregations draw individuals from a relatively small and local area. The spawning site can be reached through a migration of a few hours or less and often lies within the home range of the participating individuals. They usually (1) occur at a specific time of day over numerous days, (2) last only a few hours or less, (3) occur daily over an often lengthy reproductive period of the year, and (4) can occur year round. A single day of spawning for an individual participating in a resident spawning aggregation represents a small fraction of that individual's annual reproductive effort.

Transient spawning aggregations draw individuals from a relatively large area. Individuals must travel days or weeks to reach the aggregation site. Transient spawning aggregations often (1) occur during a very specific portion of 1 or 2 months of the year; (2) persist for a period of days or at most a few weeks and (3) do not occur year round. A single transient spawning aggregation may represent the total reproductive effort for participating individuals.

Claydon (2004) suggested that the differentiation between resident and transient spawning aggregations is artificial, with the distinction being a simple matter of scale. He argued that all spawning aggregations are resident because the aggregation site lies within the catchment area of participating individuals, and that all aggregations are transient because they are temporary. The term 'catchment area' has been increasingly used relative to spawning aggregations in reference to the total geographic region from which individuals are drawn for a specific spawning aggregation. Catchment areas are relative to the species or population that aggregates at an individual site, rather than a property of an individual animal as indicated by Claydon. Therefore, a single site may have more than one catchment area if it is a multi-species site, and a single species may have overlapping catchment areas if it uses multiple sites in a small area. Domeier and Colin (1997) stated that Resident spawning aggregations are drawn from a "relatively small and local area" while Transient aggregations "draw individuals from a relatively large area;" this language is ambiguous. Although Domeier and Colin use the term "home range" in defining Transient spawning aggregation, Nemeth (2009) clarified the distinction between Resident and Transient spawning aggregations by adding the term 'home range' to both definitions, such that individuals migrate "within or nearby" their home range for Resident aggregations and "well outside" their home range for Transient spawning aggregations. Accepting Nemeth's distinction I have modified the original definitions as follows-modifications are bolded:

Resident spawning aggregations draw individuals to a site within or nearby their adult home range. *They usually (1) occur at a specific time of day over numerous* *days, (2) last only a few hours or less, (3) occur daily over an often lengthy reproductive period of the year, and (4) can occur year round. A single day of spawning for an individual participating in a resident spawning aggregation represents a small fraction of that individual's annual reproductive effort.*

Transient spawning aggregations draw individuals to a site well outside their typical adult home range. *Transient spawning aggregations often (1) occur during a very specific portion of one or two months of the year; (2) persist for a period of days or at most a few weeks and (3) do not occur year round. A single transient spawning aggregation may represent the total reproductive effort for participating individuals.*

The difference between Resident and Transient spawning aggregations is more than just a matter of scale, there are important functional differences that can be indirectly observed but not easily described. Indirect measures of these functional differences include the fact that species that form these different types of spawning aggregations split fairly well along phylogenetic lines (Appendix, Chap. 4). Furthermore, species that form Transient aggregations are typically large and predatory while those that form Resident aggregations are generally smaller herbivores, planktivores or omnivores (exceptions do exist; e.g. humphead wrasse, *Cheilinus undulatus*). However, like any definition, these will not necessarily capture every situation and there are species that appear to fall between the two (i.e. leopard coralgrouper) – this is the nature of many definitions and as we learn more about aggregating species refinements may be introduced.

The general dichotomy between resident and transient spawning aggregations has clear management implications. For example, on the Great Barrier Reef, Australia, the leopard coralgrouper forms seasonal semi-resident (transient on a home reef, but fish do not seem to move between reefs) spawning aggregations (unusual for a grouper-Serranidae) and is the basis of an important fishery. In Australia, where one third of the GBR is protected from fishing, the (untested) assumption is that each reef contains resident spawning aggregations, and, therefore that protecting one third of the reef theoretically could protect 1/3 of the aggregations (Martin Russell personal communication). This could not be assumed for a transient spawner, since the catchment area for a single transient spawning aggregation can involve many surrounding reefs. To effectively protect a transient aggregation, an MPA would have to be very large or extremely well-placed. It is very likely that MPAs in the absence of region-wide seasonal closures will not adequately protect transient spawning aggregations. This is just one illustration of how such definitions can have practical value.

Schooling fishes are known to join conspecific schools at predictable sites and times to form both transient (e.g. jacks-Carangidae, mackerels/tuna-Scombridae, croakers-Sciaenidae) and resident (e.g. wrasses-Labridae, parrotfish-Scaridae) spawning aggregations. The movement of a single school of animals to a specific site to spawn falls under the definition of Simple Migratory Spawning (Domeier and Colin 1997). Although this phenomenon does not fall under the definition of a spawning aggregation, due to the absence of a significant density increase, if the spawning of a schooling species (such as bumphead parrotfish, *Bolbometopon muricatum*) is predictable in time and space, the management implications are similar to those of a spawning aggregation.

1.4 Egg Types and Spawning Aggregations

Most coral reef fishes release very small pelagic eggs (often near 1 mm diameter) that float to near the surface and then hatch about 24 h later (actual time can vary) (Chap. 7). There are also many examples of reef fishes that produce sinking eggs that are either brooded in the mouth or attached to the substrate. These demersal eggs often have a longer developmental period prior to hatching and are usually guarded by one or both sexes. There are also a few examples of demersal egg spawners that do not guard the eggs and a few examples of live-bearing reef fishes (brotulids). The overwhelming majority of species that aggregate to spawn release a pelagic egg. However, an increasing number of demersal spawning aggregations are being described. Demersal spawning aggregations can be further divided into those that exhibit parental care triggerfishes (Balistidae), damselfishes (Pomacentridae) and those that do not (e.g. rabbitfish). Parental care demersal spawning aggregations can also be split into groups based upon whether the nest is guarded by the male (e.g. brown puller, *Chromis hypsilepis* Gladstone 2007), the female (e.g. *Graneledone sp.* (Drazen et al. 2003) or both sexes (e.g. triggerfish)). Live bearers and mouth brooders are not known to form spawning aggregations (Chap. 4).

Past treatments of spawning aggregations have listed the relatively few demersal examples without consideration for potential functional differences between this and pelagic spawning aggregations. All species that aggregate to spawn are under selective pressure to choose spawning sites and/or times that maximize recruitment. Species that aggregate to spawn demersal eggs have additional selective pressures, since demersal spawning sites must also provide a suitable substrate for egg adhesion and an environment that facilitates keeping the eggs clean and oxygenated. There are also different selective pressures on demersal aggregating species that care for their eggs and those that do not. Parental care includes guarding the nest from predators, fanning and cleaning; for demersal aggregations without parental care, the spawning site must adequately provide these functions in the absence of the parents. It is possible that the act of spawning in an aggregation has allowed the elimination of parental care for some species that spawn demersal eggs, due to the fact that the density of eggs satiates local egg predators (Domeier and Colin 1997). Also, the absence of parental care would be more likely for species that produce eggs that hatch shortly after spawning. For example, brown puller exhibits male parental care during a 4.5 day incubation period (Gladstone 2007), while rabbitfish demonstrate no parental care and the eggs hatch in 25–32 h (Thresher 1984). Time to hatching for the siganids is similar to that of reef fishes that spawn pelagic eggs.

1.5 Spawning Aggregations in Non-reef Habitats

The growing base of literature relevant to spawning aggregations is dominated by coral reef fish examples. However, fishes in other habitats and latitudes certainly form spawning aggregations but often lack the precision of timing and location found in coral reef fishes, and are therefore harder to describe. As a theoretical example of typical precision, an aggregation of a particular species may spawn over a 5-day period on the edge of the same reef passage, at dusk on the full moon of July each and every year. In another example, fish would aggregate to spawn over a particular feature of the reef at the same time (or tidal cycle) each day over a protracted spawning season.

Chronobiological rhythms are important cues to the timing of spawning aggregations. Infradium rhythms dictate the spawning season, while circadian rhythms dictate the fine scale, daily timing of spawning. Physical cycles of temperature, salinity and tides may interact with chronobiological rhythms to dictate the timing of spawning. Tropical marine habitats are relatively stable with respect to physical cycles when compared to higher latitudes, allowing the timing of spawning aggregations to be controlled more by chronobiological cycles, thereby making them more predictable. Higher latitude spawning aggregations are susceptible to unpredictable variations in oceanographic factors, thereby making the timing of spawning aggregations less predictable than those that occur in tropical habitats. For example, temperature is an important controlling factor for the spawning of herring (Haegele and Schweigert 1985) and inter-annual variation in water temperature can influence the precise timing of spawning.

Year to year temporal variation in spawning aggregations is understandable when physical cues are taken into consideration, but the spatial precision of spawning aggregations can vary in both tropical and higher latitude habitats and this is harder to understand. Some tropical spawning aggregations and temperate anadromous spawning aggregations can be very precisely predictable with respect to location. However, there are also examples of species from several different habitats that aggregate in a general area each year, but the exact site of spawning is not always the same. Pelagic fishes and migratory coastal species appear to be more likely to exhibit less spatial precision than demersal species. Croakers, herrings/sardines (Clupeidae), mullets and flying fishes (Exocoetidae) (see Appendix; Chap. 8; Bane 1965; Parin and Lakshminaraina 1993; Stevens et al. 2003; Casazza et al. 2005) are examples of fishes that form predictable spawning aggregations on a larger spatial scale. Despite the lesser degree of spatial precision, these aggregations are still predicted, located and targeted by fisheries, making them very relevant to this discussion in both biological and fishery terms.

The overwhelming majority of documented spawning aggregations occur in coastal habitats; however, examples from offshore and deep-sea habitats are emerging. Spawning aggregations of the commercially important orange roughy (*Hoplostethus atlanticus*) are well documented to seasonally occur over specific seamounts between 700–1,000 m (Pankhurst 1988; Bell et al. 1992; Francis and Clark 1998). More recently described deep-sea spawning aggregation examples include a sculpin, (Cottidae) *Psychrolutes phrictus*, and a cephalopod (*Graneledone sp.*)

(Drazen et al. 2003). Although orange roughy produce a planktonic egg, both the sculpin and cephalopod have demersal eggs that are guarded.

1.6 Methods for Documenting a Spawning Aggregation

To properly distinguish a spawning aggregation from other forms of aggregations (e.g. feeding aggregations, shelter aggregations etc.), it is important to carefully document evidence of spawning. To this end, Colin et al. (2003) identified a suite of direct and indirect indications of spawning. The following three criteria have been used to verify directly that the fish are gathering for the purpose of spawning: (1) undisputed spawning observations, (2) females with hydrated eggs and (3) presence of post-ovulatory follicles in the ovaries of aggregating females. A fourth means of directly documenting the presence of a spawning aggregation is added here: (4) identification of very early stage eggs and larvae that can be positively associated with the aggregating species. Recent work on black marlin, *Makaira indica* (Domeier and Speare unpublished data), was able to confirm spawning through the presence of hydrated eggs, post-ovulatory follicles and the presence of larvae from 0–13 days post hatching. Plankton tows can now be considered a valuable means of verifying the act of spawning over an aggregation site, without the need to sacrifice any spawning adults.

If none of the direct signs of spawning are observed, indirect signs can be used to document new aggregations for species already proven to form spawning aggregations. Indirect signs can include behaviours or colour patterns, if these are demonstrably known to be associated only with spawning, as well as gonadosomatic index (GSI) (Chap. 9) data or the presence of swollen abdomens (indicating the presence of hydrated eggs) in a large percentage of the aggregated individuals. In the absence of witnessing the spawning event, it is not realistically possible to gather enough information to document spawning without sampling ovaries or larvae. Testes are not good indicators of the precise timing of spawning since males are running ripe prior to, and after, the actual spawning events. Sample collection should be a very high priority for all studies of spawning aggregations that involve new species, or unusual examples of species already known to aggregate (e.g. uncharacteristic site or season).

Beyond the documentation of spawning, it is also important, under the current definition, to document that spawning is occurring in densities of fish at least four times greater than that of the non-reproductive season/habitat, and that the spawning aggregation is predictably repeated in time and space. Methods for conducting underwater surveys have been described in a comprehensive methods manual (Colin et al. 2003, Chapter 9).

The scientific and grey literature on spawning aggregations contains many examples of poorly documented ‘spawning aggregations’ that are then perpetuated when they are cited, illustrating the need for rigorous field methods and for peer-review of studies that claim to document a spawning aggregation. The Society for the Conservation of Reef Fish Aggregations has compiled a global database of spawning aggregation records based on the two types of indicators, direct and indirect, (www.SCRFA.org). This database can be searched and reviewed online, but users should not cite this database as a sole source of evidence that a species forms a spawning aggregation because the database lists all reasonably documented records, including those from fisher interviews, not just those that have been reliably validated by direct observations of spawning (Sadovy de Mitcheson et al. 2008). Researchers who use the SCRFA database will be aided by the fact that SCRFA has separated aggregations that have been validated via direct means (spawning, hydrated eggs or post-ovulatory follicles) from those entered solely on the basis of indirect evidence. Those that appear with only indirect evidence need further verification before they can be cited as an aggregating species, according to the definition above. When using this database it is important to check all the references listed before deciding whether or not all criteria have been met. A decision to allow incomplete records was made by the organization to facilitate and stimulate the study and validation of species especially in cases for which more information is needed.

1.6.1 Observed Aggregations of Reef Fishes; Spawning Aggregations or Not

An observation of an unusual aggregation of reef fish is just the first step to documenting a spawning aggregation. In the absence of long-term behavioural observations and physical sampling of ovaries it is impossible to validate that a particular aggregation is for the purpose of spawning. For example, during a recent trip to the Galapagos I observed the king angelfish, *Holocanthus passer*, in an aggregation (Fig 1.2b). No spawning occurred despite watching well past sunset. In the absence of a spawning observation one must collect ovary samples in an attempt to identify hydrated eggs (can be seen with the naked eye) or post-ovulatory follicles (requires histology). The presence of well developed ovaries is not an indication of immediate spawning, since the ovaries can remain in a state of reproductive readiness for long periods of time (Chap. 9). In this example I did not have the permits to collect specimens, but after comparing the *H. passer* site to other sites, it became obvious that this was not a spawning aggregation, but rather a result of extreme currents, topography and food availability. No angelfish (Pomacanthidae) is known to form spawning aggregations (Appendix). The documentation and validation of spawning aggregations is a difficult and time-consuming task, but one that is extremely important.

1.7 The Intriguing Questions That Surround Spawning Aggregations

Although many papers have been published on spawning aggregations, several vexing questions remain: (1) why spawn in an aggregation, (2) how do spawning aggregations originally form, (3) from how large a geographic area do adults come to the spawning site, and (4) how far do the offspring disperse from the aggregation site? Although these questions may superficially appear esoteric, they are not. There are serious fisheries and economic issues that could draw immediate benefit if we knew the answers. For example, once a spawning aggregation is fished into extinction, will it ever recover? If it will recover, how long will that take? And, will it reassemble at the very same aggregation site? What if we learned so much about spawning aggregations that we could move one from an unprotected site to a protected site, or cause one to spontaneously form within a protected site, or create one from restocked animals? It would seem easier to simply protect the best spawning aggregation sites, but sometimes the necessary policy is too late or politically impossible. Chapters in this book address some of these intriguing questions in much more detail, but for the sake of argument I will highlight some of my personal favourite hypotheses.

Numerous hypotheses have been proposed to address why spawning aggregations form at all. These can be broken down into two main themes: those that address benefits to the larvae and those that address benefits to adults. Presumably the selective pressures that led to the evolution of aggregative spawning resulted in a reproductive strategy that presents a relatively high level of reproductive success for each individual. Past discussions of spawning aggregations have listed numerous hypotheses as to why marine animals aggregate to spawn; all of them extremely difficult to test (e.g. Shapiro et al. 1988; Mora and Sale 2002). When considering the overall life history of reef fishes, it is the larval phase that is most subject to mortality, creating a situation where very small benefits to the larval phase could lead to significantly more recruitment, particularly for species with high fecundity (Chap. 4). Therefore, it is likely that benefits to the larvae are an important driving force behind the evolution of the spawning aggregation; however, other hypotheses cannot be discounted.

The selection of spawning aggregation sites and times is one of the most intriguing phenomena related to this reproductive strategy. If we continue to assume, as I do, that benefits to larvae, expressed by increased recruitment, are driving the selection of spawning aggregation sites, then we are faced with two scenarios that might lead to their selection. (1) the process involves selecting sites that generate the highest level of local recruitment, and (2) a combination of certain topographic and oceanographic conditions predictably leads to increased recruitment at a very general level that is independent of spatial scale. Under the first scenario local recruitment is a significant proportion of overall recruitment generating a feedback loop to guide site selection. Under the second scenario site selection would be guided purely by instinctual/genetically determined detection of the relevant local conditions leading to a predictable choice of spawning site.

What is the catalyst for the formation of a spawning aggregation in a region where one did not previously exist? Although such a genesis has never been documented, it certainly has occurred on an evolutionary time scale. Some if not all species that aggregate to spawn are capable of spawning outside of an aggregation. If this were not true, species that form aggregations would not spawn in pairs or small numbers in captivity, and yet they do [e.g. mutton snapper, *Lutjanus analis* (Watanabe et al. 1998), mangrove red snapper, *L. argentimaculatus* (Emata 2003), Nassau grouper, *Epinephelus striatus* (Manday and Fernandez 1966; Tucker et al. 1996) camouflage grouper, *E. polyphekadion* (James et al. 1997) brown-marbled grouper, *E. fuscoguttatus*, orange-spotted grouper, *E. coioides*, Malabar grouper, *E. malabaricus*, and giant grouper *E. lanceolatus* (Pomeroy et al. 2002)]. Furthermore, some species that form spawning aggregations have been observed spawning in the field outside of an aggregation (Randall and Randall 1963; Samoilys 1997; Krajewski and Bonaldo 2005; Tuz-Sulub 2008). Nassau grouper is a much studied species that lends itself to this discussion; although Nassau grouper aggregations have been documented throughout their range, they have never been observed to aggregate in Florida, USA, or in the coastal areas of South America, regions where this species occurs in relatively low abundance (Sadovy and Eklund 1999, Yvonne Sadovy unpublished data). Spawning aggregations for other species (e.g. snappers-Lutjanidae, groupers) have been documented in Florida (Domeier et al. 1996; Coleman et al. 1996; Domeier 2004; Burton et al. 2005) so habitat may not be limiting the formation of a Nassau grouper spawning aggregation, leading to the likely conclusion that Nassau grouper spawning aggregation formation is density dependent. If spawning aggregations in general are density-dependent, this would tend to support the hypothesis that aggregations require a local recruitment-based positive feedback loop for them to form, as in the case of the bluehead wrasse, *Thalassoma bifasciatum* (see Chap. 12.14). If this were not the case we would expect individuals to be instinctually seeking out the best spawning sites regardless of density. Whether or not the formation of spawning aggregations is density-dependent is an important question to address, and furthermore, is reproduction outside of an aggregation less productive than spawning within an aggregation?

A widely cited early hypothesis suggested spawning sites were selected to promote offshore dispersal of eggs to help developing larvae avoid predation by the large number of planktivores found on coral reefs (Johannes 1978). More recent larval connectivity studies suggest, however, that the opposite may be occurring: that local larval retention may be critical for recruitment success (e.g. Colin 1992; Swearer et al. 1999; Taylor and Hellberg 2003; Paris and Cowen 2004; Domeier 2004; Almany et al. 2007). These cited studies are just a few examples of a growing body of work that support the hypothesis that the development of spawning aggregations, and the selection of aggregation sites, is facilitated by the local retention and recruitment of offspring. Following chapters of this book will explore the concept of retention in much more detail (Chaps. 6 and 7).

Without knowing precisely how or why spawning aggregations form, it is impossible to understand the impacts and implications of heavy fishing pressure upon a species that aggregates to spawn. Certainly aggregations are exceptionally susceptible to over exploitation, but does the process of spawning aggregation formation make the recovery of an overfished stock much more difficult? Can spawning aggregations rapidly recover in the absence of fishing pressure and the presence of high recruitment from a distant source? Can spawning aggregations be created by heavy stock enhancement from hatchery reared, or transplanted individuals? Although fraught with problems, stocking of an aggregating species into an area that no longer has a viable population could address some of these questions although massive restocking of the large yellow croaker, *Larimichthyes crocea*, in China did not result in restoration of viable populations in the wild (Liu and Sadovy de Mitcheson 2008).

1.8 Spawning Aggregations as Ecological and Fisheries Indicators

Spawning aggregations can be quickly disrupted and eradicated by intense fishing pressure (e.g. Sala et al. 2001; Sadovy and Domeier 2005), and once eradicated, they are not known to recover. In addition, regions of particularly heavy fishing pressure tend to be devoid of spawning aggregations (Sadovy de Mitcheson et al. 2008). The presence of a spawning aggregation at a particular site can therefore be considered an indicator of long-term ecological stability. Moreover, ecosystems that include the formation of feeding aggregations, predictable in time and space, that target the aggregating adults or mass release of eggs from a spawning aggregation may be an indication of apex ecological stability. A spawning aggregation must form and persist for some length of time before a feeding aggregation could evolve to target the aggregation. Whale sharks, *Rhincodon typus* (Heyman et al. 2001; Hoffmayer et al. 2007), and manta rays, *Manta birostris* (Lance Millbrand personal communication), have been observed to aggregate over spawning aggregations to feed on eggs of cubera snapper, *Lutjanus cyanopterus*, and convict surgeonfish, *Acanthurus triostegus*, respectively, while cephalopods, elasmobranchs, teleosts, mammals and birds prey on spawning adults at aggregation sites (Gorka et al. 2000; Smale et al. 2001; Bogetveit et al. 2008). Similarly, non-resident bull sharks, *Carcharhinus leucas*, appear to aggregate at the Great Barrier Reef during a presumed spawning aggregation of black marlin (MLD unpublished data). Since there is no way to age a spawning aggregation it is impossible to determine the length of time necessary for predatory aggregations to coincidentally form over spawning aggregations; whether it takes just a few, or thousands of generations is an important question.

Finally, spawning aggregations provide a unique opportunity to assess the fish stock. Estimating stocks of non-schooling reef fishes is very difficult, but if they form spawning aggregations they can be directly counted at the aggregation site. Landings data from these aggregations, on the other hand, may not be useful for stock assessment since the stock may overwhelm effort until it is drastically reduced through hyperstability (Sadovy and Domeier 2005, Chapter 8, 11). To illustrate this point, one need only envision a child sticking his arm into a cookie jar that he cannot see into; each reach into the jar produces a cookie until suddenly the jar is empty. Similarly, a small fleet fishing an aggregation may not show decreased catch per unit of effort (CPUE) until the aggregation is nearly extirpated. However, structured underwater visual census can produce valuable monitoring data relevant for assessing the stock independent of fishery measures. Conducting annual assessments of the size of an aggregation, even if the assessment is an index rather than an actual count, can provide valuable real-time data as long as the sampling protocol is scientifically sound and consistent (Chap. 9). Underwater visual census will reveal decreases in stock size prior to a drop in CPUE. Hydroacoustic surveys (assessing stock size via sonar) may allow for the remote monitoring of spawning aggregations in the future, but the technology and validation techniques are not yet fully realized. Unfortunately long term standardized monitoring of spawning aggregations is not widespread, particularly in the Indo-west Pacific and Indian Oceans.

1.8.1 Feeding Aggregations That Form as a Result of Spawning Aggregations

The black marlin (*Makaira indica*) is an exceptionally large pelagic predator that gathers along the Great Barrier Reef each year between the months of October and December. The aggregation is predictable, and until it was protected by the Australian government, it was once heavily targeted by the Japanese longline fleet. The aggregation remains the basis of an important recreational fishery that is now primarily catch and release (not mandated by law, but exercised by anglers). Although anecdotal information existed that suggested this aggregation was for the purpose of spawning, the presence of hydrated eggs, post-ovulatory follicles and very early stage larvae have now confirmed that this is indeed a black marlin spawning aggregation. Mature female black marlin are much larger than the males, and multiple males are frequently observed at the surface following females; presumably a prelude to spawning (MLD personal observation).

The formation of a spawning aggregation of such a large pelagic fish along a coral reef is unique. What makes this event even more interesting is the fact that large sharks seem to aggregate along the reef at the same time, possibly to prey on the black marlin. Black marlin are often attacked by sharks while being captured on rod-and-reel; furthermore, the marlin are sometimes consumed after being released by the anglers. One such event caught on film during a satellite tagging expedition documented a large group of sharks rising from deeper water to attack a black marlin that was too exhausted to escape (film by Guy Harvey). The sharks entirely consumed the 200 kg marlin in less than 60 s. To identify and track which species of sharks are responsible for preying on black marlin, a large hook was baited and dropped in the water upon releasing a black marlin. One 250 kg bull shark was immediately captured and satellite tagged. The resulting data showed this shark leaving the GBR when the marlin aggregation dispersed, traveling 500 km south to a river mouth near Townsville (MLD unpublished data) Although these observations are anecdotal, further studies may demonstrate that predatory sharks are aggregating to prey on the spawning black marlin. Although the example described here involves predators possibly gathering to feed on the spawning adults, better documented cases of the formation of feeding aggregations over spawning aggregations involve whale sharks (Fig. 1.3a) and manta rays (Fig. 1.3b) feeding on the eggs released by snappers and surgeonfish (respectively).

Fig. 1.3 Examples of the formation of feeding aggregations over spawning aggregations involve (**a**) whale sharks, *Rhincodon typus* (Photo: © Doug Perrine/SeaPics.com) (**b**) manta rays, *Manta birostris* (Photo: © Michael L. Domeier)

1.9 Summary

If done properly defining the phenomenon we call a spawning aggregation provides a framework that is both biologically meaningful while at the same time useful and practical in fisheries management discussions. Maintaining a relatively restrictive definition best serves these parallel goals. There will always exist specific examples of organisms that do not clearly fit within or outside the proposed definition; that is unavoidable, but perhaps these examples hold the most promise of insight into the mysteries of spawning aggregation formation. Although answers to many of the intriguing questions that still surround spawning aggregations seem out of reach, the capability exists to address some of them. Researchers should certainly focus on identifying both the adult and larval catchment areas; these are the geographic footprints from which the adults are drawn to an individual aggregation and in which the resulting offspring settle. Identifying these catchment areas can help guide the process of MPA designations and allow for meaningful assessment of the effect of the MPA as well as management planning generally. The information is also important for identifying suitable management units. Relocation and stock enhancement projects are possible to study the processes related to spawning aggregation formation. Granted this is much easier for small species that form resident aggregations (i.e. Warner 1988, 1990), but attempting these experiments with species that form transient aggregations could be invaluable. Finally, the threat of overfishing spawning aggregations has been apparent for decades and yet they continue to disappear. Sustainable management of spawning aggregations is slow in coming but is imperative, while MPAs alone are likely not sufficient in the absence of seasonal closures, effort controls, and strict enforcement (Chap. 11).

Acknowledgements I would like to thank Y. Sadovy de Mitcheson and P. Colin for their valuable comments on early versions of this chapter. The time devoted to this chapter would not have been possible without the generous support of the Offield Family Foundation.

References

Almany G, Berumen M, Thorrold S, Planes S, Jones G (2007) Local replenishment of coral reef fish populations in a marine reserve. Science 316:742–744

Bane G (1965) Spawning of the margined flyingfish, *Cypselurus cyanopterus* (Valenciennes), in the Gulf of Guinea. Copeia 1965:382

Bell J, Lyle J, Bulman C, Graham K, Newton G, Smith D (1992) Spatial variation in reproduction, and occurrence of non-reproductive adults, in orange roughy, *Hoplostethus atlanticus* Collet (Trachichthyidae), from southeastern Australia. J Fish Biol 40:107–122

Bogetveit R, Slotte A, Johannessen A (2008) The ability of gadoids to take advantage of a short-term high availability of forage fish: the example of spawning aggregations in Barents Sea capelin. J Fish Biol 72:1427–1449

Burton ML, Brennan KJ, Munoz RC, Parker RO (2005) Preliminary evidence of increased spawning aggregations of mutton snapper (Lutjanus analis) at Riley's Hump two years after establishment of the Tortugas South Ecological Reserve. Fish Bull 103:404–410

Casazza T, Ross S, Necaise A, Sulak K (2005) Reproduction and mating behavior of the Atlantic flyingfish. *Cheilopogon melanurus* (Excocoetidae), off North Carolina. Bull Mar Sci 77(3):365–375

Claydon J (2004) Spawning aggregations of coral reef fishes: characteristics, hypotheses, threats and management. Oceanogr Mar Biol Ann Rev 42:265–301

Coleman FC, Koenig CC, Collins LA (1996) Reproductive styles of shallow water groupers (Pisces: Serranidae) in the eastern Gulf of Mexico and the consequences of fishing spawning aggregations. Environ Biol Fish 47(2):129–141

Colin PL (1992) Reproduction of the Nassau grouper, *Epinephelus striatus* (Pisces: Serranidae) and its relationship to environmental conditions. Environ Biol Fish 34:357–377

Colin PL, Sadovy YJ, Domeier ML (2003) Manual for the study and conservation of reef fish spawning aggregations. Society for the Conservation of Reef Fish Aggregations, Special Publication 1

Domeier ML (2004) A potential recruitment pathway originating from a Florida marine protected area. Fish Oceanogr 13(5):287–294

Domeier ML, Colin PL (1997) Tropical reef fish spawning aggregations: defined and reviewed. Bull Mar Sci 60(3):698–726

Domeier ML, Koenig CC, Coleman FC (1996) Reproductive biology of the gray snapper (Lutjanidae: *Lutjanus griseus*) with notes on spawning for other western Atlantic lutjanids. In: Arreguin-Sanchez F, Munro JL, Pauly D (eds) Biology of tropical groupers and snappers. ICLARM Conference Proceedings 48, Makati City

Drazen J, Goffredi S, Schlining B, Stakes D (2003) Aggregations of egg-brooding deep-sea fish and cephalopods on the Gorda Escarpment: a reproductive hot spot. Biol Bull 205:1–7

Emata C (2003) Reproductive performance in induced and spontaneous spawning of the mangrove red snapper, *Lutjanus argentimaculatus*: a potential candidate species for sustainable aquaculture. Aqua Res 34(10):849–857

Francis R, Clark M (1998) Inferring spawning migrations of orange roughy (*Hoplostethus atlanticus*) from spawning ogives. Mar Freshw Res 49:103–108

Gladstone W (2007) Temporal patterns of spawning and hatching in a spawning aggregation of the temperate reef fish *Chromis hypsilepis* (Pomacentridae). Mar Biol 151:1143–1152

Gorka S, Petersen C, Lobel P (2000) Predator-prey relations at a spawning aggregation site of coral reef fishes. Mar Ecol Prog Ser 203:275–288

Haegele CW, Schweigert JF (1985) Distribution and characteristics of herring spawning grounds and description of spawning behaviour. Can J Fish Aquat Sci 42(S1):s39–s55

Heyman W, Graham R, Kjerfve B, Johannes R (2001) Whale sharks *Rhincodon typus* aggregate to feed on fish spawn in Belize. Mar Ecol Prog Ser 215:275–282

Hoffmayer E, Franks J, Driggers W, Oswald K, Quattro J (2007) Observations of a feeding aggregation of whale sharks, *Rhincodon typus*, in the north central Gulf of Mexico. Gulf Caribb Res 19(2):69–73

James M, Al-Thobaiti S, Rasem B, Carlos M (1997) Breeding and larval rearing of the camouflage grouper *Epinephelus polyphekadion* (Bleeker) in the hypersaline waters of the Red Sea coast of Saudi Arabia. Aqua Res 28(9):671–681

Johannes RE (1978) Reproductive strategies of coastal marine fishes in the tropics. Environ Biol Fish 3:65–84

Johannes RE (1981) Wards of the Lagoon: fishing and Marine Lore in the Palau District of Micronesia. University of California Press, Berkely

Krajewski J, Bonaldo R (2005) Spawning out of aggregations: record of a single spawning dog snapper pair at Fernando de Noronha Archipelago, equatorial western Atlantic. Bull Mar Sci 77(1):165–168

Liu M, Sadovy de Mitcheson Y (2008) Profile of a fishery collapse: why mariculture failed to save the large yellow croaker (*Larimichthys crocea*, Sciaenidae). Fish Fish 9(3):1–24

Manday D, Fernandez M (1966) Desarrollo embrionario y primeros estados larvales de la cherna criolla, *Epinephelus striatus* (Bloch) (Perciformes: Serranidae). Estudios Inst Oceanogr Habana 1:35–45

Mora C, Sale P (2002) Are populations of reef fish open or closed? Trends Ecol Evol 17(9):422–428

Nemeth RS (2009) Dynamics of reef fish and decapod crustacean spawning aggregations: underlying mechanisms, habitat linkages and trophic interactions. In: Nagelkerken I (ed) Ecological interactions among Tropical Coastal ecosystems. Springer, Netherlands

Pankhurst N (1988) Spawning dynamics of orange roughy, *Hoplostethus atlanticus*, in mid-slope waters of New Zealand. Environ Biol Fish 21:101–116

Parin N, Lakshminaraina D (1993) Flying fishes (Exocoetidae) in the coastal waters of southeastern India. J Ichthyol 33:12–25

Paris C, Cowen R (2004) Direct evidence of a biophysical retention mechanism for coral reef fish larvae. Limnol Oceanogr 49(6):1964–1979

Pomeroy R, Agbayani R, Toledo J, Sugama K, Slamet B (2002) The status of grouper culture in southeast Asia: financial feasibility analysis for grouper culture systems in the Philippines and Indonesia. In: Pomeroy R, Parks J, Balboa C (eds) Farming the reef: a state-of-the-art review of aquaculture of coral reef organisms in tropical nearshore environments. World Resources Institute, Washington, DC

Pratt HL, Carrier JC (2001) A review of elasmobranch reporducitve behaviour with a case study on the nurse shark, *Ginglymostoma cirratum*. Environ Biol Fish 60:157–188

Randall J, Randall H (1963) The spawning and early development of the Atlantic parrot fish, *Sparisoma rubripinne,* with notes on other scarid and labrid fishes. Zoology 48(2):49–60

Russell M (2001) Spawning aggregations of reef fishes on the Great Barrier Reef: implications for management. Report of the Great Barrier Reef Marine Park Authority, Townsville, Australia. 38p. http://www.gbrmpa.gov.au/__data/assets/pdf_file/0005/4100/Russell-2001.p

Sadovy de Mitcheson Y, Cornish A, Domeier ML, Colin P, Russell M, Lindeman K (2008) A global baseline for spawning aggregations of reef fishes. Conserv Biol 22(5):1233–1244

Sadovy Y, Domeier ML (2005) Are aggregation-fisheries sustainable? Reef fish fisheries as a case study. Coral Reefs 24:254–262

Sadovy Y, Eklund A-M (1999) Synopsis of biological data on the Nassau grouper, *Epinephelus striatus* (Bloch, 1792), and the Jewfish, *E. itajara* (Lichtenstein, 1822), NOAA technical report NMFS 146, Seattle, Washington, DC

Sala E, Ballesteros E, Starr R (2001) Rapid decline of Nassau grouper spawning aggregations in Belize: fishery management and conservation needs. Fish 26(10):23–30

Samoilys M (1997) Periodicity of spawning aggregations of coral trout *Plectropomus leopardus* (Pisces: Serranidae) on the northern Great Barrier Reef. Mar Ecol Prog Ser 160:149–159

Shapiro D, Hensley D, Appeldoorn R (1988) Pelagic spawning and egg transport in coral-reef fishes: a skeptical overview. Environ Biol Fish 22(1):3–14

Smale M, Sauer W, Roberts M (2001) Behavioral interactions of predators and spawning chokka squid off South Africa: towards quantification. Mar Biol 139:1095–1105

Stevens P, Bennett C, Berg J (2003) Flyingfish spawning (*Parecocoetus brachypterus*) in the northeastern Gulf of Mexico. Environ Biol Fish 67:71–76

Swearer S, Caselle J, Lea D, Warner R (1999) Larval retention and recruitment in an island population of a coral-reef fish. Nature 402:799–802

Taylor M, Hellberg M (2003) Genetic evidence for local retention of pelagic larvae in a Caribbean reef fish. Science 299:107–109

Thresher R (1984) Reproduction in Reef Fishes. Tropical Fish Hobbyist Publications, Neptune City

Tucker J, Woodward P, Sennet G (1996) Voluntary spawning of captive Nassau groupers *Epinephelus striatus* in a concrete raceway. J World Aqua Soc 27(4):373–383

Tuz-Sulub A (2008) Agregaciones de desove de mero (Serranidae: *Epinephelus sp.* y *Mycteroperca sp.*) en areas del Banco de Campeche, Yucatan, Mexico. PhD thesis, Centro Investigacion y de Estudios Avanzados del Instituto Politecnico Nacional, Unidad Merida, Departmento de Recursos del Mar

Warner R (1988) Traditionality of mating-site preference in a coral reef fish. Nature 335:719–721

Warner R (1990) Resource assessment versus tradition in mating-site determination. Am Nat 135:205–217

Watanabe W, Ellis E, Ellis S, Chaves J, Mafredi C, Hagood R, Sparsis M, Arneson S (1998) Artificial propagation of mutton snapper *Lutjanus analis*, a new candidate marine fish species for aquaculture. J World Aqua Soc 29(2):176–187

Chapter 2
Ecosystem Aspects of Species That Aggregate to Spawn

Richard S. Nemeth

Abstract A wide diversity of species form spawning aggregations and migrate from home ranges or feeding sites to specific locations for reproduction. Because most of these species comprise large carnivorous and numerous herbivorous fishes, they play a vital role in ecosystem function and fisheries economics. Nested within the functional migration area may be other spatial components including catchment area, staging area, courtship arena and the spawning aggregation site. The flux of fish biomass from feeding grounds to spawning aggregation sites as well as the energy transfer resulting from feeding, defaecation and release of propagules provides an important and largely overlooked ecological component of connectivity within marine ecosystems. Although little information exists on predator-prey dynamics at aggregation sites, a few studies suggest that some aggregating species feed along migratory pathways and at aggregation sites. Moreover, piscivores and egg predators may converge on aggregation sites to take advantage of these temporary sources of food. Multiple-species spawning aggregation sites in particular are important cross-roads of marine animal migrations and represent major nodes of biological diversity and reproductive potential. Effective management of aggregating species will require the application of ecosystem based management approaches that take into account local geophysical conditions (i.e. island shelf areas), migration patterns and key spawning habitats (i.e. promontories, reef pass channels, outer reef slopes). Most importantly, managers and fishers alike will need to acknowledge the vulnerability of aggregating species and prioritize their conservation.

R.S. Nemeth (✉)
Center for Marine and Environmental Studies, University of the Virgin Islands,
2 John Brewer's Bay, Charlotte Amalie, St. Thomas 00802, US Virgin Islands
e-mail: rnemeth@uvi.edu

Y. Sadovy de Mitcheson and P.L. Colin (eds.), *Reef Fish Spawning Aggregations: Biology, Research and Management*, Fish & Fisheries Series 35,
DOI 10.1007/978-94-007-1980-4_2, © Springer Science+Business Media B.V. 2012

2.1 Introduction

Throughout reef ecosystems many fishes utilize a reproductive strategy that requires migration to specific sites where courtship and spawning commence and fertilized eggs or larvae are released into local water masses. The reproductively mature adults that spawn at aggregation sites may represent the primary source of reproductive effort for a species (Shapiro et al. 1993). Therefore each spawning aggregation may have a strong influence on the replenishment of participating populations.

Migration and subsequent spawning by aggregating species provide an important and largely overlooked ecological component of connectivity within marine ecosystems, due to fish movements, habitat use and interspecific interactions. Since most reef-associated fishes have limited adult movements, connectivity at larger spatial scales has typically referred to the genetic exchange among local marine populations via larval dispersal (Cowen et al. 2000; Cowen 2002; Sale 2004). However, recent biophysical models of larval dispersal suggest that spatial structure of successful larval exchange for a variety of reef fish species can be as little as 10–100 km (Cowen et al. 2006). On the other hand, the adults of many species within at least five families (snooks-Centropomidae, ladyfishes-Elopidae, snappers-Lutjanidae, groupers-Serranidae, porgies-Sparidae) of reef-associated fishes annually swim these distances, or greater, when migrating from home ranges to their spawning aggregation sites (Nemeth 2009, Table 4). The extent of genetic mixing at fish spawning aggregations (FSA) among adults who have migrated from an area encompassing 100's of square kilometers is unknown but is highly relevant to understanding population structure and for management, and therefore requires greater attention when addressing issues of population connectivity. However, because information on the genetic relatedness of fish in spawning aggregations is scant (Rhodes et al. 2003), this chapter will focus primarily on spatial scales of adult connectivity in aggregating species and on the ecological interactions that occur along migration pathways and at spawning aggregation sites. Moreover, ecosystem based management (EBM) requires a good understanding of the ecology and behaviours of target species (Garcia et al. 2003), to fully explore different management scenarios.

Because aggregating species include many large carnivorous (i.e. groupers, snappers, jacks-Carangidae) and numerous herbivorous (i.e. surgeonfishes-Acanthuridae, parrotfishes-Scaridae) fishes of high commercial and ecological value, they play an important role in ecosystem function and fisheries economics. For example, the Nassau grouper (*Epinephelus striatus*) was an important commercial species until aggregation fishing nearly eliminated it from many locations throughout the Caribbean (Olsen and LaPlace 1978; Sadovy 1997; Sala et al. 2001; Aguilar-Perera 2006). Although the substantial decline in Nassau landings and subsequent loss of revenue has long been documented (Sadovy 1994, 1997; Sadovy and Eklund 1999; Claro et al. 2001), the broad ecological importance of this single species has only recently been realized. Stallings (2008) found that Nassau grouper facilitated higher rates of recruitment and maintained higher biological diversity of small reef fishes by indirectly structuring food webs through the consumption of secondary predators such as graysby (*Cephalopholis cruentata*) and coney (*C. fulvus*), which aggressively feed on newly settled fishes (Kaufman and Ebersole 1984; Nemeth 1998). Moreover, Nassau grouper as well as tiger grouper (*Mycteroperca tigris*) were found to prey on Indo-Pacific lionfish (*Pterois volitans*) (Maljković et al. 2008), which has recently invaded the Atlantic coast of North America and many locations throughout the Caribbean (Whitfield et al. 2002, Hamner et al. 2007a, Snyder and Burgess 2007). Unfortunately, Nassau grouper, tiger grouper and other aggregating groupers have been fished heavily throughout the region (Olsen and LaPlace 1978; Sadovy et al. 1994a; Sadovy 1997; Aguilar-Perera 2006), an unforeseen consequence of lack of fishery management that reduces trophic integrity and biological diversity and may be a factor allowing for the spread of Indo-Pacific lionfish and other invasive species.

This chapter examines aggregating species within a broad ecosystem context with a particular focus on the spatial scales of migration to, and movements around, fish spawning aggregation (FSA) sites, the methods used to study these movement patterns (Sect. 2.1), migration pathways and habitat linkages between home sites and spawning areas (Sect. 2.2) and the potential impacts of aggregating species on local food webs via predator-prey interactions (Sect. 2.3). This chapter also examines how these spatial aspects of migrating species can be integrated into an ecosystem approach to fisheries management (Sect. 2.4).

To better understand these various spatial and ecological aspects of FSA's, a brief description of reproduction in aggregating species is warranted. A wide diversity of species form spawning aggregations (Chap. 1, Table 1) and all migrate from home ranges or feeding sites to specific locations for reproduction. Two general categories of aggregation can be defined although some species share features of both (Chap. 1). *Resident* aggregating species such as surgeonfishes and parrotfishes can often be found in various sized groups foraging across a reef throughout the day (Ogden and Buckman 1973; Robertson 1983). These smaller species may have home ranges located only a few tens or hundreds of metres to a maximum of a few km from spawning sites. Therefore their migration distances are short. *Transient* aggregating species such as groupers and snappers usually consist of widely dispersed solitary individuals that remain within home ranges during the non-reproductive period. For these larger species spawning sites may be tens to hundreds of kilometres away, although a proportion of the spawning adults may live in close proximity to the aggregation site (Hutchinson and Rhodes 2010). At the onset of the reproductive season, various large-scale environmental cues such as water temperature or day-length, or correlates of these, may initiate migration of the spawning population (Nemeth 2009). Spawning seasons typically last 2–3 months for transient aggregating species while resident aggregating species can spawn on a monthly or even daily basis throughout much of the year (Chap. 5) (Sadovy de Mitcheson et al. 2008). The mode of spawning ranges from gamete release a few metres from the substrate between an individual male and female pair or between a male spawning with individual females within a harem, to massive group-spawning in mid-water between a female and several males (Domeier and Colin 1997).

Pair-spawning usually occurs within spawning territories defended by individual males whereas group-spawning occurs well above the substrate with no evidence of territorial defense by males.

2.2 Spatial Patterns Exhibited by Fishes That Aggregate to Spawn

During migration to spawning aggregation sites, fish may follow key landmarks or bathymetric features along particular migration pathways (Colin 1992; Mazeroll and Montgomery 1995, 1998). Although the actual migration route might not be direct, the straight-line distance from a fish's home site to the spawning aggregation site defines its *migration distance*. Individual fish or migrating groups eventually reach a specific *spawning aggregation site* and remain in the area for a few hours, days or weeks (Rhodes and Tupper 2008). After spawning, fish return to home sites along pathways that radiate outward from the spawning site. The linear distance or length of each "radial" can vary considerably among individuals within a species (Zeller 1998; Nemeth 2005; Nemeth et al. 2007). If one connects the endpoints of the longest "radials" which radiate out from the spawning site *core* (i.e. those fish whose home sites are farthest from the spawning site), the resulting polygon defines the *catchment area* of an aggregation site for a spawning population of a single species (Fig. 2.1). Throughout this chapter a spawning population is defined as all adults using a single spawning aggregation site.

As fish migrate through their catchment area and converge on the spawning site, this temporary concentration of hundreds to thousands of herbivorous or carnivorous fishes during the reproductive period provides a potentially important mechanism to interlink and possibly influence food webs. The complex biological processes and trophic interactions that may occur during migration and spawning and the mosaic of habitats through which fish may migrate within a catchment area represents the *functional migration area* (FMA) of a species (Nemeth 2009). The FMA includes migration pathways, spatial and temporal habitat use during the spawning season, all intra- and interspecific interactions and predator-prey dynamics that occur within the catchment area during the spawning season from the moment the adults depart their home ranges until the time they return. The functional migration area also takes into account the transfer of energy resulting from feeding, defaecation and release of propagules at the spawning site (Nemeth 2009). Moreover, as transient aggregating species temporarily vacate their home reefs during the reproductive season, predation pressure on their potential prey may be temporarily alleviated.

The FMA utilized may vary in size from a few hundred to several thousand square kilometres depending upon species and location (Nemeth 2009). Due to the inherent patchiness of coral reefs, not all substrates within the catchment area are suitable habitat for feeding, migration, spawning or other activities. Therefore the subset of suitable habitats represents the maximum area supporting the spawning population. To explore the possible influence of a spawning population as it moves

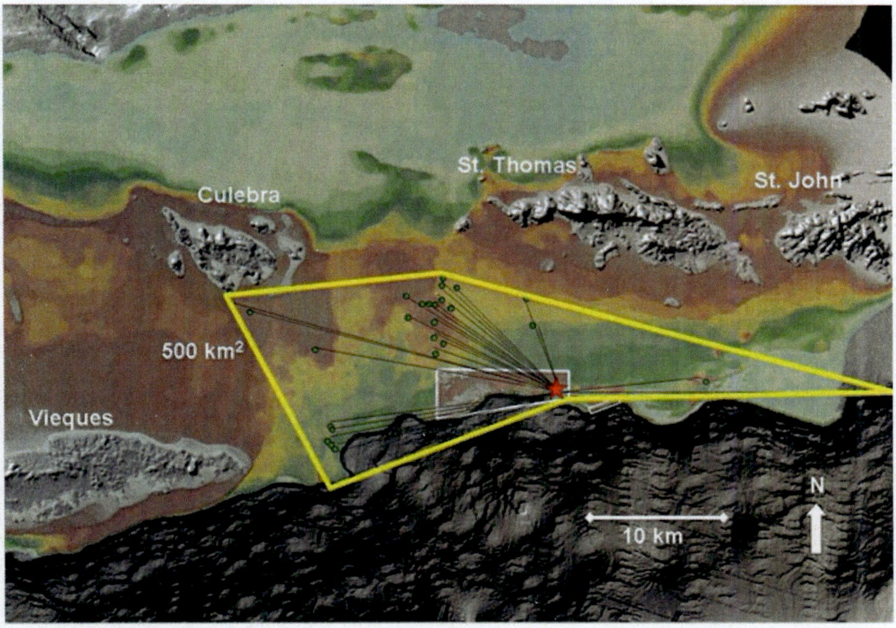

Fig. 2.1 Red hind (*Epinephelus guttatus*) functional migration area (*yellow polygon*) south of St. Thomas, US Virgin Islands calculated from a mark-recapture study (Nemeth 2005). Fish were tagged at the spawning aggregation site (*red star*) and recaptured (*dots*) by fishermen. *Straight line* migration distances (*black lines*) define the catchment area (500 km^2). Areas defined by *white lines* delimit Red Hind Bank Marine Conservation District (*large*) and Grammanik Bank (*small*)

through an ecosystem to and from a spawning site, the spatial and temporal scales at which spawning aggregations operate need to be understood.

The most effective methods for measuring movements of aggregating species are conventional mark-recapture studies and acoustic tracking (Chap. 9). Conventional tagging typically relies on the cooperation of local fishers to return tags and provide accurate information on recapture location. Fish that are recaptured on the spawning site also provide estimates of size- and gender-specific residence time, and growth rate. Fish that are recaptured away from the spawning site provide estimates of gender-specific migration distance and direction, swim speeds, resident habitats and growth rates. Although most tag returns will be from areas with the greatest fishing effort, producing gaps in spatial data, this method can be used to calculate catchment area (Fig. 2.1).

Identifying migration pathways is more difficult and tag loss and damage can reduce the effectiveness of this method (Wormald and Steele 2008). Acoustic tags, which transmit unique identification codes from individual fish, used in combination with a directional hydrophone or an array of underwater acoustic receivers (Domeier 2005), provide detailed information on location, depth, residence time, frequency and timing of migration, habitat use and small-scale movement patterns of tagged fish (Holland et al. 1993; Zeller 1999; Meyer et al. 2000; Beets et al. 2003; Rhodes and

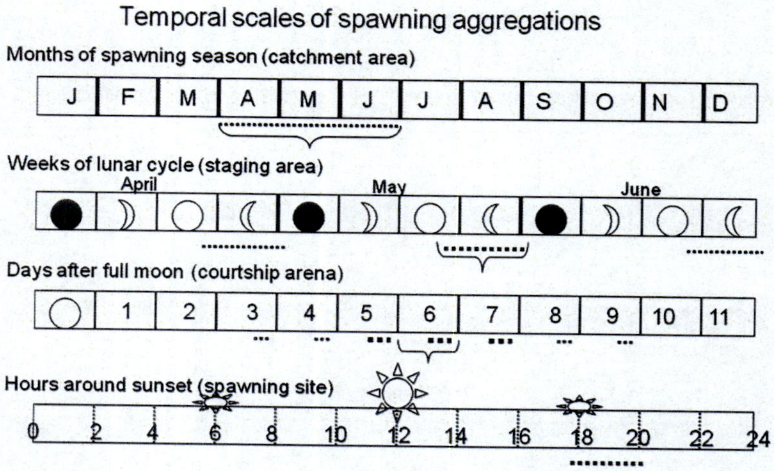

Fig. 2.2 Temporal scales of transient spawning aggregations showing characteristic timing of spawning of a Caribbean snapper (mutton snapper, *Lutjanus analis*) in the USVI. During the 3 month spawning season (April, May and June), fish only aggregate for 7–10 days following the full moon (*open circle*), with the majority spawning during 2 or 3 consecutive days and for only a few hours around sunset (• • • • • indicates periods of peak spawning activity during each time period)

Tupper 2008; Hutchinson and Rhodes 2010). However, the high cost of acoustic tags and receivers will limit sample size and the area that can be monitored which makes measurements of migration distances and estimates of catchment area difficult using this technology. Despite their respective weaknesses, these two methods used separately or in combination can provide useful information on the spatial and temporal aspects of migration which will enhance our understanding of spatial use and improve information for ecosystem based management (Rhodes and Tupper 2008).

To illustrate the relevant spatio-temporal scales of movement exhibited by species of fish that form spawning aggregations, the focus of this chapter will be transient aggregators (*sensu* Domeier and Colin 1997, Chapter 1). The FMA during spawning can be divided, for descriptive purposes, into four possible spatio-temporal scales that range from largest to smallest: (1) the *catchment area* encompasses the home ranges of all spawning adults using a single aggregation site during the annual reproductive cycle (Fig. 2.1), (2) the *staging area* is where migration pathways begin to converge and certain aggregating groupers rest, feed or visit cleaning stations during the spawning season and in the vicinity of the spawning site, (3) the *courtship arena* is where males and females begin to interact during the specific reproductive period or lunar phase, and (4) the *spawning site* is where spawning occurs over a few hours and for some species may include a *core* area, (Figs. 2.2 and 2.3). While this information is available for a few species that have been studied in detail, all four categories may not be applicable to all aggregating species with further studies required in most cases.

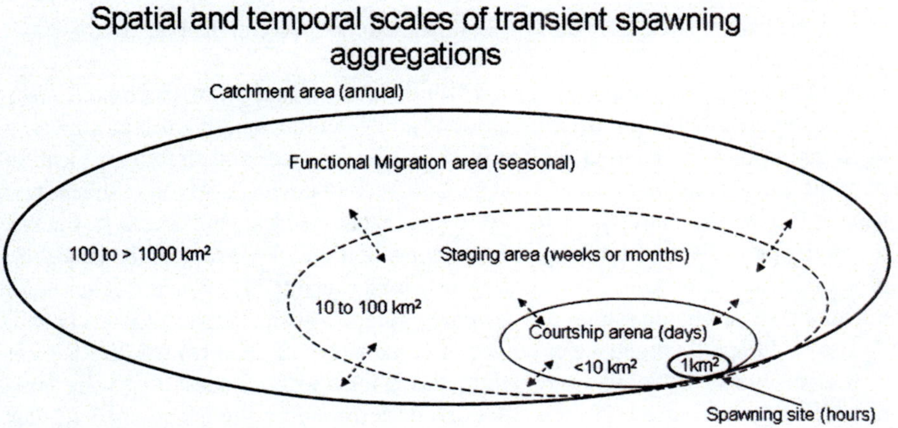

Fig. 2.3 Spatial scales of movement associated with aggregations of group-spawning species (e.g. species that do not defend spawning territories and spawn in mid-water) showing the functional migration area and the four spatio-temporal phases of the reproductive cycle. Spatial scales include the catchment area, staging area, courtship arena and spawning aggregation site

2.2.1 Catchment Area

At the largest spatial scale (100 to > 1,000 km^2) the catchment area encompasses the sum of home ranges and migration routes of a local spawning population that uses a specific aggregation site during the annual reproductive cycle (Figs. 2.1 and 2.3). The tagging methods briefly described above are useful for measuring migration distances and calculating catchment areas for FSA sites, as well as for identifying the minimum management area of a spawning population. However, migration distances and therefore catchment areas can vary greatly depending upon a number of factors. Based on limited data, the migration distance and catchment areas within a family seem to be positively correlated with fish size (Chap. 4) and also area of reef platform. For example, larger species, such as the Nassau grouper, reach maturity at 48 cm and maximum size at 94 cm or greater (Ault et al. 2008) and can migrate 110 to at least 240 km (Colin 1992; Carter et al. 1994). Nassau grouper can have catchment areas estimated at 7,500 km^2 in locations with extensive reef systems such as the Mesoamerican reef and the Bahamian archipelago (Nemeth 2009). On the other hand, on small isolated islands with deep water barriers such as the Cayman Islands and Glover's Reef, Belize, the migration distances of adult Nassau grouper may be limited to 15–70 km, with estimated catchment area of 30–100 km^2, depending upon the shelf area surrounding each island (Colin et al. 1987; Semmens et al. 2005; Starr et al. 2007). Smaller species like the red hind, *E. guttatus*, which reach maturity at 25 cm TL and maximum size at 76 cm TL (Heemstra and Randall 1993; Ault et al. 2008), have smaller catchment areas estimated to range from 90 km^2 to 500 km^2 for migration distances of 16 km to 30 km, respectively, depending upon the size of the insular shelf (Luckhurst 1998; Nemeth 2005; Nemeth et al. 2006a, 2007, Chapter 12.3).

Different tagging methodologies will influence estimation of catchment areas and can be used to answer different questions. For example tagging fish at spawning aggregation sites can be used to identify migration distances and directions, location of adult home range habitats and catchment areas (Nemeth 2005; Hutchinson and Rhodes 2010). Alternatively, adults tagged within their resident habitats can be tracked to identify previously unknown spawning aggregation sites. For example, Zeller (1998) used both these approaches to track leopard coralgrouper (*Plectropomus leopardus*) movements from home ranges to spawning aggregation sites around Lizard Island, Australia. Leopard coralgrouper tagged at home sites migrated from 0.2 to 5.2 km to four different spawning aggregation sites. Movements of these fish were contained within the narrow 20 m depth contour surrounding the island shelf (Zeller 1997, 1998; Zeller and Russ 2000), resulting in an estimated catchment area of 1.5 km^2 for fishes from the largest spawning aggregation site (e.g. Granite Head) (Nemeth 2009). Several adults were also tagged at the Granite Head spawning site and recaptured on isolated shoals up to 11 km away (Zeller 1998). These individuals increased the potential catchment area for the Granite Head FSA site from 1.5 km^2 to at least 80 km^2, a substantial increase that may have implications for future management decisions and potential MPA design for this species and shows the importance of working on a large enough sample size.

2.2.2 *Staging Area*

Within the FMA different habitats along migration routes or surrounding the spawning aggregation site may provide different services to aggregating species. As migrating adults begin to converge on the spawning aggregation site fish densities increase. For several groupers within the genera *Epinephelus*, *Mycteroperca*, and *Plectropomus*, all the adults do not pack into the spawning area for the entire spawning season but instead occupy large staging areas (Fig. 2.3) where they may congregate in groups to rest, feed or occupy cleaning stations (Samoilys 1997; Rhodes and Sadovy 2002; Nemeth et al. 2006b; Semmens et al. 2006; Robinson et al. 2008). Fish within staging areas maintain normal colouration and do not display spawning colour patterns typically found in the courtship arena or at spawning sites (see below). For example, in the Seychelles Robinson et al. (2008) observed small groups of 3–20 normal coloured brown-marbled grouper (*Epinephelus fuscoguttatus*) and camouflage grouper (*Epinephelus polyphekadion*) occupying a staging area (6,900 m^2) which surrounded the spawning site (5,750 m^2). Within the staging area grouper densities were highest about 5 days before spawning but then declined as fish moved to the spawning site to set up territories. While in staging areas, brown-marbled and camouflage groupers did not display spawning colouration, courtship behaviours or territoriality and thus temporarily resided in these locations for some other purpose (Robinson et al. 2008). Similar movement and behavioural patterns were observed for leopard coralgrouper

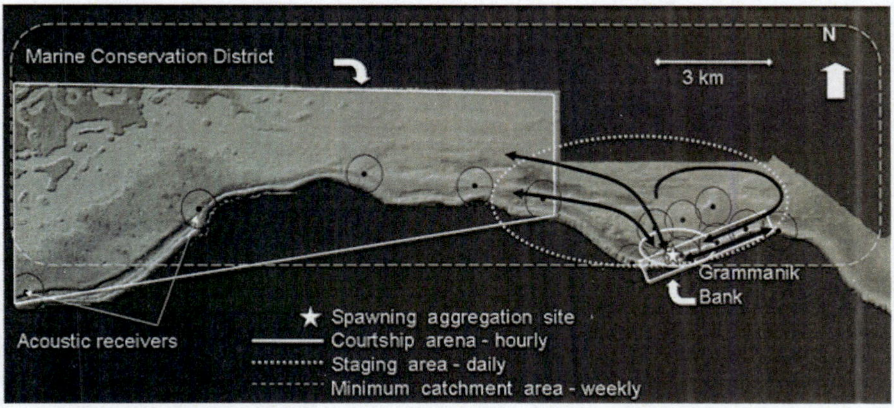

Fig. 2.4 Spatial and temporal boundaries of grouper movements around the Grammanik Bank (GB), a seasonally protected area (*rectangle*) south of St. Thomas, USVI, that includes a multi-species fish spawning aggregation site (*star*). An acoustic tagging study showed that during the spawning aggregation period Nassau (*Epinephelus striatus*) and yellowfin (*Mycteroperca venenosa*) grouper swam outside the protected area and courtship arena (*small oval*) on a daily basis. Tagged groupers moved throughout a staging area (*large oval*) and followed specific migration pathways (RSN unpublished data)

and red hind (Samoilys 1997; Nemeth 2005; Nemeth et al. 2007). In Pohnpei, the camouflage grouper spawning population size, as estimated from Rhodes and Sadovy (2002a), was at least five-fold greater than in the Seychelles (i.e. ca. 10,000 vs. 2,000 fish), and were distributed over a much larger staging area (25,000 m^2) before spawning but occupied a spawning area of similar size (ca. 5000 m^2). In the Caribbean, yellowfin grouper (*Mycteroperca venenosa*) were observed in groups of 50 or more fish briefly occupying (< week) isolated patch reefs up to 5 km away from the spawning aggregation site several weeks before spawning (RSN personal observation).

Transient aggregating species may use staging areas prior to spawning or during the several weeks between monthly spawning peaks (Nemeth 2005; Nemeth et al. 2007; Rhodes and Tupper 2008). Recent acoustic studies have shown that yellowfin grouper and Nassau grouper individuals visit the same spawning site once or twice per year, sometimes more, may actively roam 20–30 km around the spawning site within a 24 h period, and occupy a staging area of at least 15 km^2 (Starr et al. 2007, RSN unpublished data) (Fig. 2.4). The extensive swimming to and from the spawning site may represent directed movements to particular habitats for foraging or regular visits to cleaning stations (Samoilys 1997; Rhodes and Sadovy 2002b; Semmens et al. 2005, 2006; Nemeth et al. 2006b). The daily roaming behaviour of these groupers may also be associated with attracting or leading conspecific adults and/or first-time spawners to the aggregation site; similar behaviours, but at smaller spatial and temporal scales, have been reported for brown surgeonfish (*Acanthurus nigrofuscus*, Acanthuridae) (Mazeroll and Montgomery 1998). For this species certain individuals break away from a migrating group to interact with conspecifics foraging adjacent to the migration route. The conspecific acanthurids stop feeding and join the migrating group which continues to swim toward the spawning aggregation site (Mazeroll and Montgomery 1998). These behaviours may be important for establishing, learning about and maintaining traditional spawning aggregation sites, as shown in the bluehead wrasse, *Thalassoma bifasciatum* (Warner 1988), which may persist for many decades (Colin 1996).

Some species may also undertake vertical migrations during the spawning period. Detailed acoustic data from Glover's Reef Atoll, Belize showed that migrating Nassau grouper used not only shallow coral reef habitat along the shelf (15–35 m) but also deeper (range 50–255 m) outer slopes below the shelf break (Starr et al. 2007); the entire spawning population descended within a few hours to depths greater than 50 m immediately after the second, February, spawning period each year. During the 2 months of deep reef habitation, a limited number of Nassau grouper continued to migrate to and from the aggregation site but all fish remained deeper than 50 m (mean depth 72 m). In April all tagged fish ascended in synchrony within a few hours to shallow reef areas averaging 20 m depth (Starr et al. 2007). It is not clear what these Nassau grouper were doing, but the vertical depth changes may have been to feed on preferred prey, avoid predation or parasite infestations, enter different water strata (i.e. cooler, deeper waters) or to release fertilized eggs in particular ocean strata or currents which enhance larval survival or retention (Semmens et al. 2006; Starr et al. 2007; Nemeth 2009). Further studies and more environmental information are needed to understand the similarities and differences in behaviour among sites and species.

2.2.3 Courtship Arena, Spawning Site

Many investigators have identified areas immediately surrounding spawning aggregation sites where fish density and behaviours related to spawning increase dramatically in the days leading to spawning (Colin et al. 1987; Colin 1992; Carter and Perrine 1994; Rhodes and Sadovy 2002b; Whaylen et al. 2004, 2006; Heyman et al. 2005; Kadison et al. 2006; Nemeth et al. 2006b; Robinson et al. 2008). Although little is understood about these specific areas, they have been given a variety of names such as 'boundary' area, 'aggregation margins', 'areas outside the aggregation core', 'outlying areas' and 'surrounding reef areas' (Samoilys 1997; Rhodes and Sadovy 2002b; Nemeth et al. 2007; Robinson et al. 2008). In an effort to standardize this terminology, the term *courtship arena* is suggested as a potentially useful and descriptive way of identifying and characterizing the unique behaviours, spatio-temporal patterns of movement and reef habitats associated with spawning aggregations.

The courtship arena (Fig. 2.3) may be distinguished from the staging area by observed increases in fish density, courtship behaviours and colouration and/or intra/interspecific interactions which are at their most intense in close proximity to the spawning aggregation site. Several studies have observed that as fish move from the courtship arena to the core spawning site colour changes and male-male aggression may intensify and culminate with spawning rushes and gamete release (Rhodes and Sadovy 2002b; Whaylen et al. 2004; Nemeth et al. 2006b, 2007; Robinson et al. 2008). Whaylen et al. (2004) described and quantified such a shift in colouration and courtship for Nassau grouper during the crepuscular period as spawning adults moved from the reef platform to hover off the shelf edge to commence spawning. A similar pattern has also been observed for yellowfin grouper in the USVI (RSN unpublished data).

The definition of core spawning area depends upon the mode of reproduction of individual species. For species that establish spawning territories and pair spawn at aggregation sites the core area is defined as the area which consistently has the highest densities of aggregating individuals as reported for red hind, brown-marbled grouper, camouflage grouper and yellowmargin triggerfish (*Pseudobalistes flavimarginatus*) (Gladstone 1994; Rhodes and Sadovy 2002b; Nemeth et al. 2007; Robinson et al. 2008). In large aggregations multiple core spawning sites may exist (e.g. red hind, Kadison et al. 2009). For those species that spawn in large groups (i.e. group spawn – a spawning rush of three or more fish such as in Nassau grouper and snappers) the core area is the location where the majority of spawning adults ascend into the water column and spawn and this core area can change within the courtship arena even within a single evening (Sala et al. 2001; Whaylen et al. 2004; Heyman and Kjerfve 2008).

Estimates of the size of courtship arenas come mainly from studies of a few transient aggregating species (triggerfishes-Balistidae, snappers, groupers) and is currently lacking for species forming resident aggregations (surgeonfishes, parrotfishes). For transient aggregating species within snappers and groupers the courtship arena is estimated to be $<10\ km^2$ with spawning usually confined to an area $<1\ km^2$ (Fig. 2.3, see also Appendix 4.2 in Nemeth 2009). For example, Nassau and tiger groupers, and cubera and dog snappers (*Lutjanus cyanopterus* and *L. jocu*) are all transient aggregators which typically spawn along a narrow section (50–100 m) of the shelf edge (Colin 1992; Carter and Perrine 1994; Heyman et al. 2005; Nemeth et al. 2006b; Kadison et al. 2007). Heyman et al. (2005) and Kadison et al. (2006) reported that during the week of spawning cubera snapper could be found in large, highly mobile groups within a $4.5\ km^2$ area of reef (i.e. courtship arena) where they were observed performing a variety of pre-spawning behaviours. But during spawning, cubera snapper were confined to a core spawning area less than $1,000\ m^2$ where they spawned in mid-water (Heyman et al. 2005).

Spawning aggregation territories and habitats: For aggregating species that establish temporary spawning territories and pair-spawn, the courtship arena is estimated to be $0.01–0.4\ km^2$ (Appendix 4.2 in Nemeth 2009). Species known to defend territories within spawning aggregations include red hind, brown-marbled, camouflage, tiger, and leopard coralgroupers, square-tail coralgrouper (*Plectropomus areolatus*) and yellowmargin triggerfish (Shapiro et al. 1993; Gladstone 1994; Sadovy et al. 1994a, b; Samoilys and Squire 1994; Samoilys 1997; Zeller 1998; Rhodes and Sadovy 2002a, b; Nemeth 2005; Nemeth et al. 2007; Rhodes and Tupper 2008; Robinson et al. 2008). For these species, some of which include benthic spawners with a lek-like mating system (i.e. triggerfish, Gladstone 1994), the courtship

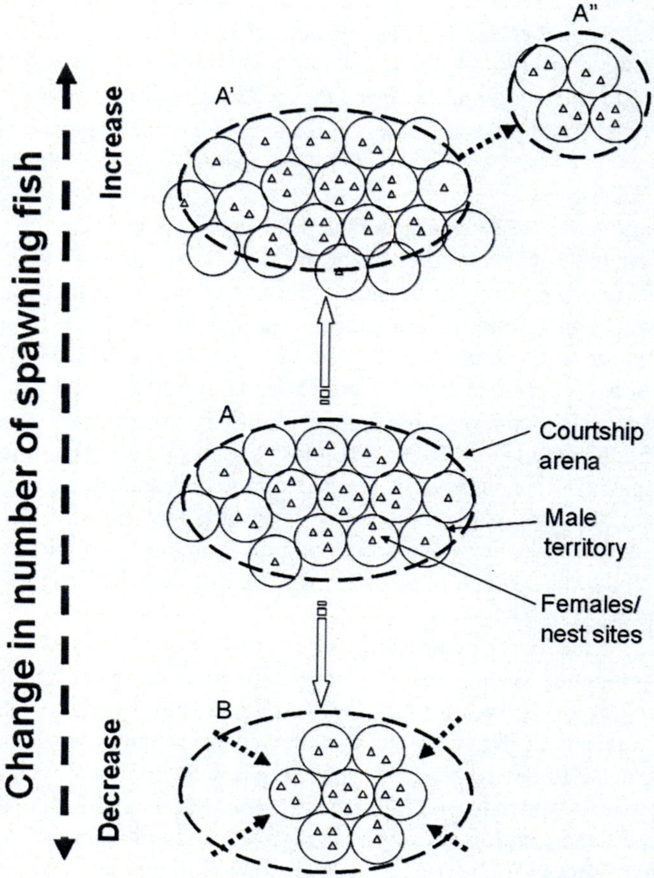

Fig. 2.5 Spatial relationship of male spawning territories (*circles*) within the courtship arena (*A*, *dashed oval*) where males compete for and defend higher quality spawning habitat (i.e. core of courtship arena). Successful males near the core will probably have greater access to females or nest sites (*triangles*) than males in marginal habitats at the edge of the courtship arena. As fish numbers decrease (*B*), from fishing mortality for example, the movement and concentration of the remaining fish into the core (*dashed arrows*) would result in a reduction in the relative size of the courtship arena compared to (*A*). Alternatively, if the number of spawning fish increases through greater recruitment or reduced fishing mortality, competition for spawning territories at the primary spawning site may increase (*A′*). In this case, subordinate males (and females) may establish a satellite or secondary spawning aggregation site nearby (*A″*)

arena and spawning area encompass multiple male spawning territories (Fig. 2.5). Males cruise along territorial boundaries, posturing, challenging, chasing and fighting rival males, displaying breeding colouration and courting females. Females either swim around to inspect nest sites (e.g. triggerfish) or remain hidden under coral heads (i.e. groupers) prior to pair spawning (Shapiro et al. 1993; Gladstone 1994; Sadovy et al. 1994a; Rhodes and Sadovy 2002b; Nemeth et al. 2006b; Pickert et al. 2006;

Nemeth et al. 2007; Rhodes and Tupper 2008; Robinson et al. 2008). Gladstone (1994) noted that dominant yellowmargin triggerfish males occupied the core of the aggregation whereas smaller males occupied apparently marginal spawning habitats on the edge of the spawning area and sometimes failed to attract a mate (Fig. 2.5).

If preferred or suitable spawning habitat is a limited resource, then as the number of fish in an aggregation increases some portion of the spawning population may occupy marginal areas or may find other suitable spawning habitat and establish satellite or 'secondary' aggregation sites (Fig. 2.5). This may explain the presence of one large (primary) spawning site and one or more smaller (secondary) spawning sites reported for red hind and leopard coralgrouper (Samoilys and Squire 1994; Samoilys 1997; Zeller 1998; Kadison et al. 2009). Alternatively, if fishing pressure at a spawning aggregation site increases, satellite aggregations may be the first to be eliminated with fish occupying marginal spawning habitat replacing those removed from the core spawning area. These replacements would maintain spawning fish density at the core of the primary aggregation and may account for a condition known as hyperstability where catch rates remain stable while the total spawning population is declining (Sadovy and Domeier 2005; Chapter 8). Finally, at multi-species spawning aggregation sites some species may have spatially distinct spawning areas whereas others can have overlapping courtship arenas and spawning sites (Fig. 2.6) (Colin and Clavijo 1988; Sala et al. 2001; Rhodes and Sadovy 2002b; Whaylen et al. 2004; Nemeth et al. 2006b; Heyman and Kjerfve 2008). Competition for limited spawning habitat may also be a factor that determines levels of inter-specific aggression and the distribution patterns observed among aggregating species at multi-species spawning sites (Rhodes and Sadovy 2002b; Nemeth et al. 2006b; Robinson et al. 2008).

2.3 Habitat Linkages During Spawning Migrations

Resident and transient aggregating species migrate from feeding sites or home ranges to spawning aggregation sites over various spatial and temporal scales. To illustrate, square-tail coralgrouper adults commonly inhabit patch reefs within lagoons, inner and outer reef slopes and fringing reefs and spawn in reef pass channels, outer reef slopes and promontories of fringing reefs (Myers 1999; Sluka 2001a, b; Pet et al. 2005; Rhodes and Tupper 2008). Seasonal habitat linkages for squaretail coralgrouper can be drawn between lagoon and fringing reefs (feeding habitats) to channels and outer reef slopes (spawning habitats). Hutchinson and Rhodes (2010), who tracked squaretail coralgrouper with acoustic telemetry, provided a specific example of habitat linkages between spawning and home sites. Squaretail coralgrouper aggregated near reef passages along reef flat, wall and slope habitats and lived the remainder of the year on either the fore reef or inside the lagoon in areas of moderate to high coral cover. To better understand the connectivity of aggregating species among habitat types, Nemeth (2009, Appendices 4.1 and 4.2) compiled a list of 13 generalized adult feeding habitats and spawning sites for 134 species of coral reef fishes known or presumed to form resident (n=42) or transient (n=92)

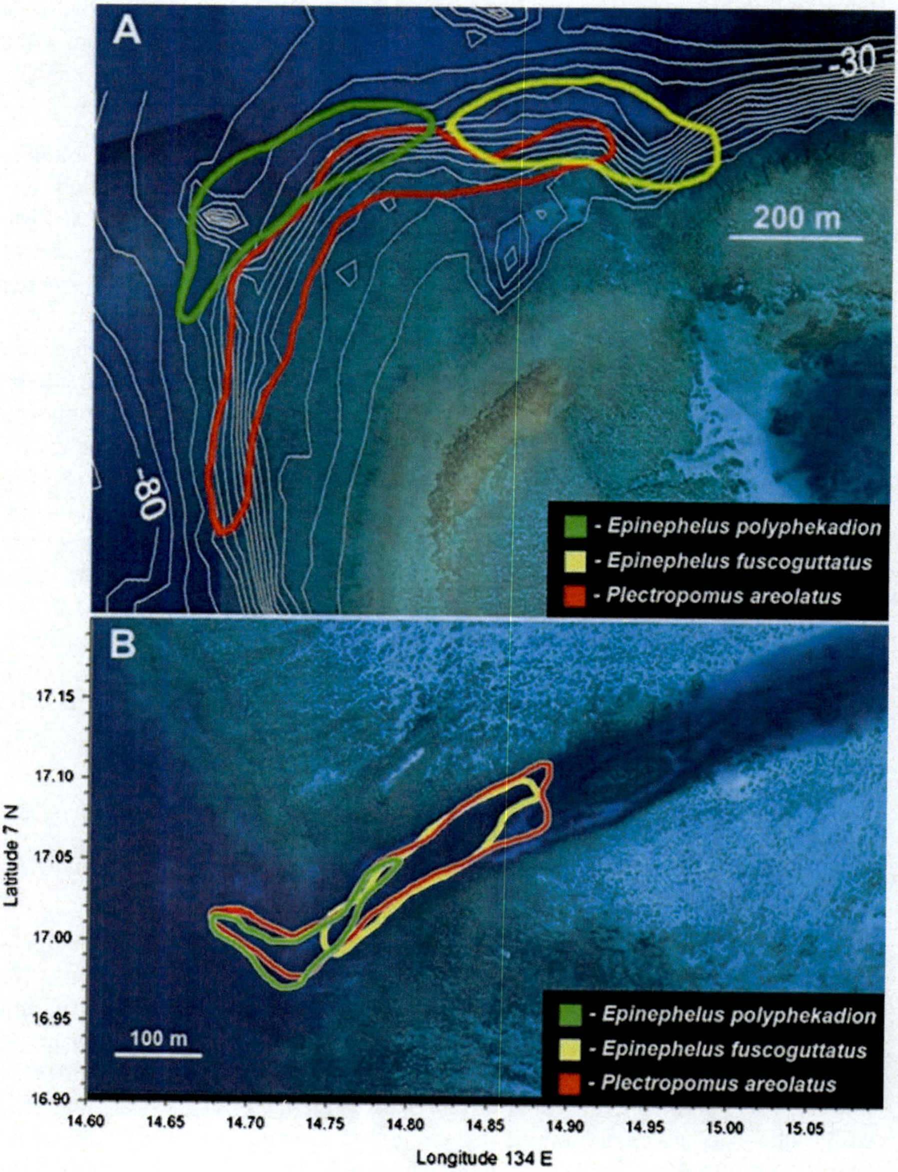

Fig. 2.6 Multiple-species spawning aggregation sites showing outer limits of spawning aggregation areas of camouflage grouper, (*Epinephelus polyphekadion*), brown-marbled grouper (*E. fuscoguttatus*) and squaretail coralgrouper (*P. areolatus*) in two channels in Palau on a single day. Channel A is broad (400 m) and deep (>30 m) while Channel B has a narrow mouth (50 m wide) and shallow (12 m). In Channel A the three groupers have space to occupy their preferred bottom type and depth while in Channel B they are more confined and have strongly overlapping distributions. The outer limits of aggregations can change day to day due to changes in numbers of fishes present (Figure courtesy of P. Colin, A. Bukurrou, S. Kiefer and Y. Sadovy de Mitcheson; Palau Conservation Society 2010)

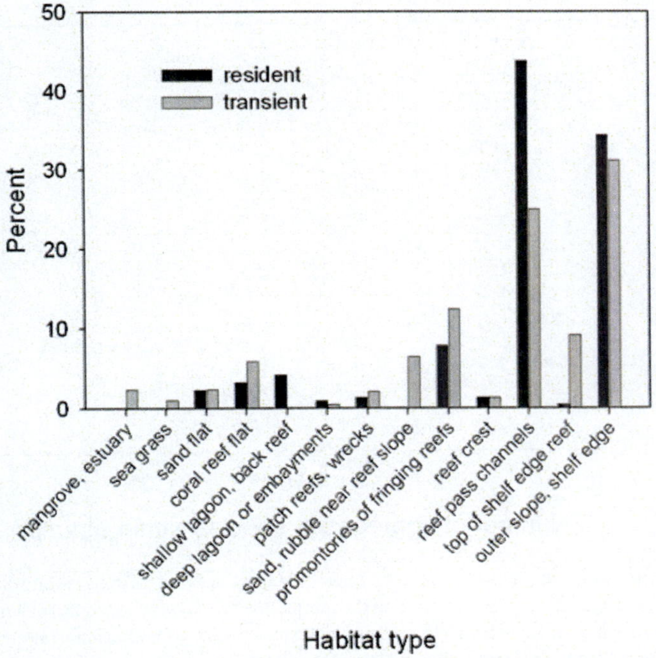

Fig. 2.7 Percent distribution of resident and transient spawning aggregations among 13 spawning habitat types arranged from shallow nearshore to deeper offshore locations (Data summarized from Nemeth (2009, Appendices 4.1 and 4.2))

spawning aggregations. A summary of these data showed that the majority of commercially important aggregating species spawn in three geomorphological types: reef pass channels, promontories along fringing reefs, and outer reef-slope drop-offs (Fig. 2.7), a result that supports the findings of Moyer (1989) and Sadovy de Mitcheson et al. (2008). Aggregating species also spawn in a diversity of other habitat types (Fig. 2.7). Aanalysis of these data showed that the number of habitat types used as spawning aggregation sites within a fish family appears positively related to species richness within the family (Fig. 2.8).

The feeding and spawning habitats described above extend from shallow near-shore to deep offshore locations. Groups of fish migrating between these habitats on a daily, monthly or annual basis for spawning create a complex matrix of movements across the seascape. Information on the intermediate habitats utilized by migrating fishes and the migration routes that are followed to reach spawning sites comes mostly from unvalidated accounts of local fishers (Johannes 1978, 1981; Garcia-Cagide et al. 2001; Claro and Lindeman 2003). A few studies have provided information on migration routes of resident aggregating species which move between feeding habitats and nearby spawning aggregation sites (Warner 1995; Colin 1996; Mazeroll and Montgomery 1998). For example, the migratory pathways of resident spawning surgeonfish occur between shallow inshore feeding areas

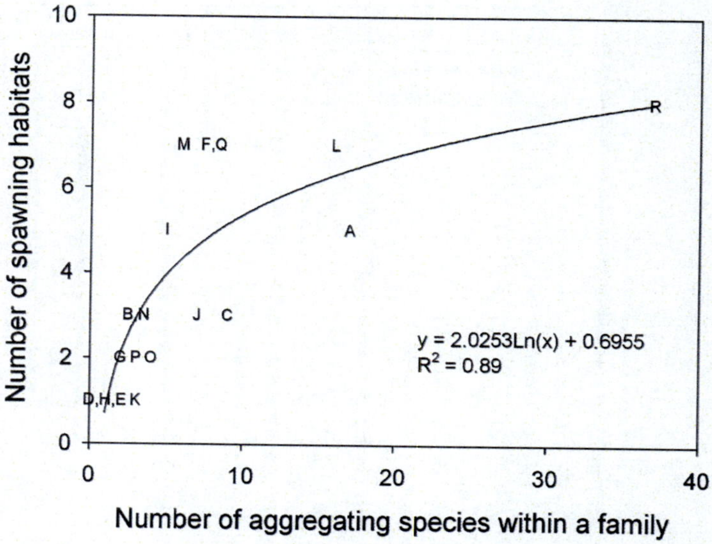

Fig. 2.8 Relationship between the number of aggregating species within 18 families known or suspected to form spawning aggregations and the number of habitat types used as spawning aggregation sites. The fish families included in this analysis were *A* Acanthuridae (surgeonfishes); *B* Balistidae (triggerfishes); *C* Carangidae (jacks); *D* Caesionidae (fusiliers); *E* Elopidae (ladyfishes); *F* Scaridae (parrotfishes); *G* Gerreidae (mojarras); *H* Centropomidae (snooks); *I* Labridae (wrasses); *J* Lethrinidae (emperors); *K* Kyphosidae (chubs); *L* Lutjanidae (snappers); *M* Mugilidae (mullets); *N* Mullidae (goatfishes); *O* Scombridae (mackerals); *P* Sparidae (porgies); *Q* Siganidae (rabbitfishes); *R* Serranidae (seabasses). *Letters* separated with a comma were overlapping. *Line* fitted with non-linear regression was significant at $p<0.0001$

and the nearest part of the reef edge, or fish travel along the reef edge to specific spawning sites (Robertson 1983; Colin 1996; Mazeroll and Montgomery 1998). In a detailed study of brown tang, *Acanthurus nigrofuscus*, Mazeroll and Montgomery (1998) found that migrating fish used at least 15 different routes at the beginning of the spawning season but only 3 routes by the end, suggesting that some socially mediated behaviours may be operating to consolidate pathways. Direct observations of spawning migrations of transient aggregating species are limited to Nassau grouper which was seen migrating in various sized groups along the edges of drop offs to spawning sites in the Bahamas, Belize and Cayman Islands (Colin 1992; Carter et al. 1994; Whaylen et al. 2004).

More recent studies have used acoustic telemetry to document generalized movement patterns of squaretail coralgrouper and red hind which are believed to migrate along similar pathways each year to spawning aggregation sites (Zeller 1998; Nemeth 2005; Rhodes and Tupper 2008). Data on habitat use during migration are more difficult to acquire but can be determined if movement patterns are coupled with detailed benthic habitat maps over the entire catchment area. For example, a recent acoustic tagging study in the USVI found that five of eight tagged yellowfin grouper (63%) and five of nine Nassau grouper (56%) consistently used two deep well-developed linear reefs as their primary migration pathways while traveling to and from their spawning aggregation site (RSN unpublished data). These preliminary data and Fig. 2.8 suggest that ecosystem based management initiatives and MPA networks should not only prioritize protection of key geomorphological reef types (i.e. reef pass channels, shelf edge reefs and promontories) but also consider the relative importance of migratory pathways and other potential spawning habitat types in maintaining high biological diversity, connectivity and ecosystem function.

2.4 Predator-Prey Dynamics and Other Ecological Processes Within Functional Migration Areas

Migrations from home ranges to spawning aggregation sites represent short-term and possibly substantial localized fluctuations in fish biomass, as judged from numbers of fish in some aggregations. Large aggregations of cubera snapper in Belize briefly attain estimated peak abundances of 4,000–10,000 individuals within the 1,000 m^2 spawning aggregation site (Heyman et al. 2005). Spawning adults in Cuba averaged about 80 cm and 8.5 kg (Claro and Garcia-Arteaga 2001) and if fish in Belize are of similar size, the aggregations represent an estimated fish biomass of 3,400–8,500 kg 100 m^{-2}. Several surgeonfish species in the Caribbean and Pacific can form even larger aggregations exceeding an estimated 20,000 individuals (Robertson 1983; Colin and Clavijo 1988). These temporary peaks in fish biomass may affect food web dynamics and energy transfer along migratory pathways and at spawning aggregation sites through feeding, defaecation, predation and reproduction. For example, Hamner et al. (2007b) found that nearly 90% of zooplankton flowing off a Palauan reef during ebb tide was composed of fish eggs from spawning aggregations of a suite of species (mostly surgeonfishes and parrotfishes but also wrasses-Labridae). High concentrations of spawning adults or fish eggs may attract a wide diversity of predators attempting to take advantage of this temporary and predictable food source.

Most available information suggests that while predation on spawning adults may be low, rates of predation on eggs after release are much higher, and can vary considerably depending upon species and location (Robertson 1983; Moyer 1987; Craig 1998; Sancho et al. 2000a; Claydon 2004). Although it is unknown what proportion of gametes contributes to the local food web, the energy gathered through the regular feeding activities of adults across a large functional migration area and exported in the form of millions of fertilized eggs could have a brief, yet important, influence on energy and nutrient dynamics during spawning periods. A comparable system is the synchronized mass spawning of corals in which the sudden increase of organic matter is rapidly assimilated into the local food web and stimulates biological activity in adjacent benthic and pelagic environments (Guest 2008). Herbivorous fishes may be especially important for the transfer of energy from benthic to pelagic environments since they consume primary algal production and convert it to eggs which are exported from the reef during daily spawning episodes (Patrick Colin personal communication).

The possible effect of aggregating species on food web dynamics can be grouped into four categories: (1) feeding by aggregating fish on animals and plants residing at spawning aggregation sites or along migration pathways; (2) predation by piscivores on migrating or spawning adults; (3) egg predation, and (4) other trophic linkages. These will each be discussed below.

2.4.1 Feeding by Aggregating Species at Spawning Aggregation Sites or Along Migration Pathways

Several differences exist between resident and transient aggregations with respect to their possible influence on food-web dynamics at spawning aggregation sites. Resident aggregations tend to be characterized by herbivorous or omnivorous species (i.e. surgeonfish, parrotfish, wrasses, Appendix) that undergo daily short-distance migrations from feeding grounds to spawning sites over prolonged seasonal cycles (Fishelson et al. 1987; Colin and Clavijo 1988). Since resident aggregating species spend only brief periods of time each day at their spawning aggregation site, most of their activity is focused on courtship and spawning and not feeding, as described for striped bristletooth and twotone tangs (*Ctenochaetus striatus* and *Zebrasoma scopas*) (Randall 1961). However, some herbivorous species (e.g. surgeonfish, parrotfish) have been observed milling and grazing on benthic algae at certain points along their spawning migration routes and during the brief periods between spawning episodes at aggregation sites (Randall and Randall 1963; Robertson 1983; Mazeroll and Montgomery 1998; Sancho et al. 2000b). These species may therefore influence the relative abundance of benthic algae and coral cover at aggregation sites and along migration routes.

By contrast, transient aggregators tend to be characterized by carnivorous and piscivorous species (i.e. snappers, groupers) that migrate 10's–100's of kilometers and spend days to weeks around their spawning aggregation sites (Garcia-Cagide et al. 2001; Rhodes and Sadovy 2002a, b; Heyman et al. 2005; Nemeth 2005; Kadison et al. 2007, 2009; Nemeth et al. 2007). During migration and aggregation fish may continue to feed on other fishes and invertebrates thus linking local food webs within their functional migration areas (McCann et al. 2005). Evidence of transient aggregating species feeding on fishes and invertebrates at aggregation sites, however, is limited to a few incidental observations of predation and gut contents of several grouper species. In Palau aggregating squaretail coralgrouper occasionally attacked schools of small fusiliers (Caesionidae) and brown-marbled grouper was observed eating a camouflage grouper which aggregated at the same site (Johannes et al. 1999). During spawning aggregations of red hind in the Virgin Islands, disgorged gut contents commonly contained true crabs, hermit crabs, juvenile lobster and small reef fishes such as the slender filefish, *Monacanthus tuckeri*, and even juvenile red hind (RSN unpublished data). Yellowfin grouper, *Mycteroperca venenosa*, stomach contents also contained a variety of reef fishes such as yellowtail snapper, *Ocyurus chrysurus*, blackfin snapper, *Lutjanus buccanella*, red hind, and Creole wrasse, *Clepticus parrae*, whereas Nassau grouper guts contained mainly crustaceans (e.g. *Justitia longimanus*) but also Creole wrasse (RSN unpublished data).

Not all species necessarily feed while aggregated but may use staging areas for maintenance. Individuals of squaretail coralgrouper in Palau, the males of which defend temporary spawning territories, become thin during the spawning season suggesting that some are not feeding or not feeding enough (Yvonne Sadovy de Mitcheson personal communication). Transient aggregating species may spend brief periods at aggregation sites but individual fish may spend several weeks to months between spawning peaks feeding in staging area habitats (Figs. 2.3 and 2.4) that surround the aggregation site (Starr et al. 2007). More detailed studies of feeding habits by aggregating species during migration and at FSA sites are needed to understand if, when or where feeding takes place and how important local prey populations are for aggregating species and their subsequent reproductive output.

2.4.2 Predation by Piscivores on Spawning Adults

Potential predators on fish in spawning aggregations are composed of a diverse group of which 22 species within 13 families attacked and occasionally consumed spawning adults (Table 2.1). Predation attempts are more commonly reported from resident aggregating species than from transient aggregators (14 vs. 2 species listed in Table 2.1) and evidently more common in the Indo-Pacific (Robertson 1983; Sancho et al. 2000a) than in the Caribbean (Colin and Clavijo 1988; Robertson et al. 1999). The higher number of observations of predation attempts on resident aggregating species probably results from their extended spawning seasons (Chap. 5) during which they spawn nearly every day during daylight hours, when visibility is good for behavioural observations. Transient aggregating species, on the other hand, typically spawn over short time periods at sunset or during the night, and are often associated with deeper water or challenging diving conditions, making direct observations by divers difficult (Colin et al. 2003). Transient aggregating species are also larger in body size making them vulnerable to only the largest of piscivores (Juanes 1994). Reports of sharks attacking transient spawning aggregations appear to be most commonly associated with fishing-induced predation where sharks and other large predators attack groupers that are hooked on fishing lines (Olsen and LaPlace 1978; Nemeth 2005; Matos-Caraballo et al. 2006; Heyman and Kjerfve 2008) (Chaps. 5 and 7).

Evidence for predation on spawning adults comes from direct observations of successful strikes and predation attempts or from an increase in the number of potential predators at FSA sites when fish are aggregated compared to non-aggregation periods. Although observations on successful predation attempts are rare (Colin and Clavijo 1988), a few studies have provided quantitative data of predation on aggregating species (Chaps. 5 and 7). These studies indicate that when predation attempts were observed, most piscivores attacked during the spawning rush, with an overall

Table 2.1 Piscivorous fishes reported to attack and occasionally consume adults at spawning aggregation sites located in the Caribbean (C) and Pacific (P)

Predators on adults	ACANTHURIDAE	*Acanthurus guttatus*	*A. lineatus*	*A. nigrofuscus*	*A. nigroris*	*A. triostegus*	*Ctenochaetus striatus*	*C. strigosus*	*Naso literatus*	*Zebrasoma flavescens*	LABRIDAE	*Thalassoma cupido*	MUGILIDAE	*Crenimugil melampygus*	SCARIDAE	*Chlorurus sordidus*	*Scarus psittacus*	*Sparisoma rubripinne*	SERRANIDAE	*Epinephelus guttatus*	*E. polyphekadion*
AULOSTOMIDAE				P																	
Aulostomus chinensis[f]												P									
BELONIDAE				P																	
CARANGIDAE																					
Caranx melampygus[a, d, g, h]		P		P	P	P	P	P		P						P	P				
Scomberomorus cavalla[c]																		C			
CARCHARHINIDAE																					
Carcharhinus amblyrhynchos[b]									P					P							
C. melanopterus[a]				P																	
LABRIDAE																					
Cheilio inermis[f]												P									
LUTJANIDAE																					
Aphareus furca[d]					P			P		P						P	P				
Lutjanus bohar[a, h]			P	P																	
L. cyanopterus[e]																				C	
MURAENIDAE																					
Gymnothorax eurostus[f]												P									
G. moringa[e]																				C	

RHINCODONTIDAE							
Ginglymostoma cirratum[e]						C	
SCOMBRIDAE							
Gymnosarda unicolor[h]	P						
SCORPAENIDAE							
Scorpaenopsis cirrhosa[f]				P			
Sebasticus marmoratus[f]				P			
SERRANIDAE			P				
Cephalopholis argus[a]		P					
Epinephelus fasciatus[f]				P			
E. fuscoguttatus[b]							P
SPHYRAENIDAE							
Sphyraena barracuda[c]					C		
SYNODONTIDAE							
Synodus ulae[f]				P			

[a]Robertson (1983)
[b]Johannes et al. (1999)
[c]Randall and Randall (1963)
[d]Sancho et al. (2000a)
[e]RSN unpublished data
[f]Moyer (1987)
[g]Sancho (2000)
[h]Craig (1998)

success rate of 2–4% (Moyer 1987; Sancho 2000; Sancho et al. 2000a). The possible impact of predation on spawning aggregations can be calculated for at least one species. At Johnston Atoll, Sancho et al. (2000a) found that piscivores (bluefin trevally *Caranx melampygus* and small jobfish *Aphareus furca*) successfully consumed spawning bullethead parrotfish (*Chlorurus sordidus*) at about 0.1 fish per hour. Since bullethead parrotfish spawns about 1 h each day (Sancho et al. 2000b), this equals a predation rate of about three fish per month. The maximum spawning population size observed by Sancho et al. (2000b) at Johnston Atoll was 300 individuals, so piscivores may consume about 1% of the bullethead parrotfish spawning population each month or 12% each year during spawning episodes.

Other non-quantitative studies report an increase in the number of predators at spawning aggregation sites during the spawning period including sharks, large pelagics (marlin-Istiophoridae, tunas and wahoo-Scombridae), large green moray eels and snappers (Nemeth 2005; Heyman and Kjerfve 2008). One recent acoustic tagging study in the USVI found that lemon sharks (*Negaprion brevirostris*) frequented red hind, yellowfin grouper and Nassau grouper spawning sites during the spawning season but were mostly absent the remainder of the year (Brad Wetherbee, Mahmood Shivji and RSN unpublished data). These observations suggest that feeding patterns of piscivores may be synchronized with the seasonal spawning activity of transient aggregations. However, the importance of spawning aggregations to the overall diet of predators that target spawning adults requires further investigation.

2.4.3 Egg Predation at Spawning Aggregations

Evidence for predation on newly released eggs or sperm at spawning aggregation sites comes from several quantitative studies, and direct observations including an observed increase in the number of potential oophagous predators. A review of the literature found that 35 species within 13 families were egg predators with the majority of observations coming from the Pacific (Table 2.2). In most cases, egg predators rapidly swim to the centre of a visible gamete cloud and pick at newly released eggs or swim through gamete clouds and ram filter feed (i.e. bigmouth mackerel *Rastrelliger kanagurta* and whale shark *Rhincodon typus*) (Colin 1976; Moyer 1987; Sancho et al. 2000a; Heyman et al. 2001).

The percent of spawning events attacked by oophagous predators was calculated for a variety of species and varied from <1% to >40% (Robertson 1983; Moyer 1987; Samoilys 1997; Sancho et al. 2000a). A number of factors contributed to this variability including spawning location, the species aggregating, spawning mode and time of day. For example, in Palau Robertson (1983) found that egg predation was similar among pair-spawning bristletooth tang and brown surgeonfish (<5%) but was considerably higher during group-spawns in these two species (27% and 42%, respectively). Egg predation rates for pair-spawning leopard coralgrouper (10–27%) and group-spawning cupid wrasse, *Thalassoma cupido*, (42%) also fell within this range (Moyer 1987; Samoilys 1997). While there are no estimates of the

Table 2.2 Planktivorous fishes reported to feed on eggs of aggregating species at spawning sites located in the Caribbean (C) and Pacific (P)

Predators on eggs	ACANTHURIDAE	*Acanthurus guttatus*	*A. lineatus*	*A. nigrofuscus*	*A. nigroris*	*A. triostegus*	*Ctenochaetus striatus*	*Zebrasoma scopas*	CAESIONIDAE	*Caesio teres*	LABRIDAE	*Thalassoma cupido*	*T. quinquevittatum*	LUTJANIDAE	*Lutjanus cyanopterus*	*L. jocu*	MULLIDAE	*Parupeneus bifasciatus*	SCARIDAE	*Chlorurus sordidus*	*Scarus psittacus*	SERRANIDAE	*Mycteroperca venenosa*	*Plectropomus leopardus*
BALISTIDAE																								
Melichthys niger					P													P		P	P			
M. vidua[d, i]			P	P	P		P	P										P		P	P			
CAESIONIDAE[h]																								P
Caesio coerulaureus[a]			P	P			P	P																
C. erythrogaster[a]			P	P			P	P																
C. lunaris[a]			P	P			P	P																
Pterocaesio chrysozonus[a]			P	P			P	P																
CARANGIDAE																								
Elagatis bipinnulata[b, c, e]															C	C							C	
DUSSUMIERIDAE[a]			P	P			P	P																
EPHIPPIDAE																								
Chaetodipterus faber[b, c]															C	C								
EXOCOETIDAE[a]			P	P			P	P																
HEMIRAMPHIDAE[g]										P														
LABRIDAE																								
Thalassoma hardwicki[a, i]			P	P			P	P																
T. amblycephalus[a]			P	P			P	P																
T. lunare							P																	

(continued)

Table 2.2 (continued)

Predators on eggs	ACANTHURIDAE	*Acanthurus guttatus*	*A. lineatus*	*A. nigrofuscus*	*A. nigroris*	*A. triostegus*	*Ctenochaetus striatus*	*Zebrasoma scopas*	CAESIONIDAE	*Caesio teres*	LABRIDAE	*Thalassoma cupido*	*T. quinquevittatum*	LUTJANIDAE	*Lutjanus cyanopterus*	*L. jocu*	MULLIDAE	*Parupeneus bifasciatus*	SCARIDAE	*Chlorurus sordidus*	*Scarus psittacus*	SERRANIDAE	*Mycteroperca venenosa*	*Plectropomus leopardus*
LUTJANIDAE																								
Lutjanus analis[e]																							C	
L. jocu[e]																							C	
Macolor niger[i, j]		P	P			P	P						P											
Ocyurus chrysurus[b, c]															C	C								
POMACANTHIDAE																								
Centropyge interruptus[f]												P												
C. tibicen[f]												P												
POMACENTRIDAE																								
Abudefduf saxatilis[a]			P	P			P	P																
Acanthochromis polyacanthus[i]							P																	
Ambliglyphidodon curacau[a, i]			P	P			P	P																
A. leucogaster[i]							P																	
Amphiprion clarkii[f]												P												
Chromis atripectoralis[a]			P	P			P	P																
C. caerulea[a]			P	P			P	P																
C. chrysura[f]												P												
C. flavomaculata[f]												P												

C. viridis[i]			P				
Dascyllus trimaculatus[f]					P		
Pomacentrus coelestis[f]					P		
P. nagasakiensis[f]					P		
RHINCODONTIDAE							
Rhincodon typus[b, c]						C	C
SCOMBRIDAE							
Rastrelliger kanagurta[a, i]	P	P	P	P			

[a]Robertson (1983)
[b]Heyman et al. (2001)
[c]Heyman et al. (2005)
[d]Sancho et al. (2000a)
[e]RSN unpublished data
[f]Moyer (1987)
[g]Bell and Colin (1986)
[h]Samoilys (1997)
[i]Claydon (2005)
[j]Craig (1998)

amount of eggs eaten from individual spawning events, Moyer (1987) observed that egg predators often remained within a gamete cloud for more than a minute and continued to feed on the remaining eggs even though other spawning clouds of *T. cupido* were released nearby. In the Caribbean, a similar suite of egg predators was reported to feed directly on gametes or increased in numbers at spawning aggregation sites, e.g. Belize, Cayman Islands, Mexico, Puerto Rico and the US Virgin Islands (Colin and Clavijo 1988; Aguilar-Perera and Aguilar-Davila 1996; Heyman et al. 2001; Whaylen et al. 2006). The majority of egg predators may opportunistically feed on newly released eggs at aggregation sites, while other species target these temporary concentrations of planktonic food. This appears to be the case for whale sharks which may undergo annual migrations on a regional scale to feed on gamete clouds of snappers, jacks and mackerels (Heyman et al. 2001, 2005; Graham and Castellanos 2005; Hoffmayer et al. 2007). The importance of these temporary food sources is unknown, but the synchronized movements of whale sharks and possibly other planktivorous species to active spawning sites highlights a possible ecosystem level component of spawning aggregations.

2.4.4 Other Trophic Linkages

Other potentially important but unstudied trophic interactions at spawning aggregation sites include host-parasite relationships, the use of cleaning stations by aggregating species and the effects of defaecation on benthic communities. During the spawning season fish spend considerable amounts of energy for migration, courtship, territoriality and spawning. Semmens et al. (2006) found that infestation by the isopod parasite *Excorallana tricornis tricornis* on Nassau grouper increased immediately following spawning. Similar infestations were also observed on red hind at an aggregation on St. Croix (RSN personal observation, Tom Daley personal communication). Semmens et al. (2006) suggested that energy expense and physiological stress during reproduction may expose Nassau grouper to higher levels of parasitism and therefore visiting cleaning stations during spawning may be particularly important for aggregating groupers (Nemeth et al. 2006b). Finally, defaecation by aggregating species is a potentially important source of nutrients into a reef system surrounding a spawning aggregation site. Faecal material may provide nutrients that enhance the growth of corals at aggregation sites (Meyer et al. 1983; Meyer and Schultz 1985) or may directly contribute to local productivity through feeding by coprophagus fishes and other detritivorous organisms (Bailey and Robertson 1982; Robertson 1982).

2.5 Approaches to Study and Management of FSA's at the Ecosystem Level

Identifying and mapping the catchment areas of spawning populations and the various ecological components within the functional migration area of aggregating species is an important step towards ecosystem based management (EBM) of multi-species tropical fisheries. Many aggregating fishes include top carnivorous and numerous herbivorous species with high commercial and ecological value, and may play an important role in ecosystem function and fisheries economics. The general concepts behind EBM are to (1) maintain environmental quality and ecosystem productivity while achieving long-term socioeconomic benefits, (2) minimize the effects of human activities which may cause irreversible change to habitat structure, species assemblages and ecosystem processes, (3) gather information on key ecological components and processes that are relevant to maintaining ecosystem integrity and resilience, and (4) adopt a precautionary approach to fisheries management when data are limited that favour conservation and sustainability principles rather than maximizing catch (Garcia et al. 2003; Pikitch et al. 2004; Appeldoorn 2008). Each of these concepts is directly relevant to the management and maintenance of species which form spawning aggregations and the aggregation sites utilized by these species.

Guidelines for EBM are meaningful for managing aggregating species so long as the relevant spatial scales of management are identified. EBM usually operates at intermediate geographic scales of a few to 1,000's of km^2 (Garcia et al. 2003), which encompasses most of the relevant biological processes and habitats used by aggregating species. The spatial parameters defined in Sect. 2.1 above may therefore provide an appropriate guide for EBM. In Fig. 2.3 the catchment area of a spawning aggregation includes the home ranges of all mature adults spawning at a single site but the trophic linkages associated with fish movements probably occur at smaller spatial scales. The area and specific habitats occupied by the greater majority of the spawning population define the Functional Migration Area whereas the long-distance migrants define the catchment area. In some locations a species may have multiple spawning aggregation sites and overlapping catchment areas whereas on small isolated islands they may have a single spawning aggregation site. In either case, replenishment of the spawning population will come from a variable supply of larvae from self-recruitment and/or larvae spawned elsewhere. Since some species are known to display strong site fidelity (Sadovy et al. 1992, 1994b; Luckhurst 1998; Zeller 1998), persistence of an aggregation under heavy fishing pressure may be related to the relative proportion of larvae from upstream sources. To increase population resilience to fishing and environmental change, EBM will need to take these factors into consideration and be applied at spatial scales appropriate to aggregating species and local geophysical constraints (i.e. island shelf areas).

Other management considerations include evaluating the role of aggregating species in ecosystem stability, inclusion of shelf edge reefs, promontories of fringing reefs and reef pass channels in MPA design and analyzing the structure of food webs and trophic linkages within functional migration areas to determine the reciprocal influence of aggregating species on ecosystem processes (Cury et al. 2003). Given the many gaps in ecological data of aggregating species, their vulnerability to overfishing and their possible inability to recover from extirpation, managers should take a precautionary approach and carefully monitor, limit or prohibit fishing on all spawning aggregation sites to minimize the risk of stock collapse (Pikitch et al. 2004; Sadovy and Domeier 2005; Sadovy de Mitcheson et al. 2008). FSA's are likely to be essential to the persistence of many fish populations and their maintenance important for managing sustainable fisheries (Nemeth 2005) (Chap. 11).

2.6 Conclusions and Future Directions

Throughout tropical reef ecosystems a wide diversity of fishes, with a range of aggregating behaviours, migrate to specific spawning aggregation sites. Some species are solitary carnivores that are widely dispersed across 100's–1,000's km^2 and undergo long distance spawning migrations only once per year. At the other extreme are herbivores that live in foraging groups and migrate short distances to spawning aggregation sites on a daily basis. Fishes that aggregate to spawn provide an important and largely overlooked ecological component of connectivity within marine ecosystems that needs to be incorporated into fishery management, EBM and MPA design.

Typically, connectivity is associated with the larval phase but aggregating species provide examples whereby adult connectivity may also be an important part of ecosystem dynamics. Spawning migrations represent short-term yet substantial fluctuations in fish biomass that increase the potential for significant trophic interactions and energy transfer from feeding grounds to spawning sites, and beyond, through the release of eggs and predator-prey encounters. Although a large body of anecdotal information exists on this topic, the most dramatic example of the influence of spawning aggregations on ecosystem-level processes is the synchronized convergence of whale sharks in Belize and other locations apparently for the purpose of feeding on fish spawn at spawning aggregation sites (Heyman et al. 2001; Hoffmayer et al. 2007). Multiple-species spawning aggregation sites in particular are important cross-roads of marine animal migrations and represent major nodes of biological diversity and reproductive potential.

The area within which species migrate represents each spawning aggregation's functional migration area and includes important migration pathways and habitats used exclusively during the spawning season. Since the functional migration area encompasses most of the relevant biological processes and habitats used by aggregating species it may also provide an appropriate guide for defining the relevant geographical boundaries necessary to implement EBM objectives. Nested within the functional migration areas are other spatially distinct components of some aggregating species that are equally relevant to fisheries management, including catchment area, staging area, courtship arena and the core spawning site. Detailed information on these spatial components is beginning to emerge for a few groupers and snappers but is lacking for most other species. Based on limited data, the catchment area encompasses the home ranges of an aggregation's spawning adults during the annual reproductive season and defines the minimum area of management for a particular species. The staging area may be important for aggregating groupers and snappers immediately before and after spawning for feeding, visiting cleaning stations or for some other purpose. Staging areas that contain healthy and biologically diverse reef systems and intact prey populations may be an important component of reproductive success for transient aggregating species which may spend weeks or months away from home sites, yet they might not be immediately obvious as associated with an FSA. Finally, protecting the areas used for courtship and spawning from habitat damage, and the direct and indirect impacts of fishing activities is critical for the long-term sustainability of spawning aggregations.

How these spatially distinct components are incorporated into management plans will depend upon management objectives, priorities and capabilities. For example, for a species whose population is being fished sustainably (either because fishing effort is low or the fishery is well-managed), protecting the core spawning site and courtship arena during the spawning season may be adequate. For species or populations that are heavily fished, incorporating the staging area into a no-fishing zone may be the most appropriate course of action. If data are available, protecting the key migration pathways, especially if these are a target of fishing, may be needed to further reduce fishing mortality in areas surrounding the FSA site. This strategic approach will also minimize the economic impact on local fishing communities. However, if the status of a local fish population is unknown or a species is overfished, a precautionary approach is needed. Establishing larger protected areas, or setting zero catch limits for multiple years may be required to help rebuild their populations. Unfortunately, detailed spatial data for most aggregating species are not available so more general and conservative approaches will need to be implemented. These may include protecting aggregating species during spawning using seasonal market closures, or broadly protecting specific habitats known to include spawning aggregations sites (i.e. reef channels, drop offs, promontories) (Chaps. 5 and 10) (Sadovy de Mitcheson et al. 2008).

Although the geographical boundaries of highly mobile marine organisms are typically difficult to determine (Pittman and McAlpine 2003), mapping the extent of fish spawning migrations and their natural boundaries is important for understanding habitat use and distribution patterns of aggregating species and would address a critical gap in knowledge for effective implementation of EBM. Because migration distances and catchment areas for aggregating species may vary depending upon the size of the insular shelf, detailed tagging and acoustic studies are needed across a broad range of species and geographical areas to identify movement patterns and habitats that are important during the reproductive season (i.e. spawning sites and migration pathways). These data should be integrated with bathymetric features and oceanographic patterns surrounding spawning aggregation sites (Colin et al. 2003; Heyman et al. 2007) to better understand the importance of adult migration, larval retention mechanisms and dispersal patterns on the connectivity of reef fish populations (Eggleston 1995; Colin et al. 1997; Dahlgren and Eggleston 2001; Whaylen et al. 2006; Nemeth et al. 2008; Cherubin et al. 2011). Moreover, incorporating these empirical data into population models can help to predict the resilience of reef fish populations exposed to increasing fishing pressure, habitat degradation or climate change, and assist in future management decisions and MPA planning (Botsford et al. 2009).

The need for more detailed information on, and effective management of, spawning aggregations is without question. Worldwide nearly half of all known FSA's are in decline or have disappeared and an equal number of aggregations have no information available as to their status (Sadovy de Mitcheson et al. 2008). Many aggregating species play an unknown, yet potentially important, role in ecosystem function. For example, the Nassau grouper was a major commercial species throughout the

Caribbean until aggregation fishing nearly eliminated it from most locations throughout its range (Olsen and LaPlace 1978; Sadovy 1997; Sala et al. 2001; Aguilar-Perera 2006). The importance of this single species for maintaining biological diversity and indirectly structuring food webs (Stallings 2008), potentially controlling invasive species such as the Indo-Pacific lionfish (Maljković et al. 2008) and (historically) in regional fisheries, highlights the need to better understand aggregating species within coral reef ecosystems and the need for their management under prevailing uncontrolled commercial exploitation.

Acknowledgements The author appreciates the fruitful discussions with the editors which greatly improved this chapter. Writing of this chapter was supported in part by the University of the Virgin Islands' Center for Marine and Environmental Studies, the NSF VI-EPSCoR program (NSF award # 346483 & 0814417) and Friday Harbor Laboratory's Helen Riaboff Whiteley Center. This is contribution # 61 for the Center for Marine and Environmental Studies at UVI.

References

Aguilar-Perera A (2006) Disappearance of a Nassau grouper spawning aggregation off southern Mexican Caribbean coast. Mar Ecol Prog Ser 327:289–296

Aguilar-Perera A, Aguilar-Davila W (1996) A spawning aggregation of Nassau grouper *Epinephelus striatus* (Pisces: Serranidae) in the Mexican Caribbean. Environ Biol Fishes 45:351–361

Appeldoorn RS (2008) Transforming reef fisheries management: application of an ecosystem-based approach in the USA Caribbean. Environ Conser 35:232–241

Ault JS, Smith SG, Luo J, Monaco ME, Appeldoorn RS (2008) Length-based assessment of sustainability benchmarks for coral reef fishes in Puerto Rico. Environ Conser 35:221–231

Bailey TG, Robertson DR (1982) Organic and caloric levels of fish feces relative to its consumption by coprophagus reef fishes. Mar Biol 69:45–50

Beets J, Muehlstein L, Haught K, Schmitges H (2003) Habitat connectivity in coastal environments: patterns and movements of Caribbean coral reef fishes with emphasis on bluestriped grunt, *Haemulon sciurus*. Gulf Caribb Res 14:29–42

Bell LJ, Colin PL (1986) Mass spawning of *Caesio teres* (Pisces: Caesionidae) at Enewetak Atoll, Marshall Islands. Environ Biol Fishes 15:69–74

Botsford LW, White JW, Coffroth M-A, Paris CB, Planes S, Shearer TL, Thorrold SR, Jones GP (2009) Connectivity and resilience of coral reef metapopulations in marine protected areas: matching empirical efforts to predictive needs. Coral Reefs 28:327–337

Carter HJ, Perrine D (1994) A spawning aggregation of dog snapper, *Lutjanus jocu*, (Pisces: Lutjanidae) in Belize, Central America. Bull Mar Sci 55:228–234

Carter JG, Marrow J, Pryor V (1994) Aspects of the ecology and reproduction of Nassau grouper, *Epinephelus striatus*, off the coast of Belize, Central America. Proc Gulf Caribb Fish Inst 43:65–111

Cherubin LM, Nemeth RS, Idrisi I (2011) Flow and transport characteristics at an *Epinephelus guttatus* (red hind grouper) spawning aggregation site in St. Thomas (US Virgin Islands). Ecological Modelling In Press (doi:10.1016/j.ecolmodel.2011.05.031)

Claro R, Garcia-Arteaga JP (2001) Growth patterns of fishes of the Cuban shelf. In: Claro R, Lindeman K, Parenti LR (eds) Ecology of the marine fishes of Cuba. Smithsonian Institution Press, Washington, DC

Claro R, Lindeman KC (2003) Spawning aggregation sites of snapper and grouper species (Lutjandidae and Serranidae) on the insular shelf of Cuba. Gulf Caribb Res 14:91–106

Claro R, Baisre JA, Lindeman KC, Garcia-Arteaga JP (2001) Cuban fisheries: historical trends and current status. pp 194–219 In: Claro R, Lindeman KC, Parenti LR, ed. Ecology of the Marine Fishes of Cuba. Washington DC, Smithsonian Institution Press

Claydon J (2004) Spawning aggregations of coral reef fishes: characteristics, hypotheses, threats and management. Oceanogr Mar Biol Annu Rev 42:265–302

Claydon JAB (2005) The structure and dynamics of spawning aggregations of coral reef fish. James Cook University, School of Marine Biology and Aquaculture. PhD dissertation, 180 pp

Colin PL (1976) Filter-feeding and predation of the eggs of *Thalassoma* sp. by the scombrid fish *Rastrelliger kanagurta*. Copeia 1976:596–597

Colin PL (1992) Reproduction of the Nassau grouper, *Epinephelus striatus*, (Pisces: Serranidae) and its relationship to environmental conditions. Environ Biol Fishes 34:357–377

Colin PL (1996) Longevity of some coral reef fish spawning aggregations. Copeia 1996:189–192

Colin PL, Clavijo IE (1988) Spawning activity of fishes producing pelagic eggs on a shelf edge coral reef, southwestern Puerto Rico. Bull Mar Sci 43:249–279

Colin PL, Shapiro DY, Weiler D (1987) Aspects of the reproduction of two groupers, *Epinephelus guttatus* and *E. striatus,* in the West Indies. Bull Mar Sci 40:220–230

Colin PL, Laroche WA, Brothers EB (1997) Ingress and settlement in the Nassau grouper, *Epinephelus striatus* (Pisces: Serranidae), with relationship to spawning occurrence. Bull Mar Sci 60:656–667

Colin PL, Sadovy YJ, Domeier ML (2003) Manual for the study and conservation of reef fish spawning aggregations. Report No. Special Publication No. 1 (Version 1.0), Society for the Conservation of Reef Fish Aggregations, www.SCRFA.org

Cowen RK (2002) Larval dispersal and retention and consequences for population connectivity. In: Sale PF (ed) Coral reef fishes: dynamics and diversity in a complex ecosystem. Academic, London

Cowen RK, Kamazima LMM, Sponaugle S, Paris CB, Olsen DB (2000) Connectivity of marine populations: open or closed. Science 287:857–859

Cowen RK, Paris CB, Srinivasan A (2006) Scaling of connectivity in marine populations. Science 311:522–527

Craig PC (1998) Temporal spawning patterns of several species of surgeonfishes and wrasses in American Samoa. Pac Sci 52:35–39

Cury P, Shannon L, Shin YJ (2003) The functioning of the marine ecosystems: a fisheries perspective. In: Valdimarsson IMSG (ed) Responsible fisheries in the marine ecosystem. FAO and CAB International, Rome/Wallingford

Dahlgren CP, Eggleston DB (2001) Spatio-temporal variability in abundance, size and microhabitat associations of early juvenile Nassau grouper *Epinephelus striatus* in an off-reef nursery system. Mar Ecol Prog Ser 217:145–156

Domeier M (2005) Methods for the deployment and maintenance of an acoustic tag tracking array: an example from California's Channel Islands. Mar Tech Soc J 39:74–80

Domeier ML, Colin PL (1997) Tropical reef fish spawning aggregations: defined and reviewed. Bull Mar Sci 60:698–726

Eggleston DB (1995) Recruitment in Nassau grouper (*Epinephelus striatus*): post-settlement abundance, microhabitat features, and ontogenetic habitat shifts. Mar Ecol Prog Ser 124:9–22

Fishelson L, Montgomery WL, Myberg AAJ (1987) Biology of surgeonfish *Acanthurus nigrofuscus* with emphasis on changeover in diet and annual gonadal cycles. Mar Ecol Prog Ser 39:37–47

Garcia S, Zerbi A, Aliaume C, Do Chi T, Lasserre G (2003) The ecosystem approach to fisheries. Issues, terminology, principles, institutional foundations, implementation and outlook. Fisheries and Agriculture Organization, Rome

Garcia-Cagide A, Claro R, Koshelev BV (2001) Reproductive patterns of fishes of the Cuban shelf. In: Claro R, Lindeman KC, Parenti LR (eds) Ecology of the marine fishes of Cuba. Smithsonian Institution Press, Washington, DC

Gladstone W (1994) Lek-like spawning, parental care and mating periodicity of the triggerfish *Pseudobalistes flavimarginatus* (Balistidae). Environ Biol Fishes 39:249–257

Graham RT, Castellanos DW (2005) Courtship and spawning of carangid species in Belize. Fish Bull 103:426–432

Guest J (2008) How reefs respond to mass coral spawning. Science 320:621–623

Hamner RM, Freshwater DW, Whitfield PE (2007a) Mitochondrial cytochrome b analysis reveals two invasive lionfish species with strong founder effects in the western Atlantic. J Fish Biol 71:214–222

Hamner WM, Colin PL, Hamner PP (2007b) Export-import dynamics of zooplankton on a coral reef in Palau. Mar Ecol Prog Ser 334:83–92

Heemstra PC, Randall JE (1993) Groupers of the world (family Serranidae, subfamily Epinephelinae). An annotated and illustrated catalogue of the grouper, rockcod, hind, coral grouper and lyretail species known to date. FAO Fisheries Synopsis, FAO species catalogue 16

Heyman WD, Kjerfve B (2008) Characterization of transient multi-species reef fish spawning aggregations at Gladden Spit, Belize. Bull Mar Sci 83:531–551

Heyman WD, Graham RT, Kjerfve B, Johannes RE (2001) Whale sharks *Rhincodon typus* aggregate to feed on fish spawn in Belize. Mar Ecol Prog Ser 215:275–282

Heyman WD, Kjerfve B, Graham RT, Rhodes KL, Garbutt L (2005) Spawning aggregations of *Lutjanus cyanopterus* (Cuvier) on the Belize Barrier Reef over a 6 year period. J Fish Biol 67:83–101

Heyman WD, Ecochard J-LB, Biasi FB (2007) Low-cost bathymetric mapping for tropical marine conservation – a focus on reef fish spawning aggregation sites. Mar Geodesy 30:37–50

Hoffmayer ER, Franks JS, Driggers WB, Oswald KJ, Quattro JM (2007) Observations of a feeding aggregation of whale sharks, *Rhincodon typus*, in the north central Gulf of Mexico. Gulf Caribb Res 19:69–73

Holland KN, Peterson JD, Lowe CG, Wetherbee BM (1993) Movements, distribution and growth rates of the white goatfish *Mulloides flavolineatus* in a fisheries conservation zone. Bull Mar Sci 52:982–992

Hutchinson N, Rhodes KL (2010) Home range estimates for squaretail coralgrouper, *Plectropomus areolatus* (Ru¨ppell 1830). Coral Reefs 29:511–519

Johannes RE (1978) Reproductive strategies of coastal marine fishes in the tropics. Environ Biol Fish 3:65–84

Johannes RE (1981) Words of the Lagoon. Fishing marine lore in the Palau district of Micronesia. University of California Press, Los Angeles

Johannes RE, Squire L, Graham T, Sadovy Y, Renguul H (1999) Spawning aggregations of groupers (Serranidae) in Palau, The Nature Conservancy Marine Research Series Publication No. 1, 144 pp

Juanes F (1994) What determines prey size selectivity in piscivorous fishes? In: Stouder DJ, Fresh KL, Feller RJ (eds) Theory and application in fish feeding ecology. University of South Carolina Press, Columbia

Kadison E, Nemeth RS, Herzlieb S, Blondeau J (2006) Temporal and spatial dynamics of *Lutjanus cyanopterus* and *L. jocu* (Pisces: Lutjanidae) spawning aggregations on a multi-species spawning site in the USVI. Rev Biol Tropic 54:69–78

Kadison E, Nemeth RS, Herzlieb S, Blondeau J (2007) Temporal and spatial dynamics of *Lutjanus cyanopterus* and *L. jocu* (Pisces: Lutjanidae) spawning aggregations on a multi-species spawning site in the USVI. Rev Biol Tropic 54:69–78

Kadison E, Nemeth RS, Blondeau JE (2009) Assessment of an unprotected red hind (*Epinephelus guttatus*) spawning aggregation on Saba Bank in the Netherlands Antilles. Bull Mar Sci 85:101–118

Kaufman LS, Ebersole JP (1984) Microtopography and the organization of two assemblages of coral reef fishes in the West Indies. J Exp Mar Biol Ecol 78:253–268

Luckhurst BE (1998) Site fidelity and return migration of tagged red hinds (*Epinephelus guttatus*) to a spawning aggregation site in Bermuda. Proc Gulf Caribb Fish Inst 50:750–763

Maljković A, Van Leeuwen TE, Cove SN (2008) Predation on the invasive red lionfish, *Pterois volitans* (Pisces: Scorpaenidae), by native groupers in the Bahamas. Coral Reefs 27:501

Matos-Caraballo D, Posada JM, Luckhurst BE (2006) Fishery-dependent evaluation of a spawning aggregation of tiger grouper (*Mycteroperca tigris*) at Vieques Island, Puerto Rico. Bull Mar Sci 79:1–16

Mazeroll AI, Montgomery WL (1995) Structure and organization of local migrations in brown surgeonfish (*Acanthurus nigrofuscus*). Ethology 99:89–106

Mazeroll AI, Montgomery WL (1998) Daily migrations of a coral reef fish in the Red Sea (Gulf of Aqaba, Israel): initiation and orientation. Copeia 1998:893–905

McCann K, Rasmussen J, Umbanhower J, Humphries M (2005) The role of space, time and variability in food web dynamics. In: de Ruiter PC, Wolters V, Moore JC (eds) Dynamic food webs. Elsevier, Amsterdam

Meyer J, Schultz ET (1985) Tissue condition and growth rate of corals associated with schooling fish. Limnol Oceanogr 30:157–166

Meyer JL, Schultz ET, Helfman G (1983) Fish schools: an asset to corals. Science 220:1047–1049

Meyer CG, Holland KN, Wetherbee BM, Lowe C (2000) Movement patterns, habitat utilization, home range size and site fidelity of whitesaddle goatfish, *Parupeneus porphyreus*, in a marine reserve. Environ Biol Fishes 59:235–242

Moyer JT (1987) Quantitative observations of predation during spawning rushes of the labrid fish, *Thalassoma cupido* at Miyake-Jima, Japan. Japan J Ichthyol 34:76–81

Moyer JT (1989) Reef channels as spawning sites for fishes on the Shiraho coral reef, Ishigaki Island, Japan. Japan J Ichthyol 36:371–375

Myers RF (1999) Micronesian reef fishes. A field guide for divers and aquarists. Coral Graphics, Guam, 216 pp

Nemeth RS (1998) The effect of natural variation in substrate architecture on the survival of juvenile bicolor damselfish. Environ Biol Fishes 53:129–141

Nemeth RS (2005) Population characteristics of a recovering US Virgin Islands red hind spawning aggregation following protection. Mar Ecol Prog Ser 286:81–97

Nemeth RS (2009) Dynamics of reef fish and decapod crustacean spawning aggregations: underlying mechanisms, habitat linkages and trophic interactions. pp 73–134 In: Nagelkerken I (ed) Ecological interactions among tropical coastal ecosystems. Springer, Dordrecht

Nemeth RS, Herzlieb S, Blondeau J (2006a) Comparison of two seasonal closures for protecting red hind spawning aggregations in the US Virgin Islands. Proc Int Coral Reef Conf 4:1306–1313

Nemeth RS, Kadison E, Herzlieb S, Blondeau J, Whiteman E (2006b) Status of a yellowfin grouper (*Mycteroperca venenosa*) spawning aggregation in the US Virgin Islands with notes on other species. Proc Gulf Caribb Fish Inst 57:543–558

Nemeth RS, Blondeau J, Herzlieb S, Kadison E (2007) Spatial and temporal patterns of movement and migration at spawning aggregations of red hind, *Epinephelus guttatus*, in the U.S. Virgin Islands. Environ Biol Fishes 78:365–381

Nemeth, RS, Kadison E, Blondeau JE, Idrisi E, Watlington R, Brown K, Smith T, Carr L (2008) Regional coupling of red hind spawning aggregations to oceanographic processes in the eastern Caribbean. In: Grober-Dunsmore R, Keller BD (eds) Caribbean connectivity: implications for marine protected area management. Marine Sanctuaries Conservation Series NMSP-08-07. U.S. Department of Commerce, NOAA, Office of National Marine Sanctuaries, Silver Spring

Ogden JC, Buckman NS (1973) Movements, foraging groups, and diurnal migrations of the striped parrotfish *Scarus croicensis* Bloch (Scaridae). Ecology 54:589–596

Olsen DA, LaPlace JA (1978) A study of Virgin Islands grouper fishery based on a breeding aggregation. Proc Gulf Caribb Fish Inst 31:130–144

Palau Conservation Society (2010) Enhanced monitoring of grouper spawning aggregation at Ebiil channel: final technical report, Palau Conservation Society and Society for the Conservation of Reef Fish Aggregations report, 31 pp, www.SCRFA.org

Pet JS, Mous PJ, Muljadi AH, Sadovy YJ, Squire L (2005) Aggregations of *Plectropomis areolatus and Epinephelus fuscoguttatus* (groupers, Serranidae) in the Komodo National Park, Indonesia: monitoring and implications for management. Environ Biol Fish 74:209–218

Pickert P, Kelly T, Nemeth RS, Kadison E (2006) Seas of change: spawning aggregations of the Virgin Islands. In: Pickert P, Kelly T (eds) DVD documentary by Friday's Films, San Francisco

Pikitch EK, Santora C, Babcock EA, Bakun A, Bonfil R, Conove DO, Dayton P, Doukakis P, Fluharty D, Heneman B et al (2004) Ecosystem-based fishery management. Science 305:346–347

Pittman SJ, McAlpine CA (2003) Movements of marine fish and decapod crustaceans: process, theory and application. Adv Mar Biol 44:205–294

Randall JE (1961) Observations on the spawning of surgeonfishes (Acanthuridae) in the Society Islands. Copeia 1961:237–238

Randall JE, Randall HA (1963) The spawning and early development of the Atlantic parrot fish, *Sparisoma rubripinne*, with notes on other scarid and labrid fishes. Zool N Y Zool Soc 48:49–60

Rhodes KL, Sadovy Y (2002a) Reproduction in the camouflage grouper (Pisces: Serranidae) in Pohnpei, Federated States of Micronesia. Bull Mar Sci 70:851–869

Rhodes KL, Sadovy Y (2002b) Temporal and spatial trends in spawning aggregations of camouflage grouper, *Epinephelus polyphekadion*, in Pohnpei, Micronesia. Environ Biol Fish 63:27–39

Rhodes KL, Tupper MH (2008) The vulnerability of reproductively active squaretail coralgrouper (*Plectropomis areolatus*) to fishing. Fish Bull 106:194–203

Rhodes KL, Lewis RI, Chapman RW, Sadovy Y (2003) Genetic structure of camouflage grouper, *Epinephelus polyphekadion* (Pisces: Serranidae) in the western central Pacific. Mar Biol 142:771–776

Robertson DR (1982) Fish feces as fish food on a Pacific coral reef. Mar Ecol Prog Ser 7:253–265

Robertson DR (1983) On the spawning behavior and spawning cycles of eight surgeonfishes (Acanthuridae) from the Indo-Pacific. Environ Biol Fish 9:192–223

Robertson DR, Swearer SE, Kaufmann K, Brothers EB (1999) Settlement vs. environmental dynamics in a pelagic-spawning reef fish at Caribbean Panama. Ecol Monogr 69:195–218

Robinson J, Aumeeruddy R, Jorgensen TL, Ohman MC (2008) Dynamics of camouflage (*Epinephelus polyphekadion*) and brown marbled grouper (*Epinephelus fuscoguttatus*) spawning aggregations at a remote reef site, Seychelles. Bull Mar Sci 83:415–431

Sadovy Y (1994) Grouper stocks of the western Atlantic: the need for management and management needs. Proc Gulf Caribb Fish Inst 43:43–65

Sadovy Y (1997) The case of the disappearing grouper: *Epinephelus striatus* (Pisces: Serranidae). Proc Gulf Caribb Fish Inst 45:5–22

Sadovy Y, Domeier M (2005) Are aggregation-fisheries sustainable? Reef fish fisheries as a case study. Coral Reefs 24:254–262

Sadovy Y, Eklund A-M (1999) Synopsis of biological data on the Nassau grouper, *Epinephelus striatus* (Bloch, 1792), and the Jewfish, *E. itajara* (Lichtenstein, 1822), U.S. Department of Commerce, NOAA Technical Report NMFS 146, FAO Fisheries Synopsis 157, Seattle, Washington

Sadovy Y, Figuerola M, Roman A (1992) Age and growth of red hind *Epinephelus guttatus* in Puerto Rico and St. Thomas. Fish Bull 90:516–528

Sadovy Y, Colin PL, Domeier ML (1994a) Aggregation and spawning in the tiger grouper, *Mycteroperca tigris* (Pisces: Serranidae). Copeia 2:511–516

Sadovy Y, Rosario A, Roman A (1994b) Reproduction in an aggregating grouper, the red hind, *Epinephelus guttatus*. Environ Biol Fishes 41:269–286

Sadovy de Mitcheson Y, Cornish A, Domeier M, Colin PL, Russell M, Lindeman K (2008) A global baseline for spawning aggregations of reef fishes. Conserv Biol 22:1233–1244

Sala E, Ballesteros E, Starr RM (2001) Rapid decline of Nassau grouper spawning aggregations in Belize: fishery management and conservation needs. Fish 26:23–30

Sale PF (2004) Connectivity, recruitment variation, and the structure of reef fish communities. Integr Comp Biol 44:390–399

Samoilys MA (1997) Periodicity of spawning aggregations of coral trout (*Plectropomous leopardus*) on the Great Barrier Reef. Mar Ecol Prog Ser 160:149–159

Samoilys MA, Squire LC (1994) Preliminary observations on the spawning behavior of coral trout, *Plectropomus leopardus* (Pisces: Serrandae), on the Great Barrier Reef. Bull Mar Sci 54:332–342

Sancho G (2000) Predatory behaviors of *Caranx melampygus* (Carangidae) feeding on spawning reef fishes: a novel ambushing strategy. Bull Mar Sci 66:487–496
Sancho G, Petersen CW, Lobel PS (2000a) Predator-prey relations at a spawning aggregation site of coral reef fishes. Mar Ecol Prog Ser 203:275–288
Sancho G, Solow AR, Lobel PS (2000b) Environmental influences on the diel timing of spawning in coral reef fishes. Mar Ecol Prog Ser 206:193–212
Semmens BX, Luke KE, Bush PG, Pattengill-Semmens C, Johnson B, McCoy C, Heppell S (2005) Investigating the reproductive migration and spatial ecology of Nassau grouper (*Epinephelus striatus*) on Little Cayman Island using acoustic tags-an overview. Proc Gulf Caribb Fish Inst 56:1–8
Semmens BX, Luke KE, Bush PG, McCoy CMR, Johnson BC (2006) Isopod infestation of postspawning Nassau grouper around Little Cayman Island. J Fish Biol 69:933–93
Shapiro DY, Sadovy Y, McGehee MA (1993) Size, composition, and spatial structure of the annual spawning aggregation of the red hind, *Epinephelus guttatus* (Pisces: Serranidae). Copeia 2:399–406
Sluka RD (2001a) Grouper and napoleon wrasse ecology in Laamu Atoll, Republic of Maldives: Part 1. Habitat, behavior, and movement patterns. Atoll Res Bull 491:1–26
Sluka RD (2001b) Grouper and napoleon wrasse ecology in Laamu Atoll, Republic of Maldives: Part 2. Timing, location, and characteristics of spawning aggregations. Atoll Res Bull 492:1–17
Snyder D, Burgess G (2007) The Indo-Pacific red lionfish, *Pterois volitans* (Pisces: Scorpaenidae), new to Bahamian ichthyofauna. Coral Reefs 26:175
Stallings CD (2008) Indirect effects of an exploited predator on recruitment of coral-reef fishes. Ecology 89:2090–2095
Starr RM, Sala E, Ballesteros E, Zabala M (2007) Spatial dynamics of the Nassau grouper *Epinephelus striatus* in a Caribbean atoll. Mar Ecol Prog Ser 343:239–249
Warner RR (1988) Traditionality of mating-site preferences in a coral reef fish. Nature 335:719–721
Warner RR (1995) Large mating aggregations and daily long-distance spawning migrations in the bluehead wrasse, *Thalassoma bifasciatum*. Environ Biol Fishes 44:337–345
Whaylen L, Pattengill-Semmens CV, Semmens BX, Bush PG, Boardman MR (2004) Observations of a Nassau grouper (*Epinephelus striatus*) spawning aggregation site in Little Cayman Island. Environ Biol Fishes 70:305–313
Whaylen L, Bush P, Johnson B, Luke KE, McCoy C, Heppell S, Semmens B, Boardman M (2006) Aggregation dynamics and lessons learned from five years of monitoring at a Nassau grouper (*Epinephelus striatus*) spawning aggregation in Little Cayman, Cayman Islands, BWI. Proc Gulf Caribb Fish Inst 57:1–14
Whitfield PE, Gardner G, Vives SP, Gilligan MR, JrWR C, Ray GC, Hare JA (2002) Biological invasion of the Indo-Pacific lionfish *Pterois volitans* along the Atlantic coast of North America. Mar Ecol Prog Ser 235:289–297
Wormald CL, Steele MA (2008) Testing assumptions of mark-recapture theory in the coral reef fish *Lutjanus apodus*. J Fish Biol 73:498–509
Zeller DC (1997) Home range and activity patterns of the coral trout *Plectropomus leopardus* (Serranidae). Mar Ecol Prog Ser 154:65–77
Zeller DC (1998) Spawning aggregations: patterns of movement of the coral trout *Plectropomus leopardus* (Serranidae) as determined by ultrasonic telemetry. Mar Ecol Prog Ser 162:253–263
Zeller DC (1999) Ultrasonic telemetry: its application to coral reef fisheries research. Fish Bull 97:1058–1065
Zeller DC, Russ GR (2000) Population estimates and size structure of *Plectropomus leopardus* (Pisces: Serranidae) in relation to no-fishing zones: mark-release-resighting and underwater visual census. Mar Freshw Res 51:221–228

Chapter 3
Why Spawn in Aggregations?

Philip Patrick Molloy, Isabelle M. Côté, and John D. Reynolds

Abstract Animals breed in groups either through necessity, when reproduction depends on a spatially limited resource, or by choice, when group-breeding conveys adaptive benefit to adults or their young. We review evidence for both hypotheses from a variety of animal taxa before turning our attention to potential explanations for fish spawning in aggregations. Spawning in aggregations may increase reproductive output by increasing mate-encounter rates, facilitating mate choice and/or improving fertilisation success. Aggregating to spawn may also dilute predation rates on adults and eggs. However, spawning in aggregations may also carry costs, such as increased conspicuousness to predators, predator-mediated behavioural changes, and increased exposure to parasites or diseases. It is also possible that spawning aggregations occur around specific, spatially rare features or resources that offer a benefit to spawning fish, irrespective of group size. Although explicit tests of most of these costs and benefits have yet to be conducted, existing evidence indicates that there is unlikely to be a single explanation for why fish spawn in aggregations. Additionally, spawning aggregations present a specific and unusual arena for sexual selection, which shapes the evolution of behaviours and life histories in group-breeding species. Consequently, elucidating the evolutionary correlates of aggregation-spawning would be a rich avenue for future research. This chapter explores the costs and benefits of aggregating to breed in a range of fish and non-fish taxa.

P.P. Molloy (✉)
Project Seahorse, Fisheries Centre, University of British Columbia, Vancouver, BC V6T 1ZA, Canada

Department of Biological Sciences, Simon Fraser University, Burnaby, BC V5A 1S6, Canada
e-mail: Philip_p_molloy@gmail.com

I.M. Côté • J.D. Reynolds
Department of Biological Sciences, Simon Fraser University, Burnaby, BC V5A 1S6, Canada
e-mail: imcote@sfu.ca; reynolds@sfu.ca

Y. Sadovy de Mitcheson and P.L. Colin (eds.), *Reef Fish Spawning Aggregations: Biology, Research and Management*, Fish & Fisheries Series 35,
DOI 10.1007/978-94-007-1980-4_3, © Springer Science+Business Media B.V. 2012

An understanding of the proximate and evolutionary causes and consequences of spawning aggregations will provide new insight into the taxonomic and geographic distribution of aggregation-spawning, as well as the potential impacts of heavy exploitation on these species.

3.1 Introduction

Two types of questions can be asked about animal behaviour: how and why. The former addresses real-time stimuli for behaviours such as hormonal or environmental cues; the latter addresses the evolutionary benefit afforded by a particular behaviour and is the stuff of behavioural ecology. In this chapter, we provide a behavioural ecological perspective of the phenomenon of aggregation-spawning.

One area of research that has captivated behavioural ecologists for a long time is why organisms form groups (Krause and Ruxton 2002). Of particular interest are those species that are solitary most of their lives but form aggregations to breed. Although fish species have been used as model systems for research into virtually all types of reproduction, fish breeding aggregations (or spawning aggregations) have received comparatively little attention. A spawning aggregation is a grouping of a single fish species that has gathered at a specific location, at a density that is at least four-times greater than normal, with the specific purpose of reproducing (Chap. 1). To form aggregations, fish may travel very long distances. For example, some Nassau groupers, *Epinephelus striatus*, travel hundreds of kilometres to spawning sites (Bolden 2000). The resulting aggregations may comprise dozens to tens of thousands of individuals and can offer a breathtaking visual spectacle.

The scarcity of behavioural research on spawning aggregations may be largely due to the logistical difficulties associated with accessing aggregations and manipulating them experimentally (Claydon 2004). However, fish spawning aggregations need not remain an unsolved mystery. The theoretical frameworks developed through studies of other group-breeding taxa allow us to cast some indirect light on spawning aggregations in fishes and derive some insight into their possible evolutionary advantages. Our objective in this chapter is to use a behavioural ecological perspective to explore the costs and benefits of aggregating to breed, to examine the mating systems and reproductive behaviours of a range of aggregation-spawning species and to draw parallels with similar mating systems in a diversity of other taxa. Finally, we formulate novel predictions tailored specifically to aggregating fishes and suggest ways in which they may be tested.

3.2 General Overview of the Adaptive Significance of Reproductive Aggregations

There are two general explanations for the evolution of reproductive aggregations in animals, irrespective of the time and place of reproduction. First, limitation of suitable space or habitat can force animals to occupy higher densities than they would

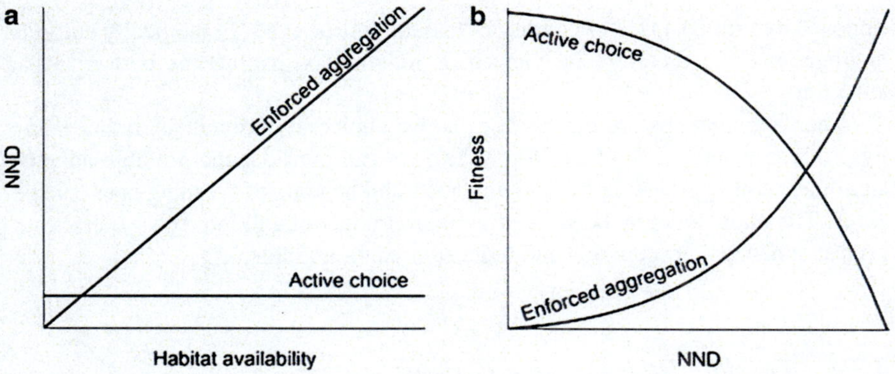

Fig. 3.1 Aggregation may form either through active choice or through enforcement by habitat availability or suitability. The predicted relationships between (**a**) habitat availability and distance between individuals (nearest-neighbour distance, NND), and (**b**) NND and fitness differ between these two ultimate causes of aggregation. NND should increase with habitat availability in species that aggregate only when habitat is limited because, in these species, fitness increases with NND. By contrast, individuals of species that choose to aggregate will seek constant proximity to others because they derive fitness benefits from doing so

at non-breeding times (Fig. 3.1a). Alternatively, aggregations may be the result of active choice by individuals for proximity to conspecifics because of benefits, for breeding adults or their young, inherent to breeding near others (Fig. 3.1b). These two hypotheses have been tested for a variety of forms of reproductive aggregations. What follows is a very brief summary of the state of knowledge to date, based on a wide variety of taxa, to set the scene for the next section, which evaluates evidence specifically for fish species.

Shortages of suitable habitat lead to reproductive aggregations in many species, with well-known examples for some birds and marine mammals that have particularly narrow breeding habitat requirements (Wittenberger 1981). Many seabirds, such as black-legged kittiwakes, *Rissa tridactyla*, for example, nest at very high densities on isolated rocky islets or steep cliffs, which are safe havens from predators (e.g. Lack 1968). Similarly, female hawksbill turtles, *Eretmochelys imbricata*, and most other species of sea turtles come ashore on beaches with specific characteristics that are required for successful nesting (Hays et al. 1995). In all of these cases, there is no evident intrinsic advantage to breeding near conspecifics. Indeed, ecologically enforced sociality may carry large costs, including increased parasitism and reduced survival of adults and young (Krause and Ruxton 2002, Fig. 3.1b). Thus when sociality is 'imposed', the extent of aggregations corresponds to the sizes of suitable breeding habitat patches. In barn swallows, *Hirundo rustica*, for example, those individuals that manage to secure very small patches can breed nearly or completely solitarily and may have the highest reproductive output because they derive all of the benefits bestowed by the good habitat and none of the costs of group-living (Shields and Crook 1987). The alternative to breeding in aggregations that are imposed by limited space or conditions is not to breed at all. Ecologically enforced aggregations can therefore be viewed as individuals making the best of a bad situation.

When aggregations are the result of active choice by individuals, rather than a by-product of habitat shortage (Fig. 3.1a), we can consider the possible adaptive significance of aggregating by studying costs and benefits of breeding near conspecifics. The main potential benefits of social reproduction fall into two general categories: enhanced reproduction and reduced predation (Table 3.1).

3.2.1 *Possible Reproductive Benefits*

A reproductive benefit of breeding in groups is increased rates of encounter with potential mates, which can facilitate mate choice (Loiselle and Barlow 1979). This benefit has been suggested particularly for lekking birds, such as great snipe, (*Galinago media*) (Höglund and Alatalo 1995). On leks, males display in small territories that are usually clumped and that provide no resources to females. Females examine a number of potential mates, mate with one (or more) and leave the mating arena to have their young (Andersson 1994; Reynolds 1996). The costs of sampling potential mates when in aggregations can also be reduced because travel time between mates is low. In barking treefrogs, *Hyla gratiosa*, for example, females position themselves near choruses of calling males so that they can detect the calls of multiple males simultaneously (Murphy and Gerhardt 2002). Individuals at aggregations can also observe and copy the choices of others (Reynolds and Gross 1990; Dugatkin and Hoglund 1992). Examples of mate choice copying in fishes have been summarized by Reynolds and Jones (1999) and include gobies (Gobiidae), damselfish (Pomacentridae) and sticklebacks (Gasterosteidae). These advantages may be particularly great for females, who are conventionally the choosier sex, and may generate strong preferences by females for aggregated over solitary males (Alatalo et al. 1992). Easier mate choice may, however, not be a benefit for females in all taxa with lek-like mating systems. In marine mammals such as elephant seals, *Mirounga angustirostris*, and in ungulates such as Uganda kob, *Kobus kob*, leks are arenas for intense male-male competition for access to mates, which reduces considerably the scope for uncoerced mate choice by females (Andersson 1994).

Breeding in dense aggregations may increase mating rates. For example, the reproductive success of small male marine iguana, *Amblyrhynchus cristatus*, increases with lek size (Wikelski et al. 1996) as did that of both male and female dance fly, *Empis borealis*, in mating swarms (Svensson and Petersson 1992). Similarly, close proximity between individuals may ensure a high fertilisation rate of the eggs in many aquatic animals that broadcast their gametes into the water column (e.g. urchins, Levitan 2004). Conversely, reproductive proximity can also facilitate extra-pair matings in pair-spawning species, which is a benefit for those who engage in it, and a cost to males that lose such matings. In colonial birds and fishes, for example, sperm competition can be intense, and individuals can lose a

Table 3.1 Potential benefits and costs of breeding in aggregations

General benefit or cost	Mechanism	Example	Evidence in spawning aggregation?
Reproduction	Increased mate encounter rate	No quantitative support, but seems intuitive	No quantitative support, but seems intuitive
	Facilitate mate comparison	Evidence for simultaneous sampling of mates by females in chorusing treefrogs (Murphy and Gerhardt 2002)	No evidence
	Reduce the cost of mate comparison	Evidence of mate choice copying in black grouse (Hoglund et al. 1995)	No evidence
	Increased fertilisation rate	Higher fertilisation rate in larger groups in green sea urchins (Gaudette et al. 2006)	No fertilisation benefits from group spawning in brown surgeonfish (Kiflawi et al. 1998) and bluehead wrasse (Petersen et al. 1992).
	Increased cost of mate guarding	Intensity of mate guarding by male purple martins increases with colony size (Davis and Brown 1999)	Territoriality becomes economically unviable at high population densities in bluehead wrasse (Warner and Hoffman 1980b)
	Increased risk of extra-pair matings	Percentage of extra-pair copulations higher in colonially than solitarily breeding birds (Møller and Birkhead 1993)	No evidence
	High sperm competition	Group-living bats (Hosken 1997) and frogs (Byrne et al. 2002) have larger testes than solitary counterparts	High investment in sperm production in many group-spawning fish (Robertson and Warner 1978; Warner and Robertson 1978; Petersen et al. 1992; Erisman et al. 2007)
Predation	Early detection of predator	Threat detection improves with size of breeding colony in Montagu's harriers (Arroyo et al. 2001). Lekking male Uganda kob detect predators earlier than solitary males (Balmford and Turyaho 1992)	No evidence
	Cooperative deterrence of predators	Mobbing by nesting Montagu's harriers (Arroyo et al. 2001)	No evidence

(continued)

Table 3.1 (continued)

General benefit or cost	Mechanism	Example	Evidence in spawning aggregation?
	Predator swamping/ satiation	Large species of African grazing mammals in large herds with precocial young have highly synchronized birth seasons (Sinclair et al. 2000)	Spawning 'epidemic' observed in several parrotfish species may satiate egg predators (Randall and Randall 1963; Colin 1978)
	Dilution effect	*Per capita* predation risk declines with lek size in Túngara frogs (Ryan et al. 1981)	No quantitative support, but seems intuitive
	Increased group detection	Lions target lekking Uganda kob (Balmford and Turyaho 1992)	Predators focus attacks on group- rather than pair-spawning fish (Sancho et al. 2000) Conspicuous and predictable nature of spawning aggregations makes them vulnerable to detection by fishers
Other	Increased risk of contagious diseases and parasites	Transmission of ectoparasites between cliff swallow chicks increases with colony size (Brown and Brown 2004)	Transmission of ectoparasites during spawning aggregations (Semmens et al. 2006; Sigura and Justine 2008)
	Lower risk of non-contagious parasites	Intensity of infection by mobile parasites decreases with increasing host group size across a variety of taxa (Côté and Poulin 1995)	No evidence
	Loss of resources	No quantitative support, but seems intuitive	No evidence
	Migration costs	Sage grouse hens move further per day when visiting leks (Gibson and Bachman 1992)	No direct evidence but long migration distances of many groupers (Nemeth et al. 2007; Starr et al. 2007) are likely to carry energetic costs
	Increased likelihood of agonistic behaviour	Agonistic encounters between Magellanic penguins rarer in areas with lower nest density (Stokes and Boersma 2000). Male frogs in denser choruses fight more (Byrne et al. 2002)	No evidence

Fig. 3.2 A spawning aggregation of longnose parrotfish, *Hipposcarus longiceps*, which has released a cloud of eggs and sperm being preyed upon by black snapper, *Macolor niger*, in Palau. By spawning in groups, parrotfish may reduce *per capita* egg predation rates. (Photo: Patrick L. Colin)

significant number of fertilisations to neighbours (Brown and Brown 1989; Jennings and Philipp 1992). In European starlings, *Sturnus vulgaris*, denser groups breed more synchronously and breeding synchrony, in turn, is linked to greater fledging success (Evans et al. 2009).

Predation may be reduced in a breeding group because of a dilution effect, whereby the likelihood of any individual being the target of a predator decreases as group size increases (Hamilton 1971). Breeding group members may detect predators earlier than solitary animals, thus increasing their probability of escape (e.g. Heard 1992). They may also be more successful in deterring or confusing predators through cooperative efforts with other group members (Wittenberger and Hunt 1985). In addition, the synchrony of reproductive activities in a group may temporarily swamp the appetite of predators, leading to lower predation *per capita* when offspring are vulnerable (e.g. Gross and MacMillan 1981) (Fig. 3.2). Although larger groups may be encountered more often by predators than smaller groups or solitary individuals (Krause and Ruxton 2002), the benefits outlined above to individual aggregation members generally outweigh the costs of increased detection rate.

A few other costs and benefits of breeding in aggregations, which do not clearly fall under the headings of predation or reproduction, can be experienced by social individuals (Table 3.1). This includes an increased risk of transmission of contagious diseases and parasites (Côté and Poulin 1995; Krause and Ruxton 2002), although the *per capita* risk of acquiring non-contagious parasites may be reduced: with more conspecifics around, the probability that any given individual will be parasitized is reduced (Côté and Gross 1993). Within aggregations, locations that confer protection from predation or yield higher mating success may be hotly contested, leading to agonistic interactions among competitors and risk of injury. Finally, temporary breeding aggregation sites are often separate from the territories or home ranges held by individuals during non-breeding periods. Travel to these breeding sites may entail additional energetic expenditure and place the traveller under higher risk of predation. Territories left unprotected while the owners are breeding can be raided or usurped.

Although the two major explanations for breeding aggregations, i.e. habitat limitation and individual choice, have been presented here as mutually exclusive alternatives, this is not necessarily the case for all reproductive groupings. Breeding in an aggregation could yield more than one benefit or entail more than one cost. A major challenge for behavioural ecologists has been, and remains, to convert all benefits and costs to a comparable currency to evaluate the 'bottom line'.

3.3 Benefits and Costs of Spawning in Fish Aggregations

We now examine evidence for or against some of the reproductive costs and benefits outlined in the previous section, focussing specifically on fish aggregations from a range of habitats. Since reproductive success is such a critical fitness component, understanding the reproductive consequences and dynamics of spawning aggregations may help to shed light on their evolutionary significance.

3.3.1 Increased Mate Encounter Rate

Mate encounter rates have never been tested explicitly in aggregative spawners; however, given that densities at spawning aggregations are higher than in non-reproductive populations, it seems likely that mate encounter rates should be higher at spawning aggregations. Increased encounter rates are likely to be particularly significant for species such as the red hind, *Epinephelus guttatus*, (Shapiro et al. 1994) and gag grouper, *Mycteroperca microlepis* (Coleman et al. 1996), which exhibit some spatial segregation of the sexes during the non-breeding season, and for species such as the blob sculpin, *Psychrolutes phrictus* (Drazen et al. 2003), and permit, *Trachinotus falcatus* (Graham and Castellanos 2005), which occur at low densities during the non-reproductive season, possibly owing to the general scarcity of non-reproductive resources such as food and/or shelter. In these species, interactions with potential mates apparently occur almost exclusively on spawning aggregations. Conversely, increased mate encounter rates are unlikely to be important in species that occur in high densities away from aggregations. For example, the blue tang, *Acanthurus coeruleus*, which forms spawning aggregations of several thousand individuals (Colin and Clavijo 1988), also forms large feeding groups during the non-reproductive season (Morgan and Kramer 2004). This is also the case for the yellow and blueback fusilier, *Caesio teres* (Bell and Colin 1986), and white trevally, *Pseudocaranx dentex* (Afonso et al. 2008a), as well as other schooling species.

Although higher mate encounter rates could be demonstrated easily by comparing the number of intersexual interactions during and away from spawning aggregations, this would not tell us much about the evolution of spawning aggregations. By definition, spawning aggregations involve high population densities (and thus, presumably, high encounter rates with mates), but it would be difficult to determine whether higher encounter rates at aggregations are actually the evolutionary driver of formation of spawning aggregations. It is possible that, because individuals that reproduced in aggregations are guaranteed to encounter a mate at that time, they are 'free' to exploit sex-specific resources during the non-breeding season. This may be the case for several shark species, such as nurse sharks, *Ginglymostoma cirratum*, scalloped hammerheads *Sphyrna lewini* and sand tiger sharks *Carcharias taurus*, which exhibit both mating in aggregations and sexual segregation out of the mating season (reviewed by Sims 2005). The evolution of spawning aggregations could therefore be a cause, rather than a consequence, of the distribution of non-spawning individuals.

3.3.2 Mate Choice Facilitation or Cost Reduction

If spawning aggregations have evolved because females prefer situations that facilitate mate choice, then we should see evidence of greater ease of choice. Females often select mates by assessing the quality of courtship displays, and courtship commonly occurs at spawning aggregations. For example, males of several epinepheline groupers exhibit intricate displays involving colour, swimming patterns, sound and physical contact (e.g. Colin 1994; Pelaprat 1999; Erisman et al. 2007). Female dusky groupers, *Epinephelus marginatus*, spawning in Corsica, France, visit or are visited by several courting males but spawning occurs with only a small number of these visitors (Pelaprat 1999), implying that mate selection occurs. Similar displays and apparent mate selection have been observed in many other aggregating taxa, including parrotfishes (Scaridae) (Clavijo 1983; Gladstone 1986), angelfishes (Pomacanthidae) (Moyer et al. 1983) and jacks (Carangidae) (Graham and Castellanos 2005) (Fig. 3.3) (also see Chaps. 12.2 and 12.18).

Courtship is also observed in non-aggregating species, so the presence of displays at aggregation alone is not strong support for mate choice facilitation. The key question is whether aggregating females are able to find, inspect and compare potential mates more easily than their non-spawning relatives. A number of observations made in relation to aggregation size or density would be consistent with this idea. With increasing aggregation size/density, the time taken by females to select a mate should decrease, but mate inspection rate should increase, as should the probability

Fig. 3.3 While aggregating, males of many species take on temporary courtship colouration; (**a**) normal coloured tiger grouper, *Mycteroperca tigris*, look very different from the (**b**) courting male (Photos: Yvonne Sadovy de Mitcheson)

that any given female will find a suitable mate. As far as we are aware, these predictions remain untested in aggregating fish species. Ultimately, if females are better able to compare mates and select partners of higher quality in larger groups, both female and offspring fitness should increase with group size, although realistically, this would be impossible to measure.

Females spawning in aggregations could save time when choosing a mate by copying other females in their choice of spawning partner; given the relatively narrow time windows for spawning in some species this could be advantageous. To our knowledge, the presence of mate-choice copying has never been tested in aggregation spawning fishes and no evidence is available to determine whether such a tactic may be used.

The benefit of mate choice facilitation should be greatest in species where females are particularly choosy. Generally, females that spawn in pairs or single-male, multiple female (i.e. harem) groups within aggregations, such as red hind (Sadovy et al. 1994) and bumphead parrotfish, *Bolbometapon muricatum* (Gladstone 1986), are likely to be choosier than those spawning in promiscuous, multiple-male groups, such as Nassau grouper (Colin 1992), striped parrotfish, *Scarus iserti* (Colin 1978), and permit (Graham and Castellanos 2005). Nevertheless, an element of mate choice may still occur in multiple-male groups. For example, females may identify a preferred male with whom they attempt to spawn, and their spawning rush is joined by other, 'less attractive' males, in a manner similar to 'hot-shot' lekking (Höglund and Alatalo 1995). However, in multi-male spawning groups, it seems more likely that females gain from spawning in aggregations due to fertilisation benefits.

3.3.3 Fertilisation Dynamics and Sexual Conflict

For males, the increase in density of conspecifics and the choosiness of females on spawning aggregations result in increased intraspecific competition. This competition could be either directly for territories and/or females, or through sperm competition for fertilisations. The investment made by males into sperm production reflects directly the intensity of sperm competition (Stockley et al. 1997), and there is ample evidence that male fish that spawn in multi-male groups invest heavily in sperm (e.g. several wrasses, Labridae, Robertson and Warner 1978, Atlantic wreckfish, *Polyprion americanus*, Peres and Klippel 2003). For example, in group-spawning groupers, such as the Nassau and leopard, *Mycteroperca rosacea*, groupers, males have very large testes relative to their body size (Nassau: ~10% body weight, Sadovy and Colin 1995; leopard: ~2.5% body weight, Erisman et al. 2007, Brice Semmens personal communication) (Fig. 3.4) (Chap. 5). In comparison, pair-spawning groupers, such as the red hind, tend to invest relatively little in sperm production (<1% body weight, Colin et al. 1987; Sadovy et al. 1994). Many aggregative spawners are known to have mixed strategies where group- and pair-spawning both occur; indeed, all group spawners studied by Colin and Clavijo (1988) also pair-spawned. Species with such mixed strategies may provide the best tests of the effects of sperm competition. Such comparisons between group- and pair-spawning individuals have been made for several wrasse (Labridae) and parrotfish species where males exhibit two colour patterns that are associated with these two mating tactics: initial-phase males are small, have drab colours or appear similar to females, and usually spawn in groups; terminal-phase males are large, brightly coloured, defend harems or resources that attract females, and spawn in pairs. The differences in sperm alloca-

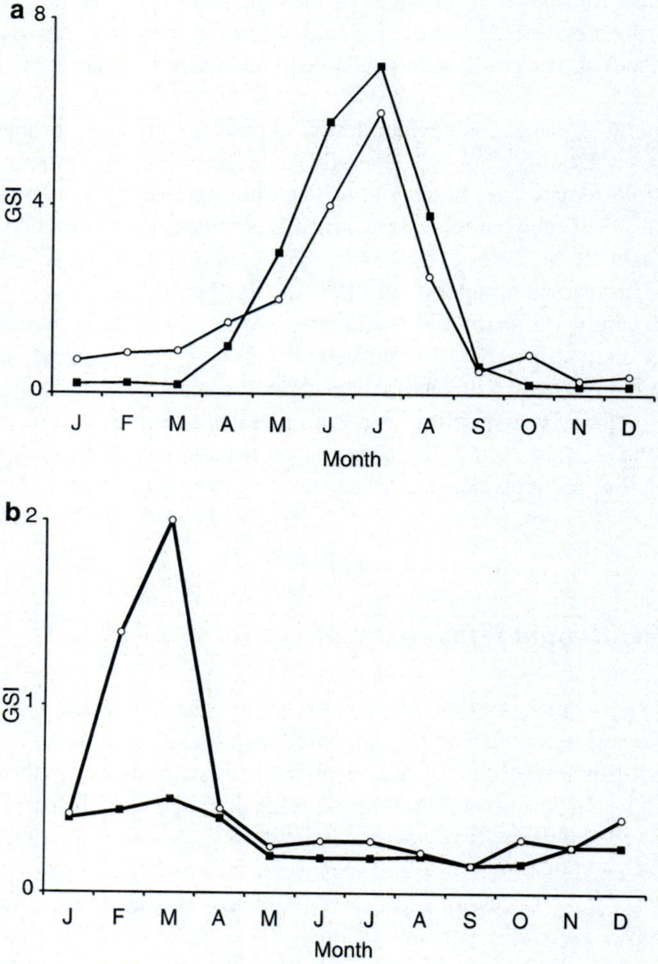

Fig. 3.4 Gonadosomatic index (*GSI*, gonad weight/body weight * 100) of species (**a**) with, brown surgeonfish (*Acanthurus nigrofuscus*), and (**b**) without, gag grouper (*Mycteroperca microlepis*). Higher GSI indicates larger gonad size relative to body weight and reflects greater exposure to sperm competition. Main spawning seasons are February and March for gag grouper and May to August for brown surgeonfish. Females *open circles*; males *closed squares* (Adapted from Fishelson et al. (1987), Hood and Schleider (1992))

tions (as indicated by gonadosomatic index, or GSI) are in keeping with sperm-competition predictions: initial-phase males produce lots of sperm while terminal-phase males only produce a little. GSI reflects the level of sperm competition experienced by the two male types: initial phase males spawn in multi-male groups where sperm competition is intense; terminal phase males spawn in pairs where sperm competition is relaxed (Robertson and Warner 1978; Warner and Robertson 1978) (Fig. 3.5). [See also Chap. 8.]

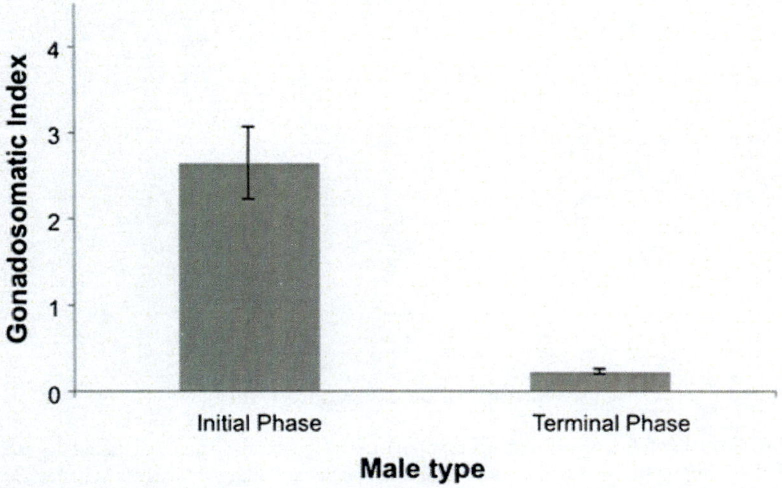

Fig. 3.5 Gonadosomatic index (see equation Fig. 3.4) ± 95% confidence intervals for non-sex-changing initial phase (*IP*) males and sex-changed terminal phase (*TP*) bluehead wrasse (*Thalassoma bifasciatum*) males. See text for details (Data derived from Warner and Robertson 1978)

Despite the strong selection for abundant sperm production when spawning in a group, sperm limitation can occur. The large amount of sperm released by males to compete for access to eggs, in addition to the potential for mating with multiple females in quick succession (particularly by the most preferred males), can lead to sperm depletion in some individuals. For example, pair-spawning, territorial male bluehead wrasse, *Thalassoma bifasciatum*, that have high spawning rates showed lower fertilisation rates than males with lower spawning rates (Warner et al. 1995, Chap. 12.14). Although these males are not aggregative spawners, this example neatly demonstrates the potential for sperm limitation in fish that spawn repeatedly. For females, spawning in groups with several males may overcome this risk of sperm limitation and instead yield increased fertilisation rates where total sperm availability is greater. Evidence for such a fertilisation benefit to females through spawning in groups is variable. In Atlantic cod (*Gadus morhua*), fertilisation rates increase with the number of spawning males (Rowe et al. 2004); however, fertilisation rates in bluehead wrasse were similar between multi-male group spawns and pair spawns (Petersen et al. 1992), despite an ~80 times increase in the amount of sperm released by group- than pair-spawning males (Fig. 12.38). In brown surgeonfish (Acanthuridae), *Acanthurus nigrofuscus*, fertilisation rates were similar between large and small spawning groups (Kiflawi et al. 1998). However, group size *per se* may not be associated with increased fertilisation rates if the number of males per female remains constant in all groups; rather, fertilisation rates should increase with male:female sex ratio. Group size will only correlate with fertilisation rates if it also correlates with sex ratio. Indeed, studies of aggregation fishing predict that sperm limitation, and thus reduced fertilisation rates, are most likely when fisheries

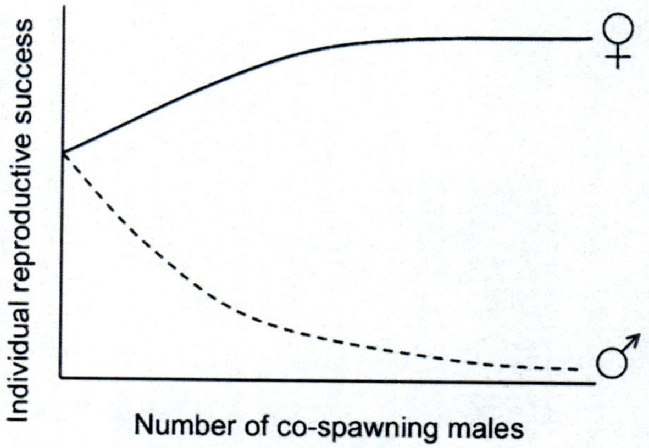

Fig. 3.6 Sexual conflict associated with number of co-spawning males and, hence, aggregation size. Female reproductive success (*solid line*) and male reproductive success (*dashed line*) are shown according to number of co-spawning males. See text for details

disproportionately kill males and drive a highly female-biased sex ratio (e.g. Vincent and Sadovy 1998 Fig. 9–10, Alonzo and Mangel 2004). Future studies on fertilisation rates in spawning aggregations should test sex-ratio effects more specifically.

If pair-spawning is associated with sperm limitation, females may suffer a cost from spawning with territorial males and a sexual conflict may arise: females prefer spawning in groups whereas territorial males prefer spawning in pairs (Fig. 3.6). Specifically, if females were free to spawn in multi-male groups they might obtain higher fertilisation rates than obtained during pair spawns; however, when males are able to defend females or resources on aggregation sites, multi-male spawning groups are not predicted to occur. As such, males may reach their maximum fertilisation potential but the lack of other spawning males may prevent females from obtaining maximum fertilisation rates. Female reproductive success increases with the number of spawning males: the more males participating in a spawn, the greater the proportion of her eggs that will be fertilised. In contrast, as more males participate in a spawn the proportion of eggs that gets fertilised by any given male (and hence, his reproductive success) declines. This creates a conflict: males should prefer to spawn in pairs, while females should prefer to spawn with many males. The resolution of this conflict depends on numerous factors such as the ability of males to monopolize and guard females, the importance of female mate choice and non-reproductive cost and benefits for females associated with reproducing in a group versus in a pair.

This sexual conflict has never been addressed in spawning aggregations and determining whether females actively encourage other males to join a spawning event may be difficult. However, since females often lead spawning rushes in group-spawning species (e.g. Moyer et al. 1983; Colin and Bell 1991; Erisman et al. 2007), one could test whether a female is more likely to initiate a spawning rush when more males are close by.

3.4 Survival Benefits and Costs

3.4.1 Dilution of Predation on Adults and Eggs

Spawning aggregations are conspicuous phenomena. In addition to the increased density of fishes, mating behaviours exhibited by the attending individuals are often elaborate and showy (Clifton and Robertson 1993). Furthermore, aggregations are spatially and temporally predictable. As such, aggregations may be particularly attractive to predators of both eggs and spawning adults. For example, whale sharks, *Rhincodon typus*, are known to aggregate around snapper spawning aggregations in Belize, where they feed on recently spawned gametes (Heyman et al. 2001). Similarly, Nassau grouper aggregations around St Thomas, BVI, attract shark predators (Olsen and LaPlace 1979), although this may have been a consequence of the fishing activity, rather than the aggregation *per se* (Chap. 5).

However, although aggregative spawning might increase detection by predators, doing so may reduce *per capita* predation rates on adult and/or egg. Although adult and egg predation by fishes at spawning aggregations has been reported (e.g. Colin 1978; Olsen and LaPlace 1979; Bell and Colin 1986; Colin et al. 1987; Aguilar-Perera 1994; Samoilys and Squire 1994; Samoilys 1997; Craig 1998; Pelaprat 1999; Sancho 2000; Heyman et al. 2005), predation *rates* are usually low (Colin and Clavijo 1988; Colin and Bell 1991; Sancho et al. 2000) and, perhaps as such, have rarely been studied in detail (Chap. 5). Sancho et al. (2000) quantified predation on a wide range of species spawning at an aggregation site at Johnston Atoll in the Central Pacific. Two predatory species, the bluefin trevally, *Caranx melampygus*, and the small-toothed jobfish, *Aphareus furca*, concentrated their attacks on individuals participating in spawning rushes. A total of 2.3% of all spawning rushes were attacked, and 4% of attacks were successful, resulting in a rate of successful predation on spawning rushes of 0.1% for 33 species combined. Species-specific rates are much lower. *Per capita* predation rates were not reported but they are likely to be extremely small because attacks were concentrated on spawning groups of four or more individuals. Similarly, despite the ubiquity of egg predation at spawning grounds and the appeal of the idea that spawning aggregations form to increase egg survivorship, *per spawner* egg predation rates have not been quantified.

To obtain such data, more detailed comparisons of predation rates on and off aggregations are required. Ideally, these data should come from observations of spawning individuals. These studies should focus on species known to spawn in groups *and* in isolated pairs. Examples include various parrotfishes and wrasses (Randall and Randall 1963; Robertson and Warner 1978; Warner and Robertson 1978; Colin and Clavijo 1988), dog snapper, *Lutjanus jocu* (Krajewski and Bonaldo 2005), and several surgeonfishes (Acanthuridae) (Colin and Clavijo 1988). Nevertheless, such studies will be hampered by how seldom spawning and predation co-occur.

It is also important to remember that although spawning aggregations may have evolved, at least in part, because of predation dilution advantages, this survivorship benefit may be negated by the intensity of modern fishing. Irrespective of 'natural' predation rates, it is likely that fishing has rendered or will render *per capita* mortality higher on exploited aggregations than during non-spawning periods. Fishing mortality may also reduce spawner numbers to levels where aggregating individuals no longer gain a dilution of *per capita* 'natural' predation rates.

3.4.2 Other Costs

Individuals may incur other less obvious costs from spawning in aggregations. Increased predation risk may cause indirect, behaviourally mediated fitness costs (Lima and Dill 1990). For example, hogfish, *Lachnolaimus maximus*, are known to abort spawning attempts when egg predator densities are too high (Colin 1982b). Similarly, group-spawning striped parrotfish temporarily reduce their spawning rate by about 25-fold following attacks by mackerel (*Scomberomorus* sp. Colin 1978), although spawning rates usually rebound within 10 min (Patrick Colin personal communication). Releasing gametes at suboptimal locations or times, or reducing the total number of gametes released as a result of predation risk, could entail reproductive costs for aggregative spawners. Such costs are more likely to be encountered in species that only spawn for a brief period each year, such as some Indo-Pacific surgeonfish and some grouper populations (Robertson 1983; Montgomery and Galzin 1993; Domeier and Colin 1997) – these species have a smaller window of opportunity to subsequently fully recoup such reproductive costs. Currently there are no studies addressing the indirect effects of risk of predation on the reproductive output of fish at spawning aggregations. Such studies could test predictions that (1) the presence of adult or egg predators reduces spawning rates, and (2) the length of time required for fish to resume spawning after predator-induced cessation increases with the number of predators or their combined attack rate.

Higher population densities may promote parasite or disease transmission. For example, spawning aggregations of the speckled blue grouper, *Epinephelus cyanopodus*, are thought to increase transmission of monogeneans, which would explain how these parasites, which prefer large hosts, are occasionally found on smaller individuals (Sigura and Justine 2008). Specific observations of isopod infestations on Cayman Island populations of Nassau grouper immediately after spawning also suggest that high spawning densities may facilitate transmission or infestation (Semmens et al. 2006). These examples are somewhat anecdotal. To demonstrate that spawning in aggregations increases susceptibility to parasitism, simultaneous comparisons of parasite loads on fish spawning in aggregations differing in densities are needed. The costs of parasite transmission are potentially important since they may form an integral part of an individual's decision to visit or avoid an aggregation to spawn (Semmens et al. 2006).

Various costs are also associated with migrating to spawning grounds. For example, migration is likely to impose energetic costs for the many species that travel considerable distances to reach spawning aggregations (Crawford et al. 1986; Zeller 1998;

Slotte and Fiksen 2000; Peres and Klippel 2003). In addition, individuals may be more conspicuous to predators while travelling to spawning grounds along migration routes, which are often highly predictable (Starr et al. 2007 and references therein). Many aggregation-spawning groupers show non-spawning home-range fidelity (e.g. Beets and Hixon 1994; Lembo et al. 1999; Kaunda-Arara and Rose 2004) and may return to the same non-spawning site after spawning (Waschkewitz and Wirtz 1990). Attending spawning aggregations can result in displacement from non-spawning territories (e.g. Nassau grouper, Patrick Colin personal communication) (Chap. 2).

3.5 Habitat Limitation as a Cause of Spawning Aggregations

The spatial rarity of spawning aggregations in many cases, along with their apparent site traditionality, might suggest that these aggregations form around specific features that are limited in availability (Chap. 5). These sites may have inherent characteristics that are beneficial to spawning fish, regardless of group size.

Hydrographic conditions that promote the swift removal of spawned eggs from the reef and its associated planktivorous predators have long been assumed, although not demonstrated, to be important determinants of spawning aggregation locations and timing (Johannes 1978; Lobel 1978). The same conditions could also enhance dispersal of larvae into new habitats (Barlow 1981; Doherty et al. 1985) or, alternatively, increase larval retention into areas that were suitable for the parents (Johannes 1978; Lobel 1978; Lobel and Robinson 1988). If so, spawning aggregations should occur at sites and times with, for example, tidal flows, non-tidal current speed and/or direction that differ in a predictable manner from random sites (Chaps. 5 and 6). Scepticism regarding these hypotheses and, more notably, the efforts taken to test them was raised over 20 years ago (Shapiro et al. 1988). Our understanding of the role of hydrographic conditions in determining where spawning aggregations are located has not improved greatly since then (Chap. 6). Comparisons of spawning to non-spawning sites have only been made for a few species. For example, spawning aggregations of brown puller damselfish, *Chromis hypsilepis*, do not coincide with tidal regimes that appear to facilitate either early survival or dispersal of larvae (Gladstone 2007b). Similarly, Colin (1992) found that currents at Nassau grouper aggregation sites do not favour offshore transport of eggs. Drazen et al. (2003) hypothesised (but did not test) that blob sculpin aggregate on seamounts to spawn due to the faster current flow across these mounts. Releasing eggs high above the substratum has also been suggested as being important, and the perception that spawning aggregations are often located around promontories appears to be consistent with this idea. Nevertheless, there is little evidence supporting this idea (Shapiro et al. 1988, Chapter 6, 7). For example, starting heights and lengths of spawning rushes varied widely among 45 Indo-Pacific species of wrasses and parrotfish at Enewetak Atoll, Marshall Islands, implying no inherent benefit associated with spawning at a particular height (Colin and Bell 1991). More detailed consideration of topographic and oceanographic features of spawning sites is given in Chap. 5.

Fish may also spawn in aggregations if certain sites are particularly safe for adults, rather than for their eggs (Johannes 1978; Shapiro et al. 1988). This could happen if there are particularly few predators in an area, or if the physical structure of a site protects spawners as they release their gametes, often at the apex of an upward spawning rush during which they are most vulnerable to predation (Sancho et al. 2000). Several studies have noted that aggregations occur at sites that have more relief compared to nearby non-spawning areas. For example, red hind aggregate to spawn around highly complex and rare habitats (Beets and Friedlander 1998), benthic-spawning brown puller damselfish select spawning sites that are significantly more rugose than neighbouring non-spawning sites, and Sancho et al. (2000) found that several group-spawning species aggregated at particularly complex sites to spawn. In all these examples, the authors hypothesise that reef complexity serves to reduce predation risk (Beets and Friedlander 1998; Sancho et al. 2000; Gladstone 2007a).

The ultimate experiment to test whether specific physical features are sought by spawning fish would be to remove all fish from an area where they aggregate, replace them with site-naïve fish, and see whether these new fish aggregate to spawn at the same location as the original population. This extraordinary experiment was conducted by Robert Warner using the wrasse as a model system (Warner 1988). Bluehead wrasse spawn either in pairs or groups, and the groups form and spawn daily at predictable times and locations (Figs. 12.38, 12.39). Some mating sites have remained in use for over 12 years without changing location (Warner 1988). Experimental replacement of entire populations from small patch reefs led to the use of sites for spawning that differed from those used by the original populations. Thus mating site locations in bluehead wrasse are not solely the result of individual assessment of current resource quality, but rather represent culturally transmitted traditions. Further experiments in which either the male or the female portion of each population was replaced demonstrated that transmission of spatial traditions occurred via females (Warner 1990). Males simply go where females aggregate to spawn.

At this point, it is unknown whether these results can be extrapolated to other species that spawn in aggregations. It would be extremely difficult to carry out such an experiment with larger species, such as groupers or snappers. Unfortunately, the first part of this experiment is effectively taking place through overfishing of spawning aggregations in many regions of the world (Sala et al. 2001; Sadovy and Domeier 2005a). Should protection and management result in recovery of populations to former levels, it will be of interest to observe whether spawning aggregations re-form in locations from which they were extirpated, which would support the idea that relatively static, beneficial environmental features exist at these sites. By contrast, the formation of aggregations at novel locations would indicate that spawning sites are not chosen on the basis of beneficial features, that many places have beneficial features but not all are used, or that the physical features that prompted site choice by previous generations are no longer present.

3.6 Sexual Selection in Spawning Aggregations

Many aggregative spawners also spawn in pairs outside of aggregations (Aguilar-Perera 1994). For example, the two male phases (initial and terminal) of many parrotfishes and wrasses (Fig. 3.5) show contrasting spawning behaviours: initial-phase males often spawn in aggregations and have group spawning (Fig. 12.45) whereas terminal-phase males usually spawn in pairs outside of aggregations and in defended territories. In the bluehead wrasse, which similarly has two male spawning strategies (Figs. 12.38, 12.39), the success of the two male types, and hence of the prevailing mating system, depends largely on population density: when it is economically viable to defend a territory, terminal-phase males do well and female- or resource-defence polygyny prevails, but when population density is high, territoriality is too costly and initial-phase group-spawning males predominate (Warner and Hoffman 1980a, b). Terminal phase queen parrotfish, *Scarus vetula*, also spawn in pairs and groups. In this species, males usually defend large permanent territories containing several females, with whom they pair-spawn (i.e. female-defence polygyny) (Robertson and Warner 1978). But Clavijo (1983) also observed lek polygyny in a Puerto Rican population of queen parrotfish: males aggregated along the shelf edge where they defended small, closely packed territories. They courted visiting females, which inspected several males before selecting one with which they spawned. The underlying mechanism for this mating-system shift was not investigated.

Selection pressures acting on males will differ according to whether males spawn in multi- or single-male spawns. In contrast to males that spawn in multi-male groups, males that spawn alone with a female compete physically and benefit from being large. Therefore, among aggregation spawners (and more generally among all fishes), we should expect to see a larger ratio of male:female sizes in species that pair spawn than in species that spawn in multi-male groups. Intraspecific evidence for this prediction from parrotfish is strong: group-spawning males are similar in size to females while pair-spawning males are larger (e.g. Randall and Randall 1963; Choat and Robertson 1975; Robertson and Warner 1978). Interspecific evidence is less convincing but also in the expected direction. For example, a qualitative comparison of the degree of sexual dimorphism in group- and pair-spawning groupers offers support for the prediction: generally, group-spawning Nassau grouper (Sadovy and Colin 1995) show complete overlap in size distribution between the sexes, while males of pair-spawning species such as the tiger grouper, *Mycteroperca tigris* (White et al. 2002), red grouper, *Epinephelus morio* (Brulé et al. 1999; note: the species is not a confirmed aggregation spawner), and the red hind (Sadovy et al. 1994; Nemeth 2005) are larger, on average, than females. Similarly, pair-spawning, terminal-phase male wrasses are larger than females while conspecific group-spawning, initial-phase, males are similar in size to females (Warner and Robertson 1978).

When males gain a greater reproductive advantage with growth than females, female-first sex-change is predicted (Warner 1975; Leigh et al. 1976; Charnov 1982).

This is expected in pair-spawners but not multi-male group-spawners (Molloy et al. 2007). It is noteworthy, then, that the only groupers for which there is evidence that some individuals are gonochoristic (i.e. they do not change sex and only ever function as one sex) are group-spawning Nassau grouper (Sadovy and Colin 1995), leopard grouper (Erisman et al. 2007) and kelp bass, *Paralabrax clathratus* (Sadovy and Domeier 2005b), where selection for large male size is relaxed. Further comparative analyses should reveal that female-first sex change is more common among those species that spawn in pairs or single-male groups than those that spawn in multi-male groups.

Highly tuned mate choice by females and intense male-male competition, particularly in pair-spawning species, lead to highly skewed male reproductive success. Not only do the largest males often defend the best resources or spawning locations, they are also disproportionately selected as mates (e.g. threadfin wrasse *Cirrhilabrus temminckii*, Kohda et al. 2005; Mediterranean parrotfish *Sparisoma cretense*, Afonso et al. 2008b; Andersson 1994). Skewed male reproductive success has rarely been quantified in aggregation spawners. One exception is an example of the Atlantic cod (North Sea population). Using microsatellite-based parentage analysis, Bekkevold et al. (2002) showed that male reproductive success followed a U-shaped distribution, indicating that most males achieved either very high or very low reproductive success. A subsequent study showed that 80% of the offspring were sired by only 2–7 males out of groups of 52–93 individuals (Rowe et al. 2008). Furthermore, larger males with longer fins used in courtship had the highest reproductive success, implying strong sexual selection in this species. Similar analyses have not been performed on other aggregation spawners but based on the discussion above relating to pair- and group-spawning males, one can predict that male reproductive success will be more heavily skewed in pair- than group-spawners (see also Chap. 8).

As more species and aggregations are studied, it is likely that within-species variation in mating systems and tendency to spawn in aggregations will be recorded in many other species. Such flexibility may prove important in relation to their ability to cope with heavy fishing pressure (Colin 1982a): if heavy fishing mortality at aggregations makes it unprofitable for individuals to spawn in groups, non-aggregated resource- or female- defence mating systems may prevail. Other species without such phenotypic plasticity may be more vulnerable to fisheries-mediated extirpation. Whether or not this is the case, and whether fish have such behavioural flexibility, are pertinent questions for fisheries conservation. As far as we are aware, no studies have attempted to test such issues. These questions could be explored using modelling techniques to predict relative vulnerability of species with and without different spawning tactics, or with varying degrees of behavioural plasticity. More empirical evidence could be obtained by monitoring relative frequency of different spawning tactics on and away from known spawning sites. Many of the methods that would be involved in the design and implementation of the fieldwork required to test the predictions described above are detailed in Chaps. 9 and 10.

3.7 Conclusions and Future Directions

A few fish species that spawn in aggregations have been very well studied. For these species, it appears that there is no consistent pattern pointing to a single explanation for the evolution of spawning in aggregations, e.g. social benefits of aggregating versus space limitation. This is not surprising, because there is considerable variation among species in social systems, travel costs, and the nature of the aggregations themselves. The phylogenetic diversity of aggregation-spawning fishes (Domeier and Colin 1997, Chapter 4) also makes it unlikely that there should be a single evolutionary explanation for aggregation spawning. Most studies lack experimental manipulations, which has prevented definitive conclusions. Although predictions from competing hypotheses are not fundamentally impossible to test, they are harder to tackle in some taxa than in others. For example, studies of birds breeding in leks and in colonies have gone much farther than studies of fishes because birds are often easier to mark for individual recognition, the parentage of the young is easier to determine, they often lend themselves better to experimentation, and unexploited populations are more widely accessible for study.

To move forward in our understanding of fish aggregations, it would be helpful to study more model systems where individuals can be tracked and their reproductive success recorded. For example, studies of groupers (e.g. Nemeth et al. 2007; Starr et al. 2007) have been very informative about travel costs and individual variation in behaviour, and studies of Atlantic cod in large tanks have been helpful for measuring individual differences in behaviour and reproductive success (Rowe et al. 2008). There is very little direct evidence concerning predation dilution benefits of spawning aggregations. However, as most fisheries around the world are causing rapid depletion of top predators, perhaps progress can be made by comparing the timing, density, and behaviour of fish in spawning aggregations inside and outside of marine protected areas.

Many spawning aggregations are themselves targeted heavily by fisheries (Chaps. 8 and 11). It would be interesting to study changes in spawning behaviour as traditional aggregations become depleted or recover from over-fishing. For example, it would be useful to test for Allee effects, also known as depensation in the fisheries literature, whereby components of individual fitness decline with decreasing group sizes (Courchamp et al. 2008) (Chap. 8). Again, comparisons of aggregations inside and outside of marine protected areas or across fishing gradients would be helpful here. Such studies would both help elucidate costs and benefits of spawning in aggregations, and they might also shed some light on the specific management needs and the effectiveness of management actions taken to protect species that exhibit this remarkable behaviour.

Acknowledgements This is a contribution from Project Seahorse at University of British Columbia, and the Earth2Ocean group at Simon Fraser University. PPM was supported by a Leverhulme Study Abroad Studentship # 2/SAS/2006/0057 and a Canadian Bureau for International Education postdoctoral research fellowship. JDR and IMC were supported by NSERC of Canada Discovery Grants. Thanks to Yvonne Sadovy and Patrick Colin for useful comments on previous drafts of this chapter.

References

Afonso P, Fontes J, Morato T et al (2008a) Reproduction and spawning habitat of white trevally, *Pseudocaranx dentex*, in the Azores, central north Atlantic. Sci Mar 72:373–381

Afonso P, Morato T, Santos RS (2008b) Spatial patterns in reproductive traits of the temperate parrotfish *Sparisoma cretense*. Fish Res 90:92–99

Aguilar-Perera A (1994) Preliminary observations of the spawning aggregation of Nassau grouper, *Epinephelus* striatus, at Mahahual, Quintana Roo, Mexico. Proc Gulf Caribb Fish Inst 43:112–122

Alatalo RV, Höglund J, Lundberg A et al (1992) Evolution of black grouse leks – female preferenced benefit males in larger leks. Behav Ecol 3:53–59

Alonzo SH, Mangel M (2004) The effects of size-selective fisheries on the stock dynamics of and sperm limitation in sex-changing fish. Fish Bull 102:1–13

Andersson M (1994) Sexual selection. Princeton University Press, Princeton

Arroyo B, Mougeot F, Bretagnolle V (2001) Colonial breeding and nest defence in Montagu's harrier (*Circus pygargus*). Behav Ecol Sociobiol 50:109–115

Balmford A, Turyaho M (1992) Predation risk and Lek-breeding in Uganda Kob. Anim Behav 44:117–127

Barlow GW (1981) Patterns of parental investment, dispersal and size among coral reef fishes. Environ Biol Fish 6:65–85

Beets J, Friedlander A (1998) Evaluation of a conservation strategy: a spawning aggregation closure for red hind, *Epinephelus guttatus*, in the U.S. Virgin Islands. Environ Biol Fish 55:91–98

Beets J, Hixon MA (1994) Distribution, persistence, and growth of groupers (Pisces, Serranidae) on artificial and natural patch reefs in the Virgin Islands. Bull Mar Sci 55:470–483

Bekkevold D, Hansen MM, Loeschcke V (2002) Male reproductive competition in spawning aggregations of cod (*Gadus morhua, L.*). Mol Ecol 11:91–102

Bell LJ, Colin PL (1986) Mass spawning of *Caesio teres* (Pisces: Caesionidae) at Enewetak Atoll, Marshall Islands. Environ Biol Fish 15:69–74

Bolden SK (2000) Long-distance movement of a Nassau grouper (*Epinephelus striatus*) to a spawning aggregation in central Bahamas. Fish B-NOAA 98:642–645

Brown CR, Brown MB (1989) Behavioral dynamics of intraspecific brood parasitism in colonial cliff swallows. Anim Behav 37:777–796

Brown CR, Brown MB (2004) Empirical measurement of parasite transmission between groups in a colonial bird. Ecology 85:1619–1626

Brulé T, Déniel C, Colás-Marrufo T, Sánchez-Crespo M (1999) Red grouper reproduction in the southern Gulf of Mexico. Trans Am Fish Soc 128:385–402

Byrne PG, Roberts JD, Simmons LW (2002) Sperm competition selects for increased testes mass in Australian frogs. J Evol Biol 15:347–355

Charnov EL (1982) The theory of sex allocation. Princeton University Press, Princeton

Choat JH, Robertson DR (1975) Protogynous hermaphroditism in fishes of the family Scaridae. In: Reinboth R (ed) Intersexuality in the animal kingdom. Springer, New York

Clavijo IE (1983) Pair spawning and formation of a lek-like mating system in the parrotfish *Scarus vetula*. Copeia 1983:253–256

Claydon J (2004) Spawning aggregations of coral reef fishes: characteristics, hypotheses, threats and management. Oceanogr Mar Biol Ann Rev 42:265–302

Clifton KE, Robertson DR (1993) Risks of alternative mating tactics. Nature 366:520

Coleman FC, Koenig CC, Collins LA (1996) Reproductive styles of shallow-water groupers (Pisces: Serranidae) in the eastern Gulf of Mexico and the consequences of fishing spawning aggregations. Environ Biol Fishes 47:129–141

Colin PL (1978) Daily and summer-winter variation in mass spawning of the striped parrotfish, *Scarus croicensis*. Fish B-NOAA 76:117–124

Colin PL (1982a) Aspects of the spawning of western Atlantic reef fishes. In: Huntsman GR, Nicholson WR, Fix Jnr WW (eds) Workshop on biological bases for reef fish management. NOAA, St. Thomas

Colin PL (1982b) Spawning and larval development of the hogfish, *Lachnolaimus maximus* (Pisces: Labridae). Fish B-NOAA 8:853–862

Colin PL (1992) Reproduction of the Nassau grouper, *Epinephelus striatus* (Pisces: Serranidae) and its relationship to environmental conditions. Environ Biol Fish 34:357–377

Colin PL (1994) Preliminary investigations of reproductive activity of the jewfish, *Epinephelus itajara* (Pisces: Serranidae). Proc Gulf Caribb Fish Inst 43:138–147

Colin PL, Bell LJ (1991) Aspects of the spawning of labrid and scarid fishes (Pisces: Labroidei) at Enewetak Atoll, Marshall Islands with notes on other families. Environ Biol Fishes 31:229–260

Colin PL, Clavijo IE (1988) Spawning activity of fishes producing pelagic eggs on a shelf edge coral reef, southwestern Puerto Rico. Bull Mar Sci 43:149–179

Colin PL, Shapiro DY, Weiler D (1987) Aspects of the reproduction of two groupers, *Epinephelus guttatus and E. striatus* in the West Indies. Bull Mar Sci 40:220–230

Côté IM, Gross MR (1993) Reduced disease in offspring: a benefit of coloniality in sunfish. Behav Ecol Sociobiol 33:269–274

Côté IM, Poulin R (1995) Parasitism and group size in social animals: a meta-analysis. Behav Ecol 6:159–165

Courchamp F, Berec L, Gascoigne J (2008) Allee effect in ecology and conservation. Oxford University Press, Oxford

Craig PC (1998) Temporal spawning patterns of several surgeonfishes and wrasses in the American Samoa. Pac Sci 52:35–39

Crawford RH, Cusack RR, Parlee TR (1986) Lipid-content and energy-expenditure in the spawning migration of alewife (*Alosa pseudoharengus*) and blueback herring (*Alosa aestivalis*). Can J Zool 64:1902–1907

Davis JA, Brown CR (1999) Costs of coloniality and the effect of colony size on reproductive success in Purple Martins. Condor 101:737–745

Doherty PJ, Williams DM, Sale PF (1985) The Adaptive significance of larval dispersal in coral reef fishes. Environ Biol Fishes 12:81–90

Domeier ML, Colin PL (1997) Tropical reef fish spawning aggregations: defined and reviewed. Bull Mar Sci 60:698–726

Drazen JC, Goffredi SK, Schlining B et al (2003) Aggregations of egg-brooding deep-sea fish and cephalopods on the Gorda Escarpment: a reproductive hot-spot. Biol Bull 205:1–7

Dugatkin LA, Hoglund J (1992) Delayed breeding and the evolution of mate copying in lekking species. J Theor Biol 174:261–267

Erisman BE, Buckhorn ML, Hastings PA (2007) Spawning patterns in the leopard grouper, *Mycteroperca rosacea*, in comparison with other aggregating groupers. Mar Biol 151:1849–1861

Evans LE, Ardia DR, Flux JEC (2009) Breeding synchrony through social stimulation in a spatially segregated population of European starlings. Anim Behav 78:671–675

Fishelson L, Montgomery LW, Myrberg AA (1987) Biology of surgeonfish *Acanthurus nigrofuscus* with emphasis on changeover in diet and annual gonadal cycles. Mar Ecol Prog Ser 39:37–47

Gaudette J, Wahle RA, Himmelman JH (2006) Spawning events in small and large populations of the green sea urchin *Strongylocentrotus droebachiensis* as recorded using fertilization assays. Limnol Oceanogr 51:1485–1496

Gibson RM, Bachman GC (1992) The cost of female choice in a lekking bird. Behav Ecol 3:300–309

Gladstone W (1986) Spawning behaviour of the bumphead parrotfish *Bolmometopon muricatum* at Yonge Reef, Great Barrier Reef. Jpn J Ichthyol 33:326–328

Gladstone W (2007a) Selection of a spawning aggregation site by *Chromis hypsilepis* (Pisces: Pomacentridae): habitat structure, transport potential, and food availability. Mar Ecol Prog Ser 351:235–247

Gladstone W (2007b) Temporal patterns of spawning and hatching in a spawning aggregation of the temperate reef fish *Chromis hypsilepis* (Pomacentridae). Mar Biol 151:1143–1152
Graham RT, Castellanos DW (2005) Courtship and spawning behaviors of carangid species in Belize. Fish B-NOAA 103:426–432
Gross MR, MacMillan AM (1981) Predation and the evolution of colonial nesting in bluegill sunfish (*Lepomis macrochirus*). Behav Ecol Sociobiol 8:163–174
Hamilton WD (1971) Geometry of the selfish herd. J Theor Biol 31:295–311
Hays GC, Mackay A, Adams CR et al (1995) Nest site selection by sea turtles. J Mar Biol Assoc UK 75:667–674
Heard DC (1992) The effect of wolf predation and snow cover on musk ox group size. Am Nat 139:190–204
Heyman WD, Graham RT, Kjerfve B et al (2001) Whale sharks *Rhincodon typus* aggregate to feed on fish spawn in Belize. Mar Ecol Prog Ser 215:275–282
Heyman WD, Kjerfve B, Graham RT et al (2005) Spawning aggregations of *Lutjanus cyanopterus* (Cuvier) on the Belize Barrier Reef over a 6 year period. J Fish Biol 67:83–101
Höglund J, Alatalo RV (1995) Leks. Princeton University Press, Princeton
Hoglund J, Alatalo RV, Gibson RM et al (1995) Mate-choice copying in black grouse. Anim Behav 49:1627–1633
Hood PB, Schleider RA (1992) Age, growth, and reproduction of gag, *Mycteroperca microlepis* (Pisces: Serranidae), in the eastern Gulf of Mexico. Bull Mar Sci 51:337–352
Hosken DJ (1997) Sperm competition in bats. Proc R Soc Lond B Biol Sci 264:385–392
Jennings MJ, Philipp DP (1992) Female choice and male competition in longear sunfish. Behav Ecol 3:84–94
Johannes RE (1978) Reproductive strategies of coastal marine fishes in the tropics. Environ Biol Fishes 3:65–84
Kaunda-Arara B, Rose GA (2004) Homing and site fidelity in the greasy grouper *Epinephelus tauvina* (Serranidae) within a marine protected area in coastal Kenya. Mar Ecol Prog Ser 277:245–251
Kiflawi M, Côté I, Mazeroll AI, Goulet D (1998) Does mass spawning enhance fertilization in coral reef fish? A case study of the brown surgeonfish. Mar Ecol Prog Ser 172:107–114
Kohda M, Sasaki K, Sakai Y et al (2005) Preliminary study on female choice of mates versus sites in a wrasse, *Cirrhilabrus temminckii*. Ichthyol Res 52:406–409
Krajewski JP, Bonaldo RM (2005) Spawning out of aggregations: record of a single spawning dog snapper pair at Fernando De Noronha archipelago, equatorial Western Atlantic. Bull Mar Sci 77:165–167
Krause J, Ruxton G (2002) Living in groups. Oxford University Press, Oxford
Lack D (1968) Ecological adaptations for breeding in birds. Methuen, London
Leigh EGJ, Charnov EL, Warner RR (1976) Sex ratio, sex change and natural selection. Proc Natl Acad Sci USA 73:3656–3660
Lembo G, Fleming IA, Øklund F et al (1999) Site fidelity of the dusky grouper *Epinephelus marginatus* (Lowe, 1834) studied by acoustic telemetry. Mar Life 9:37–43
Levitan DR (2004) Density-dependent sexual selection in external fertilizers: Variances in male and female fertilization success along the continuum from sperm limitation to sexual conflict in the sea urchin *Strongylocentrotus franciscanus*. Am Nat 64:298–309
Lima SL, Dill LM (1990) Behavioral decisions made under the risk of predation: a review and prospectus. Can J Zool 68:619–639
Lobel PS (1978) Diel, lunar, and seasonal periodicity in the reproductive behaviour of the pomacanthid fish, *Centropyge potteri*, and some other reef fishes in Hawaii. Pac Sci 32:193–207
Lobel PS, Robinson AR (1988) Larval fishes and zooplankton in a cyclonic eddy in Hawaiian waters. J Plankton Res 10:1209–1223
Loiselle PV, Barlow GW (1979) Do fishes lek like birds? In: Reese ES, Lighter FJ (eds) Contrasts in behavior. Adaptations in the aquatic and terrestrial environments. Wiley, Toronto

Møller AP, Birkhead TR (1993) Cuckoldry and sociality: a comparative study of birds. Am Nat 1142:118–140

Molloy PP, Goodwin NB, Côté IM et al (2007) Sperm competition and sex change: a comparative analysis across fishes. Evolution 61:640–652

Montgomery WL, Galzin R (1993) Seasonality in gonads, fat deposits and condition of tropical surgeonfishes (Teleostei: Acanthuridae). Mar Biol 115:529–536

Morgan IE, Kramer DL (2004) The social organization of adult blue tangs, *Acanthurus coeruleus*, on a fringing reef, Barbados, West Indies. Environ Biol Fishes 71:261–273

Moyer JT, Thresher RE, Colin PL (1983) Courtship, spawning and inferred social organization of American angelfishes (Genera *Pomacanthus, Holacanthus,* and *Centryopyge*; Pomacanthidae). Environ Biol Fishes 9:25–39

Murphy CG, Gerhardt HC (2002) Mate sampling by female barking treefrogs (*Hyla gratiosa*). Behav Ecol 13:472–480

Nemeth RS (2005) Population characteristics of a recovering US Virgin Islands red hind spawning aggregation following protection. Mar Ecol Prog Ser 286:81–97

Nemeth RS, Blondeau J, Herzlieb S et al (2007) Spatial and temporal patterns of movement and migration at spawning aggregations of red hind, *Epinephelus guttatus*, in the US Virgin Islands. Environ Biol Fishes 78:365–381

Olsen DA, LaPlace JA (1979) A study of a Virgin Island grouper fishery based on a breeding aggregation. Proc Gulf Caribb Fish Inst 31:130–144

Pelaprat C (1999) Observations on the spawning behaviour of the dusky grouper *Epinephelus marginatus* (Lowe, 1834) in the North of Corsica (France). Mar Life 9:59–65

Peres MB, Klippel S (2003) Reproductive biology of southwestern Atlantic wreckfish, *Polyprion americanus* (Teleostei: Polyprionidae). Environ Biol Fishes 68:163–173

Petersen CW, Warner RR, Cohen S et al (1992) Variable pelagic fertilization success: implications for mate choice and spatial patterns of mating. Ecology 73:391–401

Randall JE, Randall HA (1963) The spawning and early development of the Atlantic Parrotfish, *Sparisoma rubripinne*, with notes on other scarid and labrid fishes. Zoologica 48:49–63

Reynolds JD (1996) Animal breeding systems. Trends Ecol Evol 11:A68–A72

Reynolds JD, Gross MR (1990) Costs and benefits of female mate choice – is there a lek paradox? Am Nat 136:230–243

Reynolds JD, Jones JC (1999) Female preference for preferred males is reversed under low oxygen conditions in the common goby (*Pomatoschistus microps*). Behav Ecol 10:149–154

Robertson DR (1983) On the spawning behavior and spawning cycles of 8 surgeonfishes (Acanthuridae) from the Indo-Pacific. Environ Biol Fishes 9:93–223

Robertson DR, Warner RR (1978) Sexual patterns in the labroid fishes of the western Caribbean, II: the parrotfish (Scaridae). Smithson Contrib Zool 255:1–26

Rowe S, Hutchings JA, Bekkevold D et al (2004) Depensation, probability of fertilization, and the mating system of Atlantic cod (*Gadus morhua* L.). ICES J Mar Sci 61:1144–1150

Rowe S, Hutchings JA, Skjaeraasen JE et al (2008) Morphological and behavioural correlates of reproductive success in Atlantic cod *Gadus morhua*. Mar Ecol Prog Ser 354:257–265

Ryan MJ, Tuttle MD, Taft LK (1981) The costs and benefits of Frog chorusing behavior. Behav Ecol Sociobiol 8:273–278

Sadovy Y, Colin PL (1995) Sexual development and sexuality in the Nassau grouper. J Fish Biol 46:961–976

Sadovy Y, Domeier M (2005a) Are aggregation-fisheries sustainable? Reef fish fisheries as a case study. Coral Reefs 24:254–262

Sadovy Y, Domeier ML (2005b) Perplexing problems of sexual patterns in the fish genus *Paralabrax* (Serranidae, Serraninae). J Zool 267:121–133

Sadovy Y, Rosario A, Román A (1994) Reproduction in an aggregating grouper, the red hind, *Epinephelus guttatus*. Environ Biol Fishes 41:269–286

Sala E, Ballesteros E, Starr RM (2001) Rapid decline of Nassau grouper spawning aggregations in Belize: fishery management and conservation needs. Fish 26:23–30

Samoilys MA (1997) Periodicity of spawning aggregations of coral trout *Plectropomus leopardus* (Pisces: Serranidae) on the northern Great Barrier Reef. Mar Ecol Prog Ser 160:143–159

Samoilys MA, Squire LC (1994) Preliminary observations on the spawning behavior of coral trout, *Plectropomus leopardus* (Pisces: Serrandiae), on the Great Barrier Reef. Bull Mar Sci 54:332–342

Sancho G (2000) Predatory behaviors of *Caranx melampygus* (Carangidae) feeding on spawning reef fishes: a novel ambush strategy. Bull Mar Sci 66:487–496

Sancho G, Petersen CW, Lobel PS (2000) Predator-prey relations at a spawning aggregation site of coral reef fishes. Mar Ecol Prog Ser 203:275–288

Semmens BX, Luke KE, Bush PG et al (2006) Isopod infestation of postspawning Nassau grouper around little Cayman Island. J Fish Biol 69:933–937

Shapiro DY, Hensley DA, Appeldoorn RS (1988) Pelagic spawning and egg transport in coral-reef fishes - a skeptical overview. Environ Biol Fishes 22:3–14

Shapiro DY, Garcia-Moliner G, Sadovy Y (1994) Social system of an inshore stock of the red hind grouper, *Epinephelus guttatus* (Pisces: Serranidae). Environ Biol Fishes 41:415–422

Shields WM, Crook JR (1987) Barn swallow coloniality – a net cost for group breeding in the Adirondacks. Ecology 68:1373–1386

Sigura A, Justine JL (2008) Monogeneans of the speckled blue grouper, *Epinephelus cyanopodus* (Perciformes, Serranidae), from off New Caledonia, with a description of four new species of *Pseudorhabdosynochus* and one new species of *Laticola* (Monogenea: Diplectanidae), and evidence of monogenean faunal changes according to the size of fish. Zootaxa 1695:1–44

Sims DW (2005) Differences in habitat selection and reproductive strategies of male and female sharks. In: Ruckstuhl KE, Neuhaus P (eds) Sexual segregation in vertebrates: ecology of the two sexes. Cambridge University Press, Cambridge

Sinclair ARE, Mduma SAR, Acese P (2000) What determines phenology and synchrony of ungulate breeding in Serengeti? Ecology 81:2100–2111

Slotte A, Fiksen O (2000) State-dependent spawning migration in Norwegian spring-spawning herring. J Fish Biol 56:138–162

Starr RM, Sala E, Ballesteros E et al (2007) Spatial dynamics of the Nassau grouper *Epinephelus striatus* in a Caribbean atoll. Mar Ecol Prog Ser 343:239–249

Stockley P, Gage MJG, Parker GA et al (1997) Sperm competition in fishes: the evolution of testis size and ejaculate characteristics. Am Nat 149:933–954

Stokes DL, Boersma PD (2000) Nesting density and reproductive success in a colonial seabird, the magellanic penguin. Ecology 81:2878–2891

Svensson BG, Petersson E (1992) Why insects swarm – testing the models for lek mating systems on swarming *Empis broealis* females. Behav Ecol Sociobiol 31:253–261

Vincent ACJ, Sadovy Y (1998) Reproductive ecology and the conservation and management of fishes. In: Caro TM (ed) Behavioral ecology and conservation biology. Oxford University Press, Oxford

Warner RR (1975) The adaptive significance of sequential hermaphroditism in animals. Am Nat 109:61–82

Warner RR (1988) Traditionality of mating-site preferences in a coral-reef fish. Nature 335:719–721

Warner RR (1990) Male versus female influences on mating-site determination in a coral-reef fish. Anim Behav 39:540–548

Warner RR, Hoffman SG (1980a) Local population size as a determinant of mating system and sexual composition in two tropical marine fishes (*Thalassoma* spp.). Evolution 34:508–518

Warner RR, Hoffman SG (1980b) Population density and the economics of territoriality in a coral reef fish. Ecology 61:772–780

Warner RR, Robertson DR (1978) Sexual patterns in the labroid fishes of the western Caribbean, I: the Wrasses (Labroidae). Smithson Contrib Zool 254:1–27

Warner RR, Shapiro DY, Marconato A et al (1995) Sexual conflict: males with highest mating success convey the lowest fertilization benefits to females. Proc R Soc Lond B Biol Sci 262:135–139

Waschkewitz R, Wirtz P (1990) Annual migration and return to the same site by an individual grouper, *Epinephelus alexandrinus* (Pisces, Serranidae). J Fish Biol 36:781–782
White DB, Wyanski DM, Eleby BM et al (2002) Tiger grouper (*Mycteroperca tigris*): profile of a spawning aggregation. Bull Mar Sci 70:233–240
Wikelski M, Carbone C, Trillmich F (1996) Lekking in marine iguanas: female grouping and male reproductive strategies. Anim Behav 52:581–596
Wittenberger JF (1981) Animal social behavior. Duxton Press, Buston
Wittenberger JF, Hunt GL Jr (1985) The adaptive significance of coloniality in birds. Avian Biol 8:1–78
Zeller D (1998) Spawning aggregations: patterns of movement of the coral trout *Plectropomus leopardus* (Serranidae) as determined by ultrasonic telemetry. Mar Ecol Prog Ser 162:253–263

Chapter 4
Spawning Aggregations in Reef Fishes; Ecological and Evolutionary Processes

John Howard Choat

Abstract What factors have been important in the evolution of reef fish spawning aggregations? Surprisingly, basic biological features such as size, trophic ecology and anatomy are more predictive than life history features. As long as the different groups (Resident and Transient aggregators) shared basic properties of body size, nutritional ecology and anatomy they manifest similar spawning behaviours regardless of whether they are protogynous or gonochoristic, exhibit short or long generation times or have slow or fast population turnover rates. A critical element in the evolution of spawning aggregations is proposed to be the rapid advection of eggs and larvae away from the reef environment. In addition the timing of spawning episodes may be linked to specific seasonal and climatic features of the ocean environment, a variant of the match/mismatch hypothesis developed to explain spawning patterns in clupeoid fishes. Neither larval retention nor broad dispersal are seen as critical elements in the evolution of spawning aggregations. It is hypothesized that differences in aggregate spawning patterns and their underlying processes will occur in the Pacific and Atlantic oceans, a reflection of the different histories, oceanic environments and habitat structures of these two ocean basins.

4.1 Introduction

Spawning aggregations occur in a wide range of fish species distributed over a variety of marine habitats. Marine fishes are diverse, (Table 4.1) comprising over half the known species of vertebrates (Nelson 2006) and occur in a wide range of marine

J.H. Choat (✉)
School of Marine and Tropical Biology, James Cook University, Townsville, Qld 4811, Australia
e-mail: john.choat@jcu.edu.au

Y. Sadovy de Mitcheson and P.L. Colin (eds.), *Reef Fish Spawning Aggregations: Biology, Research and Management*, Fish & Fisheries Series 35,
DOI 10.1007/978-94-007-1980-4_4, © Springer Science+Business Media B.V. 2012

Table 4.1 Estimates of fish species number: fish species richness; global patterns

Category	Global (marine + freshwater)	Marine only	Coastal marine only	Coral Reef only
	31,319	18,227	11,752–12,214	5,508
Source	Eschmeyer and Fong (2010) Helfman et al. (2009)	Helfman et al. (2009)	Helfman et al. (2009) Tittensor et al. (2010)	Allen (2007)

Table 4.2 Estimates of fish species number: fish species richness; Coral reef faunas by region

Region	Indo-west Pacific	Tropical Atlantic	East Pacific	Total
	3,919	1,240	349	5,508
Source	Allen (2007)	Floeter et al. (2008) Osmar J. Luiz (2010)	Zapata and Robertson (2007)	–

environments, many of which support different forms of spawning aggregation (Table 4.2). For this reason it is unlikely that any one class of explanations will accommodate the known instances of aggregate spawning. Given the scope of this book, Chap. 4 will be restricted to those fishes that occur on coral reefs and which manifest resident and transient spawning behaviours (Nemeth 2008, Chap. 1). The emphasis is on the ecological and evolutionary processes associated with reef fish reproduction and the development of different aggregate spawning patterns.

Even this conservative approach to develop general explanations for spawning aggregations proved to be a difficult task for four reasons. Firstly, coral reefs support a very high diversity of fishes (Helfman et al. 2009). Secondly, reef ecosystems have a complex geological history especially during the Tertiary (Montaggioni and Braithwaite 2009). This has resulted in marked differences in reef systems and their fish faunas between ocean basins (Spalding et al. 2001, 2007; Bellwood and Wainwright 2002). Third, the reefs that provide an adult habitat are widely dispersed over oligotrophic oceans and connected by pelagic larvae that are associated with variation in recruitment and adult numbers over time. Fourth, dispersive larvae are disseminated over wide areas frequently recruiting to reef habitats that differ from the parent reef; this can result in major ecological and behavioural changes between parent and offspring generations. Not only is there a high diversity of species, often with different life-history features, but the fishes themselves frequently display high levels of ecological and behavioural plasticity (Warner 1997; Peterson and Warner 2002).

The reef fishes that spawn in aggregations have characteristic patterns of size structure, morphology and nutritional ecology (Chap. 2). Can these features be used to help classify the various types of spawning behaviour? Do the distinctive growth dynamics, longevities and patterns of sexual ontogeny that distinguish the different lineages of reef fish provide additional help in classifying spawning behaviours? At present there is no clear answer to these questions. However, if the different spawning modes are found to consist of mixtures of lineages showing a variety of

Table 4.3 Estimates of fish species number: reef fish families confirmed to form spawning aggregation (From www.SCRFA.org, accessed 2009)

Taxon	Number of species	Aggregation confirmed
Perciformes		
Acanthuridae	82	12
Caesionidae	22	2
Labridae (Wrasses)	440	11
Lethrinidae	40	4
Lutjanidae	107	10
Mugilidae	73	2
Mullidae	77	1
Pomacentridae	345	6
Labridae (Scarines)	98	8
Serranidae (Epinephelines)	167	18
Siganidae	32	2
Tetradontiformes		
Balistidae	41	3
Totals	**1,352**	**79**

life history features this implies that the formation of spawning aggregations is based on very general biological processes such as those that enhance the reproductive success of individuals or result in reductions in juvenile mortality. If so, this suggests that growth rates, longevity or sexual ontogeny have little influence on how aggregations form. This question will be a primary consideration in the development of this chapter.

The material for the study of aggregations, especially that relating to commercially important species is listed in the SCRFA database (SCRFA 2009 www.SCRFA.org) and is summarized in the initial chapters of this book (Chap. 1, Appendix). These record the number of species in which spawning aggregations have been proposed either on the basis of direct evidence (spawning observation, presence of hydrated eggs) or indirectly (seasonal peaks in reproductive activity, aggregating behaviour), and provide a basis for assessing the ecological and evolutionary variety within the different spawning modes, resident and transient. The record to date demonstrates that aggregations are dominated by relatively few lineages of reef-associated fishes. The majority of these species release pelagic eggs and most spawn on the outer margins of reefs including on reef promontories and outer slopes. The 2010 database lists 12 families for which spawning aggregations have been confirmed (others have been reported but not confirmed) (Table 4.3, Appendix). This list is dominated by four families of perciform fishes; groupers (Serranidae), snappers (Lutjanidae), surgeonfishes (Acanthuridae) and wrasses (Labridae) (Sadovy de Mitcheson et al. 2008).

On-going work suggests that this list will be rapidly augmented by the inclusion of a number of additional reef-associated species. The database is partitioned with respect to spawning mode; all confirmed records of groupers and snappers classified these as transient spawners; all surgeonfishes and wrasses were classified as resident spawners. Predictably there are a large number for which the spawning classification was unknown (Table 4.3, Appendix).

The focus of this chapter will therefore be the demographic and evolutionary relationships of the above four perciform families and the extent to which these help explain the phenomenon of aggregation spawning. Two types of ecological factors are considered (1) basic biological features such as body size, feeding mode and functional anatomy and (2) life history information, including longevity, age structure, growth rates and patterns of sexual development such as sex-reversal. A major conclusion is that the capacity of reef fishes to form different types of spawning aggregations is governed by basic rules of size, trophic ecology, anatomy and physiology. Groups that made up the different spawning modes shared common properties of size, nutritional ecology and anatomy. Whether they were protogynous or gonochoristic or had short or long generation times was a secondary consideration.

Advection of eggs and larvae away from reef habitats dominated by plankton feeders is hypothesized to be an important aspect of the development of aggregate spawning. Regardless of whether the propagules are retained near the reef or dispersed into oceanic waters the important consequence is that the initial stages of development occur in waters away from the reef habitats where there are high predation rates on plankton (Chap. 7). Advection may be achieved by spawning in areas subject to reef (Heyman and Kjervfe 2008) or tidal current systems (tidal jets associated with reef passes) (Chap. 9). Spawning in areas of enhanced water movement is common to most species and does not imply either retention or broad dispersal. Larval retention is simply one potential consequence after initial advection from reef habitats (Chap. 6).

A further conclusion concerns the linkage of periodic spawning episodes to particular oceanic conditions. This may be accomplished by restricting spawning to particular windows in the annual cycle so as to link spawning to oceanic regimes that result in successful larval development and growth. This is a variant of the match/mismatch hypothesis developed to explain spawning patterns in clupeoid fishes (Sinclair 1988). This contrasts with the patterns shown by some species that spawn over the entire annual cycle which exposes the annual output of propagules to a potentially greater range of environmental conditions (see also Chap. 6).

Detailed descriptions of reproduction and spawning behaviour have been compiled for only a fraction of the reef fish fauna; a more comprehensive set of examples may modify these conclusions. Moreover the terms *resident* and *transient* are not mutually exclusive and may simply represent a continuum of reproductive behaviours (Fig. 4.1). Whether we classify spawning aggregations as resident or transient may depend on ocean-specific patterns of reef structure, a consequence of the different geological and evolutionary histories (Montaggioni and Braithwaite 2009).

A number of additional hypotheses seek to explain spawning aggregations, many compiled by Claydon (2005). However, given the accumulation of biological detail relevant to this topic it is an appropriate time to focus on the basic biological attributes that are important for the development and maintenance of spawning aggregations and the processes that underlie the evolution of spawning aggregations.

Fig. 4.1 (**a**) Resident aggregation of the humphead wrasse, *Cheilinus undulatus*, in Palau. Small numbers of males and females travel frequently to specific sites for extended periods in the year and spawn in pairs during the day according to the tides. (**b**) Transient aggregation of twin-spot snapper, *Lutjanus bohar,* in Palau. Large numbers of fish gather for brief periods for a short time each year and spawn in groups; picture shows fish returning to substrate after releasing eggs and sperm (Photos: Mandy Etpison, Nial McCarthy)

4.2 Coral Reef Fishes, Diversity, Habitat Structure and Biogeography

Fishes are the most diverse group of vertebrates with a total of 31,319 species (Table 4.1) (Eschmeyer and Fong 2010). Nearly 70% of the world's marine fish species occur in only 8% of the ocean habitat and, more dramatically, coral reefs with a

Table 4.4 Ocean basin and reef habitat estimates: ocean basin data

	Pacific	Atlantic	Indian
Metric			
Ocean area km^2.10^6	152.6	81.7	67.5
Shelf area km^2.10^6	23.6	20.7	6.7
Coastline km. 10^3	135.7	111.9	66.5
Sources	https://www.cia.gov/library/publications Wright and Rothenbury (1998)		

Table 4.5 Ocean basin and reef habitat estimates: coral reef areas by region

Region	Pacific	Indian	Red Sea	Atlantic	Indian
Reef area km^2.10^3	176	36	17	24.2	0.1
Total IWP	232	–	–	–	–
Total global	256.3	–	–	–	–
Source	Spalding and Grenfell (1997)				

Table 4.6 Ocean basin and reef habitat estimates: coral reef areas as a function of shelf metrics

Metric	Pacific	Atlantic	Indian
Reef area/shelf area	0.98	0.11	0.54

total global area of 256,500 km^2 support ~6,000 species of fishes (Tables 4.2 and 4.3) (Spalding and Grenfell 1997; Spalding et al. 2001; Allen 2007) in habitats that contribute only 0.08% of the ocean by area; it is anticipated that reef areas will increase when deeper systems are incorporated. The diversity of reef fishes varies by ocean basin with the Pacific dominating. The Indo-west Pacific supports 71% of these species with the distributions strongly biased towards the western regions of the ocean and the archipelagoes lying between 10° North, 10° S and 110° to 130° East. This area supports ~2,000 species such that an area that comprises 3% of the Indo-west and central Pacific supports 60% of the reef fish species (Randall 1998; Allen 2007). In contrast the tropical Atlantic has a relatively sparse fauna with a total of 1,240 fishes (Floeter et al. 2008; Osmar J. Luiz, 2010) (Table 4.2).

The distribution and extent of coral reefs varies between ocean basins. Most reefs occur in the Pacific Ocean (176 × 103 km^2) with the majority of these on the western margin (156.2 × 10^3 km^2) which includes Southeast Asia, Papua New Guinea, Solomon Is., Australia and the adjacent archipelagos; Indonesia alone supports 51.0 × 10^3 km^2 of reef habitat (Tables 4.4–4.7) (Spalding et al. 2001). In marked contrast the east Pacific supports only one significant coral reef, the 4 km^2 reef of Clipperton Atoll (Allen 2007). The Indian Ocean shows relatively little reef development (a total of 36.0 × 10^3 km^2) especially in northwestern and eastern areas, with the western extension of this ocean, the Red Sea, having an area of 17.0 × 10^3 km^2. The tropical Atlantic supports only 24.2 × 10^3 km^2 with the great majority of this located in the western region, the Caribbean. The eastern Atlantic mirrors the eastern

Table 4.7 Ocean basin and reef habitat estimates: coral reef areas as a function of coastal metrics

Metric[a]	Western Pacific	Caribbean
Coastal length/land area	2.65 ± 2.21	0.22 ± 0.18
Reef area/coastal length	1.79 ± 0.52	0.51 ± 0.12

[a]Data. Estimates of land area and coast length Google Earth Pro; Western Pacific: Micronesia Federated States, Marshall Is, Solomons, PNG, Vanuatu, Fiji, Philippines, Caribbean: Belize, Honduras, Venezuala, Dominican Republic, Cuba, Jamaica, Haiti, Trinidad-Tobago, Virgin Is, Bahamas

Pacific with only fractional reef development with a total of <100 km^2 (Spalding and Grenfell 1997; Spalding et al. 2001). Given the major differences in reef habitat metrics among ocean basins it is possible that the dynamics of spawning aggregations will also vary on a biogeographic scale, by the number and size of aggregations present in different oceans. A greater cover of coral reef habitat and their greater structural complexity, for example, as opposed to simply hard bottom habitats would be associated with larger numbers of relatively small aggregations.

4.3 Spawning Modes and Reproductive Behaviour in Reef Fishes

Spawning aggregations in reef fishes can be classified as either *resident* or *transient* (Domeier and Colin 1997, Chap. 1). These are nested within a broader classification of reef fish reproductive biology. A basic reproductive trait is mode of egg development. This can be demersal, where fertilized eggs develop in nests or attached to the substratum, or pelagic, where eggs are fertilized and develop in open water following broadcast spawning (Chap. 7). Demersal egg development followed by pelagic larval development occurs mainly in small species (Thresher 1984; Munday and Jones 1998) the majority of which are not aggregate spawners. Reproduction is usually accomplished by pair- spawning, occurs within localized areas in the normal foraging habitat and may involve parental care. There are exceptions to the condition of small size and non-migratory spawning behaviour in species with demersal egg development. Two groups, triggerfishes (Balistidae), and the rabbitfishes (Siganidae), contain relatively large and mobile species that produce demersal eggs yet undergo aggregative spawning (Table 4.8).

Triggerfishes (mean length of 42 cm, range of 15–100 fork length cm), are predominantly pair-spawning (Kawase 1998) with varying degrees of parental care. Residential spawning aggregations have been recorded in *Pseudobalistes* (55–60 cm) and *Balistoides* (75 mm) spp. These include the yellowmargin triggerfish, *Pseudobalistes flavimarginatus*, in which spawning occurs in a lek-like system (Gladstone 1994). Aggregation has also been observed in the titan trigger fish, *Balistoides viridescens*, where it co-occurs at the same aggregation site as a large squaretail coralgrouper, *Plectropomus areolatus*, aggregation in the Solomon islands (Alec Hughes personal communication), as well as with the brown-marbled grouper, *Epinephelus fuscoguttatus*, and the camouflage grouper, *E. polyphekadion*, in Palau (Yvonne Sadovy, 2010, Chap. 12.23). Up to 500 individuals may occur in an area 800 m^2 from 6 to 25 m depth. Pair-spawning is suggested by numerous observations (Alec Hughes personal communication). It is noteworthy that the largest triggerfish aggregate at the same site as transient spawning groupers adjacent to reef passes on the outer reef slope (Alec Hughes personal communication). Colin (Chap. 12.23) reports a similar reproductive pattern for the yellowmargin triggerfish in Palau. Rabbitfishes are moderately sized (mean 28, range 18–39 cm) with aggregations reported for nine species (SCRFA 2009 www.SCRFA.org). However at least one species *Siganus argenteus* produces dispersive eggs while others (*S. luridus, S, rivulatus, S. guttatus*) produce demersal adhesive eggs (Popper et al. 1979; Juario et al. 1985). The mode of spawning is unclear although both pair- and group-spawning in small groups within the aggregation have been inferred. Aggregations of *Siganus fuscescens* have been described in Palau (Chap. 12.22).

4.3.1 Pelagic Egg Development

Pelagic egg development is the dominant mode in reef fishes of intermediate (19–38 cm) and large (>38 cm) size (Chap. 7). The former size class includes wrasses, surgeonfishes, butterflyfishes (Chaetodontidae), angelfishes (Pomacanthidae), anthiids (Serranidae; Anthiinae), and sandperches (Pinguipedidae) and the latter size group includes groupers, snappers, jacks (Carangidae), wrasses, emperors (Lethrinidae), grunts (Haemulidae) and large wrasses. Although small wrasses, particularly the bluehead wrasse, *Thalassoma bifasciatum*, participate in resident spawning migrations (Warner 1995) most species at or below this size range spawn within the normal foraging area (Robertson 1972; Walker and McCormick 2009; McCormick et al. 2010). The smaller (<19 cm) representatives, especially wrasses, sandperches and some surgeonfishes and *scarine labrids* [old name Scaridae – parrotfishes and hereafter referred to as *parrotfishes*], undergo localized pair spawning, frequently in a haremic social structure without migration within the reef systems (Robertson 1983; Kuwamura et al. 2009; Walker and McCormick 2009). Non-aggregative pair spawning is prevalent in butterflyfishes (Colin 1989; Lobel 1989) which establish long-term pair relationships, while angelfishes are predominantly haremic spawners (Allen et al. 1998) as are anthiids and sandperches (Walker and McCormick 2009). The major reproductive and biological characteristics of demersal and pelagic spawning species are summarized in Table 4.8. This predictably relates small size to high levels of species diversity and increased abundances. Despite the apparent ubiquity of pelagic spawning it is noteworthy that on coral reefs the majority of species and individuals are demersal pair-spawners.

Table 4.8 Comparison of reproductive and ecological characteristics of 22 taxa representative of main reproductive modes in reef fishes (T=transient; R=Resident)

Taxon	Egg development mode	Mean size cm FL	Species richness	Diet; % carnivory	Abundance estimate Individual/ 10 m²	Spawning mode Pair	Group	Migration
Gobiidae	Demersal	3.8	~1,600	32	75.5	x	–	No
Pomacentridae[a]	Demersal	7.3	340	40	98.6	x	–	No
Blenniidae	Demersal	5.4	310	25	9	x	–	No
Tripterygidae	Demersal	4.4	90	85	3.5	x	–	No
Apogonidae	Mouth brooders	8.1	290	100	6.5	x	–	No
Siganidae[a]	Some Demersal	28	32	6	0.3	x	–	R
Balistidae[a]	Demersal	42	41	72	0.2	x	–	R
Monacanthidae	Demersal	32	69	65	0.15	x	–	?
Chaetodontidae	Pelagic	12.1	125	85	1.6	x	–	No
Pomacanthidae	Pelagic	19.2	84	65	0.54	x	–	No
Labridae[a] (wrasses)	Pelagic	18.5	420	98	6	x	x	R
Labridae[a] (Scarines)	Pelagic	42.7	98	1	3.8	x	x	R
Acanthuridae[a]	Pelagic	38.2	82	15	5.2	x	x	R
Anthiinae	Pelagic	13	204	98	5.5	x	–	?
Cirrhitidae	Pelagic	11	34	94	3	x	–	No
Pinguipedidae	Pelagic	17	76	96	5	x	–	No
Caesionidae[a]	Pelagic	29	22	96	1.5	?	–	No
Lethrinidae[a]	Pelagic	44.5	40	98	0.21	?	–	T
Lutjanidae[a]	Pelagic	45.2	107	100	0.36	x	x	T
Serranidae[a] (Epinephelines)	Pelagic	72	167	100	0.28	x	x	T

(continued)

Table 4.8 (continued)

Taxon	Egg development mode	Mean size cm FL	Species richness	Diet; % carnivory	Abundance estimate Individual/10 m^2	Spawning mode Pair	Group	Migration
Carangidae	Pelagic	73.5	130	100	0.31	?	–	T
Haemulidae	Pelagic	50	204	95	0.35	?	–	?

Data sources: *Size* Author Froese and Pauly (2010); *Species richness* Eschmeyer and Fong (2010); *Abundance* Ackerman and Bellwood (2000), Author Unpublished; *Spawning mode* SCRFA data base (2009). *Diet analyses* Hiatt and Strasberg (1960), Randall (1967), Suyehiro (1942). Estimates for Pomacentrids and Wrasses adjusted for tropical species only

Note: Haemulid size estimates reflect the very large size of Indo-Pacific species of *Plectorhynchus*

[a]Families that contain species that aggregate to spawn (Table 4.3)

4.3.2 Variation in Reproductive Behaviour

Any classification of reef fish reproductive biology must acknowledge the intraspecific variation that occurs in population biology and reproductive parameters (Warner 1984; Clifton 1995; Gust 2004; Robertson et al. 2005; Ruttenberg et al. 2005; Trip et al. 2008; McCormick et al. 2010). Variation in local densities is an important driver of reproductive behaviour and population biology (Peterson and Warner 2002; Gust et al. 2003; McCormick et al. 2010), and may also influence patterns of male recruitment (Munday et al. 2006b). As just one example the plasticity of reproductive biology in surgeonfishes is high, ranging from pair-spawning to large aggregations and group-spawning within a single species at the same locality (Robertson 1983; Myrberg et al. 1988, Chap. 12.20).

Robertson (1983) recorded a variety of spawning behaviours in two species, brown surgeonfish, *Acanthurus nigrofuscus*, and striated surgeonfish, *Ctenochaetus striatus*, in the western Indian Ocean ranging from localized pair spawning to large resident aggregations comprising thousands of fish and group-spawning. Spawning generally occurred on ebb tides leaving a mass of fertilized eggs to be transported seawards by the tidal flow, a process well described by Hamner et al. (2007). For both brown and striated surgeonfishes, localized group spawning and pair spawning were recorded. In brown surgeonfish a lead female and groups of 6–15 males per episode resulted in polyandrous matings. Large males set up feeding territories on reef flats and pair spawned with individual females within the territory. Males may also migrate short distances to set up lekking systems and pair-spawn with adjacent females. In contrast group-spawning in large aggregations was achieved by migrations from reef flat and crest feeding sites to outer reef slopes 0.3–0.4 km from inshore habitats. Pair-spawning occurred within feeding territories. Streak or interference spawning by males from adjacent territories was observed. Although the tropical Atlantic supports only three abundant species of surgeonfish, two species at least, ocean surgeonfish, *A. bahianus*, and blue tang, *A. coeruleus,* maintain resident spawning aggregations of thousands of individuals (Colin and Clavijo 1988, Chap. 12.20, Domeier and Colin 1997). Both species have complex foraging and feeding behaviour patterns and display the same levels of reproductive complexity as their Indo-Pacific congeners.

Scarine labrids also exhibit intra-specific variation in spawning patterns (Kuwamura et al. 2009). The tropical Atlantic labrids, the redfin parrotfish, *Sparisoma rubrippine* (Randall and Randall 1963), stoplight parrotfish, *Sparisoma viride* (Van Rooij et al. 1996), and striped parrotfish, *Scarus iserti* (Colin 1978) also display aggregative spawning and complex reproductive behaviour. In the stoplight parrotfish spawning behaviour varied from pair-spawning in haremic territories to multiple male and female spawning groups.

Spawning usually occurred in deeper parts of the habitat at specific sites although the extent to which local densities were enhanced by aggregation was unclear. However in the case of the redfin parrotfish, large resident spawning aggregations were formed (Randall and Randall 1963). Localized complexity in reproductive tactics and spawning is also reported for the Caribbean resident spawner, the bluehead wrasse (Warner 1995). Kuwamura et al. (2009) provided details of spawning behaviour in 14 species of Indo-Pacific parrotfish including members of the genera *Calotomus, Chlorurus* and *Scarus* at a single site in Okinawa. Pair-spawning was observed in all 14 species and invariably involved a single female and a terminal phase male; streaking behaviour by competing males was observed in seven species and group-spawning in seven species. However the frequency of group spawning was represented disproportionately by individuals of the two most abundant species daisy parrotfish, *Chlorurus sordidus*, and rivulated parrotfish, *Scarus rivulatus*.

Similar information on locality-specific population and reproductive biology is not available for the larger transient spawning species. There are fewer aggregations compared with resident spawning species, spawning may be at dusk or nocturnal (Heyman and Kjerfve 2008; Colin 2010a) and there is often a lack of age-specific life history information, especially for many serranid species, including the well studied Nassau grouper, *Epinephelus striatus*. A feature of the comparison between resident and transient spawning is the sensitivity of resident spawning species to variations in density (Warner 1995; Petersen and Warner 2002). Depending on the local environment surgeonfishes show a range of behaviours from localized pair-spawning to very large resident aggregations on the same reef. The present information suggests that species that display classical transient spawning aggregations are more conservative with respect to reproductive behaviour. However this aspect of transient spawning needs to be further examined through a comparison of tropical West Atlantic and Indo-west Pacific snappers and groupers as reproductive behaviour in these groups may have a strong biogeographical element.

4.4 Ecological Features Associated with Resident and Transient Spawning Modes

4.4.1 The Influence of Size and Trophic Ecology on the Classification of Resident and Transient Spawning Modes

This section examines the ecological variables that might predict spawning modes in reef fishes and is divided into two sections. The first considers 20 families of reef fishes and examines two basic features, the size of fishes in the different families and their trophic biology. Large size (here expressed as fork length in cm) is an important characteristic of transient spawning migrations (Domeier and Colin 1997; Nemeth 2008, Chaps. 1 and 9), as aquatic transport costs scale with size (Denny 1993) and specific metabolic demand decreases with increasing size (Shuter and Post 1990, Chap. 2). Predation on reef fishes may also influence the type of spawning mode. Recent studies have emphasized the importance of predators on reef environments arguing that on undisturbed reef habitats predators will have dramatic impacts on the size structure and demography of prey species (Sandin et al. 2008). Predation may also have impacts on spawning individuals (Sancho et al. 2000a, b) especially for male participants (Clifton and Robertson 1993; Gust et al. 2003, Chap. 5). The impact of predation decreases with size (Hixon and Beets 1993; Sogard 1997; Sandin et al. 2008; Holmes and McCormick 2009). For this reason movement out of the normal foraging range to exposed reef areas may be restricted to species that achieve large size.

As size and density are negatively correlated (Ackerman and Bellwood 2000) aggregations of large individuals would need to draw on a broad habitat catchment area to achieve the numbers observed in transient aggregations. This would require the extensive migrations recorded for some larger aggregating species. We should expect a strong relationship between size, the distance covered in spawning migrations and the extent to which individuals move beyond their normal foraging range. For example Bolden (2000) recorded relatively long migrations in one of the larger groupers, the Nassau grouper, which contrasts with a medium-sized grouper, the red hind, *Epinephelus guttatus*, which migrates shorter distances – up to 33 km (Chaps. 2 and 12.3).

Trophic ecology will influence the daily activity, feeding and digestive processes in the various families. The greatest contrasts are between those species that graze a benthic biota including filamentous algae, detritus and meiofauna, and carnivores which actively forage for large mobile prey organisms. Herbivores and detritivores display continuous feeding activity and high food processing rates (Choat and Clements 1998; Choat et al. 2004). Most grazing species have consistent feeding sites reached by local migrations and determined by structural and biological features of the reef habitat (Montgomery et al. 1989; Claisse et al. 2009). Piscivores and larger invertebrate carnivores forage and feed more opportunistically (Hobson 1974, Bshary et al. 2006). The capacity to undergo extensive migration episodes beyond the normal foraging reef habitat may be dependent on the extent to which daily feeding requirements tie particular groups to specific reef habitats (Chap. 2).

In a preliminary analysis, 20 groups of reef fishes (Table 4.8) were subject to an ordination (PCA) analysis to examine the relationship between fish size, abundance, and nutritional ecology in the context of spawning behaviour. Two trends were observed (Fig. 4.2). PC1, which accounted for 65.8% of the variance, is shown on the horizontal axis. This was positively correlated with mean size and negatively correlated with abundance. PC2 accounted for 32.1% of the variance and was dominated by a trend on the vertical axis reflecting the trophic status of the families ranging from predominantly carnivorous groups to herbivorous and detrital feeders. Mean body size was positively correlated with PC1 while abundance was predictably negatively correlated. Both abundance and percent carnivory in the diet were negatively correlated with PC2.

The median size group (11–19.8 cm total length [TL]) (wrasses, anthiids, butterflyfishes and angelfishes) occupied an intermediate position in the size spectrum with the grazing groups, parrotfishes, rabbitfishes and surgeonfishes (28–37.8 cm TL) slightly larger. The largest families, including groupers, jacks, snappers and emperors all of which are carnivores, were grouped in the right lower quadrant of

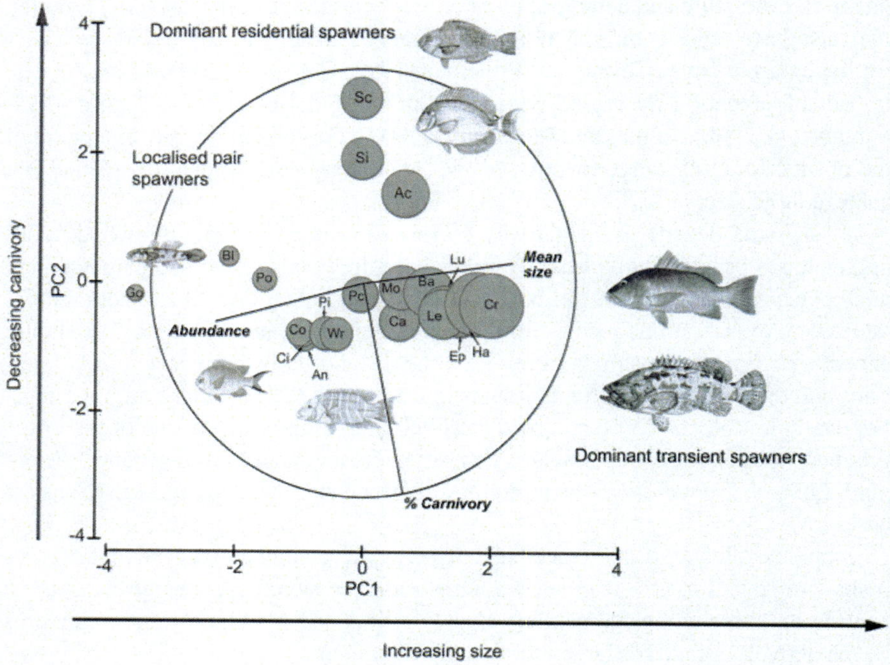

Fig. 4.2 Ordination of three variables; Mean size, Abundance and Percent carnivory using Principal Component Analysis (*PCA*). The plot shows the direction (eigenvectors) and contribution (relative length of eigenvectors) of the three variables. Bubble size represents gradient from lowest to highest values of mean size. Twenty reef fish taxa, Gobiidae Go, Blenniidae Bl, Pomacenridae Po, Chaetodontidae Co, Anthiinae An, Cirrhitidae Ci, Pinguipedidae Pi, Labridae (Wrasses) Wr, Pomacanthidae Pc, Caesionidae Ca, Acanthuridae Ac, Siganidae Si, Labridae (Scarines) Sc, Monacanthidae Mo, Balistidae Ba, Lethrinidae Le, Lutjanidae Lu, Epinephelidae Ep, Haemulidae Ha, Carangidae Cr were included in the analysis (Data shown in Table 4.8)

the PCA plot and include most of the verified instances of transient spawning aggregations. The negative correlation of abundance with PC1 confirms that the larger species are relatively rare suggesting that large transient spawning aggregations will be drawn from a large catchment area. The species grouped to the left quadrant of the plot are characterized by localized movements, small mean size, high abundances and turnover rates (Depczynski et al. 2007) and were usually pair-spawners. PC2 was negatively correlated with trophic biology and the vector shows the displacement of the three main herbivorous/detritivorous groups relative to carnivorous groups.

A number of species attain exceptionally large size but are not transient spawners. As they are all wrasses this has a strong phylogenetic component. The main species occur in the basal area of the labrid phylogeny (Westneat and Alfaro 2005) including the hypsigenyine, cheiline and scarine wrasses (Choat et al. 2006). For the two largest, the humphead, Napoleon or Maori wrasse, *Cheilinus undulatus,* and the bumphead, *Bolbometopon muricatum,* spawning occurs in groups on reef fronts and passes (Gladstone 1986; Hamilton 2004, Chaps. 9, 12.13 and 12.15) and involves pair spawning (Colin 2010a). In the case of humphead wrasse, the increases in local density are a reproductive phenomenon and represent migrations of ~5 km to the spawning site Colin (2010a). However for bumphead parrotfish there is no evidence to date that schools are augmented by migrants during reproduction. In summary, spawning behaviour was predicted by size, abundance and nutritional ecology with transient spawners represented by large carnivores and resident spawners by intermediate sized species, usually benthic grazers. Large wrasses are an exception and prompted an additional analysis to determine whether spawning behaviour might be better explained by demographic and life-history variables.

4.4.2 The Influence of Morphology, Demography and Life History on the Classification of Resident and Transient Spawning Modes

The analysis described in this section focused on the four families of reef fish, surgeonfish and wrasses (resident) and snapper and grouper (transient) fishes which make up 78% of the verified aggregate spawners. Surgeonfishes and parrotfishes show a high degree of similarity in terms of habitat, foraging behaviour, size and abundance profiles (Choat and Bellwood 1985; Choat et al. 2002, 2004). Snappers and groupers are more diverse ecologically but also share properties of size and trophic biology. Although the families within each spawning group share properties of size and trophic biology they differ significantly in life history features. Between surgeonfishes and parrotfishes (resident) there are differences in growth patterns and maximum ages in each group. Small surgeonfishes <20 cm FL may achieve maximum ages of 45 years. Larger species frequently achieve ages in excess of 40 years (Choat and Axe 1996; Choat and Robertson 2002). Most tropical labrids (especially parrotfishes) have shorter life spans including very large species. For example the humphead wrasse achieves maximum sizes an order of magnitude greater than most acanthurids but frequently reach only half the life-span of this group (Choat and Robertson 2002; Choat et al. 2006). They also display different growth curves, indeterminate in wrasses, and highly asymptotic in surgeonfishes (Choat and Robertson 2002).

A similar set of contrasts is seen in groupers and snappers (transient). Species of *Plectropomus* (the squaretail coralgrouper, the leopard coralgrouper, *P. leopardus,* and blacksaddle coralgrouper *P. laevis)* reach lengths in excess of 70 cm but maximum ages of usually 12–15 years with a maximum of 18 (Ferreira and Russ 1994; Williams et al. 2008; Heupel et al. 2010). Relatively small snappers, such as five-lined snapper, *L. quinquelineatus*, with a maximum size of <25 cm, achieve ages in excess of 30 years (Newman et al. 1996) while large species may exceed 50 years (Marriott et al. 2007). Each of the transient spawning groups also display the same differentiation of growth curves; groupers tend to have indeterminate growth curves while snappers usually display the same highly asymptotic growth curves as seen in surgeonfishes. There are also differences in reproductive patterns. Wrasses are predominantly protogynous hermaphrodites (Sadovy de Mitcheson and Lui 2008), while surgeonfishes are gonochorists with a distinctive gonad morphology (Montgomery and Galzin 1993; Montgomery et al. 1999). A similar pattern is seen in transient spawners. Groupers display complex patterns of sexual ontogeny with protogyny being the dominant mode; snappers appear to be gonochorists as there are no records of protogyny or protandry in this well-studied group (Sadovy de Mitcheson and Liu 2008; Erisman et al. 2009).

To identify ecological and life history features that might predict spawning mode, representatives of these families were examined by ordination analysis (Fig. 4.3). The first analysis considered maximum size and morphological variables (body cross sectional ratio and gape). Cross-sectional ratio is a proxy for internal body dimensions, habitat association and foraging mode (Fig. 4.3a). Gape was estimated as the ratio of the functional length of the upper jaw to head length. The use of external

Fig. 4.3 (**a**) Ordination of three morphological variables maximum length (Lmax), Body cross sectional ratio (BW/Bd) and Functional Gape (HL/M) using Principal Component Analysis (*PCA*) showing the direction (eigenvectors) and contribution (relative length of eigenvectors) of the three variables. Bubble size represents gradient from lowest to highest values of maximum size. Twenty eight taxa with records of aggregate spawning from four perciform families were included in the analysis. Acanthuridae *Acanthurus lineatus* Alin, *A.nigrofuscus* Anig, *A.blochii* Ablo, *A.bahianus* Abah, *Ctenochaetus striatus* Cstr, *Zebrasoma flavescens* Zfla. Labridae (Scarines) *Chlorurus microrhinos* Cmic, *C.sordidus* Csor, *Scarus psittacus* Spsi, *S.rivulatus* Sriv, *Sparisoma viride* Svir. Lutjanidae *Lutjanus apodus* Lapo, *L.adetti* Lade, *L.argentimaculatus* Larg, *L.bohar* Lboh, *L.campechanus* Lcam, *L.griseus* Lgri, *L.synagris* Lsyn, *L.fulvus* Lful. Serranidae *Plectropomus leopardus* Pleo, *P.laevis* Plae, *P.areolatus* Pare, *Epinephelus fuscoguttatus* Efus, *E.polyphekadion* Epol, *E.striatus* Estr, *E.adscensionus* Eads, *Mycteroperca bonaci* Mbon, *M.phenax* Mphe. Body cross-sectional ratios were estimated using the negative relationship between body depth and body cross-sectional ratios (Fulton 2005, Fig. 2.7). Fulton (2005) estimated body cross sectional ratios as body width/body depth so that species with high body planes and lateral compression, for example acanthurids, generated low values for cross sectional ratios (0.15–0.25). Species with fusiform bodies, for example serranids, generated higher values (0.4–0.5). HL is distance from snout tip to posterior margin of operculum; M length of the maxillary margin of jaw. Morphometric data extracted from digital images using Image J software. (**b**) Ordination of three demographic variables Maximum age (Tmax), Proportion of life span remaining when 50% of Linf is achieved (50% Linf), Proportion of primary males (%1° males). Principal Component Analysis (PCA) showing the direction (eigenvectors) and contribution (relative length of eigenvectors) of the three variables. Bubble size represents gradient from lowest to highest values of maximum age. Twenty eight taxa as in Fig. 4.2a. The metric used to compare growth profiles among the groups was an estimate of the proportion of the life span remaining after a specified size/developmental stage was reached in the different species. Although the size/age at which sexual maturity occurs would be the most appropriate reference point from which to estimate the remaining proportion of the life span this information was not generally available from the literature. As Von Bertalanffy Growth Functions were available for the relevant species a proxy value of the proportion of the life span remaining after 50% of *Linf* was achieved was used as an alternative.

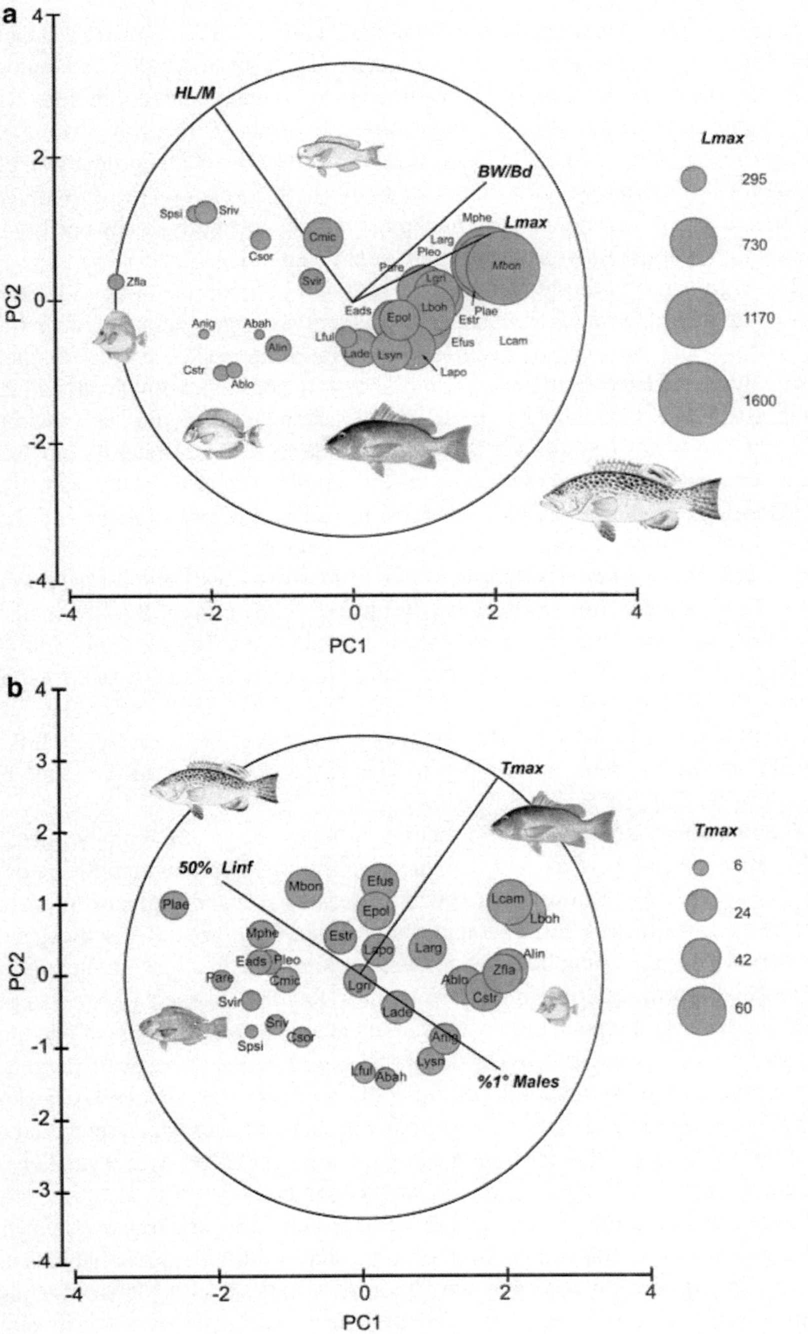
a
HL/M
BW/Bd
Lmax
PC1
PC2
Lmax
295
730
1170
1600
Spsi
Sriv
Cmic
Csor
Svir
Zfla
Anig
Abah
Alin
Cstr
Ablo
Eads
Lful
Lade
Lsyn
Epol
Lboh
Lgri
Pare
Pleo
Larg
Mphe
Mbon
Plae
Estr
Efus
Lcam
Lapo
b
Tmax
50% Linf
%1° Males
PC1
PC2
Tmax
6
24
42
60
Plae
Mbon
Efus
Epol
Lcam
Lboh
Mphe
Estr
Lapo
Larg
Alin
Eads
Pleo
Rare
Cmic
Lgri
Ablo
Zfla
Cstr
Svir
Lade
Sriv
Spsi
Csor
Anig
Lysn
Lful
Abah

body dimensions such as gape, body depth and body length as a proxy for locomotory, foraging and feeding patterns has been verified by more detailed studies on perciform fishes that have related body dimensions to functional morphology (Collar and Wainwright 2009). Body dimensions also influence reproductive biology. Sadovy (1996, Fig. 2.6) established a relationship between reproductive output and body form demonstrating that fishes with high levels of lateral compression (estimated as cross sectional ratio) had lower relative fecundity at any one time than those with fusiform body profiles at comparable lengths.

The argument that body architecture, size and internal anatomy will constrain ovary size and be associated with more frequent spawning episodes with smaller egg releases per episode has been suggested by a number of authors (Robertson 1991; Choat and Bellwood 1991; Sadovy 1996). In addition, body depth and lateral compression are linked to habitat associations, maneuverability and swimming speed in reef fishes (Fulton 2007). Fishes with high levels of lateral compression are associated with reef crests, may swim rapidly, primarily with pectoral fins (labriform) and feed on benthic resources. In contrast, representatives of fish with lower levels of lateral compression (snappers and groupers) are associated with deeper and more sheltered waters and swim primarily by body and caudal fin movements (sub-carangiform) (Fulton and Bellwood 2005; Fulton 2007). Small gape size in the surgeonfishes and parrotfishes is associated with rapid and continuous feeding and is associated with modified jaw architecture and the capacity to deliver a more powerful bite (Wainright et al. 2004; Konow et al. 2008; Price et al. 2010). The combination of a more elongate body and wider gape has been clearly linked to a predatory feeding mode with those species at the upper ends of the range being piscivores (Collar and Wainwright 2009).

The first analysis in this section focuses on morphology and biology relevant to aggregative spawning; size, body architecture, foraging and swimming mode and trophic biology. PC 1 was associated with increasing size and explained most of the observed pattern (Fig. 4.3a). The analysis retrieved three groups; (1) snappers and groupers where increasing body length was associated with a loss of laterally compressed body form resulting in fusiform shapes. Smaller snappers (Dory snapper *L. fulviflamma*) were more laterally compressed than the larger members of this family. The greatest sizes were achieved by the groupers with a characteristically large gape; (2) parrotfishes and (3) surgeonfishes; these share a small gape relative to head length with the degree of lateral compression greater in surgeonfishes which separated from the parrotfishes along this axis. Reduced gape was associated with a smaller body size and a tendency for higher levels of lateral compression.

These patterns identify differences in feeding, swimming and foraging behaviour. Species with high lateral compression occur in shallow turbulent reef habitats using pectoral fin swimming modes to achieve high swimming speeds and maneuverability (Fulton 2007). Snappers and groupers with a more fusiform body occur in habitats with less water movement and swim at slower speeds using a sub-carangiform swimming mode (Fulton 2005, 2007; Fulton and Bellwood 2005). However it is unclear how the different body forms and swimming modes perform under conditions of extended episodes of migratory swimming as seen in many transient spawning species. Although pectoral fin (labriform) swimming modes are the most efficient in environments characterized by high and variable water flow (Fulton and Bellwood 2005) experimental evidence (Korsmeyer et al. 2002) suggests that parrotfishes may efficiently employ rigid-body, median or paired-fin swimming for long periods, shifting to caudal fin swimming to achieve higher speeds. However this mode of swimming was energetically more costly and could not be maintained for extended periods of migratory swimming.

The dominant groups of resident spawners, surgeonfishes and parrotfishes, occupy shallow habitats where feeding occurs over the whole daily cycle and is achieved through rapid bites (Wainwright and Bellwood 2002; Choat et al. 2002, 2004). This is associated with a high investment in visceral anatomy and swim bladders which limit the space available for gonad development (Choat and Bellwood 1991; Sadovy 1996). The development of large ovaries (relative to body weight), as seen in transient spawning serranids (as reflected by high gonadosomatic indices, or GSIs) cannot occur in surgeonfishes and parrotfishes. Therefore, transient spawning species should show higher gonad (ovary) indices at peak spawning times than resident spawning species which have a more continuous pattern of reproductive output. In a preliminary analysis from the literature peak female GSIs for the four groups were found to be as follows; snappers 9.98 ± 1.8; large groupers 10.75 ± 1.5; surgeonfish 6.02 ± 0.7; parrotfishes 4.07 ± 0.5. In this context, gonad indices of the resident spawning large wrasses, the humphead wrasse and the bumphead parrotfish are of interest. For individuals of the former sampled from a spawning site the peak GSI obtained was 2.4% (Choat et al. 2006). For the latter peak ovarian GSI ranged from 5.9% to 6.6% (Hamilton et al. 2008a). Although these trends are in the predicted direction it is unclear whether all of these represent peak values as would occur in aggregations prior to spawning, for example once hydration has occurred. A more comprehensive analysis based exclusively on ovaries collected from spawning aggregations is required.

In addition to different capacities for gonad size and egg production patterns, resident and transient spawning groups should have differing capacities for energy storage. Transient spawning demands a capacity for *capital breeding* (reliance on stored energy for oocyte production) (Warner 1995; Stephens et al. 2009) due to the need to store resources to cover the costs of transport associated with migratory episodes and the restriction of reproduction to a limited number of episodes during the year. The ability of tropical snappers and groupers to store resources to fuel future episodes of growth and reproduction is unknown. However recent work has shown that the yellowfin grouper, *Mycteroperca venenosa*, an aggregate spawning species, may switch from growth to storage regimes using the liver as a repository for lipid storage (Stallings et al. 2010). Moreover increased storage of lipids does not appear to affect swimming performance in teleost fishes (Brix et al. 2009).

Nutritional ecology will have a major influence on the capacity to undertake long-distance migratory episodes (Chap. 2). Grazing species display continuous feeding at specific sites and a high turnover of ingested material (Choat et al. 2002, 2004). It is unlikely that these groups could defer feeding for extended periods and still maintain the costs of transport and high levels of reproductive output. In contrast snappers and groupers are predominantly carnivores on benthic invertebrates and on fishes, with a shift to increased piscivory with increase in size (Wainwright and Bellwood 2002). These groups are highly opportunistic feeders foraging over a wide range of habitats, feeding modes that are more likely to be sustained during migratory episodes than grazing on a shallow benthic biota. Thus maximum size, body architecture and trophic biology interact to constrain the migratory behaviour and episodic reproduction of transient spawning species to the larger carnivorous groups. In this context the influence of predation on the formation of spawning aggregations is unclear and requires additional study of size-dependent rates of predation on fish that make up spawning aggregations.

Is a classification of resident and transient spawners based on size and nutritional ecology linked to particular demographic and life history variables? In a second analysis the parameters maximum life span, the form of the growth curve (indeterminate or asymptotic) and reproductive ontogeny (gonochorism or protogynous hermaphroditism) were investigated (Fig. 4.3b). Fishes with asymptotic growth patterns will achieve maximum size early in life while those with indeterminate curves continue to grow over most of their life span. The different growth curves will require different patterns of investment in somatic and reproductive growth and could be an important driver of spawning patterns. The proportion of primary males in the population was used as an estimate of the importance of two distinctive patterns of sexual ontogeny, gonochorism and protogynous hermaphroditism. Protogynous species (most parrotfishes and groupers) would have a relatively low proportion of primary males compared to gonochorists.

Unlike the previous analysis which separated groups based on size, foraging and trophic variables the analysis based on demographic and life history variables retrieved different and more diffuse groupings (Fig. 4.3b). The assemblage was partitioned primarily by the pattern of sexual ontogeny with the relatively small surgeonfishes grouping with snappers and the gonochoristic Nassau grouper, while a second grouping consisted of protogynous groupers and parrotfishes. The species that grouped by a protogynous sexual ontogeny also exhibited indeterminate growth patterns. Most of the gonochoristic species, including the snappers and surgeonfishes, displayed asymptotic growth curves reaching the maximum size early in life and with limited growth following sexual maturity. There was a trend of increasing life span with the greatest ages achieved by the smaller surgeonfishes and the largest snappers and groupers (both *Epinephelus* and *Mycteroperca* spp.), and low maximum ages in parrotfishes and groupers of the genus *Plectropomus*. The main conclusion of the analysis was that basic demographic and life history features do not reflect the different spawning modes.

In summary the capacity of reef fishes to form large spawning aggregations appears to be governed by basic rules of size, trophic ecology and anatomy. The resultant spawning patterns emerge as combinations of lineages that have little in common with respect to demography and life histories. As long as the groups shared basic properties of size, nutritional ecology and anatomy they manifested the same spawning behaviours regardless of whether they were protogynous or gonochoristic, or with short or long generation times. The ecological analysis also suggested different patterns of diversification among two of the main lineages. Wrasses, which appear to be exclusively resident spawners, have a relatively small mean size, although a minority of basal species are very large (Choat et al. 2006) and show a great deal of morphological diversification (Westneat and Alfaro 2005). Groupers, which contain most of the recorded transient spawners, are more uniform morphologically with the primary axis of morphological diversification being an increase in size (Wainwright and Bellwood 2002; Craig and Hastings 2007).

4.5 Evolutionary Processes and the Development of Aggregative Spawning

The factors that drive the evolution of spawning aggregations remain unresolved. However important processes will be those that enhance survival of offspring relative to those produced by other individuals. These include two fundamental aspects of reef fish reproduction. The first is that spawning occurs at sites which result in advection of propagules to open water (Colin 2010a, b). The second is that anisogamy dictates that ova are the limiting reproductive resource and this will profoundly influence the reproductive behaviour of both sexes (Trivers 1972; Munday et al. 2006a). Given the disproportionate investment of resources in eggs relative to sperm, individual females will seek to maximize zygote survival. In this context egg release in a hydrodynamic and biological environment favouring zygote and larval survival is critical. If aggregation sites represent such environments and females migrate there to spawn, anisogamy ensures that males will follow.

This section discusses advection of eggs and zygotes from the reef following gamete release with the following sequence. (1) The issue of advection with respect to the scale of movement of propagules and the separate but related concepts of larval retention and dispersal. (2) The case for local advection as a process with general evolutionary implications including the significance of mortality early in teleost life histories and the ecological and evolutionary consequences of dispersal. (3) The factors that may restrict transient spawning to particular reef fish lineages and to a limited number of sites and spawning times (Chap. 6).

Debates concerning the capacity of currents to move passive propagules away from the reef usually focus on the extent to which they are retained in the natal area or are dispersed into the oceanic pelagic environment (Colin 2010a). Both scenarios involve the movement of propagules tens to thousands of kilometres. Advection as argued here involves a much smaller scale, tens to hundreds of metres. The high densities of benthic filter-feeding organisms on reef edge habitats and the numerous planktivorous fishes in the adjacent waters form a narrow predatory field where eggs and larvae are subject to high mortalities (Hobson and Chess 1978; Holt et al. 1985). Advection refers to transition through this narrow but intense inner predatory field to the pelagic environment (Hamner et al. 2007, Chap. 6).

The case for advection driving aggregate spawning behaviour is based on three observations. Firstly, most aggregation sites are adjacent to open water on reef fronts, passes or promontories (Sadovy de Mitcheson et al. 2008; Colin 2010a). Secondly, many of these sites are influenced by oceanic current systems, especially in the western Atlantic (Heyman and Kjerfve 2008), or by outward flowing tidal currents in the Indo-Pacific (Colin 2010a, Chap. 6). Thirdly, both types of current systems have the capacity to move propagules away from the immediate spawning site and in some cases entrain them into eddies adjacent to the reef but retain them in an open water environment beyond the predatory field. It is also noteworthy that many of the larger species that produce abundant propagules appear to spawn at dusk or nocturnally which would reduce mortality from visual predators (Colin 1992; Samoilys 1997; Heyman and Kjerfve 2008).

Teleost fishes experience exceptionally high mortality rates in the larval and juvenile phases, approximately five times those experienced by terrestrial groups (Perez and Munch 2010). Although mortality decreases with increase in age and size, only ~0.01–0.1 fish in a given cohort survive the first year of life (Claisse et al. 2009; Perez and Munch 2010). Reduction in the initial mortality rate would occur if propagules moved rapidly through the predatory field adjacent to the reef and into an open pelagic environment. Any tendency to ensure pelagic propagules passed rapidly beyond the near-reef predatory field would be favoured by selection.

Larval retention and return of propagules to a natal reef will benefit groups of reef fishes with specialized habitat requirements (Swearer et al. 2002; Almany et al. 2007; Jones et al. 2009). A successful parental generation indicates that they occupy an appropriate habitat for the next generation of recruits. While retention may benefit such species the case for those with wide distributions and general habitat requirements, including the four groups that dominate aggregate spawning, is less clear. Retention and self recruitment may only be one of a range of dispersal strategies seen in reef fishes as a number of species consistently exhibit very wide dispersal of larvae (Craig et al. 2007; Horne et al. 2008; Gaither et al. 2010). Similar arguments are made in the case of widely distributed tropical gastropods with differing levels of stability in habitats occupied by adults (Crandall et al. 2010). The case for wide dispersal in a number of reef fish groups is strengthened by two sorts of observations. Firstly, many reef species with a supposedly short planktonic larval duration can disperse over very wide distances including the East Pacific Barrier (Lessios and Robertson 2006). Secondly, transport into open-ocean and nutrient-poor larval environments does not mean that these larvae are disadvantaged relative to those retained in local nutrient rich waters. Although larval growth during the pelagic phase may be reduced, compensatory growth prior to settlement and relatively high survival compared to those retained in near-shore environments have been recorded (Hamilton et al. 2008b). The case for larval retention as a major reason for the development of spawning aggregations is not supported by the evidence of larval biology and phylogeography (see also Chaps. 6 and 7).

Although resident and transient spawners rely on advection to enhance egg survival they have different temporal patterns of spawning. This raises the question as to how the costs of reproduction are met in the two groups. The alternatives are capital breeding and *income breeding* (reliance on concurrent intake) (Warner 1995; Stephens et al. 2009). Income breeding in which daily reproduction is directly subsidized by continued feeding would be expected for many resident spawning species. In contrast the less predictable nature of food sources for transient spawning carnivores suggests that they may be capital breeders relying on reserves accumulated over periods prior to reproduction, especially as their body form and anatomy allows the development of a very large ovarian mass compared to body mass. Partitioning reproduction into a few large episodes requires the ability to store the nutrients necessary to nourish the development of large numbers of hydrated oocytes (the last stage before spawning). Resident spawners with extended seasonal reproductive periods (Ross 1983; Robertson 1991; Craig 1998; Bushnell et al. 2010, Chap. 12.20) that cover a range of seasonal and oceanic environments contrast strongly with transient spawners which reproduce at a limited number of times and places. Focusing reproductive outputs in this way may allow females to better anticipate the future larval environment. Indeed, van Woesik (2010) has identified such a proxy linking mass spawning in corals to a specific environmental feature, wind fields. In the case of reef fish aggregations a similar linkage between timing of transient spawning events and a general environmental signal has yet to be identified. However in temperate environments matching of spawning to a seasonally driven cycle of productivity and hydrodynamic features is well-established (Sinclair 1988).

4.6 Conclusions

The main conclusion of this chapter is that spawning aggregations have most probably evolved to enhance egg and larval survival through rapid advection of reproductive products away from the adult environment. The ecological distinctions between resident and transient spawners reflect differences in size and nutritional ecology but not life histories. Size, structural and nutritional constraints limit the acquisition of a transient spawning phenology to larger species, primarily those with a carnivorous nutritional ecology and the capacity to develop very large ovarian masses (i.e. high GSIs) fueled through an investment in stored energy. In addition the differences in timing and frequency of reproductive episodes in resident and transient spawners represent the spreading of recruitment risk in the former and matching of spawning to favourable larval environments in the latter. By limiting the spawning periodicity to a restricted number of episodes some tropical reef fish appear to be converging towards their cold-temperate counterparts where spawning is restricted to periods that match predictable environmental cycles. These arguments implicitly reject the idea that larval retention and self recruitment are the result of selective pressures applicable to reef fishes generally. In some species, especially those with highly specific recruitment sites and adult environments, retention of individuals would be advantageous. In others, wide dispersal and colonizing of new environments may be beneficial.

A number of predictions follow. Transient spawning will be restricted to those species above a size threshold of ~40 cm FL with an alimentary anatomy and body form that provides for the development of a large ovarian mass. There is a phyloge-

Table 4.9 Body size structure for transient spawning Epinepheline serranids and Lutjanids by geographical region (Data are mean maximum size FL (cm); maximum size estimates from Froese and Pauly (2010); author data; angling and spearfishing records)

Region	Global	IWP	Atlantic	East Pacific
Taxon Serranidae: *Epinephelus, Cephalopholis, Mycteroperca, Plectropomus, Dermatolepis*				
All species	72.0±3.5	60.5±3.5	99.4±7.5	109.1±15
Aggregating	–	82.2±5.7	109.25±10.3	121.0±25.9
Taxon Lutjanidae: *Lutjanus, Symphorichthys*				
All species	61.7±3.9	53.7±4.9	81.3±8.1	74.5±11.9
Aggregating	–	73.3±33.3	94.1±12.7	97.3±26.3

netic element in that large wrasses exceed the size threshold but appear to be resident spawners. However both body form (most large wrasses have a high degree of lateral compression) and nutritional ecology appear to preclude the development of a large ovarian mass. In this context, large wrasses, i.e. labrids such as the hogfish, *Lachnolaimus maximus*, are predicted to be confirmed as resident spawners. Transient spawning is predicted to be more prevalent in the tropical Atlantic than the western Pacific for two reasons. Firstly, reef extent and structure differ between these regions such that suitable spawning sites are restricted in the Atlantic thereby requiring fish to travel greater distances to aggregate (Tables 4.4–4.7). Secondly, the mean size of the dominant transient spawning groups, snapper and groupers is greater in the tropical Atlantic than in the Pacific (Table 4.9). The argument that the amount of reef habitat available will influence the degree of transient spawning observed also predicts that large transient spawning aggregations will be more prevalent in the Indian as opposed to the Pacific Ocean.

4.7 Future Directions

4.7.1 Atlantic and Pacific Reef Structures

Although many of the recorded Caribbean spawning aggregations are unequivocally transient in nature, it is unclear to what extent many Indo-Pacific aggregate spawning species meet the criteria of migratory episodes to form aggregations, and restriction of spawning to specific periods of the annual cycle. The western Pacific generally supports more suitable aggregation sites and possibly higher densities of aggregating species per unit length of coastline than the Atlantic although more comprehensive information on the densities of groupers from Pacific and Atlantic Ocean reef systems is a priority. Not only is there a greater area of reef habitats in the Pacific, the reef systems are more complex in the western Pacific than the western Atlantic with the length of coastline per unit area of land mass being 12 times greater in the western Pacific, a reflection of the different geological histories of each region. Moreover the amount of reef habitat per kilometre of coast is 3.5 times greater in the western Pacific than in the Caribbean (Tables 4.4–4.7). Inter-ocean differences in spawning behaviour will be important in groupers such as members of the genus *Plectropomus* which occur at higher densities than any species of Caribbean grouper (Chiappone et al. 2000; Pears 2005). If so, the catchment area from which aggregations are drawn and the distances to suitable aggregation sites will both be smaller in the western Pacific. This predicts that (a) the average size of undisturbed aggregations, and (b) the magnitude of trans-habitat migrations will be smaller in the western Pacific and the number of potential aggregation sites per unit of coastal habitat will be greater. In this context, details of spawning aggregations in leopard coralgrouper support the predictions of numerous small sites with resident aggregation (Chap. 12.9).

Moreover, if the association between mass spawning events and clear regional environmental cues (van Woesik 2010) is a general phenomenon, then the timing and frequency of aggregations is also expected to be a geographically specific feature. In addition, Colin (Chap. 9) drew attention to the differences in tidal amplitude between the oceans and this approach needs to be extended to seasonal wind fields that may influence the pelagic environment and the extent of deeper reef habitats that may provide additional spawning sites.

4.7.2 Regional Differences in Body Size and Transient Spawning

The list of aggregate spawning reef fishes comprises at least 11 families (78 species). With respect to transient spawning two lineages, groupers and snappers, dominate the present examples with the greatest diversity for these groups occurring in the Indo-Pacific. However for both groups the majority of confirmed records are from the Atlantic and East Pacific. The PCA analysis (Fig. 4.2) demonstrated an association between mean size and spawning mode. A more comprehensive analysis of size structure for each group was carried out to determine (1) the mean size of the total species pool in each group, (2) the geographical distribution of size within each group and (3) mean size of aggregating species in each group (Table 4.9). Analyses of size structure for entire lineages and of the currently recognized aggregating species demonstrated a consistent trend within groupers and snappers of a greater mean size in the Atlantic and with transient spawning species dominated by large Atlantic and East Pacific species. The geographical and reproductive size trends observed in the groupers and snappers can be interpreted with respect to the evolutionary history of each group. For the diverse genus *Epinephelus* the phylogenetic analysis of Craig and Hastings (2007) revealed a basal group of very large and geographically widespread species with a mean size of 105 cm TL and a derived clade consisting of small species (44 cm TL) confined to the IWP. The clade of very large groupers, presently included in the genus *Mycteroperca,* is basal to *Epinephelus* and confined to the Atlantic and East Pacific. Although a comprehensive phylogenetic analysis is not available for *Lutjanus* spp. the most derived clade is a complex of relatively small species also confined to the IWP (Miller and Cribb 2007). However a more informed analysis of transient spawning in each ocean requires a better database on the age-specific dynamics and abundance estimates (densities) of the main species, especially the larger groupers, including the Nassau grouper. Although there are a number of studies on the behaviour and spawning patterns (Sadovy and Colin 1995; Sadovy and Eklund 1999; Erisman et al. 2009) we are largely ignorant of the life span and age specific reproductive schedules of this iconic species.

4.7.3 Lack of Information

Our capacity to understand the biological basis of spawning aggregations is compromised by the paucity of information on the anatomy and physiological processes associated with the annual reproductive cycles especially of the larger groupers and snappers. This includes information on the costs of reproduction in both resident and transient spawning groups in the context of income or capital breeding systems.

There are a number of management and conservation implications in this approach. If, as suggested, Indo-Pacific species and especially groupers are mainly resident spawners with a smaller mean size than those of the tropical Atlantic then protection of individual reefs (i.e. spatial approaches) may be an appropriate management response. This is especially true for the leopard coralgrouper, one of the most important species in the live reef fish trade which attain high abundances compared to the larger Atlantic species and appear to form numerous small resident spawning aggregations (see Chap. 8). For tropical Atlantic groupers which achieve large size, occur in habitats with relatively small areas of reef environment and exhibit transient spawning, more extensive areas, possibly in addition to other management measures, would need to be protected (Chap. 11).

A final caveat concerns our coverage of the reproductive biology of reef fishes. As Tables 4.1–4.3 demonstrates the examples for which we have verified examples of spawning behaviour cover a tiny fraction of the reef fish fauna. However, regardless of the coverage we require better information on the trophic ecology, reproductive cycles and the wet machinery of internal nutrient storage and allocation. These are the areas currently neglected in reef fish ecology. An appropriate end point is a quotation from Munday et al. (2006a) in a discussion on the adaptive significance of sex change. "This will require detailed information about sex-specific fecundity, growth, mortality and movement patterns at the individual level. Only then can we really assess how reproductive value is affected by the different breeding tactics (male, female or non-breeder) that individuals can use." This also applies to the study of spawning aggregations.

Acknowledgements I thank Y. Sadovy and P. Colin for their valuable comments on several earlier versions of this chapter. The ideas were discussed with P. Munday, G. Russ and R. Robertson. Research support was provided by The National Geographic Society, The Queensland Government/ Smithsonian Institution Collaborative Research Program on Reef Fishes. R. Robertson provided major support in associated field work. Thanks are also due to R.Hamilton and A.Hughes for information on spawning aggregations in the Solomon Islands. Alex Vail, Lizard Island Research Station, provided additional information on the spawning of *Plectropomus leopardus,* E. Trip assisted with the analyses and V. Ward and K.Clements with the figure preparation.

References

Ackerman JL, Bellwood DR (2000) Reef fish assemblages: a re-evaluation using enclosed rotenone stations. Mar Ecol Prog Ser 206:227–237

Ackerman JL, Bellwood DR, Brown JH (2004) The contribution of small individuals to density-body size relationships: examination of energetic equivalence in reef fishes. Oecologia 139:568–571

Allen GR (2007) Conservation hotspots of biodiversity and endemism for Indo-Pacific coral reef fishes. Aquat Conserv Mar Freshw Ecosyst. doi:10.1002/aqc.880

Allen GR, Steene R, Allen M (1998) A guide to angelfishes and butterflyfishes. Odyssey Publishing/ Tropical Reef Research, Perth

Almany GR, Berumen ML, Thorrold SR, Planes S, Jones GP (2007) Local replenishment of coral reef fish populations in a marine reserve. Science 316:742–744

Bellwood DR, Wainwright PC (2002) The history and biogeography of fishes on coral reefs. In: Sale PF (ed) Coral reef fishes. Dynamics and diversity in a complex ecosystem. Academic, San Diego

Bolden SK (2000) Long-distance movement of a Nassau grouper (*Epinephelus striatus*) to a spawning aggregation in the central Bahamas. Fish Bull 98:642–645

Brix O, Gruner R, Ronnestad I, Gemballa S (2009) Whether depositing fat or losing weight, fish maintain a balance. Proc Biol Sci 276:3777–3782

Bshary R, Hohner A, Ait-El-Djoudi K, Fricke H (2006) Interspecific communicative and coordinated hunting between groupers and giant moray eels in the Red Sea. PLoS Biol 4:2393–2398

Bushnell ME, Claisse JT, Laidley CW (2010) Lunar and seasonal patterns in fecundity of an indeterminate, multiple-spawning surgeonfish, the yellow tang *Zebrasoma flavescens*. J Fish Biol 76:1343–1361

Chiappone M, Sluka R, Sealy KS (2000) Groupers (Pisces: Serranidae) in fished and protected areas of the Florida Keys, Bahamas and northern Caribbean. Mar Ecol Prog Ser 198:261–272

Choat JH, Axe LM (1996) Growth and longevity in acanthurid fishes; an analysis of otolith increments. Mar Ecol Prog Ser 134:15–26

Choat JH, Bellwood DR (1985) Interactions amongst herbivorous fishes on a coral reef: influence of spatial variation. Mar Biol 89:221–234

Choat JH, Bellwood DR (1991) Reef fishes: their history and evolution. In: Sale PF (ed) The ecology of coral reef fishes. Academic Press San Diego

Choat JH, Clements KD (1998) Vertebrate herbivores in marine and terrestrial environments: a nutritional ecology perspective. Annu Rev Ecol Syst 29:375–403

Choat JH, Robertson DR (2002) Age based studies. In: Sale PF (ed) Coral reef fishes diversity and dynamics in a complex system. Academic Press San Diego

Choat JH, Clements KD, Robbins WD (2002) The trophic status of herbivorous fishes on coral reefs. I. Dietary analyses. Mar Biol 140:613–623

Choat JH, Robbins WD, Clements KD (2004) The trophic status of herbivorous fishes on coral reefs II: food processing modes and trophodynamics. Mar Biol 145:445–454

Choat JH, Davies CR, Ackerman JL, Mapstone BM (2006) Demography of a large teleost, *Cheilinus undulatus* with a review of size distribution in labrid fishes. Mar Ecol Prog Ser 318:237–246

Claisse JT, Kienzle M, Bushnell ME, Shafer DJ, Parrish JD (2009) Habitat- and sex-specific life history patterns of yellow tang *Zebrasoma flavescens* in Hawaii, USA. Mar Ecol Prog Ser 389:245–255

Claydon J (2005) Spawning aggregations of coral reef fishes: characteristics, hypotheses, threats and management. Oceanogr Mar Biol Annu Rev 42:265–301

Clifton KE (1995) Asynchronous food availability on neighboring Caribbean coral reefs determines seasonal patterns of growth and reproduction for the herbivorous parrotfish *Scarus iserti*. Mar Ecol Prog Ser 116:39–46

Clifton KE, Robertson DR (1993) Risks of alternative mating tactics. Nature 366:520

Colin PL (1978) Daily and summer-winter variation in mass spawning of the striped parrotfish *Scarus croicensis*. Fish Bull 76:117–124

Colin PL (1989) Aspects of the spawning of western Atlantic butterflyfishes (Pisces: Chaetodontidae). Environ Biol Fish 25:131–141

Colin PL (1992) Reproduction in the Nassau grouper *Epinephelus striatus* (Pisces: Serranidae) and its relationship to environmental conditions. Environ Biol Fish 34:357–377

Colin PL (2010a) Aggregation and spawning of the humphead wrasse *Cheilinus undulatus* (Pisces: Labridae): general aspects of spawning behaviour. J Fish Biol 76:987–1007

Colin PL (2010b) Marine environments of Palau. Coral Reef Research Foundation. Indo-Pacific Press, San Diego

Colin PL, Clavijo IE (1988) Spawning activity of fishes producing pelagic eggs on a shelf edge coral reef, southwestern Puerto Rico. Bull Mar Sci 43(2):249–279

Collar C, Wainwright PC (2009) Ecomorphology of centrarchid fishes. In: Cooke S, Philipp DR (eds) Centrarchid fishes: diversity, biology and conservation. Wiley-Blackwell, Chichester

Craig PC (1998) Temporal spawning patterns of several surgeonfishes and wrasses in American Samoa. Pac Sci 52:35–39

Craig MT, Hastings PA (2007) A molecular phylogeny of the groupers of the subfamily Epinephelinae (Serranidae) with a revised classification of the Epinephelini. Ichthyol Res 54:1–17

Craig MT, Eble JA, Bowen BW, Robertson DR (2007) High genetic connectivity across the Indian and Pacific Oceans in the reef fish *Myripristis berndti* (Holocentridae). Mar Ecol Prog Ser 334:245–254

Crandall ED, Taffel JR, Barber PH (2010) High gene flow due to pelagic larval dispersal among South Pacific archipelagos in two amphidromous gastropods (Neritomorpha: Neritidae). Heredity 104:563–572

Denny MW (1993) Air and water: the biology and physics of life's media. Princeton University Press, Princeton

Depczynski M, Fulton CJ, MarnaneMJ BDR (2007) Life history patterns shape energy allocation among fishes on coral reefs. Oecologia 153:111–120

Domeier ML, Colin PL (1997) Tropical reef fish spawning aggregations: defined and reviewed. Bull Mar Sci 60:698–726

Erisman BE, Craig MT, Hastings PA (2009) A phylogenetic test of the size-advantage model: evolutionary changes in mating behavior influence the loss of sex change in a fish lineage. Am Nat 174:E83–E99

Eschmeyer WN, Fong JD (2010) Species of fishes by family/subfamily. Online version. http://research.calacademy.org/research/ichthyology/catalog/SpeciesByFamily.asp

Ferreira BP, Russ GR (1994) Age validation and estimation of growth rate of the coral trout, *Plectropomus leopardus* (Lacepede 1802), from Lizard Island, Northern Great Barrier Reef. Fish Bull 92:46–57

Floeter SR, Rocha LA, Robertson DR, Joyeux JC, Smith-Vaniz WF, Wirtz P, Edwards AJ, Barreiros JP, Ferreira CEL, Gasparini JL, Brito A, Falcon JM, Bowen BW, Bernardi G (2008) Atlantic reef fish biogeography and evolution. J Biogeogr 35:22–47

Froese R, Pauly D (eds) (2010) FishBase. World Wide Web electronic publication. www.fishbase.org. version (07/2010)

Fulton CJ (2005) Wave energy and the role of swimming in reef fish ecology. PhD thesis. James Cook University, Townsville, Australia

Fulton CJ (2007) Swimming speed performance in coral reef fishes: field validations reveal distinct functional groups. Coral Reefs 26:217–228

Fulton CJ, Bellwood DR (2005) Wave-induced water motion and the functional implications for coral reef fish assemblages. Limnol Oceanogr 50:255–264

Gaither MR, Toonen RJ, Robertson DR, Planes S, Bowen BW (2010) Genetic evaluation of marine biogeographical barriers: perspectives from two widespread Indo-Pacific snappers (*Lutjanus kasmira* and *Lutjanus fulvus*). J Biogeogr 37:133–147

Gladstone W (1986) Spawning behaviour of bumphead parrotfish *Bolbometopon muricatum* at Yonge Reef, Great Barrier Reef. Japan J Ichthyol 33:326–328

Gladstone W (1994) Lek-like spawning, parental care and mating periodicity of the triggerfish *Pseudobalistes flavimarginatus* (Balistidae). Environ Biol Fish 39:249–257

Gust N (2004) Variation in the population biology of protogynous coral reef fishes over tens of kilometres. Can J Fish Aquat Sci 61:205–218

Gust N, Choat JH, Ackerman J (2003) Demographic plasticity in tropical reef fishes. Mar Biol 140:1039–1051

Hamilton RJ (2004) The demographics of bumphead parrotfish (*Bolbometopon muricatum*) in lightly and heavily fished regions of the Western Solomon Islands. PhD thesis, University of Otago, Dunedin, New Zealand

Hamilton RJ, Adams S, Choat JH (2008a) Sexual development and reproductive demography of the green humphead parrotfish (*Bolbometopon muricatum*) in the Solomon Islands. Coral Reefs 27:153–163

Hamilton SL, Regetz J, Warner RR (2008b) Postsettlement survival linked to larval life in a marine fish. Proc Natl Acad Sci USA 105:1561–1566

Hamner WM, Colin PL, Hamner PP (2007) Export–import dynamics of zooplankton on a coral reef in Palau. Mar Ecol Prog Ser 334:83–92

Helfman GS, Collette BB, Facey DE, Bowen BW (2009) The diversity of fishes, 2nd edn. Wiley-Blackwell, Oxford

Heupel MR, Williams AJ, Welch DJ, Davies CR, Adams S, Carlos G, Mapstone BD (2010) Demography of a large exploited grouper, *Plectropomus laevis*: implications for fisheries management. Mar Freshw Res 61:184–195

Heyman WD, Kjerfve B (2008) Characterization of transient multi-species reef fish spawning aggregations at Gladden Spit, Belize. Bull Mar Sci 83:531–551

Hiatt RW, Strasberg DW (1960) Ecological relationships of the fish fauna of the coral reefs of the Marshall Islands. Ecol Monogr 30:65–127

Hixon MA, Beets JP (1993) Predation, prey refuges, and the structure of coral-reef fish assemblages. Ecol Monogr 63:77–101

Hobson ES (1974) Feeding relationships of teleostean fishes on coral reefs in Kona, Hawaii. Fish Bull NOAA 72:915–1031

Hobson ES, Chess JR (1978) Trophic relationships among fishes and plankton at Enewetak Atoll, Marshall Islands. Fish Bull NOAA 76:133–153

Holmes TH, McCormick MI (2009) Influence of prey body characteristics and performance on predator selection. Oecologia 159:401–413

Holt GJ, Holt SA, Arnold CR (1985) Diel periodicity of spawning in sciaenids. Mar Ecol Prog Ser 27:1–7

Horne JB, van Herwerden L, Choat JH, Robertson DR (2008) High population connectivity across the Indo-Pacific: congruent lack of phylogeographic structure in three reef fish congeners. Mol Phylogenet Evol 49:629–638

Jones GP, Almany GR, Russ GR et al (2009) Larval retention and connectivity among populations of corals and reef fishes: history, advances and challenges. Coral Reefs 28:307–325

Juario JV, Duray MN, Dwray VM, Nacario JF Almendras JME (1985) Breeding and larval rearing of the rabbitfish, *Siganus guttatus* (Bloch). Aquaculture 44:91–101

Kawase H (1998) Reproductive behavior and evolution of triggerfish (Balistidae) and filefish (Monacanthidae). Japan J Ichthyol 45:1–19

Konow N, Bellwood DR, Wainwright PC, Kerr AM (2008) Evolution of novel jaw joints promote trophic diversity in coral reef fishes. Biol J Linn Soc 93:545–555

Korsmeyer KE, Steffensen JF, Herskin J (2002) Energetics of median and paired fin swimming, body and caudal fin swimming, and gait transition in parrotfish (*Scarus schlegeli*) and triggerfish (*Rhinecanthus*). J Exp Biol 205:1253–1263

Kuwamura T, Sagawa T, Suzuki S (2009) Interspecific variation in spawning time and male mating tactics of the parrotfishes on a fringing coral reef at Iriomote Island, Okinawa. Ichthyol Res 56:354–362

Lessios HA, Robertson DR (2006) Crossing the impassable: genetic connections in 20 reef fishes across the eastern Pacific barrier. Proc Biol Sci 273:2201–2208

Lobel PS (1989) Spawning behavior of *Chaetodon multicinctus* (Chaetodontidae); pairs and intruders. Environ Biol Fish 25:125–130

Marriott RJ, Mapstone BD, Begg GA (2007) Age-specific demographic parameters, and their implications for management of the red bass, *Lutjanus bohar* (Forsskal 1775): a large, long-lived reef fish. Fish Res 83:204–215

McCormick MI, Ryen CA, Munday PL, Walker SP (2010) Differing mechanisms underlie sexual size-dimorphism in two populations of a sex-changing fish. PLoS One 5(5):e10616

Miller TL, Cribb TH (2007) Phylogenetic relationships of some common Indo-Pacific snappers (Perciformes: Lutjanidae) based on mitochondrial DNA sequences, with comments on the taxonomic position of the Caesioninae. Mol Phylogenet Evol 44:450–460

Montaggioni LF, Braithwaite CJ (2009) Quaternary coral reef systems, vol 5, History, development processes and controlling factors. Elsevier Science, Amsterdam

Montgomery WL, Myrberg AA, Fishelson L (1989) Feeding ecology of surgeonfishes (Acanthuridae) in the northern Red Sea, with particular reference to *Acanthurus nigrofuscus* (Forsskal). J Exp Mar Biol Ecol 132:179–207

Montgomery WL, Umino T, Nakagawa H, Vaughn I, Shibuno T (1999) Lipid storage and composition in tropical surgeonfishes (Teleostei: Acanthuridae). Mar Biol 133:137–144

Montgomery WI, Galzin R (1993) Seasonality in gonads, fat deposits and condition of tropical surgeonfishes (Teleostei, Acanthuridae). Mar Biol 115:529–536

Munday PL, Jones GP (1998) The ecological implications of small body size among coral reef fishes. Oceanogr Mar Biol Annu Rev 36:373–411

Munday PL, Buston PM, Warner RR (2006a) Diversity and flexibility of sex-change strategies in animals. Trends Ecol Evol 21:89–95

Munday PL, White JW, Warner RR (2006b) A social basis for the development of primary males in a sex-changing fish. Proc Biol Sci 273:2845–2851

Myrberg AA Jr, Montgomery WL, Fishelson L (1988) The reproductive behavior of *Acanthurus nigrofuscus* (Forsskal) and other surgeonfishes (Fam. Acanthuridae) off Eilat, Israel (Gulf of Aqaba, Red Sea). Ethology 79:31–61

Nelson JS (2006) Fishes of the world, 4th edn. Wiley, Hoboken

Nemeth RE (2008) Dynamics of reef fish and decapod crustacean spawning aggregations: underlying mechanisms, habitat linkages, and trophic interactions. In: Nagelkerken I (ed) Ecological connectivity among tropical coastal ecosystems. Springer, Dordrecht

Newman SJ, Williams DM, Russ GR (1996) Age validation, growth and mortality rates of the tropical snappers (Pisces: Lutjanidae) *Lutjanus adetii* (Castelnau, 1873) and *L. quinquelineatus* (Bloch, 1790) from the central Great Barrier Reef. Mar Freshw Res 47:575–584

Pears RJ (2005) Comparative demography and assemblage structure of serranid fishes: implications for conservation and fisheries management. PhD thesis James Cook University, Townsville, Australia

Perez KO, Munch SB (2010) Extreme selection on size in the early lives of fishes. Evolution 64:2450–2457

Petersen CW, Warner RR (2002) The ecological context of reproductive behaviour. In: Sale P (ed) Coral reef fishes dynamics and diversity in a complex system. Academic, Amsterdam

Popper D, Pitt R, Zohar Y (1979) Experiments on the propagation of Red Sea siganids and some notes on their reproduction in nature. Aqua 16:177–181

Price SA, Wainwright PC, Bellwood DR, Kazancioglu E, Collar DC, Thomas J, Near TJ (2010) Functional innovations and morphological diversification in parrotfishes. Evolution 64:3057–3068

Randall JE (1967) Food habits of reef fishes in the West Indies. Stud Trop Oceanogr Miami 5:665–847

Randall JE (1998) Zoogeography of the shore fishes of the Indo-Pacific Region. Zool Stud 37:227–268

Randall JE, Randall HA (1963) The spawning and early development of the Atlantic parrotfish, *Sparisoma rubripinne*, with notes on other scarid and labrid fishes. Zool NY Zool Soc 48:49–60

Robertson DR (1972) Social control of sex-reversal in a coral reef fish. Science 177:1007–1009

Robertson DR (1983) On the spawning behavior and spawning cycles of eight surgeonfishes (Acanthuridae) from the Indo-Pacific. Environ Biol Fish 9:192–223

Robertson DR (1991) The role of adult biology in the timing of spawning of tropical reef fish. In: Sale PF (ed) The ecology of fishes on coral reefs. Academic, San Diego

Robertson DR, Ackerman JL, Choat JH, Posada J, Pitt J (2005) Ocean surgeonfish, *Acanthurus bahianus*. I. The geography of demography. Mar Ecol Prog Ser 295:229–244

Ross RM (1983) Annual, semilunar, and diel reproductive rhythms in the Hawaiian labrid *Thalassoma duperrey*. Mar Biol 72:311–318

Ruttenberg BI, Haupt AJ, Chiriboga AI, Warner RR (2005) Patterns, causes and consequences of regional variation in the ecology and life history of a reef fish. Oecologia 145:394–403

Sadovy Y (1996) Reproduction of reef fishery species. In: Polunin NV, Roberts CM (eds) Reef fisheries. Chapman & Hall, London

Sadovy de Mitcheson Y, Liu M (2008) Functional hermaphroditism in teleosts. Fish Fish 9:1–43

Sadovy de Mitcheson Y, Cornish A, Domeier M, Colin PL, Russell LKC (2008) Reef fish spawning aggregations: a global baseline. Conserv Biol 22(5):1233–1244

Sadovy Y, Colin PL (1995) Sexual development and sexuality in the Nassau grouper, *Epinephelus striatus* (Bloch) (Pisces: Serranidae). J Fish Biol 46:961–976

Sadovy Y, Eklund AM (1999) Synopsis of biological data on the Nassau grouper, *Epinephelus striatus* (Bloch, 1792), and the Jewfish, *E. itajara* (Lichtenstein, 1822). Seattle, WA, NOAA/ National Marine Fisheries Service, NOAA Technical Report NMFS, 146

Samoilys MA (1997) Periodicity of spawning aggregations of coral trout *Plectropomus leopardus* (Pisces: Serranidae) on the northern Great Barrier Reef. Mar Ecol Prog Ser 160:149–159

Sancho G, Solow AR, Lobel PS (2000a) Environmental influences on the diel timing of spawning in coral reef fishes. Mar Ecol Prog Ser 206:193–212

Sancho G, Petersen CW, Lobel PS (2000b) Predator-prey relations at a spawning aggregation site of coral reef fishes. Mar Ecol Prog Ser 203:275–288

Sandin SA, Smith JE, DeMartini EE, Dinsdale EA, Donner SD, Friedlander AM, Konotchick T, Malay M, Maragos JE, Obura D, Pantos O, Paulay G, Richie M, Rohwer F, Schroeder RE, Walsh S, Jackson JBC, Knowlton N, Sala E (2008) Baselines and degradation of coral reefs in the northern Line Islands. PLoS One 3:1548

SCRFA database (2009) www.SCRFA.org

Shuter BJ, Post JR (1990) Climate, population viability, and the zoogeography of temperate fishes. Trans Am Fish Soc 119:314–336

Sinclair M (1988) Marine populations Washington Sea Grant Program. University of Washington Press, Seattle, USA

Sogard SM (1997) Size-selective mortality in the juvenile stage of teleost fishes: a review. Bull Mar Sci 60:1129–1157

Spalding MD, Grenfell AM (1997) New estimates of global and regional coral reef areas. Coral Reefs 16:225–230

Spalding MD, Ravilious C, Green EP (2001) World atlas of coral reefs. University of California Press, Berkeley

Spalding MD, Fox HE, Allen GR, Davidson N, Ferdana ZA, Finlayson M, Halpern BS, Jorge MA, Lombana AL, Lourie SA, Martin KD, McManus E, Molnar J, Recchia CA, Robertson J (2007) Marine ecoregions of the world: a bioregionalization of coastal and shelf areas. Bioscience 57:573–583

Stallings CD, Coleman FC, Koenig CC, Markiewicz DA (2010) Energy allocation in juveniles of a warm-temperate reef fish. Environ Biol Fish 88:389–398

Stephens PA, Boyd IL, Mcnamara JM, Houston AI (2009) Capital breeding and income breeding: their meaning, measurement, and worth. Ecology 90:2057–2067

Suyehiro Y (1942) A study of the digestive system and feeding habitats of fish. Japan J Zool 10:1–303

Swearer SE, Shima JS, Hellberg ME, Thorrold SR, Jones GP, Robertson DR, Morgan SG, Selkoe KA, Ruiz GM, Warner RR (2002) Evidence of self-recruitment in demersal marine populations. Bull Mar Sci 70:251–271

Thresher RE (1984) Reproduction in reef fishes. TFH Publications, Neptune City

Tittensor DP, Mora C, Jetz W, Lotze HK, Daniel Ricard D, Berghe EV, Worm B (2010) Global patterns and predictors of marine biodiversity across taxa. Nature 466:1098–1101

Trip EL, Choat JH, Wilson DT, Robertson DR (2008) An inter-oceanic analysis of demographic variation in a widely distributed Indo-Pacific coral reef fish. Mar Ecol Prog Ser 373:97–109

Trivers RL (1972) Parental investment and sexual selection. In: Campbell B (ed) Sexual selection and the descent of man. Aldine, Chicago, pp 1871–1971

van Rooij JM, Kroon FJ, Videler JJ (1996) The social and mating system of the herbivorous reef fish *Sparisoma viride*: One-male versus multi-male groups. Environ Biol Fish 47:353–378

van Woesik R (2010) Calm before the spawn: global coral spawning patterns are explained by regional wind fields. Proc R Soc Lond Ser B Biol Sci 277:715–722

Wainwright PC, Bellwood DR (2002) Ecomorphology of feeding in coral reef fishes. In: Sale PF (ed) Coral reef fishes. Dynamics and diversity in a complex ecosystem. Academic, San Diego

Wainwright PC, Bellwood DR, Westneat MW, Grubich JR, Hoey A (2004) A functional morphospace for the skull of labrid fishes: patterns of diversity in a complex biomechanical system. Biol J Linn Soc 82:1–25

Walker SP, McCormick MI (2009) Sexual selection explains sex-specific growth plasticity and positive allometry for sexual size dimorphism in a reef fish. Proc Biol Sci 276:3335–3343

Warner RR (1984) Mating behavior and hermaphroditism in coral reef fishes. Am Sci 72:128–136

Warner RR (1995) Large mating aggregations and daily long-distance spawning migrations in the bluehead wrasse *Thalassoma bifasciatum*. Environ Biol Fish 44:337–345

Warner RR (1997) Evolutionary ecology: how to reconcile pelagic dispersal with local adaptation. Coral Reefs 16(Suppl):S115–S120

Westneat MW, Alfaro ME (2005) Phylogenetic relationships and evolutionary history of the reef fish family Labridae. Mol Phylogenet Evol 36:370–390

Williams AJ, Currey L, Begg GA, Murchie CD, Ballagh AC (2008) Population biology of coral trout species in the eastern Torres Strait: implications for fishery management. Cont Shelf Res 28:2129–2142

Wright J, Rothery DA (1998) The ocean basins: their structure and evolution Butterworth-Heinemann Oxford in association with The Open University

Zapata FA, Robertson DR (2007) How many species of shore fishes are there in the Tropical Eastern Pacific? J Biogeogr 34:38–51

Chapter 5
Timing and Location of Aggregation and Spawning in Reef Fishes

Patrick L. Colin

Abstract Fishes that engage in aggregation spawning do so at specific sites and times and have a wide diversity of spawning strategies. There are some clear distinctions in locations and timing between transient aggregations (TA) and resident aggregations (RA). TAs have larger predatory fishes, occur infrequently, but seasonally largely on or near shelf edge areas. RAs have herbivores and omnivores, are more numerous and occur in both shelf edge areas and inshore regions. Aggregations differ between the Indo-west Pacific (IWP) and tropical western Atlantic (TWA); probably due to current dominated regimes in the IWP due to the higher tidal amplitudes and barrier reef/channel geomorphology. Migration patterns are related to the frequency of spawning. Nearly all TA (and some RA) sites are used by multiple species, either sequentially or simultaneously. Many aggregation sites are stable in location over decades with only slight variation. Spawning at some TAs is now known to occur during periods of low current speed. The entrainment of TA propagules into oceanic circulation after spawning is uncertain with tendencies at some TA sites for retention of propagules. Spawn from RAs is less likely to become entrained into oceanic circulation. Water temperature regimes may be an important determinant of seasonality of spawning and early life history success. Most aggregations occur over a limited temperature range. The daily and lunar timing of aggregation spawning may be related to needs of pelagic life history. Predation on spawning adults is rare while predation of released eggs is common, but neither factor is believed to limit or structure aggregations.

P.L. Colin (✉)
Coral Reef Research Foundation, P.O. Box 1765, Palau 96940, Koror
e-mail: crrf@palaunet.com

Y. Sadovy de Mitcheson and P.L. Colin (eds.), *Reef Fish Spawning Aggregations: Biology, Research and Management*, Fish & Fisheries Series 35,
DOI 10.1007/978-94-007-1980-4_5, © Springer Science+Business Media B.V. 2012

5.1 Introduction

Our understanding of why spawning aggregations of tropical and temperate marine fishes occur at particular places ("where") and times ("when") is only in its preliminary stages. Spawning aggregations were first noted by fishers, probably discovering them initially by chance. By returning later to the same locations, particularly at the same daily time, lunar phase or season, they gradually learned that aggregations occur on a lunar or annual schedule and used this knowledge to increase their ability to capture food. Such knowledge is typical of human populations dependent on direct exploitation of the natural world for survival. The scope and limits of the locations and timing for the species involved would become what we know today as traditional ecological knowledge (TEK, also known as local ecological knowledge see Chap. 10).

Nearly all reef fishes have a pelagic larval life history stage, usually starting either with the hatching of demersal eggs into free-swimming larvae or (in the case of most aggregators) the release and fertilization of planktonic eggs. For both aggregating and non-aggregating species, spawning is often concentrated into a relatively narrow time frame. High mortality during the pelagic phase is normal with few "propagules" surviving to settlement. The time and location of spawning aggregations are intimately involved with the behaviour and ecology of the spawning adults and the pelagic life history of larvae through oceanographic and climatic interactions.

Questions immediately arise for scientists regarding "why this site" and "why this time" are utilized by a particular species and, if multiple species use the same site simultaneously or sequentially, "is this site special for some reason"? Despite recent attention, scientists are still wondering about the ultimate "why" of location and timing. The science of reef fish spawning aggregations is still largely in the basic "discovery" phase with the information available coming from a limited geographic scope and suite of species, often restricted to fisher interviews regarding TEK. Few researchers conduct field work on the science of aggregations and it is understandable that a clear view of spawning aggregations has yet to emerge. Increased attention from the conservation advocacy community has resulted in redirection of support to conservation efforts in which detailed scientific study of aggregations is generally regarded as unnecessary for specific conservation agendas. For example, the promotion of Marine Protected Areas (MPAs), which may or may not include aggregation areas, tends to discourage new research with the rationale that once designated such areas are thereby protected, hence there is little need to monitor or understand aggregations. It also sidesteps other possible management approaches, such as seasonal and market closures. Efforts to obtain "hard" knowledge on spawning aggregations often become subservient to political or social considerations.

With around 100 species of tropical and temperate reef fishes found in all oceans presently reported to aggregate for spawning (Chaps. 4 and 8, Appendix, Sadovy de Mitcheson et al. 2008), and perhaps many more species doing so without our knowledge, there is potentially a high diversity of reproductive strategies.

Subsequently in this chapter the following abbreviations will be used: tropical western Atlantic (TWA), tropical Indo-west Pacific (IWP) and eastern Pacific (EP). The TWA includes both the North Atlantic (Caribbean, Gulf of Mexico, Bahamas-Turks and Caicos archipelago, US east coast and Bermuda), and South Atlantic (Brazil and Atlantic Islands). The IWP includes the islands of the central and western North and South Pacific, much of Southeast Asia, the tropical Indian Ocean to the east coast of Africa and the Red Sea.

If there is a single, or even a few, overriding reason(s) for the location and timing of spawning, it is likely scientists would have identified it (them) by now. For every generality or hypothesis put forward about spawning aggregations, there is some example that argues exactly the opposite. Also, what occurs in one location does not necessarily follow in another and the potential for unknown flexibility in spawning ecology is always present. Most studies have tended to focus on one or at most a few sites, and often put forward generalities based on those locations. While documenting and analyzing what happens at individual sites is necessary and important, it may not provide answers to the overall questions of benefits or limitations of sites. The locations and timing of aggregations are not necessarily "adaptive" for reproductive success at all times (Chaps. 1, 2, 4). Use of a site may promote success some of the time while another site may be advantageous at other times (Colin 1995). While we have a limited ability to monitor the numbers of fish present and occasionally their success in spawning, scientists are usually unable to determine whether such spawning actually results in life history success through to the settlement stage, and beyond.

Most likely the factors influencing spawning aggregations will be an amalgam combining aspects of the oceanographic and atmospheric environments (currents, topography, seasonal temperature, salinity and weather) with the biological (species specific needs with regard to feeding, social structure, predation, physiology and genetics of larval and juvenile-adult stages) (Chap. 4). Also part of the mix are historical aspects of reef ecology and geomorphology as reef environments have shifted from steep escarpments to broad reef expanses in the last 20,000 years since the start of the Holocene transgression (Paulay 1990).

The known scope and patterns of the "where and when" of aggregation spawning reveal a wide diversity of strategies. Species and behaviour range from small resident aggregation (RA) species, typified by *Thalassoma bifasciatum,* the bluehead wrasse in the TWA and a variety of small wrasses (Labridae) and parrotfish (Scaridae) in the IWP, to large transient aggregation (TA) species, which reach their epitome in the TWA with *Epinephelus striatus*, the Nassau grouper and *Lutjanus cyanopterus*, the cubera snapper (Chap. 1). While not as well known as these TWA species, new knowledge indicates some IWP snappers (Labridae) (e.g. blackfin snapper – *Lutjanus fulvus* and blue-lined sea bream – *Symphorichthys spilurus* – Chaps. 12.10 and 12.11) have large TA's comparable to those of the TWA, but more information on reproductive status through specimen collection is needed. Those IWP groupers with documented aggregations tend to have many (to multiple 1,000s) individuals spread over a relatively broad area at densities lower than those found in some TWA groupers (PLC unpublished data, Figs. 2.6, 9.1, 9.9).

A few aggregating species have benthic eggs (10 out of 169 "confirmed" to "possible" aggregating species – Appendix) with somewhat different reproductive strategies but are not considered in detail in this chapter. For the rabbitfishes (Siganidae) our knowledge is rudimentary; some may only pair-spawn (PS) and not all may have negatively buoyant adhesive eggs (Chap. 12.22). For the triggerfishes (Balistidae), which excavate nests in sandy areas with demersal (but not attached) eggs, knowledge is still fragmentary (Chap. 12.23).

Much can be learned about spawning sites from the study of both TA and RA sites, as well as from non-aggregating species (pair and haremic spawers) that may spawn at or near aggregation sites. Nearly all known TA sites have multiple species using them, either simultaneously or sequentially. TA sites are generally fewer in number that RA sites in a given area, but whether this provides any relative advantage in promoting reproductive success is not known. Comparative studies between TA, RA and non-aggregation sites may identify those elements (if any) found at TA or RA sites that benefit reproductive success. RA sites offer some advantages in studying aggregations, being more numerous and often occurring over longer periods of the year usually with spawning during daylight hours.

5.2 Stability of Aggregation Site Location

Long-term use of specific aggregation sites is found among reef fishes throughout the world. TEK for TAs indicates consistent occurrence for some at particular spots over multiple human generations (Craig 1969; Johannes 1978, 1981; Sadovy de Mitcheson et al. 2008). For RAs, consistent use of sites is also documented for periods from several years (Heyman et al. 2005; Colin 2010) to decades (Colin and Clavijo 1978; Colin 1996). Small year to year changes in locations (10s to a few 100s of meters) have been reported for some large TAs (Colin 1992; Whaylen et al. 2004; Kadison et al. 2006), possibly resulting from "nomadic" (not tied to any particular feature along a section of homogeneous reef) behaviour of fish (Kadison et al. 2006), temporary changes in current direction or confusion of "staging areas", occupied prior to spawning (Chap. 2), with the actual nearby site of spawning.

Highly accurate data on aggregation locations have been collected only over a relatively short time-frame but continued monitoring of locations over the long-term is important. Documentation of aggregation site locations and the area occupied by the aggregation, determined by Global Positioning System (GPS), will allow determination of whether sites are moving over time. Mapping of site features and bathymetry relative to the overall aggregation is also a high priority (Chap. 9).

5.3 Location and Geomorphology of Sites

For most of the last 120,000 years sea-level has been significantly lower than at present, with its lowest (about −120 m) just 20,000 years ago, and arriving at today's level only a few thousand years ago (Paulay 1990). When sea level was lower, what

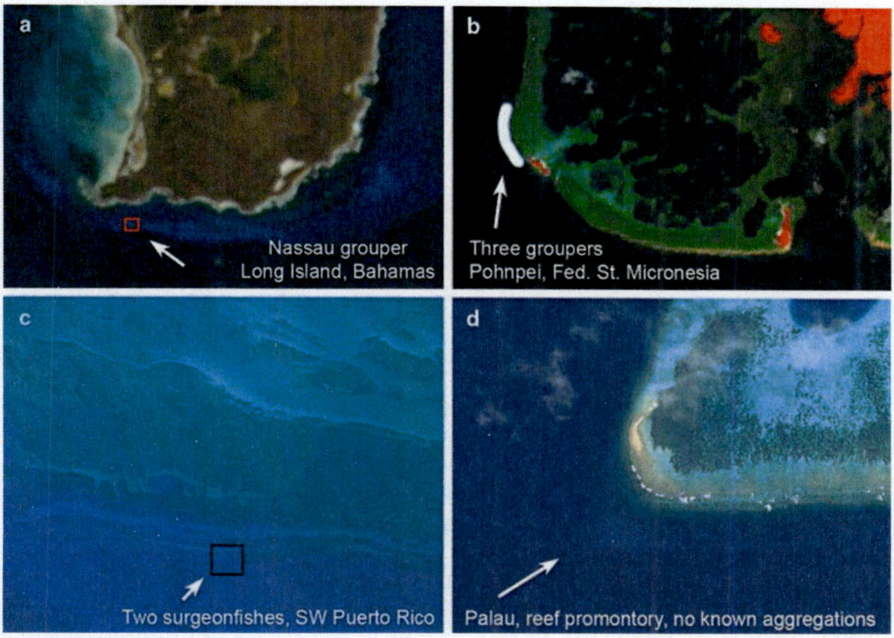

Fig. 5.1 Examples of areas with and without spawning aggregation sites located on outer reefs adjacent to deep ocean. (**a**) Nassau grouper, *Epinephelus striatus*, transient aggregation (*TA*) site without sharp projections at the southern end of Long Island, Bahamas. (**b**) TA area for three species of groupers, Kehpara Marine Reserve in Pohnpei, Caroline Islands, is an arching section of reef (**c**) resident aggregation site for two surgeonfishes on the shelf edge reef off SW Puerto Rico; an area without any major projecting reef features. (**d**) A promontory at Agupelu Reef, Palau, similar in appearance of other aggregation areas, without any known spawning aggregations

today are "drop off" areas were steep escarpments at the water's surface dropping away quickly. Shallow lagoon or fore reef areas would have been uncommon, and during glacial low sea levels no other habitat was available for aggregation and spawning. Once sea level rose to present levels, if spawning on drop-off areas is selectively neutral or somewhat advantageous, and traditionally transmitted sites important for encountering mates, there would be no selective pressure for fishes to change where they aggregate (e.g. move inshore or elsewhere).

Outer reef areas facing the open ocean at the edge of insular and continental shelves are a common location for spawning aggregations of many commercially important species (Fig. 5.1) (Sadovy de Mitcheson et al. 2008, www.SCRFA.org). The "drop offs" where the bottom slopes away into deep water are a nearly ubiquitous geomorphologic feature of outer reefs varying from gently sloping to vertical, with the width of the reef comprising the slope duly influenced. In the TWA nearly all TAs are found at or very near shelf edge sites. RAs in the TWA occur both on the shelf edge and also in areas far removed. The limited migration distance RA species can cover each day may limit the areas where they aggregate. Areas are probably selected due to certain features (Randall and Randall 1963) (Chap. 4), but continued use of sites may be traditional and learned (Warner 1988, 1995).

Fig. 5.2 The projection of reef at Reef Bay, St. John, US Virgin Islands (*white arrow*), where Randall and Randall (1963) studied the spawning of the redfin parrotfish, *Sparisoma rubripinne*, in 1960–1961 is nearly 15 km from the insular shelf edge

Despite some superficial similarities the oceanography environment is quite different between outer shelf TA sites and inner shelf RA sites. For example, the redtail parrotfish, *Sparisoma rubripinne*, aggregation site of Randall and Randall (1963) is superficially a reef "promontory" and might be considered equivalent to an outer reef promontory. However the projection of reef extends out onto a flat bottom of sand and rubble located nearly 15 km inside the actual insular shelf edge south of St. John, US Virgin Islands (Fig. 5.2). The currents would differ greatly between inner shelf areas and outer reef environments. It is probably difficult for eggs and early larvae from the inner shelf areas to be transported off the insular shelf (Appeldoorn et al. 1994; Hensley et al. 1994), and as a result larvae might spend their entire planktonic life over the insular shelf.

On TWA insular shelf edges TAs can occur along sections of reef without any prominent seaward projections, at areas with projections (promontories), and at the ends or corners of islands and atolls. It is normal for an outer reef of more than a few km long to have some areas that to some extent project seaward from an otherwise relatively straight general line of reef. Their existence does not imply the presence of either RAs or TAs. That some aggregations occur at promontories does not mean that all aggregations occur at promontories, nor that all promontories have aggregations. The limited evidence indicates that, among numerous apparently equivalent promontories, only a few of those might actually have TAs (Colin 1992) (Fig. 5.3).

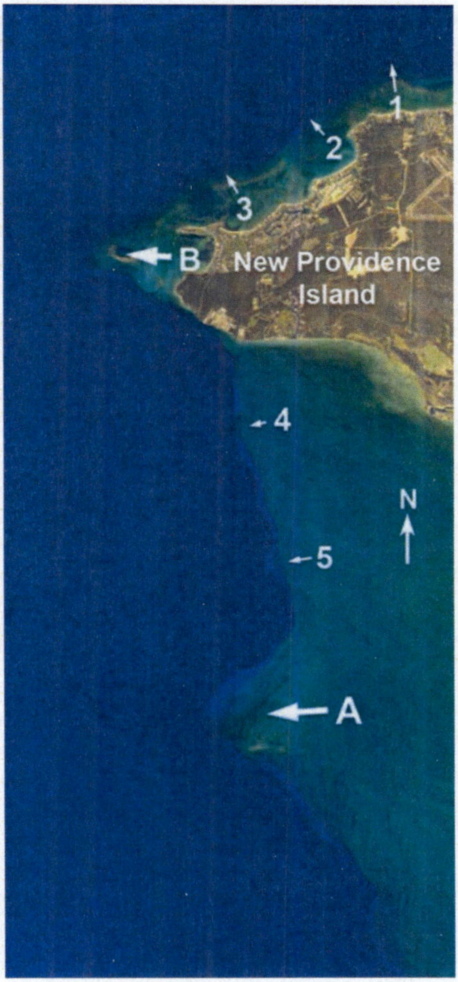

Fig. 5.3 The promontories along the shelf edge reefs of the Bahamas (western end of New Providence Island shown) are common and of various sizes. The view shows two major ("*large scale*") promontories; (**a**) is a historically known Nassau grouper aggregation site, now fished out, and (**b**) is not known as an aggregation site. Five lesser ("*small scale*") promontories (1–5) are indicated: none are known as aggregation sites

The IWP is somewhat different, with TAs not only on the outer slope, but also often at the mouth and inner reaches of channels through barrier and other types of reefs. Such channels are rare in the TWA. RAs occur across a wider range of habitats including many locations on broad insular shelves and inner reefs; not just outer reef faces (Chap. 2). Site selection may be determined more by the limitations of migration distance than by other factors.

To evaluate geomorphology across geographically distinct sites, a standardized terminology is needed. The terms used previously are often vague, confusing and poorly defined. Ideally in this volume these terms are used consistently in the hope that some general definitions can be developed and applied to future usage. Reefs are variable in shape with ends, curves, and bends along their lengths, projections at various locations, and basins among possible features. The deeper and shallower

Fig. 5.4 The projection of outer reef can characterized by the angle formed by *tangential lines* drawn along the reef faces either side of the projection. "Promontories" can have an angle less than 90°, while "corners" are 90° or greater. "Bends" have angles greater than a 90° angle, but also have a very rounded outer profile. Examples shown for Landsat 7 image of Little Inagua Island, Bahamas

portions of reefs may not have the same profile; a linear section of shallow reef may have deeper projections or incised areas with steep faces. Features, such as projections, can be large, medium or small scale and often the same terms have been applied to features at all scales (Fig. 5.3). *Promontory* is a term often used (and misused) to describe a projection of reef relative to spawning aggregations, including redundant terms for a promontory, such as "salient promontory" (Heyman and Kjerfve 2008) and "a prominent reef promontory" (Rhodes 2003). Promontory has not been defined or characterized by most researchers. Features termed promontories range from major reef structures several km across to small submerged projections on a reef, about two orders of magnitude different in scale. The general definition of a promontory as "a high point of land or rock projecting into a body of water" is applicable in some cases (Heyman et al. 2005). Others features, such as a bend in the reef, have also been called promontories (Fig. 5.1b).

The following general guidelines are offered as one way to standardize terms. If tangential lines are drawn along the general line of a reef face, where the reef changes direction, these lines form an angle (Fig. 5.4). For present purposes, the tangential lines for a "promontory" would ascribe an angle of less than 90°. Angles

Fig. 5.5 Large to medium scale reef promontories in northern Exuma Sound, Bahamas, show structures that are (**a**) "cusp-like" (**b**), relatively projecting or (**c**) or rounded at its outer reaches

greater than 90° are considered to represent "corners" or if greatly rounded "bends". When a promontory is a large-scale feature, it may appear acute (less than 90°) in a satellite or aerial image, but when viewed from the water's surface or underwater it appears to be a gently rounded feature (Promontory "A" in Fig. 5.3).

The grouping of reef projections at different scale levels is somewhat arbitrary, but is a useful starting point for comparative purposes. Large-scale promontories project a kilometre or more from the general reef profile or form the terminal end of a large area of reef or island several km or more in length (Figs. 5.3 and 5.5) and can be cusp-like, with a pointed projection formed by the intersection of two arcs (Fig. 5.5a). Such projections can be found in many areas of the Bahamas (Figs. 5.3 and 5.5) and some (but not all) are reported to be or have been Nassau grouper aggregation sites. Other projections are rounded, not really qualifying as promontories, but still project out to sea. Some of these are reported as grouper aggregation sites by fishermen, but not verified by site visits. Other areas of the Bahamas that are not in any sense projections are confirmed aggregation sites (Fig. 5.1a).

Medium-scale projections are measured in 100s of metres of either extension or width where it emerges from the general reef profile (Figs. 5.3 and 5.6). Divers can, by swimming over a broad area, see the feature as a promontory. Medium-scale features, such as Blue Corner (Fig. 5.6a) and Ngeraul reef (Fig. 5.6b) in Palau are known aggregation sites. Small-scale features can be measured in tens of metres and can be seen easily by a diver as a recognizable feature (Fig. 5.7). Such areas are relatively common along barrier and outer fringing reefs. On Palau's outer reefs there are probably 100 or more projections similar to those illustrated in Fig. 5.7. In general, small-scale projections are not known to have either TA or RA spawning aggregations in the IWP, but few have been examined in sufficiently to completely rule that out.

Narrow islands with fringing reefs (e.g. Little Cayman, Cayman Brac and Roatan, Fig. 5.8a, b) and narrow atolls (e.g. Layang Layang, Fig. 5.8c), have acuminate ends generally with similarly shaped projecting outer reefs. While such ends superficially appear sharply pointed they are often a relatively straight reef face that only gradually curves along the area where an aggregation occurs (Fig. 5.9). The oceanographic environment at a narrow island end would probably be very different from

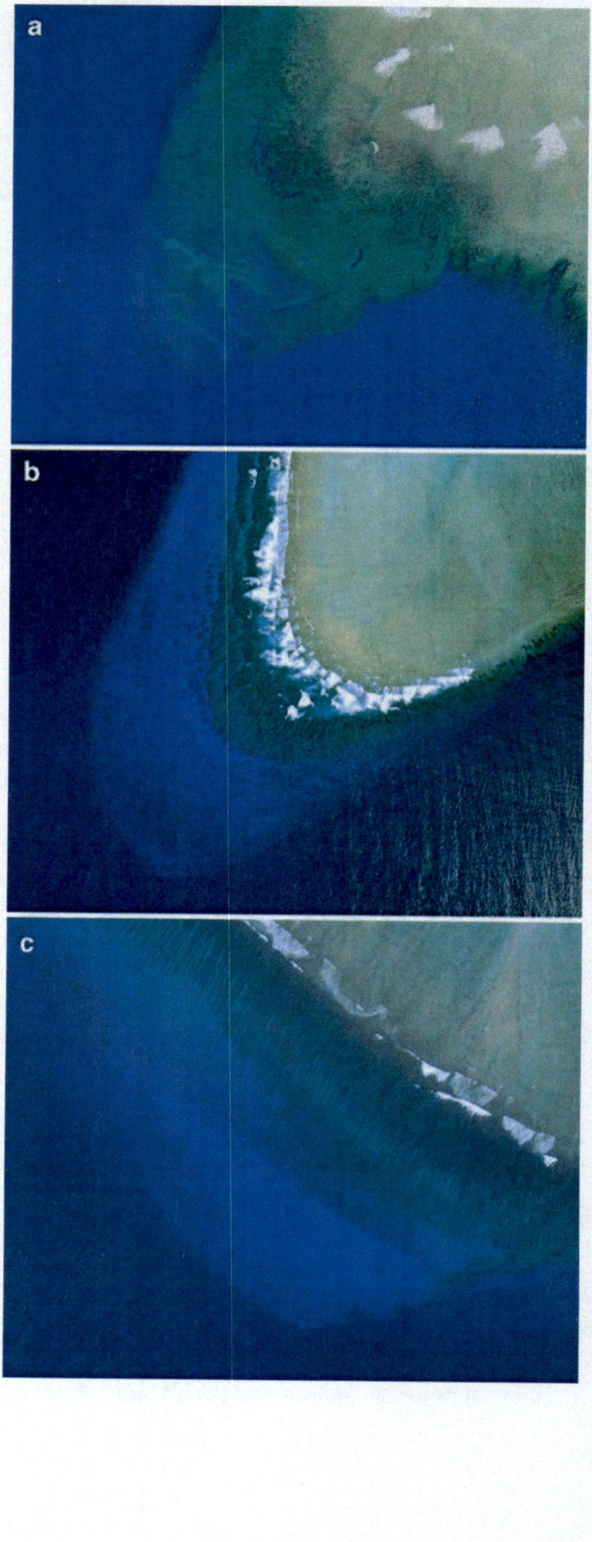

Fig. 5.6 Promontories are variable and include a range shown in these medium-scale (100's of meters across the feature) reef projections in Palau. They include (**a**) a sharp and pronounced promontory at "Blue Corner" (renowned for large fishes), (**b**) an acute angled reef, NW tip of the main reef and (**c**) a blunter projection of reef at the north end of Kayangel Atoll

Fig. 5.7 Small-scale promontories (tens of metres across the feature) on outer reef faces (northern barrier reef of Palau shown) occur in many areas and are readily apparent to divers as a projection of reef to seaward. Such small features are not known to be major locations for spawning aggregations, but few have been examined in detail

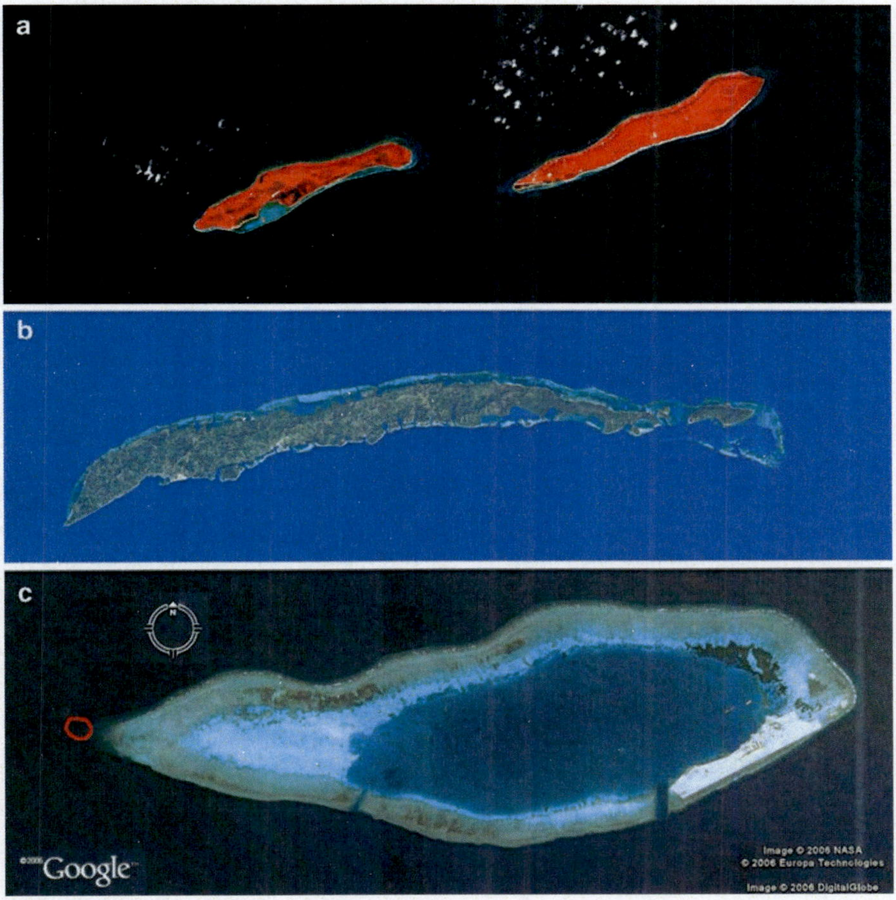

Fig. 5.8 Narrow atolls and islands often have acuminate projections of reef at both ends; sometimes with spawning aggregation sites. (**a**) Little Cayman (*left*) and Cayman Brac were historically known to have had Nassau grouper aggregations at both ends of the islands. Today an aggregation is reported only on the western end of Little Cayman. Down-current eddies at the tips which might help in retaining eggs and larvae in their vicinity (see Fig. 5.10a). (**b**) Roatan, Bay Islands, Honduras, is narrow with acuminate reefs at its tips with unverified reports of spawning aggregations at the ends of this island. (**c**) Layang Layang atoll, South China Sea, had a single aggregation of humphead wrasse (*Cheilinus undulatus*) at its western end (*white circle*), which is now fished out. No aggregations are known from the eastern end of the atoll

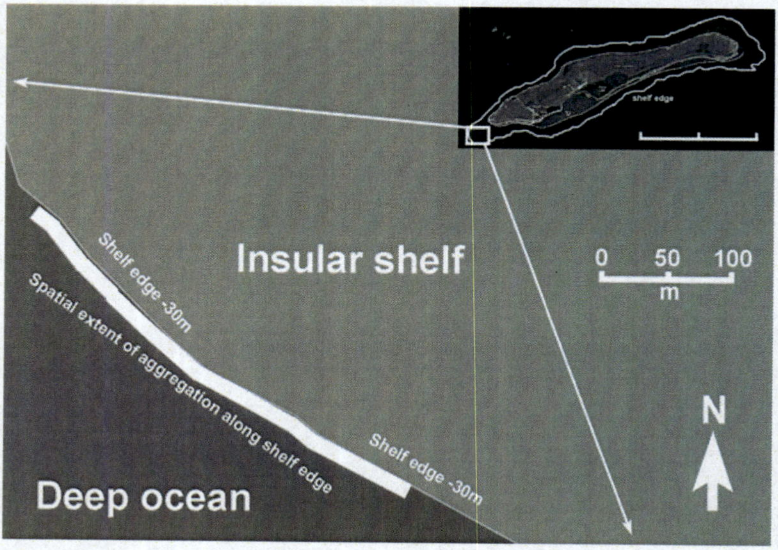

Fig. 5.9 When viewed close up many promontory areas have little projection or geomorphological features identifying it as the end of an island reef; this example (Adapted from Whaylen et al. (2006), shows a Nassau grouper aggregation site at Little Cayman Island (Fig. 5.8a))

a promontory along a barrier reef, having water flowing down both sides of the island merging into a single stream at the end of the island reef. If the axis is oriented parallel to oceanic currents, it is almost like a vessel cleaving the water, leaving relatively little turbulent flow in their down-current "wake". In most situations oceanic currents hit islands at an angle to the island axis of the island, with currents forming eddy fields at the island's downcurrent end and eddying circulation induced on the "back" side of the island (Hamner and Hauri 1981; Rissik et al. 1997). The eddies produced in the wake of a small island may assist in retaining eggs and larvae nearby at times (Fig. 5.10a), but often are of limited extent compared to a larger island, reducing the time propagules bearing water might be retained (Fig. 5.10b).

"Bank reefs", where the reef does not reach close to the surface and is exposed to current on all sides, are another geomorphologic configuration associated with spawning aggregations. Two such areas (Fig. 5.11a, b) in the TWA region, Saba Bank (red hind, *Epinephelus guttatus*, Nemeth et al. 2008) and "Riley's Hump", Dry Tortugas (mutton snappers, *Lutjanus analis*, Burton et al. 2005) can have water flow across the bank area. There is less tendency for bank reefs to have "island effect" circulation on their down-current side and lower prospects for retention of eggs and larvae. Domeier (2004) used drifter vials to examine potential dispersal of larvae from Riley's Hump and found rapid transport of drifters to SE Florida over time periods similar to larval development of snappers. If the end of an island reef, perhaps as a shallower reef area, is far enough away from the island, its environment

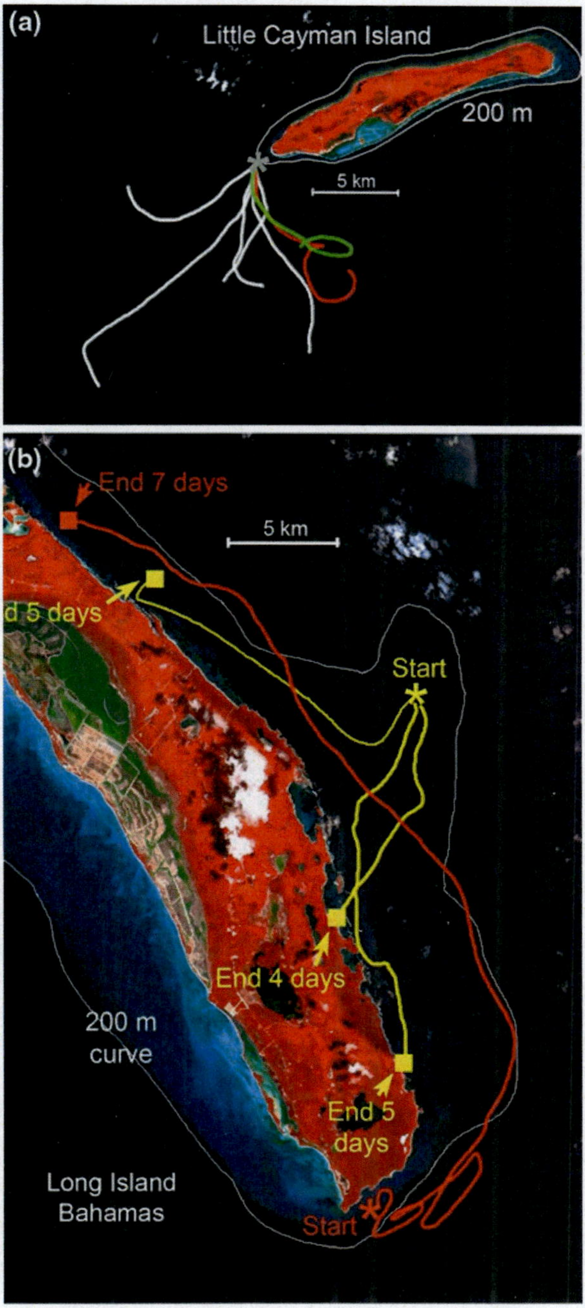

Fig. 5.10 Depending on geomorphology, location and weather conditions spawn from aggregations has different prospects for becoming part of oceanic circulation. (**a**) Current drifters started at a Nassau grouper spawning site, western end Little Cayman quickly moved into oceanic waters (Adapted from Grouper Moon Project, unpublished). (**b**) Current drifters launched at spawning sites in south Long Island, Bahamas remained over or close to the insular shelf for several days (Adapted from Colin 1992)

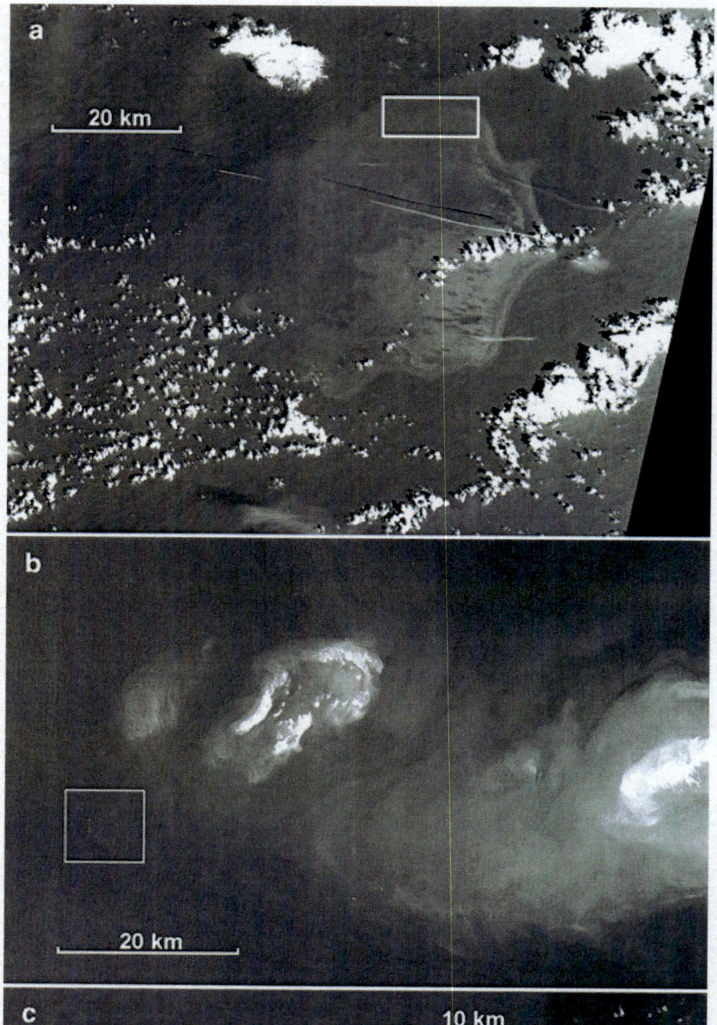

Fig. 5.11 Bank reefs can also have spawning aggregations and their hydrodynamic environment is very different from that of shallow reefs and islands. (**a**) Saba Bank, a sunken atoll, has a red hind aggregation within the box indicated, as well as a variety of promontories and other features visible in this satellite image. (**b**) "Riley's Hump" (*white box*) at the SW corner of the Dry Tortugas is a known aggregation site for snappers and potentially other fishes. (**c**) Lang Bank (*arrow*), at the eastern end of St. Croix is a red hind aggregation site. It is more like an isolated bank, than part of an insular shelf, due to its distance from any land

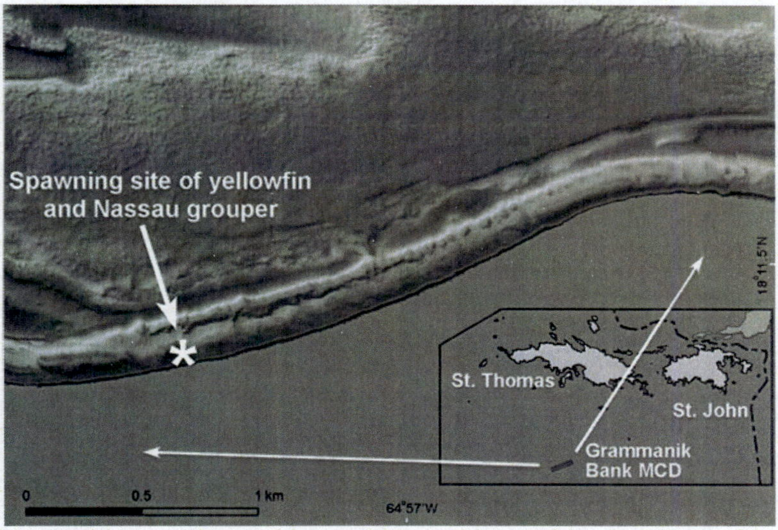

Fig. 5.12 Transient aggregations of Nassau grouper (*Epinephelus striatus*) and yellow fin groupers (*Mycteroperca venenosa*), occur on a relatively straight uniform section of insular shelf away from the ends of islands or promontories south of St. Thomas, US Virgin Islands (Image courtesy E. Kadison, from Rothenberger et al. 2008)

resembles that of a bank reef. Lang Bank, at the eastern end of the St. Croix insular shelf (Fig. 5.11c), is an aggregation area for red hinds and other fishes (Nemeth et al. 2007). A broad shelf deeper than the bank aggregation area separates it from the island 20 km away, so the bank is openly exposed to general oceanic currents and not generating island effects itself.

Some TWA species with TA, such as the Nassau grouper, utilize both promontory and non-promontory sites for aggregation. Detailed information on geomorphology of these sites is available and these include outer reef slopes (Figs. 5.12 and 5.1a) without projections or promontories (Virgin Islands – Rothenberger et al. 2008; Bahamas – Smith 1972; Colin 1992; Sadovy and Eklund 1999), projections off a long barrier reef (Belize – Carter et al. 1994; Paz and Truly 2007), ends or points on narrow islands or atolls (Belize – Heyman et al. 2001, Starr et al. 2007; Cayman Islands – Whaylen et al. 2004; Aguilar-Perera 2006), and non-promontory areas of atolls (Mexico – Aguilar-Perera and Aguilar-Davila 1996). Colin (1992) pointed out that in southern Long Island, Bahamas promontories existed within 10–15 km of a south Long Island aggregation site, which itself is not located at a promontory, and that fishers reported never having seen any Nassau grouper aggregations at these promontories. Despite variation in geomorphology of Nassau grouper sites, they all share the characteristic of being on or very close to the outer slope. A similar level of geomorphologic variability (promontory and non-promontory) may exist for sites of other large TA species, such as cubera snapper, but more detail is needed before such a comparison can be made.

5.4 Remote Sensing to Identify Aggregation Sites

Attempts to discover unknown aggregation sites using remote sensing rely on characterizing the geomorphology of known sites, then identifying similar areas as potential sites (Kobara and Heyman 2008). While appealing in principle, the variability of TA site geomorphology would alone make it unlikely such would be effective in practice. Even so, the limitations of the commonly used satellite images, often low resolution Landsat coverage, render such attempts often futile. Many TWA TAs occur at depths below those visible in satellite images (Colin 1992; Sadovy et al. 1994; Nemeth et al. 2007). When the bottom cannot be seen, then there is no ability to discern geomorphology as an indication of possible aggregation occurrence. For example, for the south Puerto Rico-Virgin Islands shelf, a "low tech" marine chart shows the basic geomorphology of the area fairly clearly (Fig. 5.13a) but a multibeam sonar map has much more detail (Fig. 5.13b). A Landsat 7 image, however, shows no bottom visible near any aggregation sites (Fig. 5.13c) and a map interpretation (Reefbase.org), based on the satellite image, also does not indicate shallow bottom in the area. The same problem occurs for many Nassau grouper aggregations in the Bahamas. These limitations argue against using satellite images to identify aggregation sites even if geomorphology alone were a valid indicator (which it is not). This is discussed further in Chap. 9.

5.5 Migration to and from Aggregation Sites

To aggregate, individual fishes have to undergo migrations of varying distances unless among the small number already resident at the aggregation site. TA migrations are believed to occur at lunar or seasonal time frames while those of RA may occur daily for periods of weeks or months. Why do fishes migrate to TAs? Why do RA fishes migrate at all?

Migration was discussed in more detail in Chap. 2 (also see Nemeth 2009). The migration distances for fishes with RAs are probably limited to a few km, since they must make the round trip from resting (and possibly feeding) sites to the aggregation site daily. Data are lacking for most species, but brown surgeonfish, *Acanthurus nigrofuscus*, in the Red Sea have daily migrations between feeding and spawning sites (Mazeroll and Montgomery 1998; Myrberg et al. 1988) for much of the year, following discrete pathways in an organized manner.

Different patterns of migration to RAs may occur on a single reef, the distance increasing with larger fish size. For example, on a shallow reef front in Palau two patterns of migration were found. In the first, the small bullethead parrotfish, *Chlorurus sordidus,* striated surgeonfish, *Ctenochaetus striatus,* and brown surgeonfish aggregate in many areas on the reef front to spawn, migrating no more than 100–200 m daily (Fig. 5.14a). They form an almost continuous, band-like, distribution of spawning fish on the reef front (PLC unpublished data, Robertson 1983).

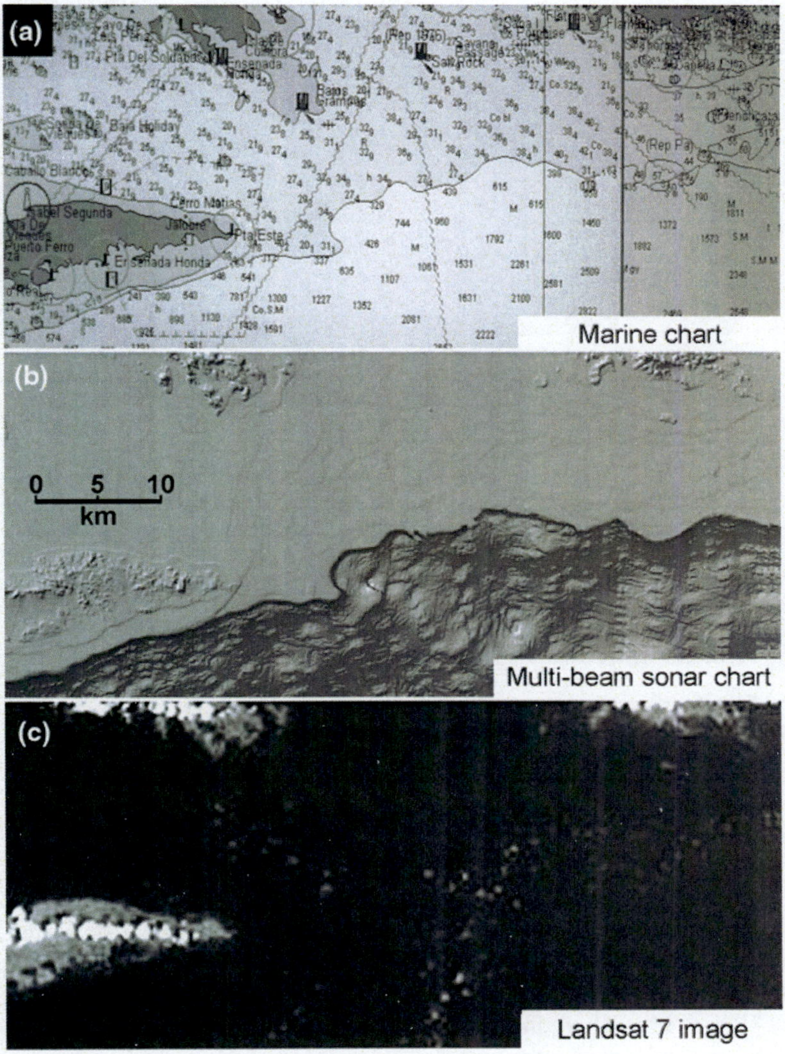

Fig. 5.13 When comparing geomorphology and aggregation sites visible in the same area along the southern shelf of the Virgin Island and Puerto Rico using (**a**) conventional bathymetric maps, (**b**) multi-beam sonar and (**c**) satellite images, each provides different information. Traditional charts (**a**), particularly with accurate position information, can provide an overall perspective on aggregation areas relative to other habitats. Multi-beam sonar (**b**) can produce an unequaled image of the bottom, showing details of geomorphology that not apparent with any other method. Satellite images (**c**), such as this Landsat 7 image, can provide benthic habitat information if the water is shallow and clear, particularly if there is strong contrast, but in many area where aggregations occur neither the insular shelf or its edge are visible. See Fig. 9.12 for aggregation location

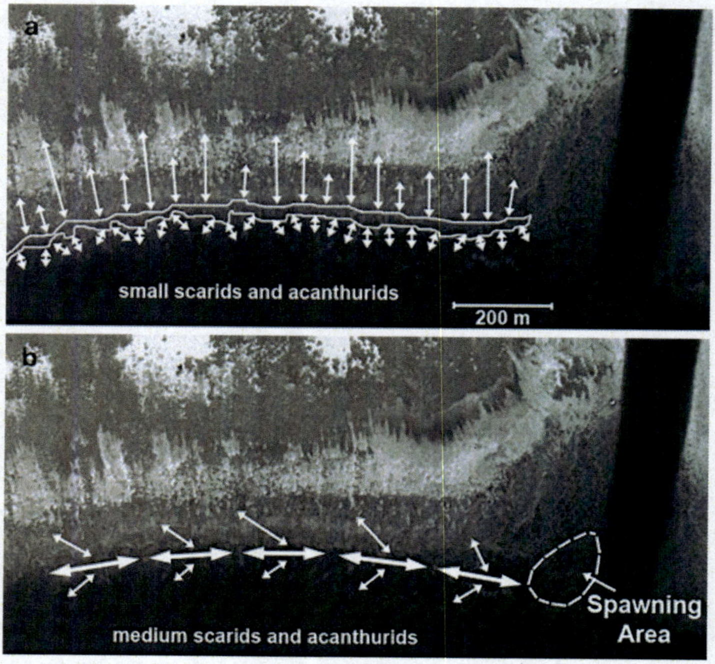

Fig. 5.14 Daily migration patterns of fishes which form resident aggregations (*RA*) on shallow reefs can differ among fishes. For Lighthouse Reef, Koror, Palau, (**a**) small parrotfishes and surgeonfishes migrate only short distances to form a band of spawning fishes on the seaward margin of the reef, while (**b**) somewhat larger species migrate at least a few km along the shallow reef to spawn on the margin of the deep tidal channel (Photo and unpublished data: Patrick L. Colin)

The larger greenthroat parrotfish (*Scarus prasignathos*), tan-faced parrotfish (*Chlorurus frontalis*) and ringtail surgeonfish (*Acanthurus blochii*) migrate along pathways through the aggregation areas of the smaller species (and while the smaller fishes are spawning) to an area near the edge of a deep tidal channel at least 1.5 km distant (Fig. 5.14b) where they spawn. An hour later they return along the same lines, moving along the depth contours (3–6 m deep) of the reef front to their home ranges. The humphead wrasse, *Cheilinus undulatus*, is the largest known RA species and its maximal daily migrations, in Palau on the order of 5–10 km (Colin 2010) and about 10 km at Layang Layang atoll, South China Sea; that distance limited by the atoll size (Nicholas Pilcher 2005) may represent near an upper limit to RA species daily migration distances.

TA species migrate to and from sites usually for a specific lunar or seasonal period and tend to be predatory species able to forego feeding for prolonged periods (Chap. 2). Nassau grouper have documented migration distances to spawning aggregations of 150–250 km along unbroken reef tracts (Bahamas – Colin 1992; Bolden 2000; Belize-Yucatan - Carter et al. 1994) and lesser distances among islands and reefs isolated by deep waters (Belize, Glover's Reef 35 km – Starr et al. 2007; US Virgin Islands, St. Thomas, 6–33 km, St. Croix, 5–18 km – Nemeth et al. 2007).

These data support the belief that large TA aggregators do not migrate between reefs, islands or atolls separated by deep (several hundred meters or more) water. However, at locations where reefs exist on a contiguous insular or continental shelf aggregating fishes may easily migrate "between reefs" (Sadovy et al. 1994; Zeller 1998).

How do fishes with large TAs migrating long distances find the sites? It has been shown in some smaller RA species younger fishes learn routes of migration and location from older fish (Mazeroll and Montgomery 1998; Warner 1988) and it is likely some ability exists in this regard for larger TA species. Similar to RA species, it is likely TA species would use reef features as a migration guide. Observations of probable TA migrants swimming along shelf edge reefs (Colin 1992; Carter et al. 1994; Whaylen et al. 2006; Starr et al. 2007) indicate that larger fishes may use shelf edge, or drop off, contours as "highways" to reach transient aggregation sites. Shelf edges are essentially a linear environment and at the shelf edge a migrating fish can swim one of two ways along this feature. The possibility of an innate navigational compass or ability to guide migrations also exists (Helfman and Shultz 1984), but has not been demonstrated for aggregating fishes.

Earlier suggestions the fish might locate aggregation sites by swimming either upcurrent (Colin et al. 1987) or down-current (Carter 1986) along the shelf edge to the end of an island or to a promontory were overly simplistic. Currents are likely too variable daily and seasonally to be of great importance in locating aggregations (Carter et al. 1994; Colin 1992). However, Nemeth et al. (2008) did find that most red hinds in the Virgin Islands migrated in an up-current direction to reach aggregation sites. In areas where currents are directionally consistent, a correlation of current and migration directions might exist, however the current alone is probably not the sole means to locate aggregations. Acoustic tagging studies, in which the migration direction can be determined by successive detection of fish by receivers (Starr et al. 2007), hold promise for providing information on migration, particularly if combined with current measurements.

Given the longevity of many large TA species, an individual fish could potentially migrate to the same or multiple aggregations for a decade or more. For Nassau grouper, with sexual maturity at 4–7 years and a maximum age of nearly 30 years (Sadovy and Eklund 1999), most participating individuals may well have migrated to a site numerous times, assuming a yearly migration. With heavy fishing pressure the average age of fish in aggregations decreases and could reduce the number of fish from which younger ones might learn the locations of aggregations.

5.6 Oceanography of Spawning Sites

The oceanographic conditions at sites have generally been proposed as a critical factor attracting individuals and promoting site use over time. Physical factors such as temperature, current speed and direction, light, water clarity, water column structure, waves, seasonal weather or productivity differences could play a role.

Sites might be beneficial for adults, perhaps easy to locate during migrations and traditionally used, or for propagules, through reduced predation on released eggs or increased survival of eggs and larvae to the recruitment stage. Our knowledge of these factors is only preliminary and any hypotheses should be tested empirically.

A major geographic difference exists between the TWA and IWP with the former generally lacking major tidally driven currents while they are common in the latter. This comes largely from differences in tidal amplitudes. Most IWP reef areas have ranges of 1–2 m, inducing strong tidal currents between closely spaced islands and reefs, while TWA locations have only about one half metre range and broad deep (15–20 m) insular shelves without nearby shallow areas and strong tidal currents (see also Chap. 4). A few areas in the TWA, such as the Exuma chain (Bahamas) and the Florida Keys, have large shallow bank areas where strong tidal currents course through gaps between islands, however these are exceptions.

Most offshore currents will run parallel to an island shore or relatively straight barrier reef, but where a physical structure, such as a promontory, interrupts the parallel flow the current will be forced to deviate in direction (Chap. 6). This produces eddying in the offshore flow which may help to retain or disperse larvae. Tidal channels have currents as tidal jets extending out to sea while shear flow on their edges will produce eddies on the sides of the channel. Where barrier reefs exist water flows from lagoon to ocean across the reef on falling tides as a lens, and can extend a short distance out to sea from the reef as a discrete water mass (see Fig. 6.12).

The question of whether the oceanography of spawning sites promotes the dispersal and/or retention (or neither) of eggs and resulting larvae can be approached in many ways. Currents measured at sites and times of aggregation spawning provide robust data on chances of dispersal versus retention. The concept of dispersal kernals (Chap. 6) is useful in visualizing the distribution and movement of masses of propagules. Recent studies of self-recruitment have clearly established that some larvae can settle near to where they were spawned; "near" being a relative term (Swearer et al. 1999; Jones et al. 1999, 2005), but dispersal should ensure that most larvae settle variable distances from where they were spawned (Cowen 2002; Mora and Sale 2002).

In the TWA minimal currents ("quiet currents" of Whaylen et al. 2006) have been measured or observed at large TA sites during dusk spawning. For Nassau grouper, Whaylen et al. (2006) reported spawning bursts to occur when currents "were slack or negligible". For red hind, Nemeth et al. (2007) found currents at a shelf edge aggregation site of red hinds at their annual and lunar minima during the winter spawning season with the week preceding the full moon phase when aggregation and spawning is occurring having currents of only 2.5–3.5 cm s^{-1} during spawning. Heyman and Kjerfve (2008) reported variable currents at a snapper-grouper TA site, with a mean modest speed of only 8 cm s^{-1}. A new analysis of the current data at a Nassau grouper site (Colin 1992) indicates during the late afternoon and early evening period (means 3–5 cm s^{-1}) currents were minimal on the full moons of December and January (Fig. 5.15) and currents milder at the aggregation site than at two sites 5 and 6.5 km distant in opposite directions. There are also a number of qualitative reports of mild currents at spawning sites and times

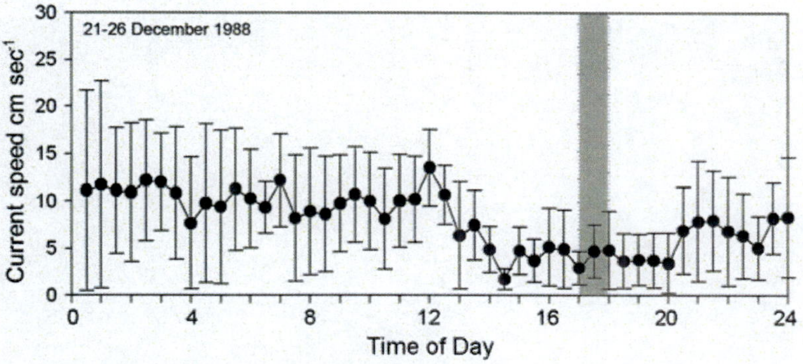

Fig. 5.15 Currents at a Nassau grouper aggregation site at Long Island, Bahamas were found to be at a minimum in the late afternoon when the groupers are spawning (Data from 21–26 December 1988, mean is the *black circle* with ±1 SD for each 30 min period shown). Spawning was observed between 17:00 and 18:00 h, indicated by the *grey bar*.

(Colin et al. 1987; Colin 1992; Sadovy et al. 1994). For RAs in the TWA currents seems to play a relatively small role in spawning as many RA species in the region aggregate and spawn in the late afternoon, without exact relationship to the currents (Randall and Randall 1963; Colin and Clavijo 1978, 1988; Warner 1995).

The relationship of TAs and currents is different in the IWP. Three groupers (brown-marbled grouper – *Epinephelus. fuscoguttatus*, camouflage grouper – *E. polyphekadion* and squaretail coralgrouper – *Plectropomus. areolatus*) commonly aggregate simultaneously near the ocean end of channels where tidal currents exist (Sadovy 2005). In Palau the currents were examined at two barrier reef channels with these aggregations (PLC unpublished data) and they had flow from lagoon to ocean about 60% of the time, with 40% inflow to the lagoon for a net export of water through them (Fig. 9.3). The camouflage grouper was observed spawning just after dawn as the current starting moving offshore (Chap. 12.5), while spawning time of the other two species is not known. Drifters released at the aggregation site at different times starting after high tide indicate eggs were carried in a tidal jet no more than about 1–2 km (Fig. 5.16) off the reef and then carried in the direction of the along reef current.

Other IWP large TA's are on outer reef faces, rather than in or near tidal channels, and the currents relative to spawning are unknown. The FSA's of blue-lined sea bream, as well as the possible FSA's of twin-spot snapper, *Lutjanus bohar*, occur on outer reef slopes close to promontories, exposed to along-reef currents, which are occasionally quite strong (1 m s^{-1} or more). Sakuae et al. (Chap. 12) indicate a close relationship between current direction and spawning in blue-lined sea bream with eggs potentially being transported into open water. Spawning by twin-spot snapper occurred near the same site, but with nearly opposite current conditions that still appeared to carry eggs into open ocean (Chap. 12). Samoilys (1997) reported little current at dusk during periods when leopard coralgrouper, *Plectropomus leopardus*, were aggregated to spawn on continental shelf reefs of the Great Barrier Reef.

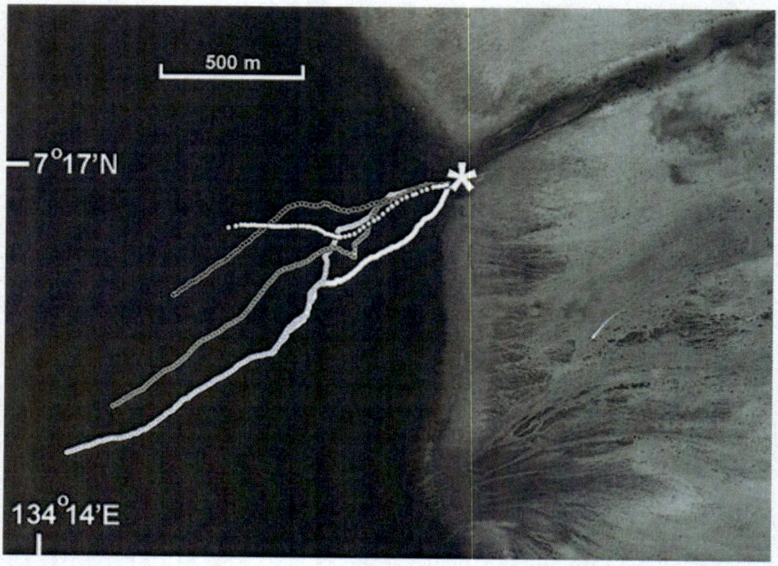

Fig. 5.16 Current-following drifters started at 30 min intervals after high tide at a grouper aggregation site at Ulong Channel (Ngerumakaol), Palau during aggregation were carried 1–1.5 km out to sea by tidal jets created by falling tides. After low tide they stalled offshore as tidal currents came to a stop

In the IWP some RA species (as well as various non-aggregating fishes) spawn along reef fronts during the few hours after high tide, migrating (as discussed previously) either to the seaward front or along the reef to channel edges (Fig. 5.14), and start spawning as the tide begins to turn and move slowly off the reef (see Fig. 6.12). In Palau such spawning locations and times did not always ensure offshore dispersal of eggs; a portion of these moved back towards their natal reefs after low tide with some being carried across the reef into lagoon environments (Hamner et al. 2007, PLC unpublished data). For other IWP fishes aggregating in tidal channels of barrier reefs and atolls, spawning by various RA species occurs after high tide when currents start moving out towards the open ocean (Bell and Colin 1986; Moyer 1989; Craig 1998).

Three factors affect the movement of eggs after spawning; the vertical movement of eggs in the water column, diffusion of eggs as particles and the advection of water containing eggs. These are discussed in more detail in Chaps. 6 and 7, but it is important to establish here that nearly all aggregating species have positively buoyant eggs that both ascend slowly and diffuse outward from their release point. Currents producing horizontal transport (advection) in reef areas where aggregations occur are produced by a number of mechanisms. Weather and sea swell effects can also dominate shallow transport of eggs and larvae in lagoons, channel areas and nearshore outer reefs (Sancho et al. 2000a).

The entrainment of eggs and larvae into oceanic circulation is uncertain and appears somewhat location-specific. Certainly there is a mix of entrainment potential among spawning sites; those larvae that become entrained in larger current systems have a higher chance of being widely dispersed (Domeier 2004, “Grouper Moon project” – unpublished data), while those not entrained may be retained more locally (Colin 1992; Heyman et al. 2005). In areas with extensive shallow reefs or large coastlines it is often not easy for reef fish eggs from both aggregating and non-aggregating species to become entrained into oceanic circulation as currents do not necessarily favour transport off the shelf edge (Appeldoorn et al. 1994; Hensley et al. 1994), however oceanic entrainment may be common for small islands with narrow shelves and isolated reefs due to differences in water circulation near them (Fig. 5.10).

5.7 Temperature Regimes and Other Physical Parameters

Water temperature determines the rates of many of the metabolic functions of fishes. Most reef fishes live in a fairly narrow range of annual temperatures with water temperatures in the low latitude tropics typically quite stable over the year, with a range of only about 1–3°C in the IWP and 3–4°C in the TWA. At higher latitudes with reefs (e.g. Great Barrier Reef, Hong Kong, southern Japan, Bermuda) summer high temperatures are similar to equatorial regions but winters are considerably cooler for greater annual variation.

Many RA species in the equatorial tropics spawn year-round throughout the limited annual temperature range, but little is known for areas with greater yearly variation. In most places, TAs are reported to occur seasonally within a particular temperature regime, usually at locations where there is an annual variation of at least a few °C (Domeier and Colin 1997). In areas of the IWP where there is little annual temperature variation, TA species may have lengthy aggregation periods (Fig. 5.17) to as much as year-round (or nearly so), but always with a lunar periodicity (Johannes et al. 1999, Chap. 12). If a species with a broad latitudinal distribution has its spawning season varying with latitude, this would potentially indicate a close relationship between temperature and spawning. For Nassau grouper, spawning occurs at a fairly consistent temperature (see below). Other species may not have such a close correlation, but often truly definitive data are lacking. Camouflage grouper aggregation, for example, is reported at different temperatures (between about 24°C and 29.5°C) over its range (Chap. 12).

Temperature determines how fast embryos develop in the egg, when eggs will hatch, when feeding can begin, the time window a larva has to initiate feeding before it starves, and how fast it can grow under otherwise suitable conditions. At hatching, pelagic larvae lack a mouth, formed fins and pigmented eyes, and must live off their yolk for a limited time. After development of functional eyes, fins and a complete digestive system with mouth and anus, the yolk is nearly absorbed. A “critical period” of limited duration (a few hours?) occurs at this point during which the larvae must start feeding or starve (Leggett and Dublois 1994; Yoseda et al. 2006) (Chap. 7).

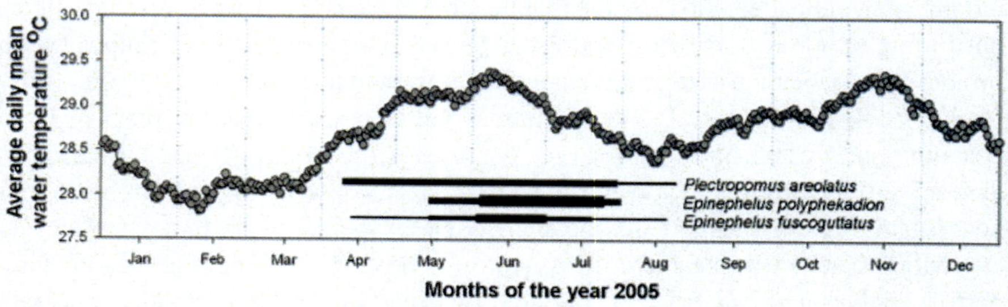

Fig. 5.17 In low latitude tropical areas water temperature may not vary much during the year. At Ulong Channel (Ngerumakaol), Palau, daily mean temperatures were found to vary by about 1.7°C during 2005. The presence of grouper aggregations is shown relative to water temperatures and the relative numbers of three groupers are shown by the thickness of the bars for each species (Photo and unpublished data: Patrick L. Colin)

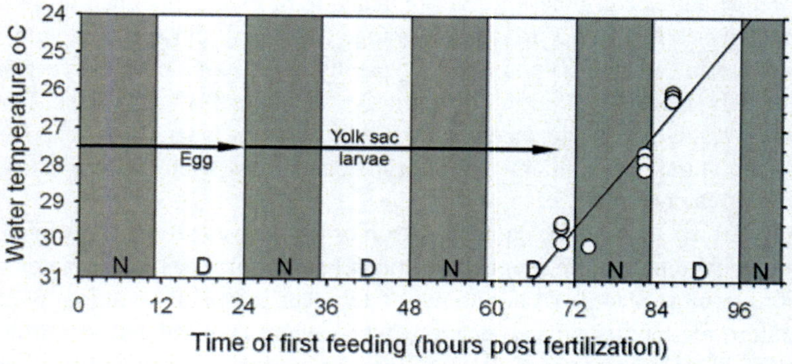

Fig. 5.18 The time between fertilization and first feeding for Nassau grouper larvae (includes egg and yolk sac larva) is dependent on temperature. Eggs/larvae reared at 26°C are ready to begin feeding on the fourth morning after spawning. *Grey/white* segments are *D*-day and *N*-night. Larvae reared at higher temperatures are ready to begin feeding at sunset or during the night and less likely to successfully initiate feeding (Data from Watanabe et al. 1995)

For the Nassau grouper temperature may be a critical factor in early larval success. They are known to spawn within a relatively narrow temperature range (about 25.5–26.5°C) with timing varying geographically with water temperatures (Carter 1986; Colin 1992; Sadovy and Eklund 1999; Whaylen et al. 2004, 2006; Starr et al. 2007). For the central Bahamas the annual temperature range is about 5°C, with lowest temperatures (24.0–24.5°C) in late February and early March, with 25.5–26.5°C occurring during the known spawning season of December–January (Colin 1992). Watanabe et al. (1995), in hatchery studies on Bahamian Nassau grouper, found that the incubation time for eggs and first day survival post hatching was inversely related to temperature. Time from fertilization to first feeding was also found to decrease linearly with rising temperature, from 86 h at 26°C to 71 h at 30°C (Fig. 5.18). Interpolating from Watanabe et al. (1995), at 25°C it should be about 91 h and 88 h at 25.5°C. In the field spawning occurred around sunset with the day-night length in the Bahamas approximately 11–13 h during December–January (Colin 1992). Dawn after the fourth night post-spawning would occur around 85 h after fertilization. At 26°C larvae would be ready to begin feeding an hour or two after sunrise on that morning (Fig. 5.18) and would have nearly a full day in which to initiate feeding, helping to ensure that the larvae will survive the upcoming night when feeding may not be possible. Larvae would be ready to feed in the middle of night if reared at 28°C and at sunset on the fourth night at 30°C (Fig. 5.18).

Is temperature similarly important among other TA species? Unfortunately few data on time to first feeding versus temperature, or indeed in relation to spawning, are available for other reef fishes. Sheaves (2006) suggested that early life history stages of sparid fishes (porgies) might be intolerant of higher water temperatures and that spawning (not necessarily aggregation-spawning) during cool periods by those found in warmer water might be a tactic for survival in such waters. Samoilys (1997) found leopard coralgrouper aggregations on the Great Barrier Reef to occur at temperatures between 24.25°C and 28.5°C, within an overall annual range of about 23–30°C. Ulong Channel in Palau had an annual water temperature range of 1.5°C (28.0–29.5°C) with two groupers aggregated during limited temperature ranges (28.5–29.5°C for brown-marbled grouper, *E. fuscoguttatus* and squaretail coralgrouper) while a third had a more in limited temperature range (29.0–29.5°C for camouflage grouper, *Epinephelus polyphekadion*) (Fig. 5.17). In Fiji the same species aggregated at temperatures of approximately 24°C with an annual temperature range at aggregation site is about 24–28.5°C (Yvonne Sadovy de Mitcheson unpublished data). Perhaps the Fijian populations have different rates of development through the yolk sac larvae, or perhaps, if rates are the same, have the time between spawning and first feeding lengthened by one day in the cooler water temperatures of Fiji, still allowing first feeding to start early in the day. Alternatively, they appear to spawn at the lowest point of the annual temperature cycle so this may be a cue for adults. Kadison et al. (2006) found cubera snapper spawning in the US Virgin Islands at temperatures above 26.9°C to nearly 28°C, but unfortunately comparative data are not available for this species from other areas. If data are gathered on annual temperature regimes, as well as studies of egg and early larval development conducted related to temperature (often from aquaculture), it will become possible to test ideas of the possible importance of temperature in early life history and whether this is a general principle related to reef fish aggregation. The documentation of temperature at spawning sites is a simple, inexpensive standard activity and should be regularly conducted (Chap. 9).

Drop-off areas where most TA fishes aggregate generally have cooler water at depths not very far below aggregation areas. Thus fishes can influence their temperature regime by simply varying the depth at which they reside. Starr et al. (2007) documented a fascinating instance where Nassau grouper at Lighthouse Reef, Belize, moved from the normal shallow water (less than 30 m) habitats to deeper (60–70 m depth) water after their January spawning aggregation. Several possible reasons for this change can be suggested, selection for a preferred temperature being one of them.

Among other factors, salinity may play a role in where reef fishes aggregate because high salinities (over 30–32 ppt) may be necessary for reef fish eggs to be positively buoyant in order to rise towards the surface rather than drifting towards the bottom (Colin et al. 1996; Ellis et al. 1997). It seems unlikely that any reef fishes will migrate to low salinity areas to spawn (Chap. 2). Unverified information, such as that of Craig (1969) reporting goliath grouper, *Epinephelus itajara*, to migrate to mangrove channels in Belize to spawn (which may have somewhat lower salinity than reef areas), has not been confirmed by subsequent work which indicates the species migrates offshore to spawn (Chap. 12.4). It is believed oxygen will not normally be a factor influencing locations and timing of spawning, as waters where aggregations occur are generally near saturation. Most aggregation areas on reefs have relatively clear water, although the turbidity can vary over a normal range for reef areas where RA species spawn.

5.8 Spatio-Temporal Factors

Aggregations exist in several different time frames: annual, seasonal, lunar month and diel. The time spent on direct courtship and spawning is very low for large TA species, on the order of 0.1–0.2% per annum for adult Nassau grouper, but if migration is included might represent 5–10% of total time. RA species spend more time engaged in courting and spawning, perhaps as much as 3% for humphead wrasse (Colin 2010), and if daily migration time is included, the value is similar to that for TA species. Many RA species, despite their regular spawning, can maintain activities such as feeding when *en route* during migration and while at aggregation sites (Chap. 2). TA species are often benthic predators, and while migrating or aggregating the fish may either go for periods without feeding or range out from aggregation areas to feed elsewhere at times (Chap. 2, Nemeth et al. 2007). Noticeably skinny squaretail coralgrouper have been seen in markets at the end of the spawning season in Palau and other locations, suggesting that their feeding does not keep up energy requirements during the spawning season (Yvonne Sadovy de Mitcheson 2009).

5.8.1 Time of Day for Spawning

The time of day when spawning in reef fishes occurs varies with species and conditions (Thresher 1984) and activity patterns vary among aggregating species. At the equator there is no difference in day length throughout the year, with day and night being approximately equal at all times. As one moves away from the equator, the lengths of day and night change during the year. For example, at the latitude of Palau (7°N), day length is 11:41 h on the winter solstice, 12:34 on the summer solstice; a change of only 53 min or 7.5% annually. At the northern limits of coral reefs, the differences are marked. In Bermuda (32°N) the winter solstice day is 10:01 h long and the summer solstice 14:17 h, a 40% difference of 4:16 h.

All known RA fishes are diurnally active species. For TA species, many are active both day and night (groupers, snappers). For largely nocturnal fishes (e.g. squirrelfishes – Holocentridae, bigeyes – Priacanthidae) it is not known how and when they spawn, but aggregation spawning is possible. In the TWA large TA fishes, for which there is information, spawn from very late in the afternoon to sunset. Nassau grouper spawn in a narrow time window from about 20 min before to 20 min after sunset (Colin 1992; Sadovy et al. 1994; Whaylen et al. 2004). Other TWA groupers are similar (Colin et al. 1987) while cubera snapper spawn in the hour before sunset (Heyman et al. 2005; Kadison et al. 2006). In the EP Sala et al. (2003) found the snapper *L. argentiventris* to spawn within 1 h of sunset in Gulf of California.

Other species may not be so tied to a particular time of day. In the EP Sala et al. (2003) found Pacific dog snapper, *Lutjanus novemfasciatus*, spawned at 2 pm along reef walls while bigeye trevally, *Caranx sexfasciatus*, spawned as pairs out of an aggregating school during the day at no specific times (Chap. 12.18). Less is known for IWP TA species, but some IWP snappers are now known to spawn in the early morning (blue-lined sea bream and twin-spot snapper) on outer reef faces (Chap. 12.11). Many attempts to observe spawning by IWP TA groupers have been unsuccessful, but camouflage grouper have been observed spawning just after dawn (Chap. 12.5). Some of these species may be spawning in the middle of the night (Rhodes and Sadovy 2002). The only IWP grouper with numerous sunset spawning observations is leopard coralgrouper and its behaviour appears to be intermediate between a RA and TA species (Chap. 12.9). Hamilton (2005) reports on reliable observations of the spawning by the longfin emperor, *Lethrinus erythrocles*, near midnight.

RA fishes in the IWP in areas with moderate to high tidal amplitudes have a different pattern with nearly all spawning starting after high tide and lasting for 1–2 h. Along the lengthy barrier reef of Palau, RA species as well as more numerous harem and pair spawners (PS), spawn on the seaward edge of the reef crest after the high tide (Robertson 1983, PLC unpublished data) at times ranging from near dawn to sunset. Newly shed and fertilized RA and harem/pair-spawned eggs are essentially the same (size, transparency, buoyancy) and have a similar pattern of advection off the reef (Hamner et al. 2007). While spawning starts just after high tide, in some areas there appears a sequence to the spawning by 20–40 species (only 2–3 of those RA spawners) most over the period of 60–90 min after high tide. The humphead wrasse is the final species to spawn, starting until 2–2.5 h after high tide (Colin 2010). By the time it is finished for the day, no other RA or PS fishes remain spawning.

A few RA species are believed to spawn only in the morning. The bumphead parrotfish, *Bolbometopon muricatum*, spawns as pairs out of larger schools from just after dawn (Gladstone 1986, PLC unpublished data) to late morning (Mandy Etpison personal observation), but not during the afternoon (Chap. 12.15). Some other large, but non-aggregating, IWP parrotfish are also only known to spawn in the morning (Colin and Bell 1991). The bullethead parrotfish, a small RA fish, group-spawned after high tides only from early morning to early afternoon in

Palau, while other species did so throughout the day depending on the time of high tide (Chap. 12.21, PLC unpublished data).

5.8.2 *Lunar Aspects of Aggregations*

Specific lunar phases are associated with many life history events in reef fishes and delineating the lunar components of aggregation and spawning is of special importance. Tidal amplitudes are greatest (spring tides) and tidal currents strongest during full and new moon periods, but these periods differ greatly in the ambient light levels at night. The full moon provides significant nocturnal light in the shallow reef environment and most fishes (and even humans) are able to see quite well with it. Just prior to the full moon, the moon will already be above the horizon at dusk and remain in the sky most of the night. In the days after the full moon, moonrise occurs an hour or two after sunset and it is initially dark. After the moon has risen it is still nearly full and provides light for the remainder of the night. Those fishes making long (multi-day) migrations to aggregations might, if they travelling at night, can presumably migrate more easily around the time of the full moon with its illumination; while species with shorter migrations (which may not be moving at night) would not be tied to a particular lunar phase. In the TWA some species known to have lengthy migrations (Nassau grouper, red hind) aggregate on the full moon, but additional information is needed for other species before this can be considered a possible general principle. Conversely, on the new moon, it is dark all night long. The days around the new moon provide essentially no lunar illumination.

The length of the lunar (synodic) month is approximately 29.5 days, and differs from the length of Julian calendar months (28–31 days). Because of this disparity, averaging about 0.9 days per month, the dates of the lunar phases change each year on the Julian calendar. This makes direct comparison of aggregation/spawning times on specific lunar phases of a given month between years problematic. The dates of the lunar phases are about 10–11 days earlier in each successive year (Fig. 5.19) so every 3 years the dates of lunar phases in a calendar month will have shifted back to near the same position relative to the calendar month. Hence in Palau the full moon of January 2001 was on the 10th, that of 2004 on 8th. The full moon in January 2002 was the 28th, but the full moon of December 2001 (30 December) is actually closer to the 10 January 2001 moon. Whaylen et al. (2006) attempted to compensate for full moons that occur earlier or later in a calendar month by using the winter solstice (Dec 21 or 22) as a benchmark which changes little year to year. This is useful and this (or similar) measure(s) should be used when comparing times of aggregation and spawning between years.

The lunar timing of spawning may also determine the timing of pelagic phase recruitment to the bottom. For Nassau grouper in the Bahamas, settlement of pelagic juveniles occurred about 37–42 days after spawning and juveniles are believed to recruit at night (Shenker et al. 1993; Colin et al. 1997). Based on the larval life length, this will occur near the time of the new moon, when moonlight is minimal, and may benefit recruits in reducing predation as they settle. If a species has a

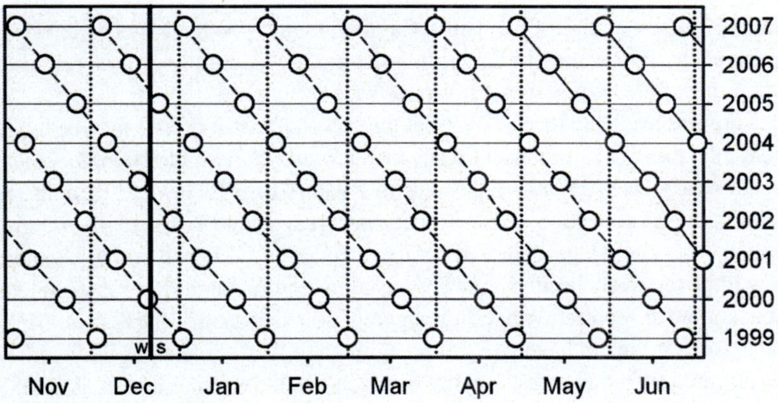

Fig. 5.19 The date of the full moon, indicated by the *open circle* (as well as other moon phases) shifts approximately 9–10 days earlier with each successive year. Every three years, the calendar dates of given moon phases are back to being close in alignment (*vertical dotted lines*). Often it is more useful to compare lunar phases or the time of events by the number of days after the winter solstice (*black vertical line* labelled "ws") or similar astronomical event, rather than strictly considering calendar months

shorter or longer pelagic life history stage, a new moon spawning could also result in recruitment on the new moon one or more months after spawning. Samoilys (1997) found leopard coralgrouper to spawn most often during the new moon while Doherty et al. (1994) found coral trout recruiting on the new moon, based on numbers of daily otolith increments, at an age of 25 days after spawning. Further studies of larval life length, based on otolith ages, are needed for large TA species to see whether spawning times typically produces recruitment around the new moon.

5.8.3 Seasonal Periodicity

Major astronomical markers each year include the solstices and the equinoxes. In the northern hemisphere the winter solstice is December 21 or 22 and the summer solstice June 21 or 22. They are reversed in the southern hemisphere. The equinoxes occur on March 20 or 21 (vernal in north hemisphere) and September 22 or 23 (autumnal in same). On these dates day and night length are equal everywhere.

Spawning aggregations are usually associated with specific seasonal periods and lunar phases. In the TWA nearly all known TA species have their spawning in the winter or spring (Domeier and Colin 1997, Chap. 2), however in Bermuda some groupers aggregate in June and July (Bardach et al. 1958; Luckhurst 1998) possibly due to overall lower water temperatures there. The goliath grouper, *Epinephelus itajara*, may be an exception, with probable (but not confirmed) spawning aggregations occurring during the summer (Bullock et al. 1992; Colin 1994, Chap. 12.4). Within the Caribbean portion of the TWA, individual species that aggregate seem to have the same seasonal timing across the region. Cubera snapper have the same spawning seasonality throughout the Caribbean (US Virgin

Islands – Kadison et al. 2006 –, Little Cayman Island – Whaylen et al. 2004; Belize – Heyman et al. 2005) and similar timing occurs for Nassau groupers, red hind, and yellowfin groupers, *Mycteroperca venenosa*.

Data are less available for the IWP, but the seasonal extent of TAs may be longer and different than the TWA. The blackfin snapper, *Lutjanus fulvus*, has been documented to aggregate nearly every month of the year in Palau (Chap. 12.10) and there are reports from traditional knowledge of TAs occurring year round, particularly in groupers (Hamilton et al. 2005) e.g. squaretail coralgrouper (Chap. 12.8). It would be interesting to know if the relatively limited annual water temperature ranges in equatorial locations might account for reported variation in aggregation seasonality. TEK from Melanesia indicates some groupers have year-round aggregations in some areas while others had seasons limited to 3–4 months, varying between localities (Hamilton et al. 2005).

Validating differences in patterns of aggregation timing and of fish numbers is not simple. Often the timing and fish number peaks of aggregation for any month and over fairly long seasonal periods are not easily documented (Chap. 9). Peak fish numbers in an aggregation may occur for only 1 day during a lunar month, and not necessarily on the same day of the lunar cycle each year, with surveys on days either side of the peak having dramatic decreases in fish numbers. Unless surveys cover several days in a row, the peak period may be missed and a sub-peak day mistakenly assumed to represent the peak fish numbers (e.g. Fig. 12.77). The diversity of methods employed to gather data on numbers of fish in aggregations also introduces another set of challenges to comparing results. For example, three species of groupers aggregating at Ulong Channel in Palau have reported durations or repeated aggregation formation over 2.5–6 months (spring and summer), with some fishes present continuously at the site. Johannes et al. (1999) reported, during 1994–1996 peak aggregation months, as February-March for squaretail coralgrouper, June for camouflage grouper and May-June for brown-marbled grouper. Colin (unpublished data) using a different and more accurate GPS-based method at the same site (Chap. 9) obtained different results, finding the peaks in 2005 to be June–July for squaretail coralgrouper, July for camouflage grouper and June for brown-marbled grouper. Figure 12.15 shows peaks at a nearby site. It is not known whether differences between the studies are artefacts or real. Johannes et al. (1999) also reported different months for peak aggregation for three different channels in Palau. It is clear that surveys need to be conducted simultaneously in different areas to address such questions. The value of quantitative and repeatable methods for determining fish distribution and abundance in aggregations as well as gathering physical data is highlighted by this example (Chap. 9).

It has been suggested that the seasonality and locations of reef fish spawning may be related to the cycles of primary productivity, with fishes spawning during periods when there is adequate zooplankton, supported by phytoplankton production, for larval feeding (Johannes 1981; Heyman et al. 2005). This might be the case in areas with clear seasonality of primary production, such as the EP (Erisman et al. 2007) but is less likely to occur in the tropics which potentially have relatively stable, but low, zooplankton populations (Chaps. 6 and 7). Evidence indicates that adequate early feeding of larvae is important for ultimate survival of the pelagic stage (Bergenius et al. 2002).

5.8.4 Location and Height Above the Bottom of Spawning

Nearly all fishes with planktonic eggs release their gametes some distance above the bottom and the eggs are positively buoyant. Exceptions among aggregation spawners are rabbitfishes, triggerfishes and damselfishes (Pomacentridae). Some TA species, such as cubera snappers (Heyman et al. 2005; Kadison et al. 2006) ascend, often almost to the surface. Among RA species the humphead wrasse spawns as pairs near the surface (Colin 2010) while the yellow and blueback fusilier, *Caesio teres,* spawned as a large mass close to the surface (Bell and Colin 1986). Various surgeonfishes and parrotfishes also spawn very close to the surface, but ascend from only a few metres depth (Robertson 1983). The ascent of fishes (including most aggregation spawners) above the bottom causes expansion of the swim bladder and, as originally suggested by Randall and Randall (1963), this may aid in expulsion of the gametes at the top of the ascent. This expansion may also limit how high fishes can rise above the bottom, ascending only to a depth where expansion aids spawning, but not to the point of endangering the fish through loss of buoyancy control or internal damage from overinflation.

For most species, benthic-based egg predators are believed to be the most important factor favouring spawning at some height above the bottom. Thresher (1984) found that both large and small species spawning during the day released their eggs at the same height above the bottom, but for those spawning at dusk, larger species spawned higher in the water column. Due to positive buoyancy, it is likely most pelagic reef fish eggs ascend up into the upper few metres of the water column before hatching, even if released at 20–30 m depth. The egg ascent rates measured have been 8.4–18.6 cm min^{-1} for cubera snapper (Heyman et al. 2001) and 11 cm min^{-1} for coney, *Epinephelus fulvus* (Colin 1992). At these rates eggs spawned at 10–20 m would reach the surface within a few hours, well before hatching. Spiralling of rising and spawning fish may produce a vortex that helps the ascent of eggs towards the surface (Heyman et al. 2005). Advective mechanisms also tend to transport eggs away from reefs and benthic based predators (usually at speeds much faster than ascent rates), although changing tidal currents can later bring them back towards natal areas (Hamner et al. 2007). Other aspects of the fate of eggs and larvae just after spawning are discussed in Chaps. 6 and 7.

5.8.5 Multi-species Aggregations and Sites

The multi-species use of aggregation sites was first reported by Smith (1972) who noted in the Bahamas, based on fisher information, that black groupers, *Mycteroperca bonaci,* also used the same aggregation site (but at a different time) as Nassau groupers. Subsequently Olsen and LaPlace (1979), Colin (1992), Carter (1989), Carter et al. (1994), Whaylen et al. (2006), Sala et al. (2001), Starr et al. (2007), and Sadovy et al. (1994) reported use of aggregation sites by multiple grouper species for the TWA. In the 1990's conservation entities began trumpeting the "discovery" of multi-species use of sites as a means to publicize the need for conservation protection for such sites. However, virtually all known large TA sites worldwide are used by multiple TA species; exceptions to this generality may reflect nothing more than a lack of adequate study to detect other species using the site (some red hind sites, Colin et al. 1987; Nemeth et al. 2007) or local extirpation of some species. Sites where goliath grouper may be aggregating to spawn (spawning has never been seen) are not yet known to be sites shared with other TA species; again this may simply be a lack of study on a year-round basis since most aggregation sites are only studied during aggregation periods.

Multi-species use of a site can be simultaneous, sequential, or in most cases both. Often there are slight differences in the locations where different species aggregate at sites (Heyman and Kjerfve 2008; Nemeth et al. 2007, PLC unpublished data) and in the timing of their reproduction; ample reason to support the detailed mapping of sites (Chap. 9). In the IWP the camouflage grouper, brown-marbled grouper and squaretail coralgrouper use the same sites across a wide geographic range, often simultaneously, despite some inter- and intra-specific aggression (also see Rhodes and Sadovy 2002). At Ulong Channel, Palau, a relatively narrow and shallow channel, the three species have simultaneous, largely overlapping, distributions while in another much wider and deeper channel mouth 30 km away, the three species overlap much less (Fig. 2.6). In Fiji, at a reef channel aggregation site for the three species, the camouflage grouper is more concentrated on one side of the channel than the other, while the brown-marbled grouper only occurs on the opposite side of the channel and close to a fourth species, the blacksaddle coralgrouper, *Plectropomus laevis*, aggregation (Yvonne Sadovy de Mitcheson unpublished data).

Less attention has been directed toward multi-species use of RA sites. Reef tops, particularly seaward margins of barrier reefs, are locations for spawning by both resident IWP aggregators (typically brown surgeonfish, striated surgeonfish, bullet-head parrotfish) and non-aggregating fishes with spawning occurring on daytime falling tides, often in a sequence (Robertson 1983, PLC unpublished data). In the TWA, several examples of RA multi-species use are known, including for the redfin parrotfish, *Sparisoma rubripinne* and spotted goatfish, *Pseudupeneus maculatus,* (US Virgin Islands – Randall and Randall 1963; Colin and Clavijo 1978) and two surgeonfishes in Puerto Rico (Colin and Clavijo 1988).

5.8.6 Predation on Spawning Adults and Eggs

Predation attempts by piscivores on spawning reef fishes, including those forming aggregations, appear to be uncommon, despite limited quantitative data. Chapter 2 details many of these (Table 2.1) and suggests that reports of attacks are more common for RA than TA and more common in the IWP than TWA. In general reports for tropical reefs indicate roughly one attack (often not successful) per

100–1,000 spawns although potential predators are often present near areas where reef fishes are spawning (Colin 1978; Colin and Clavijo 1988; Clifton and Robertson 1993; Craig 1998; Sancho et al. 2000b). Some authors have failed to observe, despite the presence of potential predators (sharks and large fishes) any attacks on aggregated fishes (Colin and Clavijo 1978; Colin and Bell 1991; Colin 1992; Kadison et al. 2006) during many hours of observation. Some reliable reports (e.g. Jim Forrest 2010) do indicate occasional predation by sharks on aggregated or spawning fishes. However, it is very difficult to observe such events, as they are probably rare, and just the presence of a human observer may disturb the activity. It has been suggested reef fishes often take on a "stupor"-like condition to allow spawning to continue despite possible risks due to the presence of and attacks by predators (Johannes 1981), but there is little to support this conjecture and much that argues against it (see following section).

Despite its occurrence, it seems likely that predation on spawning adults plays only a minor role in structuring aggregation time and location. In nearly all cases spawning fishes are very aware of their surroundings and the sequence of events leading to spawning can be easily interrupted in its preliminary phases. For example, Colin and Clavijo (1978) saw no predation attempts on aggregated goatfishes in the Virgin Islands, but the simple appearance (no attack) of a large mackerel (*Scomberomorus* sp.) caused the entire spawning population (100's of fish) to hide and not reappear for several minutes. If attacked, the group may be disturbed for a short period and cease spawning, but quickly resume once the possible danger of predators has passed. Most potential predators are highly mobile and active at nearly all times of day and if aggregations shifted to different areas or times, perhaps to reduce predation risks, predators would be easily able to follow such changes and the risk of predation would remain unchanged. Rather it appears that individual and collective behaviour and awareness is the observed defence against adult predators.

Fishing activity on TA's changes the natural relationship between spawning fish and potential predators. It is believed such activity on TA's often draws in predators, such as sharks, that might otherwise not occur there (Olsen and LaPlace 1979; Colin 1992) and may produce the impression that predators are more prevalent than they would be in an undisturbed situation. Heyman and Kjerfve (2008) report various predators present at the Gladden Spit aggregation site, Belize, with "sharks and bottlenose dolphins preying on aggregating snappers and groupers, the former, most commonly after aggregating fish were hooked on a fishing line".

Predation on eggs after their release occurs relatively often, orders of magnitude more frequently than predation on spawning adults, but a wide range of egg predation occurrence is documented, making it premature to generalize based on only a single or few sites (Chap. 2, Table 2.2). For some RA species a high percentage of spawns is attacked, probably due to predictability of spawning occurrence and limited height above the bottom where eggs are released. This risk almost certainly influences exactly where and when gametes are released (Colin and Bell 1991), but would serve to reduce egg predation rather than prevent it. For aggregations egg

Fig. 5.20 "Cloud predators" rush into areas where concentrated planktonic gametes have been released by both aggregation and pair spawning reef fishes and quickly start ingesting eggs (and sperm?) from the spawn. (**a**) A group of longnose parrotfish, *Hipposcarus longiceps*, releasing eggs and sperm attracting black snappers, *Macolor niger*, which rush in to feed on the spawn. (**b**) Within seconds snappers are at the spawn site sucking in water containing eggs and sperm by buccal pumping and will continue to feed from the cloud for a few more seconds, then move on and await subsequent spawns

numbers are extremely large and overall it is believed that egg predators may only take a small percent of the egg output (Hamner et al. 2007).

Egg predators are typically benthic-based zooplanktivores that normally select particulate items in mid-water above reefs functioning as "cloud predators", that attack aggressively masses of just-released gametes (Colin and Bell 1991) spawned by aggregating and non-aggregating fishes. The predators are not specialized for feeding solely on fish eggs, but most are aware when fish are preparing to spawn and rush in within seconds after gamete release to feed. While most pick individual eggs from the cloud of eggs and sperm, the black snapper, *Macolor niger*, suctions up masses of water which it expels through its gills through buccal pumping, a specialized type of plankton feeding (Fig. 5.20) while the striped mackerel, *Rastrelliger kanagurta*, filters eggs by ram filtering (Colin 1976). These fish eggs, rich in lipids, probably represent a supplemental source of food and egg predators can be small (such as *Praeneus* silversides in Palau) to large. Whale sharks are believed to specifically arrive at reefs to prey on spawn of benthic snappers (Heyman et al. 2001) as well as on eggs of pelagic little tunny (*Euthynnus alletteratus*) in the open sea (Hoffmayer et al. 2007). Other general zooplanktivores can feed on pelagic eggs, either through generalized particulate feeding (eggs are still visible, but not concentrated) or by filter feeding.

For large TA species the level of attacks by egg predators seems somewhat lower compared to RA species, particularly given the probable millions of eggs spawned daily in a short time. Colin (1992) reported only a few instances of Nassau grouper egg predation by yellowtail snapper, *Lutjanus chrysurus*, out of many (circa 100) spawns. Heyman et al. (2005) reported at least three species of fishes (in addition to whale sharks) to prey on newly released eggs of cubera snapper in Belize.

5.8.7 *The Spawning Stupor*

The prime, and oft-cited, example of a "spawning stupor" is from the observation that schooling mullets were easily dip-netted at night from the side of a ship tied to a dock while numerous bright lights were directed at the water, while sharks were also attacking the group (Helfrich and Allen 1975). Johannes (1978) was present at this site and dip-netted numbers of large fringelip mullet (*Crenimugil crenilabris*), normally a difficult fish to approach, from what he considered a spawning aggregation without significantly disturbing the remaining fish. Black tip reef sharks were reported to be feeding on the mullet also, the prey fish making "only sluggish and ineffectual efforts to avoid" the sharks. It was believed the fish were in a stupor-like condition allowing spawning (the fish were ripe, but not actually spawning?) to proceed without interruption despite the dangers from predators. However, "night lighting" is a common technique used to dazzle and attract fishes to a boat so they can easily be netted and the influence of the bright lights and possibly other conditions present mean that this should not be considered normal behaviour. Helfrich and Allen (1975) reported the shark attacks to occur early in their observations (after 20:30 h), while spawning (at some distance from the vessel) was believed to occur only after midnight, with no instances of shark predation noted during that probable spawning event.

In support of spawning stupors, Johannes et al. (1999) cited the observation by Robertson (1983) of a 1.3 m long carcharinid shark (not identified) attacking a single spawning pulse of several hundred brown surgeonfish at Lizard Island, Australia. The outcome of the attack was uncertain, but probably unsuccessful and did not disrupt the surgeonfish spawning. However, this can interpreted as the spawning fish were sufficiently aware during the attack that those individuals close to the shark avoided being bitten, while the others continued spawning. Several attacks on spawning brown surgeonfish in Palau failed to interrupt spawning behaviour for more than a few seconds (PLC unpublished data). In a final example Johannes et al. (1999) cited the fact that divers could swim among spawning cubera snapper without disturbing them as indicative they were in a stupor state. It is more reasonable to suppose that the fish were aware of events around them but had no reason to fear or flee from divers, as divers regularly visit these aggregations so fish become habituated to their non-threatening presence (Heyman et al. 2001, 2005).

There is no doubt the behaviour of many fishes is altered when in spawning aggregations, such as letting divers approach closer than normal (Hamilton 2005, Kadison et al. 2007), but some reports have interpreted such changes as "spawning stupors" (Burton et al. 2005; Gladstone 1996). The existence of a "spawning stupor" which overwhelms all self-preservation behaviour, is at best an uncertain (and in a sense unfortunate) concept, which like many appealing myths in biology gains credence with each incorrect repetition, misapplication or misinterpretation.

5.9 Discussion

As this brief review indicates, our understanding of the reasons for the location and timing of aggregations is at best preliminary. Only a few species have been examined in any detail, and at only in a few locations. Interestingly, two extremes of the aggregation spectrum, the Nassau grouper (a large TA species with previously massive aggregations) and the bluehead wrasse (a small RA species often with small aggregations) are perhaps the best-known species. New information is slowly becoming available; assumptions and conclusions regarding species and the phenomenon of spawning aggregations in general are likely to undergo revision.

What is known at present indicates some general working ideas:

- There are some clear distinctions in locations and timing between TAs and RAs. The former occurring infrequently, but almost exclusively, on shelf edge areas while the latter are more numerous and occur at a higher diversity of locations. They are characterized by different types of fishes; TAs with larger predatory species and RAs with herbivores and omnivores. They have different protection and conservation needs.
- The TAs and RAs of the IWP are different from those of the TWA. Differences can possibly be attributed to the higher tidal amplitudes and geomorphology of barrier reefs and their channels of the IWP with resultant effects on the oceanic environment.
- TA fishes have longer, less frequent migrations to aggregations while RA species have shorter, more frequent migrations (Chap. 2). These are related to the frequency of spawning and duration of spawning season.
- Aggregation sites are stable in location over decades with only slight unexplained variation in some species year to year.
- There are no known differences between eggs and larvae of aggregating and non-aggregating species within families (Chap. 7).
- The currents occurring at TA and RA sites during spawning are just beginning to be elucidated. Several examples with low current speed, termed "quiet currents", are now known at the time of spawning for Nassau grouper, red hind and cubera snapper. Quiet currents tend of reduce chances of offshore dispersal. Currents during spawning at RA sites are also often weak and would not cause advection of eggs far off the reef.
- The entrainment of propagules into oceanic circulation after spawning is not a given, and at some TA sites there appears a tendency for propagules to be retained nearshore. Spawn from RAs has a harder time becoming entrained into oceanic circulation. Geomorphology of the reef and spawning sites is important in this regard (Chap. 6).
- Water temperature regimes may be an important determinant of seasonality of spawning and early life history success. Aggregations often occur over a limited temperature range.

- The daily and lunar timing of aggregation spawning may be related to needs of pelagic life history.
- Predation on spawning adults is rare while predation of released eggs is common. Neither factor is believed to limit or structure aggregations.

The hypotheses about the "where and when" of aggregations have tended to focus on whether aggregations produce dispersal or retention of propagules, essentially following the same arc as the theories about reef fish populations in general. Previously, it was thought that reef fish populations were selected to be "open"; to spread their progeny across areas as wide as possible to ensure that at least some of the population finds conditions suitable for survival. Similarly it was initially proposed (Johannes 1978, 1981; Barlow 1981) that aggregations formed in areas that promoted the offshore dispersal of eggs and larvae into oceanic waters to either be dispersed widely or potentially returned by mesoscale eddies after lengthy periods at sea, rotating on a schedule in synch with larval development times, to their natal region (Johannes 1978, 1981; Lobel and Robinson 1986). While innately appealing, these hypotheses had little supporting data, but this did not prevent them being widely interpreted as solid facts in the literature and not subjected to the rigorous testing necessary to fully evaluate (Shapiro et al. 1988). More recently, focus has shifted to ideas of larval populations being retained near their sources (Swearer et al. 1999; Jones et al. 1999, 2005). Aggregations may play a role in retention of eggs and larvae, rather than dispersal, with locations and times of spawning preventing or at least limiting eggs from entering oceanic circulation (Colin 1992; Whaylen et al. 2004; Hamner et al. 2007). Arguments have ranged back and forth, and much of the time results are inconclusive. Black or white arguments may be futile in the end, since populations are most likely to exhibit characteristics of both strategies and are likely to be quite sensitive to locality differences.

New data argue that at least some, if not all, TA spawning occurs at times and at sites which do not favour dispersal of propagules (Colin 1992; Whaylen et al. 2004; Nemeth et al. 2007; Heyman and Kjerfve 2008; Kadison et al. 2009). It previously assumed that spawning, particularly for TA species, would occur at times of relatively strong currents potentially favouring offshore dispersal of eggs (Johannes 1978, 1981; Barlow 1981). The realization in the last decade that a number of TA fishes spawn at times when currents are minimal, rather than maximal, has been important in understanding that TAs may help to retain larvae, rather than disperse them. Certain spawning sites, particularly bank reefs and narrow islands with small shelves, do favour dispersal and entrainment of propagules into oceanic circulation (Domeier 2004, Grouper Moon Project unpublished data), so it is clear both dispersal and retention mechanisms are operating, often at different sites and potentially in the same area at different times (Colin 1995).

A review of the diversity of locations, timing and conditions between TA and RA spawning aggregations, as well as comparison to non-aggregation spawning sites, provides some insight into the factors likely to be influencing aggregation-spawning. Nearly all TA species either spawn on the reef drop-off or in areas closely connected (channels) with outer reefs. RA species also spawn in such areas, but additionally utilize more inshore areas. RA species generally seem to occur on the same sort of geomorphologic features, such as seaward facing reef slopes, associated with TA aggregations, but are less specific in their aggregation sites. Pair and haremic spawning species with planktonic eggs have an even greater range of spawning locations.

Where temperatures vary seasonally, selection of a certain temperature may be a pivotal factor in determining when spawning is occurring and its potential success. First, it may well be critical in the proper of maturation of gametes, and second it may ensure proper timing for initiation of feeding (Chap. 7). An entire cohort of larvae might be lost in a few hours if they are unable to start feeding. Much of the strategy of aggregation-spawning may be structured around selecting temperatures important for success; with other factors still being important, but less crucial to overall success or failure. That some fishes, both aggregation and non-aggregating species, spawn only during morning hours may be related to this need. The role of temperature in early life history and possible effects from changing climates needs to be carefully examined (Rombought 2007).

While strictly speculative at this point, it is possible that much of the dynamics behind aggregating behaviour may have evolved to optimize the quality of the large number of eggs spawned, particularly for TA species. For TA species, if studies from reef fishes with demersal eggs can be projected (Chap. 7), the seasonal short duration of spawning, near simultaneous maturation all ova, hormonal and nutritional factors associated with building towards a single or few episode(s) of spawning, and intra-specific behaviour and interactions stimulating hormone production resulting in eggs and larvae with potential for faster growth may lead to improved feeding and higher survival of larvae to the settlement stage. It can be suggested that evolution of aggregation spawning has resulted in its most specialized form (the large transient aggregation), in the production of vast numbers of optimal eggs to be spawned at the optimal time for the highest chance of success. While this "spawn all your eggs from one location" strategy may have occasional massive failures, the long lives of most TA fishes give their populations high likelihood of reproductive success over time.

References

Aguilar-Perera A (2006) Disappearance of a Nassau grouper spawning aggregation off the southern Mexican Caribbean coast. Mar Ecol Prog Ser 327:289–296

Aguilar-Perera A, Aguilar-Davila A (1996) A spawning aggregation of Nassau grouper *Epinephelus striatus* Pisces: Serranidae in the Mexican Caribbean. Environ Biol Fish 45:351–361

Appeldoorn RS, Hensley DA, Shapiro DY, Kioroglou S, Anderson BG (1994) Egg dispersal in a Caribbean coral reef fish. *Thalassoma bifasciatum*. II Dispersal off the reef platform. Bull Mar Sci 54:271–280

Bardach JE, Smith CL, Menzel DW (1958) Bermuda fisheries research program final report. Bermuda Trade Development Board, Hamilton

Barlow GW (1981) Patterns of parental investment, dispersal and size amongst coral-reef fishes. Environ Biol Fish 6:65–85

Bell LJ, Colin PL (1986) Mass spawning of *Caesio teres* (Pisces: Caesionidae) at Enewetak Atoll, Marshall Islands. Environ Biol Fish 15:69–74

Bergenius MAJ, Meeken MG, Robertson DR, McCormick ML (2002) Larval growth predicts recruitment success of a coral reef fish. Oecologia 131:512–525

Bolden SK (2000) Long-distance movement of a Nassau grouper (*Epinephelus striatus*) to a spawning aggregation in the central Bahamas. Fish Bull 98:642–645

Bullock LH, Murphy MD, Godcharles MR, Mitchell ME (1992) Age, growth and reproduction of jewfish, *Epinephelus itajara*, in the eastern Gulf of Mexico. Fish Bull 90:243–249

Burton ML, Brennan KJ, Muñoz RC Jr (2005) Preliminary evidence of increased spawning aggregations of mutton snapper (*Lutjanus analis*) at Rileys Hump two years after establishment of the Tortugas South Ecological Reserve. Fish Bull 103:404–410

Carter HJ (1986) Moonlight mating of the multitudes. Anim King 89(6):63–71

Carter HJ (1989) Grouper sex in Belize. Nat Hist Oct 1989:60–69

Carter HJ, Marrow GJ, Pryor V (1994) Aspects of the ecology and reproduction of Nassau grouper (*Epinephelus striatus*) off the coast of Belize, Central America. Proc Gulf Caribb Fish Inst 1990:65–111

Clifton KE, Robertson DR (1993) Risks of alternative mating strategies. Nature 366:520

Colin PL (1976) Filter feeding and predation on the eggs of *Thallasoma* sp. by the scombrid fish *Rastrelliger kanagurta*. Copeia 1976(3):596–597

Colin PL (1978) Daily and summer-winter variation in mass spawning of the striped parrotfish, *Scarus croicensis*. Fish Bull 76(1):117–124

Colin PL (1992) Reproduction of the Nassau grouper, *Epinephelus striatus*, (Pisces: Serranidae) and its relationship to environmental conditions. Environ Biol Fish 34:357–377

Colin PL (1994) Preliminary investigations of reproductive activity of the jewfish, *Epinephelus itajara* (Pisces: Serranidae). Proc Gulf Caribb Fish Inst 43:138–147

Colin PL (1995) Surface currents in Exuma Sound, Bahamas and adjacent areas with reference to potential larval transport. Bull Mar Sci 56(1):48–57

Colin PL (1996) Longevity of some coral reef fish spawning aggregations. Copeia 1996:189–192

Colin PL (2010) Aggregation and spawning of the humphead wrasse *Cheilinus undulatus* (Pisces: Labridae): general aspects of spawning behaviour. J Fish Biol 76:987–1007

Colin PL, Bell LJ (1991) Aspects of the spawning of labrid and scarid fishes (Pisces: Labroidei) at Enewetak Atoll, Marshall Islands with notes on other families. Environ Biol Fish 31:229–260

Colin PL, Clavijo IE (1978) Mass spawning by the spotted goatfish, *Pseudupeneus maculatus* (Bloch) (Pisces: Mullidae). Bull Mar Sci 28:780–782

Colin PL, Clavijo IE (1988) Spawning activity of fishes producing pelagic eggs on a shelf edge coral reef, southwestern Puerto Rico. Bull Mar Sci 43:249–279

Colin PL, Shapiro DY, Weiler D (1987) Aspects of the reproduction of two groupers, *Epinephelus guttatus* and *E. striatus*, in the West Indies. Bull Mar Sci 40:220–230

Colin PL, Laroche WA, Koenig CC (1996) Development from egg to juvenile of the red grouper, *Epinephelus morio* (Pisces: Serranidae) in the laboratory. In: Arreguin-Sanchez F, Munro JL, Balgos MC, Pauly D (eds) Biology, fisheries and culture of tropical groupers and snappers, ICLARM conference proceedings. ICLARM, Manila

Colin PL, Laroche WA, Brothers ED (1997) Timing of ingress and settlement in the Nassau grouper, *Epinephelus striatus*, (Pisces: Serranidae), with relationship to spawning occurrence. Bull Mar Sci 60(3):656–667

Cowen RK (2002) Oceanographic influences on larval dispersal and retention and their consequences for population connectivity. In: Sale PF (ed) Coral reef fishes: new insights into their ecology. Academic, London

Craig AK (1969) The grouper fishery of Cay Glory, British Honduras. Annu Assoc Am Geogr 59:252–263

Craig PC (1998) Temporal spawning patterns of several species of surgeonfishes and wrasses in American Samoa. Pac Sci 52:35–39

Doherty PJ, Fowler AJ, Samoilys MA, Harris DA (1994) Monitoring the replenishment of coral trout (Pisces: Serranidae) populations. Bull Mar Sci 54(1):343–355

Domeier ML (2004) A potential larval recruitment pathway originating from a Florida marine protected area. Fish Oceanogr 13:287–294
Domeier ML, Colin PL (1997) Tropical reef fish spawning aggregations: defined and reviewed. Bull Mar Sci 60(3):698–726
Ellis EP, Watanabe WO, Ellis SC, Ginzoa J, Moriwake A (1997) Effect turbulence, salinity and light intensity on hatching rate and survival of larval Nassau grouper, *Epinephelus striatus*. J Appl Aquacult 7(3):33–43
Erisman BE, Buckhorn ML, Hastings PA (2007) Spawning patterns in the leopard grouper, *Mycteroperca rosacea*, in comparison with other aggregating groupers. Mar Biol 151:1849–1861
Gladstone W (1986) Spawning behavior of the bumphead parrotfish *Bolbometopon muricatum* at Yonge Reef, Great Barrier Reef. Jpn J Ichthol 33:326–328
Gladstone W (1996) Unique annual aggregation of longnose parrotfish (*Hipposcarus harid*) at Farasan Island (Saudi Arabia, Red Sea). Copeia 1996(2):483–485
Hamilton RJ (2005) Indigenous ecological knowledge (IEK) of the aggregating and nocturnal spawning behaviour of the longfin emperor, *Lethrinus erythropterus*. SPC Tradit Mar Resour Manag Knowl Inf Bull 18:9–17
Hamilton RJ, Matawai M, Potuku T, Kama W, Lahu P, Wark J, Smith AJ (2005) Applying local knowledge and science to the management of grouper aggregation sites in Melanesia. South Pac Community Live Reef Fish Inf Bull 14:7–19
Hamner WM, Hauri IR (1981) Effects of island mass: water flow and plankton pattern around a reef in the Great Barrier Reef lagoon, Australia. Limnol Oceanogr 26(6):1084–1102
Hamner WM, Colin PL, Hamner PP (2007) Export-import dynamics of zooplankton on a coral reef in Palau. Mar Ecol Prog Ser 334:83–92
Helfman GS, Shultz ET (1984) Social transmission of behavioral traditions in a coral reef fish. Anim Behav 30:379–384
Helfrich P, Allen PM (1975) Observations on the spawning of mullet *Crenimugil crenilabus* (Forsskål), at Enewetak, Marshall Islands. Micrones 11:219–225
Hensley DA, Appeldoorn RS, Shapiro DY, Ray M, Turingan RG (1994) Egg dispersal in a Caribbean coral reef fish. *Thalassoma bifasciatum*. I. Dispersal over the reef platform. Bull Mar Sci 54:256–270
Heyman WD, Kjerfve B (2008) Characterization of multi-species reef fish spawning aggregations at Gladden Spit, Belize. Bull Mar Sci 83(3):531–551
Heyman WD, Graham RT, Kjerfve B, Johannes RE (2001) Whale sharks *Rhincodon typus* aggregate to feed on fish spawn in Belize. Mar Ecol Prog Ser 215:275–282
Heyman WD, Kjerfve B, Graham RT, Rhodes KL, Garbutt L (2005) Spawning aggregations of *Lutjanus cyanopterus* (Cuvier) on the Belize Barrier Reef over a 6 year period. J Fish Biol 67:83–101
Hoffmayer E, Franks J, Driggers W, Oswald K, Quattro J (2007) Observations of a feeding aggregation of whale sharks, *Rhincodon typus*, in the north central Gulf of Mexico. Gulf Caribb Res 19(2):69–73
Johannes RE (1978) Reproductive strategies of coastal marine fishes in the tropics. Environ Biol Fish 3:65–84
Johannes RE (1981) Words of the Lagoon. Fishing marine lore in the Palau district of Micronesia. University of California Press, Los Angeles, USA
Johannes RE, Squire L, Graham T, Sadovy Y, Renguul H (1999) Spawning aggregations of groupers (Serranidae) in Palau. The Nature Conservancy, Arlington, USA
Jones GP, Milicich MJ, Emslie MJ, Lunow C (1999) Self-recruitment in a coral reef fish population. Nature 402:802–804
Jones GP, Planes S, Thorrold SR (2005) Coral reef fish larvae settle close to home. Curr Biol 15:1314–1318
Kadison E, Nemeth RS, Herzlieb S, Blondeau J (2006) Temporal and spatial dynamics of *Lutjanus cyanopterus* and *L. jocu* (Pisces: Lutjanidae) spawning aggregations on a multi-species spawning site in the USVI. Rev Biol Trop 54(suppl 3):69–78

Kadison E, Nemeth RS, Blondeau JE (2009) Assessment of an unprotected red hind (*Epinephelus guttatus*) spawning aggregation on the Saba Bank in the Netherland Antilles. Bull Mar Sci 85(1):101–118

Kobara S, Heyman WD (2008) Geomorphometric patterns of Nassau grouper (*Epinephelus striatus*) spawning aggregation sites in the Cayman Islands. Mar Geod 31:231–245

Leggett WC, Dublois E (1994) Recruitment in marine fishes: is it regulated by starvation and predation in the egg and larval stages? Neth J Sea Res 32:119–134

Lobel PS, Robinson AR (1986) Transport and entrapment of fish larvae by ocean mesoscale eddies and currents in Hawaiian waters. Deep Sea Res 33(4):483–500

Luckhurst BE (1998) Site fidelity and return migration of tagged red hinds (*Epinephelus guttatus*) to a spawning aggregation site in Bermuda. Proc Gulf Caribb Fish Inst 50:750–763

Mazeroll AI, Montgomery WL (1998) Daily migrations of a coral reef fish in the Red Sea (Gulf of Aqaba, Israel): initiation and orientation. Copeia 1998:893–905

Mora C, Sale PF (2002) Are populations of coral reef fish open or closed? Trends Ecol Evol 17:422–428

Moyer JT (1989) Reef channels as spawning sites for fishes on the Shirahao Coral Reef Ishigaki Island, Japan. Jpn J Ichthyol 36:371–375

Myrberg AA, Montgomery WL, Fishelson L (1988) The reproductive behavior of *Acanthurus nigrofuscus* (Forskal) and other surgeonfishes (Fam. Acanthuridae) off Eilat, Israel (Gulf of Aqaba, Red Sea). Ethology 79:31–61

Nemeth RS (2009) Dynamics of reef fish and decapod crustacean spawning aggregations: underlying mechanisms, habitat linkages and trophic interactions. In: Nagelkerken I (ed) Ecological interactions among tropical coastal ecosystems. Springer, Berlin

Nemeth RS, Blondeau J, Herzlieb S, Kadison E (2007) Spatial and temporal patterns of movement and migration at spawning aggregations of red hind, *Epinephelus guttatus*, in the U.S.Virgin Islands. Environ Biol Fish 78:365–381

Nemeth RS, Kadison E, Blondeau JE, Idrisi N, Watlington R, Brown K, Smith T, Carr O (2008) Regional coupling of red hind spawning aggregations to oceanographic processes in the eastern Caribbean. In: Grober-Dunsmore R, Keller BD (eds) Caribbean connectivity: implications for marine protected area management, Marine Sanctuary Conservation Series ONMS-08-07. NOAA, Silver Spring

Olsen DA, LaPlace JA (1979) A study of Virgin Islands grouper fishery based on a breeding aggregation. Proc Gulf Caribb Fish Inst 31:130–144

Paulay G (1990) Effects of late Cenozoic sea-level fluctuations on the bivalve faunas of tropical oceanic islands. Paleobiology 16(4):415–434

Paz GE, Truly E (2007) The Nassau Grouper spawning aggregation at Caye Glory, Belize: a brief history. Report to the Nature Conservancy, Arlington

Randall JE, Randall HA (1963) The spawning and early development of the Atlantic parrot fish, *Sparisoma rubripinne*, with notes on other scarid and labrid fishes. Zoologica 48:49–60

Rhodes KL (2003) SCRFA spawning aggregation survey: federated States of Micronesia. Western Pacific fisher survey series, vol 2. Society for the Conservation of Reef Fish Aggregations

Rhodes KL, Sadovy Y (2002) Temporal and spatial trends in spawning aggregations of camouflage grouper, *Epinephelus polyphekadion*, in Pohnpei, Micronesia. Environ Biol Fish 63:27–39

Rissik D, Suthers IM, Taggart CT (1997) Enhanced zooplankton abundance in the lee of an isolated reef in the south Coral Sea: the role of flow disturbance. J Plankton Res 19(9):1347–1368

Robertson DR (1983) On the spawning behavior and spawning cycles of eight surgeonfishes (Acanthuridae) from the Indo-Pacific. Environ Biol Fish 9:192–223

Rombought PJ (2007) The effects of temperature on embryonic development and larval development. In: Wood C, McDonald M (eds) Global warming: implications for freshwater and marine fishes. Cambridge University Press, Cambridge

Rothenberger P et al (2008) The state of coral reef ecosystems of the US Virgin Islands. In: Waddell JE, Clarke AM (eds) The state of coral reef ecosystems of the United States and Pacific freely associated states, 2008. NOAA Technical Memorandum NOS NCCOS 73, Silver Spring

Sadovy Y (2005) Troubled times for the trysting trio: three aggregating groupers in the live reef food-fish trade. South Pac Community Live Reef Fish Inf Bull 14:3–6

Sadovy Y, Eklund A-M (1999) Synopsis of biological data on the Nassau grouper, *Epinephelus striatus* (Bloch, 1792), and the Jewfish, *E. itajara* (Lichtenstein, 1822). NOAA Technical Report NMFS 146, Seattle, Washington

Sadovy Y, Colin PL, Domeier ML (1994) Aggregation and spawning in the tiger grouper, *Mycteroperca tigris* (Pisces: Serranidae). Copeia 2:511–516

Sadovy de Mitcheson Y, Cornish A, Domeier M, Colin PL, Russell LKC (2008) Reef fish spawning aggregations: a global baseline. Conserv Biol 22(5):1233–1244

Sala E, Ballesteros E, Starr RM (2001) Rapid decline of Nassau grouper spawning aggregations in Belize: fishery management and conservation needs. Fisheries 26:23–30

Sala E, Aburto-Oropeza O, Paredes G, Thompson G (2003) Spawning aggregations and reproductive behavior of reef fishes in the Gulf of California. Bull Mar Sci 72(1):103–121

Samoilys MA (1997) Periodicity of spawning aggregations of coral trout *Plectropomus leopardus* (Pisces: Serranidae) on the northern Great Barrier Reef. Mar Ecol Prog Ser 160:149–159

Sancho G, Petersen CW, Lobel PS (2000a) Predator-prey relations at a spawning aggregation site of coral reef fishes. Mar Ecol Prog Ser 203:275–288

Sancho G, Solow AR, Lobel PS (2000b) Environmental influences on the diet timing of spawning in coral reef fishes. Mar Ecol Prog Ser 206:193–212

Shapiro D, Hensley D, Appeldoorn R (1988) Pelagic spawning and egg transport in coral-reef fishes: a skeptical overview. Environ Biol Fish 22(1):3–14

Sheaves M (2006) Is the timing of spawning in sparid fishes a response to sea temperature regimes? Coral Reefs 25:655–669

Shenker J, Maddox ED, Wishinski E, Pearl A, Thorrold SR, Smith N (1993) Onshore transport of settlement stage Nassau grouper (*Epinephelus striatus*) and other fishes in Exuma Sound, Bahamas. Mar Ecol Prog Ser 98:31–43

Smith CL (1972) A spawning aggregation of Nassau grouper, *Epinephelus striatus* (Bloch). Trans Am Fish Soc 101:257–261

Starr RM, Sala E, Ballesteros E, Zabala M (2007) Spatial dynamics of the Nassau grouper *Epinephelus striatus* in a Caribbean atoll. Mar Ecol Prog Ser 343:239–249

Swearer SE, Caselle JE, Lea DW, Warner RR (1999) Larval retention and recruitment in an island population of coral-reef fish. Nature 402:709–802

Thresher RE (1984) Reproduction in reef fishes. Tropical Fish Hobbyist Publications, Neptune City, USA

Warner RR (1988) Traditionality of mating-site preferences in a coral reef fish. Nature 335:719–721

Warner RR (1995) Large mating aggregations and daily long-distance spawning migrations in the bluehead wrasse, *Thalassoma bifasciatum*. Environ Biol Fish 44(4):337–345

Watanabe WO, Lee CS, Ellis SC, Ellis EP (1995) Hatchery study of the effect of temperature on eggs and yolksac larvae of the Nassau grouper *Epinephelus striatus*. Aquaculture 136:141–147

Whaylen L, Pattengill-Semmens CV, Semmens BX, Bush PG, Boardman MR (2004) Observations of a Nassau grouper (*Epinephelus striatus*) spawning aggregation site in Little Cayman Island. Environ Biol Fish 70:305–313

Whaylen L, Bush P, Johnson B, Luke K, McCroy C, Heppell S, Semmens B, Boardman MR (2006) Aggregation dynamics and lessons learned from five years of monitoring at a Nassau grouper (*Epinephelus striatus*) spawning aggregation in Little Cayman, Cayman Islands, BWI. Proc Gulf Caribb Fish Inst 59:479–487

Yoseda K, Dan S, Sugaya T, Yokogi K, Tanaka M, Tawada S (2006) Effects of temperature and delayed initial feeding on the growth of Malabar grouper (*Epinephelus malabaricus*) larvae. Aquaculture 256(1–4):192–200

Zeller DC (1998) Spawning aggregations: patterns of movement of the coral trout *Plectropomus leopardus* (Serranidae) as determined by ultrasonic telemetry. Mar Ecol Prog Ser 162:253–263

Chapter 6
Oceanography of the Planktonic Stages of Aggregation Spawning Reef Fishes

William Marion Hamner and John Louis Largier

Abstract Aggregation spawning reef fish have planktonic (pre-flexion) and nektonic (post-flexion) larval phases. Pelagic eggs hatch in about 24 h into non-motile, yolk-sac larvae and several days later into pre-flexion larvae which feed actively and control depth despite incomplete fin development and limited motility. Thus, eggs, yolk-sac and pre-flexion larvae are all passive, planktonic drifters, yet they are not necessarily flushed away by coastal currents into the open sea. Here we investigate how the physics of near-shore flow features, such as diffusion of the initial spawning cloud, tidal advection in coastal boundary currents and entrainment into tidal eddies, interact with the complexities of reef topography to reduce offshore dispersion. We use scale modelling to estimate dispersal via turbulence of gamete clouds for the first few days post-spawning. Thereafter, we emphasize empirical information from tracked drifters released at spawning sites to evaluate alongshore tidal advection and entrainment into tidal eddies. Larval behaviour is of great importance not only to trajectories, but also because subtle depth adjustments by planktonic pre-flexion larvae permit contact with concentrated foods on density discontinuities or at convergent fronts. This food source is critical for transformation from pre-flexion plankton into post-flexion nektonic larvae.

W.M. Hamner (✉)
Department Ecology and Evolutionary Biology, University California, Los Angeles, USA
e-mail: whamner@ucla.edu

J.L. Largier
Bodega Marine Laboratory, University of California, Davis, USA
e-mail: jlargier@ucdavis.edu

Y. Sadovy de Mitcheson and P.L. Colin (eds.), *Reef Fish Spawning Aggregations: Biology, Research and Management*, Fish & Fisheries Series 35,
DOI 10.1007/978-94-007-1980-4_6, © Springer Science+Business Media B.V. 2012

6.1 Introduction

A pelagic larval stage is found in all species of reef fishes known to aggregate for spawning (Appendix). These larvae are the main dispersal stage, transported via water movement and active larval swimming (Leis 1991, 2006). The distance pelagic larvae can travel typically exceeds the migration of adults to/from aggregation sites (although adults can migrate tens to hundreds of kilometres). Pelagic larvae are the only life history stage during which most resident reef fishes can or will cross water barriers, often including deep channels through the reef, and therefore dispersion of larvae rather than of adults is primarily responsible for population structure of many reef fishes. Strathman et al. (2002) suggest that an early planktonic life history is particularly advantageous because planktonic larvae can feed readily in productive inshore waters, avoid reef predators, and break cycles of parasitism. They note that the benthic parental habitat is clearly a favourable location for recruitment of juveniles, with natural selection favouring recruitment to natal reefs.

The early life history (ELH) of aggregating reef fishes from spawning to settlement requires from about 2 weeks to a few months. It starts with passive pelagic eggs that take 24 h or less to hatch. After hatching from pelagic eggs, larvae go through a relatively non-motile yolk-sac stage of a few days, then transition to larvae which can actively feed but still have limited motility due to incomplete fin development (pre-flexion larvae). Near the midpoint of the planktonic phase the larvae have a greater sensory capacity (see Chap. 7) and develop supporting elements for the caudal fin (post-flexion) as well as well-defined medial fins with supporting elements, resulting in a major increase in horizontal swimming ability. Thus pelagic larvae of reef fishes have both an initial planktonic (pre-flexion) stage and a subsequent nektonic (post-flexion) phase. As plankton, the subject of this chapter, eggs and larvae are initially passive drifters, but they soon develop limited motility that permits them to swim sufficiently well to feed and control their vertical location in the water column. As nekton, larval fishes can swim well and hold their own against weaker horizontal ocean currents, as well as potentially sensing the presence of distant reefs and navigating directionally in the pelagic waters around coral reefs in order to return to and settle on the coral reef.

The release of millions to billions of propagules within a short time and limited volume of water during aggregation-spawning poses questions about the potential disadvantages or advantages (costs versus benefits) of this life history strategy compared with other methods of spawning. In this chapter we examine aspects of the physical and biological ocean environment that influence the survival and distribution of the planktonic phase of those reef fishes that originate from spawning aggregations. Physical mechanisms such as tidal currents and eddies are important in initial transport and in the short term (a few days) a significant portion of eggs and early larvae can be returned back to their initial spawning sites (Hamner et al. 2007). After these initial days, behavioural mechanisms, such as vertical migration and reaction to frontal conditions, are also important in determining where larvae will go during their first several weeks of life. Later, after the larvae become more active, horizontal swimming, navigation or schooling behaviours are sufficient to potentially enable late larval and juvenile stages to detect and move towards suitable settlement habitat (see Chap. 7).

We address the first few days after spawning using scale modelling to obtain estimates of dispersal of the initially highly concentrated cloud of sperm and eggs via turbulence in the upper water column above a spawning aggregation. We address later dispersal and advection of primarily passive, pre-flexion, larvae noting the importance of well-known near-shore flow features that reduce the extent of advection into offshore far-field currents. It is important to note that our knowledge of nearshore coral reef oceanography is incomplete, particularly for windward reefs, and much of this chapter is directed at what is expected rather than what has been proved.

6.2 The Start of Early Life History-Modelling Dispersion of Fertilized Eggs from Aggregation Sites

Population continuity following spawning is a function of (1) spawning location, time of spawning, and numbers of eggs spawned, (2) the dispersion (i.e., transport and mixing) and survival of propagules from a single or multiple spawning sites that determine the distribution of potential settlers, and (3) the presence of suitable settlement habitat and the density of settlers (Botsford et al. 2009). Among aggregation spawning species, the first two components, in particular, may differ from other reef fishes (see also Chap. 7). The role of aggregation-spawning in dispersal outcomes has been largely ignored. Do aggregations, for example, produce large cohorts recruiting at the same time, as is considered in Chap. 7, or do these aggregations, rather, confer benefits on eggs and especially early stage larvae (this Chapter)?

Since the specifics of spawning, the starting point of ELH, determine where and when larvae develop at sea, the physical factors impacting the eggs and early stage larvae, are important. The geomorphology, timing and local environmental conditions at aggregation spawning sites have been described in Chap. 5, but it is important also to understand how rapidly recently fertilized eggs become part of the planktonic ecosystem and how they are thereafter dispersed by patterns of water flow interacting with larval behaviour. Rather than thinking of the eggs from aggregation-spawning as a point source (except in a broad geographic sense), a more comprehensive view of dispersal can be conceptualised as a *dispersal kernel*, which expresses the probability that a larva released from a particular location will be transported to, and successfully settle at, other specific locations where adult habitat is available (Largier 2003). Dispersal kernels describe a continuum of dispersal, which is comprised of dispersal patterns from a variety of sources each of which typically approximates a bell-shaped curve of larval distributions as a function of distance from the spawning source (Fig. 6.1). While diffusion (due to variable currents) accounts for this spreading out of larvae from a common source, advection (due to the mean current) will result in an offset of this bell-shaped curve from the

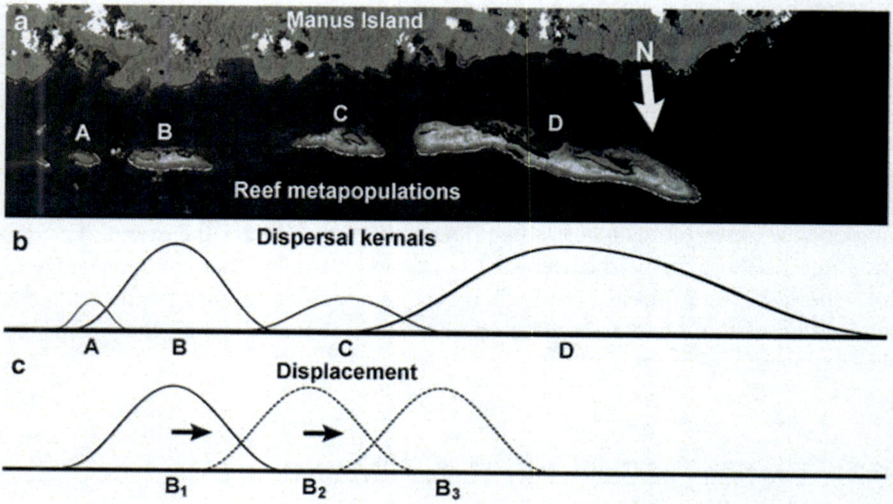

Fig. 6.1 Hypothetical dispersal kernals originating from a series of reefs along the north coast of Manus Island, Papua New Guinea. (**a**) Four offshore barrier reefs of varying sizes, separated by openings of varying distances. (**b**) In a hypothetical example, the density of spawn of fishes might be found spatially in individual bell-shaped dispersal kernals reflecting the density and distribution of source organisms. (**c**) Advective water movement may retain the basic form of a dispersal kernal, but transport (displace) it some distance from its origin (Based on Botsford et al. 2009)

spawning origin (Fig. 6.1c, also see Largier 2003). Different oceanographic conditions and population attributes influence the shape and movement of the dispersal kernel (Botsford et al. 2009) with individual kernels more advective (greater average transport) or diffusive (broader spread) with both spatial and temporal variations, depending on the conditions where they are formed (Fig. 6.1). While a single dispersal event, or even a single dispersal season, may yield patchy settlement that reflects the specific circulation events during the dispersal period (e.g. Siegel et al. 2008), in general it is expected that when one aggregates over population-relevant space and time scales the dispersal pattern will approximate a Gaussian-like pattern (i.e. higher values in the centre and increasingly lower values at greater distances from the centre).

Given the many potential pathways between multiple origins and multiple destinations, dispersal kernels are typically obtained from oceanographic models with few attempts to assess dispersal kernels in the field for aggregation spawners. Holm (2004) modelled dispersal of cubera snapper, *Lutjanus cyanopterus*, eggs in Belize, and also included short-term drifter studies for comparison. For Nassau grouper, *Epinephelus striatus*, Colin (1992) and the "Grouper Moon Project" (www.reef.org/data/groupermoon.html) deployed drifters at spawning sites and obtained very different results; one set of drifters was retained for up to a week on an insular shelf edge while the drifter set at the second location was taken rapidly out to the open ocean. If anything, this example shows that it is inappropriate to generalize about population-scale dispersal outcomes from a single aggregation site or short time period (as discussed above). Determination of typical "recruitment pathways" from an aggregation site requires consideration of many realizations and has most often been theoretically examined through biophysical modelling (e.g. Paris et al. 2005). While it is possible in principle to measure actual dispersal from an aggregation spawning site over a time period consistent with the entire pelagic stage of a fish (Domeier 2004), the method of determining such dispersal is subject to numerous unknowns which reduce the specificity of the data and one would need to repeat this study numerous times to obtain a typical recruitment pathway.

6.2.1 A Simple Model of Dispersion

Here we outline the scale of the initial dispersion of fertilized eggs, and relate it to aggregation-spawning, with a view to better understanding the possible costs and benefits of this reproductive strategy. Once released, planktonic eggs are carried by the moving water and mixed outward as a dispersing cloud. Each egg is a "particle" which follows an unpredictable path due to the influence of multiple flow features on different scales and the ubiquitous presence of turbulence; however, there are probable dispersion patterns related to the overall nature of the fluid flow.

Given the stochastic nature of propagule dispersion, it is best to model the process as a "random walk" (e.g. Siegel et al. 2008), which allows the variability in outcomes to be seen. However, this requires knowledge of the three-dimensional, time-varying flow field at the time of each spawning event. In reality, we have only rudimentary knowledge of water flows during spawning and can only make limited scale-estimates of the probable patterns of dispersion. For this purpose, one averages over multiple spawning events (in space and time) to develop an advection-diffusion view of dispersion at the scale of the population. In this approach, the typical or mean flow over the time and spatial scale of interest is estimated and referred to as "advection" (in oceanography the horizontal movement of water) and the zero-mean variability in flows during this time of interest is estimated and referred to as "diffusion" – net outward movement of particles from an area of higher to lower concentration (e.g. Largier 2003). Note that in this simplified approach "shear dispersion" due to mixing across shear in the mean flow is subsumed in the along-flow diffusion term. Also, when referring specifically to the transport of those propagules that successfully recruit in adult habitat, the process is called "dispersal". This advection-diffusion approach is not intended to provide single-burst predictions, but rather to represent a probable outcome for numerous releases of eggs (or even over repeated spawning by the same population cohort) for a given spawning location and time period.

The advection-diffusion equation is an expression of "mass balance", i.e., a method of tracking the mass of planktonic particles and how they will spread out in time, with concentration decreasing as the patch or cloud expands (i.e. "dilution"). The governing equation, in one dimension, is as follows:

$$\partial_t C = -\partial_x (u.C - K_x.\partial_x C)$$

where C(x,t) is concentration of particles at given place and time; u is the mean "advection" velocity in the x-direction, K_x is the "eddy diffusivity" describing the level of mixing or "diffusion" due to flow variability in the x-direction. Ideally a three-dimensional equation is needed to fully describe dispersion, but for shallow water with a well-mixed water column, typical of most fish aggregation sites, one only need track concentration (per surface area) in the two horizontal dimensions – usually expressed as alongshore and cross-shore in coastal environments.

In the context of aggregation spawning, three phases of dispersion are of interest: the first is the dispersion of gametes immediately after fertilization (tens of seconds to a few minutes), the second is the initial dispersion of fertilized eggs/larvae when concentrations are greater than they would be in the absence of aggregated spawning (days), and the third, the later dispersion of early phase planktotrophic larvae over subsequent days and weeks that must result in delivery to suitable recruitment habitat for successful completion of dispersal. It is believed that eggs and yolk-sac larvae behave as passive drifters for the first two phases (Leis 2006), while in the later stages of the third phase transport is significantly affected by larval behaviour (and thus not addressed by this simple model).

6.2.2 Formation of the Initial Cloud of Gametes

At spawning the need to bring about contact between eggs and sperm is crucial so that fertilization can occur. The high rates of egg fertilization that are observed (Kiflawi et al. 1998; Colin unpublished data) suggest that contact rates are typically high enough to ensure near total fertilization in aggregation spawning, presumably due to the close proximity of so many gametes and/or due to the significant small-scale turbulence produced by the rapid swimming action of spawning fish (Fig. 6.2, Bell and Colin 1986). Whether aggregation spawning fertilization rates differ from non-aggregation spawning rates has not been accurately assessed, and, given the numbers of species involved and diversity of spawning behaviour, is not really a useful comparison. For most aggregation spawning fertilization rates are 90% or higher. The dimensions and concentration of this initial cloud of gametes depend on the number of spawning adults, their fecundity and the nature of currents during spawning, factors which often cannot be determined. We can assume that the initial shape and size of the cloud are determined by the distribution of spawning adults and their numbers and sizes, and then altered by the effect of currents.

While vigorous large-scale mixing due to energetic flow structures is undesirable at this time, as it will tear the gamete clouds apart too quickly, small-scale turbulent mixing may assist in a high rate of contact between these "particles" (Petersen et al. 1992; Kiflawi et al. 1998). If aggregations spawn in locations or times of weak flow (see Chap. 5), large-scale mixing can be precluded or minimized and the gamete clouds may persist in high concentrations for several minutes, often with subsequent spawning releases occurring within the same water volume as earlier ones, perhaps increasing already high fertilization rates (Kiflawi et al. 1998).

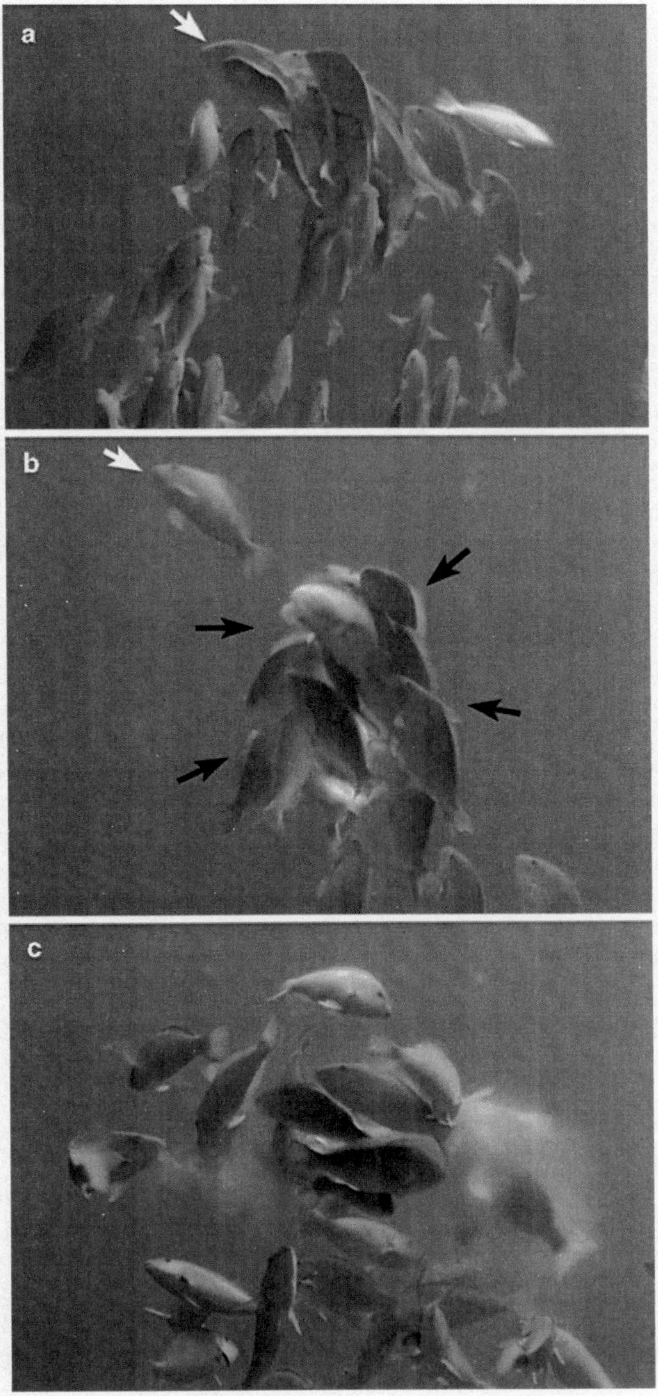

Fig. 6.2 (**a**) The spawning rush of longnose parrotfish is usually led by a single probable female (*white arrow*), believed to be female and (**b**) after initial release other tightly clustered individuals (males and females) release additional gametes (*black arrows*). (**c**) As the clustered fish break apart slight turbulence from their movements promotes contact between eggs and sperm (*visible white cloud*) and rapid fertilization

A large transient aggregation, such as for Nassau grouper, will occupy an area perhaps 30–50 m in radius, but when actually spawning will occupy a much smaller volume of water (as little as 1–2 m across, see Chaps. 1, 5 and Glossary have definitions of aggregation types). Some spawning aggregations are elongate, as adults spawn along an escarpment (e.g. humphead wrasse – *Cheilinus undulatus* Colin 2010, or various parrotfish (Scaridae) and surgeonfish (Acanthuridae), see Fig. 5.13, Chap. 5). Given that aggregation spawning on any given day usually occurs for a period of an hour or less, even if currents are relatively weak (<0.05 m/s), the gamete clouds may stretch over a distance of 100 m or more due to advection. In a sense, successive spawnings from an aggregation which itself does not move significantly can be thought of as "puffs of smoke" produced in an area where the wind is blowing them away, to be replaced by the next puff. The result is either a single mass of gametes or a succession of "puffs" in a "line" determined by the current speed and direction combined with the timing of spawning. Typical concentrations are obtained by dividing the number of propagules by the volume of the overall cloud, providing information on egg density prior to the continual process of dispersion after spawning.

It seems reasonable to assume that the spawning process creates a vertically mixed cloud near the ocean surface due to the innate buoyancy of eggs and that the subsequent cloud dispersion is largely due to horizontal mixing processes. Even if spawned at some depth, the buoyancy of eggs would bring them close to the surface prior to hatching (Chap. 5) and the same horizontal mixing processes would cause the egg cloud to expand outward as it ascends. Away from the immediate effect of small-scale shear and eddies due to the interaction of flow with the roughness of the reef (which are minimized by spawning well above the bottom), Richardson (1926), Okubo (1980) and many later studies (e.g. Stacey et al. 2000 in nearshore waters) have shown that horizontal diffusivity scales with $L^{4/3}$ where L is the cloud size – with values of K of order 0.01 m^2/s for a cloud of order 30 m and K of order 0.001 m^2/s for a cloud of order 5 m (Fischer et al. 1979; $L \sim 3\sigma$ where σ is variance of particle positions). The time scale for mixing can be scaled as $t \sim a.L^2/K$ where $a \sim 0.05$. A 5 m diameter cloud would mix on a time scale of $t \sim 1{,}250$ s (21 min) while a 30 m diameter cloud would mix on a time scale of $t \sim 4{,}500$ s (75 min).

The above Okubo-based estimates of cloud dispersion rates should be considered a minimum, corresponding to dispersion in the absence of local flow-topography interactions (see Sect. 6.4). There are few quantitative reports on the actual development of gamete clouds. Heyman et al. (2005) qualitatively report clouds of cubera snapper eggs of 64 m^3 some 15 s after spawning (order 5 m horizontal scale) expanding to 1,800 m^3 after 1 min (order 30 m horizontal scale), although there is some question that this expansion rate is overstated (Patrick L. Colin personal communication 2010). This rate of mixing is much faster than expected from diffusive effects alone ($K \sim \Delta L^2/18.\Delta t \sim 1$ m^2/s), which either requires that small-scale (1–10 m scale) shear and eddies in the flow field greatly dominate the background eddy diffusion effects (cf. Clarke et al. 2007) or suggest that the field estimates are in error.

The rapid growth of the cloud observed by Heyman et al. (2005) (concentration diluted 28-fold in less than a minute) underscores the importance of spawning at times and places where flow speeds are weak. The time it takes before gamete contact rate declines enough to affect fertilization rate depends on the initial concentration and the strength of turbulence in the gamete cloud and the duration of egg viability after release. However, initial concentrations may be extremely high so that gamete contact rates are sufficient even after dilutions of orders of magnitude – as would happen in this case within the 8 min when the cloud was visually evident. Intuitively there are major differences between the types of gamete clouds produced by large transient aggregations and the larval dispersal of gametes released during pair-spawning by smaller reef fishes. But at present it is not possible to accurately compare costs and benefits between these different modes of spawning.

Irrespective of the ambient flows at the time of spawning, the fish themselves ensure small-scale turbulent mixing (Colin 1992; Heyman et al. 2005) through vigorous swimming motions (e.g. Chap. 12), which have been observed in both "burst" and "broad launch" modes (Chap. 7). This suggests that broad-launch spawning may work to place gametes in mid-water shear layers (typical in a stratified water column). In such a shear layer, shear dispersion will rapidly bring about contact between "particles" across this vertically confined horizontal layer, at the same time as the shear spreads out the cloud horizontally (which may be useful for broader final dispersal outcomes).

6.2.3 Dispersion of Gamete Cloud

It is expected that eggs and yolk-sac larvae act as passive particles, normally slightly buoyant but readily mixed downward from the surface under all but the calmest conditions (e.g. Ellis et al. 1997; Hamner et al. 2007). In water depths of 10 m or less, one can expect neutrally buoyant particles to mix throughout the water column within a few hours, from $t \sim H^2/K_z$ where K_z is vertical diffusivity (and assuming $K_z \sim 0.01$ m^2/s from above, or from $K_z \sim \kappa.u_*.H/6$ where κ is von Karman constant and u_* is friction velocity ~0.01 m/s). For slightly buoyant particles, a quasi-steady vertical distribution with concentration decreasing with depth (if spawned at or near the surface) will be attained in a similar amount of time. We thus continue to assume that cloud dispersion is dominated by 2-dimensional horizontal mixing.

Recent studies of water flow over and around coral reefs and associated modelling of flows promise credible computer simulations of dispersing larval clouds at selected locations (for review see Monismith 2007). However, the need to characterize the dispersal of egg and larval clouds in general, and the overall lack of either detailed observations or models of water flow at almost all locations, lead to the emphasis here on relative scale estimates of this problem. Continuing to use scale expressions for diffusive mixing (as above), one can note that diffusivity $K \sim \Delta L^2/18.\Delta t$ and thus the time for a cloud of size L_1 to mix out to L_2 can be estimated by $\Delta t \sim 0.05(L_2^2 - L_1^2)/K$. Since K increases with the size of the cloud (see above) this problem is best solved numerically. Nevertheless one can obtain an idea of the rate of mixing by considering stages of mixing. A 30 m cloud ($K \sim 0.01$ m^2/s) mixes to

60 m in 3.7 h and corresponds to fourfold dilution of egg/larval concentration; in turn, the 60 m cloud (K ~ 0.025 m^2/s) mixes to 120 m in 6 h and dilution is now 16-fold; and so on – 120–240 m in 9.5 h; 240–480 m in 15 h; and 480–960 m in 24 h – yielding a total dilution of 1,024 and patch size of order 1 km in 58 h (2.4 days). Thus, if adult spawners move from within a 1 km long reef to a point when they aggregate to spawn a cloud 30 m in diameter, rather than spawning as a distributed population creating a 1 km cloud of eggs/larvae, then it will take a few days for the aggregation spawned cloud to approach the size and concentration of the non-aggregated cloud (which would have grown somewhat over the same period to about 3 km in size). A key factor here is the spatial scale from which spawners aggregate; it will take significantly more time for larval densities to approach those of non-aggregation-spawned larvae if spawners come together from distances of 10 or 100 km or more. The densities of larvae produced from an average 100 km distance migration (probably typical of Nassau grouper) would still be more concentrated by the time of first feeding than if those same larvae were spawned over the entire 100 km distance. The additional larvae of many other fishes present in the water invariably greatly outnumber Nassau (and other) grouper larvae. Serranid larvae are always very low in relative numbers compared to other reef fishes in plankton tows, light trap and crest net catches (Leis 1991; Dufour and Glazin 1993; Sponaugle et al. 2005), so that instead of competing among themselves, the grouper larvae are in reality competing more with larvae of other species, i.e. the costs of aggregating in terms of higher grouper larval concentrations are moot.

Typically, mixing in the direction of advection is dominated by shear dispersion effects (e.g. Clarke et al. 2007), including the role of wakes and eddies associated with small-scale topography (cf., Largier 2004 for discussion of along-stream dispersion at larger scales). As suggested by data in Heyman et al. (2005) and dispersion studies in other environments (e.g. Stacey et al. 2000; Clarke et al. 2007), observed nearshore dispersion rates are better represented by K ~ 1–10 m^2/s, even at smaller cloud sizes. During advective transport, if along-stream mixing can be represented by K ~ 1 m^2/s, then the 30 m cloud would mix to 1 km and 1,000-fold dilution in about ½ day (or faster for larger K). This is consistent with the few observations collected by Appeldoorn et al. (1994) at spawning sites of bluehead wrasse on inshore reefs off Puerto Rico. They found that water parcels tracked from different sites tended to merge during the 24 h period of observations, i.e. the clouds from different aggregation sites had merged within a day of release, yielding a larger, more dilute, cloud that can be expected to be similar to one which would have resulted from spawning without aggregation. However, in the case of cubera snapper, where spawning may yield 10^{10} eggs in a cloud of order 30 m, a 1,000-fold dilution still corresponds to high plankton concentrations of order 10,000 eggs/m^2 – corresponding to 1,000 eggs/m^3 over a 10 m near-surface layer. Clearly, subsequent mixing is probably required to preclude competition among young larvae for food. The second 1,000-fold dilution, which would reduce concentrations to order 1 larva/m^3, can be expected to take a bit longer than a few days – although non-zero larval mortality may decrease larval concentrations more quickly.

The above discussion considers a patch moving along the reef, but the situation is notably different at a reef passage or at the sudden end of a topographic reef feature, where flow may separate from the shoreline with some of the larva-laden water being exported offshore while other parts of the larval cloud are retained nearshore. If one considers the associated jet and wake flow features of scale >100 m and flow speeds >0.1 m/s as part of the dispersion process, then values of K would approach 100 m^2/s, similar to those values used for representing tidal dispersion in similarly scaled estuary flows. In this scenario, one would see the small 100–1,000 m cloud of larvae rapidly ripped apart into swirls and whirls that could be represented by a cloud of order 10 km or larger. In most spawning situations there are strong topographic features (Colin 1992; Heyman et al. 2005; Hamner et al. 2007) and the larval cloud is likely to reach such a separation point within hours or a day after being spawned – suggesting that the dense concentrations of eggs/larvae produced by aggregation spawning on a reef will be diluted down to more typical non-aggregation concentrations within less than a few days. Later herein we discuss such oceanographic features and note that some features also present opportunities for retention of a larval cloud.

The above discussion has focused on diffusion – the spreading out and dilution of a cloud of eggs/larvae due to zero-mean flow variability, whether due to smaller scale turbulence or larger scale flow features like eddies. This diffusion is superimposed on a background mean flow that advects the cloud along the reef – and may elongate the cloud formed during prolonged spawning (e.g. Nassau grouper spawning for 30 min will yield a cloud 100–200 m long in an advective flow of 0.05–0.1 m/s). The phenomena of advection and diffusion are related because stronger flow over the reef and past small-scale topography will yield shear and eddies that account for stronger diffusion, exceeding background values expected from the "4/3-law" values. Further, the comparison between reef length and advection velocity gives an estimate of how long the cloud diffuses in the vicinity of the reef and when it will reach a topographic break (due to its advection along the reef). At the break the associated flow separation, through shear effects, can literally tear the cohesive cloud apart, causing the cloud to become more dispersed. These events, which destroy the continuity of the cloud, can occur rapidly after spawning (such as along the edge of a tidal channel with outflowing current) or take longer (cloud is slowly moved as a discrete, slowly diffusing, mass along a continuous outer reef face).

Over time scales of a few hours, tidal flows act as advection (moving the eggs or larvae horizontally). However, dispersion over time scales of days or multiple events (on different days) will see tidal flows as zero-mean flow variability and a key contributor to the total diffusion. For multiple spawn bursts over less than an hour (e.g. cubera snapper, Heyman et al. 2005), or for multiple spawning events over the reproductive time scale of a given population, it is the ensemble average of spawning outcomes that is of interest and tidal flows will be quasi-random time-varying flows that affect dispersal variance (i.e. a diffusion effect). For example, Hensley et al. (1994) observed that bluehead wrasse tended to group spawn at sites that favour off-reef transport of eggs, and although the offshore tidal flow was short-lived, it enhanced along-reef dispersion through inducing shear dispersion. Further, Holm (2004) obtained field data and model results for the movement of fertilized eggs of cubera snapper off Belize until the time of hatching (17–20 h after spawning) which indicate a wide variety of possible larval trajectories (depending on tide/wind/currents), presenting a broad cloud of larval destinations at the scale of population dispersal.

In the above section, we have taken the broad-brush approach of scaling length and time scales from an advection-diffusion approach. In reality, dispersal will be more complex in space and time, but in the absence of better data and understanding this provides a rough estimate of the problem. There is much to be discovered relating to dispersal of early life stages of aggregating reef fishes, and we expect that field investigations will yield plenty of surprises in this regard.

6.3 Nearshore Oceanography: Dispersion and Advection of Early Larvae in Marginal Shallow Waters

Existing marine populations have obviously evolved successful larval strategies, with both local and distant recruitment occurring within a single population. Looking at spawning and flow across reefs one sees the origins of this bimodal outcome early in the dispersion of the egg/larvae cloud – where flow separates from topography, leading to the offshore export of some propagules (which may return to the same reef later or easily be advected across deep offshore waters to a distant reef), while others remain in the flow attached to the topography (greatly increasing the chances of being retained in a wake or other retention zone long enough to ensure recruitment back to a nearby reef).

A wide diversity of oceanographic features results from the interaction of flow with topography (e.g. structure of the shoreline, variability of benthic topography and bottom depth) on coral reefs. We expect that three flow patterns, *coastal boundary layers* (CBL), *lateral trapping in embayments*, and *eddies* (gyres) are important in the transport outcomes for aggregation spawned propagules. Resident and transient spawning aggregations do not form everywhere but only at quite specific locations on the coral reef and only at particular times of day, tide, and time of year (see Chap. 5).

Resident spawning aggregations on coral reefs are often composed of many species of small herbivorous fishes. In Palau, a suite of 30–40 species of surgeonfishes, parrotfishes, and wrasses spawn every day for a few hours after high tide along the forereef throughout the year (Patrick L. Colin personal observation). In particular, spawning aggregations of the bullethead parrotfish, *Chlorurus sordidus,* and two surgeonfishes, the brown surgeonfish, *Acanthurus nigrofuscus,* and striped bristletooth, *Ctenochaetus striatus*, occur at intervals of about 50 m along the reef face of the eastern barrier reef (Colin and Bell 1991; Domeier and Colin 1997; Hamner et al. 2007, Patrick L. Colin personal observation), aggregating to spawn each day at shallow

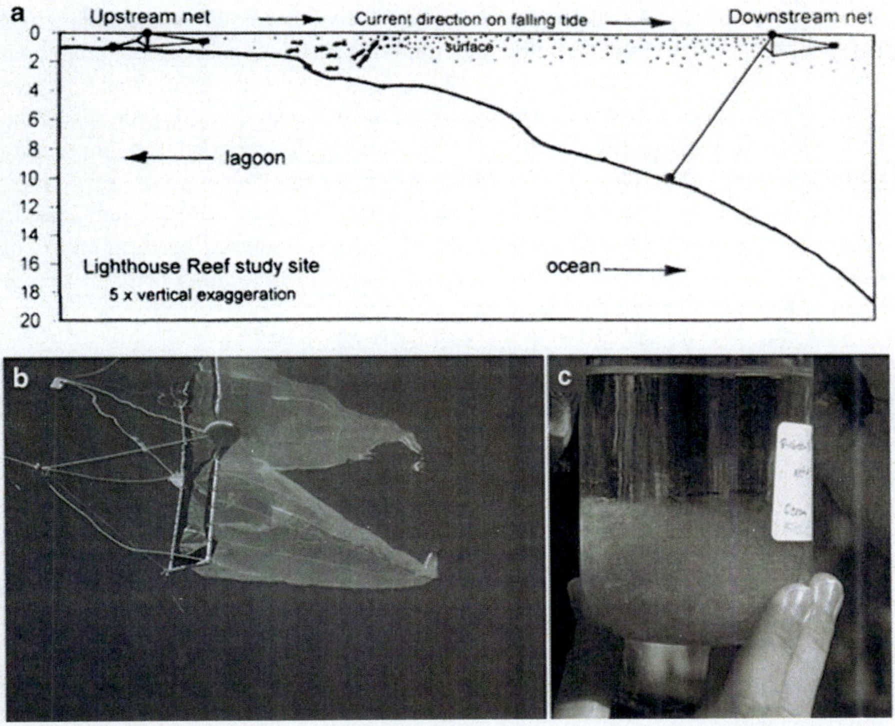

Fig. 6.3 (**a**) Resident aggregations of parrotfish and surgeonfish spawn on the ebb tide flowing off the reef into the coastal boundary layer east of Lighthouse Reef, Palau. (**b**) Moored neuston nets capture zooplankton flowing off the reef on ebbing tides. (**c**) Some plankton samples consist largely (99%) of fish eggs captured downstream of spawning during the first 2 h of ebb tide (Redrawn from Hamner et al. 2007)

coral outcroppings that are both close to their normal home ranges and adjacent to the nearest deep water on the seaward edge of the barrier reef. Hamner et al. (2007) investigated the export of eggs from these aggregations into the gentle alongshore flow that forms a coastal boundary current along the face of the barrier reef. Resident fishes aggregated and spawned after high tide, releasing fertilized eggs into tidal currents that flowed across the reef from the lagoon to the offshore edge of the reef, thus placing spawn in the coastal boundary layer flow along the outer reef face (Fig. 6.3). Drifters released at these spawning sites indicated that most of the developing eggs released during these spawns were retained in the CBL for least one tidal cycle and probably longer. This layer is a discrete water mass that is advected gently back and forth every day along front of the eastern barrier reef by linear, reversing tidal currents without significant offshore export due to eddies. This dispersal outcome may be typical of many resident aggregation spawners.

In contrast, transient aggregation spawners are typically much larger, predatory fishes, such as groupers (Serranidae) and snappers (Lutjanidae). These species often migrate many kilometres to spawning sites, usually to the shelf edge in areas adjacent to deep, often oceanic water. They generally have two or more short spawning periods per year, with a strong lunar component, implying a critical role of tidal phase and amplitude (Colin 1992; Heyman et al. 2005). They usually spawn at dusk (and in a few cases at night or dawn) at 10–30 m depth near topographic projections of the reef. If their propagules drift along the reef in the alongshore current, they are usually entrained into topographic eddies generated at sudden changes in reef form. Alternatively, or even simultaneously, other propagules may be advected into oceanic waters via a separated jet. Thus, the fates of the eggs spawned by these transient aggregators often quickly diverge and appear variable, alternating between export to the ocean and retention near the reef.

6.3.1 Resident Aggregations Spawn into Coastal Boundary Layers

Csanady (1972) and Largier (2002, 2003, 2004) used the phrase *coastal boundary layer* for the steady, directional flow of water that moves along an unbroken beach or straight rocky shoreline. A key difference between a coastal boundary layer and eddies (discussed later) is that CBL flow is primarily linear and tidally reversing whereas flow in topographic eddies is rotational. Largier (2003) noted that coastal boundary currents are both slow and retentive, and argued that most larvae spawned into CBLs will take several days to disperse offshore. For many populations, larvae may spend a major portion of their early life history in the CBL (Largier 2003).

Although coastal boundary currents are characteristic of linear coastlines, most coral reefs include *embayments* of various sizes and shapes; typically semi-enclosed bodies of water that are fully open on only one of four sides, bounded by headlands or promontories on the reef face, or segmented by passes through the coral reef or by rivers or estuaries that preclude through-flow of water. Coastal boundary currents in embayments are usually dominated by reversing tidal flows (indeed, mean currents are often near-zero and more attention must be given to tidal boundary layers along the edge of coral reefs; Hamner et al. 2007). The extent to which tidal dispersion contributes to the *flushing* of an embayment depends on the ratio of the typical length scale of topographic variation to the *tidal excursion distance* (Geyer and Signell 1992). However, viewed at a larger spatial scale (in which bays are seen as bumps along the coast), the zero/slow mean along-coast flow through the bay may be viewed as a region where the CBL is extra wide (order 10 km rather than more usually order 1 km) (Largier 2004).

Larval retention zones in tropical reef environments are not well-known (Largier 2003), but at least two examples have since been described from Palau associated with resident spawning aggregations of small herbivorous fishes. Both of these retention areas occur in deeply incised embayments in the barrier reef (Fig. 6.4), with larval retention zones defined by movements of drifters released when resident fishes aggregated for spawning (Fig. 6.5). The east side retention zone had tidal currents which moved water and propagules off and onshore (returning toward or

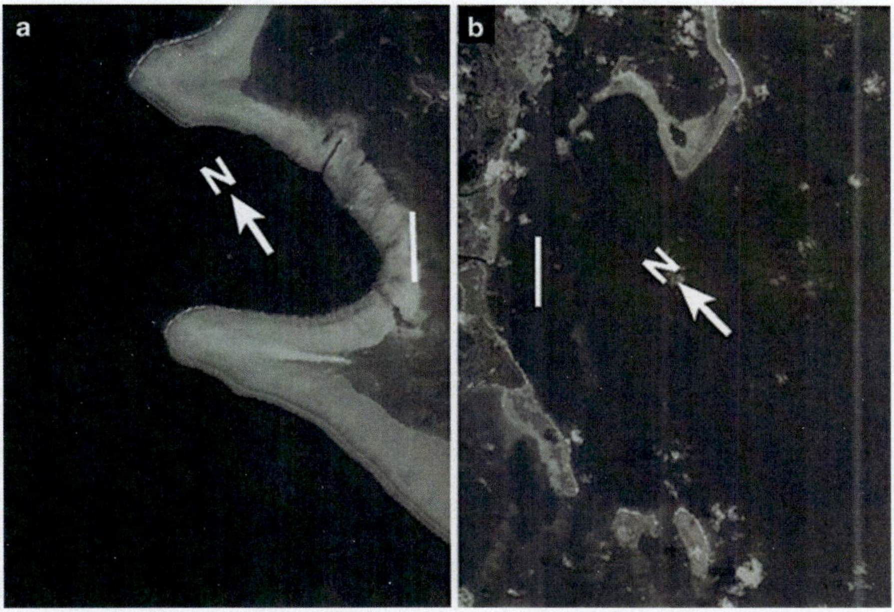

Fig. 6.4 Long-shore current retention zones in Palau have three sides enclosed by land or reef, with one side open to offshore waters. (**a**) Ulong Reef and associated embayment on western side of Palau lagoon. (**b**) Lighthouse Reef region on the east side of Palau. *Vertical white bars* equal 2 km. *White arrows* indicate north

across the reef on flood tides), as well as CBL currents that moved both north and south along the face of the barrier reef (Hamner et al. 2007). One east-side drifter (out of nearly 20) did not remain in the CBL, but ran east on both ebb and flood tides toward the open sea, illustrating the "leaky" nature of retention zones. Drifters released in a similar embayment off the western barrier reef (Fig. 6.5, unpublished data) moved a short distance seaward from the reef on the falling tide, then moved in whatever direction the CBL was going, remaining relatively close to the barrier reef throughout ebb tide. Most drifters reversed direction after 6 h, returning toward the barrier reef or crossing the reef into the lagoon on flood tide. We expect that rugged and sinuous coral reefs and tropical islands similarly will have many retention zones, wherever the length scale of coral reef embayments exceeds the 6-h alongshore tidal excursion (Geyer and Signell 1992).

While one may consider relatively small embayments as part of the coastal boundary layer, these bays also sequester a significant volume of water, plankton and fish larvae from the overall alongshore flow in a process called *lateral trapping* (Wolanski 2001). Most recently, Jessopp et al. (2008) investigated eight bays of various size and configuration along the Irish coast and demonstrated that the largest, deepest bays had the slowest flushing rates, the longest retention times, and the greatest larval diversities. The potential for detrainment from the coastal current and

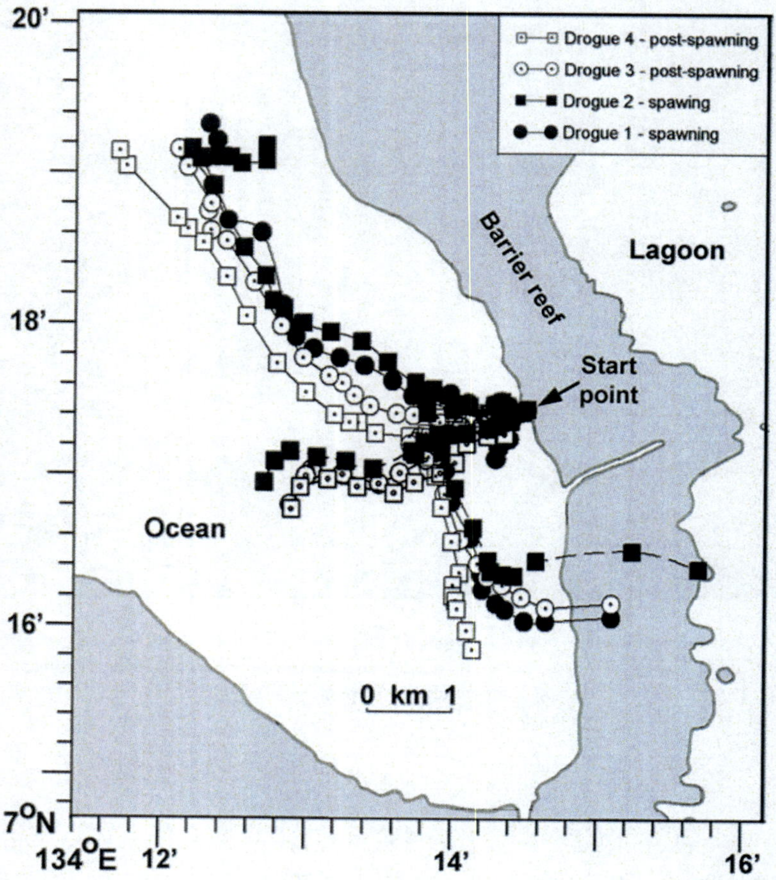

Fig. 6.5 Movement of current drifters on three different days released at a resident spawning aggregation site in the Ulong Reef embayment. Palau, at 30 min intervals. Drifters took somewhat different paths each day. Some became entrained within the coastal boundary layer while others reversed direction on the flood tide and advected east towards and across the barrier reef (Colin and Hamner, unpublished data)

retention in a bay occurs when water and plankton are trapped on the inshore side of the long shore flow, and it is one of the most important topographic mechanisms for retention of materials such as pollutants or larvae near reefs (Wolanski 2001). Lateral trapping can be seen clearly at Cid Island in the Whitsunday Island, Australia (Fig. 6.6, redrawn from Hamner and Hauri 1977). Archambault et al. (1998) demonstrated lateral trapping for small embayments, as have McCulloch and Shanks (2003) and Shanks et al. (2003).

Cross shore *tidal jets* disrupt coastal boundary currents but their disturbance often does not completely interrupt alongshore flow unless the jet is particularly

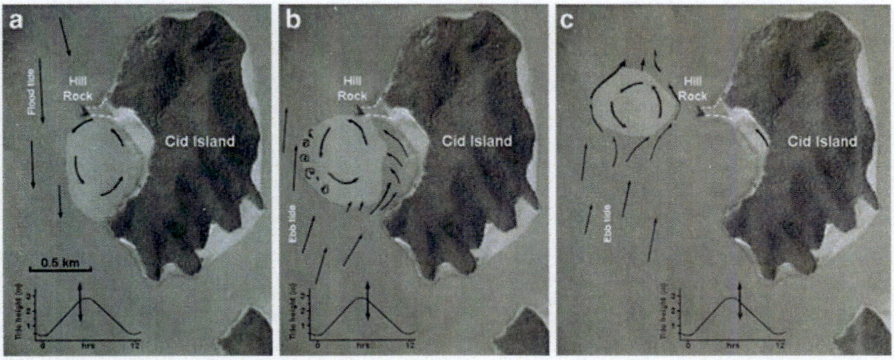

Fig. 6.6 Photo-drawings of the sediment-laden (stippled) gyre to the west of Cid Island, Australia, during various phases of the tidal cycle. Sequences (**a–c**) drawn from low altitude photographs taken obliquely from the south; Cid Island image taken by satellite. Vertical bars on the idealized tidal cycles indicate time of photograph (Redrawn from Hamner and Hauri 1977)

large, such as at the mouth of a large bay or river. Tidal jets through passes in coral reefs are generally narrow, fast-flowing features, typically ejected from narrow passages or where flow separates from the shore (Wolanski and Hamner 1988). Many aggregations are believed to spawn eggs into tidal jets (see Figs. 2.6 and 5.13, Chap. 12.10). Channels through the reef with associated tidal jets enhance offshore flow during ebb tide and also augment flow back into the lagoon during flood tides, often producing mid-channel axial convergent fronts during the return (Wolanski and Hamner 1988). Barrier reef channels in Palau are 15–70 m deep (Colin 2009), they have strong flows up to about 10–12 m/s, and a number of species of fishes occur in transient spawning aggregations along the sides or bottoms of these channels. Tidal jets in Palau (Fig. 6.7) move water from the CBL up to 2–3 km offshore during ebb tides, but most drifters released into the eastern embayment during ebb tide either returned toward the reef on flood tide or drifted along the reef in the CBL as tidal currents slacked.

Thus, from the limited amount of information available (mostly from Palau), it appears that resident spawning aggregations of coral reef fishes release their spawn at ebb tide each day into the nearest deep water, either a seaward facing reef front or tidal channel. Associated with these topographic features is a tidal, coastal boundary current that flows gently back and forth along the face of the reef. The length of the reversing tidal excursion defines the temporal stability of the parcels of larvae-rich water that move alongshore in the coastal boundary layer (i.e. early larvae of resident aggregation spawning fishes are entrained for several days or longer). The topography of the fore-reef face, benthic rugosity, depth and linear sinuosity of the reef determine the shape of the embayments along the reef margin as well as the magnitude of lateral trapping, sequestration, residence times and flushing rates of water, plankton and fish larvae. Tidal jets clearly disrupt alongshore flow and enhance cross-shore dispersion but generally these do not completely destroy the integrity of the coastal boundary layer.

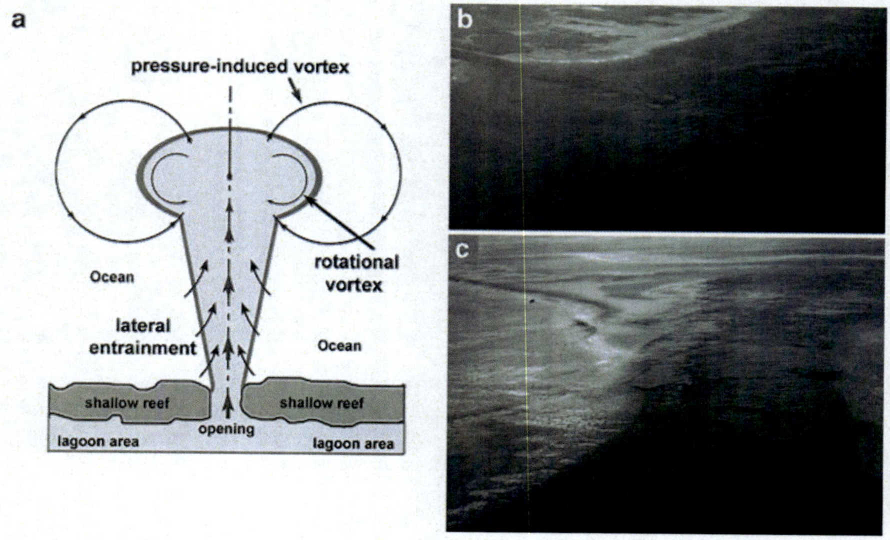

Fig. 6.7 (**a**) Tidal jet current fields emanating from a reef channel on a falling tide. The jet produces vortices at its outer end and a series of small eddies that entrain water along its axis. (**b**) Aerial view of a tidal jet, Lighthouse Reef channel, Palau, on a falling tide. (**c**) Aerial view of a tidal jet at West Channel, Palau, looking from lagoon toward the open sea with turbulent flow along the sides of the channel persisting after the jet has passed through the channel

It will be interesting in the future to learn how small herbivorous coral reef fishes manage to reproduce in situations other than as described above for Palau, on a barrier reef subjected to strong tides and modest winds. Since small herbivorous fishes occur everywhere on all coral reefs, they undoubtedly will exhibit different patterns of spawning elsewhere. For example, we do not know how these fishes spawn on windward reefs subjected to strong trade winds and continuous wave pounding. Nor is it known how small herbivorous fishes on small coral outcrops in the centre of coral reef lagoons spawn, yet they must spawn into the surrounding lagoon because small reef fishes will not leave the shelter of the coral to swim alone over deep water. If the reef flat is exceptionally wide, will these small fishes remain close to their home range or will they migrate further to spawn? Clearly, more research is needed on the behaviour of fishes that form resident spawning aggregations and on the fate and distribution of their eggs and larvae after becoming planktonic.

6.3.2 *Transient Aggregation Spawning and Topographic Eddies*

The fates of larvae spawned at the shelf edge by transient spawning aggregation fishes are often different from those spawned by resident aggregations. Transient spawners are large, carnivorous fishes that migrate considerable distances to aggregate, and they generally spawn at dusk at the shelf edge near topographic projections of the reef adjacent to deep water. Transient spawning aggregations form only several days a year at a particular phase of the lunar cycle, often on consecutive months (Colin 1992; Heyman et al. 2005). After spawning, the fertilized eggs may initially drift along the reef in the alongshore coastal boundary current, but may also be quickly entrained into topographic eddies generated by nearby reef promontories. Tidal eddies that form downstream of headlands, may be shed from the headland source and move offshore or they may be reconstituted with reversed rotation on the opposite side of the promontory when the tide reverses. Because these tidal eddies persist briefly, they have been called transient eddies (Signell and Geyer 1991; Geyer and Signell 1992). Retention of both water and particles (eggs/larvae) in eddies depends on vorticity dynamics. Vorticity may lead to tidal eddies dissipating over the subsequent tidal cycle, reforming on the other side of the headland, or detaching and drifting into deeper offshore waters, carrying a discrete body of entrained water and materials such as sediments (Hamner and Hauri 1977) and fish larvae (Kingsford et al. 1991; Burgess et al. 2007). In contrast, eddies due to persistent mean flow will remain attached to the headland and may accumulate and retain larvae for as long as the flow and thus the eddy persist.

Two transient aggregations of the Nassau grouper provide an example of the diversity of influence of the coastal boundary layer and island eddies on eggs and early larvae. Drifters started at the Bahamas Nassau grouper spawning aggregation sites on Long Island indicated that eggs and early larvae were retained close to the island and over the shelf for many days. Drifters followed tidal trajectories with directional reversals consistent with the behaviour of transient eddies formed by promontories along the edge of Long Island (Fig. 5.9, Chap. 5; Colin 1992). On the other hand, at Little Cayman Island (a much smaller island with a narrow island shelf) drifters released at Nassau grouper spawning sites indicated entrainment of larvae within a few hours into the general Caribbean circulation ("Grouper Moon project" – unpublished data). These larvae would thereafter spend much of their time in the large, offshore, mesoscale eddies which predominate south of Cuba. These two studies point out the divergent prospects of aggregations of the same species, assuming that drifters reasonably mimic the general movement of eggs and larvae. At one site larvae were retained by small transient, tidal eddies, while at a second site eggs were advected rapidly away from shore into meso-scale eddies that circulated hundreds of kilometres over deep oceanic water.

Sometimes divergent sites have similar potential larval fates. For example, Nemeth et al. (2008), using stationary current meters, found that the average current direction for three red hind aggregation sites separated by 200 km all exhibited transport of fertilized eggs onto the island shelf or banks (St. Thomas, US Virgin Islands – island shelf; St. Croix, USVI – bank reef; Saba Bank, Netherlands Antilles – bank reef). In this study, broad-scale conditions appeared to foster general retention of eggs and larvae.

Clearly an understanding of small-scale and mesoscale eddies is exceptionally important for evaluating the oceanographic processes that affect larval survivorship for both resident and transient aggregators. The locations where transient aggregators spawn greatly increase the possibility that larvae will eventually be entrained into topographically generated eddies. Unfortunately, the literature on flow interactions with reef and island topography is incomplete and it is difficult to generalize about the fate of spawn in generic topographic locations – indeed, Bakun (2006) concurs that this literature is incomplete, sparse, incomprehensible and confusing. It is incomplete because fine-scale eddies are not addressed by most physical oceanographers, who work preferentially with meso-scale and larger oceanic flows. It is sparse because topographic eddies are not relevant to most fisheries biologists who investigate pelagic fishes that often live far from shore. It is often incomprehensible to biologists poorly trained in the mathematics of fluid mechanics, and it is confusing due to the inconsistent use of terms that are vague or imprecise.

Eddies are ubiquitous in the ocean, at all locations and at all scales. Every obstacle in the ocean – be it island, reef, or submerged bank, a whale or a copepod – sets up a downstream eddy field. Usually the fine-scale oceanography of a particular area is not well known when field sampling of spawning fishes is contemplated and it is difficult to tailor the sampling regime for eddy effects. There are, however, some general, well known rules associated with flow of fluids around objects; those rules apply to how water flows around coral reefs (Wolanski 2001). For a qualitative appreciation of eddies and vorticity around coral reefs and their retentive potential there are five factors of immediate concern: *velocity* and *direction* of the current, the *size* and *shape* of the reef, and the *depth* of the water. Tides alter velocity and direction of flow around a reef every day, with range varying throughout the lunar month. Within any 6-h tidal period flow around a reef might be (a) smooth, or (b) turbulent with stable eddies, or (c) turbulent with an elongate wake (Fig. 6.8). Flow around an island, headland or reef will differ during flood and ebb tide, depending on topography, tidal strength and wind direction. Thus, when a current encounters an abrupt change in bathymetry such as a submerged bank, a coral reef or an island, it diverges to flow around the obstruction, and eddies form in the lee. Shallow water eddies in reversing tidal flows typically last only about 2–3 h and they transport particles around/through the reef several times over a period of days, effectively trapping them in the immediate vicinity of the reef and preventing them being swept away by far-field currents (Black et al. 1990, 1991; Kingsford et al. 1991; Burgess et al. 2007). Tidal eddies have been described at islands (Wolanski et al. 1984), headlands (Alldredge and Hamner 1980), and coral reefs of the GBR (Hamner and Hauri 1981; Wolanski et al. 1984; Young et al. 1993; Suthers et al. 2004; White and Wolanski 2007).

It is not easy to determine whether transiently aggregating fishes have spawned into a linear coastal boundary layer or into a transient, tidal, topographic eddy unless one simultaneously deploys drogues or obtains aerial images at that specific site because quite different flow patterns look exactly the same to an observer at the edge of the reef. A range of methods can be used to examine water movements. For small bodies of water fluorescent dyes are often used, but in many locations it is possible to visualize flow from aircraft because sediments are suspended by strong tidal currents, as in the Whitsunday Islands, Australia (Hamner and Hauri 1977). Satellite images also show differences in water temperatures and chlorophyll. In addition, current meters, acoustic doppler current profilers, and drifters are

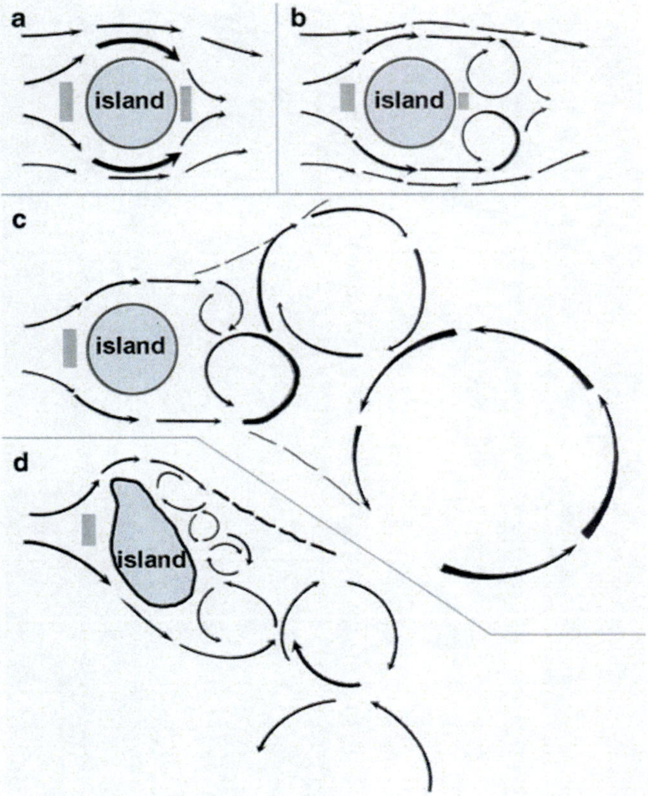

Fig. 6.8 Flow around islands (*top* 3 figures) varies according to the velocity of flow and shape of the island. If a perfectly round island **(a)** if flow is gentle, it separates upstream and merges smoothly downstream. **(b)** If flow is faster paired stable eddies are generated in the lee, with upstream and downstream upwelling. **(c)** Faster flow causes downstream eddies are shed (not retained) and drift free into an extended, expanding island wake with eddies of alternating vorticity. **(d)** Island shape affects downstream flow and small, new eddies form continuously and expand as they drift down the lee-side of the island

important for ground truth data on local flows. Current-following drifters are by far the cheapest, quickest and easiest ways to measure local flow, and over the short term (hours to a few days) they are believed to adequately mimic egg and larval movements. Some methods to examine nearby flow patterns of water are included in Chap. 9.

Pandora Reef is a useful example of how clouds of eggs/larvae from potential spawning sites near headlands can be transported from the reef into the open sea in a smoothly flowing CBL without vorticity or spawned into the vorticity of either an entrained or free eddy field (Hamner and Hauri 1981). This reef (Fig. 6.9) is subject to strong tidal currents within the Great Barrier Reef lagoon. Three different potential spawning sites on a flood tide are considered: one close to a headland within a

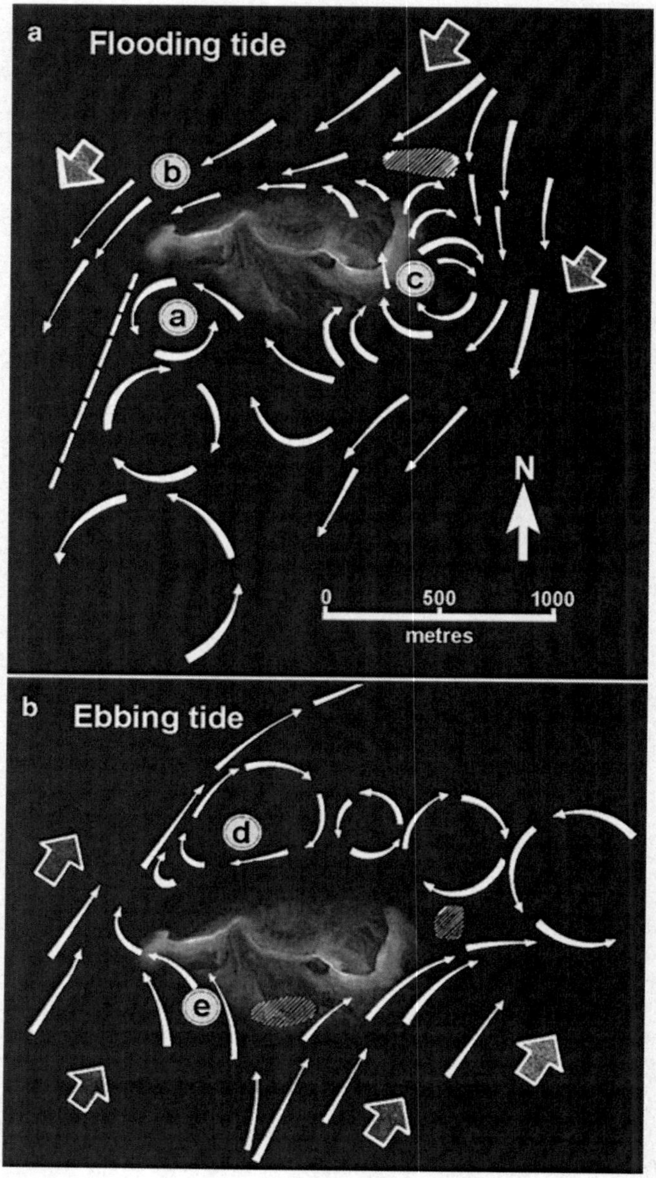

Fig. 6.9 Pandora Reef, Australia, has strong tidal currents and larvae spawned on flood or ebb tides at different locations and have different patterns of advection, dispersal, and vorticity. (**a**) On flood tides a site close to a headland within a topographic eddy if the current sufficiently strong or the reef sufficiently small or asymmetric, the eddy may be released from reef attachment, and in this case the eddy and larvae will drift downstream to the south (site A). A flood tide site within an alongshore coastal boundary layer larvae will have drift westward past the end of the reef, merge smoothly into the flood tide far-field current (site B). A site where water is entrained but not released into the current that flows around the end of the reef, has an eddy with central upwelling but no net horizontal advection (site C). (**b**) On ebb tides, eggs spawned into an eddy entrained on the north side of the western headland will rotate and larvae will remain close to the reef (site D). Spawning on the island's south side near the western headland (site E) will cause eggs to advect in an alongshore boundary current with no vorticity, first advecting west, then north

topographic eddy (site a), a second within an alongshore coastal boundary layer immediately next to the reef (site b), and one where water is entrained but not released into the current that flows around the eastern end of Pandora Reef (c). At the first site, if the current impinging on the reef is sufficiently strong or if the reef sufficiently small or asymmetric, the eddy may have sufficient vorticity to be released from reef attachment, and the eddy and larvae will drift to the south, downstream, in a "von Karman vortex street" (also Fig. 5 in Hamner and Hauri 1981). For the second site, the larvae will drift westward past the end of the reef, merge smoothly into the flood tide far-field current, and then drift south. The shear-zone between the main current and the island-entrained eddy is impenetrable to passive particles, so larvae will not be entrained into the downstream reef-generated eddy field (Wolanski 2001) but drift free of topographic influence. The concentration of larvae will decrease slowly and smoothly according to normal dispersal and natural mortality, and they will advect without interruption until the far-field current changes direction or velocity. At the third site (c), the eddy has upwelling at the center but no net horizontal advection, so the larval cloud will remain within the eddy relatively close to the reef as long as the current flowing past the eastern end of the reef does not change direction or velocity.

During ebb tide, eggs spawned into an eddy entrained on the north side of the western headland (site d in Fig. 6.9) will rotate clockwise and the cloud of larvae will remain close to the reef and the larvae may become more concentrated on eddy-generated fronts. Later, however, when the tide turns this eddy will shed downstream but this time it will be a free-eddy. If fish spawn on the south side of the island near the western headland (site e) during ebb tide, their eggs will be released into an alongshore boundary current with no vorticity, they will first advect west and then north. Consequently, larvae spawned on a flood or an ebb tide at any one of three different locations on the reef, all near headlands, can end up with a variety of different patterns of advection, dispersal, and vorticity.

At Bowden Reef, Kingsford et al. (1991) investigated meroplankton and larval fishes in slicks of tidal, topographically eddy-induced fronts and also in Langmuir circulation convergent fronts (Fig. 6.10), modelling flow around the reef mathematically. They sampled for meroplankton on surface slicks using a small plankton purse-seine, and found that presettlement reef-fishes and zooplankton were exceptionally abundant in frontal slicks. The waters around Bowden Reef, on the Great Barrier Reef shelf, are vertically well mixed (Wolanski 2001) and flows around these reefs are barotropic. It is interesting that the tidal fronts northwest and west of Bowden Reef are sufficiently robust that slicks of 1–2 km length retain their integrity and oscillate in position over several or more tidal cycles.

6.4 Topographic Meso-scale Eddies: Oceanic Islands and Atolls

Shallow water wakes for the Great Barrier Reef are ubiquitous in the vertically well-mixed, barotropic waters over the continental shelf (Hamner and Hauri 1981; Wolanski et al. 1984; Wolanski 2001), but benthic friction generates a secondary

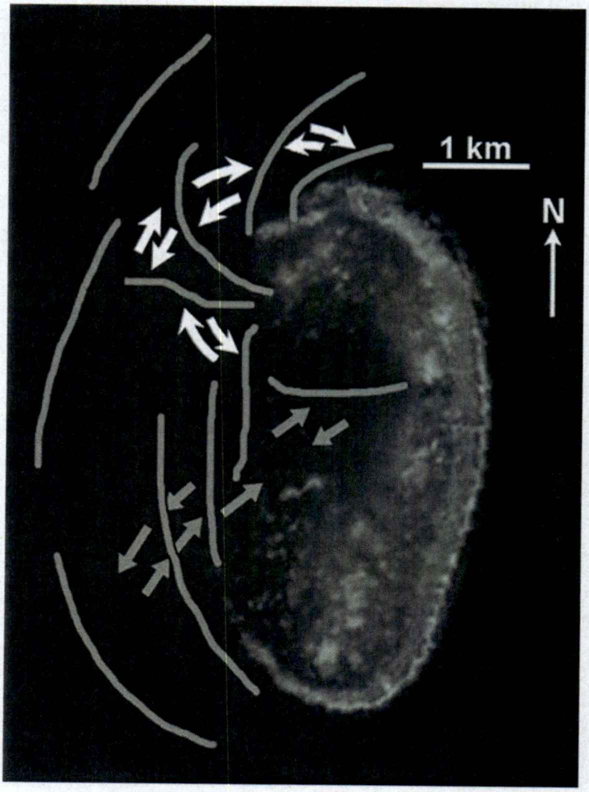

Fig. 6.10 At Bowden Reef, Great Barrier Reef, surface slicks (*thin grey curved lines*) changed location with spring tide (*white arrows*), and parts advected into the lagoon, while grey arrows indicate predicted movements of slicks. A topographically controlled front (slick) off the northwest corner of the reef rotated north or south with the tide (Redrawn from Kingsford et al. 1991)

circulation such that bottom water flows radially toward the eddy centre with upwelling and downwelling (Wolanski et al. 1984; Tomczak 1988; Wolanski and Hamner 1988; Wolanski 2001; Suthers et al. 2004; White and Wolanski 2007), giving shallow-water wakes 3-dimensional structure. It is now clear that 3-D wakes behind shallow-water continental reefs are all qualitatively different from the mostly 2-D wakes that occur downstream of oceanic reefs and islands (Tomczak 1988), although there is also gentle upwelling and downwelling in large mesoscale oceanic gyres. All oceanic islands and atolls also produce wakes, but in deep oceanic waters there is no bottom friction to generate rapid recurrent 3 dimensional flow. However, from a satellite 2-D and 3-D wakes look much the same (Teinturier et al. 2010), with smooth flow around the reefs when ambient far-field currents are gentle, paired stable eddies as flow increases, and downstream vortex streets when currents are strong. Dong et al. (2009) recently modelled wakes for islands surrounded by deep water, including the relationships between island size, wake instability, coherent vortex formation, and mesoscale and sub-mesoscale eddy activity. The presence of sloping island sides, as opposed to vertical sides, introduces an additional set of variables affecting the structure of the downstream wake. Wakes of deep water islands and reefs cannot entrain and retain densely concentrated zooplankton or larval fishes as rapidly as shallow flows around islands and reefs on continental and island shelves.

Tropical oceanic islands come in all shapes and sizes but lack a continental shelf and usually have only narrow fringing reefs. Some are isolated (e.g. Bermuda, Kosrae, Mauritius), while others are close to similar islands and other types of reefs (e.g. Lesser Antilles, Moorea). Most have constant or slowly varying directional oceanic flow patterns. Islands, oceanic reefs and atolls in deep water, all deflect currents according to the size and the shape of the island or reef and the speed of the far-field current. Tidal currents around oceanic reefs do generate oscillatory flow and retention in the coastal boundary layers close to shore (e.g. larval squid are retained in a tidal coastal boundary layer along the west side of Catalina Island, Southern California; Zeidberg and Hamner 2002), but further offshore oceanic currents are unidirectional, generally stronger than tidal currents, and net transport around oceanic islands and coral reefs is primarily the result of far-field current dynamics. Exactly how reefs around oceanic islands maintain populations of animals with pelagic larvae in spite of directional advection is not known, but island wakes are expected to play a key role. Downstream wakes in the open ocean can be stable and similar over long periods of time (Barkley 1972; Lobel and Robinson 1986; Dong et al. 2009). Barkley (1972) described large eddies for the Johnston Atoll wake, with alternating vorticity and alternating upwelling and downwelling, which were formed and shed regularly into a continuously expanding vortex street. Long-term stable island wakes with paired-eddies that do not shed are purported to exist, and these presumably could retain larvae from the island or reef, potentially abetting subsequent recruitment and enhancing local fishing. Such a stable eddy system in the wake of Tobi Island, south of Palau was described qualitatively by Johannes (1981), based on information from native fishers. However more recent knowledge based on oceanographic modelling (Heron et al. 2006) makes its existence uncertain.

Where oceanic islands are not separated by great distances it is likely that eddies with larvae from one island will impinge on other islands during the larval life, allowing larvae to reach settlement habitat on the second island. Where islands are crowded, a mix of overlapping eddies could occur (e.g. Lesser Antilles) and produce a complex mixture of genomes from distant sources (Cowen and Castro 1994; Oxenford et al. 2008).

6.5 Physical Oceanography and the Trophic Environment of Early-Stage Larvae

There are two physical features in which larval food is aggregated in sufficient concentrations for feeding by early-stage fish larvae: high vertical gradients, as in the thermocline, and high horizontal gradients, as in fronts. The importance of larval first-feeding and chlorophyll layers is now widely recognized in fisheries, and although storms and deep-mixing of surface waters can quickly obliterate the chlorophyll maximum, less has been written about the atmospheric or hydrographic criteria associated with its formation. Since Hjort (1914) it has been believed that variations in year-class strength of Atlantic herring, *Clupea harengus*, are largely

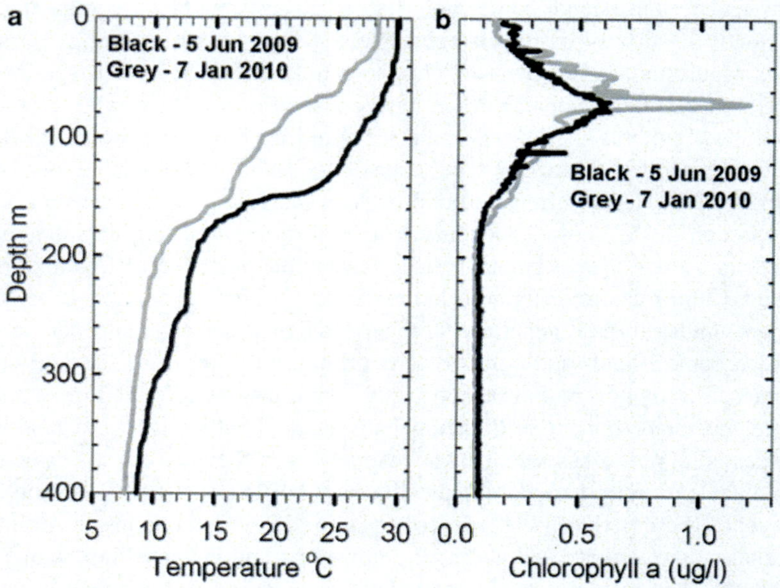

Fig. 6.11 An example of vertical distribution of temperature and chlorophyll in the tropical oceanic water column close to Palau's coral reef areas showing change as a result of ENSO conditions with "normal" (*black line*) and a strong El Niño (*grey line*) periods

due to differential mortality of the larvae related to the availability of food for the first-feeding larvae. Laboratory rearing studies with California anchovy larvae (Lasker et al. 1970; Lasker 1975) indicated the density of larval food must be higher than that usually found at sea to obtain even moderate larval growth and survival. Off southern California anchovy larvae were able to feed extensively on water from the chlorophyll maximum layer, which contained phytoplankton, but they did not survive if fed phytoplankton deficient surface waters. Recently, there has been a resurgence of interest in the horizontal phytoplankton layers concentrated at the thermocline (or halocline), called "thin layers." Sullivan et al. (2010) reported that fine-scale, dense patches of organisms are "ubiquitous features" in the ocean and the term "thin layer" (vertical extents from centimetres to a few meters) describes horizontally concentrated patches of organisms, or particles that can extend horizontally for many kilometres and persist for weeks.

Reef fish larvae, including those of aggregation spawners, are believed to feed on small zooplankters that graze on phytoplankton, hence concentrations of phytoplankton in the tropics could have some correlation with the occurrence of aggregation spawned larvae and their food. In the tropical western Pacific, for example, the thermocline is deep, often below 100 m, but so far only one study indicates that aggregation reef fish larvae occur in this depth range (Oxenford et al. 2008). In Palau, during El Niño conditions, however, thermoclines become much shallower (Fig. 6.11) resulting in different amounts of phytoplankton in the water column at different depths. However, the effects on reef fish larval feeding and survival are largely unknown.

Fronts are formed at sea when water masses of different density converge, with one sinking below the other. Where water masses meet they have on or near the surface a concentrated line of organisms and materials that float or are able to swim to stay at or near the surface. Fronts attract a variety of nektonic animals. Recently, Genin et al. (2005) confirmed, using 3D acoustic imaging in the Red Sea, that the accumulation of high concentrations of zooplankton along a frontal zone between convergent water masses was due to their directional swimming behaviour (up to tens of body lengths per second) against upwelling and down-welling currents.

The importance of fronts and eddies in the overall economy of the sea has been reemphasized recently by Kai et al. (2009). The importance of fronts to early stage larvae of reef fishes is virtually unknown, but there is strong evidence they have important influences on pre-settlement fishes (Kingsford et al. 1991). Larvae that are entrained at a front, of course, are well-placed to find food because they need only maintain position and wait while a conveyor belt of concentrated food moves continuously toward them. Fronts are surprisingly common around coral reefs because almost any given coral head can generate topographic eddies (Hamner and Hauri 1977; Wolanski and Hamner 1988; Wolanski 2001) and because all eddies generate fronts along the outside edges of the vortex (White and Wolanski 2007). Hydrographic structure, such as the slicks of internal waves, tidally generated eddies, jets, fronts, thermal plumes and Langmuir circulations, may influence both distribution patterns of fishes and zooplankton, as well as their direction of movement (Alldredge and Hamner 1980; Sakamoto and Tanaka 1986; Wolanski and Hamner 1988; Kingsford et al. 1991; Wolanski 2001). Whether fronts play any greater role for aggregation spawned larvae than for any other type of fish larvae is not known, but needs exploration.

Convergence of water masses also generates directional horizontal flow along the front that can transport entrained organisms back toward shore or into a coral reef lagoon (Largier 1993). Eggleston et al. (1998) state that since "fronts constrain cross-frontal flow, they serve to deflect incident flow resulting in strong along-frontal and down-welling flows that transport larvae collected at the front." Strong swimming larvae in surface waters, such as late stage larval fishes, may use fronts as 'larval conduits' to intersect adequate settlement habitat.

There are also fronts associated with the coastal boundary layer since these currents retain their integrity as they flow along an unbroken coastline, often with surface waters slightly warmer than offshore waters on calm warm afternoons. In Palau, on ebbing tides, warmer water from the lagoon flows out across the reef flat and it slides seaward over slightly cooler oceanic water (Fig. 6.12). The frontal edge of the buoyant sheet of lagoon water flowing off the reef carries most of the eggs spawned on this reef every day at ebb tide by resident spawning aggregations of surgeonfishes and parrotfishes. Given the importance of fronts for larval foraging, it is surprising that fronts have seldom been carefully sampled.

Fig. 6.12 Flow from lagoon to ocean on falling tides across the western barrier reef of Palau produces a relatively gentle current which moves seaward across the reef, generating a front visible in this aerial photograph. When conditions are calm, lagoon water, slightly less dense than ocean water, floats like a surface lens over oceanic water

6.6 Summary

Aggregation-spawning reef fishes produce eggs, yolk-sac and pre-flexion larvae that are planktonic but which are not necessarily advected away into the open sea. Diffusion dynamics of spawning clouds, tidally limited advection by coastal boundary currents, entrainment within tidal eddies, and lateral trapping by reef topography reduce offshore dispersion. Particularly instructive have been investigations in which drifters were released at spawning sites to evaluate long-shore advection and eddy entrainment.

The role of aggregation-spawning for gamete cloud dispersal has been largely ignored. Do aggregations produce large cohorts that recruit at the same time, or do aggregations confer direct benefits on eggs and early stage larvae? The types of gamete clouds produced by small, herbivorous resident-fish aggregations and by large, carnivorous transient fish aggregations are different, but we cannot yet compare their relative costs and benefits. Comparison between reef length and advection velocity gives an estimate of how long the spawning cloud diffuses in the vicinity of the reef and when it reaches a topographic break. Propagules that initially remain in the flow attached to reef topography, greatly increase their chance of local recruitment. Coastal boundary currents, lateral trapping in embayments, and eddies are flow features associated with specific reef topography that act to retain aggregation-spawned propagules. Resident and transient spawning aggregations do not form everywhere but only at specific locations and particular times of day, tide, and year.

Resident fish aggregators spawn at ebb tide each day into the nearshore waters that flow tidally back and forth along the face of the reef. Tides flush embayments in the reef, but embayments also sequester water, plankton and fish larvae from larger scale circulation by lateral trapping. Resident aggregation spawning fishes are mostly small herbivores that are common and occur everywhere on all coral reefs, and they spawn frequently (Chap. 4).

Most transient fish aggregations are composed of large carnivorous fishes that are less abundant than resident species, often migrate long distances to form spawning aggregations, but only spawn on ebb tides at dusk at the edge of the continental shelf near topographic projections of the reef and then only for several days a year at a particular phase of one or several lunar cycles (Chaps. 7 and 12). While some larvae are ejected into large-scale offshore flows, other larvae are entrained into and retained by topographic tidal eddies generated by reef promontories. Understanding eddy dynamics is critical for evaluating larval survivorship for both resident and transient aggregators; the precise location of fish spawning aggregations on a given reef affects both retention and dispersal of larvae.

It is difficult to tell *a priori* if fish have spawned into a linear, coastal boundary current or into a rotating, topographic eddy. Without additional information from drifters or aerial images, these different flow regimens look exactly alike to a swimmer on the edge of the reef. While wakes behind shallow-water continental reefs are different from those downstream of oceanic reefs and islands, they both aggregate plankton for first-feeding larvae via horizontal layering (thermoclines, chlorophyll maxima, thin-layers) and vertical discontinuities (fronts and slicks).

Acknowledgements We have been aided greatly in our review by two books, *Marine Populations… An Essay on Population Regulation and Speciation* by Michael Sinclair (1988), and *Physical Oceanographic Processes of the Great Barrier Reef* by Eric Wolanski (2001), and by various reviews cited herein, particularly those by Sponaugle et al. (2002) and Cowen and Sponaugle (2009). Data for Fig. 6.11 of water column temperature and chlorophyll concentrations were provided by Dan Rudnick of Scripps Institution of Oceanography through Office of Naval Research grants N00014-06-1-0776 and N00014-10-1-0273. Pat Colin drafted most of the figures, provided editorial assistance and contributed significantly to the chapter.

References

Alldredge AL, Hamner WM (1980) Recurring aggregation of zooplankton by the tidal current. Estuar Coast Mar Sci 10:31–37

Appeldoorn RS, Hensley DA, Shapiro DY et al (1994) Egg dispersal in a Caribbean coral reef fish. *Thalassoma bifasciatum*. II Dispersal off the reef platform. Bull Mar Sci 54:271–280

Archambault P, Roff JC, Bourget E et al (1998) Nearshore abundance of zooplankton in relation to shoreline configuration and mechanisms involved. J Plankton Res 20:671–690

Bakun A (2006) Fronts and eddies as key structures in the habitat of marine fish larvae: opportunity, adaptive response and competitive advantage. Sci Mar 70S2:105–122

Barkley RA (1972) Johnston Atoll's wake. J Mar Res 30:201–216

Bell LJ, Colin PL (1986) Mass spawning of *Caesio teres* (Pisces: Caesionidae) at Enewetak Atoll, Marshall Islands. Environ Biol Fish 15:69–74

Black KP, Gay SL, Andrews JC (1990) Residence times of neutrally-buoyant matter such as larvae, sewage or nutrients on coral reefs. Coral Reefs 9:105–114

Black KP, Moran PJ, Hammond LS (1991) Numerical models show coral reefs can be self-seeding. Mar Ecol Prog Ser 74:1–11

Botsford LW, White JW, Coffroth MA et al (2009) Connectivity and resilience of coral reef metapopulations in marine protected areas: matching empirical efforts to predictive needs. Coral Reefs 28:327–337

Burgess SC, Kingsford MJ, Black KP (2007) Influence of tidal eddies and wind on the distribution of presettlement fishes around One Tree Island, Great Barrier Reef. Mar Ecol Prog Ser 341:233–242

Clarke LB, Ackerman D, Largier J (2007) Dye dispersion in the surf zone: measurements and simple models. Cont Shelf Res 27:650–669

Colin PL (1992) Reproduction of the Nassau grouper, *Epinephelus striatus* (Pisces: Serranidae) and its relationship to environmental conditions. Environ Biol Fish 34:357–377

Colin PL (2009) Marine environments of Palau. Indo-Pacific Press, San Diego

Colin PL (2010) Aggregation and spawning of the humphead wrasse, *Cheilinus undulatus,* (Pisces: Labridae): general aspects of spawning behaviour. J Fish Biol 76:987–1007

Colin PL, Bell LJ (1991) Aspects of the spawning of labrid and scarid fishes (Pisces: Labroidei) at Eniwetak Atoll, Marshall Islands with notes on other families. Environ Biol Fish 31:229–260

Cowen RK, Castro LR (1994) Relation of coral reef fish larval distributions to island scale circulation around Barbados, West Indies. Bull Mar Sci 54:228–244

Cowen RK, Sponaugle S (2009) Larval dispersal and marine population connectivity. Annu Rev Mar Sci 1:443–466

Csanady GT (1972) The coastal boundary layer in Lake Ontario: Part II. The summer-fall regime. J Phys Oceanogr 2:168–176

Domeier ML (2004) A potential larval recruitment pathway originating from a Florida marine protected area. Fish Oceanogr 13:287–294

Domeier ML, Colin PL (1997) Tropical reef fish spawning aggregations: defined and reviewed. Bull Mar Sci 60:698–726

Dong C, Idica EY, McWilliams JC (2009) Circulation and multiple-scale variability in the Southern California Bight. Prog Oceanogr 82:168–190

Dufour V, Glazin R (1993) Colonization patterns of reef fish larvae to the lagoon at Moorea Island, French Polynesia. Mar Ecol Prog Ser 102:143–152

Eggleston DB, Armstrong DA, Elis WE et al (1998) Estuarine fronts as conduits for larval transport: hydrodynamics and spatial distribution of Dungeness crab postlarvae. Mar Ecol Prog Ser 164:73–82

Ellis EP, Watanabe WO, Ellis SC et al (1997) Effect of turbulence, salinity and light intensity on hatching rate and survival of larval Nassau grouper, *Epinephelus striatus*. J Appl Aquac 7(3):33–43

Fischer HB, List EJ, Koh RCY, Imberger J et al (1979) Mixing in inland and coastal waters. London: Academic Press

Genin A, Jaffe JS, Reef R et al (2005) Swimming against the flow: a mechanism of zooplankton aggregation. Science 308:860–862

Geyer WR, Signell RP (1992) A reassessment of the role of tidal dispersion in estuaries and bays. Estuaries 15:97–108

Hamner WM, Hauri IR (1977) Fine-scale currents in the Whitsunday Islands, Queensland, Australia: effect of tide and topography. Aust J Mar Freshw Res 28:333–359

Hamner WM, Hauri IR (1981) Effects of island mass: water flow and plankton pattern around a reef in the Great Barrier Reef lagoon, Australia. Limnol Oceanogr 26:1084–1102

Hamner WM, Colin PL, Hamner PP (2007) Export-import dynamics of zooplankton on a coral reef in Palau. Mar Ecol Prog Ser 334:83–92

Hensley DA, Appeldoorn RS, Shapiro DY et al (1994) Egg dispersal in a Caribbean coral reef fish. *Thalassoma bifasciatum*. I. Dispersal over the reef platform. Bull Mar Sci 54:256–270

Heron SF, Metzger EJ, Skirving WJ (2006) Seasonal variations of the ocean surface circulation in the vicinity of Palau. J Oceanogr 62:413–426

Heyman WD, Kjerfve B, Graham RT et al (2005) Spawning aggregations of *Lutjanus cyanopterus* (Cuvier) on the Belize Barrier Reef over a 6 year period. J Fish Biol 67:83–101

Hjort J (1914) Fluctuations in the great fisheries of northern Europe viewed in the light of biological research. Rapp P-V Reun Cons Int Explor Mer 20:1–228

Holm HE (2004) Horizontal dispersion of eggs of cubera snapper, *Lutjanus cyanopterus*, at Gladden Spit, Belize. MS thesis, University of South Carolina, Columbia

Jessopp M, Mulholland OR, McAllen R et al (2008) Coastline configuration as a determinant of structure in larval assemblages. Mar Ecol Prog Ser 352:67–75

Johannes RE (1981) Words of the Lagoon: fishing marine lore in the Palau district of Micronesia. University of California Press, Los Angeles

Kai ET, Rossib V, Sudreb J et al (2009) Top marine predators track Lagrangian coherent structures. Proc Natl Acad Sci USA 106:8245–8250

Kiflawi M, Mazeroll AI, Goulet D (1998) Does mass spawning enhance fertilization success in coral reef fish? A case study of the brown surgeonfish. Mar Ecol Prog Ser 172:107–114

Kingsford MJ, Wolanski E, Choat JH (1991) Influence of tidally induced fronts and Langmuir circulations on distribution and movement of presettlement fishes around a coral reef. Mar Biol 109:167–180

Largier JL (1993) Estuarine fronts: how important are they. Estuaries 16(1):1–11

Largier JL (2002) Linking oceanography and nearshore ecology: perspectives and challenges. In: Castilla JC, Largier JL (eds) The oceanography and ecology of the nearshore and bays in Chile. Ediciones Universidad Católica de Chile, Santiago

Largier JL (2003) Considerations in estimating larval dispersal distances from oceanographic data. Ecol Appl Suppl 13:71–89

Largier JL (2004) The importance of retention zones in the dispersal of larvae. Am Fish Soc 42:105–122

Lasker R (1975) Field criteria for survival of anchovy larvae: the relation between inshore chlorophyll maximum layers and successful first feeding. Fish Bull 73:453–462

Lasker R, Feder HM, Theilacker GH et al (1970) Feeding, growth and survival of *Engraulis mordax* larvae reared in the laboratory. Mar Biol 5:345–353

Leis JM (1991) The pelagic phase of coral reef fishes: larval biology of coral reef fishes. In: Sale PF (ed) The ecology of fishes on coral reefs. Academic, San Diego

Leis JM (2006) Are larvae of demersal fishes plankton or nekton? Adv Mar Biol 51:59–141

Lobel PS, Robinson AR (1986) Transport and entrapment of fish larvae by ocean mesoscale eddies and currents in Hawaiian waters. Deep Sea Res 33:483–500

McCulloch A, Shanks AL (2003) Topographically generated fronts, very nearshore oceanography and the distribution and settlement of mussel larvae and barnacle cyprids. J Plankton Res 25:1427–1439

Monismith SG (2007) Hydrodynamics of coral reefs. Annu Rev Fluid Mech 39:37–55

Nemeth RS, Kadison E, Blondeau JE et al (2008) Regional coupling of red hind spawning aggregations to oceanographic processes in the eastern Caribbean. In: Grober-Dunsmore R, Keller BD (eds) Caribbean connectivity: implications for marine protected area management, Marine Sanctuary Conservation Series ONMS-08-07. NOAA, Silver Spring, MD, USA

Okubo A (1980) Diffusion and ecological problems: mathematical models. Springer, Berlin

Oxenford HA, Fanning P, Cowen RK (2008) Spatial distribution of surgeonfish (Acanthuridae) pelagic larvae in the eastern Caribbean. In: Grober-Dunsmore R, Keller BD (eds) Caribbean connectivity: implications for marine protected area management, Marine Sanctuary Conservation Series ONMS-08-07. NOAA, Silver Spring, MD, USA

Paris CB, Cowen RK, Claro R et al (2005) Larval transport pathways from Cuban snapper (Lutjanidae) spawning aggregations based on biophysical modeling. Mar Ecol Prog Ser 296:93–106

Petersen CW, Warner RR, Cohen S et al (1992) Variable pelagic fertilization success: implications for mate choice and spatial patterns of mating. Ecology 73(2):391–401

Richardson LF (1926) Atmospheric diffusion shown on a distance-neighbour graph. Proc R Soc Lond A A110:709–737

Sakamoto W, Tanaka Y (1986) Water temperature patterns and distribution of fish eggs and larvae in the vicinity of a shallow sea front. Bull Jpn Soc Sci Fish 52:767–776

Shanks AL, McCulloch A, Miller J (2003) Topographically generated fronts, very nearshore oceanography and the distribution of larval invertebrates and holoplankters. J Plankton Res 25:1251–1277

Siegel DA, Mitarai S, Costello CJ et al (2008) The stochastic nature of larval connectivity among nearshore marine populations. Proc Natl Acad Sci USA 105:8974–8979

Signell RP, Geyer WR (1991) Transient eddy formation around headlands. J Geophys Res 96:2561–2575

Sinclair M (1988) Marine populations Washington Sea Grant Program. University of Washington Press, Seattle, USA

Sponaugle S, Cowen RK, Shanks A et al (2002) Predicting self-recruitment in marine populations: biophysical correlates and mechanisms. Bull Mar Sci 70:341–375

Sponaugle S, Lee T, Kourafalou V, Pinkard D (2005) Florida Current frontal eddies and the settlement of coral reef fishes. Limnol Oceanogr 50:1033–1048

Stacey MT, Cowen EA, Powell TM et al (2000) Plume dispersion in a stratified, near-coastal flow: measurements and modeling. Cont Shelf Res 20:637–663

Strathmann RR, Hughes TP et al (2002) Evolution of local recruitment and its consequences for marine populations. Bull Mar Sci 70(Suppl 1):377–396

Sullivan JM, Donaghay PL, Rines JEB (2010) Coastal thin layer dynamics: consequences to biology and optics. Cont Shelf Res 30(1):50–65

Suthers IM, Taggart CT, Kelley D et al (2004) Entrainment and advection in an island's tidal wake, as revealed by light attenuance, zooplankton and ichthyoplankton. Limnol Oceanogr 49(1):283–296

Teinturier S, Stegnera A, Didelleb H et al (2010) Small-scale instabilities of an island wake flow in a rotating shallow-water layer. Dyn Atmos Oceans 49:1–24

Tomczak M (1988) Island wakes in deep and shallow water. J Geophys Res 93:5153–5154

White L, Wolanski E (2007) Flow separation and vertical motions in a tidal flow interacting with a shallow-water island. Estuar Coast Shelf Sci 77:457–466

Wolanski E (2001) Physical oceanographic processes of the Great Barrier Reef. CRC Press, Boca Raton

Wolanski E, Hamner WM (1988) Topographically controlled fronts in the ocean and their biological significance. Science 241:177–181

Wolanski E, Imberger J, Heron ML (1984) Island wakes in shallow coastal waters. J Geophys Res 89:10553–10569

Young IR, Black KP, Heron ML (1993) Circulation in the ribbon reef region of the Great Barrier Reef. Cont Shelf Res 14:117–142

Zeidberg LD, Hamner WM (2002) Distribution of squid paralarvae, *Loligo opalescens* (Cephalopoda: Myopsida), in the Southern California Bight in the three years following the 1997 El Niño. Mar Biol 141:111–122

Chapter 7
Aggregation Spawning: Biological Aspects of the Early Life History

Patrick L. Colin

Abstract Most reef fishes have bipartite life histories, separate pelagic-oceanic (egg/larvae) and benthic (juvenile/adult) periods. The several-week pelagic period has early planktonic (egg, yolk sac and preflexion larva) and later nektonic components (post-flexion larva to settlement); the plankton-nekton transition timing is variable. For aggregating species, larvae are weak swimmers early in life, but late stages are often strong swimmers able to perhaps influence their settlement locations. No obvious differences were found between larval stages of aggregating and non-aggregating species and both types of spawning are found within single families, and even within a species. There are no egg types, morphologies, feeding strategies or special structures exclusive to aggregating species. Initial dispersal is determined by location and time of spawning. Pelagic eggs are buoyant, keeping them in near-surface waters and away from benthic predators. The larvae go through a series of stages (egg, yolk sac larvae, pre- and post-flexion larvae, pelagic juvenile), becoming larger and more capable over time. Critical periods occur and can cause major mortality of a cohort. Ocean conditions during the early egg and yolk sac stage are critical to survival followed by initiation of feeding as a second critical event. During pelagic life larvae must survive in open water, find appropriate food as larvae and avoid predators. Cohorts from aggregations can recruit as a large pulse, but other fishes may also have such pulses. The mass spawnings of reef invertebrates, such as stony corals, are generally not comparable to those of fishes, while crustaceans (spiny lobsters, marine crabs, terrestrial crabs) have some similarities. There is a need for fisheries oceanography research on aggregation spawning, as well as more work on laboratory culture. The question of potential maternal benefits to larvae needs careful attention.

P.L. Colin (✉)
Coral Reef Research Foundation, P.O. Box 1765, Palau 96940, Koror
e-mail: crrf@palaunet.com

Y. Sadovy de Mitcheson and P.L. Colin (eds.), *Reef Fish Spawning Aggregations: Biology, Research and Management*, Fish & Fisheries Series 35,
DOI 10.1007/978-94-007-1980-4_7, © Springer Science+Business Media B.V. 2012

7.1 Introduction

During their early life history (ELH), which covers the period between spawning and settlement as juvenile fishes from the water column, reef fish larvae must survive and grow in the pelagic environment in order to successfully recruit to the benthic stage. Chap. 6 examined the physical and biological environment impacting the dispersal and retention of reef fish larvae, emphasising those with aggregation spawning. This chapter examines the biology of reef fish larvae from spawning through metamorphosis to try to (1) discern any intrinsic differences between aggregation spawned and other reef fish larvae, and (2) determine the common features of larval development across those families with at least some members engaged in aggregation spawning. Basic information is first provided on the morphology of larvae, the probable critical periods in ELH, then examines what differences and similarities exist between species in their spawning behaviour and initial dispersal, and how predators might affect success of spawning. Comparisons are made with potentially similar broadcast spawners and future research directions are indicated.

The larvae and the juveniles/adults of reef fishes occupy very different habitats, the former generally in open water "pelagic" environments and the latter in benthic or near benthic habitats (Fig. 7.1). On an individual basis the ELH of aggregation spawning reef fishes seems essentially identical to those reef fishes lacking aggregation spawning. The same determinants of survival (feeding, predation, transport, water quality) operate on all larvae. If differences do exist, they might provide some insight into the reasons why some fishes engage in aggregation spawning. While considerable progress has been made in understanding the early life history of reef fishes in general (reviews include Cowen and Sponaugle 2009; Leis and McCormick 2002; Cowen 2002), attention has been focused more on species with demersal eggs and short larval lives. Settlement-stage fishes can be captured in good condition using light traps and other means. Some studies utilize cultured larvae, again mostly from families with demersal eggs, which are easier to culture. The understandable attention paid towards the more easily studied species has resulted in a tendency to consider all larvae to behave similarly and to not appreciate the vast differences between early and late stage larvae.

While it is often feasible to investigate ELH at its start (spawning or hatching) and its conclusion (settlement), when fishes are near reefs or other shelf areas, the period of weeks between is not easily examined. The numbers of larvae in the open ocean are generally low relative to the water volume, and deciphering the pelagic life of reef fishes is not easy, usually requiring costly shipboard work followed by tedious sorting and identification. Our rudimentary knowledge severely limits the understanding of what affects success of a given cohort of fishes.

Questions about whether eggs/larvae from a large mass spawn stay together for extended periods of time (until at least time of first feeding) have been examined in the previous chapter, and the initial answer seems to be a qualified "no". The physical mechanisms operating at these early stages generally favour gradual dispersal away from the location of spawning, with perhaps only the slight buoyancy and later vertical movement ability of yolk sac larvae to promote concentration along oceanic fronts.

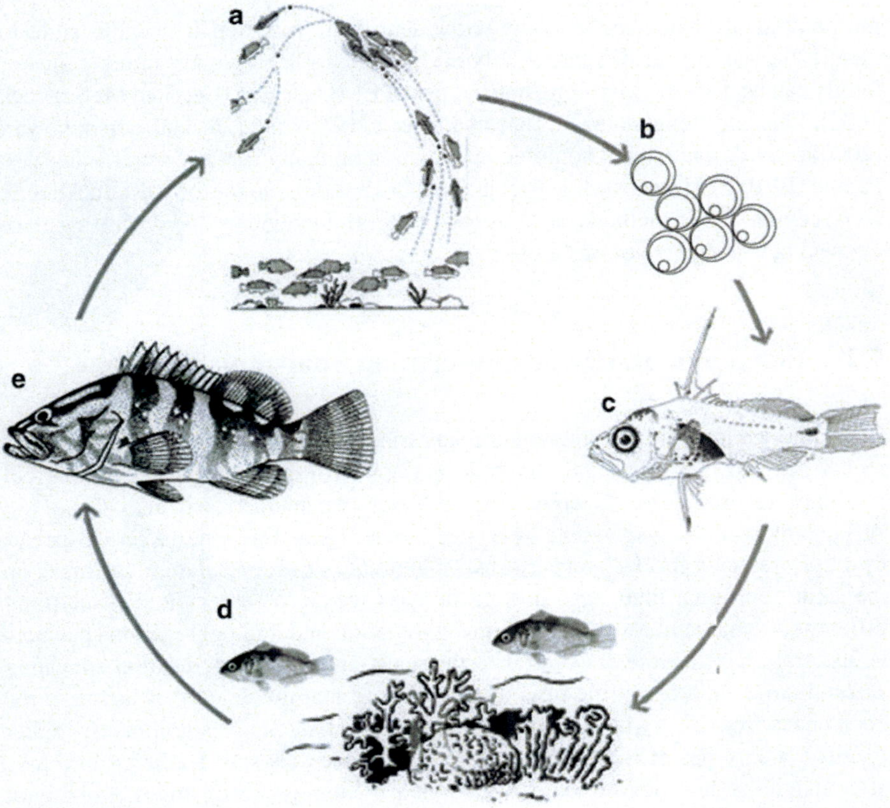

Fig. 7.1 The bipartite life cycle (pelagic eggs and larvae with benthic juveniles and adults) of most reef fishes is shown here, using the Nassau grouper as an example. (**a**) Spawning, in this case from an aggregation, produces (**b**) small fertile eggs which drift away with the current. (**c**) Larvae develop over several weeks in the pelagic environment and (**d**) eventually settle to take up benthic life as juveniles. (**e**) Over time, usually a number of years, fish grow to become mature adults which spawn to start the cycle again (Figure: Yvonne Sadovy de Mitcheson with permission)

Aggregation propagules do not, apparently, form a cohesive group that would remain so in the pelagic environment. So does aggregation spawning provide any benefit to early larvae? That question is still open, but may be examined by seeing whether or not the conditions at times of aggregation spawning might promote success of larvae through recruitment, particularly if other larval components are present which might compete with fish larvae.

Reef fish eggs and larvae initially act as drifting plankton, and at a later stage some become capable of behaving more as free-swimming nektonic organisms (Leis 2006). The exact stage of transition between the plankton-nekton, when larvae go from being passive drifters to active swimmers, is variable. The extent to which late stage larvae use sensory means to navigate or seek out settlement locations is an area of considerable research, but also of much uncertainty. Larvae normally differ greatly in morphology from juveniles/adults of the same species and have fewer morphological characters to assist with identification. Often it is impossible to identify larvae to species (particularly early stages). Placing them into a genus or family can be difficult, inhibiting many types of ELH research (Leis and McCormick 2002). Leis and Rennis (1983), Leis and Trnski (1989) and Leis and Carson-Ewart (2000) provide available taxonomic information on many shallow water Indo-west Pacific (IWP) ELH stages while Richards (2006) has done so for the western Atlantic (WA) region. Other methods, such as use of otolith increments and chemistry, have allowed new perspectives on ELH.

7.2 Taxonomic Range of Aggregation Spawning

Both aggregation and non-aggregation spawning occur within various families of reef fishes and their eggs and larvae do not appear to differ markedly among co-familials according to spawning strategy (see Appendix) (see also Chap. 4). Aggregation- and pair-spawning by related species (and even within a single species in some wrasses-Labridae and parrotfishes-Scaridae) often occur simultaneously on the same reef area, their eggs mixing in open water. Whether there is anything different or beneficial for ELH between aggregation and non-aggregation spawners is uncertain but nonetheless central to the question of why aggregation spawning exists at all. It appears at the time of first feeding that the density of larvae in the open ocean from a very large aggregation spawning may not be significantly higher than for other types of spawning (see Chap. 6). Once they start feeding a few days after hatching, how larvae can be sustained by the generally low zooplankton populations found in most tropical waters is crucial. Chapter 6 indicated some biophysical mechanisms that potentially concentrate the larvae, the food items they need and potential predators. Other mechanisms might reduce the density of larvae through predation and dispersal so they become less concentrated.

7.3 Early Life History Stages of Coral Reef Fishes

Most fishes with spawning aggregations have planktonic eggs. A limited group of aggregating species has eggs which develop on the bottom (see Appendix) and whose dispersal and pelagic life does not start until hatching. Knowledge of larval stages and behaviour usually comes from capture of specimens using plankton nets, light traps and other techniques, as well as rearing of eggs in culture. Some *in situ* studies are being undertaken, mostly with settlement stage fishes (Leis et al. 1996, 2003; Paris et al. 2008). The ELH stages of aggregating fishes are typical of other reef fishes, and those stages can be broken down as follows:

Eggs (pelagic) – These can be of variable shape, but most often nearly spherical; typically ranging from about 0.6 to 1.2 mm diameter; larger for eels (up to 4 mm), and some non-perciform groups like *Fistularia* and tetradontiforms. Some parrotfishes have distinctive spindle-shaped eggs up to about 2.5 mm long and 0.5 mm wide. Most free-spawned eggs are slightly buoyant, nearly transparent, with yolk and oil globule(s), as well as a hard outer chorion. The yolk provides food for the embryo in the egg and yolk sac larva stages. The oil globule may contain materials, such as triacylglycerol (TAG) lipids, important for larval nutrition. Most pelagic reef fish eggs hatch within 24 h to yolk-sac larvae. Species with negatively buoyant, adhesive, small eggs (triggerfishes-Balistidae, rabbitfishes-Siganidae) also hatch quickly. Eggs attached to the bottom or other objects (damselfishes-Pomacentridae) are generally larger, not spherical, take longer to hatch and are ready to feed within hours of hatching.

Yolk sac larvae – Pelagic eggs at hatching are essentially embryos attached to the large yolk sac, often with slight positive buoyancy, and have very limited swimming ability. Over a few days they develop functional eyes, a complete gut with mouth and anus, rudimentary fins and nervous system. The benthic eggs of triggerfishes and rabbitfishes quickly hatch into larvae similar to those with pelagic eggs. Others with demersal eggs (e.g. damselfishes) go through the yolk sac stage in the egg capsule for some days and at hatching are capable of weak swimming and ready to feed quickly. Most fishes with demersal eggs do not aggregate for spawning. Otoliths (ear bones) begin to develop in the yolk sac stage.

Early stage larvae (preflexion) – Pelagic larvae initiate feeding on external foods once the systems necessary (complete gut, pigmented eyes, ability to orientate) have been developed during the yolk sac stage. Larvae possess a notochord without any caudal fin supporting structures. The nearly transparent bodies of early larvae, an adaptation to pelagic life, have the body musculature divided into myomeres, which have a near one-to-one correspondence with the eventual number of vertebrae. Pigment is often limited to the eyes, over the gut area and as scattered chromatophores or melanophores on the body. There is a medial fringe of thin tissue, the fin fold, around the posterior end which becomes differentiated into the medial fins. Rudimentary pectoral fins are usually present. A swim bladder may form and usually inflates some days after the larvae start to feed, reducing negative buoyancy. The otoliths develop with rings being laid down which allow for age determination, if the rings are formed daily or at another known interval, since otolith formation.

Late stage larvae (post-flexion) – As larvae grow, the posterior end of the notochord flexes upward in the process called "flexion" establishing the location where the supporting elements of the caudal fin will develop. Once flexion is complete, the larva is in the "post-flexion" stage. Formation of caudal fin supporting elements occurs over time and the vertebrae develop. Larvae often become more pigmented, and develop specialized structures in some families (some may start prior to flexion). Swimming ability is greatly increased once the caudal fin is developed. The larval stage is considered to end when all external meristic characters (fin spines and rays, etc.) develop and temporary special features for pelagic life are lost.

The pelagic juvenile stage – Still living in open water, they have nearly the full morphology of the benthic stages, but usually are partially transparent and lack the benthic juvenile colour pattern. The pelagic juvenile stage (also called "pre-settlement") may persist for some time if a suitable settlement environment is not found. Pelagic juveniles are usually ready to settle from the plankton and take up benthic life.

The benthic juvenile stage – The young fish takes up life on the bottom and quickly acquires normal juvenile coloration. Any adaptations for pelagic life are quickly lost. This stage is also termed "post-settlement".

7.4 Dynamics of Life History Stages and Events

Aside from the developmental stages, there are a number of critical life history events during pelagic life, and failure to transition through these results in high mortality of a given larval cohort. ELH for all reef fishes can potentially be divided into early, middle and late stages. The early starts at fertilization, ends with the initiation of notochord flexion and can be characterized by multiple critical events, including first feeding and swim bladder inflation. During the middle stage the caudal and other fin supporting elements become complete, some specialized structures (which potentially enhance survival) become fully developed, and significant vertical movements in pelagic environments may occur. Rapid growth during this stage appears to be important in eventual success in recruitment and early benthic life. The late stage (also termed the "settlement stage") starts at the transition to a pelagic juvenile, loss of specialized structures, enhancement of swimming ability and the potential to use sound, olfactory or other cues to navigate towards suitable settlement habitats, followed by settlement to a benthic existence. During all stages larvae are moving, either by drifting or by actively swimming (and navigating?), and may be selecting a given depth, perhaps on a diel schedule. Once ready to feed, they need to do so continually (with interruptions at night?). The negative consequences of failing to do so are probably more important in the early stages. They must avoid predators at all times.

7.4.1 The Start of Pelagic Life – Aspects of the Early Stages

After spawning, egg and yolk sac larva mortality can potentially be high if conditions of turbulence, salinity, temperature and light are unsuitable, as demonstrated in culture studies. Ellis et al (1997) reported that low turbulence, high light and high salinity were optimal for survival of larval Nassau grouper, *Epinephelus striatus*, from hatching to first feeding. Increased survival under high versus low (to zero) light intensity points to possible relationships between light intensity (affecting behaviour through phototaxis and swimming activity), diel timing, and size at first feeding, which may extend throughout ELH. Finally, larvae in completely static water had a less than 1% hatching success suggesting that exceptionally calm seas could have a negative effect on hatching. While this study and others (Akatsu et al. 1983; Watanabe et al. 1995, 1996, 1998), have been conducted in rearing tanks rather than under natural conditions, they demonstrate that such factors may influence survival in the field. Unpredictable weather conditions may result in spawning under less than optimal conditions and result in failure of a cohort spawned in an aggregation. Certainly cohort size can vary considerably in aggregating species. For example, in the leopard coralgrouper, *Plectropomus leopardus*, a single cohort was found to dominate in areas closed to fishing on central the Great Barrier Reef, Australia (Russ et al. 1996).

First feeding – If physical conditions are acceptable, the critical transition from yolk feeding to capturing food can occur. Larvae ready to begin feeding lack a caudal supporting structure and other well-developed fins; however they can swim in a coordinated fashion to orient, approach and strike at potential food items. They are not capable of swimming well enough for migration purposes, have limited ability to control their depth and often lack an inflated swim bladder. It seems likely that early feeding of aggregation-spawned reef fish larvae will be dependent on vision for feeding and limited to periods of adequate daylight. At first feeding the larvae capture food items usually by launching themselves at the food item and ingesting it. Within few days feeding ability develops so they can feed by striking forward using the pectoral fins or an "S" or "C" shaped bend in the body, and engulfing prey with, by now, protrusible jaws.

If the correct type and size of food organisms are present and feeding is initiated (a "match") survival may be high, or if the larval state and food available are incorrect (a "mismatch") the larvae face starvation in a very short time period. This "match-mismatch" hypothesis (Sinclair 1988) has been widely accepted from studies of temperate ichthyoplankton, and is undergoing some revisions to thinking as more information becomes available (Leis and McCormick 2002). Almost certainly the limited time between exhaustion of the yolk and starting to feed externally is a "critical period" but the time limits are poorly known for reef fish larvae. The times of hatching and exhaustion of the yolk after spawning are dependent on temperature; higher temperatures producing shorter time spans. For larvae of cultured Malabar grouper, *Epinephelus malabaricus*, for example, Yoseda et al. (2006) found the volume of yolk remaining, representing endogenous energy reserves at time of opening of the mouth and the onset of feeding, to be higher at 25°C than at 28° and 31°C. By delaying the start of feeding by 6–24 h after mouth opening, they found only a short period where this delay did not affect survival. Overall the higher the energy reserves remaining when feeding can begin, the greater the chance of successfully transitioning to exogenous food. Some results seem contrary, however. For example, Sugama et al. (2004) reported highest survival of highfin grouper, *Cromileptes altivelis* (Serranidae), larvae at 28°C (versus 25° and 31°C) although larvae, fed rotifers (density of 5–10 per litre), held at 31°C ingested more rotifers and had higher growth rates.

Inflation of the swim bladder – Most reef fishes with pelagic eggs, including aggregation species, have swim bladders as juveniles/adults which develop during larval life.

The point at which the bladder becomes functional (gas-filled) is considered "inflation" and can represent another "critical period". For some larvae it is believed they must rise to the surface to ingest air (Czesny et al. 2005). In some living larvae the swim bladder is visible as a silvery structure over the gut due to their generally transparent nature; however the swim bladder is often masked by melanophores and not always visible externally. Once inflation has occurred, the larvae may change from negatively to neutrally buoyant, this change apparent in the angle at which the body is held in the water. Data on time of inflation are sparse, but, for example, in the red grouper, *Epinephelus morio,* it was observed at 12 days post-hatching (Colin et al. 1996) and Drass et al. (2000) observed it 4–6 days post-hatching in 2.4 mm notochord length (NL) reared larvae of the red snapper, *Lutjanus campechanus.*

Flexion – Swimming of sufficient speed and duration to match ambient currents is nearly impossible for early larvae. Prior to flexion, swimming is adequate for food capture and perhaps to change depths to a limited degree, but not to migrate any distance. During their early stages larvae might be concentrated by mechanisms, such as by maintaining a given depth along a descending front, but may not actively seek such mechanisms. Their swimming ability might allow them to remain, once encountered, with high concentrations of food. While flexion is easily noted in larval specimens, the age of its occurrence is often not known for field-caught specimens. Rearing studies provide some indication that it may occur roughly half way (or more) through the larval life. For example Clarke et al. (1997) reported notochord flexion in several species of snappers (Lutjanidae) to start at 11–12 days post-hatching and to be complete at 16–18 days, while metamorphosis occurred at ages of 22–33 days. For red grouper Colin et al. (1996) found flexion to occur at about 16 days, with a total pelagic life of 35 days.

7.4.2 Are There Differences in ELH Between Aggregation and Non-aggregation Spawners?

Fish families with aggregation spawning also have non-aggregating species. Within these families the larvae of the two spawning types do not appear different and the requirements to promote survival to recruitment are probably similar. Are there, however, possible differences in egg type or size for fishes with different strategies? Are all propagules equivalent? For egg types, it seems unlikely that differences exist. If a family has two egg morphologies (e.g. spherical versus spindle in parrotfishes), they are found for both types of spawning. Where dual modes of spawning (pair- and group-spawning) exist in one species (e.g. parrotfishes and wrasses), the eggs produced are also identical, or nearly so, and individual females may alternate between pair and group spawning. Robertson (1996) did report a 5% higher egg volume in group-spawned versus pair-spawned eggs of *T. bifasciatum*. This represents a difference of less than 2% in diameter and may represent a slightly different degree of hydration.

However, egg size differences may be uncorrelated with adult fish size and vary geographically or seasonally. For spherical eggs, small differences in egg diameter produce greater differences in egg volume which consists largely of yolk. A 26% increase in diameter results in a doubling of egg, and presumably yolk, volume. Some large reef fishes have small eggs; the humphead wrasse, *Cheilinus undulatus*, and bumphead parrotfish, *Bolbometopon muricatum*, the largest members of their families, have hydrated eggs averaging only 0.66 and 0.65 mm in diameter, respectively (Colin 2010; Bell and Colin 1986). The Micronesian wrasse, *Labropsis micronesica*, a non-aggregating wrasse reaching less than 100 mm SL, has a 0.80 mm diameter egg, twice the volume of a humphead wrasse egg. In the Atlantic its largest wrasse, the hogfish *Lachnolaimus maximus* (a possible, but unverified aggregator), has eggs 1.2 mm in diameter (Colin 1983) while eggs of the bluehead wrasse, *Thalassoma bifasciatum* (reaching about 150 mm SL), with dual spawning modes, are about 0.56 mm in diameter (Holt and Riley 1999). While the egg of the hogfish is quite large, the eggs of most other Atlantic species seem similar to comparable Pacific species (Richards 2006).

For the eggs of a single species, the influence of maternal age and condition may determine the quality of eggs and subsequent survival/growth of larvae. The "provisioning" of larvae with energy-rich TAG lipids (found largely in the oil globule) increased with female age in black rockfish, *Sebastes melanops* (Scorpaenidae), and correlated with subsequent growth and survival (Berkeley et al. 2004). For reef fishes the importance of egg quality is known for some damselfishes with demersal eggs (reviewed by Leis and McCormick 2002), although little is known for those with pelagic eggs. In damselfishes maternal hormones and nutritional components incorporated into egg yolk, the size of eggs and the size of larvae at hatch are affected by feeding conditions of the female and behavioural interactions with conspecifics. The larvae resulting from such "superior" eggs grow and develop faster reducing time spent in the pelagic environment, thereby reducing the potential for mortality of a cohort. It has been shown for species of surgeonfish (Bergenius et al. 2002) and wrasse (Grorud-Colvert and Sponaugle 2006), both species with aggregation spawning, that fast early growth can have positive effects on recruitment success.

The role of social factors in spawning aggregations (with the possible exception of sex change) is unstudied. It is conceivable that aggregations provide abundant opportunity for adults, particularly females, to be influenced by their interactions with conspecifics through hormonal actions and increased "quality" of their eggs, similar to the effect identified in damselfishes or through sexual selection and mate choice (McCormick 1998, 1999a, b; McCormick and Nechaev 2002, Chap. 3). For transient aggregators (TAs), who often have their reproduction limited to a few short periods each year, if social interactions and hormonal effects can improve "quality" of eggs produced, their relatively short spawning season may also help to maintain egg size and quality (as opposed to long spawning seasons when egg quality may decline over time, Leis and McCormick 2002). It would be important to know if the nutritional fitness of the ova is determined by feeding activities prior to any migration, which might take several days during which feeding may not occur, to evaluate the benefits to egg quality when a transient aggregator then undertakes a significant migration prior to spawning. Groupers at transient aggregations are known to feed between lunar spawning periods, perhaps to maintain nutritional fitness and egg quality (see Chap. 2).

Social factors, perhaps increasing the survival of transient spawned larvae, might provide reasons for the evolution and maintenance of aggregating behaviour. If the larvae of transient aggregators, whose individuals often go through long migrations and elaborate preparations for spawning, are inherently superior compared to non-aggregating species, it might help explain why aggregation spawning has evolved in such species. The "stage duration" hypothesis "that larvae that grow and develop faster have higher survival and enhanced recruitment" (Leis and McCormick 2002) has support from studies of temperate and tropical species (Bergenius et al. 2002; Sponaugle and Pinkard 2004; Sponaugle et al. 2006). Studies of hormone and nutrition levels in adult females and in ova produced, both during ovulation and after spawning, are needed of both aggregating and non-aggregating species, perhaps from single families, to test whether aggregating species show particular benefits.

7.4.3 *Spawning and Gamete Release*

While maternal condition may increase growth and survival, potential benefits at the start of ELH may accrue from aggregation spawning when the environmental conditions are beneficial to the larvae. This might involve release into patches of water having specific properties conducive to larval survival (food in the form of zooplankton?), dispersal and retention. What little is known about the ocean dynamics related to the release of large numbers of propagules into a limited volume of water over short time periods was considered in Chap. 6. Some information on the biologically mediated dynamics of aggregations (food, predation) is provided here.

The overall movements of fishes engaged in the release of eggs and sperm in "free-spawning" reef fishes has been given various descriptive terms – rush, burst, dash, ascent, spawning run – and is the start of the process of early life history. Fishes present in aggregations spawn in three manners; as individual pairs (pair-spawn), as small groups that briefly break away from a larger aggregation (group-spawn), or in a single group comprising most individuals in the aggregation (mass spawn). Fertilization occurs quickly (within 1 min or so), unlike in many broadcast spawning invertebrates, and in most cases the vent areas of males and females are in very close proximity when gametes are released (Colin 2010), presumably increasing egg-sperm contact. Available data indicate that fertilization rates are generally quite high (90–100%) for many pair- and group-spawning species (Kiflawi et al. 1998). The massive sperm clouds released by some transient spawners (such as Nassau grouper, cubera snapper – *Lutjanus cyanopterus* and leopard grouper – *M. rosacea*) and high associated gonadosomatic indices of their males suggest in such cases that virtually all healthy eggs get fertilized (Colin 1992; Sala et al. 2003; Heyman et al. 2005). This may not be the case for some smaller pair spawning reef fishes (Petersen et al. 1992) and sperm limitation may occur (Petersen et al. 2001).

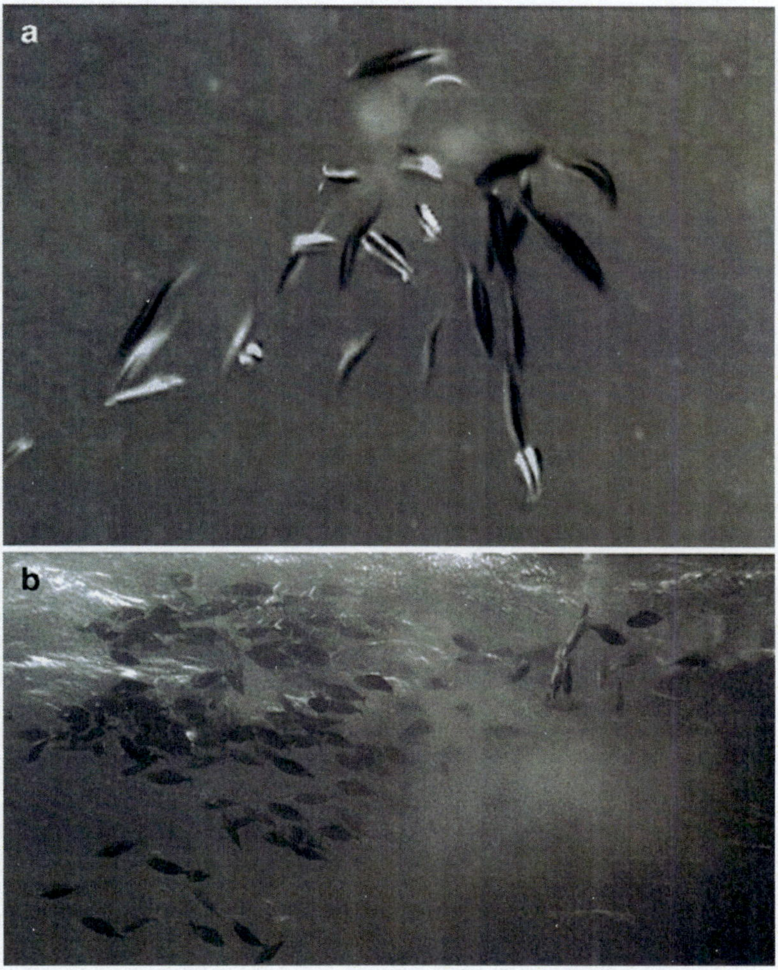

Fig. 7.2 (**a**) A relatively small gamete cloud is released by blunthead wrasse, *Thalassoma amblycephalum*. (**b**) Aggregation spawning by yellow and blueback fusilier, *Caesio teres,* produced a large gamete cloud easily visible due to the large amount of sperm released (Photos: Patrick L. Colin)

The nature of gamete release is related to depth of water over the reef. Where the depth is 5–10 m or more, the fish ascend to release eggs and sperm as a "burst", often spherical in shape. The group of fish follow a single ascending female (Figs. 7.2 and 7.3c), some taking inward radial paths, and sometimes spiral as a group, to release the gametes at a central point or along the spiral. Heyman et al. (2005) suggested that this spiralling might assist in propelling the eggs towards the surface after release. After release spawning fish radiate outward, usually descending towards the bottom (Fig. 7.2a). Spawn ascents that start with only a few fish are often joined by additional individuals within seconds. If the later arrivals are somewhat

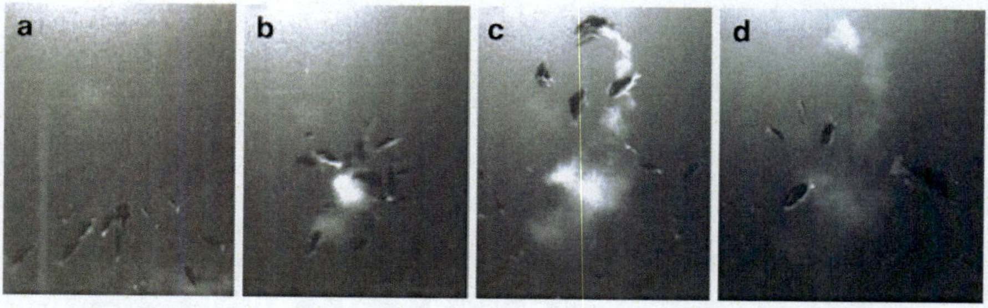

Fig. 7.3 The Nassau grouper, *Epinephelus striatus*, produces a highly visible gamete cloud when spawning, largely due to massive amounts of sperm being released. (**a**) A mix of males and females follow a single leading female prior to release. (**b**) The fish come together, release gametes in a limited volume of water, and then radiate outward. (**c**) In this case, a pair of Nassau grouper swam through the area of the gamete burst, continued upward and released gametes above the initial cloud. The male was streaming sperm, leaving a visible track of its path upward prior to spawning. (**d**) A few seconds after spawning the fish have all moved outward from the site of gamete release and the cloud is gradually starting to become less visible as it dissipates (Photos: Lori J. Colin)

behind the first group, their gamete release may not be in precisely in the same location, but close enough to the already present gametes that this usually serves to expand the overall size of the gamete cloud for a single spawning event. In some larger species, such as Nassau grouper (Fig. 7.3) and cubera snapper, there can be a succession of several gamete releases as fish rush in to join an initial group and start spiralling upward towards the surface, not breaking off quickly after the initial release (Heyman et al. 2005). In mass spawning, the entire mass of fish ascend, often slowly, then at the peak start swirling and releasing gametes, continuing for many seconds (Fig. 7.2b), and is seen in the spawning of yellow and blueback fusilier, *Caesio teres* (Bell and Colin 1986).

There can be variations on this general sequence. In one instance a group-spawn by Nassau groupers was followed just a few seconds later by a single pair (apparently male and female as only one fish was spewing sperm) which arched through where the group had already released and continued in an upward trajectory (Fig. 7.3). While unusual (it was the only such spawning I observed, photographed or recorded on video) it does indicate that more than one female might participate in a given spawning rush. For some large transient aggregators, when there are a series of individuals spawning near simultaneous bursts over the area of the aggregation, the individual spawns appear to merge to form a single large massive gamete cloud.

It is usually difficult to see the gametes when released. Eggs are nearly transparent, and only when copious are the sperm visible. For most pair and small group-spawning fishes, the gametes appear as a faint burst of particles in the water at the apex of a spawning rush which quickly diffuses and disappears (Fig. 7.2a, Colin 2010). Males of some of the larger transient aggregating species have large testes and their sperm can be seen as milky clouds in the water for a few minutes presumably because they release so much sperm at one time (Fig. 7.3).

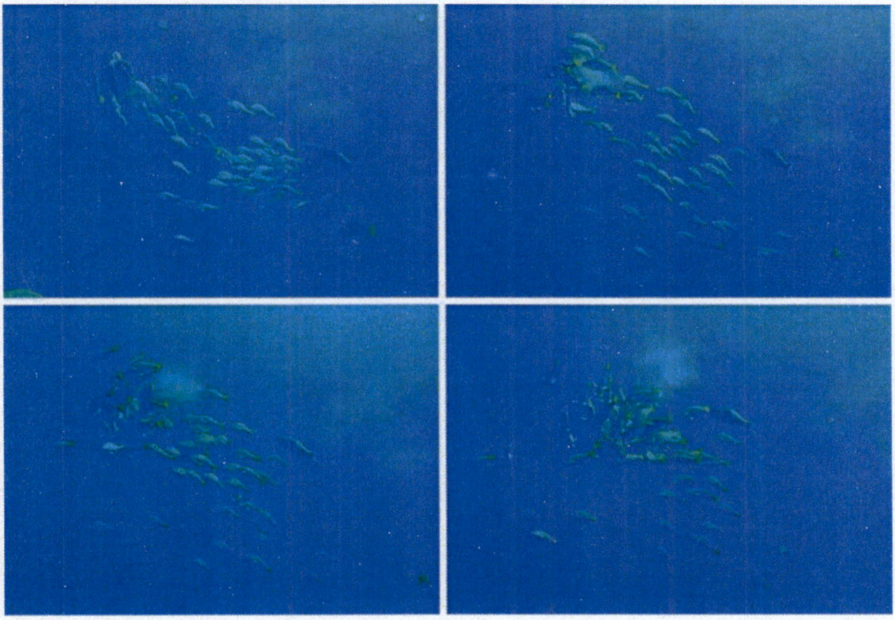

Fig. 7.4 Aggregation spawning by the longnose parrotfish, *Hipposcarus longiceps*, on the western barrier reef of Palau. Elapsed time is shown in the *upper right corner* of each photo. A single male redlip parrotfish, *Scarus rubroviolaceus*, is seen on the *right side* and had no involvement in the spawning activity (Photo: Patrick L. Colin)

If water depths are shallow (only a few metres) and the expanse of an aggregation large compared to the depth (many 10's of metres across), multiple groups will make short spawning ascents from over the entire aggregation area, producing a flurry of spawning which leaves a large area of water with a visible gamete cloud. This is particularly seen in those IWP resident aggregators, typically surgeonfishes and parrotfishes, spawning on lengthy shallow fore reefs just after high tide (Hamner et al. 2007). Such a "broad launch" is often seen in surgeonfishes and parrotfishes (Fig. 7.4a), and can result in a band of gametes being carried off the shallow reef, rather than single gamete clouds, with current from the falling tide for roughly an hour (Hamner et al. 2007).

Other resident aggregators, such as bluehead wrasse, outside of tidally influenced areas, may have a series of individual spawning bursts transported sequentially away from the spawning site by whatever currents exist. Appeldoorn et al. (1994) tracked water masses found at multiple spawning sites of bluehead wrasse, and also at similar non-spawning sites on the same reef. They found transport away from the start site to be greater for releases at non-spawning sites, while there were no differences in broadscale transport within the reef tract. They found that patches from sites tended to merge during the 24 h period of observations. In the same area

Hensley et al. (1994) found that bluehead wrasse tended to group-spawn at sites that potentially offer advantages for off-reef transport of eggs, however, the benefits were only evident over the short term.

Nearly all pelagic reef fish eggs are very slightly positively buoyant after fertilization (see Chap. 5) and this buoyancy is helpful in keeping them in the water column where they can be dispersed and advected away from predators. Under extremely calm conditions eggs might rise and rest upon the water surface, something that happens in aquaria, and potentially exposes them to incident UV radiation. Some temperate fish eggs are known to have UV blocking compounds (Chioccara et al. 1980; Plack et al. 1981), but these have not been investigated for tropical reef fish pelagic eggs. Hamner et al. (2007) found that, for pelagic eggs (largely surgeonfishes, parrotfishes and wrasses) spawned near the surface at a site 80 m downcurrent and about 10 min after spawning, 90% of eggs were in the top 4 m of the water column with 60% of these in the top metre. On windy days with surface waves, fish eggs and zooplankton flowing off the reef would mix deeper, with greater under-sampling if plankton nets are moored at the surface.

7.5 Life in the Pelagic Environment – Aspects of the Middle Stage

The middle phase of ELH is characterized by an increase in swimming ability, which allows easier control of vertical distribution, necessary for selection of depth during a daily cycle, and food capture to maintain rapid growth to increase survival. Armsworth (2001) points out that, if possible, it is much more efficient to use vertical changes in position to move into water layers where advection will carry a larva towards a "desired point" than to actually swim the entire distance toward that point. The increased swimming ability is a result of swim bladder inflation, development of caudal supporting elements, and increasing development and strength of fins. There is also greater ability to avoid predators through growth of anti-predator structures (fin spines, head and body spination) and better swimming. Finding enough food consistently is a critical factor, both for fast growth and to prevent starvation as food reserves are slight. Schooling is unlikely. There is ample evidence of shifts in vertical distribution of larvae between day and night, with nocturnal periods characterized by shallower distributions. Vertical distributions can also modify dispersal trajectories.

7.5.1 Development of Specialized Structures

Specialized structures in larval reef fishes include elongate dorsal, anal and pelvic fin spines, elaborate head spination, and fin-rays with bulbous growths, with such structures limited to a single or a few families. Specialized growth includes highly compressed, high dorsal-ventral aspect (deep-bodied) and very elongate, larval form (leptocephalus) and often such larvae have been given distinctive names, sometimes based on initially erroneous descriptions of them as separate genera (acronurus, tholichthys). Specialized larvae and structures are found in many families with aggregation, in both aggregating and non-aggregating species of those families. Structures, such as spines (epinepheline groupers – Colin and Koenig 1996, surgeonfishes – Randall 1961), may serve in predator deterrence, but the functional use of these features is usually not apparent from studies of preserved specimens. For example, the dorsal and pelvic fin spines of epinepheline groupers form early (starting as soon as 7 days post-hatching at a preflexion notochord length of only 3 mm) and grow rapidly, reaching their maximum length in another 7–10 days at 6–7 mm NL (Colin and Koenig 1996). The spines are quite motile, can be erected nearly vertical and their ends clearly marked with dark melanophores. Their normal positions and angles combined with the location of the snout form a tetrahedron, presumably presenting potential predators with a greatly increased effective size of the larvae. Other examples of specialized structures occur in surgeonfishes; a highly compressed deep body, nearly as deep as long, with formidable dorsal and anal spines which increase its effective size. Once the caudal peduncle spine forms (at about 17 mm TL) the larvae is called an "acronurus" until settlement and metamorphosis.

Body form, without spines or other hard structures, may be modified. The leptocephalus is an elongate flattened leaf-like larva found in eels (order Anguilliformes) and a few primitive bony fishes, such as bonefishes and tarpon (order Elopiformes). Some elopiform fishes may spawn in aggregations. Bonefishes (Albulidae), for example, are reported to migrate through channels to spawn on outer reef slopes, although they may engage in simple migratory spawning rather than aggregate. At least one moray eel aggregates near the reef to spawn (Thresher 1984; Ferraris 1985).

7.5.2 *Growth and Feeding of Mid-stage Larvae*

While much of the larval life between the transport immediately after spawning away from the site and eventual appearance as recruits weeks later is largely unknown, some recent techniques provide tools to examine this life period. The occurrence of rings laid down, near daily, in otoliths of larvae has allowed precise ageing of many species, while the chemistry of otoliths, particularly the presence of stable isotopes, can indicate whether larvae have grown up in oceanic or near shore environments (Patterson et al. 2005). Hamilton et al (2008) used the occurrence of lead isotopes in otoliths of *T. bifasciatum* to distinguish larvae which developed in offshore (about 55%) and nearshore (about 45%) waters and followed their survival during the first month of benthic life which suggested different larval histories suffered differential mortality. Patterson and Swearer (2007) used elemental signatures in otoliths to attempt to identify natal locations for an endemic and a widespread species of wrasse with some success.

Llopiz and Cowen (2009a, b) found in the Straits of Florida high incidences of food in guts of larval fishes, including many from families with aggregation spawning, and a surprising diversity of feeding strategies, based on gut contents. Rapid gut evacuation and no evidence of nocturnal feeding was prevalent in scombroid larvae (based on sampling larvae after sunset) and it is likely reef fish larvae from these waters do the same. However the possibility exists that some reef fish larvae might be able to feed if in shallow water on nights with a near full moon as this provide considerable illumination in shallow water.

Interestingly Miller (2009) suggests that leptocephali larvae probably feed on particulate matter in the water column, such as marine snow, faecal pellets and discarded apendicularian house and that such larvae would not compete for food with those of perciform fishes, which are believed to feed on crustacean zooplankton. Other fish larvae from aggregation families utilize additional types of planktonic organisms for food which may be abundant. Sampey et al. (2007) found that the one surgeonfish larva and nemipterid (sea bream) larvae had eaten appendicularians (larvaceans), while Llopiz and Cowen (2009a) found appendicularians to be important food items of surgeonfish larvae and snapper larvae. Recognition of the importance of these other planktonic groups, such as appendicularians, compared to those traditionally considered primary food items (copepods crustaceans) for larval reef fishes, also complicates the examination of zooplankton density and larval feeding.

7.6 Life in the Pelagic Environment – Late Stages

Once fins and other structures are completely developed, the larva enters its end form, the settlement stage (also known as the "pelagic juvenile"). There may some persistence of "larval" characters (Quere and Leis 2010), but these are eventually lost after settlement. Consistent feeding is probably still important, but fish can survive short periods without feeding, and for some it appears they do not feed during metamorphosis. They develop the capability to potentially move towards preferred settlement environments as swimming ability is sufficient to match or exceed most current regimes encountered and any sensory systems can detect sonic, olfactory and other cues. The fish settle from the pelagic environment into benthic habitats.

The literature concerning the biology of settlement stage coral reef fishes is far larger than that dealing with earlier life history stages. Settlement stage fishes can be captured in a number of manners, identified while alive, used in experiments, and have large otoliths with information on life history recorded in them. Species with spawning aggregations are often among those captured by light traps and nets, and some are quite abundant compared to other species in overall samples. The bluehead wrasse and surgeonfishes are among the most common larvae captured and are resident aggregators with long spawning seasons. Others, typically transient aggregators, such as most of the groupers, are not so common, and make it difficult for

Fig. 7.5 Pelagic juvenile Nassau grouper, *Epinephelus striatus*, photographed after capture in a channel net at Lee Stocking Island, Bahamas. The fish is about 15 mm in standard length and has the transparent body characteristic of fish in the pelagic environment. The gut is covered by silver reflective pigment. Within a few days the fish acquire the pigment typical of juvenile (Photo: Patrick L. Colin)

extensive studies to be done using them that relate directly to the outcomes of aggregations. However, there are a few reports of very large settlement pulses of aggregating species (see Sect. 7.6.3). Some specialized sampling, such as channel nets, can effectively capture some species (Colin et al. 1997; Keener et al. 1988). Crest nets (established on tops of reef using waves and tidal currents to capture fishes) are an alternative method to sample incoming reef fishes, but have not been particularly effective at sampling aggregation spawned fishes with the only family contributing substantially to catches being the surgeonfishes (Dufour and Galzin 1993, Hair et al. 2000).

7.6.1 Settlement and Metamorphosis

The larvae of most species transition into the juvenile stage while still in mid-water and may be able to position themselves in the vicinity of reefs during this stage as many are strong swimmers in both speed and duration (Leis and McCormick 2002; Leis and Carson-Ewart 1997). Stobutzki and Bellwood (1997) found unfed settlement stage individuals of three families (surgeonfishes, snappers, emperors – Lethrinidae) with spawning aggregations to have the highest swimming endurance (time capable of maintaining 13.5 cm s^{-1}) among nine families. If settlement stage fish are fed while swimming, their endurance increases several-fold and some were capable of continuous swimming at 13.5 cm s^{-1}. The settlement stage fishes of many aggregation spawning families tend to be large and capable (Doherty et al. 1994), putting them among those species most likely to be able to move towards settlement habitat. In studies where Nassau grouper pelagic juveniles (Fig. 7.5) were captured while recruiting in channel nets (Shenker et al. 1993; Colin et al. 1997) they were by far the largest and most robust of the many pelagic stages captured (Thorrold et al. 1994), and one of the few species which consistently survived the trauma of net capture and subsequent handling.

Settling fish may "try out" a reef and then quickly return to the water column, not remaining on the first reef they encounter. Once a suitable settlement location is found, the juvenile undergoes metamorphosis, in which it changes its body structure and physiology, losing structures that contribute to survival in the pelagic (transparency, spination), and acquire characters useful in the benthic, environment (pigmention, changes in mouth orientation, changes in alimentary tract for a different diet). Some evidence suggests that fishes stop feeding during metamorphosis, using internal reserves to support themselves for a period of several days.

The settlement behaviour and selection of settlement habitat of some aggregation spawning predatory fishes have been examined (Eggleston 1995; Wright et al. 2005; Nakamura et al. 2009). Quere and Leis (2010) found the Spanish flag snapper, *Lutjanus caponotatus,* to have complex settlement behaviour with strong swimming capabilities and the ability to interact with a variety of reef residents. Leis and Carson-Ewart (1999) reported similar settlement behaviour in the squaretail coralgrouper, *Plectropomus areolatus*.

In the absence of a suitable reef, or any reef at all, how long can pelagic juveniles persist in the plankton? It appears that different species have different tolerances for extension of pelagic life (McCormick 1999b; Leis and McCormick 2002). For some aggregating species they seem to have fairly high limits, if the largest size of pelagic stages is any indication. Among groupers, Smith (1971: 77) reported two transparent "late larvae", one a Nassau grouper 45.6 mm long and a mutton hamlet, *Epinephelus afer*, 37.5 mm long (whether standard or total length not indicated), taken 8 miles from land in Bermuda and kept in aquaria until they transformed. Randall (1956) noted pelagic surgeonfish larvae (acronurus) as large as 60 mm (average size for settlement about 26 mm SL). Lara et al (2009) provided evidence of delayed metamorphosis in goliath grouper, *E. itajara*, with larval life, based on otolith increments and settlement marks, ranging from a low of around 30 days to as high as 60–72 days. Experimental evidence of delayed metamorphosis is uncommon. However, McCormick (1999a, b) caught settlement-ready convict surgeonfish, *Acanthurus triostegus*, acronuri at night and held them at 3–6 m depth over a 50 m deep bottom inside moored monofilament net cages which delayed their metamorphosis. Control fish, held in benthic cages, metamorphosed within 5 days.

The larval biology of freckled goatfish, *Upeneus tragula*, a widespread IWP goatfish, may be of particular relevance, since it has pelagic eggs. While it has not been observed to aggregate for spawning (some other goatfishes do), it is possible it will be found to aggregate, or that its ELH is similar to aggregating goatfishes. McCormick and Molony (1995) found in an experimental study that larvae of freckled goatfish settled earlier in warmer water with shorter body length and reduced body depth than those reared in cooler water.

7.6.2 Swimming Capabilities and ELH of Reef Fishes

Fisher (2005) indicated that larval reef fishes for several families would be able to "substantially influence their dispersal pattern" through swimming for the last half of their planktonic lives. Four of the families with the highest average swimming speeds have aggregating species (surgeonfishes, rabbitfishes, snappers, emperors). Despite this capability late in larval life, larvae of these same families are weak swimmers early in larval life. For example, Fisher (2005) examined swimming ability in 10 species of reef fishes, but only one, the brown surgeonfish, *Acanthurus nigrofuscus*, had a pelagic egg (also an aggregation spawner). The brown surgeonfish at hatching were the slowest swimmers with a value near zero (Fisher 2005, Fig. 2) reflecting the feeble ability of yolk sac larvae to distribute themselves or control their depth. Overall she found that larval reef fishes (minus brown surgeonfish) increase their swimming speed in an essentially linear relationship with age (for the species examined) and that during the latter half of larval life they swim sufficiently well to be able to influence their location in relation to average currents near reefs. The vast increase in swimming capability with growth needs to be put into the perspective of the entire ELH in that larvae with pelagic eggs will spend roughly the first half of their larval life without the ability to swim against average currents. Once they have the capability to swim robustly, their location in the ocean could be quite far from any shallow water habitat.

7.6.3 Do Aggregation-Spawning Species Recruit as Large Cohorts?

If large numbers of propagules are started simultaneously from a limited number of locations via aggregation spawning, does this produce a corresponding pulse of recruitment by a large settlement-ready cohort at the end of pelagic life? It would seem the prospects for large recruitment pulses from aggregation spawning, particularly from TAs, would be high. For resident aggregators, which often have lengthy spawning seasons, it seems less likely recruitment would be pulsed, unless survival of some cohorts are enhanced by processes during pelagic life. There is some limited evidence to support such a conclusion, but the connections between aggregation spawning and the occurrence of large cohorts is far from certain. Large pulses of recruitment might be produced by non-aggregation spawners which have a short spawning season. Such would probably be indistinguishable from aggregation spawning recruitment pulses. Differential survival in the pelagic stage among fishes which spawn would also tend to produce recruitment in large cohorts. Interestingly, some capture-based aquaculture fisheries are based on catching large pulses of settlement phase larvae (Sect. 8.2.3).

There are records of large recruitment pulses for reef fishes; many from aggregation spawning species or families. Such are known in surgeonfishes, groupers, wrasses, tetraodontiform fishes, bigeyes (Priacanthidae), Moorish idols (Zanclidae) and rabbitfishes (Fig. 7.6). In a large recruitment pulse, high numbers of settlement-ready fishes appear over just a few days and, based on locations where such pulses have been sampled, recruit over reef crests (Dufour and Galzin 1993) in the Indo-Pacific, through tidal channels from offshore waters (Keener et al 1988; Shenker et al. 1993;

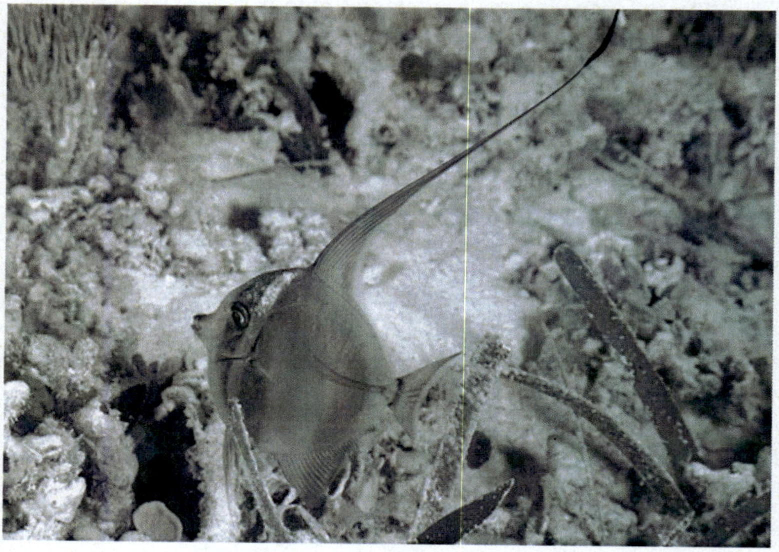

Fig. 7.6 Pelagic juvenile Moorish idol, *Zanclus cornutus*, at the time of recruitment. Photographed at night, the fish has begun to acquire the coloration typical of juvenile fish (copyright Mandy T. Etpison)

Colin et al. 1997) and simply by the appearance of vast numbers of juveniles (Fig. 7.7) where previously there were few or none (Letourneur et al. 1998; Chabanet et al 2005, PLC personal observation).

Instances of massive recruitment by surgeonfishes are well documented from a number of species known to be resident aggregation spawners (Randall 1961, Doherty et al. 2004, Robertson 1983). While not all family members are known to aggregate, it seems likely many others will be added as observations increase. The specialized acronurus larva is useful in that it is easily identified in plankton samples (Oxenford et al. 2008) and is highly adapted for pelagic life. Convict surgeonfish is known to have large fluxes ("regular broad peaks") of larvae across the reef (Dufour and Galzin 1993, Randall 1961). Planes et al. (1993) reported successive pulses in recruitment of convict surgeonfish in Tahiti, saying they "may reflect reproductive periodicity or result from physical oceanic processes". Sancho et al. (1997) documented a recruitment pulse of spotted surgeonfish, *Ctenochaetus strigosus*, at Johnston Island. Similar pulses have been observed in Palau on several occasions, most recently in 2009 (Fig. 7.7). They appear as large schools of small individuals on both outer reef slopes and inshore areas, which they would have had to enter through small tidal channels. Such groups are usually subject to very high predation, as often there is not sufficient shelter for fishes (Doherty et al. 2004). The schools formed by the massive recruitment of two species of surgeonfishes in Palau in 2009 (Fig. 7.7) were subject to heavy predation by jacks (Carangidae) and other piscivores. Over a few weeks the size of school decreased and eventually disappeared, the fish presumably either eaten or having successfully taken up residence

Fig. 7.7 School of newly recruited *Ctenochaetus* sp. surgeonfishes in the lagoon inside the western barrier reef of Palau. The fish were present over an open sandy *bottom*. If approached the entire group would swim away from the snorkeler and in general are extremely easily disturbed (copyright Mandy T. Etpison)

elsewhere on the reef. Whether such large recruitment pulses originate from spawning in Palau is not known, but some of the species do aggregate for spawning in Palau. Other species can recruit directly into the reef in vast numbers, and may be subject to predator pressures, but their disappearance is not so easily detected (Fig. 7.8).

There are instances where recruitment occurrence is tied back to reproductive seasons by inference from knowledge of spawning season and recruitment periods. Such may occur for both aggregation and non-aggregating species, although in many cases whether or not aggregation spawning occurs is not known. Spawning by a large population of fish outside of aggregations over a short period of time, given that oceanographic mechanisms may serve to concentrate eggs or larvae, may be indistinguishable from aggregation spawning. Lunar effects may also be hard to see in such cases, although it is possible to back-date otoliths to determine time of spawning.

Pulse recruitment of the Nassau groupers in the Exuma Cays in the Bahamas, directly tied to aggregation spawning, has been documented using channel nets for a period of a few days about 1.5 months after spawning (Shenker et al. 1993; Colin et al. 1997). Additional recruitment may occur in areas other than those fed through tidal channels, but is harder to quantify as such recruitment is not so easily monitored (Colin et al. 1997). Doherty et al. (1994) documented pulse recruitment in the squaretail coralgrouper on the Great Barrier Reef. Doherty and McIllwain (1996)

Fig. 7.8 Cloud of newly recruited surgeonfishes, with a few damselfishes in the foreground, on the barrier reef of Palau, April 2009 (copyright Mandy T. Etpison)

found possible pulses of recruitment of some aggregation spawners (snappers) using channel nets at Ningaloo Reef, Western Australia, but for other families where pulses have been seen elsewhere, such as surgeonfishes, few settlement stage larvae were captured.

Relationships between spawning season and recruitment are known in other groupers. Keener et al (1988) showed a clear connection between aggregation spawning of the gag grouper, *Mycteroperca microlepis*, with appearance of juveniles over a few days in nets fished in a tidal inlet. Recruitment of some smaller groupers may not necessarily be a result of aggregation spawning. The coney, *E. fulvus*, spawns in small haremic groups without lunar timing during the winter in the Bahamas and new recruits appear at appropriate times where their otolith increments back-date to known spawning times (PLC unpublished data) so the settlement is seasonal, but does not appear to be pulsed. For other western Atlantic groupers believed to have limited spawning seasons there is a loose relationship between the known months of spawning and the appearance of juveniles 1–2 months later (PLC unpublished data)

The same is possibly true for some small IWP groupers. Dufour et al. (1996) found simultaneously large fluxes of honeycomb grouper, *Epinephelus merra*, for a few days across a number of seaward reef flats in Moorea after the new moon in February and March. Letourneur et al. (1998) and Chabanet et al (2005) reported massive recruitment of honeycomb grouper following storms at Reunion Island,

Indian Ocean with up to over 400 individuals per 20 m^2 area and populations falling to near normal levels (only a few percent of those originally seen) within a few weeks, either through predation on or migration of juvenile groupers. While not verified as an aggregation spawner, there is some evidence this grouper may indeed use that reproductive strategy (Chap. 5).

For wrasses large recruitment pulses are known in bluehead wrasse (Masterson et al. 1997) and California sheephead, *Semicossyphus pulcher* (Cowen 1985). Masterson et al. (1997) reported a large synchronous pulse of recruitment of bluehead wrasse to sites on 3 US Virgin Islands in 1 year, but not in 2 other years, while the spawning seasonality of the species varies across its range.

In the tetradontiform fishes (triggerfishes and pufferfishes-Monacanthidae) there are several records of massive and intermittent recruitment, although it is unknown if many species in this group have aggregation spawning. Robertson (1988) reported large-scale recruitment of queen triggerfish, *Balistes vetula*, in Caribbean Panama. Stimson (2005) documented occurrence of large episodic recruitments of the fantail filefish, *Pervagor spilosoma*, to reefs in Hawaii, but its spawning mode is not known. Rabbitfishes are also known to have aggregation spawning and massive recruitments. In Guam, Kami and Ikehara (1978) reported scribbled rabbitfish, *Siganus spinus*, and forktail rabbitfish, *S. argenteus*, juveniles to appear on the reef flats a few days before or after the last quarter moon in April and May. There are reports of "balls" of siganid juveniles occurring in deep reef waters. Aggregation spawning and large recruitment pulses are known for rabbitfish in Palau (Chap. 12.22).

Fishes in some other families, such as Hawaiian bigeye, *Priacanthus meeki*, and Moorish idol, *Zanclus cornutus*, have had bursts of abundance on Hawaiian reefs with subsequent disappearance or dropping back to normal levels (Stimson 2005). It seems likely the Moorish idol may well have aggregation spawning in at least some areas (Palau, PLC personal observation), but nothing is known about bigeye spawning (Chap. 12.19).

7.6.4 Is Mass Spawning of Marine Invertebrates Comparable to Reef Fishes and Does It Provide Any Insight into Reef Fish Pelagic Life?

Many marine invertebrates have a dispersive larval stage, and some spawn at discrete times and places *en masse* (Babcock et al. 1992). While a review of this subject is beyond the scope of this chapter, a few comparisons are useful, particularly when invertebrate spawning has been compared to fish aggregation spawning (see also Chap. 3).

The mass spawning of stony corals (Scleractinia) has been documented from many locations throughout the tropics (Baird et al. 2009), and is often limited in seasonal and lunar periods. Van Woesik (2010), reporting that corals spawn at times of calm wind conditions in different areas, suggested that "spawning during regionally calm periods may also be applicable to other marine organisms, such as fishes" citing Johannes (1981). While this might seem a useful comparison to fish aggregation spawning, the similarities are limited. Many species of stony corals release sperm and egg bundles over a period ranging from a few minutes to as much as an hour relying on eventual mixing and fertilization in the water column. Corals do not, in the sense of reef fishes, aggregation spawn, as they can not migrate to a spawning area, but release gametes over their entire range within a reef area. The sex products are quite buoyant, compared to pelagic fish eggs, and quickly form slicks on the surface that are carried by currents or waves. Wolanski and Hamner (1988) point out that "coral eggs are larger and more buoyant than most other eggs from animals that have a planktonic stage and will tend to float even in areas of downwelling" and that waves disperse larvae across the water's surface. Coral spawn can be thought of more as an oil slick on the surface, than the discrete fertile eggs of fishes which might be distributed in the upper few metres of water (Hamner et al. 2007). Only later (many hours to a few days) do the fertile eggs transform into larvae which leave the surface and swim in the water column for a short larval life, in some only a few days. It might also be relevant that no reef fish is known to synchronize its spawning with that of stony corals.

Some crabs and lobsters are probably the invertebrates whose reproductive and larval lives are most comparable to those of reef fishes. This similarity was also noted by Arvedlund and Kavanagh (2009) who stated "almost all demersal tropical teleost fishes have pelagic larvae that may disperse, in common with most tropical marine decapod larvae". For crabs and lobsters, their already fertilized eggs hatch to swimming larvae, often *en masse*, live in the water column, and have to feed and find settlement habitat at the end of the larval life; all comparable to reef fishes. In Palau, Hamner et al. (2007) found on some days very large numbers of crab larvae among zooplankton normally dominated by planktonic fish eggs largely from aggregation spawning. The mass larval release of terrestrial coconut and Coenobitide hermit crabs are similar to aggregation spawning of reef fishes. As detailed by Fletcher (1993) ovigerous coconut crabs with already fertilized eggs move to shore. "Spawning" takes place after sunset when high tide corresponds with dusk, usually on a semi-lunar pattern with the first and last quarters of the moon. Coconut crabs go through 4–5 larval stages requiring 15–28 days, depending on temperature, and the final larval stage must find a terrestrial environment into which to recruit and emerge from the ocean. Marine crabs are similar, however, although they may release their larvae anywhere in the ocean. The oceanographic mechanisms that might tend to concentrate or enhance survival of fish larvae also affect crab and spiny lobster larvae (Eggleston et al. 1998).

7.7 What Do We Know About ELH for Aggregation Spawning Fishes?

While knowledge of the larval biology of coral reef fishes has advanced significantly in the last decade, much of it concerns species with demersal eggs, such as damselfishes, or small fishes with pelagic eggs (e.g. bluehead wrasse) with little subsistence or commercial value, other than as aquarium trade fishes in some cases. We remain largely ignorant of much of the ELH of those larger reef fishes with aggregations. As Quere and Leis (2010) point out "larger species of importance to commercial and recreational fisheries (e.g. groupers, snappers, grunts-Haemulidae) are particularly poorly represented in studies of larval reef-fish behaviour". There is fragmentary information on a few aggregating species, such as Nassau grouper, but along with concerns regarding human impacts on aggregations themselves, there is a pressing need for studies of these (albeit difficult to work on) species of conservation and fisheries importance which are depleted or near extinction in many areas.

At present, there do not seem to be major differences in or benefits documented for ELH from aggregation versus other spawning strategies. Egg sizes or types, based on the limited data available, do not appear to be correlated with aggregation occurrence. By the time larvae are ready to begin feeding, the densities of reef fish larvae with aggregation spawning are unlikely to be much different from other reef fishes (Chap. 6). Behavioural or physical processes may help to concentrate or disperse larvae, but are not documented for the larvae of aggregating fishes. With the designation of large numbers of marine protected areas in low latitude regions, often with the intention to protect reef fisheries, understanding the larval fish ecology in such regions is becoming increasingly important, but not receiving sufficient attention.

In many respects knowledge of ELH of aggregating species is best known for the early life stages. Many of these fishes are of aquaculture interest, and detailed studies looking at the effect of variables such as temperature, turbulence, light, and salinity, have shown that these can have major effects on survival of a cohort. When the results of culture studies, such as specificity of temperature for survival of Nassau grouper through the yolk sac stage, correspond to the conditions observed in the field during the limited spawning season, this implies that culture work can provide important insight into reasons for recruitment changes or failure. Coleman et al (1999) point out "because spawning and recruitment in aggregating spawners are episodic events, they are subject to the vagaries of environmental conditions, and unfavourable conditions such as low water temperature during planktonic stages can drastically reduce year-class strength".

The existence of critical periods, part of the match/mismatch hypothesis, does seem to occur in ELH. As long as conditions are suitable for life through the yolk sac stage, the initiation of feeding is the first major bottleneck in life history. Once ready to feed the time window to begin feeding is relatively short (Yoseda et al. 2006). Being ready to start feeding at sunset is probably not beneficial to early larvae, which are also unlikely to be able to feed at night, and having to wait 12 h before being able to feed will probably ensure that most larvae will soon starve.

The eggs and yolk sac larvae are 'drifters' initially not active in promoting their retention or dispersal. These are largely determined by where and when they are spawned and it is likely drifters or dye studies could provide reasonable estimates of transport. At the end of the yolk sac stage, swimming ability is limited and directed towards feeding and perhaps making diel vertical migrations in response to light. Passive drifting raises the problem of "wash out", in which larvae would be increasingly dispersed from their natal region and some mechanisms described in Chap. 6 which serve to limit this. Once feeding is successfully initiated, for reef fish larvae in general, under normal conditions starvation may not be an overriding factor in mortality (Llopiz and Cowen 2009a, b). Potentially, predation may have equal or greater importance in larval mortality.

The relative importance of physical versus biological mechanisms of retention will likely change as presettlement fishes develop (Leis and McCormick 2002). Once established in the pelagic environment early stage larvae transition within 2 weeks or less to post-flexion larvae, capable of more extensive swimming and daily vertical movements. Post-flexion larvae may be able to seek out and stay with concentrations of food items, perhaps seeking out convergent fronts with more food items, but also potentially more larval predators. Structures, such as large fin spines in grouper/snapper larvae and spines in acanthurids, may develop that reduce the potential for predation and enhance survival in shallow oceanic waters. Larvae remain nearly transparent, reflective pigment found over the gut area and only those areas necessary to be pigmented, such as the eyes, are easily visible.

While middle and late-stage larval fishes from aggregations are able to swim capably (Stobutzki and Bellwood 1997), there is, as yet, no strong evidence they school (Leis 2006) or utilize the epibenthos during their ELH. Leis (2006) points out "where occupancy of the epibenthic boundary layer has been looked for, it has generally been found, but only in a minority of the species" (e.g. gobiids, sciaenids). For most reef fishes available evidence indicates ELH stages occurring well above the epibenthic. Also, while the possibility of active navigation to settlement habitat through sensory means (hearing, olfaction, wave motion detection, rheotaxis, magnetic sense) has been suggested by numerous authors (Tolimieri et al. 2000; Leis and Carson-Ewart 2003; Simpson et al. 2004; Leis and Lockett 2005; Montgomery et al. 2006; Gerlach et al. 2007; Dixson et al. 2008), there are numerous caveats evident in any application of such information in a broad oceanographic sense. For example, Heenan et al. (2008) commented that all earlier studies on the response of settlement stage larvae using acoustic playback to assess auditory attraction had been done in the same location, often with the single same reef sound recording, and use of a reef sound recording from a different area, did not attract fishes. Most studies have relied on a small suite of families, usually fishes with demersal or orally-brooded eggs, with experiments being carried out either in small flumes or in reef waters on broad continental shelves (Great Barrier Reef lagoon), and are considered applicable only in situations where reefs are from a few 100 of metres to perhaps 1 km away from settlement-stage larvae.

Whether the sensory abilities of settlement stage pelagic larvae are capable of finding reefs from many miles at sea has not been conclusively demonstrated. New efforts, such as the "orientation with no frame of reference" effort (Paris et al. 2008) hold promise of providing detailed information on the behaviour of larvae at sea and how they use their sensory capabilities to locate settlement habitat. Retention mechanisms may help to keep late stage larvae within recruitment range of some reefs, but the occurrence of settlement stage fish far at sea shows the "leaky" nature of the system. For most aggregation spawned fishes "self recruitment" is unlikely, however for some there is a reasonable chance some larvae might recruit into an area that constitutes the catchment for migration to their natal aggregation site. Detailed work on the genetics of aggregation fishes, particularly those with TAs, may illustrate the presence of meta-populations in these fishes.

7.8 Future Work

There is need for more field work documenting spawning output from aggregations, and the short term transport and entrainment of propagules following through to recruitment. There are some outstanding examples of projects that are doing an excellent job of providing at least part of the picture of the importance of aggregations. Work in US Virgin Islands (Beets and Friedlander 1998; Nemeth et al. 2007, 2008) on groupers and snappers has provided a wealth of new information and is perhaps the best example of how protection of aggregations can allow recovery to occur. They have not, as yet, expanded the work to include most of ELH and recruitment. The project in the Cayman Islands looking at Nassau grouper ("The Grouper Moon") project, a relatively small-scale effort with modest funding, is an outstanding example of the type of work that needs to be done, as it has attempted to link aggregation and spawning with the pelagic stage ideally through to benthic recruitment and is being conducted over a multi-year period (Whaylen et al. 2004, 2007). That study takes advantage of an existing, protected aggregation to learn what cannot otherwise be learned, but is applicable elsewhere. Similarly, work on spawning aggregations in Belize, although effort has been variable year to year, has provided much information (Heyman et al. 2005; Starr et al. 2007; Sala et al. 2001) but needs to be continued and make the transition to include studies of the fate of larvae at sea. Research on snappers in areas like Riley's Hump in Florida (Burton et al. 2005) could be expanded in seasonal duration, geographic scope (to more aggregation locations) and continued on into the pelagic stage. Domeier (2004) released drift vials at Riley's Hump at the projected time of spawning to demonstrate the high probability of recruits from that aggregation reaching much of South Florida, yet his effort has not been continued by later researchers documenting aggregation at the site. For so many species, we know bits and pieces about their aggregation and spawning, but the whole picture of their ELH remains unclear. These "whole picture" efforts are needed to include as many locations as possible where aggregations still exist.

Aggregations potentially can be approached from a fisheries oceanography standpoint, by targeting studies during the periods when the largest aggregators are spawning. Methods for doing this are well known from fisheries work in more temperate latitudes and could be applied with some modifications for the tropics. In an area like Palau, while there is a relatively substantial amount known about aggregations, there has been no work done on ichthyoplankton or fisheries oceanography.

While large TAs have a mystique that makes them appealing phenomena to investigate, the smaller and more numerous RA provide many more opportunities to increase understanding of aggregations. RAs occur regularly in many locations and are often easily accessed from field stations. They allow collection of spawning fishes, their gonads and released eggs plus investigations of the relationship between environment and reproduction. RAs are often still found in areas where TAs have largely been eliminated by overfishing. Research methods can often be developed first on RA and then applied to TA at a later time.

The interest of the aquaculture industry in rearing larger reef fish species, many with aggregations, will provide valuable information on those factors important in ELH, as long as those results are published in the open literature. The changes in survival during culture through the yolk sac stage brought about by minor differences in temperature, light, salinity and turbulence (Watanabe et al. 1995, 1996, 1998; Ellis et al. 1997; Sugama et al. 2004; Yoseda et al. 2006) imply those same factors are acting in the ocean and provide a backdrop for assessing environmental conditions when aggregation and spawning occurs. They certainly point to the first week of life being the most sensitive for reef fish with pelagic eggs. A vigorous assessment of the role of temperature regimes in relation to timing of spawning over geographic ranges is needed; there are few data that cover the entire seasonal temperature regime where aggregation spawning occurs.

There is need to expand studies of recruitment by larger, commercially important reef fishes. Light traps are useful for some, but not all, species. Methods such as channel nets, reef crest nets or other passively fishing nets have proven useful. The success of capturing settlement stage Nassau grouper using channel nets in the Bahamas (Shenker et al 1993; Colin et al. 1997) as well as South Carolina (Keener et al. 1988) has not been replicated (attempted) elsewhere. Given the relatively small settlement window known for the species, it would be relatively easy to replicate this effort in other regions with tidal currents during the expected (and other) times of settlement.

The role of maternal nutrition and egg quality (McCormick 1998; Green and McCormick 2005; Green 2008) merits careful attention among aggregating fishes. A benefit derived from aggregation spawning (particularly for transient species) may result from increases in egg quality, through nutritional factors, limited spawning season, and intraspecific behaviour. This could be tested in a number of ways, from comparing nutritional and hormonal status of ovaries of captured fishes, capture of eggs after spawning (Chap. 9), to experimental work. Since egg quality in some reef fishes with demersal eggs can increase ELH survival and growth of larvae, as well as reduce the time in the pelagic phase, this may be of major importance in the evolution and maintenance of aggregation spawning.

References

Akatsu S, Al-Abdul-Elah KM, Teng SK (1983) Effects of salinity and water temperature on the survival and growth of brown spotted grouper larvae (*Epinephelus tauvina*, Seranidae). J World Maric Soc 14:624–635

Appeldoorn RS, Hensley DA, Shapiro DY, Kioroglou S, Anderson BG (1994) Egg dispersal in a Caribbean coral reef fish. *Thalassoma bifasciatum*. II Dispersal off the reef platform. Bull Mar Sci 54:271–280

Armsworth P (2001) Directed motion in the sea: efficient swimming by reef fish larvae. J Theor Biol 210:81–91

Arvedlund M, Kavanagh K (2009) The senses and environmental cues used by marine larvae of fish and decapod crustaceans to find tropical coastal ecosystems. In: Nagelkerken I (ed) Ecological connectivity among tropical coastal ecosystems. Springer Netherlands

Babcock R, Mundy C, Keesing J, Oliver J (1992) Predictable and unpredictable spawning events: in situ behavioural data from free-spawning coral reef invertebrates. Invertebr Reprod Dev 22(1–3):213–228

Baird AH, Guest JR, Willis BL (2009) Systematic and biogeographical patterns in the reproductive biology of scleractinian corals. Annu Rev Ecol Evol Syst 40:551–571

Beets J, Friedlander A (1998) Evaluation of a conservation strategy: a spawning aggregation closure for red hind, *Epinephelus guttatus*, in the U.S. Virgin Islands. Environ Biol Fish 55:91–98

Bell LJ, Colin PL (1986) Mass spawning of *Caesio teres* (Pisces: Caesionidae) at Enewetak Atoll, Marshall Islands. Environ Biol Fish 15:69–74

Bergenius MAJ, Meeken MG, Robertson DR, McCormick ML (2002) Larval growth predicts recruitment success of a coral reef fish. Oecologia 131:512–525

Berkeley SA, Chapman C, Sogard SG (2004) Maternal age as a determinant of larval growth and survival in a marine fish, *Sebastes melanops*. Ecology 85(4):1258–1264

Burton ML, Brennan KJ, Muñoz RC Jr (2005) Preliminary evidence of increased spawning aggregations of mutton snapper (*Lutjanus analis*) at Rileys Hump two years after establishment of the Tortugas South Ecological Reserve. Fish Bull 103:404–410

Chabanet P, Moyne-Picard M, Pothin K (2005) Cyclones as mass settlement vehicles for groupers. Coral Reefs 24(1):138

Chioccara F, Della Gala A, de Rosa M, Novellino E, Prota G (1980) Mycosporine amino acids and related compounds from the eggs of fishes. Bull Soc Chim Belge 89:1101–1106

Clarke ME, Domeier ML, Laroche WA (1997) Development of larvae and juveniles of the mutton snapper (*Lutjanus analis*), lane snapper (*Lutjanus synagris*) and yellowtail snapper (*Lutjanus chrysurus*). Bull Mar Sci 61(3):511–538

Coleman FC, Koenig CC, Eklund AM, Grimes CB (1999) Management and conservation of temperate reef fishes in the grouper-snapper complex of the southeastern United States. Am Fish Soc Symp 23:233–242

Colin PL (1983) Spawning and larval development of the hogfish, *Lachnolaimus maximus* (Pisces: Labridae). Fish Bull 80(4):853–862

Colin PL (1992) Reproduction of the Nassau grouper, *Epinephelus striatus*, (Pisces: Serranidae) and its relationship to environmental conditions. Environ Biol Fish 34:357–377

Colin PL (2010) Aggregation and spawning of the humphead wrasse, *Cheilinus undulatus* (Pisces: Labridae): general aspects of spawning behaviour. J Fish Biol 76:987–1007

Colin PL, Koenig CC (1996) Spines in larval red grouper, *Epinephelus morio*: development and function. Proc Gulf Caribb Fish Inst 44:31–38

Colin PL, Laroche WA, Koenig CC (1996) Development from egg to juvenile of the red grouper, *Epinephelus morio* (Pisces: Serranidae) in the laboratory. In: Arreguin-Sanchez F, Munro JL, Balgos MC, Pauly D (eds) Biology, fisheries and culture of tropical groupers and snappers, ICLARM conference proceedings 48. International Center for Living Aquatic Resources, Manila.

Colin PL, Laroche WA, Brothers ED (1997) Timing of ingress and settlement in the Nassau grouper, *Epinephelus striatus*, (Pisces: Serranidae), with relationship to spawning occurrence. Bull Mar Sci 60(3):656–667

Cowen RK (1985) Large-scale pattern of recruitment by the labrid *Semicossyphsus pulcher*: causes and implications. J Mar Res 43:719–742

Cowen RK (2002) Larval dispersal and retention and consequences for population connectivity. In: Sale PF (ed) Coral reef fishes: dynamics and diversity in a complex ecosystem. Academic, New York

Cowen RK, Sponaugle S (2009) Larval dispersal and marine population connectivity. Annu Rev Mar Sci 1:443–466

Czesny SJ, Graeb BDS, Dettmers JM (2005) Ecological consequences of swim bladder noninflation for larval yellow perch. Trans Am Fish Soc 134:1011–1020
Dixson DL, Jones GP, Munday PL, Planes S, Pratchett MS, Srinivasan M, Syms C, Thorrold SR (2008) Coral reef fish smell leaves to find island homes. Proc R Soc B 275:2831–2839
Doherty P, McIllwain J (1996) Monitoring larval fluxes through the surf zones of Australian coral reefs. Mar Freshw Res 47:383–390
Doherty PJ, Dufour V, Galzin R, Hixon MA, Meekan MG, Planes S (2004) High mortality during settlement is a population bottleneck for a tropical surgeonfish. Ecology 85:2422–2428
Doherty PJ, Fowler AJ, Samoilys MA, Harris DA (1994) Monitoring the replenishment of coral trout (Pisces: Serranidae) populations. Bull Mar Sci 54(1):343–355
Domeier, ML (2004) A potential larval recruitment pathway originating from a Florida marine protected area. Fish Oceanog 13:287–294
Drass DM, Bootes KL, Lyczkowski-Shultz J, Comyns BH, Holt GJ, Riley CM, Phelps RP (2000) Larval development of red snapper, *Lutjanus campechanus*, and comparisons with co-occurring snapper species. Fish Bull 98:507–527
Dufour V, Galzin R (1993) Colonization patterns of reef fish larvae to the lagoon at Moorea Island, French Polynesia. Mar Ecol Prog Ser 143–152
Dufour V, Riclet E, Lo-Yat A (1996) Colonization of reef fishes at Moorea Island, French Polynesia: temporal and spatial variation of the larval flux. Mar Freshw Res 47:413–422
Eggleston DB (1995) Recruitment in Nassau grouper *Epinephelus striatus*: post-settlement abundance, microhabitat features, and ontogenetic habitat shift. Mar Ecol Prog Ser 124:9–22
Eggleston DB, Armstrong DA, Elis WE, Patton WS (1998) Estuarine fronts as conduits for larval transport: hydrodynamics and spatial distribution of Dungeness crab postlarvae. Mar Ecol Prog Ser 164:73–82
Ellis EP, Watanabe WO, Ellis SC, Ginzoa J, Moriwake A (1997) Effect turbulence, salinity and light intensity on hatching rate and survival of larval Nassau grouper, *Epinephelus striatus*. J Appl Aquac 7(3):33–43
Ferraris CJ (1985) Redescription and spawning behavior of the muraenid eel *Gymnothorax herrei*. Copeia 1985:518–520
Fisher R (2005) Swimming speeds of larval coral reef fishes: impacts on self-recruitment and dispersal. Mar Ecol Prog Ser 285:223–232
Fletcher WJ (1993) Coconut crabs. In: Wright A, Hill L (eds) Nearshore marine resources of the South Pacific. Institute of Pacific Studies, University of the South Pacific, Suva, pp 643–681
Gerlach G, Atema J, Kingsford MJ, Black KP, Miller-Sims V (2007) Smelling home can prevent dispersal of reef fish larvae. Proc Natl Acad Sci USA 104:858–863
Green BS (2008) Maternal effects in fish populations. Adv Mar Biol 54:1–105
Green BS, McCormick MI (2005) Maternal and paternal effects determine size, growth and performance in larvae of a tropical reef fish. Mar Ecol Prog Ser 289:263–272
Grorud-Colvert K, Sponaugle S (2006) Influence of condition on behavior and survival potential of a newly settled coral reef fish, the bluehead wrasse *Thalassoma bifasciatum*. Mar Ecol Prog Ser 347:279–288
Hair CA, Doherty PJ, Bell JD, Lam M (2000) Capture and culture of presettlement coral reef fishes in Solomon Islands. In: Proceedings of the International Coral Reef Symposium, vol 2, Bali, pp 819–827
Hamilton SL, Regetz J, Warner RR (2008) Postsettlement survival linked to larval life in a marine fish. Proc Natl Acad Sci USA 105:1561–1566
Hamner WM, Colin PL, Hamner PP (2007) Export-import dynamics of zooplankton on a coral reef in Palau. Mar Ecol Prog Ser 334:83–92
Heenan A, Simpson SD, Braithwaite VA (2008) Testing the generality of acoustic cue use at settlement in larval coral reef fish. In: Proceedings of the International Coral Reef Symposium, Ft Lauderdale, pp 554–558
Hensley DA, Appeldoorn RS, Shapiro DY, Ray M, Turingan RG (1994) Egg dispersal in a Caribbean coral reef fish. *Thalassoma bifasciatum*. I. Dispersal over the reef platform. Bull Mar Sci 54:256–270

Heyman WD, Kjerfve B, Graham RT, Rhodes KL, Garbutt L (2005) Spawning aggregations of *Lutjanus cyanopterus* (Cuvier) on the Belize Barrier Reef over a 6 year period. J Fish Biol 67:83–101

Holt J, Riley C (1999) Laboratory spawning of coral reef fishes-effects of temperature and photoperiod. U.S.-Japan Coop Program in Natural Resources (UJNR) Technical Report 28, pp 33–38

Johannes RE (1981) Words of the Lagoon. Fishing marine lore in the Palau district of Micronesia. University of California Press, Los Angeles

Kami HT, Ikehara II (1978) Notes on the annual juvenile siganid harvest in Guam. Micronesica 12:323–325

Keener P, Johnson GD, Stender BS, Brothers EB (1988) Ingress of postlarval gag, *Mycterperca microlepis* (Pisces: Serranidae) through a South Carolina barrier island inlet. Bull Mar Sci 42:376–396

Kiflawi M, Mazeroll AI, Goulet D (1998) Does mass spawning enhance fertilization success in coral reef fish? A case study of the brown surgeonfish. Mar Ecol Prog Ser 172:107–114

Lara MR, Schull J, Jones DL, Allman R (2009) Early life history stages of goliath grouper *Epinephelus itajara* (Pisces: Epinephelidae) from Ten Thousand Islands, Florida. Endanger Species Res 7:221–228

Leis JM (2006) Are larvae of demersal fishes plankton or nekton? Adv Mar Biol 51:57–141

Leis JM, Carson-Ewart BM (1997) Swimming speeds of the late stage larvae of some coral reef fishes. Mar Ecol Prog Ser 159:165–174

Leis JM, Carson-Ewart BM (1999) In situ swimming and settlement behaviour of larvae of an Indo-Pacific coral-reef fish, the Coral Trout (Pisces, Serranidae, *Plectropomus leopardus*). Mar Biol 134:51–64

Leis JM, Carson-Ewart BM (eds) (2000) The larvae of Indo-Pacific coastal fishes, vol 2, Fauna Malesiana handbook. Brill, Leiden/Boston/Koln

Leis JM, Carson-Ewart BM (2003) Orientation of pelagic larvae of coral-reef fishes in the ocean. Mar Ecol Prog Ser 252:239–253

Leis JM, Lockett MM (2005) Localization of reef sounds by settlement-stage larvae of coral-reef fishes (Pomacentridae). Bull Mar Sci 76:715–724

Leis JM, McCormick MI (2002) The biology, behavior, and ecology of the pelagic, larval stage of coral-reef fishes. In: Sale PF (ed) Coral reef fishes: dynamics and diversity in a complex ecosystem. Academic, New York

Leis JM, Rennis DS (1983) The larvae of Indo-Pacific coral reef fishes. New South Wales University Press/Sydney and University of Hawaii Press, Honolulu

Leis JM, Trnski T (1989) The larvae of Indo-Pacific shorefishes. New South Wales University Press, Sydney

Leis JM, Sweatman HPA, Reader SE (1996) What the pelagic stages of coral reef fishes are doing out in blue water: daytime field observations of larval behavioural capabilities. Mar Freshw Res 47:401–441

Leis JM, Carson-Ewart BM, Hay AC, Cato DH (2003) Coral reef sounds enable nocturnal navigation by some reef fish larvae in some places at some times. J Fish Biol 63:724–737

Letourneur Y, Chabanet P, Vigliola L, Harmelin-Vivien M (1998) Mass settlement and post-settlement mortality of *Epinephelus merra* (Pisces: Serranidae) on Réunion coral reefs. J Mar Biol Assoc United Kingd 78:307–319

Llopiz JK, Cowen RK (2009a) The successful and selective feeding of larval fishes in the low-latitude open ocean: is starvation an insignificant source of mortality? ICES ICES CM 2009/T:14

Llopiz JK, Cowen RK (2009b) Variability in the trophic role of coral reef fish larvae in the oceanic plankton. Mar Ecol Prog Ser 381:259–272

Masterson CF, Danilowicz BS, Sale PF (1997) Yearly and inter-island variation in the recruitment dynamics of the bluehead wrasse (*Thalassoma bifasciatum*, Bloch). J Exp Mar Biol Ecol 214:149–166

McCormick MI (1998) Behaviorally induced maternal stress in a fish influences progeny quality by a hormonal mechanism. Ecology 79(6):1873–1883

McCormick MI (1999a) Experimental test of the effect of maternal hormones on larval quality of a coral reef. Oecologia 118:412–422
McCormick MI (1999b) Delayed metamorphosis of a tropical reef fish *(Acanthurus triostegus)*: a field experiment. Mar Ecol Prog Ser 176:25–38
McCormick MI, Molony BW (1995) Influence of water temperature during the larval stage on size, age and body condition of a tropical reef fish at settlement. Mar Ecol Prog Ser 118:59–68
McCormick MI, Nechaev IV (2002) Influence of cortisol on developmental rhythms during embryogenesis in a tropical damselfish. J Exp Biol 293(5):456–466
Miller MJ (2009) Ecology of anguilliform leptocephali: remarkable transparent fish larvae of the ocean surface layer. Aqua BioSci Monogr 2(4):1–94. doi:10.5047/absm.2009.00204.0001
Montgomery JC, Jeffs A, Simpson SD, Meekan M, Tindle C (2006) Sounds as an orientation cue for the pelagic larvae of reef fishes and decapod crustaceans. Adv Mar Biol 51:143–196
Nakamura Y, Shibuno T, Lecchini D, Watanabe Y (2009) Habitat selection by emperor fish larvae. Aquat Biol 6:61–65
Nemeth RS, Blondeau J, Herzlieb S, Kadison E (2007) Spatial and temporal patterns of movement and migration at spawning aggregations of red hind, *Epinephelus guttatus*, in the U.S.Virgin Islands. Environ Biol Fish 78:365–381
Nemeth RS, Kadison E, Blondeau JE, Idrisi N, Watlington R, Brown K, Smith T, Carr O (2008) Regional coupling of red hind spawning aggregations to oceanographic processes in the eastern Caribbean. In: Grober-Dunsmore R, Keller BD (eds) Caribbean connectivity: implications for marine protected area management, Marine Sanctuary Conservation Series ONMS-08-07. U.S. Department of Commerce, National Oceanic and Atmospheric Administration, Office of National Marine Sanctuaries, Silver Spring
Oxenford, HA, Fanning P, Cowen RK. (2008) Spatial distribution of surgeonfish (Acanthuridae) pelagic larvae in the eastern Caribbean. In: Grober-Dunsmore R, Keller BD (eds) Caribbean connectivity: implications for marine protected area management, Marine Sanctuaries Conservation Series ONMS-08-07. U.S. Department of Commerce, National Oceanic and Atmospheric Administration, Office of National Marine Sanctuaries, Silver Spring
Paris CB, Guigand CM, Irisson J-O, Fisher R, D'Alessandro E (2008) Orientation with no frame of reference (OWNFOR): a novel system to observe and quantify orientation in reef fish larvae. In: Grober-Dunsmore R, Keller BD (eds) Caribbean connectivity: implications for marine protected area management, Marine Sanctuaries Conservation Series ONMS-08-07. U.S. Department of Commerce, National Oceanic and Atmospheric Administration, Office of National Marine Sanctuaries, Silver Spring
Patterson HM, Swearer SE (2007) Long-distance dispersal and local retention of larvae as mechanisms of recruitment in an island population of a coral reef fish. Austral Ecol 32(2):122–130
Patterson HM, Kingsford MJ, McCulloch MT (2005) Resolution of the early life history of a coral reef fish using otolith chemistry. Coral Reefs 24(2):222–229
Petersen CW, Warner RR, Cohen S, Hess HC, Sewell AT (1992) Variable pelagic fertilization success: implications for mate choice and spatial patterns of mating. Ecology 73(2):391–401
Petersen CW, Warner RR, Shapiro DY, Marconato A (2001) Components of fertilization success in the bluehead wrasse, *Thalassoma bifasciatum*. Behav Ecol 12(2):237–245
Plack PA, Fraser NW, Grant PT, Middleton C, Mitchell AI, Thomson RH (1981) Gadusol, an enolic derivative of cyclohexane-1,3-dione present in the roes of cod and other marine fish. Biochem J 199:741–747
Planes S, Lefevre A, Legendre P, Galzin R (1993) Spatio-temporal variability in fish recruitment on a coral reef (Moorea, French Polynesia). Coral Reefs 12(2):105–113
Quere G, Leis JM (2010) Settlement behaviour of larvae of the Stripey Snapper, *Lutjanus carponotatus* (Teleostei: Lutjanidae). Environ Biol Fish 88(3):227–238
Randall JE (1956) A revision of the surgeonfish genus *Acanthurus*. Pac Sci 10(2):159–235
Randall JE (1961) A contribution to the biology of the convict surgeonfish of the Hawaiian Islands, *Acanthurus triostegus sandoicensis*. Pac Sci 15:215–272

Richards WJ (ed) (2006) Early stages of Atlantic fishes: an identification guide for the Western Central North Atlantic, vol 2. CRC Press, Boca Raton

Robertson DR (1983) On the spawning behavior and spawning cycles of eight surgeonfishes (Acanthuridae) from the Indo-Pacific. Environ Biol Fish 9:192–223

Robertson DR (1988) Extreme variation in the settlement of the Caribbean triggerfish *Balistes vetula* in Panama. Copeia 1988(3):698–703

Robertson DR (1996) Egg size in relation to fertilization dynamics in free-spawning tropical reef fishes. Oecologia 108:95–104

Russ GR, Lou DC, Ferreira BP (1996) Temporal tracking of a strong cohort in the population of a coral reef fish, the coral trout, *Plectropomus leopardus* (Serranidae: Epinephelinae), in the central Great Barrier Reef, Australia. Can J Fish Aquat Sci 53:2745–2751

Sala E, Ballesteros E, Starr RM (2001) Rapid decline of Nassau grouper spawning aggregations in Belize: fishery management and conservation needs. Fisheries 26(10):23–30

Sala E, Aburto-Oropeza O, Paredes G, Thompson G (2003) Spawning aggregations and reproductive behavior of reef fishes in the Gulf of California. Bull Mar Sci 72:103–121

Sampey A, McKinnon AD, Meekan MG, McCormick MI (2007) Glimpse into guts: overview of the feeding of larvae of tropical shorefishes. Mar Ecol Prog Ser 339:243–257

Sancho GD, Ma D, Lobel PS (1997) Behavioural observations of an upcurrent reef colonization event by larval surgeonfish *Ctenochaetus strigosus* (Acanthuridae). Mar Ecol Prog Ser 153:311–315

Shenker J, Maddox ED, Wishinski E, Pearl A, Thorrold SR, Smith N (1993) Onshore transport of settlement stage Nassau grouper (*Epinephelus striatus*) and other fishes in Exuma Sound, Bahamas. Mar Ecol Prog Ser 98:31–43

Simpson SD, Meekan MG, McCauley RD, Jeffs A (2004) Attraction of settlement-stage coral reef fishes to reef noise. Mar Ecol Prog Ser 276:263–268

Sinclair M (1988) Marine populations. An essay on population regulation and speciation. Washington Sea Grant, Seattle, WA (USA)

Smith CL (1971) A revision of the American groupers: *Epinephelus* and allied genera. Bull Am Mus Nat Hist 146(2):67–242

Sponaugle S, Pinkard DR (2004) Impact of variable pelagic environments on natural larval growth and recruitment of the reef fish *Thalassoma bifasciatum*. J Fish Biol 64:34–54

Sponaugle S, Grorud-Colvert K, Pinkard D (2006) Temperature-mediated variation in early life history traits and recruitment success of the coral reef fish *Thalassoma bifasciatum* in the Florida Keys. Mar Ecol Prog Ser 308:1–15

Starr RM, Sala E, Ballesteros E, Zabala M (2007) Spatial dynamics of the Nassau grouper *Epinephelus striatus* in a Caribbean atoll. Mar Ecol Prog Ser 343:239–249

Stimson J (2005) Archipelago-wide episodic recruitment of the file fish *Pervagor spilosoma* in the Hawaiian Islands as revealed in long-term records. Environ Biol Fish 72:19–31

Stobutzki IC, Bellwood DR (1997) Sustained swimming abilities of the late pelagic stages of coral reef fishes. Mar Ecol Prog Ser 149:35–41

Sugama K, Trijoko, Ismi S, Maha Setiawati K (2004) Effect of water temperature on growth, survival and feeding rate of humpback grouper (*Cromileptes altivelis*) larvae. In: Rimmer MA, McBride S, Williams KC (eds) Advances in grouper aquaculture, ACIAR Monograph Series 110:61–66

Thorrold SR, Shenker JM, Maddox ED, Mojica R, Wishinski E (1994) Larval supply of shore-fishes to nursery habitats around Lee Stocking Island, Bahamas. II. Lunar and oceanographic influences. Mar Biol 118:567–578

Thresher RE (1984) Reproduction in reef fishes. Tropical Fish Hobbyist Publications, Neptune

Tolimieri N, Jeffs A, Montgomery JC (2000) Ambient sound as a cue for navigation by the pelagic larvae of reef fishes. Mar Ecol Prog Ser 207:219–224

Van Woesik R (2010) Calm before the spawn: global coral spawning patterns are explained by regional wind fields. Proc R Soc B 277:715–722

Watanabe WO, Lee C-S, Ellis SC, Ellis EP (1995) Hatchery study of the effects of temperature on eggs and yolksac larvae of the Nassau grouper *Epinephelus striatus*. Aquaculture 136:141–147

Watanabe WO, Ellis SC, Ellis EP, Lopez VG (1996) Evaluation of first-feeding regimens for larval Nassau grouper *(Epinephelus striatus)* and preliminary, pilot-scale culture through metamorphosis. J World Aquac Soc 27:323–331

Watanabe WO, Ellis EP, Ellis SC, Chaves J, Manfredi C (1998) Artificial propagation of mutton snapper, *Lutjanus analis*, a new candidate marine fish species for aquaculture. J World Aquac Soc 29:176–187

Whaylen L, Pattengill-Semmens CV, Semmens BX, Bush PG, Boardman MR (2004) Observations of a Nassau grouper (*Epinephelus striatus*) spawning aggregation site in Little Cayman Island. Environ Biol Fish 70:305–313

Whaylen L, Bush P, Johnson B, Luke K, McCroy C, Heppell S, Semmens B, Boardman MR (2007) Aggregation dynamics and lessons learned from five years of monitoring at a Nassau grouper (*Epinephelus striatus*) spawning aggregation in Little Cayman, Cayman Islands, BWI. Proc Gulf Caribb Fish Inst 59:479–487

Wolanski E, Hamner WM (1988) Topographical controlled fronts in the ocean and their biological influences. Science 241:177–181

Wright KJ, Higgs DM, Belanger AJ, Leis JM (2005) Auditory and olfactory abilities of pre-settlement larvae and post-settlement juveniles of a coral reef damselfish (Pisces: Pomacentridae). Mar Biol 147:142

Yoseda K, Dan S, Sugaya T, Yokogi K, Tanaka M, Tawada S (2006) Effects of temperature and delayed initial feeding on the growth of Malabar grouper (*Epinephelus malabaricus*) larvae. Aquaculture 256(1–4):192–200

Chapter 8
Fishery and Biological Implications of Fishing Spawning Aggregations, and the Social and Economic Importance of Aggregating Fishes

Yvonne Sadovy de Mitcheson and Brad Erisman

Abstract This chapter explores the fishery and biological implications of exploiting aggregating marine fishes, their general importance to subsistence, commercial, and recreational fisheries and the possible consequences of losing them. We synthesize and examine empirical data from a wide range of taxa to determine whether, when and why fish spawning aggregations need to be targets of management. We examine the socioeconomic importance, costs and benefits of exploiting reef fish aggregations for both extractive and non-extractive (e.g. tourism and reproductive output) purposes. We provide recommendations and guidance for future research, education, management and conservation planning for aggregating species.

8.1 Introduction

This chapter explores the fishery and biological implications of exploiting aggregating marine fishes, their general importance to subsistence, commercial, and recreational fisheries and the possible consequences of losing them. We synthesize and examine empirical data from a wide diversity of species range of taxa to determine whether, when and why fish spawning aggregations need to be targets of management. Although we focus on reef fishes, we extensively use relevant or important examples from non-reef species, where informative, because many share the same problems of perception and lack of management; particularly illustrative examples

Y. Sadovy de Mitcheson (✉)
Division of Ecology & Biodiversity, University of Hong Kong, Pok Fu Lam Road, Hong Kong, China
e-mail: yjsadovy@hku.hk

B. Erisman
Marine Biology Research Division, Scripps Institution of Oceanography, University of California at San Diego, La Jolla, CA 92093-0202, USA
e-mail: braderisman@gmail.com

Y. Sadovy de Mitcheson and P.L. Colin (eds.), *Reef Fish Spawning Aggregations: Biology, Research and Management*, Fish & Fisheries Series 35,
DOI 10.1007/978-94-007-1980-4_8, © Springer Science+Business Media B.V. 2012

are given in more detail. We examine the socioeconomic importance, costs and benefits of exploiting reef fish aggregations for both extractive and non-extractive (e.g. tourism and reproductive output) purposes. Against this background, we provide recommendations and guidance for future research, education, management and conservation for aggregating species. We use the definition of spawning aggregation (transient and resident) as elaborated in Chap. 1.

Fishes that aggregate to spawn form an important component of marine fisheries and ecosystems globally. While most such species are caught throughout the year, it is commonly during the spawning season that landings volumes are greatest. Such aggregations naturally offer a welcome opportunity for large catches and easy earnings often representing a seasonal bounty to be shared by whole communities. They are targeted for recreational and tourist fishing charters, or may be important for cultural events. While most aggregations are exploited for fish flesh, for some species the more valuable target might be ripe gonads or swimbladders (Fig. 8.1). In some species the millions of young produced briefly each year by the concentrated spawning of hundreds to thousands of assembled adults support capture-based aquaculture (CBA) for which large numbers of newly settling or settled juveniles are collected *en masse* and grown out in captivity prior to sale. For these species, aggregations are the only known time that reproduction occurs and are thus important for maintaining their populations and the fisheries these support.

Unfortunately, many fishes that form spawning aggregations have undergone marked declines in their fisheries and these reproductive events are increasingly targeted (Sect. 8.2). Numerous documented examples, ranging from reefs to estuarine environments (Table 8.1), to the pelagic realm and deepwater seamounts, demonstrate unequivocally that aggregations can be substantially reduced by excess fishing effort over just a few spawning seasons leading to collapse of the fisheries they sustain, with important biological, social and economic implications.

Despite widespread declines globally, spawning aggregations of most species that form them are typically unmanaged. Among key unanswered questions that may account for such lack of attention, are (1) whether aggregations themselves need to be a focus of management as opposed to applying more conventional management (such as annual quotas or gear controls) to the fish populations of interest (Chap. 11); (2) whether reduced or extirpated aggregation sites can recover following management intervention and how long might be needed for recovery; and (3) the importance of healthy aggregations to reproduction, and hence the implications of aggregation declines and losses for harvested species. The strong appeal of aggregations as targets for fishing, their importance in many seasonal fisheries, the apparent abundance when aggregation catches are witnessed, and general lack of data on catches, together with the three above noted factors, make their management particularly challenging.

The preservation of sufficient spawning (reproductive) biomass in exploited populations is a core guiding principle in fisheries science for ensuring sustainable resource use. However, how this principle is best applied to aggregating species is unclear for two reasons. First, the extent to which declining trends in fisheries of aggregating species are due to exploitation of aggregations *per se* is uncertain: many

Fig. 8.1 Catches from spawning migrations or aggregations are valued for (**a**) ripe gonads (unknown mullet species) sold in Taipei market, Taiwan (Photo: Yvonne Sadovy de Mitcheson), (**b**) grouper flesh (*Epinephelus fuscoguttatus*) in Fiji (Photo: Randy Thaman), (**c**) croaker swimbladders (*Cynoscion othonopterus*) in Mexico (Photo: © Ismael Mascarenas)

aggregations, for example, occur in fish species that are naturally vulnerable to unmanaged fishing due to other life history attributes, such as longevity and late sexual maturation (Reynolds et al. 2005). Additionally, most species are also exploited during non-aggregation periods. Second, the potential impact of removing or disturbing ripe adults from aggregations prior to spawning is not known. In an attempt to better understand the implications of fishing on aggregations, and identify key data gaps, we explore the biological, social, and economic implications of

Table 8.1 Commercial examples of fisheries on aggregating species fished at aggregations – reef fishes and other examples

Species (common and latin) names	Family	Gear types	Country/Region	Fishery history, management, and conservation status	References
Nassau grouper *Epinephelus striatus*	Epinephelidae	Handline, longline, fish traps, speargun, gillnet	Caribbean-wide	One third of aggregations have been eliminated or reduced to negligible numbers. Disappearances have been documented in the Bahamas, Florida USA, Puerto Rico, US Virgin Islands, Honduras, Cuba, Belize, and Mexico	Sadovy and Eklund (1999), Chap. 12.6
Nassau grouper *Epinephelus striatus*	Epinephelidae	Handlines, spearguns, gill nets	Mahahual, Quintana Roo, Mexico	Aggregations of up to 15,000 fish formed each year at the same site, but due to increased fishing pressure in the 1990s, aggregations have not formed since 1996. Management not enforced	Aguilar-Perera (2007), Chap. 12.6
Nassau grouper *Epinephelus striatus*	Epinephelidae	Fish traps	Cuba	Targeted almost exclusively during aggregation periods; 20 of 21 historical aggregation sites no longer form. Management measures increasing but recovery not recorded	Claro et al. (2009), Chap. 12.6
Nassau grouper *Epinephelus striatus*	Epinephelidae	Handline, speargun, fish traps	Belize	Eighty percent decline in last 25 years (15,000 down to 3,000 fish) in aggregation size at Glover's Reef; Only 2 of 9 known aggregation sites remained by 2001; reduced from 30,000 to 1-5,000 fish. All known aggregation sites underwent dramatic declines in the abundance of spawning fish over the last 2 decades. Fishing particularly intense on spawning aggregations. Current aggregation protection does not appear to be restoring this species and a minimum capture size was recently introduced	Sala et al. (2001), Carter et al. (1994), Heyman and Requena (2002), Janet Gibson 2010, Chap. 12.6

Nassau grouper *Epinephelus striatus*	Epinephelidae	Handlines, traps, spearguns	Bermuda	Fished exclusively during aggregation periods. Commercial landings dropped from 75,000 tons in 1975 to less than 10,000 tonnes in 1981. The four known historical aggregation sites have ceased to form	Bannerot et al. (1987), Luckhurst (1996), Chap. 12.6
Goliath grouper *Epinephelus itajara*	Epinephelidae	Hook and line; spearfishing, trawl and longline bycatch	Gulf of Mexico	Aggregations of 100–150 fish on deepwater wrecks declined to 0–10 fish by 1989	Sadovy and Eklund (1999), Chap. 12.4
Goliath grouper *Epinephelus itajara*	Epinephelidae	Hook and line, spearfishing, trawl and longline bycatch	Florida USA	Fishery closed in 1990 due to rapid declines in catch and CPUE. Aggregations of several dozen fish were common off the East Coast of Florida in the 1950s and 1960s but were not observed anywhere in the 1990s. The population increased after the closure and the species is no longer considered to be of concern in US waters. A sevenfold decrease in the CPUE of Goliath grouper occurred between 1965 and 1969 in south Florida sport-fishery	Sadovy and Eklund (1999), McClenachan (2009), Chap. 12.4
Goliath grouper *Epinephelus itajara*	Epinephelidae	Hook and line, spearfishing,	Yucatan Peninsula, Mexico	Surveys from 2004 to 2006 revealed that Goliath grouper represent a negligible commercial fishery in the region. The absence of landed fish is attributed to widespread overfishing of this species from its spawning aggregations	Aguilar-Perera et al. (2009), Chap. 12.4
Goliath grouper *Epinephelus itajara*	Epinephelidae	Handlines, spearguns, traps,setlines, longlines, and drumlins	Belize	Adults are now rarely caught and the commercial fishery is dominated by juveniles (90% of catch)	Graham et al. (2009), Chap. 12.4

(continued)

Table 8.1 (continued)

Species (common and latin) names	Family	Gear types	Country/Region	Fishery history, management, and conservation status	References
Pacific goliath grouper *Epinephelus quinque fasciatus* (= *E. itajara*)	Epinephelidae	Spearguns	Gulf of California, Mexico	Landings dropped from 750 mt to less than 10 tonnes between 1980 and 1988. Aggregations have not been observed since 1995. Historically, fish were harvested almost exclusively during aggregations. However, in recent years spearfishers removed solitary fish throughout the year	Kira (1999) and Sala et al. (2004)
Gag grouper *Mycteroperca microlepis*	Epinephelidae	Trawl	Florida USA	Many aggregations disappeared between the1970s and 1980s; fished during and following spawning aggregations; Sex ratios shifted from 6:1 females to males in the 1970s under light fishing pressure to 30:1 in the 1990s; Commercial sales ban during spawning season implemented in 2000; In 2009 final rule outlines measures for recreational bag limits, commercial quotas, and closed season for all fishing from January to April for shallow water groupers (gag, black, red, scamp, red hind, rock hind, coney, graysby, yellowfin, yellowmouth, tiger). Dehooking tools required as necessary	Coleman et al. (1996), Koenig et al. (2000), Southeast Fishery Bulletin 2009 National Marine Fishery Service, USA, Chap. 12.7
Scamp *Mycteroperca phenax*	Epinephelidae	Trawl	Florida USA	Declines in landings and average size of landed fish; Percentage of large males decreased from 34% to 21%, median length decreased from 610 to 570 mm TL, and the percentage of fish 10 years or older decreased from 17% to 7% between 1979 and 1997; aggregations protected by recreational and commercial harvest ban during spawning season of shallow water groupers in 2009	Coleman et al. (1996), Harris et al. (2002); Southeast Fishery Bulletin 2009 National Marine Fishery Service, USA

Tiger grouper *Mycteroperca tigris*	Epinephelidae	Fish trap, spearfishing, handline	Puerto Rico	Commercial landings from aggregation site off Vieques Island dropped from 5 tonnes in 1995 to 2 mt in 1998 (60% decline). Sex-selective spearing focused on males to protect eggs. Site not protected despite declines	Matos-Caraballo et al. (2006) and Sadovy et al. (1994a, b)
Red Hind *Epinephelus guttatus*	Epinephelidae	Fish trap, spearfishing, handline	US Virgin Islands	Reduction in catch, skewed sex ratios (15 females:1 male), reduced size of aggregating fish between the 1970s and 1990s; fished primarily during aggregations; closure of the aggregation fishery resulted in increase in size of fish, shift in sex ratio (4:1) and increased size of landed fish on fishing grounds outside protected site. Average density and biomass of spawning fish increased by 60% in five years, and maximum density doubled	Beets and Friedlander (1999), Nemeth (2005), Chap. 12.3
Mutton snapper *Lutjanus analis*	Lutjanidae	–	Belize	Commercial fishers target aggregations in spring months. Major declines in CPUE (59%), mean landings per boat (22%), and median fish size reported over the last decade. Widespread population declines due to overfishing of aggregations reported as early as the 1940s. There is little effective protection	Graham et al. (2008), Rachel Graham 2010
Camouflage grouper *Epinephelus polyphekadion*	Epinephelidae	Hook and line, spear	Fiji	The species is intensively fished on a number of spawning aggregations. According to interviews, CPUE decreased significantly at four sites from about 300 kg in the 1990s to 100 kg in 2002-3 per trip. Some aggregation sites are no longer fished because numbers have declined substantially. The government programme to open ice plants in more remote areas could encourage aggregation fishing and legislation is in draft form to protect spawning aggregations of groupers	YS, unpublished data, Chap. 12.5

(continued)

Table 8.1 (continued)

Species (common and latin) names	Family	Gear types	Country/Region	Fishery history, management, and conservation status	References
Camouflage grouper, *Epinephelus polyphekadion*, brown-marbled grouper, *E. fuscoguttatus* and squaretail coralgrouper *Plectropomus areolatus*	Epinephelidae	Mainly hook and line	Solomon Islands, Western Province	These three species often aggregate together and are targeted for the live reef food fish export trade. The live reef fish trade started in 1994 with aggregations heavily targeted. Interviews suggested that after about two years most fishers noted declines in sizes of fish and numbers caught. There was also concern that ripe females taken for the live trade had high mortality rates. The export business stopped due to such concerns but restarted later	Johannes and Lam (1999), Chap. 12.2, 12.5, 12.8
Camouflage grouper, *Epinephelus polyphekadion*, brown-marbled grouper, *E. fuscoguttatus* and squaretail coralgrouper *P. areolatus*	Epinephelidae	Spear, hook and line	Palau, western Pacific	Several aggregations with these three species have been heavily exploited in central and southern Palau since at least the mid 1900s. Interviews with patriarch fishers suggest that landings have dropped significantly from over 1 tonne per trip to less than 100 kg on aggregations. The aggregations are protected from April to July each year when they cannot be fished or marketed but aggregating still occurs in August; there is little enforcement and substantial poaching	Sadovy (2007), YSM personal observation, Chaps. 12.2, 12.5, 12.8

Brown-marbled grouper and squaretail coralgrouper *Epinephelus fuscoguttatus* and *Plectropomus areolatus*	Epinephelidae	Spearfishing, hook and line	Southern Manus, PNG.	Fisher knowledge indicated declines in aggregations and in 2004 led the community to institute a lunar-based ban on spearfishing and commercial fishing at the aggregation site in the10 days leading up to and including the new moon in every month of the year. Only subsistence hook and line fishing was allowed. In early 2007 when the results of monitoring indicated that aggregation numbers had not improved a 1 year ban on all fishing on the site was introduced in the 10 days of each month of the year during which aggregations formed	Hamilton et al. (2004), Chaps. 12.2, 12.8
Gulf grouper *Mycteroperca jordani*	Epinephelidae	Gill nets, hook and line, spearguns	Gulf of California, Mexico	Comprised 45% of finfish production of Baja California Sur in 1960, but dropped to 6% by 1972 and to <1% by 2004. There has been a tenfold reduction in daily landings and CPUE between 1950s and 2000s, and entire fishery is comprised of juvenile fish. In 1960s, a fleet of six boats would land up to 63 tonnes/month from one site during peak aggregation season. Recent underwater surveys reveal a population of 3 adult Gulf grouper at the site. No targeted management related to protection of aggregations; no commercial limits or restrictions; daily catch limit for recreational fishers is not clear and not enforced	Saenz-Arroyo et al. (2005), Sala et al. (2003), Aburto-Oropeza et al. (2008)

(continued)

Table 8.1 (continued)

Species (common and latin) names	Family	Gear types	Country/Region	Fishery history, management, and conservation status	References
Leopard grouper *Mycteroperca rosacea*	Epinephelidae	Gill nets, speargun, hook and line	Gulf of California, Mexico	The most commercially important grouper in terms of landings and market value in the region. Increased fishing pressure on feeding and breeding aggregations correlated with widespread declines in landings and average size of landed fish. No targeted management related to protection of aggregations; no commercial limits or restrictions; daily catch limit for recreational fishers is not clear and not enforced	Sala et al. (2003, 2004), Aburto-Oropeza et al. (2008)
Leopard coralgrouper *Plectropomus leopardus*	Epinephelidae	Hook and line	Australia	*P. leopardus* is the most commercially important of several coral trout species in Australia. It is nowadays mainly taken for the high value live reef food fish export trade. In the last 20 years, annual commercial landings on the Great Barrier Reef (GBR) varied between 900 and 2,500 mt and the relatively small spawning aggregations are apparently not a specific fishing target. However, collapse of one spawning site was attributed to fishing. Management is by minimum sizes and short term seasonal closures at spawning time	Chap. 12.9

Squaretail coralgrouper *Plectropomus areolatus*	Epinephelidae	Hook and line	Kiribati, western Pacific	In parts of Kiribati, targeting spawning aggregations was traditionally practiced that became intensive in the 1980s when an Outer island Fisheries Project started to buy fish from the outer islands to sell in Tarawa, the capital. At the peak season, an estimated catch of 2 mt of fish (about 1,200 individuals) per day was normal. Underwater surveys showed that mean densities and sizes in the area declined from 0.13 individuals per 100 sq.m and a mean total length (TL) of 40 cm to 0.04 individuals per 100 sq.m and mean TL of 33 cm by 2004	Being Yeeting 2009, Chap. 12.8
Grey mackerel *Scomberomorus semifasciatus*	Scombridae	Gill net	Australia	This fishery historically concentrated on a peak spawning aggregation time and location. Since 1988 about 65% of the annual catch is taken in October from a single spawning site. While the fishery-dependent CPUE remains stable, this may be a case of hyperstability because anecdotal information suggests that the aggregation used to be much larger and that there were once similar aggregations throughout the GBR, which have now ceased to exist or have diminished substantially. In 2008, no grey mackerel schools were found at all and landings have fallen markedly in the last few years. In 2009 size limits and total allowable catch was introduced in Queensland east	Martin Russell David Cook, personal observations 2010

(continued)

Table 8.1 (continued)

Species (common and latin) names	Family	Gear types	Country/Region	Fishery history, management, and conservation status	References
Dusky rabbitfish *Siganus fuscescens*	Siganidae	Throw-nets (mainly), throw spears, surround nets, and spear guns while night fishing with flashlights	Palau	In Palau at least 38 locations are known as pre-spawning aggregation sites for this species. Fishermen position themselves along the migration path as the fish head to the outer reef and fish with throw nets. In the early 1990s fishers became concerned about the declines in the number and sizes of the sites and production has declined. The reasons for declines are not clear but heavy fishing of spawning migrations is very likely an important factor. In 1994 legislation ("Marine Protection Act") banned fishing for *S. fuscescens* from March 1 to May 31 during the supposed peak spawning season but in 2006 the period was reduced to 2 months to allow increase in landings. Despite protection, commercial landings are still reported throughout the March-May closed season	Chap. 12.22
Milkfish *Chanos chanos*	Channidae	Corralling of fish using superlights and dynamite	Philippines, central, Mactan Island	The species is heavily fished as adults during migrations in the spawning season each April for 1–3 weeks when many vessels come to Mactan Island to join local fishers to catch the fish, as well as caught during the settlement stage; 'bangus' fry have long been taken for mariculture growout. CPUE of bangus have declined markedly in the last two decades according to interviews. Philippines Fisheries Code of 1998 could provide some protection but is not enforced. Declines in bangus numbers could partly be due to heavy aggregation fishing	Amores (2003) report in www.SCRFA.org

Flathead grey mullet *Mugil cephalus*	Mugilidae	Hook and line and a wide range of net types; commercially gill and cast nets	Florida Gulf Coast, USA	Large numbers of mullet are taken during the migration to spawning grounds offshore. These fish are prized for their flesh and their roe is part of a large international trade. The Flathead grey mullet is marketed fresh, dried, salted, and frozen with the roe sold fresh or smoked. It is a very important commercial fish in many other parts of the world. During the autumn and winter months, adult mullet migrate far offshore in large aggregations to spawn. A net ban has been in effect in the state of Florida since 1995. Prior to the net ban amendment, mullet were severely overfished throughout the state. Currently, mullet are on the road to full recovery	http://www.flmnh.ufl.edu/fish/gallery/descript/stripedmullet/stripedmullet.html
Other examples					
Giant seabass *Stereolepis gigas*	Polyprionidae	Hook and line, spearfishing; gill net; longline	Baja California, Mexico and Southern California, USA	Commercial landings peaked at about 360 mt in the 1940s, dropped below about 90 mt in 1964, and collapsed by 1980s; recreational fishery went from 800 fish/year to less than 20 per year in 1980s; fished primarily during summer spawning aggregations that formed in kelp beds. No targeted management of aggregations, commercial limits or restrictions; daily catch limit for recreational fishers not clear and not enforced	Domeier (2001), Pondella and Allen (2008)

(continued)

Table 8.1 (continued)

Species (common and latin) names	Family	Gear types	Country/Region	Fishery history, management, and conservation status	References
White seabass *Atractoscion nobilis*	Sciaenidae	Gill nets	Southern California USA	Commercial landings fluctuated between 100 and 400 tonnes/year until the 1980s when landings plummeted by 90% by the 1980s. Commercial fisheries targeted spawning aggregations that formed from March to July; Commercial fishing ban during first half of spawning season (March to May), banning of nearshore gillnets (within 3 miles of coast), and 1 fish per day recreational limits implemented to protect spawning aggregations	Allen et al. (2007), Pondella and Allen (2008)
Gulf corvina *Cynoscion othonopterus*	Sciaenidae	Gill nets	Mouth of Colorado River Delta, northern Gulf of California, Mexico	Aggregation fishery is main source of income for the nearby town of Santa Clara for 2 months each year and involves the entire community. Generates 3,000 to 4,000 tonnes and c. $1 million USD in revenue from February to April each year and is main source of fish for the Easter Season in Mexico. Spawning aggregation site protected via a no fishing zone, but is not consistently enforced	Roman-Rodriguez (2000), Brad Erisman, upublished data
Totoaba *Totoaba macdonaldi*	Sciaenidae	Gill nets	Northern Gulf of California, Mexico	Commercial landings dropped from 2,500 tonnes in 1942 to 59 tonnes in 1975 when fishery closed; stocks have not recovered; fishery targeted winter/spring spawning aggregations that form near the mouth of the Colorado River Delta and fish migrating along coast to reach spawning sites. Listed on CITES Appendix I in 1975	Lercari and Chavez (2007); Cisneros-Mata et al. (1995)

Giant yellow croaker *Bahaba taipingensis*	Sciaenidae	Gill nets	Estuaries of Hong Kong and southern Mainland China.	Targeted mainly on seasonal aggregations in estuaries for its highly valuable swimbladder. The species is likely to be close to extinction and the rare catch nowadays of a single large individual often makes news headlines. Landings plummeted in the middle of the twentieth century. Endangered on the IUCN Red List	Sadovy and Cheung (2003)
Large yellow croaker *Larimichthyes crocea*	Sciaenidae	Trawl-dragged seine nets	Estuarine and offshore areas, mainly of Mainland China and Hong Kong	Once a very important fishery in coastal China and heavily fished since the 1950s with little effective management. Heavily fished during the spawning seasons and in deeper overwintering areas. Marked declines occurred in the 1980s and catches of wild fish are now uncommon	Liu and Sadovy (2008)

their exploitation. Specifically, we need to know: (1) the degree to which a fishery depends on aggregation catches and how the increased ease of capture at such times can promote overfishing; (2) the extent to which the subtleties of mating behaviour and other reproductive parameters are or might be affected by fishing activities on aggregations, (3) the social and economic challenges and implications of managing, versus losing, aggregations, and (4) the contribution of healthy aggregations to the overall productivity of the fishery.

Determining the appropriate management approaches for aggregating species is important because of the obvious appeal and attraction of fishing them, especially if this is a traditional or highly lucrative activity. The historical absence of controls on aggregation-fishing or collective experience of marked declines, however, makes their management uniquely challenging, as we shall illustrate. The need for steady market supplies may make a seasonal break in fishing, as could occur with management, unpopular. Combine these factors with the prevailing lack of information on the specific biological and economic impacts of fishing on aggregations *per se* and it becomes easier to appreciate why so few are adequately managed and protected. Even conservation-focused measures, such as marine protected areas, do not routinely consider aggregation protection except when species are already threatened with extinction and thus merit conservative management. It has also become clear that public perspectives and education about aggregations require considerably more attention.

8.2 Types of Extractive Exploitation

Many fish species taken in coastal fisheries of the tropics and subtropics, whether for subsistence, recreational, cultural or commercial purposes, aggregate at predictable times and places to spawn. Not all aggregating species are fished at their spawning aggregations, however. Some, like certain rabbitfishes (Siganidae), mullets (Mugilidae) and bonefishes (Albulidae) are specifically targeted during spawning migrations. Intensive fisheries for CBA on the massive larval settlement pulses of certain mullet, grouper (Serranidae), milkfish (Channidae) or rabbitfish species are partly possible due to aggregation-spawning.

This section provides a general overview of the subsistence, recreational and commercial uses of spawning aggregations with indications of management (covered more in Chap. 11) and landings trends. While some smaller aggregating species, like the blacktail snapper, *Lutjanus fulvus* (Chap. 12.10), are not specifically targeted (and therefore do not currently represent management challenges) we particularly focus on those species targeted heavily at or travelling to their aggregations, rather than on aggregating species in general, because: (1) aggregations or pre-spawning migrations of ripe fish are particularly vulnerable to overfishing and often make up a significant proportion of the annual catch; (2) aggregations or pre-spawning migrations are increasingly a specific target for fishing and can be very valuable; and (3) managing aggregation-fishing and conserving aggregations pose a unique challenge, both for conventional management as well as for marine protected area designation.

8.2.1 Subsistence and Traditional Fisheries

Native fishers have a long history of exploiting aggregating species for subsistence and cultural purposes. In the Pacific, subsistence use is best known from anecdotal accounts and a growing number of published studies based on fisher interviews (e.g. Johannes 1981; Hamilton 2005a, b; Sadovy 2007). Indeed, an annual cycle of seasonally plentiful resources typifies fisheries in many places with fishing pressure concentrated sequentially on different species for brief periods (Johannes 1981; Hickey 2006). In Niue and Fiji, Pacific, goatfish (Mullidae) are seasonally plentiful with eggs and associated with specific community activities (Vunisea 2005, Loraini Sivo personal communication 2010). Accounts of subsistence use in the Caribbean include the Nassau grouper, *Epinephelus striatus*, in southwestern Puerto Rico, which was once caught in massive numbers while aggregating and then salted to last for many months thereafter (Sadovy 1993).

Species that predictably migrate in large numbers close to shore *en route* to spawning areas, such as mullet, rabbitfish and bonefish, are often important for seasonal fishery events in which whole communities take part. In Manus, Papua New Guinea, the mullet *Crenimugil crenilabris* was regularly targeted for its roe and flesh. Communities gather during its pre-spawning migrations and catch large numbers using gill and hand nets, even dynamite (Hamilton 2003, 2005a, b). In Tarawa, Kiribati, bonefish have been heavily fished by gill nets on their spawning runs and of several runs, all but one yielded smaller and fewer fish between 1977 and the late 1990s. These changes were attributed to fishing and causeway construction that impeded coastal migrations (Johannes and Yeeting 2000). Regulations introduced in 1994 prohibited fishing during the 3 days before and after the full moon migration period and restricted fishing methods; some recovery in numbers was noted and there is now both recreational and subsistence use of the species (Being Yeeting personal communication 2009).

Much subsistence fishing of aggregations nowadays involves a commercial component and the line between the two activities is increasingly blurred. While subsistence fishing is often poorly documented, and volumes, species and other purely subsistence fishing activities are rarely well understood, indications are that catches can be substantial (Sect. 8.2.1). In one study in Fiji, for example, many major species for both subsistence and commercial use are fished from aggregations. In Pohnpei, Micronesia, subsistence fishing of aggregations is largely ignored by local legislation, which focuses primarily on commercial fishing. However, the removal of reproductively active fish for subsistence may equal or even exceed that of commercial catches (Rhodes et al. 2005). In Palau, subsistence fishers have shifted to more modern, typically less selective, fishing gears and practices, with heavy illegal use of SCUBA and spearing, including on aggregations. The many fast boats that now exist in the country enable fishing to occur throughout the Palauan archipelago, and exploitation is shifting from a focus on subsistence and 'custom' (traditional use) to commercial use, including for the tourism sector. While the fishery is no longer regularly monitored, many aggregations of squaretail coralgrouper, *P. areolatus,* brown-marbled grouper, *E. fuscoguttatus,* camouflage grouper, *E. polyphekadion*, bluespine unicornfish, *Naso unicornis,* twin-spot snapper, *Lutjanus bohar,* and long-face emperor, *Lethrinus olivaceus*, long used for subsistence, have declined over the last decade or so according to fisher perceptions (Johannes 1981; Sadovy 2007).

8.2.2 Recreational and Sport Fisheries

Recreational fishing on spawning aggregations and migrations in tropical and subtropical coastal species is attracting management attention because of associated high economic value and growing popularity. Examples come from Australia, the Pacific and Indian Oceans, and the Caribbean. Some of the fishes involved are considered good eating fish, like seabreams (Sparidae) and croakers (Sciaenidae), while others, such as tarpon or bonefish, are mostly valued for their sport appeal and fighting ability. Recreational fishing tends to involve selective gears such as hook and line and speargun and may involve tag (or capture) and release.

Relatively well-studied coastal recreational fisheries occur in Australia and the USA. In western Australia, there is high recreational interest in the silver seabream (snapper or silver bream) (Sparidae; *Pagrus auratus* = *Chrysophrys aurata*, Chap. 12.12), mulloway and black jewfish (Sciaenidae: *Argyrosomus hololepidotus* and *Protonibea diacanthus*), and in dhufish (Glaucosomatidae: *Glaucosoma hebraicum*), among other species, much of it involving catch and release. In Shark Bay, for example, a major recreational fishery on the silver seabream occurs in winter, when sea conditions are most conducive to fishing from small boats and the fish aggregate to spawn. By the mid-1990s, there was serious concern that recreational catches for this species had reached unsustainable levels, largely due to aggregation-fishing (Marshall and Moore 2000). In 2003, a total allowable catch, and other management measures, was agreed for each snapper stock, an unusual situation for such a small and mostly recreational fishery in Australia (Chap. 12.12). The black jewfish in Australia's Northern Territory is also taken by several fishing sectors but predominantly by recreational fishermen with heavy targeting of aggregations (Phelan et al. 2008). Dhufish and mulloway are highly regarded recreationally but long-lived and fished on aggregations with concerns about overfishing for both species (Mackie et al. 2009).

In parts of the Caribbean and SE USA, tarpon, *Megalops atlanticus*, is highly valued recreationally and taken in large numbers during pre- or post-spawning migrations. The fisheries are now heavily regulated because of marked declines noted in the past from overfishing and loss of habitat, with commercial take no longer permitted in Florida. Nowadays, the tarpon fishery in the region is largely one of tag and release (http://www.tarbone.org/about-btt/then-and-now.html) and among the most lucrative fisheries in the state of Florida.

Recreational fisheries that involve aggregating species occur in many parts of the Pacific. In Kiribati, for example, bonefish, *Albula vulpes*, is taken for subsistence (see above) and, since the early 1980s, as part of a high-value fishery. For many years, its spawning migrations were subjected to fishing, but due to strong support from the tourism sector and interest in long-term sustainability, fishing is now

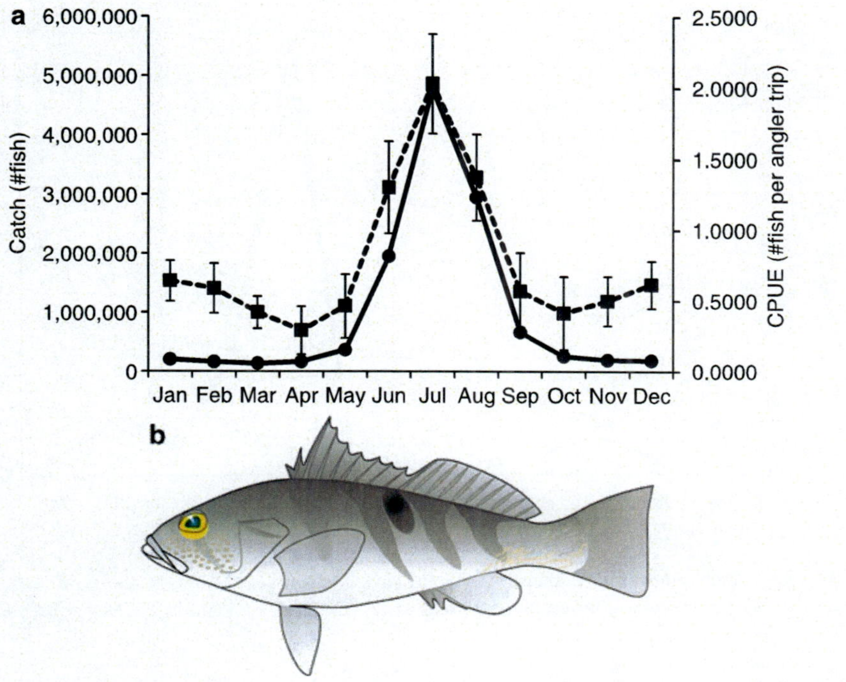

Fig. 8.2 (**a**, **b**) Barred Sand Bass, *Paralabrax nebulifer*, have consistently ranked as one of the top two recreational fisheries of Southern California for more than three decades (Love et al. 1996; Dotson and Charter 2003). The species forms massive spawning aggregations over offshore sand-flats from May to August, with a peak in July (Turner et al. 1969; Love et al. 1996). While all commercial catch has been banned since WWII, its aggregations are not managed. However, the recreational fishery is managed by a 12-inch minimum size limit and a 10-fish per day bag limit (combined daily limit for the 3 *Paralabrax* spp. that co-occur). Catch-*circle* and *solid line* (Drawing: © Larry Allen)

prohibited during spawning times (Being Yeeting personal communication 2009). Where this recreational fishery developed, there were conflicts with locals wanting to maintain it as a food source. However, after public consultation, more than 90% of the community agreed that it was a worthwhile resource to protect and use as a basis for tourism. In the Gulf of California, the endemic Gulf grouper, *Mycteroperca jordani*, is a prized target of recreational anglers and spearfishers, and sportfishing tournaments are still held during the spawning season despite dwindling numbers of fish (Sala et al. 2003; Saenz-Arroyo et al. 2005, BE personal observation). Kelp Bass (*Paralabrax clathratus*) and Barred Sand Bass (*P. nebulifer*) have consistently ranked top among species targeted by recreational fisheries of southern California, USA, over the past three decades (Dotson and Charter 2003), with 70–80% of annual landings and the highest fishing effort occurring during the summer months when both species form spawning aggregations. Although banned from commercial harvest, no specific management of aggregation sites or periods exist for the recreational fishery of either species, and landings of both have declined significantly over the last 5–10 years (Fig. 8.2).

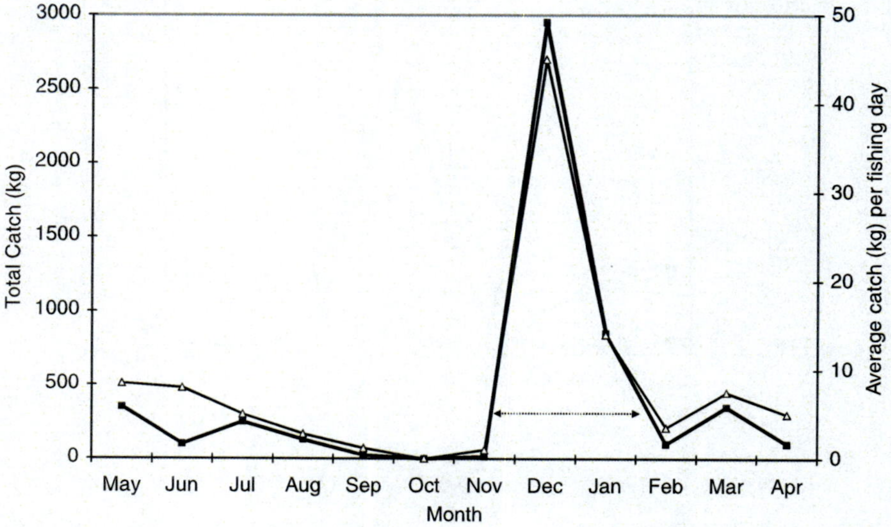

Fig. 8.3 Catch trends (total and average) in *E. fuscoguttatus* in eastern Australia show clear peaks during the limited spawning season (indicated by *horizontal line*) (Pears et al. 2007, Chap. 12.2)

8.2.3 Commercial Fisheries

Spawning aggregations are increasingly the target of commercial fisheries, using a wide range of gears and for both local and international markets. Many such fisheries are lucrative, yet few are managed and many show evidence of declines probably associated with aggregation-fishing (Sadovy de Mitcheson et al. 2008, Table 8.1). Species range from those taken exclusively at aggregations, either live or dead, for their flesh to those taken for their roe or swimbladder. Inshore coastal migrations of rabbitfishes, tarpon and mullets have also long attracted seasonal fishing operations, because they are highly predictable and many have declined or no longer form. Some species are mainly or exclusively taken in significant numbers when they aggregate, such as the brown-marbled grouper in Australia (Chap. 12.2) (Fig. 8.3) and Palau, and several deepwater species, such as orange roughy, *Hoplostethus atlanticus*. In the southern Gulf of California in Mexico, of the ten top commercially important reef fishes in terms of landings, eight are aggregation-spawners (Erisman et al. 2010; Figs. 8.4 and 8.5). Commercial fishing gears range from bottom trawls, purse seines and gill nets to hook and line, fish traps, spearguns with or without the use of compressed air diving, and even explosives and cyanide (Sect. 8.3).

In extreme cases species are threatened with extinction by commercial aggregation-fishing. The giant yellow croaker (Chinese bahaba), *Bahaba taipingensis*, a southern China endemic, and the totoaba, *Totoaba macdonaldi*, a similarly large croaker endemic to the Gulf of California, had fisheries targeting both principally for their high-value swimbladders. The large yellow croaker, *Larimichthyes (=Pseudosciaena) crocea*, is now considered to be endangered (Wang et al. 2009) and the totoaba was

Fig. 8.4 (**a**, **b**) Leopard grouper, *Mycteroperca rosacea*, spawn within aggregations from March to June in the southern Gulf of California, Mexico (Erisman et al. 2007), with peak spawning in April–May. Commercial fishers target aggregations using gill nets, nighttime spearfishing and hookah, and hook and line. Data show mean monthly commercial landings for 1999–2007 (mean and SE) (Photo: © Octavio Aburto/iLCP)

the first marine fish listing on a CITES (Convention on International Trade in Endangered Species) Appendix I in 1975 (Musick et al. 2000). Many of the larger species of groupers (Serranidae) are threatened primarily because of overfishing, especially when focused on aggregations (Tuuli 2010; Sect. 8.3.1).

Seasonal concentrations of spawning adults can form the basis for specialised fisheries. For example, herring, such as the Pacific herring, *Clupea pallesi*, eggs are

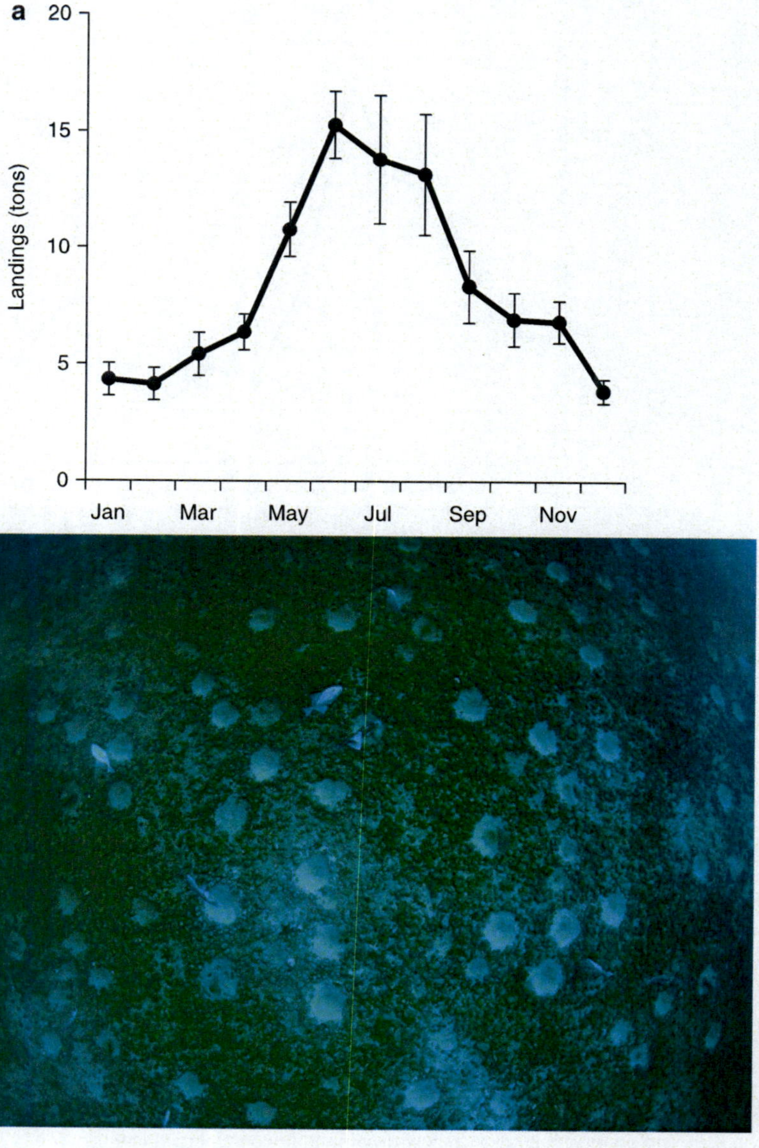

Fig. 8.5 (**a**, **b**) Finescale triggerfish, *Balistes polylepis*, aggregate to spawn and nest (*light circles* in photo) from May to September in the southern Gulf of California, Mexico (Sanchez-Velasco et al. 2009, BE personal observation). Commercial fishers target these aggregations by nighttime hookah fishing and gillnets. Data show that commercial landings for 1999–2007 (mean and SE) peak during this period (Photo: © Octavio Aburto/iLCP)

valued in a lucrative United States export market for herring roe, and for 'kazunoko kombu' (roe-on-kelp), a delicacy in Japan. In San Francisco Bay, California giant kelp is suspended in the spawning area of the species as spawning substrate, then prepared and exported to Japan. Several important fisheries target the enormous larval pulses that result from the large spawning events associated with aggregations for capture based aquaculture (CBA). Reef-associated fishes used for CBA include certain rabbitfishes, mullets, milkfish and groupers for which seasonal fisheries on post-settlement phase fish target the massive and concentrated numbers of settling recruits (Lovatelli and Holthus 2008). As one example, over a few months of the year in many Pacific islands the young of several species of rabbitfish, barely 2 months old, regularly recruit as a massive balls of tiny fish on shallow seagrass areas, mangroves, reef flats or beach areas. They are collected in their millions by push nets, cast nets, seine nets, and lift nets and grown to market size under captive conditions. They are so abundant that they have been referred to as ‘endless’ (Teitelbaum et al. 2008).

Over the last two decades, a particularly valuable fishery for living reef fish, destined for the Chinese seafood market and known as the live reef food fish trade (LRFT), has placed pressure on grouper populations in parts of the Indo-Pacific (Sadovy et al. 2003). The apparently insatiable consumer demand for these fishes drives an intensive fishery to fill air and sea consignments on a frequent and regular basis from many source countries. Aggregations are obvious targets for several of the preferred species, particularly brown-marbled grouper, camouflage grouper and squaretail coralgrouper, that form them, leading to declines in catches in the Solomon Islands, Papua New Guinea, Palau and the Seychelles (Johannes and Riepen 1995, Hamilton and Matawai 2006; Aumeeruddy and Robinson 2006). In Papua New Guinea (PNG) a Hong-Kong based trial LRFT operation claimed that it was not economically viable if it could not target aggregations and the operation ceased when aggregation fishing was not permitted (Leban Gisawa, PNG fisheries 2008). Fiji became involved in the LRFT in the late 1990s but stipulated no fishing during spawning or aggregation periods (Ledua Ovasisi, Fiji Fisheries, 2008). Aside from the large numbers of live fish removed from aggregations, which can represent a high proportion of assembled fish, catches of fully ripe females can lead to significant wastage because gravid fish tend to undergo high mortality shortly after capture.

8.3 Fishery Implications of Fishing Spawning Aggregations

A major challenge is to determine the extent to which exploitation of spawning aggregations contributes to declines in marine fish populations. Few of the 15,500 known marine fish species have both the necessary biology and fishery information for such detailed assessment. Nonetheless, this task is critically important from a management perspective, since many of the world’s major fisheries support or once supported aggregation-fishing. The massive landings that once characterized many highly productive fisheries involved seasonal gatherings of fish at magnitudes far greater than are seen today (Roberts 2007). Presumably, aggregation-spawning evolved, for whatever proximate reason, because, ultimately, it resulted in greater reproductive success compared to a non-aggregating habit. Unfortunately, the very biological characteristic, aggregation-spawning, that evidently contributes to making such species so productive may also prove to be their ‘Achilles Heel’ when

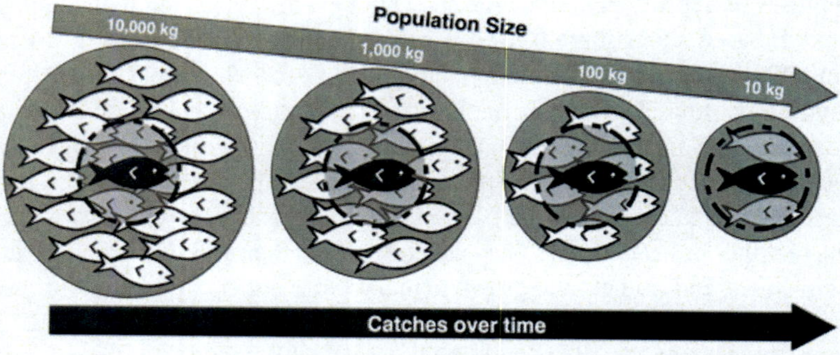

Fig. 8.6 Hyperstability refers to a phenomenon in which an observed index of stock abundance (e.g. catch per unit of effort or CPUE) remains stable (represented by *black fish in dotted circle*) while the abundance (population size) of the stock in question is actually declining (*black arrow* denotes past to the *left* and now to the *right*). The figure represents how CPUE can remain stable over time even as total fish numbers (*white fish and grey arrow*) are declining because fish are still aggregating to spawn; this will occur when the fishing effort is not so high that it removes all the fish at once but gives the illusion that fish numbers are not changing (Drawing: Octavio Aburto)

exploitation is uncontrolled. That is to say, those species that have evolved to spawn in massive aggregations may have done so because it is a particularly successful reproductive strategy, but however, the habit makes them very susceptible to overfishing.

To understand the impact of fishing on spawning also aggregations, two factors must be considered, the high catchability of fish while gathered for spawning, and the possible *direct* and *indirect* effects of fishing on aggregated and reproductively active fish. Aggregating behaviour can make fish particularly easy to catch (i.e. high catchability and hyperstability) during the spawning season yet very difficult to monitor (Fig. 8.6, Chap. 11) or manage relative to non-aggregating species. Exploitation of aggregated fish may directly or indirectly compromise reproductive function or output by disrupting the mating process or due to possible Allee effects (see Sect. 8.4.2) at low population levels. While standard fishery modelling approaches incorporate sex ratios, fecundity and spawning biomass into stock assessments, they typically do not factor in subtleties of reproductive biology (Hilborn and Walters 1992; Vincent and Sadovy 1998). Yet literature on other vertebrate taxa clearly shows that details of social and reproductive behaviour can be important components in the management of wild populations (Caro 1998). We explore each of these points.

Although aggregations may be subjected to intense fishing pressure, it is typically difficult to determine whether aggregation fishing is a major causative factor of population declines because most species with this habit are also fished outside of aggregations. This is a major impediment to promoting the need for the specific management of aggregations. An alternative explanation could be that a general failure to manage these fisheries throughout the year is the major cause of any declines noted, rather than aggregation-fishing *per se*. To tease apart these possible explanations, we consider the extent to which unmanaged exploitation (extrinsic factor) of aggregations could be a specific driver of population declines (see Sect. 8.3.1) and examine the role of fishing technology (Sect. 8.3.2). In Sect. 8.4 we go on to explore the possible biological (intrinsic) factors affecting responses to aggregation fishing. We then consider whether aggregations in general, or perhaps certain types of aggregations, need specific management attention, and address management options.

8.3.1 *Specific Effects of Aggregation Versus Non-aggregation Fishing*

In the management of aggregating species, understanding the specific impacts of fishing their aggregations, as distinct from the overall impacts of total fishing pressure on the target population throughout the year is important, albeit challenging. Natural resource managers must decide where funds and capacity are best directed, what kind of management is most appropriate and socially acceptable, and the economic implications for fishing communities of actions such as aggregation closures, or, conversely, their losses. Since detailed fishery and biological information is unavailable for most exploited reef fishes, and because few datasets allow for in-depth examination of their fisheries, teasing out cause and effect of aggregation versus non-aggregation fishing is problematic. The challenge is further compounded by the fact that other life-history characteristics, such as longevity or late sexual maturation, can strongly influence responses to fishing, and many aggregating species exhibit these characteristics (Reynolds et al. 2005). Moreover, making comparisons across very different taxa is complicated by phylogenetic differences, while analyses that treat species values as statistically independent points are questionable because closely related species may share traits through common descent rather than through independent evolution (Felsenstein 1985).

At least four approaches allow examination of possible fishery implications of fishing spawning aggregations. The first is to use information qualitatively across a wide taxonomic range of species and determine whether those in which aggregations are specifically targeted have shown the most marked changes. The second is to compare the fisheries status of several aggregating species within a single multi-species fishery, in which all species are exposed to similar fishing profiles and show general biological similarities yet exhibit different degrees of aggregating behaviour and extent of targeting of aggregations. A third approach, and one that addresses phylogenetic issues, is to select a single lineage that has both aggregating and non-aggregating species and examine fishery or conservation status by species. The fourth involves a comparison of the fishing history of a single aggregating species among locations or populations, ranging from situations where its aggregations are targeted and make a significant contribution to annual landings to those where aggregations contribute little to the overall fishery. While none of these four approaches entirely avoids possible confounding factors of life history differences, relatedness or the relative importance of non-aggregation fishing, examining the question from different perspectives at least provides a semi-quantitative means of evaluating the importance of aggregation-exploitation *per se* on fisheries. It also helps to identify the kinds of information needed in the future to address the question in more detail.

8.3.1.1 Approach One – Spawning Aggregation-Fishing as a Possible Threat Factor Across Taxa

Many aggregating species produce higher catch per unit of effort (CPUE) or higher percentages of annual landings when taken from aggregations than during other times of the year. Indeed, many of those species that have suffered serious declines or are otherwise considered to be of conservation concern, such as those listed as threatened under International Union for Conservation of Nature (IUCN) criteria, or depleted by FAO, are mainly or exclusively fished during aggregation periods.

The FAO database of world fisheries for 2004 indicates that of stocks for which information is available (c. 441 stocks), 52% are fully exploited and a further 24% overexploited or depleted (FAO 2005). Of the latter group many involve exploitation of spawning aggregations. Examples include European plaice *Pleuronectes platessa,* Atlantic cod, *Gadus morhua,* haddock, *Melanogrammus aeglefinus*, Atlantic herring, *Clupea harengus,* whiting, *Merlangius merlangus,* Argentine hake, *Merluccius hubbsi,* Geelbeck croaker, *Atractoscion aequidens,* Red steenbras *Petrus rupestris,*, icefish, *Champsocephalus spp.* Atlantic bonito, *Sarda sarda*, Atlantic halibut, *Hippoglossus hippoglossus*, and Pacific halibut *Hippoglossus stenolepis* (King 1985; Smale 1988; Griffiths and Hecht 1995; Smedbol and Wroblewski 1997; Hutchings et al. 1999; Parkes 2000; Hoarau et al. 2005; Pajaro et al. 2005; Zengin and Cincer 2006; Loher and Seitz 2008; Tobin et al. 2010). Particularly noteworthy are species like the Atlantic and southern bluefin tunas, *Thunnus thynnus* and *T. maccoyii*. These two tunas show aggregating behaviour (Farley and Davis 1998; Fromentin and Powers 2005) that appears to be more concentrated than for any other pelagic fishes and both have undergone some of the most marked declines among pelagic species, with heavy fishing focused on their aggregations. Other examples are provided below in the sections on groupers, South African sea breams (Sparidae) and croakers. Among reef fishes, species of many taxa typically form aggregations, and 60% of known tropical aggregations have undergone declines (Sadovy de Mitcheson et al. 2008).

Fuzzy logic (a form of mathematical logic in which truth can assume a continuum of values between 0 and 1) is one approach to explore possible factors associated with intrinsic extinction vulnerability to fishing. Many aggregating species have other characteristics, in addition to concentrated spawning, associated with high vulnerability to fishing or threat factors, such as large body size and longevity (e.g., Reynolds et al. 2005). For reef fishes in Fiji, Cheung et al. (2005) considered maximum body size, natural mortality and spawning aggregations, among other variables, and found, using fuzzy logic, that the incorporation of aggregations in this novel analysis greatly increased the goodness-of-fit between the estimated vulnerabilities and the empirical population trends, suggesting that aggregating species are more threatened, regardless of other characteristics, all else being equal. In a global analysis of more than 14,000 species that used a similar fuzzy logic approach, seamount aggregating fishes were shown to have a higher intrinsic vulnerability to fishing in comparison with other groups of commercially exploited marine fishes, due to a suite of life history characteristics such as long lifespan, late sexual maturation, slow growth, and low natural mortality (Morato et al. 2004).

8.3.1.2 Approach Two – Single Fishery, Multiple Species That Vary in Duration and Predictability of Aggregations

An unusually detailed and long-term fishery database from Cuba for six commercially important aggregating snappers (cubera snapper, *Lutjanus cyanopterus,* mutton snapper, *L. analis,* grey snapper, *L.griseus,* lane snapper, *L. synagris,* yellowtail snapper, *Ocyurus chrysurus*, and the Nassau grouper) was examined for species-specific trends (Claro et al. 2009). The species are reported to all share similar coastal water habitats, be part of the same multi-species fishery, have all been exposed to similar social, economic and management factors over a 45-year period, and evidently show different intensity of aggregation behaviour in relation to reproduction. They are also among the larger and longest-lived species in the fishery. In all six species, more than 50% of annual landings were taken during the spawning season, and all declined over time. Although cause and effect cannot be established, aggregation predictability and targeted fishing were likely major factors in declines, because the most marked declines occurred among those species that exhibit the briefest (fewest months) and most highly predictable (fish highly concentrated at relatively few sites) aggregations (Fig. 8.7). Marked declines were also observed in mullets, which are often targeted on or moving to spawning aggregations in Cuba. Other groups of fishes (jacks-Carangidae, mojarras-Gerreidae, and grunts-Haemulidae), not knowingly targeted in association with aggregations, involved in the same multi-species fishery have evidently not undergone similar declines (Claro et al. 2009).

8.3.1.3 Approach Three – Spawning Aggregations Within a Single Taxon

This approach compares trends within three very different taxonomic groups, groupers, sea breams and croakers, each taxon with both aggregating and non-aggregating species. The groupers (Family Epinephelidae, subfamily Epinephelinae) were recently assessed for conservation status according to IUCN Red List categories and criteria; threatened (T), near threatened (NT), least concern (LC) and data deficient (DD). The rich sea bream assemblage of South Africa has a relatively well documented fishery history, while many large croakers are fished on their aggregations with declines noted in some.

A comparative analysis of 163 groupers indicates that, among species of known reproductive strategy, spawning aggregation formation is associated with higher extinction risk (Fig. 8.8); in all cases aggregations are exploited. Since larger species

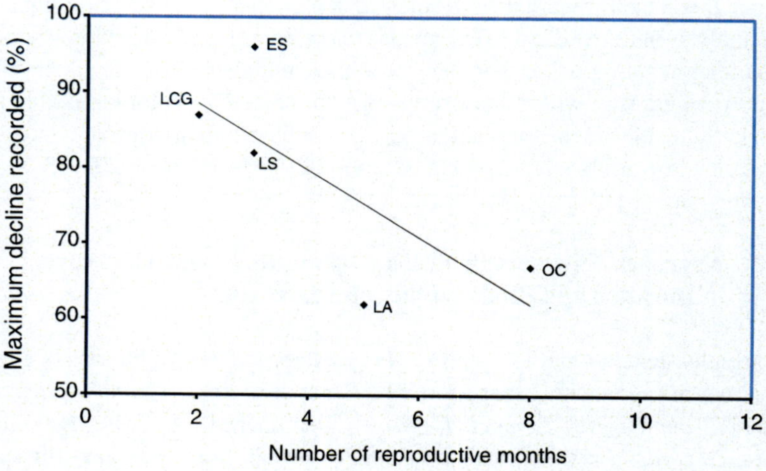

Fig. 8.7 Six commercially important coral reef species (two lumped), of maximum body sizes of 50 cm or more, taken in the coastal multi-species fishery in Cuba from a 45-year dataset with monthly landings data resolution. Y-axis indicates the largest overall drop within the 45-year period examined expressed for the 6 species according to the number of months that each species is known to be reproductive months. The trend shows greatest declines in those species that have the shortest and generally more concentrated spawning period (r= −0.74). *Lutjanus cyanopterus* and *L. griseus* data are lumped (LCG) *Epinephelus striatus* (ES), *L. synagris* (LS), *L. analis* (LA), *Ocyurus chrysurus* (OC) (Claro et al. 2009)

in a taxon are more likely to spawn in aggregations than related smaller ones (Domeier and Colin 1997, Chap. 4) body size is potentially a confounding factor and further consideration is warranted; nonetheless, the data tend to support the conclusion that aggregating species are more vulnerable. For example, comparing similar-sized and sympatric species pairs in the western Atlantic (red grouper, *Epinephelus morio*-near threatened, no indication of strong aggregating habit, paired with Nassau grouper–threatened, transient aggregator, hundreds to tens of thousands of fish in an aggregation) and Indo-Pacific (leopard coralgrouper, *Plectropomus leopardus*–near threatened, resident aggregator, tens to a few hundred fish at a single aggregation paired with squaretail coralgrouper-threatened, transient aggregator, hundreds to thousands of fish at a single aggregation) suggests that declines in populations, as judged by landings trends in the fishery over time, have been greatest in transient species that aggregate at the fewest known sites and most predictably. This is clearly suggested by marked declines in Nassau grouper and squaretail coralgrouper and is consistent with the general and perhaps intuitive observation that, all else being equal, transient spawners are likely to be more susceptible to overfishing than resident spawners.

Of 42 species of sea bream (family Sparidae) found in South African waters (30 are endemic) the ‘seventy-four’, *Polysteganus undulosus*, has undergone the greatest overall declines in this comparatively well-monitored coastal fishery. It is one of the only sea breams in the region, of the 42 present, that was once fished intensively at specific spawning aggregation sites. In the early 1900s the species

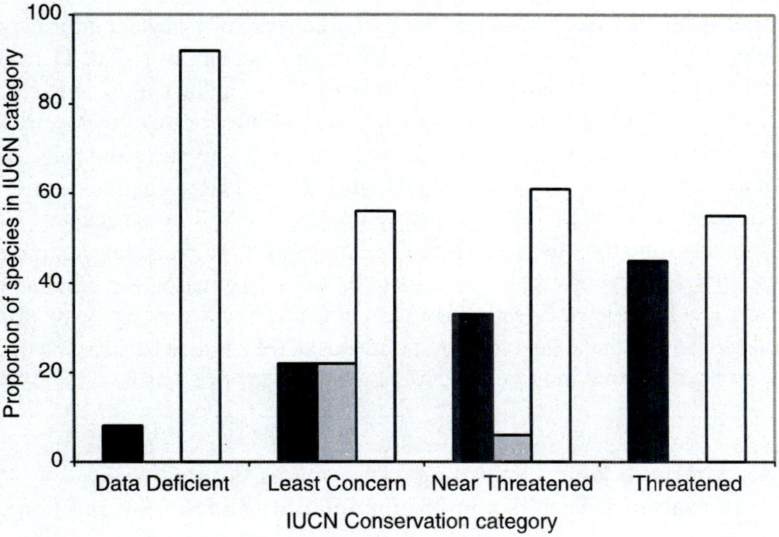

Fig. 8.8 Conservation status of 163 groupers of the family Epinephelidae according to IUCN Red List criteria (www.iucnredlist.org), data deficient, least concern, near-threatened and threatened. For each conservation category the numbers of species known to aggregate (*black bar*) known not to aggregate (*grey bar*) and of unknown reproductive strategy (*white bar*) are represented as a proportion of the category. The graph suggests that those species known to aggregate are more likely to be threatened or near-threatened (Craig et al. 2011; Sadovy de Mitcheson et al. in press)

contributed over 50% of landed catch from line-boats out of Durban harbour (Garratt 1996; Penney et al. 1999). By the late 1960s, catches in the main fishing area, KwaZulu-Natal (KZN), had collapsed and after several failed management interventions during the 1980s, a moratorium was placed on the species in 1998 (Mann 2007).

Although large (up to 120 cm TL), the seventy-four is not the largest sea bream in S. Africa. However, it has one of the shortest and most predictable spawning seasons of the South African sea bream assemblage, appears to be the most threatened sea bream and its collapse is one of the best documented examples of severe overfishing in S. African waters (reviewed in Mann 2007). Its large spawning aggregations once lasted from July to November and were heavily fished leading to marked declines in catches and, ultimately, the need for management. However, the species showed little recovery despite a fishing moratorium. While there are positive indications of larger adults in KZN and increased abundance of juveniles in the Eastern Cape, poaching continues and the high price paid for this desirable, and now rare, fish means the greatest threat to its recovery is the lack of effective enforcement (Mann 2007). Another sea bream, the red steenbras (*Petrus rupestris*) has suffered a similar fate. Although its spawning aggregations are less well known; experienced fishers strongly indicate that this species aggregates during its spawning periods (August-November) when the largest catches are made (Bruce Mann personal communication 2009).

The croakers include many highly valued species aggregating to spawn, typically in coastal or estuarine areas and which are targeted at this time by commercial or recreational fishers (Sects. 8.2.2 and 8.2.3). Of 28 species reviewed, information on aggregation behaviour was available for 10, mainly the largest (>1 m TL), species that had the shortest reproductive seasons (Tuuli 2010). In the Colorado River Delta in the northern Gulf of California the totoaba and Gulf corvina (*Cynoscion othonopterus*) have undergone complete fishery collapses and were the only species known to form massive, dense, brief and highly localized aggregations at just a few sites (Cisneros-Mata et al. 1995; Roman-Rodriguez 2000). The totoaba was the first commercial marine food fish to be listed on Appendix I (1975) and is yet to recover. In Australia, aggregation-fishing appears to be the major reason for declines in the mulloway and black jewfish (see above), and in China it was very likely the principal factor leading to the threatened conservation status of the Chinese bahaba, and the large yellow croaker (Sadovy and Cheung 2003; Liu and Sadovy de Mitcheson 2008).

8.3.1.4 Approach Four – Spawning Aggregations Within a Species, Geographic Variation in Fishing Intensity and Stock Condition

Fishery status can be compared for a single species among places or populations where aggregations are subjected to different levels of fishing pressure. To do this we need (1) assessments of population status (either from fishery stock assessment or using some other indicator such as trends in body size or in CPUE) and (2) monthly landings data or information that distinguish catches between spawning and non-spawning seasons. Although more quantitative data are needed to identify the effects of fishing on aggregations versus impacts of fishing on the stock as a whole, available qualitative and quantitative information for the Nassau grouper strongly suggests declines in fish sizes and numbers associated specifically with aggregation-fishing. For example, where aggregations have been protected for a number of years and monitored, numbers are substantial or have stabilized (e.g. Cayman Islands; Philippe Bush 2009), whereas where there has been little effective protection aggregations are much reduced or no longer form (e.g. Belize, Cuba, Puerto Rico and parts of Mexico) (Sadovy 1993; Aguilar-Perera 1994, 2007; Sala et al. 2001; Heyman and Wade 2007; Claro et al. 2009).

Since it is unlikely that more detailed assessments are forthcoming in the near term, we propose that the indications from different Nassau grouper stocks, combined with the analyses in approaches 1–3 above, unequivocally call for a conservative approach to managing aggregation-fisheries.

8.3.2 *Effects of Gear Types and New Technologies*

Access to and exploitation of spawning aggregations has become easier due to the emergence of new technologies and more effective gear types. Devices such as detailed fishing charts, Geographic Information Systems (GIS), bathymetric maps, sonar, and Global Positioning Systems (GPS) make aggregations less difficult to find, characterize, exploit, relocate, and communicate to others (Sadovy and

Domeier 2005). Satellite imagery (e.g. Google Earth) provides a bird's eye view of fishing areas, increasing the likelihood of locating the spawning grounds of target species once the physical attributes of such sites are better understood. The live fish trade industry has used helicopters and planes to seek concentrations of fishing activity at aggregations (Leban Gisawa personal communication 2008, Bob Johannes personal communication 2000). Increased range and capacity of large fishing vessels, including those with live wells or viviers that can carry 15–20 mt, greatly increase interest in commercial aggregation fisheries (Johannes and Riepen 1995; Johannes 1997). Finally, the widespread introduction of ice plants in the Pacific to encourage and facilitate market linkages is a favoured development tool often tied to tuna licenses; iceplants greatly increase access to urban markets by remote fishing communities to markets by providing temporary storage of fish until cargo vessels can take them to market, thereby opening up remote aggregations to commercial exploitation (Being Yeeting personal communication 2009).

Certain gear types and fishing practices have negative impacts on aggregation sites and fishes. Gill nets or bottom trawls damage important habitat features of aggregation sites (Koslow et al. 1997; Hamilton et al. 2005). In the Gulf of Mexico, sites with high densities of *Oculina* coral hold high abundances of gag, *Mycteroperca microlepis*, and scamp, *Mycteroperca phenax*, groupers; commercial trawl fisheries damage these spawning habitats, thereby accelerating population declines (Koenig et al. 2000). Dynamite is regularly applied to take milkfish, *Chanos chanos*, in the Philippines (Table 8.1). Discards of sub-legal size fish may comprise a substantial proportion of the recreational or commercial catch at aggregation sites but are often associated with high post-release mortality due to barotrauma, gut-hooking, stress and fatigue from prolonged handling times, or predation following release (Burns et al. 2002; Rudershausen et al. 2007).

Expansion of the SCUBA diving tourist industry, especially the increase in availability of inexpensive equipment (tanks, fins, underwater lights, etc.) has greatly contributed to the rise in the number of commercial, recreational, and artisanal fishers and divers who utilize this method to harvest fishes from aggregations for sport or profit (e.g. Paz and Truly 2007). Night-time spearfishing, which involves compressed air supplied via SCUBA or hookah, allows divers to easily harvest large numbers of resting fishes at aggregation sites with minimal effort. This activity is widely illegal but its prohibition is rarely enforced, and it is considered to be the major cause of aggregation declines in the Gulf of California and throughout the Pacific islands (Gillett and Moy 2006, www.seawatch.org) (Fig. 8.9). Even the simple technology of a reliable d-cell battery torch (Toshiba and others), allowed night fishing while snorkelling in many areas of the Pacific (Pat Colin personal communication 2010).

8.4 Biological Effects of Fishing Spawning Aggregations

Animals congregate for numerous reasons. Conservation biologists and fishery managers, however, rarely concern themselves with the implications of grouping for conservation action (Reed and Dobson 1993). The one exception is in relation to the

Fig. 8.9 (**a**, **b**) Illegal nighttime fishing for leopard grouper, *Mycteroperca rosacea* in the Gulf of California, Mexico (Photos: © Octavio Aburto/iLCP (fish) and © Seawatch.org (divers))

Allee Effect (Allee 1931) whereby reduced populations show unexpectedly low rates of recovery (see Sect. 8.4.2).

Spawning aggregations have evolved for reasons that have yet to be fully understood and that may well vary among species and lineages (Chaps. 2 and 4). Such reasons are important to consider, not just for academic interest but for understanding the impact of fishing. Possible adaptive advantages to spawning in large temporary groups include conditions important for fertilization, encountering mates and synchronizing spawning, sexual selection and related behaviours such as stimulation of spawning by conspecifics and anti-predator reasons. The specific times and locations where gatherings of ripe fish occur may provide selective benefits related to larval dispersal, conditions associated with egg/larval survival, or habitat factors such as appropriate substrate for nesting (e.g. in triggerfish, Chap. 12.23) or high substrate complexity where large numbers of adults can temporarily find shelter. However, the significance of mate choice or of specific habitats for reproducing in large groups has been little examined in fishes (Chap. 3). Therefore, we must keep in mind the various possible adaptive reasons for aggregation-formation when exploring the possible direct and indirect biological impact(s) of exploiting spawning aggregations.

When fisheries heavily target spawning aggregations, the most obvious sign of change to the fishery manager is likely to be declines in landings volumes or reduced numbers of fish at the aggregation site. Depending on the level of fishing effort, such declines may take a long time to become apparent because of *hyperstability* (Fig. 8.6, Chap. 11). Yet significant negative changes to the population may occur before landings decline. Many reef fishes have complex social and mating systems that may be disrupted by fishing activities. Fishing could directly alter population structure, via shifts in adult sex ratio or mean body size, in ways that affect reproduction. Indirectly, declines in fish numbers due to fishing could affect reproduction by impairment of visual, olfactory or auditory stimuli, social cue transmission or by influencing other interactions such as courtship or sexual selection that only occur during brief aggregation periods. For those species that live in deep-water or habitats or are normally solitary, the spawning aggregation may be the only time fish encounter conspecifics for mating, or are exposed to the population sex ratio or other aspect of the demographic profile (see Chap. 12.7).

The possible biological effects of fishing on aggregations will, therefore, depend on the way(s) in which fishing is conducted, the absolute and relative volumes of fish taken, the selectivity of fishing, differential movements of sexes in and out of aggregations and the adaptive significance for the species of the aggregating habit itself. The combination of direct and indirect effects of fishing could ultimately influence the reproductive output from aggregations with short and long-term implications for the fishery as well as for the targeted population. These issues are addressed below and considered in light of fishery management implications and future information needs.

8.4.1 Abundance, Sex Ratio, and Body Size

Unmanaged harvesting from fish spawning aggregations by commercial, recreational, or even artisanal fisheries is often associated with declines in aggregation size (number of fish) over time. This is not surprising, given that heavy fishing can remove large proportions of aggregations in just a few days (e.g. 10–20%, Nassau grouper, Sala et al. 2001; 20–30%, camouflage grouper, Rhodes and Sadovy 2002a, b), and that heavy exploitation of aggregations might reflect high fishing mortality in the fishery in general. In extreme cases, entire aggregations have been removed

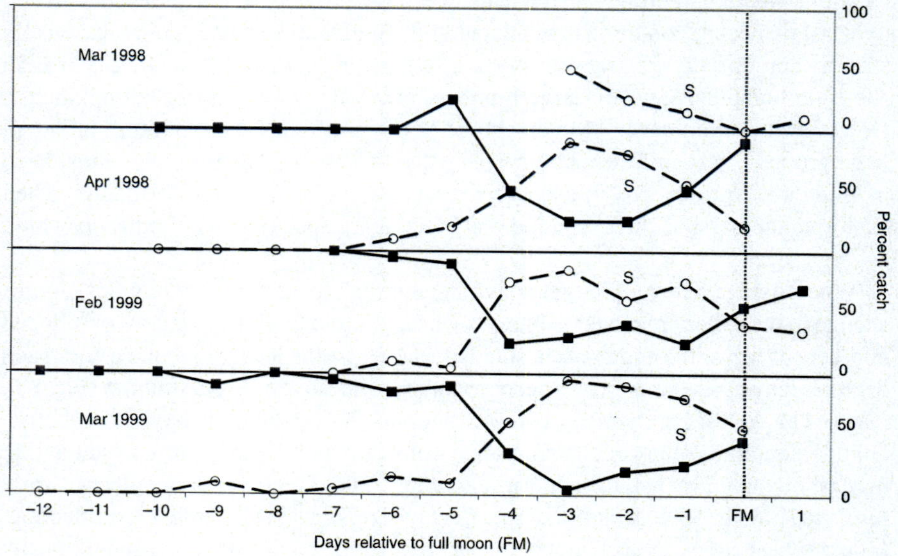

Fig. 8.10 Percent of males (*square and solid line*) and females (*circle and dotted line*) of camouflage grouper, *Epinephelus polyphekadion*, in aggregation catches in the 12 days leading up to full moon (*FM*) and 1 day after FM in Pohnpei during one of several annual spawning months from the Kehpara Island Marine Sanctuary (February 1998–March 1999). *Graph* shows arrival of males followed by females several days later and the resulting shift in adult sex ratio at the site (Spawning = *S*) (Rhodes and Sadovy 2002a, b)

within a single or a few spawning seasons (Johannes and Riepen 1995; Hamilton et al. 2005). Serial depletions from fishing are likely a major contributor to the complete disappearance of a target species from larger regions (seventy-four Chale-Matsau et al. 2001, orange roughy Clark 2001, large yellow croaker Liu and Sadovy de Mitcheson 2008, *Epinephelus quinquefasciatus = itajara,* Sala et al. 2004, Nassau grouper Sadovy and Eklund 1999; Sala et al. 2001; Aguilar-Perera 2007; Claro et al. 2009).

Sex-specific differences in the temporal movement patterns of fish into, out from, and around aggregation sites can interact with fishing in ways that can alter adult sex ratios. In red hind (*Epinephelus guttatus*), camouflage grouper and Atlantic cod males arrive earlier and remain longer at aggregation sites than females, presumably to establish and maintain mating territories (Nemeth et al. 2007, Rhodes and Sadovy 2002a, b; Robichaud and Rose 2003, Windle and Rose 2007, Rhodes and Tupper 2008) (Fig. 8.10). Conversely, females of these species may be more transient, moving onto spawning grounds to spawn and then leaving or migrating to nearby areas between mating periods to ripen new batches of eggs and/or to feed (Morgan and Trippel 1996; Nemeth et al. 2007, Chap. 2). In such cases, fishing only on the spawning grounds or only on adjacent areas can selectively remove one or the other sex, thereby altering reproductive sex ratios with potential to reduce reproductive output (Beets and Friedlander 1992; Shapiro et al. 1993; Rowe and Hutchings 2003). Selective fishing could have explained the highly male-skewed sex ratio of squaretail coarlgrouper at Ulong Channel, Palau (Johannes et al. 1999) (see below).

Sex ratio effects are also relevant to sex-changing species, which tend to have naturally male-biased (protandrous) or female-biased (protogynous) adult sex ratios. In certain protogynous groupers that form spawning aggregations (e.g. red hind, Shapiro et al. 1993, gag grouper Koenig et al. 1996), sex ratio assessment during aggregation periods may be an important cue for sex change, since this is the only time when adult males and females are known to come together in significant numbers; in all sex-changing species studied, the incidence and timing of sex change is mediated at the level of social groups in response to behavioural cues (Muñoz and Warner 2003; Munday et al. 2006). Particularities in fisher behaviour can also skew sex ratios. In an exploited aggregation of tiger grouper, *Mycteroperca tigris*, sex ratios in catches were skewed towards males because spearfishers actively selected males to "protect" females and the eggs they bear (Sadovy et al. 1994a, b; Matos-Caraballo et al. 2006). The selective male removals could potentially cause problems with mate choice, mate encounter rates, or other reproductive behaviours, depending on the mating system. In another example, a male-biased sex ratio in an aggregation of squaretail coralgrouper in Palau was associated with greater harassment (i.e. chasing) of the relatively smaller number of ripe females moving around the aggregation site in Palau compared to sites with less male bias (Johannes et al. 1999).

Heavy fishing pressure on spawning aggregations could be a major factor in reducing the average length and size range of fish, although this may be a general effect of fishing rather than one specifically related to aggregation fishing. In leopard coralgrouper in Australia (Adams et al. 2000), Nassau grouper in Belize and Mexico (Carter et al. 1994; Aguilar-Perera 2007), red hind in the United States Virgin Islands (USVI) (Beets and Friedlander 1999), and in leopard grouper, *M. rosacea* in the eastern Pacific (Sala et al. 2003) sizes of fish taken from aggregations have declined relative to past baselines (Figs. 8.11, 8.12). In a few species, such as Atlantic cod, stripey seaperch (*Lutjanus carponatus*), gag grouper and scamp, reduced body size is associated with reductions in ages of sexual maturity and in sex change, or in decreases in egg size, larval survivorship, and batch fecundity of females (Coleman et al. 1996; Olsen et al. 2004, 2005; Evans et al. 2008). This is a major concern for fisheries, given the relative reproductive value of large females due to the association of high fecundity with large size and, in some species, high egg and larval quality (Berkeley et al. 2004; Birkeland and Dayton 2005). While declines in body size have several possible causes, aggregation-fishing could be a major contributing factor if it is particularly intensive or size-selective. Conversely, reduction of fishing pressure can result in larger and more plentiful fish (Nemeth 2005).

8.4.2 Allee and Other Mating Behaviour Effects

Allee Effects, expressed by positive relationships between various fitness components (e.g. number of matings or eggs fertilized) and population densities, are related to mate-finding factors that can be influenced by fishing (Allee 1931; Stephens et al. 1999). In birds, the once abundant passenger pigeon, *Ectopistes migratorius,* became extinct, at least in part, because pairs no longer bred once colony sizes were reduced

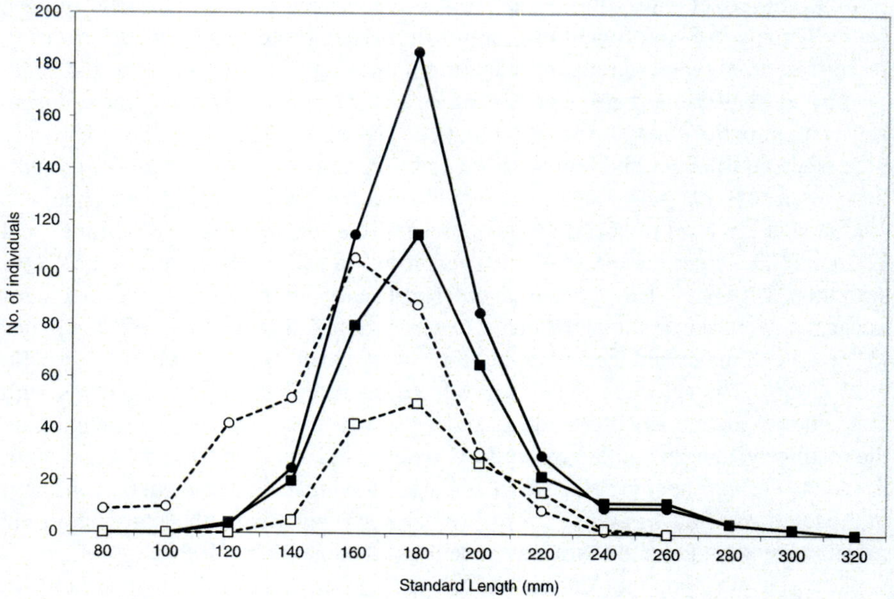

Fig. 8.11 Size frequency distribution by sex (female *circle*, male *square*) data for Nassau grouper, *Epinephelus striatus*, from catches taken at a (formerly) non-exploited aggregation site (Northern Two Caye, *solid symbols*) and an exploited aggregation site (Cay Glory, *open symbols*) in Belize. Data show that both sexes are smaller at the exploited site (Carter et al. 1994)

by hunting and habitat losses (Blockstein and Tordoff 1985). Allee Effects may be important if they constrain population growth and recovery in overfished fish species. Mate-finding Allee Effects may be important in aggregating species if the primary purpose of aggregation formation is to find mates and avoid the problem of low density (which may occur in large species that disperse widely during non-aggregation times), or if there is some positive reproductive effect of groups of animals coming together such as mate choice and fertilization rate (see Chap. 3). Important factors might involve mate encounter rates and mate attraction, gamete density and sperm limitation, physiological stimulation of courtship, female choice, sexual selection, and reproductive investment (Gascoigne et al. 2009). It is relevant to consider whether there is evidence that such factors might be associated with aggregation-spawning, or with different types of mating systems (i.e. transient versus resident aggregations, group- versus pair-spawning) within aggregations.

Species for which Allee Effects have been demonstrated or hypothesized range from Pacific sardine, Downs herring, scallops, abalone, sea urchins, and giant clams, to muskrats, flour beetles, sheep ticks, and passenger pigeons (Chap. 3, Frank and Brickman 2000). Examples of Allee Effects include difficulty in finding a mate, and a breakdown in social structure and migration patterns. Species that have characteristic social behaviour such as group-mating, group defense, and schooling may be at increased risk of extirpations under heavy exploitation due to Allee Effects (Frank and Brickman 2000). While there are few examples of clear Allee

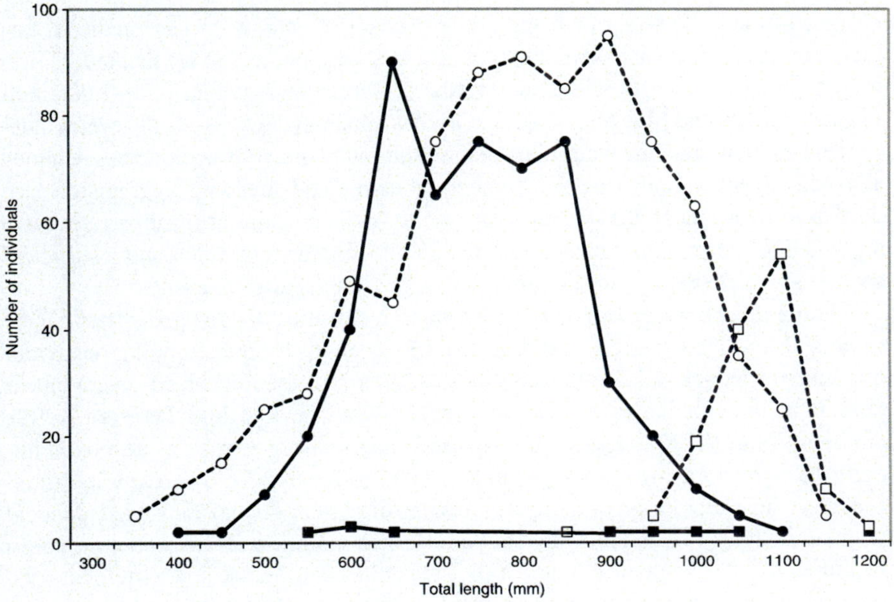

Fig. 8.12 Size frequency distribution of males (*squares*) and females (*circles*) gag grouper, *Mycteroperca microlepis*, from the Gulf of Mexico, USA. Historical data from 1977 to 1980 (*open symbols and dotted line*, Hood and Schlieder 1992), and from 1991 to 1992 (*closed symbols and solid line*, Koenig et al. 1996). Sex ratios of males to females dropped from 17% to 2%, and mean sizes of both sexes declined over the time period

Effects in commercial fishery species, data at small population sizes are few and the possibility cannot be ignored (Myers et al. 1995).

If one of the adaptive values of congregating to spawn is to bring together males and females within a complex mating system, or for mate selection, major deviations from the natural abundances, body sizes, or adult sex ratios within spawning aggregations may be important. Extensive disruption of mating behaviour due to fishing is also possible. Such changes could potentially result in significant declines in reproductive output that translate into disproportionately lowered population growth rates, i.e. an Allee Effect. For example, queen conch (*Strombus gigas*) normally aggregates to mate but spawning ceases below a threshold density of 48 individuals ha^{-1} (Stoner and Ray-Culp 2000). Densities are known to be important in determining the prevalence of group- versus pair-spawning in an area, for example in the bluehead wrasse (Chap. 12.14).

A decrease in the number of breeding males per female and reduction in sperm concentration could result in lowered numbers of fertilizations and a reduced population size. Sperm limitation may be a significant barrier to the recovery of overfished populations of gag grouper in Florida; in 1977–1982 males represented 17% of ripe fish at aggregation sites in the Atlantic and Gulf of Mexico, which declined to 1–3% by 1991–1994 when large numbers of gravid but 'unspawned' females were noted to be present at aggregations (Coleman et al. 1996) (Fig. 8.12). Model simulations designed to test the effect of fishing on reproductive activity of gag predicted that mean fertilization rate and the number of fertilized eggs per recruit are markedly lower in small mating groups (<50 individuals) than in larger aggregations (100–1,000 individuals, Alonzo and Mangel 2004). A study that examined the results of several independent captive-breeding studies in Atlantic cod found that fertilization rates declined and had higher variance as the number of males and breeding aggregation size decreased (Rowe et al. 2004). For both species, reductions in fertilization rates were attributed to sperm limitation associated with a low number of males and insufficient sperm to fertilize the eggs of all females during group-spawning events.

Mobile invertebrate examples show that high densities of eggs and sperm can be critical to fertilization success and presumably account for the clumping behaviour of adults for gamete release in some species. Sperm limitation can be severe unless numerous individuals spawn simultaneously (e.g. Levitan and Petersen 1995), although Yund (2000) suggests that sperm limitation may not be as severe as initially thought in marine free-spawners. Many commercially over-exploited non-sedentary invertebrate species are recruitment-limited or display density-dependent population dynamics and the effect of possible sperm limitation has sometimes been included in fishery models (Yund 2000). In the bluehead wrasse sperm numbers and fertilization rates are lower in single-male matings than during multiple-male matings (Shapiro et al. 1994; Marconato et al. 1997) (Chap. 3).

Fishing spawning aggregations could reduce reproductive output through effects on courtship and mate choice. Assortative mating, whereby individuals tend to mate with fish of similar size, is found in several aggregating species (bluehead wrasse Shapiro et al. 1994, bucktooth parrotfish *Sparisoma radians* Marconato and Shapiro 1996, Atlantic cod Rowe and Hutchings 2003, leopard grouper Erisman et al. 2007), while in others, females show preferences for larger males (e.g. Chap. 3 Rasotto et al. 2010). If preferred mate phenotypes are removed during or just prior to spawning periods by fishing, the choosier sex may respond by releasing fewer gametes or performing fewer spawning rushes, thereby reducing reproductive output (Shapiro et al. 1994, Marconato et al. 1997).

Fishing could potentially lower mating frequency or fertilization rates through disruption of courtship in other ways. For example, sound production plays an important role for attracting individuals at aggregation sites, in mate competition, during courtship, and for stimulating or synchronizing gamete release or maturation in several species of croakers (spotted weakfish *Cynoscion nebulosus* Gilmore 2003, white weakfish *Atractoscion nobilis* Aalbers and Drawbridge 2008, the goliath grouper Mann et al. 2008) and in the Atlantic cod (Rowe and Hutchings 2006). At reduced aggregation sizes, acoustic intensity may be insufficient to attract all individuals to spawning sites or may decrease fertilization rates due to reduced synchronization of gamete release (Rowe and Hutchings 2003).

Stress can affect reproduction in fishes although there is little evidence to suggest that this ultimately results in reduced annual reproductive output. In the common snook, *Centropomus undecimalis*, silver seabream and red gurnard, *Chelidonichthyes kumu*, stress in captivity can cause changes in hormone levels, fecundity, egg size and development, and egg survival (e.g. Morgan et al. 1999). However, the extent to which fishes may be stressed in exploited or disturbed spawning aggregations is not known.

Reductions in aggregation numbers could affect social cues that stimulate short or long-term changes in reproductive condition or are otherwise associated with mating. In many aggregating species, males show temporary courtship colour changes (Colin et al. 2003). In some groupers, courtship activity and colour pattern changes associated with spawning are less intense (qualitative observations include less activity by males in female pursuit and less frequent courtship) at sites with smaller, dispersed aggregations when compared to sites with large, dense aggregations (Nassau grouper Colin 1992, leopard grouper Erisman et al. 2007). Several groupers and the Atlantic cod form dominance hierarchies with male defence of breeding territories (e.g. scamp and gag grouper Gilmore and Jones 1992, tiger grouper Sadovy et al. 1994a, b, Atlantic cod Hutchings et al. 1999). Removal of fish during aggregation and breeding periods could affect these hierarchies, with unknown impacts on reproductive output. Since ovulation periods of aggregating species are often highly synchronized to coincide with short courtship and spawning periods, delayed or disrupted courtship could potentially cause eggs to over-ripen, thereby reducing egg viability or developmental success of eggs and larvae (Rowe and Hutchings 2003). The possible importance of aggregations for social cues in relation to sex change was addressed in Sect. 8.4.

One of the few specific examples of clear evidence that fishing on spawning aggregations can directly and negatively impact reproductive potential comes from the Atlantic cod (Rose et al. 2008). A spawning ground in Bar Haven, Placentia Bay, Newfoundland, was opened in 1997 and from 1998 to 2000 was heavily fished at spawning time. Between 33% and 40% of the total annual catch came from this area followed by a stock decline and less recovery in spawning biomass than predicted by fishery models. The data from the resulting model suggest not only that there was weak compensation in survival at low stock size, but that the mortality rate of the Bar Haven spawning fish was considerably higher than that in the general population, and that a decline in egg production and then recruitment were the results of aggregation-fishing.

Some aggregating species show high site fidelity, whereby fish return each year to spawn at the same sites; disruption of factors enabling such returns could potentially influence reproduction. Aggregations can persist at the same sites for many years, even decades, suggesting a degree of traditionality (Colin 1996; Domeier and Colin 1997). Examples of site fidelity across years exist for a range of species (*Epinephelus alexandrinus* = *Mycteroperca fusca* Waschkewitz and Wirtz 1990, groupers and wrasses Domeier and Colin 1997, Nassau grouper Starr et al. 2007, European plaice Hunter et al. 2003, camouflage grouper Rhodes and Tupper 2008). There is also good evidence in several species that social learning and tradition play a role in the repeated formation of aggregations at specific sites, and that younger fish learn to use and find sites from older, experienced fish (bluehead wrasse Warner 1988, 1990, Atlantic cod Rose 1993, European plaice Arnold and Metcalfe 1995, brown surgeonfish *Acanthurus nigrofuscus* Mazeroll and Montgomery 1998). Indeed, the learning component of fish migratory behaviour may be particularly important in those species comprised of multiple age groups, providing the opportunity for social transmission of migration routes (Dodson 1988). The removal of older, larger adults may therefore affect spawning and aggregation-formation indirectly through loss of knowledge of spawning site locations and migration routes and could seriously compromise recovery initiatives such as restocking using hatchery-produced fish in such species.

8.5 Socioeconomic Importance of Aggregation-Fishing and Aggregating Species

From a commercial perspective, aggregations represent excellent sources of large numbers of fish that can be taken quickly, predictably and efficiently, thereby saving both time and operation costs. These qualities, conversely, make them particularly challenging to manage. For valuable and high volume niche fisheries, such as those for live fish (e.g. groupers) and roe (e.g. mullet), and for species only accessible at aggregations, these are especially appealing, and, indeed, possibly the only (in the case of roe), time to quickly obtain high catches. Aggregations are also economically attractive for recreational fisheries, providing the opportunity for private-boat fishers to reach high catch limits in single trips, or, in the case of tourist fishers, allowing them to maximize the number of fish caught during daily fishing charters. They can also be very appealing to sports divers for viewing. We explore the socioeconomic implications of aggregation protection and exploitation.

8.5.1 Gluts, Prices, and Fisher Behaviour

The economics of having large gluts of fish filling local markets over a short spawning season and the potential for fish wastage have received little attention but likely offset some of the more obvious advantages associated with the commercial exploitation of aggregations. The flooding of a large number of fish into a limited market, i.e. where supply temporarily and massively exceeds demand, can result in the price per unit weight declining such that fishers get less cash per fish than at other times. In extreme cases, fish are wasted because there are too many for small local markets, or because they cannot be stored until market prices improve. During the peak fishing periods for Gulf corvina in the northern Gulf of California, Mexico, as much as 1,000 tonnes of fish may be caught and sold to local buyers over just 3–4 days. This flood of fish causes market prices to plummet, often over the course of hours. When the supply of fish exceeds the capacity of local processing and export plants, buyers close their markets completely, and several hundred tonnes of fish meat may be abandoned on the beach or thrown into the local dump by fishermen who can no longer sell their catch (Brad Erisman unpublished data) (Fig. 8.13).

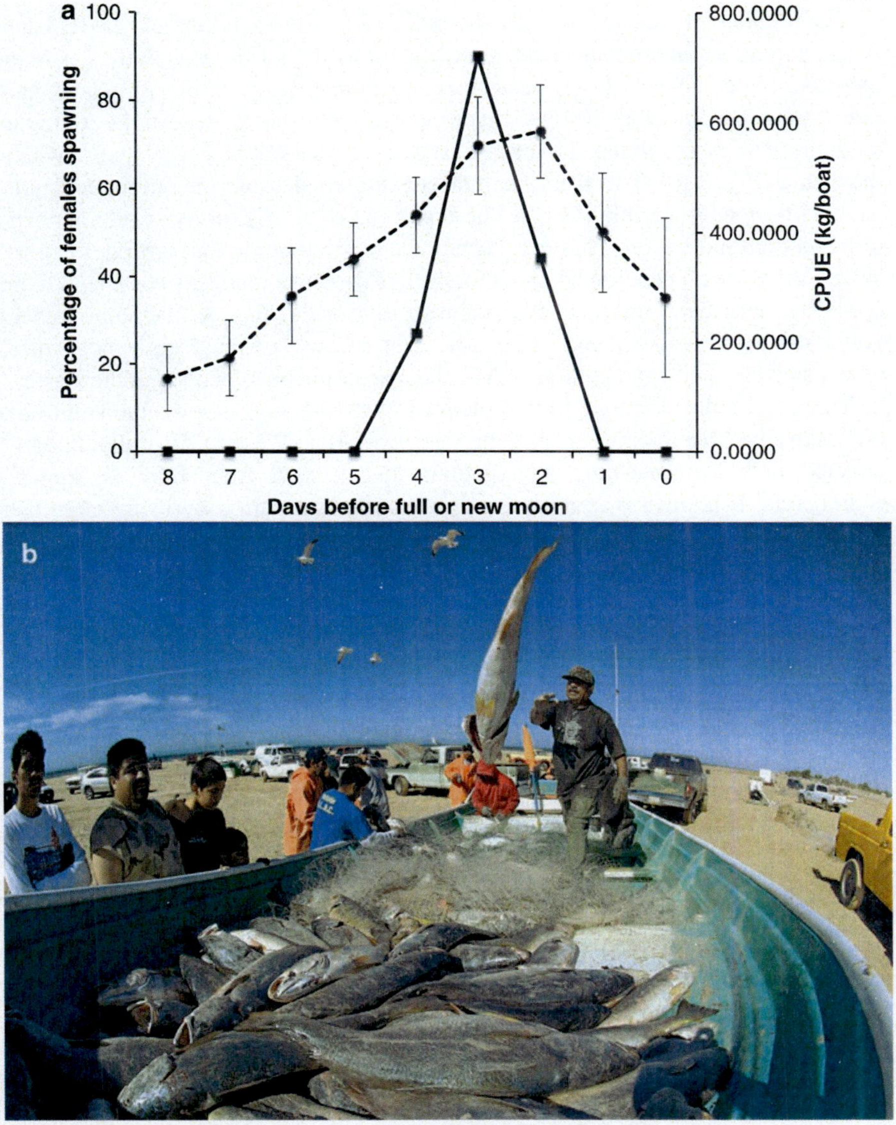

Fig. 8.13 **(a, b)** Gulf corvina, *Cynoscion othonopterus,* aggregations are fished with 12.5 cm mesh gillnets deployed from small boats that harvest up to 1 tonne of fish in a single net. Fisheries regulations allow only 1 net per boat. The commercial fishery only takes fish during the time periods when corvina migrate into the area to form aggregations and spawn in and near the estuaries during the week preceding the new and full moons from late February to early May each year (i.e. 4–5 times); spawning months have been combined for the graph. Each fishing period lasts 4–8 days, with the majority of landings occurring at the exact days of spawning when fish are inside the estuary. During these days, the landings and CPUE (*dotted line*) are highest; percent females spawning solid line (Photo: © Octavio Aburto/iLCP)

The problem of wastage may also be associated with the taking of ripe females. Aggregations are an obvious focus of interest for the LRFT because of the desire by fishers or traders to quickly accumulate large volumes ready for export in large carrier vessels (Sadovy et al. 2003). Females in many transient aggregations are found on or close to the spawning site for at least a few days until they spawn, after which they leave (Fig. 8.10). This means that most adult females taken at or around aggregation sites tend to be full of eggs. These gravid fish do not survive well after capture and high mortalities can occur during the holding phase that precedes export. Where there are no cost implications for traders (because mortality costs tend to be borne by the fishers) there is little incentive to cease fishing. While some traders prefer not to accept such fish from fishermen because of these high mortalities, others actively seek aggregations to fish and appear unconcerned about the waste.

Three examples illustrate the problem of gluts and suggest possible solutions. First, the silver seabream snapper fishery in Shark Bay, Western Australia, initially targeted only the spawning aggregations that formed from May to August. Commercial fishermen maximized catches during that period by using large fish traps (Chap. 12.12), which resulted in a glut of relatively low-priced landings during a short period flooding local markets. The fishery was placed under strict management in 1987 with entry limited to 51 tradable permits. In 2004 only 23 vessels were targeting snapper. Since 2001 these dedicated finfish boats were managed with an annual quota. These changes have caused fishers to think about ways of increasing the value of their individual catch, rather than ways of catching more fish. Nowadays the catch is landed throughout the year, avoiding the annual glut which occurred in the spawning season, and resulting in much better quality fish being taken to market (Gary Jackson personal communication 2010).

In the second and third examples, quite different, fisheries show possible responses to the problem of gluts. In Fiji several coastal species are fished heavily on aggregations. In some places, the price paid to the fisher, per fish or weight, at such times is lower (10–50%) for certain species when landings increase during aggregation time (Nanise Kuridrani Fiji Fisheries unpublished data). Concerns over the long-term implications of aggregation-fishing led the Fijian Fisheries Department to prohibit the capture of fish, destined for export, during the aggregation period. (Aisake Batibasaga Head of Fisheries Research Fiji Fisheries). In the third example, record low prices for the Atlantic wreckfish, *Polyprion americanus*, during spawning aggregations spurred the creation of an individual transferable quota (ITQ) system that resulted in catches being more evenly spread throughout the year (Gauvin et al. 1994).

While restrictions on sales or catches during the aggregation period may address problems associated with price declines due to gluts of target species, successful management will also depend heavily on both the responses of fishers to management and on market forces. For example, in the international LRFT, fish supply cannot fill demand and so all fish will get good prices even when seasonally plentiful. Moreover, as one supply area dries up, another will be opened such that the Hong Kong trade centre for the LRFT brings in fish from a wide range of, and ever changing, supply countries (Sadovy et al. 2003). In this case, and possibly for other international markets where demand is far higher than supply, local short-term gluts are unlikely to affect retail prices, although they can affect prices paid locally to fishers.

At the national level, examples of fisher responses to aggregation protection are instructive. In the Gulf of Mexico, USA, fishers for gag grouper, in response to protection from fishing during its peak aggregation month, changed their behaviour and intensified activity prior to and after protection; note that, initially, only one spawning month was protected with all spawning months protected subsequently. Economic analyses show that gag aggregation protection did not result in reduced fishing pressure on the target stock, or in increases in biomass, probably because of this behaviour (Smith et al. 2008). In other words, fisher response was so marked that it evidently offset the biological benefits from the temporary fishing restriction. The study indicates that seasonal restrictions cannot necessarily operate alone to reduce fishing pressure in an open-access fishery and highlights the importance of factoring in fisher behaviour and economics to biological models for management. In Pohnpei, during the closed season for grouper aggregations, the pressure on other reef fishes increased, in particular for parrotfishes (Scaridae), goatfishes and emperors, (Lethrinidae), to maintain overall sales (Rhodes and Tupper 2007; Rhodes et al. 2008). Loss of Nassau grouper stocks in the USVI and elsewhere due to lack of management and overfishing in the late 1970s led to fishers targeting smaller, less frequently marketed, groupers (e.g. Munro and Blok 2005; Karras and Agar 2009).

8.5.2 Costs and Benefits of Aggregation vs. Non-aggregation Fishing

Large and predictable fish gatherings are an obvious target for fishing, but long-term sustainability in the absence of sufficient controls may only apply for subsistence, or occasional cultural, uses. When commercial pressures come into play, without effective management (Sect. 8.5.1) this activity is considerably less appealing as a long-term good use of natural resources. A summary of the costs and benefits of aggregation exploitation versus total protection, highlights areas where further analyses are warranted to determine best overall and long-term monetary value from aggregating species (Table 8.2). The middle ground of managed aggregation fishing is also a possibility, with elements of both benefits and costs from the more extreme examples presented (see also Chap. 11).

While aggregation closures may greatly benefit stocks, associated costs and other implications can have serious detrimental effects on communities that may depend significantly on aggregations for their livelihood, and all possible management options should be considered. For example, the Colorado River Delta region located in the northernmost section of the Gulf of California is one of the most productive fishery regions in Mexico. A major source of this productivity is the seasonal occurrence of several commercially important fishes such as Gulf corvina, totoaba, Gulf croaker, *Micropogonias megalops*, sierra mackerel, *Scomberomorus sierra*, and several species of sharks and rays that migrate there to feed on sardines and shrimp and to spawn within massive aggregations (Cudney and Turk 1998; Rodriguez-Quiroz 2010). The local communities that surround the Delta rely on these seasonally plentiful resources as their main source of revenue, and the closure of their

Table 8.2 Economic and social benefits and costs of aggregation fishing versus total aggregation protection if aggregations supply a significant component of the annual catch of the target species population

Aggregation fishing permitted	Aggregation completely protected from fishing
Benefits	
– Reduced search time – Cost savings on fuel – Large, predictable catches can increase short-term profits – May temporarily reduce pressure on other reef resources – A ready periodic larder for social/cultural needs – Regular seasonal event for occasional community or customary activities	– Higher catch at non-aggregation times and over the long-term – Protection relatively easy to enforce as limited in time and space – Possible use for recreational diving benefits from aggregations – May produce more stable, higher market prices throughout the year by avoiding market gluts and ensuring larger fish populations, assuming that population is not overfished at non-aggregation times
Costs	
– Market gluts of fish that can lower price per fish and produce uneven supply of fish to market over the year – Taking risks in bad weather because competitors are fishing – Lack of consideration for making best long-term economic use of the fishery overall – Gluts can lead to wastage of unsold fish or those that cannot be stored for long time periods – Inequality of resource access, if some people can reach aggregation and some cannot – Possibility of overfishing resulting in long-term loss of resource for all and throughout year (Sect. 8.2)	– Loss of revenue to commercial/artisanal/recreational fisheries that have traditionally depended on fishing of aggregations – Possible increase in pressure on other reef resources temporarily – High incentive for poaching – Benefits of aggregation protection could be reduced if non-aggregation fishing intensifies and is unmanaged – Could be expensive to manage an aggregation and enforce protective regulation – Could lead to irregular/low supply of target species at certain times of the year – Could disrupt social or cultural activities

Note that the expense of monitoring is not included since this is necessary for both options

fisheries would likely result in the socioeconomic collapse of these areas. On the other hand, uncontrolled aggregation fishing could ultimately lead to loss of these valuable resources so careful management is clearly needed. Indeed, total protection of aggregations may, in the long term, enhance catches at non-reproductive times of the year and avoid waste of resources due to gluts with greater overall societal benefits. Economic analyses and fisher behaviour, therefore, in addition to biological considerations, are needed for understanding the effects of fishing on aggregating and non-aggregating animals and the implications of aggregation protection and to highlight the consequences of no protection. Such analyses might lead to solutions like ITQs or catch controls. For economic benefits, there is one further element that needs attention – the recreational value. This includes both recreational fisheries on aggregations and aggregation viewing by divers.

8.5.3 Economic Benefits from Recreational Use of Spawning Aggregations

In some places there may be considerable economic benefit in protecting aggregations entirely from commercial exploitation, substituting sport diving for extractive use, or introducing catch and release recreational fisheries. For example, the value of an aggregation of grouper in Belize was estimated at 20 times that of the extracted fish (per annum) if protected and used solely for diving tourism (Sala et al. 2001). This calculation focused only on the recreational diver benefit without factoring in the (additional) long-term benefit/value to the non-aggregation fishery of allowing fish to spawn. The extent to which displaced fishers might directly benefit in such cases is not clear, but work in Belize has successfully retrained displaced fishers as tour guides (Janet Gibson personal communication 2010). In Komodo National Park, Indonesia, the estimated value to the fishery of protecting aggregations within the park, allowing fish to fulfil their reproductive function and for their diving recreational value was significant (Ruitenbeek 2001). In the western Atlantic Turks and Caicos Islands, the Nassau grouper is a high-profile species in the dive tourism industry and divers are evidently willing to pay more for dive packages on which they observe more and/or larger fish (Rudd and Tupper 2002). Lack of effective management of such fisheries may thus impose significant economic externalities on the dive tourism industry (Rudd and Tupper 2002). Recreational fisheries on aggregations can be economically valuable and are covered in Sect. 8.2.2.

8.6 Perceptions and Education Needs

Despite a growing understanding of the aggregating habit in fishes, the food and commercial importance of aggregating species general, and recognition of the need for management, this is rarely a high fishery or conservation priority. Reasons for this include: (1) a lack of understanding, or even awareness of the presence of exploited aggregations in an area or that known aggregations may need management; (2) the presence of other management measures in the fishery which are assumed to provide sufficient protection in the absence of aggregation management; (3) the attractiveness of fishing seasonal high abundances and hence strong resistance to management; (4) the existence of roe fisheries that depend specifically on spawning fish; (5) the long-term traditional use of aggregations but lack of any previous experience of their vulnerability because, historically, levels of use were too low to be a problem or because of a shifted baseline with younger fishers unaware of changes; (6) a desire or need to maintain regular supplies of target fish to markets; (7) the need to increase fish landings to satisfy local or international markets (Sect. 8.1); (8) a lack of knowledge about the various possible direct and indirect effects of fishing on aggregations that are, as a consequence, not incorporated into fishery models (see research needs Sect. 8.7) and; (9) a lack of regular inclusion of targeted aggregation sites or of habitat often associated with aggregations (such as outer reef slope/shelf drop-off areas) into marine protected area planning (Sadovy de Mitcheson et al. 2008).

Challenges also persist in achieving effective aggregation management because of misperceptions or lack of understanding that make it difficult to gain public and political support. It is particularly in such areas that greater awareness and better understanding of these fisheries, and the economic and social implications of losing them, are needed. An obvious example is that the many fish taken over a short, often intense, period of time from aggregations give the illusion that there are plenty of fish, with no hint that a high proportion of a stock or population might be involved and hence that management might be advisable. As just one example, orange roughy in the 1980s seemed to be a boundless resource. "On one occasion 54 tonnes were caught in only 20 minutes' trawling – but this apparent abundance was an illusion, as the fish had been concentrated in spawning schools." http://www.teara.govt.nz/en/fishing-industry/5 (Robertson 1991). In managed fisheries, there may also be the belief that conventional management obviates the need for specific attention to be paid to aggregation management (e.g. leopard coralgrouper trout in Australia Chaps. 11, 12.9).

From a technical fishery monitoring perspective the seasonal change in behaviour of both fish and fishermen associated with aggregations make fish abundance difficult to estimate because catchability (i.e. the proportion of the total stock caught by one unit of fishing effort) changes over the year with the result that CPUE (widely used as an indicator of fish abundance) is not proportional to fish abundance. High catchability means that fish will continue to be caught in large numbers in aggregations even as the population as a whole declines, making aggregation CPUE a poor indicator of fishery status (Fig. 8.6, Chap. 11). For this reason, much care is needed if only fishery-dependent data are available to assess aggregation status, and every effort made to periodically monitor aggregations by fishery-independent means, ideally including visual assessment (e.g. Colin et al. 2003).

The value of protecting spawning adults to the fishery at non-aggregation times needs to be considered in fishery models to assist managers and political and community leaders in better appreciating the practical and economic implications of management, or lack thereof, in the long term. It is unlikely that dive tourism alone will bring higher economic benefits, especially if there is heavy social or economic pressure to fish in the region. The switch from viewing spawning aggregations as a target of fishing to a focus of protection will require a major shift in understanding and perspective that fishery analyses can provide. It is of value to note that many fishers intuitively understand the importance of allowing females full of eggs to spawn and can readily accept why aggregations might need to be managed (YSM and BE personal observation). Indeed, in some countries (e.g. New Zealand, Australia, USA, Bahamas) berried lobsters (females with external fertilized eggs) have variously been protected in fisheries for as much as 100 years; the same thinking has not yet been applied to female fish.

Finally, we recognize the problem of shifting baselines (Pauly 1995). Most exploited aggregations are not managed, may have already declined substantially or there is no community experience of reduced landings in aggregations when these are newly exploited. This means that much of the potential production from an aggregation may be lost before action is understood to be needed. Since even small numbers of fish concentrated briefly in one small area can appear remarkable, even scientific workers newly exposed to aggregations may have little idea about former abundances. As one specific example in reef fisheries, young researchers today in the Caribbean conducting UVC are often impressed to see a few hundred Nassau grouper gathered in one aggregation at sites that just three or four decades ago held tens of thousands of fish (YSM personal observation). Old film footage and verbal accounts of divers, biologists, or fishermen who have seen such spectacular biological events are particularly valuable for highlighting such changes.

8.7 Data Gaps

Based on the above review, we identify key data gaps which, if addressed, could substantially improve our ability to effectively monitor, manage and assess aggregations. Focusing on the gaps would help answer important questions such as how many fish can be removed safely from a fished aggregation, the economic value of management, whether full aggregation protection may result in a significant improvement of the associated, non-aggregation, fishery, and could explain why some, such as Atlantic cod, are not recovering as predicted by fisheries models (Hutchings and Reynolds 2004).

8.7.1 *Empirical Data on Effects of Fishing on Reproductive Output and Fishery Modelling*

Biological studies and theoretical considerations suggest that certain biological attributes of fishes could and sometimes do predispose some species to negative impacts on reproductive output brought about by aggregation-fishing or by related disturbances. What little evidence there is, however, while suggestive, is inconclusive and focused studies are needed to test hypotheses regarding possible impacts.

What is clear is that a precautionary approach is needed because so little is known about the subtleties of long-term effects of fishing on reproductive output and how such effect(s) might be mediated. We have only to consider the other animal taxa and a few of the examples presented in this section to acknowledge the possibilities. It is clearly a failing that fishery models and management thinking rarely take such actual and potential complexities of life history into account (Sadovy and Vincent 2002, Sect. 8.4). For example, direct, observational and experimental studies on the relationship between aggregation size, density, and reproductive activity would be particularly useful for determining whether minimum aggregation sizes are necessary to maintain viable populations. While reproduction may well continue even in much reduced populations, whether such could maintain a substantial fishery at historical levels seems unlikely based on what we currently know.

8.7.2 Fisheries Effects of Various Management Policies

It is important to document or model how various forms of management or protection of aggregations can generate increased revenue to fisheries in the long term, to track successes and failures following management, and to understand fisher responses following protection. Nemeth (2005) demonstrated that closure of spawning aggregations for red hind led to increased aggregation densities as well as increased the number and average length of red hind in adjacent fisheries. Likewise, the banning of commercial gillnets on the spawning grounds of white seabass is correlated with a recent increase in CPUE and catch of commercial fisheries in California (Allen et al. 2007).

Other possible management models could also be explored, such as pulse fishing or rotational closures (e.g. Graham 2001). However, while in principle complex management approaches might make sense on paper, the reality of management capacity usually precludes such approaches and calls for more conservative and simple measures.

8.7.3 Fisheries-Independent Surveys and Underwater Monitoring of Aggregations

Direct surveys of aggregation sites are necessary to validate the existence of spawning aggregations, estimate fish abundance and monitor the timing and duration of aggregations. The evaluation of these and other characteristics (e.g. catchment areas, larval connectivity) of aggregations is necessary because fisheries landings and CPUE data often do not accurately reflect the status of stocks for aggregating species due to the problem of hyperstability (Chap. 11, Sect. 8.3). Certain aggregation sites pose serious challenges for monitoring due to their remote locations, poor visibility for diving, strong currents, or other factors. Increased utilization of multibeam hydroacoustic surveys and other new technologies show considerable promise for surveying aggregation sites. However, hydroacoustic surveys are unreliable unless carefully ground-truthed (Chap. 9).

8.7.4 The Economics of Aggregation Fisheries

Economic assessments of aggregation fisheries and alternative uses of aggregations are important because economics drive natural resource use practices and policies, in which gains or losses in revenues rather than ecosystem health impart the greatest influence on decision-makers. Appreciation of the overall and long-term economic value of the fishery production supported specifically by aggregations, as well as economic implications of their losses, can come through fishery modelling combined with economic data and biological information. This requires data on the numbers, sizes and catchment areas of aggregating fish, market prices of fish, biological impacts of overexploitation of reproductive fish on the overall fishery, basic fishery parameters and percentage of annual catch from aggregations, fishing costs, profitability, general market conditions of supply and demand under different scenarios, and differences in catchability between spawning and non-spawning periods. The value and characteristics of the non-aggregation fishery component with or without aggregation management also need assessment.

8.7.5 Advantages of Fully Protected Aggregations and Implications

If certain aggregation sites are closed to fishing, other human use activities will usually need to be implemented to mitigate economic losses from fishing restrictions. These may include the introduction of diving charts to view aggregations or training for catch-and-release sportfishing of aggregations. Before such activities can be offered as alternative uses, however, it must be considered whether the alternatives could themselves potentially have negative impacts on reproductive activity requiring specific management steps. Therefore, studies which focus on understanding how the presence of divers, the use of various fishing gears, and catch-and-release fishing activities affect mating behaviour or survivorship are needed. While divers can evidently approach some species without apparent impact (as in the case of spawning cubera snapper in Belize), or species may be habituated to divers, little attention has been paid to this subject to date. In the case of the humphead wrasse (or Napoleon fish) *Cheilinus undulatus*, for example, the presence of divers interrupts courtship and spawning (Patrick Colin personal observation). Wider implications of protection also warrant examination, such as the impacts of shifting fishing effort from aggregations to other reef resources during periods of aggregation protection.

8.7.6 Information on Importance of Aggregations to Fisheries and Ecosystem Health

Due to widespread declines in harvested fish populations and the pervasive degradation of marine ecosystems worldwide over the past several decades, fisheries management is undergoing a theoretical and sometime actual shift away from single-species management and towards more ecosystem-based management approaches. To integrate spawning aggregations into this new management framework, their role(s) in ecosystem trophic and community structure, food web dynamics, and other ecosystem processes and parameters need to be appreciated (Chap. 2).

Unfortunately, research specifically aimed at understanding the effects of decreased aggregation sizes, or the extirpations of spawning aggregations, on the structures and processes of marine ecosystems, or on the target fisheries at non-aggregation times, is in its infancy (Chaps. 2 and 11). Certainly, there is anecdotal evidence that aggregations can attract large biomasses of megafauna and that some have long been in existence (Chap. 1).

8.7.7 Inclusion of Aggregation Sites as Key Biodiversity Areas or 'Indicators' of Reef Fishery Condition

While aggregations are not routinely incorporated into marine protected area design, their possible importance for 'congregatory' species that help to identify important areas for biodiversity, is being evaluated. For example, the selection of 'key biodiversity areas' as a conservation tool was initially applied to terrestrial species and the criteria used for evaluation address the vulnerability and irreplaceability of such sites (Eken et al. 2004). For congregatory species, sites that hold large proportions of the global population of a species on a regular basis and at specific times might be considered irreplaceable. This has already been applied in the case of birds for which the threshold is provisionally set at 1% of the global population of the species. Such a criterion could apply to fish species with relatively few transient aggregations. Considerably more information is needed to better understand the importance of individual aggregation sites for this criterion to be applied, however. The problem with site-only based protection of exploited species is that it does not, alone, address the problem of overexploitation so could only be one tool used in the case of most commercial fishes that are fished on their aggregations.

Aggregations have considerable potential to be used as markers of ecosystem, or as indicators of general fishery health (Chaps. 1 and 10), which would both highlight their importance and attract greater attention to them. While there are both advantages and disadvantages of using aggregations as indicators, on balance information to date suggests that they appear to be a good indication of the condition of related fisheries: where there is heavy overfishing, aggregations are small or non-existent, whereas in relatively unfished or managed areas, they persist (Sadovy de Mitcheson et al. 2008, SCRFA Newsletter No. 12, May 2009 www.SCRFA.org).

8.7.8 Perception and Education

Understanding fisher perceptions of the success, or otherwise, of protected sites and seasons is a crucial part of overall long-term acceptance and effectiveness of management measures. In one of the relatively few studies to directly address this, fishers from St. Croix, USVI, gave mixed responses as to whether they felt that a newly expanded area more effectively protected an aggregation site than existing area/season protection and gave them benefits. The general sense was that they did not fully understand the purpose of the additional protection and also felt that poaching was occurring because of insufficient enforcement and unclear boundaries to the protected area. There was also lack of clarity over perceived benefits of the seasonal closure, possibly due to slow recovery, and concern over economic hardship from the measures (Karras and Agar 2009). This study, and other experiences following management, demonstrates the importance of effective enforcement, communication with, and involvement of, fishers in planning, and monitoring of management outcomes. More studies on fisher perceptions are needed.

8.8 Concluding Comments

The habit of congregatory spawning is found in many species of fish from different ecosystems, in addition to coral reefs, and is common among commercial species. Indeed, it may be that the aggregating habit characterizes many exploited species expressly because it is a reproductive strategy associated with high productivity. If so, it is perhaps ironic that it is the aggregating habit that makes many such species particularly vulnerable to uncontrolled fishing.

Understanding the implications of overfishing or of eliminating aggregations is frustrated by insufficient information and by our lack of understanding regarding why aggregations have evolved. Empirical data from aggregations at different stages of exploitation and in relation to the populations they support are needed. From a biological perspective, there is much opportunity to incorporate elements of fish mating systems into fishery models to generate different scenarios against which to assess observations and make predictions or conduct comparative assessments. For example, if sperm competition is important for group-spawners and is negatively affected by declining fish numbers or sex-selective fishing, then one might predict that such species might show more marked declines than non group-spawning species. What are the effects of reduced aggregation sizes on reproductive activity and is there a minimum threshold size below which spawning ceases to occur? We might predict that transient aggregations are more vulnerable to targeted aggregation fishing than resident aggregators. From an economic perspective, are there more long-term benefits from fishing outside of aggregations and avoiding the possible waste and reduced prices associated with fishing gluts of spawning fish?

Turning to management, answers to the preceding questions should point the way to better determining the most effective and acceptable approach(es) to managing different types of aggregating species, and whether management of aggregations is necessary, in addition to management measures on target stocks (Chap. 11). Furthermore, valuable insights gained as we learn more of these species point to other important considerations. As just one example, it is clear that many transient spawning species aggregate at outer shelf edge areas, including drop-offs and reef passages. These are habitats that could be incorporated into marine protected areas since they currently receive little management or protection.

While in principle it should be possible to control fishing activity on aggregations to within sustainable levels, in practice this is unlikely to occur. Moreover, we still understand little of the implications of overfishing or of selective fishing on the long-term persistence of populations of aggregating species. And, even when management is in place, compliance is often low and enforcement typically falls short of being sufficient. For these reasons, it is clear that the most practical and precautionary approach is to avoid anything other than subsistence aggregation-fishing, at least in the case of transient aggregations, and to treat them as critical sources of fish that require protection for the benefit of the fishery. This situation calls for extreme caution in widely revealing the locations of little-known aggregations since it could easily expose them to greater fishing pressure. It also calls for consideration of the future of fisheries that depend exclusively on spawning fish, such as for roe or because the reproductive season is the only time that ripe fish can be caught.

More effort is needed to directly survey and monitor aggregations, since fishery-dependent data may not always reflect the status of aggregations (due to hyperstability), and to assess the economic value under scenarios ranging from unmanaged to fully protected aggregations. While new technologies may be applied or developed to survey aggregations at locations in which diver surveys are not possible, care is needed to assess their reliability, as in the case of hydroacoustic methods. Economic analyses are needed of aggregating fisheries, alternative use activities, and of assessments of the importance of aggregations to ecosystem structure and health. Finally, the benefits of managing and protecting aggregations need to be communicated to stakeholder groups, policy makers, and the general public so that consensus, support and compliance for policies are obtained.

Underpinning progress in management is a need for a shift in understanding about the importance of aggregations for both the fishery and the fish themselves. In particular, the 'illusion of plenty' that a healthy aggregation will inevitably produce represents a particular challenge for aggregation-fisheries. In many situations, it will be necessary to undergo a major paradigm shift, from seeing spawning aggregations as special opportunities for fishing to understanding them as particularly important times at which fish need to be protected, to ensure that they produce the next generation.

References

Aalbers SA, Drawbridge MA (2008) White seabass spawning behavior and sound production. Trans Am Fish Soc 137:542–550

Aburto-Oropeza O, Erisman B, Valdez-Ornelas V et al (2008) Commercially important serranid fishes from the Gulf of California: ecology, fisheries, and conservation. Cienc Marina Conserv 1:1–44

Adams S, Mapstone BD, Russ GR et al (2000) Geographic variation in the sex ratio, sex specific size, and age structure of *Plectropomus leopardus* (Serranidae) between reefs open and closed to fishing on the Great Barrier Reef. Can J Fish Aquat Sci 57:1448–1458

Aguilar-Perera A (1994) Preliminary observations of the spawning aggregation of Nassau grouper, *Epinephelus striatus*, at Mahahual, Quintana Roo, Mexico. Proc Gulf Caribb Fish Inst 43:112–122

Aguilar-Perera A (2007) Disappearance of a Nassau grouper spawning aggregation off the southern Mexican Caribbean coast. Mar Ecol Prog Ser 327:289–296

Aguilar-Perera A, Gonzalez-Salas C, Tuz-Sulub A et al (2009) Fishery of the Goliath grouper, *Epinephelus itajara* (Teleostei:Epinephelidae) based on local knowledge and fishery records in Yucatan, Mexico. Rev Biol Trop 57:557–566

Allee WC (1931) Animal aggregations: a study in general sociology. University of Chicago Press, Chicago

Allen LG, Pondella DJ II, Shane MA (2007) Fisheries independent assessment of a returning fishery: abundance of juvenile white seabass (*Atractoscion nobilis*) in the shallow nearshore waters of the Southern California Bight, 1995-2005. Fish Res 88:24–32

Alonzo SH, Mangel M (2004) The effects of size-selective fisheries on the stock dynamics of and sperm limitation in sex-changing fish. Fish Bull 102:1–13

Amores DAY (2003) The milkfish spawning aggregregation of Mactan Island, Central Philippines. Ocean Care Advocates, Inc. Report, April 2003

Arnold GP, Metcalfe JD (1995) Seasonal migrations of plaice (*Pleuronectes platessa*) through the Dover Strait. Mar Biol 127:151–160

Aumeeruddy R, Robinson J (2006) Closure of the live reef food fish fishery in Seychelles. Secr Pac Community Live Reef Fish Inf Bull 16:3–9

Bannerot SP, Fox WW Jr, Powers JE (1987) Reproductive strategies and the management of snappers and groupers in the Gulf of Mexico and Caribbean. In: Polovina JJ, Ralston S (eds) Tropical snappers and groupers: biology and fisheries management. Westview Press, Boulder

Beets J, Friedlander A (1992) Stock analysis and management strategies for red hind, *Epinephelus guttatus*, in the U.S. Virgin Islands. Proc Gulf Caribb Fish Inst 42:66–80

Beets J, Friedlander A (1999) Evaluation of a conservation strategy: a spawning aggregation closure for red hind, *Epinephelus guttatus*, in the US Virgin Islands. Environ Biol Fish 55:91–98

Berkeley SA, Chapman C, Sogard SM (2004) Maternal age as a determinant of larval growth and survival in a marine fish, *Sebastes melanops*. Ecology 85:1258–1264

Birkeland C, Dayton PK (2005) The importance in fishery management of leaving the big ones. Trends Ecol Evol 20:356–358

Blockstein DE, Tordoff HB (1985) Gone forever, a contemporary look at the extinction of the *Passenger Pigeon*. Am Birds 39:845–851

Burns KM, Koenig CC, Coleman FC (2002) Evaluation of multiple factors involved in release mortality of undersized red grouper, gag, red snapper and vermillion snapper. Mote Marine Laboratory Technical Report 790

Caro T (1998) Behavioral ecology and conservation biology. Oxford University Press, New York

Carter J, Marrow GJ, Pryor V (1994) Aspects of the ecology and reproduction of Nassau grouper, *Epinephelus striatus*, off the coast of Belize, Central America. Proc Gulf Caribb Fish Inst 43:65–111

Chale-Matsau JR, Govender A, Beckley LE (2001) Age, growth and retrospective stock assessment of an economically extinct sparid fish, *Polysteganus undulosus*, from South Africa. Fish Res 51:87–92

Cheung WWL, Pitcher TJ, Pauly D (2005) A fuzzy logic expert system to estimate intrinsic extinction vulnerabilities of marine fishes to fishing. Biol Conserv 124:97–111

Cisneros-Mata MA, Montemayor-Lopez G, Roman-Rodriguez MJ (1995) Life history and conservation of *Totoaba macdonaldi*. Conserv Biol 9:806–814

Clark M (2001) Are deepwater fisheries sustainable? The example of orange roughy (*Hoplostethus atlanticus*) in New Zealand. Fish Res 51:123–135

Claro R, Sadovy de Mitcheson Y, Lindeman KC et al (2009) Historical analysis of Cuban commercial fishing effort and the effects of management interventions on important reef fishes from 1960–2005. Fish Res 99:7–16

Coleman FC, Koenig CC, Collins A (1996) Reproductive styles of shallow-water groupers (Pisces: Serranidae) in the eastern Gulf of Mexico and the consequences of fishing spawning aggregations. Environ Biol Fish 47:129–141

Colin PL (1992) Reproduction of the Nassau grouper, *Epinephelus striatus* (Pisces: Serranidae) and its relationship to environmental conditions. Environ Biol Fish 34:357–377

Colin PL (1996) Longevity of some coral reef fish spawning aggregations. Copeia 1996:189–191

Colin PL, Sadovy YJ, Domeier ML (2003) Manual for the study and conservation of reef fish aggregations, Special Publications 1. Society for the Conservation of Reef Fish Aggregations, San Diego, California, pp 1–98

Craig MT, Sadovy de Mitcheson YJ, Heemstra PC (2011) Groupers of the world; a field and market guide. NICS digital and Print Publisher, South Africa and IUCN, Gland

Cudney R, Turk P (1998) Pescando entre mareas del Alto Golfo de California. Centro intercultural de estudio de desiertos y océanos, Puerto Peñasco

Dodson J (1988) The nature and role of learning in the orientation and migratory behavior of fishes. Environ Biol Fish 23:161–183

Domeier ML (2001) Giant sea bass. In: Leet WS, Dewees CM, Klingbeil R, Larson EJ (eds) California's living marine resources: a status report. Calif Fish Game, Sacramento

Domeier ML, Colin PL (1997) Tropical reef fish spawning aggregations: defined and reviewed. Bull Mar Sci 60:698–726

Dotson RC, Charter RL (2003) Trends in the Southern California sport fishery. Calif Coop Ocean Fish Investig Rep 44:94–106

Eken G, Bennun L, Brooks TM et al (2004) Key biodiversity areas as site conservation targets. BioScience 54:1110–1118

Erisman BE, Buckhorn ML, Hastings PA (2007) Spawning patterns in the leopard grouper, *Mycteroperca rosacea*, in comparison with other aggregating groupers. Mar Biol 151: 1849–1861

Erisman BE, Mascarenas I, Paredes G, Aburto-Oropeza O, Hastings PA (2010) Seasonal, annual, and long-term trends for commercial fisheries of aggregating reef fishes in the Gulf of California, Mexico. Fish Res 106:279–288

Evans RD, Russ GR, Kritzer JP (2008) Batch fecundity of *Lutjanus carponatus* (Lutjanidae) and implications of no-take marine reserves on the Great Barrier Reef, Australia. Coral Reefs 27: 179–189

FAO (2005) Review of the state of world marine fishery resources, Fisheries Technical Paper 457. Food and Agriculture Organization of the United Nations, Rome

Farley JH, Davis TLO (1998) Reproductive dynamics of southern bluefin tuna, *Thunnus maccoyii*. Fish Bull 96:223–236

Felsenstein J (1985) Phylogenies and the comparative method. Am Nat 125:1–15

Frank KT, Brickman D (2000) Allee effects and compensatory population dynamicds within a stock complex. Can J Fish Aquat Sci 57:513–517

Fromentin J, Powers JE (2005) Atlantic bluefin tuna: population dynamics, ecology, fisheries, and management. Fish Fish 6:281–306

Garratt PA (1996) Threatened fishes of the world: *Polysteganus undulosus* Regan, 1908 (Sparidae). Environ Biol Fish 45:362

Gascoigne J, Berec L, Gregory S et al (2009) Dangerously few liaisons: a review of mate-finding Allee effects. Popul Ecol 51:355–372

Gauvin JR, Ward JM, Burgess EE (1994) Description and evaluation of the wreckfish (I) fishery under the individual transferable quotas. Mar Res Econ 9:99–118

Gillett R, Moy W (2006) Spearfishing in the Pacific Islands. Current status and management issues. FAO/FishCode Review. No.19. FAO, Rome

Gilmore RG Jr (2003) Sound production and communication in the spotted seatrout. In: Bortone SA (ed) Biology of the spotted seatrout. CRC Press, Boca Raton

Gilmore GR, Jones RS (1992) Color variation and associated behavior in the epinepheline groupers, *Mycteroperca microlepis* (Goode and Bean) and *M. phenax* Jordan and Swain. Bull Mar Sci 51:83–103

Graham TR (2001) Advantages of pulse fishing in live product fisheries. Secr Pac Community Live Reef Fish Inf Bull 9:5–10

Graham RT, Carcamo R, Rhodes KL, Roberts CM, Requena N (2008) Historical and contemporary evidence of a mutton snapper (*Lutjanus analis* Cuvier, 1828) spawning aggregation fishery in decline. Coral Reefs 27:311–319

Graham RT, Rhodes KL, Castellanos D (2009) Characterization of the goliath grouper *Epinephelus itajara* (Lichtenstein, 1822) fishery of southern Belize for conservation planning. Endanger Species Res 7:195–204

Griffiths MH, Hecht T (1995) On the life-history of *Atractoscion aequidens*, a migratory sciaenid off the east coast of southern Africa. J Fish Biol 47:962–985

Hamilton R (2003) A report on the current status of exploited reef fish aggregations in the Solomon Islands and Papua New Guinea – Choiseul, Ysabel, Bougainville and Manus Provinces. Western Pacific Fisher Survey Series, Society for the Conservation of Reef Fish Aggregation vol 1

Hamilton RJ (2005a) Indigenous Ecological Knowledge (IEK) of the aggregating and nocturnal spawning behaviour of the longfin emperor *Lethrinus erythropterus*. SPC Tradit Mar Resour Manag Knowl Bull 18:9–17

Hamilton R (2005b) Papua New Guinea. SCRFA Newsl 8:11–13

Hamilton RJ, Matawai M (2006) Live reef food fish trade causes rapid declines in abundance of squaretail coralgrouper (*Plectropomus areolatus*) at a spawning aggregation site in Manus, Papua New Guinea. Secr Pac Community Live Reef Fish Inf Bull 16:13–18

Hamilton RJ, Matawai M, Potuku T (2004) Spawning aggregations of coral reef fish in New Ireland and Manus Provinces, Papua New Guinea: local knowledge field survey report (unrestricted access version). Report prepared for the Pacific Island Countries Coastal Marine Program, The Nature Conservancy. TNC Pacific Island Countries Report No. 4/04

Hamilton RJ, Matawai M, Potuku T et al (2005) Applying local knowledge and science to the management of grouper aggregation sites in Melanesia. Secr Pac Community Live Reef Fish Inf Bull 14:7–19

Harris PJ, Wyanski DM, White DB et al (2002) Age, growth, and reproduction of scamp, *Mycteroperca phenax*, in the southwestern north Atlantic, 1979—1997. Bull Mar Sci 70: 113–132

Heyman W, Requena N (2002) Status of multi-species spawning aggregations in Belize. The Nature Conservancy, Punta Gorda

Heyman W, Wade B (2007) Status of reef fish spawning aggregations in Belize. Proc Gulf Caribb Fish Inst 58:301–306

Hickey FR (2006) Traditional marine resource management in Vanuatu: acknowledging, supporting and strengthening indigenous management systems. SPC Tradit Mar Resour Manag Knowl Inf Bull 20:11–23

Hilborn R, Walters CJ (1992) Quantitative fisheries stock assessment: choice, dynamics and uncertainty. Chapman & Hall, New York

Hoarau G, Boon E, Jongma DN et al (2005) Low effective population size and evidence for inbreeding in an overexploited flatfish, plaice (*Pleuronectes platessa* L.). Proc R Soc B 272: 497–503

Hood PB, Schlieder RA (1992) Age, growth, and reproduction of gag, *Mycteroperca microlepis* (Pisces: Serranidae), in the eastern Gulf of Mexico. Bull Mar Sci 51:337–352

Hunter E, Metcalfe JD, Reynolds JD (2003) Migration route and spawning area fidelity by North Sea plaice. Proc R Soc B 270:2097–2103

Hutchings JA, Bishop TD, McGregor-Shaw CR (1999) Spawning behaviour of Atlantic cod, *Gadus morhua*: evidence of mate competition and mate choice in a broadcast spawner. Can J Fish Aquat Sci 56:97–104

Hutchings JA, Reynolds JD (2004) Marine fish population collapses: consequences for recovery and extinction risk. Bioscience 54:297–309

Johannes RE (1981) Words of the lagoon: fishing and marine lore in the Palau District of Micronesia. University of California Press, Berkeley

Johannes RE (1997) Grouper spawning aggregations need protection. Secr Pac Community Live Reef Fish Inf Bull 3:13–14

Johannes RE, Lam M (1999) The life reef food fish trade in the Solomon Islands. Secr Pac Community Live Reef Fish Inf Bull 5:8–15

Johannes RE, Riepen M (1995) Environmental, economic, and social implications of the live reef fish trade in Asia and the Western Pacific. The Nature Conservancy and the South Pacific Forum Fisheries Agency, Honolulu

Johannes RE, Squire L, Graham T, Sadovy Y, Renguul H (1999) Spawning aggregations of groupers (Serranidae) in Palau. The nature conservancy marine conservation research report #l, The Nature Conservancy and the Forum Fisheries Agency. 144 p

Johannes RE, Yeeting B (2000) I-Kiribati knowledge and management of Tarawa's lagoon resources. Atoll Res Bull 498:1–24

Karras C, Agar JJ (2009) Cruzan fishers' perspectives on the performance of the Buck Island Reef National Monument and the red hind seasonal closure. Ocean Coast Manag 52:578–585

King DPF (1985) Morphological and meristic differences among spawning aggregations of north-east Atlantic herring, *Clupea harengus* L. J Fish Biol 26:591–607

Kira GS (1999) The unforgettable sea of Cortez. Baja California's golden age 1947–1977: the life and writings of Ray Cannon. Cortez Publications, Torrance

Koenig CC, Coleman FC, Collins LA et al (1996) Reproduction in gag (*Mycteroperca microlepis*) in the eastern Gulf of Mexico and the consequences of fishing spawning aggregations. In: Arreguin-Sanchez F, Munro JL, Balgos MC, Pauly D (eds) Biology, fisheries and culture of tropical groupers and snappers, ICLARM Conference Proceedings 48. ICLARM, Manila

Koenig CC, Coleman FC, Grimes CB et al (2000) Protection of fish spawning habitat for the conservation of warm-temperate reef-fish fisheries of shelf-edge reefs of Florida. Bull Mar Sci 66:593–616

Koslow JA, Bax NJ, Bulman CM, Kloser RJ, Smith ADM, Williams A (1997) Managing the fish-down of the Australian orange roughy resource. In: Hancock DA, Smith DC, Grant A, Beumer JP (eds) Developing and sustaining world fisheries resources: the state of science and management. 2nd World Fisheries Congress Proceedings. CSIRO Publishing, Collingwood

Lercari D, Chavez EA (2007) Possible causes related to historic stock depletion of the totoaba, *Totoaba macdonaldi* (Perciformes: Sciaenidae), endemic to the Gulf of California. Fish Res 86:136–142

Levitan DR, Petersen C (1995) Sperm limitation in the sea. Trends Ecol Evol 10:228–231

Liu M, Sadovy de Mitcheson Y (2008) Profile of a fishery collapse: why mariculture failed to save the large yellow croaker (*Larimichthys crocea*, Sciaenidae). Fish Fish 9(3):1–24

Loher T, Seitz A (2008) Characterization of active spawning season and depth for eastern Pacific halibut (*Hippoglossus stenolepis*), and evidence of probable skipped spawning. J Northwest Atl Fish Sci 41:23–36

Lovatelli A, Holthus PF (2008) Capture-based aquaculture, global overview, FAO Fish Technical Paper 508. FAO, Rome, pp 1–298

Love MS, Brooks A, Busatto D et al (1996) Aspects of the life histories of the kelp bass, *Paralabrax clathratus*, and barred sand bass, *P. nebulifer*, from the southern California Bight. Fish Bull 94:472–481

Luckhurst BE (1996) Trends in commercial fishery landings of groupers and snappers in Bermuda from 1975 to 1992 and associated management issues. In: Arreguin-Sanchez F, Munro JL, Balgos MC, Pauly D (eds) Biology, fisheries, and culture of tropical groupers and snappers. ICLARM, Manila

Mackie MC, McCauley RD, Gill HS, Gaughan DJ (2009) Management and monitoring of fish spawning aggregations within the west coast bioregion of Western Australia. Final report to Fisheries Research and Development Corporation on Project No. 2004/051. Fisheries Research Report No. 187. Department of Fisheries, Western Australia

Mann B (2007) Reassessment of the seventy-four *Polysteganus undulosus* stock after a 10-year moratorium. Oceanographic Research Institute ORI. Unpublished Report No. 244 March 2007

Mann DA, Locascio JV, Coleman FC et al (2008) Goliath grouper *Epinephelus itajara* sound production and movement patterns on aggregation sites. Endanger Species Res 7:229–236

Marconato A, Shapiro DY (1996) Sperm allocation, sperm production and fertilization rates in the bucktooth parrotfish. Anim Behav 52:971–980

Marconato A, Shapiro DY, Petersen CW et al (1997) Methodological analysis of fertilization rate in the bluehead wrasse *Thalassoma bifasciatum:* pair versus group spawns. Mar Ecol Prog Ser 161:61–70

Marshall DG, Moore SA (2000) Tragedy of the commons and the neglect of science: planning and management in the Shark Bay World Heritage Area. Environ Plan Law J 2:126–137

Matos-Caraballo DJ, Posada JM, Luckhurst B (2006) Fishery-dependent evaluation of a spawning aggregation of Tiger grouper (*Mycteroperca tigris*) at Vieques Island, Puerto Rico. Bull Mar Sci 79:1–16

Mazeroll AI, Montgomery WL (1998) Daily migration of a coral reef fish in the Red Sea (Gulf of Aqaba, Israel): initiation and orientation. Copeia 4:893–905

McClenachan L (2009) Historical declines of goliath grouper populations in South Florida, USA. Endanger Species Res 7:175–181

Morato T, Cheung WWL, Pitcher TJ (2004) Vulnerability of seamount fish to fishing: fuzzy analysis of life history attributes. In: Morato T, Pauly D (eds) Seamounts: biodiversity and fisheries. 2004 Fish Centre Research Report 12. Fisheries Centre, University of British Columbia

Morgan MJ, Trippel EA (1996) Skewed sex ratios in spawning shoals of Atlantic cod (*Gadus morhua*). ICES J Mar Sci 53:820–826

Morgan MJ, Wilson CE, Crim LW (1999) The effect of stress on reproduction in Atlantic cod. J Fish Biol 54:477–488

Munday PL, Buston PM, Warner RR (2006) Diversity and flexibility of sex-change strategies in animals. Trends Ecol Evol 21:89–95

Muñoz RC, Warner RR (2003) A new version of the size-advantage hypothesis for sex change: incorporating sperm competition and size-fecundity skew. Am Nat 161:749–761

Munro JL, Blok L (2005) The status of stocks of groupers and hinds in the northeastern Caribbean. In : Proceedings of the 56th Annual Gulf and Caribbean Fisheries Institute, Tortola, pp 283–294

Musick JA, Harbin MM, Berkeley A et al (2000) Marine, estuarine and diadromous fish stocks at risk of extinction in North America (exclusive of Pacific salmonids). Fisheries 25:6–30

Myers RA, Barrowman NJ, Hutching JA et al (1995) Population dynamics of exploited fish stocks at low population levels. Science 25:1106–1108

Nemeth RS (2005) Population characteristics of a recovering US Virgin Islands red hind spawning aggregation following protection. Mar Ecol Prog Ser 286:81–97

Nemeth RS, Blondeau J, Herzlieb S et al (2007) Spatial and temporal patterns of movement and migration at spawning aggregations of red hind, *Epinephelus guttatus*, in the U.S. Virgin Islands. Environ Biol Fish 78:365–381

Olsen EM, Knutsen H, Gjoseaeter J et al (2004) Life-history variation among local populations of Atlantic cod from the Norwegian Skagerrak coast. J Fish Biol 64:1725–1730

Olsen EM, Lilly GR, Heino M, Morgan MJ et al (2005) Assessing changes in age and size at maturation in collapsing populations of Atlantic cod (*Gadus morhua*). Can J Fish Aquat Sci 62:811–823

Pajaro M, Macchit GJ, Martos P (2005) Reproductive pattern of the Patagonian stock of Argentine hake (*Merluccius hubbsi*). Fish Res 72:97–108

Parkes GB (2000) Protecting young fish and spawning aggregations of *Champsocephalus gunnari* in subarea 48.3 (South Georgia): a review. CCAMLR Sci 7:75–86

Pauly D (1995) Anecdotes and the shifting baseline syndrome of fisheries. Trends Ecol Evol 10:430

Paz GE, Truly E (2007) The Nassau Grouper spawning aggregation at Caye Glory, Belize: a brief History. The Nature Conservancy, Belize

Pears RJ, Choat JH, Mapstone BD, Begg GA (2007) Reproductive biology of a large, aggregation-spawning serranid, *Epinephelus fuscoguttatus*: management implications. J Fish Biol 71: 795–817

Penney AJ, Mann-Lang JB, van der Elst RP et al (1999) Long-term trends in catch and effort in the KwaZulu-Natal nearshore linefisheries. S Afr J Mar Sci 21:51–76

Phelan MJ, Gribble NA, Garrett RN (2008) Fishery biology and management of *Protonibea diacanthus* aggregations in far Northern Cape York Peninsula waters. Cont Shelf Res 28: 2143–2151

Pondella DJ II, Allen LG (2008) The decline and recovery of four predatory fishes from the Southern California Bight. Mar Biol 154:307–313

Rasotto MB, Sadovy de Mitcheson Y, Mitcheson G (2010) Male body size predicts sperm number in the mandarinfish. J Zool Lond 281:161–167

Reed JM, Dobson AP (1993) Behavioural constraints and conservation biology: conspecific attraction and recruitment. Trends Ecol Evol 8:253–256

Reynolds JD, Dulvy NK, Goodwin NB et al (2005) Biology of extinction risk in marine fishes. Proc R Soc Lond B 272:2337–2344

Rhodes KL, Sadovy Y (2002a) Reproduction in the camouflage grouper (Pisces: Serranidae) in Pohnpei, Federated States of Micronesia. Bull Mar Sci 70:851–869

Rhodes KL, Sadovy Y (2002b) Temporal and spatial trends in spawning aggregations of camouflage grouper, *Epinephelus polyphekadion*, in Pohnpei, Micronesia. Environ Biol Fish 63:27–39

Rhodes KL, Tupper MH (2007) A preliminary market-based analysis of the Pohnpei, Micronesia, grouper (Serranidae: Epinephelinae) fishery reveals unsustainable fishing practices. Coral Reefs 26:335–344

Rhodes KL, Tupper MH (2008) The vulnerability of reproductively active squaretail coralgrouper (*Plectropomus areolatus*) to fishing. Fish Bull 106:194–203

Rhodes KL, Joseph E, Mathias D et al (2005) Reef fish spawning aggregation monitoring in Pohnpei, Federated States of Micronesia, in response to local management needs. Secr Pac Community Live Reef Fish Inf Bull 14:20–24

Rhodes KL, Tupper MH, Wichilmel CB (2008) Characterization and management of the commercial sector of the Pohnpei coral reef fishery, Micronesia. Coral Reefs 27:443–454

Roberts C (2007) The unnatural history of the sea. Island Press, Washington, DC

Robertson DA (1991) The New Zealand orange roughy fishery – an overview. The Southern Trawl Fisheries Conference. Issues and opportunities. Melbourne 6–9 May 1990. Bureau of Rural Resources, Australian Bureau of Agricultural and Resource Economics, Canberra, pp 38–48 (Proceeding No. 10)

Robichaud D, Rose GA (2003) Sex differences in cod residency on a spawning ground. Fish Res 60:33–43

Rodriguez-Quiroz G (2010) Artisanal fisheries in the conservation zones of the Upper Gulf of California. Rev Biol Mar Ocean 45(1):89–98

Roman-Rodriguez MJ (2000) Estudio pobacional del chano norteno, *Micropogonias megalops* y la curvina Golfina *Cynoscion othonopterus* (Gilbert) (Pisces: Sciaenidae), especies endemicas del alto Golfo California, Mexico. Instituto del Medio Ambiente y Desarrollo Sustentable del Estado de Sonora. Informe final SNIB-CONABIO proyecto No. L298, Mexico D.F.

Rose GA (1993) Cod spawning on a migration highway in the north-west Atlantic. Nature 366:458–461

Rose GA, Bradbury IR, de Young B et al (2008) Rebuilding Atlantic cod: lessons from a spawning ground in coastal Newfoundland. In: Kruse GH, Drinkwater K, Ianelli JN, Link JS, Stram DL, Wespestad V, Woodby D (eds) Resiliency of gadid stocks to fishing and climate change. Alaska Sea Grant College Program, Fairbanks

Rowe S, Hutchings JA (2003) Mating systems and the conservation of commercially exploited marine fish. Trends Ecol Evol 18:567–572

Rowe S, Hutchings JA (2006) Sound production by Atlantic cod during spawning. Trans Am Fish Soc 135:529–538

Rowe S, Hutchings JA, Bekkevold D et al (2004) Depensation, probability of fertilization, and the mating system of Atlantic cod (*Gadus morhua* L.). ICES J Mar Sci 61:1144–1150

Rudd MA, Tupper MH (2002) The impact of Nassau grouper size and abundance on scuba diver site selection and MPA economics. Coast Manag 30:133–151

Rudershausen PJ, Buckel JA, Williams EH (2007) Discard composition and release fate in the snapper and grouper commercial hook-and-line fishery in North Carolina, USA. Fish Manag Ecol 14:103–113

Ruitenbeek HJ (2001) An economic analysis of the spawning aggregation function in Komodo National Park, Indonesia. Secr Pac Community Live Reef Fish Inf Bull 9:13–16

Sadovy Y (1993) The Nassau grouper, endangered or just unlucky? Reef Encounters June 10–12

Sadovy Y (2007) Report on current status and exploitation history of reef fish spawning aggregations in Palau. Western Pacific Fishery Survey Series: Society for the Conservation of Reef Fish Aggregations, vol 3. SCRFA and the Palau Conservation Society, www.SCRFA.org

Sadovy Y, Cheung WL (2003) Near extinction of a highly fecund fish: the one that nearly got away. Fish Fish 2003(4):86–99

Sadovy Y, Domeier ML (2005) Are aggregation-fisheries sustainable? Reef fish fisheries as a case study. Coral Reefs 24:254–262

Sadovy YJ, Eklund AM (1999) Synopsis of biological data on the Nassau Grouper, *Epinephelus striatus* (Bloch, 1792) and the Jewfish, *E. itajara* (Lichtenstein. 1822). NOAA Technical Report 146. U.S. Dept. of Commerce, National Oceanic and Atmospheric Administration, National Marine Fisheries Service, Scientific Publications Office, Seattle

Sadovy YJ, Vincent ACJ (2002) Ecological issues and the trades in live reef fishes. In: Sale PF (ed) Coral reef fishes dynamics and diversity in a complex ecosystem. Academic, San Diego

Sadovy YJ, Colin PL, Domeier ML (1994a) Aggregation and spawning of the tiger grouper, *Mycteroperca tigris* (Pisces: Serranidae). Copeia 1994:511–516

Sadovy Y, Rosario A, Roman A (1994b) Reproduction in an aggregating grouper, the red hind, *Epinephelus guttatus*. Environ Biol Fish 41:269–286

Sadovy YJ, Donaldson TJ, Graham TR, McGilvray F, Muldoon GJ, Phillips MJ, Rimmer MA, Smith A, Yeeting B (2003) The live reef food fish trade while stocks last. Asian Development Bank, Manila

Sadovy de Mitcheson Y, Cornish A, Domeier M et al (2008) A global baseline for spawning aggregations of reef fishes. Conserv Biol 22:1233–1244

Sadovy de Mitcheson Y et al (in press) Fishing groupers towards extinction: a global assessment of threats and extinction risk in a billion dollar fishery. Fish Fish

Saenz-Arroyo A, Roberts CM, Torre J et al (2005) Using fishers' anecdotes, naturalists' observations and grey literature to reassess marine species at risk: the case of the Gulf Grouper in the Gulf of California, Mexico. Fish Fish 6:121–133

Sala E, Ballesteros E, Starr RM (2001) Rapid decline of Nassau grouper spawning aggregations in Belize: fishery management and conservation needs. Fisheries 26:23–30

Sala E, Aburto-Oropeza O, Paredes G et al (2003) Spawning aggregations and reproductive behaviour of reef fishes in the Gulf of California. Bull Mar Sci 72:103–121

Sala et al (2004) Fishing down coastal food webs in the Gulf of California. Fisheries 29:19–25

Sanchez-Velasco L, Lavın MF, Peguero-Icaza M et al (2009) Seasonal changes in larval fish assemblages in a semi-enclosed sea (Gulf of California). Cont Shelf Res 29:1697–1710

Shapiro DY, Sadovy Y, McGehee MA (1993) Size, composition, and spatial structure of the annual spawning aggregation of the red hind, *Epinephelus guttatus* (Pisces: Serranidae). Copeia 1993: 399–406

Shapiro DY, Marconato A, Yoshikawa T (1994) Sperm economy in a coral reef fish, *Thalassoma bifasciatum*. Ecology 75:1334–1344

Smale MJ (1988) Distribution and reproduction of the reef fish *Petrus rupestris* (Pisces: Sparidae) off the coast of South Africa. S Afr J Zool 23:272–287

Smedbol RK, Wroblewski JS (1997) Evidence for inshore spawning of northern Atlantic cod (*Gadus morhua*) in Trinity Bay, Newfoundland, 1991-1993. Can J Fish Aquat Sci 54:177–186

Smith MD, Zhang J, Coleman FC (2008) Econometric modelling of fisheries with complex life histories: avoiding biological management failures. J Environ Econ Manag 55:265–280
Starr RM, Sala E, Ballesteros E et al (2007) Spatial dynamics of the Nassau grouper *Epinephelus striatus* in a Caribbean atoll. Mar Ecol Prog Ser 343:239–249
Stephens PA, Sutherland WJ, Freckleton RP (1999) What is the Allee effect? Oikos 87:185–190
Stoner AW, Ray-Culp M (2000) Evidence for Allee effects in an over-harvested marine gastropod: density-dependent mating and egg production. Mar Ecol Prog Ser 202:297–302
Teitelbaum A, Pryor T, Legarrec F et al (2008) Rabbitfish: a candidate for aquaculture in the Pacific? SPC Fish Bull 127:40–44
Tobin D, Wright PJ, Gibb FM et al (2010) The importance of life stage to population connectivity in whiting (*Merlangius merlangus*) from the northern European shelf. Mar Biol 157:1063–1073
Turner CH, Ebert EE, Given RR (1969) Man-made Reef Ecology. Calif Dep Fish Game Fish Bull 146:1–221
Tuuli CD (2010) The croaker fishery and dried swimbladder trade in Hong Kong, and the reproductive biology of the greyfin croaker, *Pennahia anea.* M. Phil. thesis. The University of Hong Kong, Hong Kong
Vincent ACJ, Sadovy Y (1998) Reproductive ecology in the conservation and management of fishes. In: Caro T (ed) Behavioral ecology and conservation biology. Oxford University Press, New York
Vunisea A (2005) Women's changing roles in the subsistence fishing sector in Fiji. In: Novaczek I, Mitchell J, Veitayaki J (eds) Pacific voices: equity and sustainability in Pacific Island fisheries. Oceania Printers Limited, Suva
Wang Y, Hu M, Sadovy Y et al (2009) Threatened fishes of the world: *Bahaba taipingensis* Herre, 1932 (Sciaenidae). Environ Biol Fish 85:335–336
Warner RR (1988) Traditionality of mating-site preference in a coral reef fish. Nature 335:719–721
Warner RR (1990) Resource assessment versus tradition in mating-site determination. Am Nat 135:205–217
Waschkewitz R, Wirtz P (1990) Annual migration and return to the same site by an individual grouper, *Epinephelus alexandrinus* (Pisces, Serranidae). J Fish Biol 36:781–782
Windle MJS, Rose GA (2007) Do cod form spawning leks? evidence from a Newfoundland spawning ground. Mar Biol 150:671–680
Yund PO (2000) How severe is sperm limitation in natural populations of marine free-spawners? Trends Ecol Evol 15:10–13
Zengin M, Cincer AC (2006) Distribution and seasonal movement of Atlantic Bonito (*Sarda sarda*) populations in the southern Black Sea coasts. Turkish J Fish Aquat Sci 6:57–62

Chapter 9
Studying and Monitoring Aggregating Species

Patrick L. Colin

Abstract The scientific study and monitoring of spawning aggregations requires field and laboratory work using a wide variety of physical and biological methods; but field work, no matter the technological tools available, will remain of fundamental importance in studying aggregations particularly for future comparison. Methods are becoming increasingly standardized allowing for meaningful comparison between sites, times and species. Methods can be fishery-dependent (gathering data from captured fishes) or -independent (observational, instrumental physical, interview data) within the subject areas of aggregation (1) discovery, (2) composition, (3) dynamics, (4) life history parameters and (5) physical parameters. Methods for each are summarized with selected subjects explored in more detail. Particularly problematic areas for data gathering have been fishery-independent determination of numbers and/or sizes of fishes and their distribution within aggregations and the larger environment. GPS based methods of mapping aggregation location, extent and fish density provide discrete snapshots of an aggregation, allowing visualization of dynamics over days to years and are repeatable any time in the future by others. Digital imaging (still and video) allows documentation previously impossible. Acoustic tagging, particularly alongside conventional tagging, allows delineating spatial and temporal aspects of aggregations. Hydroacoustic surveys are promising, but require validation of data on species present, fish numbers and their sizes. Aggregation sites should be instrumented to record physical data (e.g. temperature, currents, light) ideally year round, rather than just during aggregation, to allow comparison of aggregation periods with the entire year, and non-aggregation sites also evaluated for comparison Detailed bathymetric mapping of sites is important and feasible and allows visualization of geomorphology in relation to aggregations.

P.L. Colin (✉)
Coral Reef Research Foundation, P.O. Box 1765, Palau 96940, Koror
email: crrf@palaunet.com

Y. Sadovy de Mitcheson and P.L. Colin (eds.), *Reef Fish Spawning Aggregations: Biology, Research and Management*, Fish & Fisheries Series 35,
DOI 10.1007/978-94-007-1980-4_9, © Springer Science+Business Media B.V. 2012

9.1 Introduction

As a biological and oceanographic phenomenon, spawning aggregations provide a window into the natural complexity of the processes that maintain populations. Their study and monitoring normally requires fieldwork and involves a wide variety of physical and biological methods. This chapter is not a complete guide on how to study fish spawning aggregations, but is intended as a summary of some methods that have been used, recently developed methods (since 2003) and what might be done in the future. The SCRFA (Society for the Conservation of Reef Fish Aggregations) Methods Manual (Colin et al. 2003), available online at www.SCRFA.org, is supplementary, more comprehensive for some techniques and is to be updated at regular intervals. This chapter touches on the challenges of studying aggregations and highlights the need to better understand and properly monitor them.

Once the existence of a spawning aggregation is established (and even this seemingly simple task requires a strictly applied methodology) the type of information desired about the aggregation varies with its intended use. Academic scientists may be interested in "hard" data to address questions such as 'how and why' while the management and conservation community may focus solely on obtaining information that tells them the status of aggregating fishes, changes over time, and how to reverse negative trends in exploited populations. Monitoring is important and needed to assess the impact of management, yet like establishing the existence of aggregations, the meaningful monitoring of large temporary gatherings of fish can often be a major challenge.

Once the information needs are identified, the appropriate methods can be selected. Choosing the wrong method often results in dubious and inconsistent data while losing the opportunity to establish baselines and squandering resources. Careful selection of what is possible (and what is not) will hopefully result in information and results that are accurate, repeatable and informative. Standardization of methods will allow comparison of aggregations across geographic areas or of those examined by different workers over time. This has often been lacking in the past, but SCRFA has been diligent (with some success) in encouraging researchers to use consistent methods. Comparability of information is improving.

Work on spawning aggregations is still largely in the "discovery phase", in places where these phenomena are just beginning to become known. For a number of species and aggregation sites, knowledge has advanced to the point that more rigorous scientific investigations can now be undertaken. This requires collection of accurate data using suitable methods so the results from individual and geographically separate sites are comparable over time. Aggregations have usually been discovered by fishers, followed in some cases by site visits by biologists for confirmation of aggregation existence as well as basic data acquisition. This still continues, particularly to validate traditional ecological knowledge (Chap. 10). Sites are beginning to be instrumented to measure physical parameters while fishes are being tagged with acoustic and conventional methods to examine adult connectivity.

Despite progress, there is still a need to increase and improve overall work on aggregations to assess effects of fishing or track the outcomes of management action. In some cases the existence of aggregations is proposed based on unclear criteria (Chap. 1). Often information from fishers is accepted as definitive regarding aggregations, migration pathways, and other aspects of biology important for management without due diligence. Such information is still extremely valuable and should be documented, but cannot be treated as scientific fact until properly validated.

Interest in spawning aggregations has increased, but attention to other aspects of life history is also necessary to understand how to manage such species. Early life histories (from fertile egg to recruiting juvenile) and oceanography, for example, receive little attention (Chaps. 6 and 7). In efforts to identify whether specific conditions apply at aggregation sites, areas without aggregations are typically not used as control or comparative sites, to see if differences exist. Moreover, sites with aggregation are often not examined outside of aggregation periods. Researchers should be concerned with the entire life history of the species within an oceanographic framework (Lindemann et al. 2000).

9.2 Types of Information Needed for Work on Spawning Aggregations

Table 9.1 indicates some of the aspects of aggregations of probable interest to researchers and managers and identifies methods likely to prove useful in examining these subjects. Subsequently, each method is briefly described with more detailed notes regarding some of the methods. Methods useful for documenting fisheries and their long term monitoring and preservation are dealt with in Chaps. 10 and 11.

9.3 Fishery Dependent and Independent Methods

Some types of aggregation work can be undertaken without collection of fish by fishers or researchers. While the uncontrolled commercial exploitation of aggregations is generally not beneficial in the long term, if fishing activities are being carried out it would be counterproductive for researchers to pass up the opportunity to sample and otherwise learn from the capture of fish. Table 9.1 has an indication of whether a particular type of study method is fishery-dependent (relies on capture of fish) or independent of fishing.

Table 9.1 Some methods for examining spawning aggregations, using both fishery-dependent (FD) or -independent (FI) methods

Subject/parameter	F/D or F/I	Method
1. Discovery of aggregations	DI	DO, AT, TN
2. Composition of aggregation		
Numbers of individuals	DI	DO, VR, SP, TC, TG, SS, HA
Sex of individuals	DI	FS, US, DO
Size at sexual maturity	D	GN
Size of gonads	D	FS, US, FM
Potential fecundity	D	FM
Actual fecundity	D	FM
Estimated size of individuals (size frequency)	I	DO
Actual size of individuals (size frequency)	D	FS
Ages of individuals	D	OT
Population structure	D	GS
3. Dynamics of aggregations		
Occurrence of aggregation-daily, lunar, seasonal	DI	DO, UV
Methods of migration	DI	DO, EM
Distance of migration	DI	TG, AT
Persistence of individuals at site	DI	AT, TG
Location of individuals between aggregations	DI	AT, TG
Timing of spawning	DI	DO, AM
Location of gametes at release	I	DY
Percentage of fertilization	I	EC
Buoyancy of gametes after spawning	I	EC, AF
Drift of gametes after spawning	I	DL
Tracking eggs and larvae	I	DL
Predation on spawners	I	DO, VR
Predation on gametes	I	DO, VR
Spawning behaviour and coloration	I	DO, VR
Behaviour of fish after spawning	I	DO, VR
4. Life history parameters		
Sex change	D	FS
Growth rate	D	OT, TG
Monitoring spawning and recruitment success	DI	DO
Food and feeding during aggregation	D	FS
5. Physical parameters		
Geomorphology of aggregation aites	I	DO, BM, TS
Rugosity of site	I	BM, DO
Bathymetry of site	I	BM
Habitat mapping of sites	I	AF, DO, US, TS
Current regime	I	IS
Currents at time of spawning	I	IS, DO
Temperature, salinity, oxygen, water clarity	I	IS, DO

(continued)

Table 9.1 (continued)

Subject/parameter	F/D or F/I	Method
Location of gametes after release	I	DR, DY, DO
Migration pathways	DI	DO, AT

Key to FD/FI and Methods-FD/FI
D fishery dependent, *I* fishery independent, *DI* both, *AF* artificial fertilization and gamete collection, *AM* acoustic monitoring, *AT* acoustic tag, *BM* bathymetric mapping, *DO* direct observation by snorkeling or diving, *DR* drifter release, *DY* dye release and water tagging, *EC* egg collection, *EM* experimental manipulation, *FM* fecundity measurements, *FS* fishery sampling from fisher catch, *GC* gut contents, *GD* GPS density survey, *GN* gonad study, *GS* genetic sampling, *HA* hydroacoustic survey, *IS* instrument at site, *LD* longterm database, *OT* sampling and analysis of otoliths, *RV* remote operating vehicle, *SC* specimen collection, *SP* still photography, *SS* stationary (tether) survey, *TC* transect count, *TG* tagging, *TN* traditional knowledge through interviews, *TS* tape and compass survey, *UV* underwater visual census, *VR* video recording, *US* ultrasound imaging

9.4 Methods Described

The basic uses and limitations of methods are described below in single paragraphs. For many topics, more detailed technical information is presented in a subsequent section and the SCRFA study manual (Colin et al. 2003, www.SCRFA.org). For some additional methods, references are provided.

AF – Artificial Fertilization and Gamete Collection

One of the simplest ways to obtain large numbers of fertile eggs from a spawning aggregation is to collect individuals through diving or fishing at the time of spawning, hand-strip and mix the gametes. Fish to be speared can sometimes be selected by sex or size (Colin et al. 1987). Running ripe males can often be obtained some time ahead, and if maintained alive sperm can be stripped as needed. The sperm of dead males remain viable in the testes for hours. Testes (either in the fish or removed) can be chilled (but not frozen) for a day or more and the sperm remains viable (Colin et al. 1996). Eggs must generally be obtained at the time of spawning either through collection of females ready to spawn or following hormone injections of fish held in aquaria (Watanabe et al. 1995). It is uncertain how long hydrated eggs can be maintained before fertilization. Standard techniques are used to mix eggs and sperm for fertilization, followed by first cleavage within an hour or less to confirm development.

AM – Acoustic Monitoring

When reef fishes spawn planktonic eggs, they may produce both hydrodynamic sounds (noise) and species-specific sounds. Potentially these sounds can be monitored by an acoustic receiving system at or near the spawning site and used to document spawning activity over time (Lobel 1992).

AT – Acoustic Tagging

Implantation or external application of an acoustic transmitter to a fish or other marine animal allows tracking of its movements either by directional hydrophone or passive acoustic receivers positioned in the area of interest (Domeier 2005). Surgical implantation is now commonplace (Zeller 1997, 1998a; Nemeth et al. 2007). Fishes can be tagged while dispersed (outside of aggregations) or during aggregation. Internal tags used for fishes larger than about 50–60 cm SL transmit for a year or more. They also can provide information on temperature, depth and other parameters. The receiving stations gather data for many months, but need to be recovered for download at intervals and battery replacement. Acoustic tagging is expensive compared to conventional tagging, with a cost of US$100–200 per implanted tag, as well as US$2,000–3,000 per receiver. Sample sizes are usually constrained by costs, but should be maximized when possible. Arrays of over 50 receivers have been deployed for single studies, as well as hundreds of fish being tagged (Domeier 2005).

BM – Bathymetric Mapping

Electronic depth sounders with an integral Global Positioning System (GPS) unit allow recording of latitude – longitude – depth (y, x, z) data at sea, an essential part of any spatial examination of aggregation sites. A survey pattern run from a boat over an aggregation site and nearby areas is used to produce customized bathymetric maps (Colin et al. 2003). Multibeam surveys are expensive and equipment intensive, but are the ultimate in bathymetric mapping (Weaver et al. 2005; Nemeth et al. 2008). More information is provided below.

DO – Direct Observation by Snorkelling or Diving

Direct observation of aggregations is usually done by swimmers (snorkelling or SCUBA) at the aggregation site. The observer is equipped with suitable diving equipment depending on conditions. An underwater watch, preferably digital, and a recording slate are essential. Care should be taken that the presence of one or more observers in the water does not alter the behaviour of aggregated fish. Fish may be skittish, and often it is difficult to observe spawning. Observers try to get as close as possible to the fish, to see details that would be missed at a distance, but fish are often disturbed by this close approach. Many areas with aggregations also have high levels of spearfishing, so many fishes are already wary of humans. Possibly using red underwater lights at night would allow observation of fish behaviour for species suspected of spawning after dark; most reef fishes can see red light, but its penetration is limited in water, perhaps reducing its disturbance effect compared to white light. More information on snorkelling and diving with aggregations is included in Colin et al. (2003).

DR – Drifter Release

Current-following (Lagrangian) drifters can track water movement over time, and assuming that eggs are incorporated into the same water mass, can be used with certain caveats to track eggs and larvae. Most drifters have a drogue of various designs positioned at the depth of water to be tracked plus a marker float, usually with electronics for locating the drifter or recording its track. Deployment times range from a few minutes to months and years with costs increasing as the tracking and electronics involved become more sophisticated. The ability of drifters to mimic the movement of eggs and larvae is probably reasonably good for a few hours to a day or more, however once larvae develop the ability to change depth and swim horizontally this changes. Physical factors, such as wind shear, waves and windage of the drifter float, also introduce errors in tracking that increase over time.

DY – Dye Release and Water Tagging

Parcels of water can be tagged with dye for short term tracking (a few hours) of water which may contain eggs or larvae. This allows an estimate of where spawned eggs might go in the first few hours after spawning, and whether they would tend to be retained locally or dispersed into distant waters. Fluorescein, which produces a fluorescent green colour in the water, is most commonly used. An observer can mark a parcel of water where eggs are released by a squirt of dye (mixed with seawater) from a small flip-top plastic bottle. Party balloons filled with dye and water can be stationed underwater at a desired location and punctured by a pin releasing a sizeable dye cloud instantly (Appeldoorn et al. 1994). The relative strength and direction of currents can be determined by observers by releasing a small amount of dye and following its movement or tracking the dye stream emanating from a pinhole in a plastic bottle moored above the reef (Atkinson 1987). Large dye releases can be used to assess movement of large water masses with eggs (Carter 1994). However, dye usually disperses quickly to the point it is not easily visible. Fluorometry can be used to track the movement at low dye concentrations. Alternative means of "tagging" or tracking water with eggs have been suggested or tried, using "egg surrogates" such as fluorescent particles, wax particles and fatty globules (Appeldoorn et al. 1988). Drift vials (Domeier 2004) and drift cards are other possible means of tracking water moving away from an aggregation site.

EC – Egg Collection

To collect relatively small numbers of eggs from a gamete cloud (to describe the characters of the egg or examine fertilization rates) a fine mesh hand net can be used by a diver to sweep the water where eggs are released. If larger numbers of eggs are needed for rearing attempts, then they can be gathered using larger hand or plankton nets. The eggs should be collected after a wait of up to 1 min after release, since premature collection of eggs results in low fertilization rates (Petersen et al. 1992; Colin et al. 2003). Release of a small quantity of dye in the area of gamete cloud is often useful for tracking the eggs for that time. For large scale rearing attempts, it is usually necessary to strip gametes from ripe fish or otherwise artificially fertilize eggs obtained from broodstock (Watanabe et al. 1995).

EM – Experimental Manipulation

Experimental manipulations are relatively uncommon in aggregation studies. Mazeroll and Montgomery (1998) manipulated potential cues for migration pathways in Red Sea surgeonfish, while Warner (1988, 1995) manipulated entire spawning groups of bluehead wrasse, *Thalassoma bifasciatum*, to examine whether locations of aggregation sites were learned or innate. The various disruptions to spawning caused by the presence of humans around the aggregation and fishing activities are uncontrolled manipulations. In some regards their effects can be tested by comparing with undisturbed populations (Johannes et al. 1999; Colin et al. 2003). Knowing these effects are perhaps most important in considering whether certain activities, such as tourist SCUBA diving, should be permitted on aggregations, and if so, how they are best conducted.

FM – Fecundity Measurements

For fishes, fecundity is the number of eggs a female can produce and spawn and can be expressed at the level of a single spawning event, annually or over a lifetime. It is sometimes important information to know in any evaluation of the role aggregations play in population maintenance but its estimation is difficult and is only relevant for certain kinds of questions. Fecundity, for example, is sometimes used in standard fish stock assessments, such as spawning stock biomass per recruit models. Given the very high mortality rates between egg production and settlement, it is becoming increasingly evident that fecundity *per se* is not often directly relevant in pelagic spawning fishes for assessing reproductive success. More important is the relevance of body size to fecundity and egg quality, as well as the levels of natural mortality associated with the egg and larval stages. The potential (total number of all oocytes) and actual (total number of advanced oocytes) fecundity of females are typically different values due to atresia and other factors (West 1990) and it is important to be very clear in describing the methodology applied in fecundity studies (Sadovy 1996). The numbers of eggs for a given female is determined by first measuring the size of the ovaries, either their volume (usually by displacement of water in a graduated cylinder) or weight. Ovaries are then subsampled to determine numbers of eggs per unit weight or volume and their state of readiness for spawning. Ultrasound imaging has been used (Whiteman et al. 2005) to measure fecundity in living fish, determining gonad size with imaging techniques, then using cannula samples for oocytes per unit volume. Using hydrated eggs is one of the best approaches to measuring batch fecundity (i.e. number of eggs likely to be released in a spawning event) and this is only possible close to spawning. For annual or lifetime fecundity further methods have to be applied to determine spawning frequency, reproductive longevity and size specific parameters in relation to egg production.

FS – Fishery Sampling from Fisher Catch

A fishery on the aggregation may give access to data on length, weight, sex ratio, gonad condition and abundance from fishes caught by the fishery. Such "hard data" on fish numbers and size are useful for evaluating the accuracy of visually acquired information and as part of an assessment of management outcomes. Researchers can work with fishers to sample their catch for information that ranges from catch per unit of effort, catch volumes, gear selectivity, and sizes and sexes of fish taken in the fishery, to species identification, etc. Sometimes fishery capture data are not representative of the overall aggregation, i.e. males may be caught more often by hook and line than females, spear fishermen may specifically target males, or fishing only occurs at certain stages of aggregation formation. It may be useful to pay for sampling (measuring, weighing) as you may be disturbing people trying to make a living. Purchasing part of the catch allows you to keep the samples for other work without impacting the income of fishers. Often samples must be processed quickly for sale so researchers need to be able to work fast. If fish are being cleaned and filleted, the carcasses may normally be discarded. These should be retained and the fisher paid for them. If intact (fillets removed only) the length can be determined, the body cavity contents (including gonads) may still be present, tissue samples taken and otoliths removed from the head. The downside of fishery sampling is the prospect that fishing is wiping out the aggregation under study.

GC – Gut Contents

Whether fishes are feeding while engaged in aggregation is poorly known, so examining gut contents is useful for answering this question as well as for broader ecosystem implications. Some species seem to leave aggregation sites for part of the day and return at the time of spawning (Chaps. 2, 12.11). They may be feeding elsewhere. The guidelines of Randall (1963) on sampling gut contents still hold. Spearing or netting are the best means to obtain unbiased gut contents, but fish taken in traps or by hook and line should be excluded from any normal gut content analysis. Nonetheless, it is useful to know if fish in an aggregation will take a baited hook, or go into a baited trap, as this may indicate how easily they can be captured on the site using different fishing methods, providing perspective on how easy it is to overfish an aggregation.

GD – GPS Density Survey

This new technique produces a series of geolocated density measurements of fish in an aggregation (or any other fish that can be visually surveyed) by running a series of transects across an aggregation while towing a position logging GPS unit. The positioned density data can be used to produce a bubble plot or contour plot of fish density; estimated numbers of fish present in the aggregation can be obtained (Fig. 9.1). This method is discussed further below.

GN – Gonad Study

Examination of gonads is important because their size, especially relative to body size, and condition can be used to determine reproductive readiness and time of spawning. The Gonadosomatic Index (GSI), the weight of gonads relative to that of the fish (typically expressed as GSI = gonad weight/body weight * 100), can cheaply and readily indicate seasonality (See Chap. 3, Fig. 3.4, Chap. 12, Fig. 12.62), size of sexual maturation (from gonads taken from adults during the spawning season), and even lunar periodicity. The presence of hydrated eggs signals imminent spawning within about 24 h. Gonad samples can often be taken from fisher catches throughout the year or by market sampling. If no fisher catches are available, investigators may

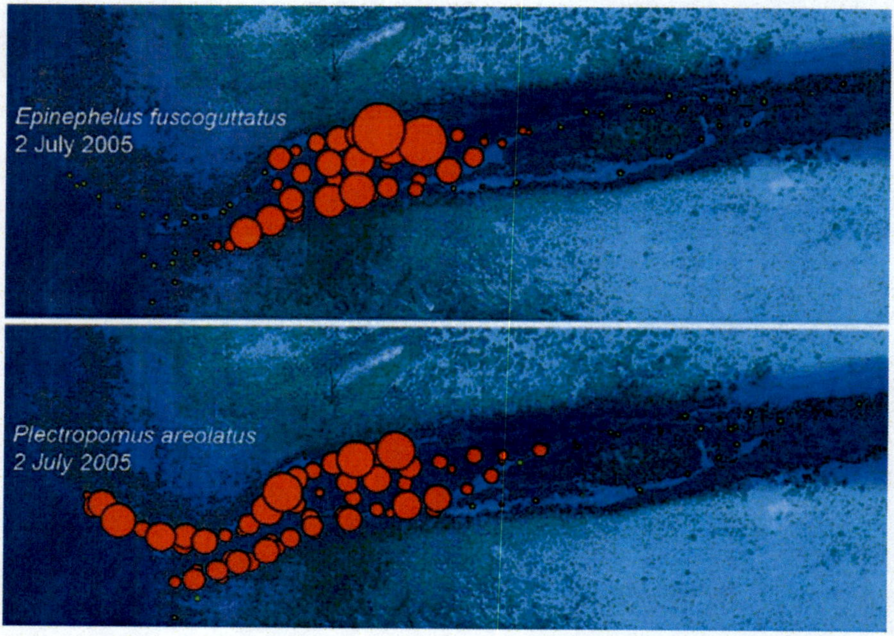

Fig. 9.1 Documentation of the distribution of aggregated fishes using the "GPS density" method provides a georeferenced measure of density which can be used to prepare bubble plots or contour plots that can be placed against an aerial image background, habitat map or bathymetric map. Locations where fish were present are indicated by large circles whose diameter increases with greater density, while locations where no fish were encountered are indicated by small, differently coloured, symbols. The distributions of different species (*top*: *Epinephelus fuscoguttatus*, *bottom*: *Plectropomus areolatus*) at the same time can be shown compared to one another, as indicated in the two panels here for the species and date

need to catch and process fish. In the absence of direct spawning observations, gonad condition, especially the presence of hydrated oocytes or post ovulatory follicles (the latter needs to be determined using histology) can be considered direct evidence of spawning (e.g. West 1990). Often gonads are discarded as fish are cleaned, but for some species they are a prized portion saved by fishers, which can make them costly or difficult to obtain. Sex can often be determined by externally, and gently, squeezing the body cavity of the fish. Milt (sperm) will often be exuded in ripe males, while if ready to spawn hydrated eggs may pour out of the genital opening of females. However, there is a period very shortly before spawning when eggs cannot easily be extruded due to tightening of the oviduct musculature at the vent. Fish can also be cannulated, if it is important not to kill them (being tagged or for conservation reasons) and sex and condition of the gonad determined. Samples may be preserved for histology, often requiring subsampling of the gonad, and preserved in Davidson's solution (or in its absence 10% formalin). Timing of spawning can be estimated from fresh gonad samples taken hourly over a period of time (Rhodes and Sadovy 2002a, b)

GS – Genetic Sampling

Fin clips placed in a suitable vial with a high concentration of ethanol is the simplest reliable technique for genetic sampling. Genetic samples are useful for looking at population structure relative to whether a single aggregation represents a single genetic stock, or whether there are multiple stocks at a single aggregation. Such information, combined with other biological and oceanographic information, is important for management decisions. If opportunity arises to collect tissues it might be important to take samples, even if not for immediate use, especially in the case of threatened species.

HA – Hydroacoustic Survey

This is a remote sensing method that combines the return sonar echoes from fishes with GPS positioning to map out the distribution and numbers of fishes over an area. It has been used for single species aggregations of commercially valuable fish like Atlantic cod, *Gadus morhua* (Lawson and Rose 2000) and orange roughy, *Hyplostethus atlanticus* (Bull et al. 2001). For reef fish aggregations, however, difficulties arise in separating echoes from a variety of fishes whose acoustic signals have been poorly characterized and may co-occur in an aggregation area. The high density of fish in some aggregations makes discrimination of individual fish problematic. Their availability for detection may vary according to their height above the substrate or their short-term movements in and out of detection range (Taylor et al. 2006). Such surveys should not be used without detailed validation because of the tendency to overestimate reef fish numbers (see later).

IS – Instrumentation at Site

Instruments installed at the aggregation site measure physical data, such as temperature, current speed and direction, tide, salinity and light. To document conditions at the aggregation site, they should be installed as close to the site as is reasonable. Similar monitoring instruments can be installed at locations other than the aggregation site to determine any differences between aggregation and non-aggregation sites.

Temperature loggers are inexpensive and should be deployed for any serious study. Current meters are useful to deploy at aggregation sites (Fig. 9.2) and many incorporate pressure (tide) and temperature sensors as well. A single instrument can now provide a suite of data for a spawning site that was almost unimaginable not very long ago. Information can either be extremely detailed (Fig. 9.3a) or show the overall picture of conditions at a site (Fig. 9.3b). Tide sensors provide important data because many aggregation areas are far from established tide stations and shelf edge sites may have very different tides. Salinity may be relatively constant at sites, but seasonal data would be useful if variation is likely. Recording light meters are useful if observations are made on a number of days of the timing of courtship and spawning activity, because day to day variation in the timing of spawning may be related to the light intensity as influenced by cloud cover. At most aggregation sites it is probably unnecessary to measure oxygen, as this is usually near saturation values

Fig. 9.2 Modern current meters provide accurate data on current speed and direction at (**a**) a single depth for the acoustic doppler current meter and (**b**) at multiple depths above (or *below*) for the acoustic doppler current profilers (*ADCP*). The units gather provide data on temperature and, if hard mounted on the bottom, the tide (Photos: Patrick L. Colin)

in areas where aggregations would occur. Theft or disturbance of instruments is a problem in many areas, but some (temperature and salinity loggers) can be hidden for extra protection if needed.

OT – Sampling and Analysis of Otoliths

There are many standard techniques for preparing and analyzing otoliths (reviewed by Campana 2001; Campana and Thorrold 2001). For tropical species that aggregate to spawn, it may be possible to identify annual growth rings in otoliths (e.g. Choat et al. 2006). For larvae and young juveniles, daily growth increments may be visible and quantifiable in otoliths, including settlement marks indicating the transition from pelagic to benthic life. Otolith chemistry (stable isotope ratios) also provides some prospects for looking at life habits of young and adult aggregating fishes (Patterson et al. 1999), or the environment in which larvae have grown up (Patterson et al. 2005).

RV – Remote Operating Vehicle (ROVs) and Submersibles

ROVs are seldom used for spawning aggregation observations, but can be useful where fish are deep or easily disturbed by a diver. Starr et al. (2007) used a ROV to observe spawning by the yellowfin grouper, *Mycteroperca venenosa*, at 40–45 m along a reef drop off at Glover's Reef, Belize, as well as observing aggregations of two other grouper species. Submersibles are usually beyond the financial capabilities of most aggregation studies, but Erisman et al. (2009) observed spawning of leather bass, *Dermatolepis dermatolepis*, from a sub at Cocos Island (east Pacific), while Gilmore and Jones (1992) observed deep grouper aggregations off Florida.

SC – Specimen Collection

Fishes from aggregations may need to be collected by researchers themselves, if there is no active fishery on the aggregation, using hook and line, fish traps, or by

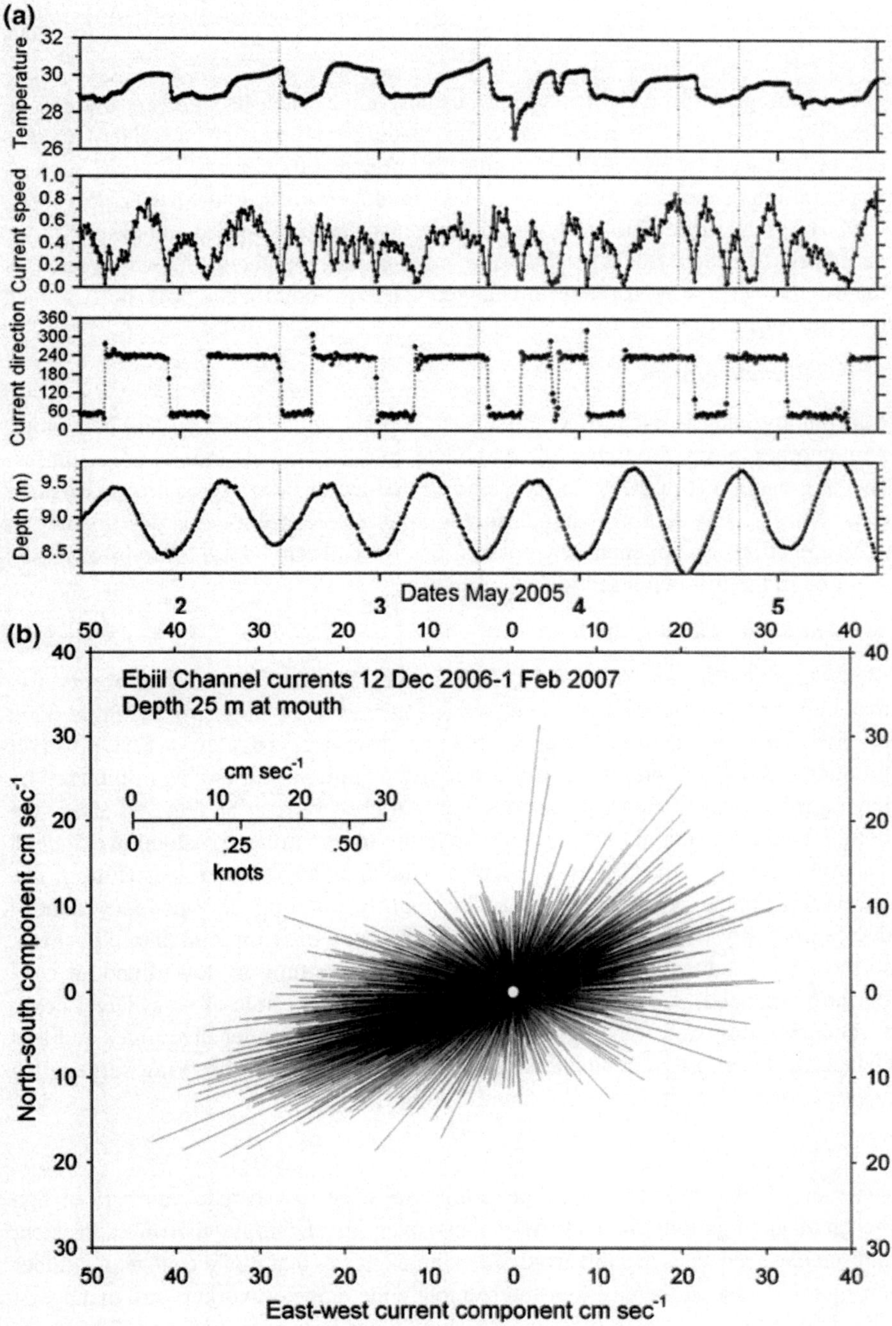

Fig. 9.3 (**a**) The types of data obtained from acoustic current meters provide a great deal of information on the physical parameters at an aggregation site and, if such data are combined with biological information regarding time of aggregation, courtship and spawning, these meters provide insights into the dynamics of fish spawning aggregations. (**b**) General current speed and directions at a site can be seen in a circulation diagram showing vectors of all currents measured during a monitoring period. This channel has a net export of water

spear. Line and trap caught fishes are usually alive when brought up and can be measured and tagged, then released. Swim bladders may need to be deflated; the gas can be released by hypodermic needle. If surgical procedures are used, such as implantation of acoustic tags, anaesthetics can reduce stress on the fish (Zeller 1997, 1998a). When returned to the water, fish are disoriented, stressed and easy prey for predators. A remote release trap (Nemeth 2005) or a weighted line which releases the fish have been used for returning tagged fish to the bottom and potentially reduce predator mortality.

SP – Still Photography

Still photography is used to document fish abundance, colouration and behaviour, among other things. Good photos can convey to others the magnitude of an impressive aggregation (Smith 1972) or provide evidence of new aggregations (Erisman et al. 2009). More recently, digital photography has revolutionized documentation of aggregations and opened new possibilities for gathering data. More information is found in the following section.

SS – Stationary (Tether) Survey

Many aggregations are reasonably stable in their location allowing an observer to make a series of observations, such as frequency of spawning, from a single point relative to the aggregation. The presence of an observer, particularly a SCUBA diver producing noise and bubbles could disturb aggregations and can be minimized by having the observer stationary and quieter. An observer can be tethered above the bottom by a line secured to the bottom; hovering like an inflated balloon over a field. The line, secured to the bottom, is clipped to the diver by a quick release fitting. The observer is made positively buoyant by slightly inflating the buoyancy control device normally worn while diving. There is no need to swim to maintain position, so the diver produces less noise and bubbles (as breathing is slowed and air consumption reduced) and the position provides an excellent field of view. Diver depth is reduced (compared to being on the bottom) allowing longer dive times and less chance of decompression sickness. Tethering was useful in observing surgeonfish aggregations in Puerto Rico (Colin and Clavijo 1988).

TC – Transect Count

A variety of different transect type counts are used to estimate numbers of fish present in aggregations and how they change over time. Counting fish on a consistent and standardized basis in short-lived, often large, groups that might change in number during the course of the day is of interest to a wide range of workers and managers. The difficulties and costs of doing so should not be underestimated. These are discussed more fully in the following section.

TG – Tagging

Conventional tagging, where an externally visible marker or branding with identifying information is attached to the fish, is still extremely valuable and when used with acoustic (sonic) tagging, the two are mutually advantageous. A number of studies have used conventional anchor type tags to assess site fidelity, movement between aggregations and home ranges for groupers (Colin 1992; Luckhurst 1998; Nemeth 2005; Starr et al. 2007) and snappers. Interestingly, the two longest distance migration records for Nassau grouper, *Epinephelus striatus,* (250 and 220 km) were the result of fortuitous recovery (when captured fish were cleaned) of implanted acoustic tags (Carter 1994; Bolden 2000). This is not surprising since external tags often only last for a few months before being lost (Shapiro et al. 1994). There are numerous considerations with regard to conventional tagging to increase tag retention and readability. Ideally external tags can have some type of numbering or coding system, such as specific placement on the body, which can be deciphered by a diver seeing a fish, so its presence can be noted without capturing it inside or outside of aggregations. Freeze branding is one method that holds promise for quickly tagging fish at an aggregation site for easy visibility by divers (Koenig and Coleman 1998). Nemeth (Chap. 2) includes additional information about considerations for conventional tagging studies. It is extremely important that a campaign to notify and inform fishing communities and to encourage/reward returns is extensively conducted in relation to tagging projects.

TN – Traditional Knowledge Through Interviews

Nearly all spawning aggregations have been discovered by fishers and in many communities fishers continue to acquire increasingly detailed knowledge of aggregation timing and locations. Fishers, particularly older individuals generally referred to as "patriarch" fishers, have a wealth of information and many are willing to share it, especially so it will not be lost when they pass away. A protocol for conducting fisher interviews was described by Johannes (1981) and adapted by others (Hamilton 2005) in which questions are asked in a manner that helps to produce accurate responses comparable to information gathered from other individuals and locations. Efforts are often made to also validate such information. Chapter 10 in this volume includes more detailed information on the use of traditional knowledge (Local Ecological Knowledge or LEK) of spawning aggregations and the importance of this approach for other types of information exchange as well as its challenges and limitations.

TS – Tape and Compass Survey

This crude method is sometimes the only alternative for producing maps showing the general relationships of features and habitats and for calculating area in a spawning aggregation. Distance and bearing between chosen points at the site are determined with measuring tape and a hand bearing compass and recorded on underwater slates (Fig. 9.4a). Any features along the survey line, such as the reef-sand interface, are sketched relative to the survey line (Fig. 9.4b) and this information is later used to prepare a rough map an area and plot the relationship of objects (Fig. 9.4c). It is useful in areas where bottom features are not visible in aerial or satellite photos and can allow mapping of home ranges of fishes and determining the outer limits of aggregation sites. The method is discussed in more detail in Colin et al. (2003).

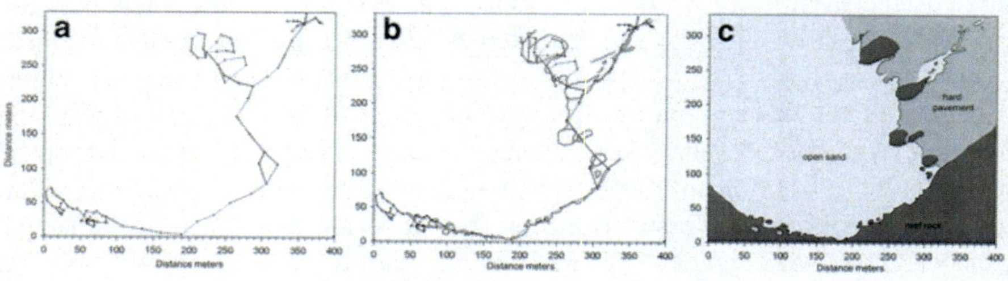

Fig. 9.4 The physical features of aggregation sites are often important to document in relation to the areas where fish aggregate and spawn. In cases where the bottom is too deep to easily prepare habitat maps from aerial or satellite images, a basic compass and measuring tape map can be prepared. The area shown here is a reef ledge area near Lee Stocking Island, Exuma Cays, Bahamas where coneys (*Epinephelus fulvus*), and blue tang (*Acanthurus coeruleus*) spawn. The three stages in preparing such a map are (**a**) the tape and compass survey along or between identifiable features on the bottom, (**b**) sketching of all visible features in relation to the survey lines and (**c**) identification and filling in of habitats on the base map

US – Ultrasound Imaging

Ultrasound imaging, using equipment designed for medical use on humans, has been used to sex and assess gonad condition of red hind, *Epinephelus guttatus* (Whiteman et al. 2005) from a spawning aggregation, as well as striped bass, *Morone saxatilis* (Will et al. 2002). Such data collection is usually part of a study in which fish are subsequently tagged and released (Nemeth 2005).

UV – Underwater Visual Census

Numerous methods have been developed for Underwater Visual Censuses (UVC) and UVC is conducted for many purposes. It is extremely important to apply an appropriate methodology for UVC surveys for these to be meaningful, repeatable and consistent. However, most standard UVC surveys of aggregations have a single, total count transect (either permanent or random) run through the aggregation area. While simple and an approach widely used for general surveys of reef fish diversity and abundance, the data generated have only limited use for aggregations because of the dynamic nature of aggregations in the short term over both time and within a small area, the often large numbers of species involved and the critical importance of timing of surveys. If single permanent transects are subdivided into segments, much more information is produced (Fig. 9.5) especially if multiple transects are used. "GPS Density Surveys", considered previously, are useful in some (but not all) cases (Fig. 9.1). If feasible, they provide the best survey information among present techniques, and are discussed further below.

VR – Video Recording

Underwater video recording, like underwater photography, is an important means of documenting behaviour of aggregating and spawning fishes. If done in a consistent and controlled manner, additional data can be derived from video recordings beyond

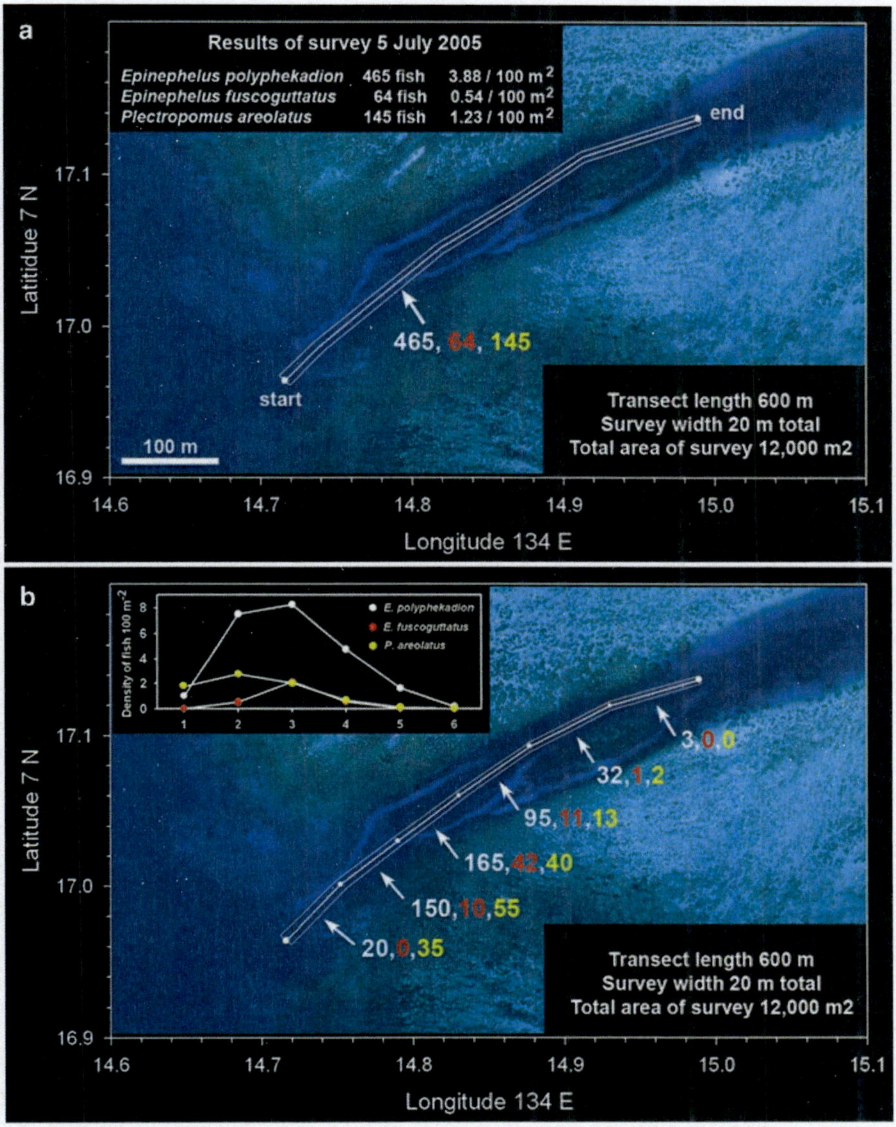

Fig. 9.5 A single transect up to several hundred meters in length is commonly used to quantify numbers of fish in spawning aggregations, generally through counting fish numbers in a swath width either side of the transect. Such surveys provide only a single measure of fish density and numbers. Single transect surveys can be easily improved by dividing the transect into a series of segments, in this case by distance, and counting fish in each segment separately, but multiple such transects are likely needed to cover an entire aggregation site (see Sect. 9.6.5)

simple documentation of behaviour (detailed in Colin et al. 2003). Video recording is also invaluable for conveying to others the large numbers of fish and dynamic nature of aggregations that cannot be obtained from any other medium. Aspects of video recording are discussed subsequently.

9.5 Additional Information on Methods: Biological, Physical and Sociological

The previous section was intended to provide only brief descriptions of available techniques, but not include detailed information on any particular technique. Additional information on some techniques is provided in the following section. Most are techniques the author believes hold promise for high quality data return with acceptable levels of effort and skills required.

9.5.1 Discovery of Aggregations

Most spawning aggregations were revealed to researchers through interviews with fishers as "traditional, or local, knowledge" passed down orally through generations of fishers, or as relatively new knowledge gained recently by commercial fishers. Traditional knowledge is common in Pacific Island cultures (Johannes 1981; Aswani and Hamilton 2004) and to a lesser degree in some Caribbean cultures (Craig 1966). Active programmes are recording and verifying this information. Techniques for interviews of fishers are discussed in Chap. 10. Older commercial fishers from "non-traditional" societies are often important sources of additional information. For example, the discovery of spawning aggregations of mutton snapper, *Lutjanus analis*, off the Dry Tortugas, Florida, USA, was made by commercial fishers, who shared their proprietary knowledge with scientists and managers (Burton et al. 2005).

Seldom do researchers actually encounter unknown or undocumented aggregations, but this can occur, particularly for species of limited commercial importance that are not searched for by fishers (Randall and Randall 1963). It may also happen in the case of dive guides who spend a lot of time in the water. The chances of success in locating unknown aggregations can be increased by searching during periods when many species are aggregating, such as the month and the lunar phase for aggregation by a given species in the region, examination of market samples to identify spawning periods, or searching in types of habitats known to harbour aggregations elsewhere. Visiting shelf edge Caribbean reefs around the full moon of December through May, for example, might reveal new aggregations of species already known to aggregate at such times and locations in the region. In Palau the humphead wrasse, *Cheilinus undulatus*, spawns on outer reef faces, both along straight sections of reefs and near promontories (Colin 2010) and several new aggregations were located in Palau by searching at times when aggregation and spawning were occurring at known sites. Towing a diver or snorkeler behind a boat might be productive as much more area can be covered compared to a normal SCUBA dive. This approach is useful for species that can vary somewhat in their aggregation area between years, such as the red hind (e.g. Kadison et al. 2006). Gathering new information about species already known to aggregate somewhere else is valuable, and helps improve the overall perspective of what aggregations mean biologically, as well as possibly allowing for comparative studies if aggregations are subjected to different levels of fishing effort. Then there is the "dumb luck" factor, which tells us that to discover something which is unknown it is best to be in the right place at the right time. The chances of this go up as more time is spent in the field looking at what is going on and, just as importantly, being able to recognize it when you see it.

The worldwide increase in diving tourism offers an opportunity for dive guides, underwater photographers and filmmakers to discover new aggregations. Expedition boats now visit the most remote areas of the ocean and if personnel on such vessels are attuned to looking for aggregations, many new discoveries could be made. The discovery of a spawning aggregation of the leather bass was made by guides from an expedition vessel (Aburto-Oropeza and Hull 2008).

Remote sensing is suggested for identifying aggregation sites (Kobara and Heyman 2007) using criteria such as the presence of reef promontories which may (or may not) be associated with them (see Chap. 5). Satellite images provide a remarkable perspective on the upper geomorphology of reef areas, but are limited in some regards. In many areas of the Caribbean region the insular shelf edge, along with many known TA sites, is not even visible due to clarity of the water in Landsat seven images (see Fig. 5.12, Chap. 5) and consequently satellite images are of little value in locating such sites. Often the location of promontories and general geomorphology is already known from other sources, such as generally available marine bathymetric charts, reducing the importance of satellite images in locating such features. Remote sensing seems an easy, high technology alternative to tedious field work to discover aggregations, but has not been applied with any success in locating any new aggregation. It is critical not to dismiss the need for field work to discover or verify the occurrence of aggregations. Moreover, hypothesis-testing for locating new aggregation sites needs to include the study of control sites not expected to be aggregation areas, and could be confounded by loss of active sites due to overfishing in the case of commercial species.

9.6 Measurement of Biological and Behavioural Aspects of Aggregations

The biological and physical dynamics of aggregations are of great interest since these are critical in the replenishment of population through recruits. Aggregations exist in a physical environment which cannot be separated from the biological, but aside from exceptional physical conditions causing reproductive failure, variation in biological aspects may be more influential in determining the outcome of reproduction (See Sect. 8.4).

9.6.1 Behavioural Aspects

Documenting the behaviour of aggregating fishes is important but often a difficult prospect. Data are obtained by observation, photographic or videographic methods, instrumentation or measurement. Interviews with fishers and experienced divers can provide useful information on behaviour and are considered further in Chap. 10. Behavioural work should not be limited to times and locations of aggregation. For some transient species, not all individuals are known to migrate to aggregation sites. Behaviour during non-aggregation periods has potential for providing added insight into changes occurring during aggregation and spawning. For example, the social structure of Nassau grouper on their normal home ranges is quite complicated, both during and outside of aggregation periods, with what appear to be dominance hierarchies, territoriality and intra-specific aggression (Chap. 12.6).

9.6.2 Tagging and Tracking

Conventional external tags are very useful for aggregation studies. Fish can be tagged while aggregated (Nemeth et al. 2007) or while dispersed (Colin 1992). Tag-recapture has also been used to estimate size of aggregations (Luckhurst et al. 2006). Nemeth et al. (2007) determined the "functional spawning migration area" of red hind aggregations in the Virgin Islands from tagging studies (Chap. 2). Distances and routing of migration relative to the aggregation site, differences between sexes and sizes of fish, can be examined if a sufficient number of fish are tagged. Having a fishery outside of the aggregation, and fishers willing to return the tags (usually with a reward and following an information campaign) is important for success. In such cases, the preparation phase for such a study is important to ensure that fishers know about the tags and where to return them as well as the reasons for the study. Planning such studies needs to take into account that recapture rates when using such tags (at least not in relation to aggregations) may be as low as 5%, so sample sizes need to be scaled accordingly (Nemeth 2005). The education value of this type of tagging is high since the fishing community is typically directly involved and can immediately understand the distances that fish are moving if informed of tagging location and reason for study.

Acoustic tagging is often intended to track "local" movements of fishes, but if large numbers of receivers are dispersed can monitor movements around entire islands or along coastal features (Domeier 2005; Nemeth et al. 2007; Starr et al. 2007). Tagging fish before migration to an unknown spawning site combined with a dispersed array of receivers could be a useful method for discovering an unknown spawning aggregation. This approach would probably be useful for species like Nassau grouper which are believed to migrate along shelf edges at specific times (Colin 1992) and along consistent pathways which can be instrumented with receivers.

External and acoustic (ultrasonic) tagging are mutually advantageous (Nemeth et al. 2007; Starr et al. 2007). External tags can provide information on short to long distance migration to and from spawning aggregations, as well as growth of individuals. Acoustic tagging provides data on occurrence and persistence of individuals at an aggregation, site fidelity over time and local movements (Zeller 1998b; Zeller and Russ 1998; Nemeth et al. 2007; Starr et al. 2007).

Fig. 9.6 Digital underwater photography now makes it simple to photograph behaviour related to spawning aggregations and spawning. In this photograph, two male camouflage grouper, *Epinephelus polyphekadion*, are engaged in a territorial dispute over areas within an aggregation area (Photo: Patrick L. Colin)

9.6.3 Photography and Videography

Still photography and videography are powerful tools for documenting aggregations (Fig. 9.6) and for gathering otherwise unobtainable data. Cameras with incredible technology are being released regularly with ever higher resolution and features, such as video clip time lapse, that were almost unthinkable previously. Cameras are now small enough to be less burdensome to carry on a regular basis. Digital still photography allows high resolution information on patterns and colours of courting and spawning fishes, with the ability to instantly assess via the view screen whether usable images of the aggregation and fishes are being obtained. It is generally found that use of additional lighting, such as high powered video lights, is disruptive to fish behaviour. Usually aggregations and spawning behaviour must be filmed using available light. Since spawning often occurs at times of low light (near sunset), the light sensitivity of the camera used is an important issue. Rapid sequences of photos (multiple frames per second) can be shot, with fast-moving behaviour recorded at resolutions much higher than video. Colin (2010) has a sequence of spawning by the humphead wrasse taken with a digital still camera. The still photo capabilities of video cameras (and video capabilities of still cameras) are also improving, to the point that usable sequences of behaviour can often be taken from either still or video sequences from either type of camera.

Planning can amplify the data return from photography and videography. Most digital cameras and video have a clock which should be set to the correct local time. A GPS receiver has a highly accurate time display once a position is acquired and

Fig. 9.7 An aggregation of many thousands of blacktail snapper, *Lutjanus fulvus*, is shown on a side of a barrier reef channel in Palau. The photograph is a panorama made up of several photos shot in just a few seconds using a digital SLR camera and covering over 120° horizontally. The thousands of snappers appear almost as a wave extending over the bottom and down the slope of the channel (Photo: Patrick L. Colin)

wristwatches and the clocks on cameras can be set to "GPS time". The accurate time allows the exact time of events filmed to be determined and for sequences of photos the time between frames is also recorded.

Digital still cameras are useful for shooting a rapid series of photos while panning across a large aggregation. The short interval between photos means the fish do not move any significant distance between photos. These can later be stitched into a single panoramic image covering a field of view of 180° or more, which otherwise could not be obtained with a single photograph (Fig. 9.7).

If digital still photography is combined with a GPS float logging positions at an interval of a few seconds (such as is done during GPS density surveys of aggregations) the location of each photo, based on the approximate position recorded at a given time, is known. Such photos are invaluable as a "hard" record of the aggregation at a given location and time. Complete photograph coverage of an aggregation with known photo positions could be done by taking a series of vertical photos (camera pointing straight down) from above the aggregation while logging positions. The numbers of fish could potentially be determined from each photograph and the photos could be combined into a single photomosaic of the aggregation.

For video cameras, many of the techniques are similar to still photography. If using a video camera to document behaviour, it is often useful to have the video camera running continuously from a known start time during a dive (assuming the recording medium has sufficient duration). Behaviour of interest often occurs unexpectedly and if the camera is already recording (rather than waiting for the lag time for recording to start once the record button is pushed) rapid action can be captured. Analogue video cameras are able to produce a useable image at lower light levels, but lack the resolution of digital video in higher light conditions. At present no single video camera is best for all conditions where it might be necessary to document behaviour through video recording.

9.6.4 Time Lapse Video and Still Photography

A stationary still or video camera positioned so it can see an aggregation site is a very useful, but underutilized, tool for examining the changes in fish abundance and behaviour over a period of time (Whaylen et al. 2006). It can be especially useful for wary fish, such as the humphead wrasse (Colin 2010). The camera is mounted on a tripod or other fixed structure so it is steady and properly oriented. Such photo and video monitoring sites should be carefully chosen, based on previous knowledge, as a slight error in locating the camera system can mean no images of value are recorded, while the activity of interest is occurring and visible just a short distance away. There is always the factor that if the aggregation moves slightly between days or years, the equipment, even with detailed prior knowledge, might not be positioned for optimal information gathering.

A modern digital camera can (at present) take thousands of high resolution photographs during a single deployment. If such a camera in an underwater housing stationed on a tripod at an aggregation site takes a photo every few minutes, a detailed record of the aggregation would be obtained over a period of days, assuming a strobe is used at night (Fig. 9.8). A simple system is limited to available light; however, for night time use a strobe system could be utilized. Some cameras have a connector for externally triggering the camera via an electrical switch and these would be easier to use for a timed deployment using a simple timer circuit and relay to trigger the camera. Other cameras, without a remote switch capability, might need a small solenoid to trigger the camera, via the shutter release button, and this would involve more battery power for such a circuit.

It might be useful to position the time lapse camera on a tripod pointed straight up at the aggregation site. Using available light, any fishes present above the camera within the angle of view of the lens should be visible and since they are against a lighter water surface, might be easily countable in each frame.

Video camera provide similar capabilities, some can be configured to take a short sequence (a few sec or more) at some regular interval (Whaylen et al. 2006). Recording on hard drives, instead of video tape, allows even more time. Probably with such equipment battery life becomes the limiting factor and can be extended using external battery power.

9.6.5 Measurement of Numbers and Density of Fish in an Aggregation

One of the most basic questions about any aggregation is how many fish are present; a seemingly simple question which is often surprisingly challenging to properly address. Seven different methods were suggested in Table 9.1 to approach this problem (and there are probably many others possible), however, none are truly accurate and in some the results cannot be cross-checked or verified. Ideally multiple

Fig. 9.8 Time lapse images taken with a high capacity digital camera system can allow the monitoring of aggregation presence/absence as well as aspects of behaviour over many day without the presence of an observer at the site. Images of a blacktail snapper (*Lutjanus fulvus*) aggregation in Palau were taken on one day at four different times. The sequence shows (**a**) fish present in the early morning in numbers similar to the rest of the day, (**b–c**) they remain at the site until at least mid-afternoon and (**d**) at sunset the they all leave the sites quickly a few minutes before sunset

approaches to estimating numbers should be used, as agreement between several methods provides some confidence that the data approach actual values. There is a need to balance simplicity of methods with reliability of data and consider planning for repeated evaluations in the long-term. Methods can be divided into "visual" and "non-visual".

Visual estimates of individual fish numbers in an aggregation can be accurate if an aggregation consists of less than 50–100 fish. An observer could feasibly count all individuals present with high accuracy if they are in clear water (so most or all can be seen at one time), the entire site is comprehensively searched, and if they do not move significantly or hide during the count. Such situations are uncommon, but do occur (Samoilys 1997, Chap. 12.9). If fish are constantly in motion or hide on observers approach, accurate counts are difficult. For small aggregations multiple successive counts, each attempting to survey all fish, should be attempted. If the numbers are consistent between counts, this provides some confidence the total aggregation has been surveyed with precision. If numbers vary, with individuals moving, hiding or behaving unnaturally, this implies reduced precision for surveys and such data viewed more sceptically. Hiding can even be sex-specific (Sadovy et al. 1994). Other factors, such as water visibility, frustrate determining fish numbers in even small aggregations.

Accuracy decreases with increasing aggregation size (i.e. number of fish), or if the species concerned is shy or otherwise difficult to survey. What sort of accuracy can be expected for various types of surveys? For fish numbers over about 100 individuals accuracy drops rapidly with increasing numbers, errors amplified by movements of individuals and the group as a whole. At some point there is a major disconnect between the ability of a human observer(s) to determine the numbers of fish present by direct observation and actual numbers.

If many thousands of fish are aggregated for spawning, how can the accuracy of estimated numbers be determined (e.g. Fig. 9.8, Chaps. 12.6, 12.9, 12.10, 12.11). At present there is no way to do that, although referring to examples from the literature on counting birds in flocks may be useful when the task is to count large numbers of massed fish. How can the differences between 5,000 versus 6,000 fish swimming in a densely packed aggregation be determined by divers in the water? It should just be accepted that using present visual methods estimates of fish numbers in large aggregations are inaccurate, that estimates of fish numbers become increasingly unreliable with increasing numbers and that this information should be treated as more qualitative than quantitative. Extreme qualitative changes (say from 1,000 to 200 fish) in fish numbers can be detected by direct observational surveys, but the actual values cannot. But by definition, estimates cannot be accurate and there is little basis to say the "accuracy of estimates" can be improved. Estimates can be made more precise through improved or multiple methods, ideally so estimated numbers can have the same level of precision day to day. However accuracy remains elusive and other factors such as inter-diver variation in counts should also be considered.

Training in estimating fish numbers can help to an extent, mostly by looking at underwater photographs or videos of fish groups (without knowing the actual number visible) and estimating the numbers present, and then comparing with the actual numbers of fish visible. In this manner the observers can get an idea of what 100, or 500 or 1,000 fish look like underwater. Using multiple observers and arriving at consensus values for fish numbers may be helpful, but the basic lack of accuracy still persists. Estimates should not be treated later as accurate data. Such qualitative data, while useful, does not identify absolute trends in population numbers in aggregations (Heyman et al. 2005) and there is still no way to assess accuracy levels, although Heyman and Kjerfve (2008) state, using two teams of observers, "trends in abundance were captured with accuracy within 10% since the maximum numbers ..... were obtained on the same day by both teams". Where there were major differences between team counts, this was attributed to one team having seen a school of fish which the other missed. Doing statistical analyses on such estimates is inappropriate.

Single line transects through aggregation areas have been used to obtain quantitative data on fish numbers although multiple, often parallel transects are most likely

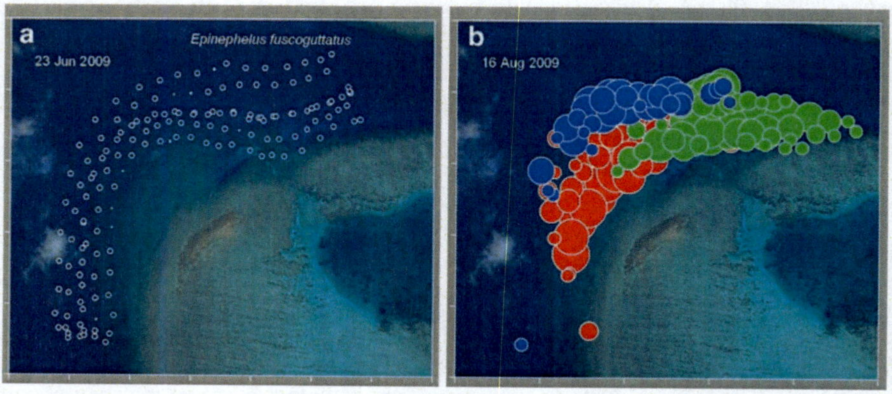

Fig. 9.9 (**a**) Five transects were set up in a channel in north Palau, the central one permanent. (**b**) The distributions of three species of groupers are shown relative to the channel on a peak aggregation day in August 2009. A single transect would not cover most of the area of the aggregation and would give a very biased view of the abundance of fishes at the site – the central transect is repeated, hence the close overlap of points (Source Palau Conservation Society 2010)

needed to cover all or the majority of an aggregation site and take into account areas of low and high fish densities and differential species distribution (Fig. 9.9). For aggregations spread over a large area permanent transects of 100 m or more in length are established at sites with observers counting the numbers of fish(es) along the transects, usually requiring about 20–30 min to do one survey. Two observers each count the number of subject fish on either side of the transect (Fig. 9.5), or, alternatively, one counts numbers while the other estimates sizes. Line surveys are used for general reef fish abundance determination, and all the factors which limit the accuracy of such line surveys need to be re-examined if used for spawning aggregations. Kulbicki (1998) points out that different species react differently to divers during line surveys, affecting the accuracy of data obtained, and such effects have not been quantified for line transects of spawning aggregations. (Johannes et al. 1999-Appendix 2) suggest some factors that could be affecting the numbers of fishes counted during transects; water clarity, current, SCUBA versus snorkel, hiding of fish, double counting, missed fish, time of day, time of lunar phase, and others. They suggested a number of arbitrary correction factors to be applied to field data to compensate for these various factors. It is probably simpler to just accept there will be "sampling error" rather than trying to compensate for unknowable errors and make efforts to increase the precision and repeatability of survey data. The main advantage to the single line transect is that is simple to set up and conduct follow up surveys, but there are a number of disadvantages, see below.

9.6.6 Towards Improved Survey Methods

Underwater survey methodology needs to improve to better quantify and characterize fish numbers, aggregation area and fish density. As more is learned about the variation of fish numbers and timing, as well as their distribution within and relative to aggregation sites, so protocols need to be modified accordingly to best characterize spawning aggregations and accommodate the different ways in which the fish aggregate and move within them or the aggregation area shifts over time. Underwater surveys typically involve transect work, using one or several transects, but this can be challenging because of prevailing conditions or because fish assemble in three dimensions. Ideally, studies should strive to design for long-term duration and in any case the methods should be described clearly enough to allow for repeatability.

The single line transect generally provides only a single total count, and if the length and swath width are known, a single density measurement (Fig. 9.5a). If the survey transect crosses the entire aggregation, then areas of higher (which may be a core area, Chap. 2) and lower density (or with no fish at all) are going to be lumped together in the single density determination. An approach of measuring density of individuals within subunits of an overall aggregation site seems to hold hope for more realistic estimates of overall numbers. If a single long permanent transect, set up using survey tape and compass, is subdivided into equal length subunits, each of which is marked separately by floats or other means. Each subunit is counted separately (but successively). This allows several measurements of density at known locations, providing increased resolution of the location and limits of the aggregation and fish within it and an average value to be calculated (Fig. 9.5b). The approach obviates the need to constantly update a single overall count of the transect (which can run to hundreds of fish). If the transect is permanent, with its location suitably quantified, and the same increments are used for each survey, the surveys obtained are repeatable at another time by another observer. Ideally, transect counts should be replicated, and possible inter-diver differences examined.

Lacking a permanent transect with subdivisions, something as simple as making fish counts at time intervals (1 to a few minutes) while maintaining a constant approximate swim rate over the bottom can provide valuable relative information on fish density and distribution within the aggregation. It is possible to use a method of measuring distance "on the fly" (Nemeth 2005) where time units (counts made on some time interval) can be transformed into distance units by towing a waterproof GPS logging positions at time intervals during the transect swims. Either method allows determination of density of fish within portions of the aggregation area, an improvement over the single transect total count.

A single permanent transect runs the risk of initially being located to represent the overall aggregation, however, slight shifts in location of aggregated fish can render a permanent transect unrepresentative. Ideally multiple permanent transects can to be set up so that if the location of the aggregation shifts, the data from individual transects will provide evidence of this and, if necessary, new transects can be added to include additional areas with aggregated fish. A SCRFA/Palau Conservation Society partnership found a single transect aggregation monitoring project to be unrepresentative and it was replaced by five permanent parallel transects which fully cover the present known aggregation of three species of groupers (Fig. 9.9). Each transect was surveyed by an observer towing a logging GPS, which recorded distance covered each minute, and numbers of fishes counted on a per minute basis. This allowed for an average density to be calculated for each species allowing for statistical treatment of data and better comparisons over time (Yvonne Sadovy de Mitcheson, Asap Bukurrou, Scott Kiefer and PC unpublished data, Palau Conservation Society 2010).

Permanent survey areas can also be set up as quadrats of equal area, which are each surveyed independently. Kadison et al. (2006) delineated an area 1.5 km long by 100 m wide of Grammanik Bank, an elongate shelf edge reef feature in the US Virgin Islands, into six equal segments each about 160 m in length. Divers swam or used Diver Propulsion Vehicles 5 m off the bottom of 30–40 m depth to survey each section of the reef in a 20 m wide transect, counting species and numbers below them. Short narrow (30 m by 2 m) transects (Nemeth 2005) were carried out by a diver swimming at a constant speed with the tape rolling out behind and using a 1 m piece of PVC pipe marked at 5 cm intervals to assist in estimating fish lengths. Shapiro et al. (1993) divided up a red hind aggregation site into large square sections using ropes and four divers simultaneously swam the entire grid.

If fish are aggregated across a broad area and not in a single tight cluster, the "GPS density" method mentioned earlier can be used to construct maps of fish density and allow repeatable surveys. A GPS receiver in a waterproof float is towed by a single or multiple observer(s) swimming tracks across the aggregation area including areas where no fish occur. The GPS records its position every 15 s to 1 min while data on numbers of fish seen within a given swath width by the observer are recorded for each minute (using a watch synchronized with the GPS time). Later the downloaded position data provide the distance surveyed in each minute, the swath width provides an area (of a rectangle) and fish numbers observed produces a density measurement within the rectangle. The central position measurement of the rectangle (at 30 s point for each min) gives the geographic centre of the rectangle. The series of geolocated density measurements, as well as the locations where no fish occur, is used to construct a map of the distribution of fish during a given survey (Fig. 9.1). This method will be described in greater detail elsewhere (Colin et al. in prep), but the basic method produces high quality information on distribution of fish within a spawning aggregation, and estimates of the total number of fish within an aggregation. The outer limits of an aggregation can also be delineated, if distinct, (Fig. 9.10a). The area occupied by an aggregation day to day can be shown as well as change occurring over time (Fig. 9.10b). This is, perhaps, the closest to an archival method for documenting the status of spawning aggregations yet available. Using the same methods it is repeatable later by another individual with real hope for comparable results. Methods for determining numbers of fish, if they are GPS based, can similarly indicate changes in locations of aggregations.

With extremely dense aggregations comprised of many hundreds to thousands of fish, such as occurs in the Nassau grouper, cubera, *Lutjanus cyanopterus*, dog snapper, *L. jocu*, or twin-spot snapper, *L. bohar*, things are different (Fig. 9.11, also see Fig. 12.24). Whaylen et al. (2004, 2006) attempted to determine the number of Nassau grouper in an aggregation by several divers estimating the dimensions of the

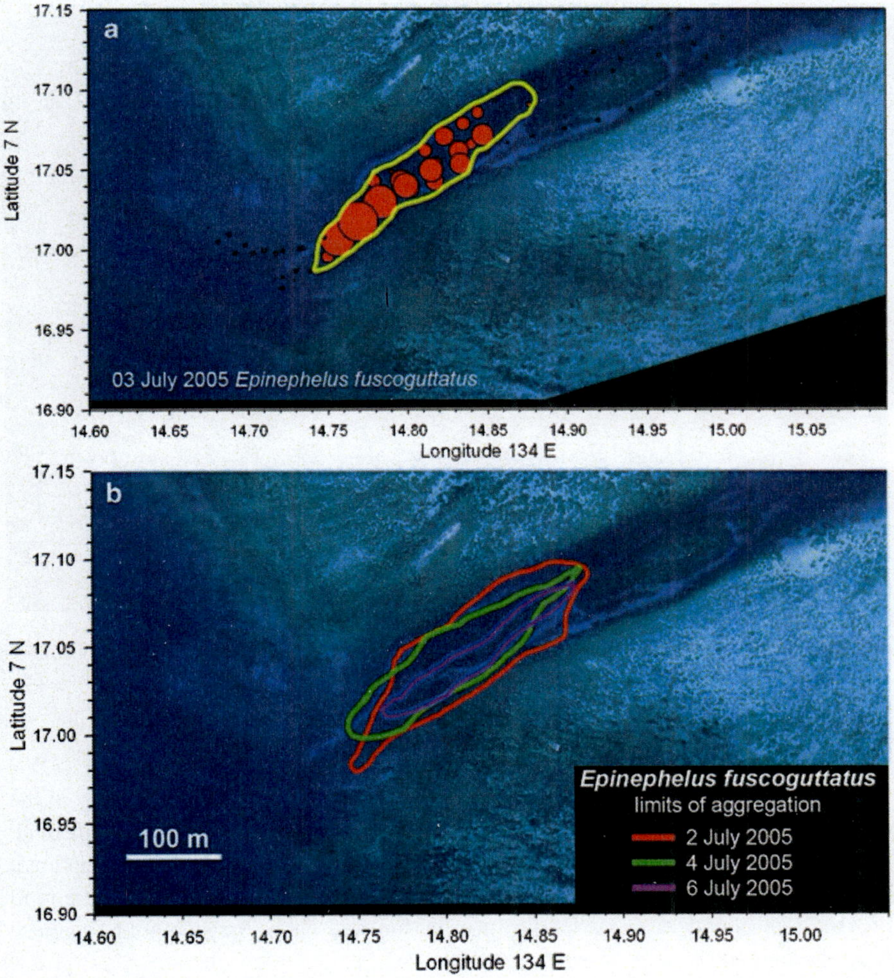

Fig. 9.10 (**a**) Using the "GPS density" survey method, the outer distributional limits of a spawning aggregation can be shown using the outer edges of areas containing fish to define the aggregation limits. (**b**) If surveys are done on successive days, the changes in the overall area and location of the aggregation can be clearly shown. This figure shows changes occurring from surveys conducted 2 days apart

aggregation using topographic reference points and marker lines at known positions, the numbers of fish within cubical volumes of the aggregation (they indicate using a $(3.3\ \text{m})^3$ cube), extrapolating to the total volume of the aggregation and averaging the estimates. During a typical survey divers conducted at least ten abundance estimates. Multiple daily counts were averaged. Video techniques, either filming compact aggregations from one point or swimming the length of the elongate aggregation with the camera aimed at the aggregation in one clip were applied and the numbers

Fig. 9.11 Photo: Filming a 'ball' of dog snapper, *Lutjanus jocu*, in a spawning aggregation in Belize. Counting the number of fish is challenging in such 'balls' (© Doug Perrine/SeaPics.com)

generated compared with visual estimates. Despite these efforts, the accuracy of the estimates are hard to assess (for alternative method with same species see Sala et al. 2001). Methodologies need to be developed to assist in counting three dimensional spaces with fish. Rand et al. (2006) used stereo images of a grouper aggregation to provide fish abundance and length information and show the potential for these types of methods.

To ensure that surveys can be repeated in future and that fish counts are meaningfully collected, UVC surveys need to include information detailing exactly what area was surveyed and how data were collected. For example, census data in Johannes et al. (1999) are an example of the problems of doing replication when transects are not clearly identified and survey methods are changed during the work. Although they identified "core areas" for each of three grouper aggregation sites in Palau, each site having three species occurring and defined them as "to include the largest proportion of the aggregation it was possible to count on a single SCUBA tank", "core areas" were not identified on maps, by geographic coordinates nor were their areas or lengths of transects identified. Of the three aggregation sites surveyed, one was small enough the entire site was covered during each survey, while the other two were too large or deep to be totally surveyed. At times the aggregation extended beyond the "core area" at two sites and fish data from outside the core area were sometimes added to counts. Fish abundance data were reported only as "number of fish counted" for each area and sampling time. Data were reported to suggest "trends", rather than total numbers of fishes in aggregations; these deficiencies made it impossible for subsequent workers to repeat their field measurements except in the case of the fully surveyed site.

9.6.7 Hydroacoustic Surveys

One way to survey fish numbers without having to enter the water is by using hydroacoustic surveys. Perhaps the best attempt to conduct a hydroacoustic survey of a reef fish spawning aggregation is that of Taylor et al. (2006) where relatively careful underwater surveys were combined with the acoustic data. They defined the extent of and numbers present (roughly 2,000 fish) in a Cayman Islands Nassau grouper aggregation, stressing the importance of "ground truthing" (i.e. combining this approach with diver surveys of numbers present to ensure that the hydroacoustic method reflects numbers of fish species of interest present). They found that hydroacoustic surveys tended to overestimate the abundance of fish, but appears to provide a good measure of the spatial extent of the aggregation (Chap. 12).

Ehrhardt and Deleveaux (2007) attempted similar work for Nassau grouper in the Bahamas without ground-truthing. Their estimate of the area covered by the aggregation off High Cay, Andros at 555,000 m^2 is much larger than any known Nassau grouper aggregation, while their estimate of fish numbers was much higher than any other that had been surveyed in recent years in the country, or indeed anywhere currently in the range of the species. They estimated a total of 10,500 fish of an average weight of 2.01–2.13 kg. This reported weight is immediately suspect as these values are below even the lower limit for aggregating Nassau grouper. Colin (1992) found Nassau grouper from aggregations at Long Island, Bahamas to weigh between 2.7 and 9.7 kg (Colin 1992), with Nassau grouper from other aggregations having similar weights. These disparities call into question the accuracy of their results, emphasizing the need for ground-truthing, and could be highly misleading if used directly for management, for example in stock assessment. Similarly Johnston et al. (2006) did broad scale acoustic surveys for red hind aggregations covering hundreds of km of the Puerto Rico insular shelf, but lacked adequate ground-truth validation. Although extensive schools of fish were recorded, it is unknown whether the data do represent spawning aggregations (or even red hind) or other fishes hovering above the bottom.

Hydroacoustic surveys remain a developing technology, with good potential for assessing spawning aggregations of reef fishes if applied rigorously. Stringent protocols for ground-truthing and assessing accuracy are essential. The unknown number of species which could produce signal returns similar to target species, as well as the attraction of predators to sites (Heyman et al. 2005; Nemeth et al. 2007) means hydroacoustic surveys without careful ground truth will not provide reliable data and will tend to overestimate fish numbers.

9.6.8 Measuring Sizes of Fishes

The size of fish in an aggregation is, after the numbers present, the most useful information to acquire. Size sampling often requires fish capture and this may not perfectly reflect the actual size distribution of fish in the aggregation because of size or sex selectivity of fishing gear. Size information can also be acquired "remotely". This might be done by pointing lasers, a set distance apart, at fishes and have divers photograph them. This will not work for all species, as Kadison et al. (2006) tried using green lasers to estimate lengths of grouper, but found that large fish had an extreme aversion to laser beams. Often observers estimate the size of fishes visually; a technique which has value, but for which the data are innately unverifiable and consequently must be treated as inaccurate A variety of methods has been proposed to train observers to estimate lengths of fishes underwater (Bell et al. 1985) which appear to increase the ability to closely estimate lengths under the controlled conditions of training. There is no guarantee similar results are obtained on fishes in the field and the skills of length estimation required regular retraining. It must be remembered that visual size estimates are not accurate. The problems of using estimated lengths in fishery and conservation considerations are similar to the estimates of fish numbers present in aggregations; the data are useful but any pretensions of accuracy are specious.

Three dimensional (3D) imagery has great promise as a means of determining sizes of fish in an aggregation (Klimley and Brown 1983). Rand et al. (2006) used 3D videography to capture images of a Nassau grouper aggregation and extract size estimates of individual fish, construct a size frequency curve for the fish and determine nearest neighbour distances. Similar methods have been applied to submersible and ROV images (Rochet et al. 2006). Some earlier efforts using stereo images have been listed elsewhere (Colin et al. 2003) and are not repeated here.

9.6.9 Working with Gametes and Eggs

If eggs are collected in the field, or fish are spawned artificially the characters of the eggs (shape, size, and presence of pigment or oil globules) can be determined by microscopic examination. For field collection, immediately after spawning the area where the gametes have been released can be marked with a small squirt of fluorescein dye from a squeeze bottle. After this delay the water near the dye marker is swept with the hand net for a short period and the contents collected in the net put into a plastic bag. Most planktonic reef fish eggs should be handled gently just after spawning, but can be easily manipulated once hardened up after fertilization. After an hour or so they are quite robust and can usually be safely washed in a sieve and gently handled.

Appeldoorn et al. (1988) provide a thorough discussion of various tracers useful in tracking the dispersal and fate of pelagic fish eggs. Kiflawi et al. (1998) and

Petersen et al. (1992) provide interesting information on collection and handling of eggs in fertilization studies. Because of the positive buoyancy of planktonic reef fish eggs the mixing of surface layers is particularly important as wind induced shears and formation of slicks can greatly affect transport of eggs. Appeldoorn et al. (1988) examined dye, surface drifters, drogues and particles as potential tracers. None was ideal and some (plastic particles mimicking fish eggs) are highly labour intensive to track. They suggest a combination of methods may be useful (using dye to track dispersal of beads, etc.). Small drifting items (drift cards, drift bottles, drift vials – see Domeier 2004) can also serve as egg and larval tracers, in those cases relying on eventual discovery of the item by other people, usually from beaches, followed by reporting of their discovery.

The spawning of a large reef fish aggregation can be a spectacular event. In the case of large transient aggregators, such as Nassau grouper and cubera snapper, many millions of fertile eggs can be produced in just a few minutes in a very small volume of water. While questions of what the concentrated spawning of aggregations might mean to early life history are addressed in Chap. 6, the huge spawnings also provide an opportunity to conduct studies that would otherwise be difficult.

Egg numbers and gamete cloud size

Beyond the number of fish present, it is useful to know the possible contribution of a spawning aggregation to the future generations of fish by knowing something about how many eggs are produced by an aggregation during the period it occurs. This information provides a "starting point" for any future analysis of life history parameters of an aggregating fish, stock assessment, etc. Batch fecundity information provides some idea of the numbers of eggs a given female can produce upon spawning, but is not a direct measure of the numbers of eggs spawned by all females in an aggregation. Sampling and quantifying the eggs released is one way to do this, but is difficult to do effectively. The eggs are tiny, nearly transparent and drift away within seconds to minutes with the current. The water where spawning occurred would have to "tagged" with a drogue or dye and then a grid around the supposed centre of eggs sampled using standard plankton sampling techniques. If spawned below the surface, time would be required for the eggs to float to near the surface (where sampling is easier). The eggs would disperse horizontally outward from the location of release with the patch expanding quickly in the hours after spawning (Chap. 6), but would be limited in their horizontal distribution by their slight buoyancy tending to keep them near the surface.

Where there are predictable currents downstream of spawning fishes, moored plankton nets can be used to quantitatively sample the eggs streaming away from the spawning. For example, Hamner et al. (2007) moored plankton nets about 80 m downcurrent of a site where both resident aggregating and non-aggregating fishes were spawning just after high tide on a reef in Palau. They found at least 50% of eggs captured came from spawning aggregations of a few species of parrotfish and surgeonfishes, and fish eggs made up to 90% of the zooplankton exported from this reef on falling tides.

Egg density and ascent rates

This information allows estimation of the time required for eggs to ascend from the spawning depth to near the surface, where water transport may be different and affect the advection of eggs from the spawning site. Most planktonic reef fish eggs are very slightly positively buoyant in seawater. Their density is somewhat less than seawater, and their buoyancy and depth of release are important in determining how fast they will rise to the surface after spawning. The small differences between the density of eggs and that of seawater is not easy to directly determine. Eggs cannot be accurately weighed, as water clings to them once they are removed from water; nor can their volume be easily measured to the accuracy required. However, their density can be quickly determined by putting fertilized eggs into containers with different densities of seawater (by diluting with fresh water) to see at what salinity the eggs are neutrally buoyant (Colin et al. 1996). Ascent rates are generally measured by releasing eggs, usually from a pipette, at the bottom of a container with seawater of the salinity found at the spawning site and then measuring their rise over time. Great care is needed to ensure that the water in the container is totally still, with no vertical water movement which would bias results.

9.6.10 Dynamics of Migration

The dynamics of migration to and from aggregations has not received much attention beyond anecdotal observations, particularly for transient aggregations. Observers have seen numbers of fish moving to (or away) from aggregation sites, but the fish usually are followed for only a few minutes (Colin 1992, Chap. 2). Acoustic tagging holds great promise for learning more about migrations (Nemeth et al. 2007; Starr et al. 2007), but for some species large acoustic deployment areas are needed to encompass the entire migration catchment for larger species. For something like Nassau grouper, this may mean covering in some manner hundreds of km of reef with receivers. Obviously for such large scale situations, receivers can only be placed at intervals, but hopefully sufficiently close that general movements can be determined and later surveys focused on the areas to which fish are migrating from aggregations.

In addition to the documentation of aggregation catchments, exactly how fishes may navigate during migration, particularly the long distance migrations of some transient aggregators (Colin 1992; Bolden 2000; Starr et al. 2007), is a matter of considerable interest. The technology to undertake such studies is now available through acoustic tagging and receiver arrays (Domeier 2005; Nemeth et al. 2007; Starr et al. 2007), but generally has not been deployed on a scale suitable for such studies. For some transient aggregators, stationary still cameras taking photos at regular intervals might provide new information on occurrence of fish and their migration behaviour, but locations for such cameras need to be carefully chosen as it is possible to completely miss a migration if the camera is placed only slightly off the migration track.

Once migration pathways have been documented, it is possible to evaluate or experimentally manipulate them to learn more of migration dynamics. This is applicable for resident aggregators, whose regular (often daily) movements allow undertaking work on a regular and extended basis. For example, Mazeroll and Montgomery (1998) experimentally tested a number of possible methods of migration navigation for brown surgeonfish, *Acanthurus nigrofuscus*, and found landmarks were important in determining the path, but that magnetic field and sun compass migration were not. Kiflawi and Mazeroll (2006) captured migrating surgeonfish, determined their sex, tagged and released them. By observing migration behaviour of tagged individuals of known sex over subsequent days they determined females most often lead migrating groups.

9.6.11 Physical Parameter Considerations

The instrumentation of aggregation sites for physical parameters (temperature, currents, tide, light), is of critical importance for understanding the dynamics of aggregations and allowing comparison of conditions across numbers of sites. Colin and Clavijo (1988), Colin (1992), Heyman et al. (2005), Nemeth et al. (2007) and Whaylen et al. (2006) provide examples of instrumented sites that have provided new insights into aggregation dynamics (see Chap. 5). Annual physical data from an aggregation site allow correlation of biological activity (timing of aggregations, spawning) within the physical framework. Measuring water movement at and near sites is important for documenting whether conditions might promote use of a certain site over others and dispersal of propagules after spawning. It is also important to instrument beyond aggregation sites if the intention is to examine differences (if any) between sites and between known aggregation sites and sites where no aggregations form.

The locations of instruments at an aggregation site depend on the parameters being measured. A current meter deployed to record currents at the actual spawning site needs to be located where the fish are spawning. Thermographs need to be located close to the aggregation site, but unless there is thermal stratification, anywhere near the aggregation (a few tens of meters) should be acceptable. The same applies to salinity measurements. Light sensors could be bottom-mounted or put on moorings at distances above the bottom.

The further removed an instrument is from the aggregation site, the less representative the data. Bolden (2000) used temperatures measured at Lee Stocking Island, Bahamas (a tidally dominated bank habitat) as representing those at a shelf edge Nassau grouper aggregation site more than 120 km south. In this case data taken far distant from the aggregation site confuse the question of water temperature regimes at the aggregation site. When the intention is to clarify the relationship of physical environment to aggregation and spawning, data must be measured at the site.

9.6.12 Currents

If fish are spawning 10 m above a 20 m deep bottom, then currents need to be measured at the 10 m level if possible. A single point current meter could be placed on a mooring above the bottom. An acoustic doppler current profiler (ADCP), while more expensive, mounted on the bottom directly below where the fish spawn provides currents in the entire water column as well as depth and tide data (Fig. 9.2). The fluid dynamics around aggregation sites can be complicated. For example, some sites are located near to or at reef promontories, which should produce eddies nearby as currents move past them.

It would be useful to have multiple current meters at an aggregation site (Colin 1992; Heyman et al. 2005). Currents in relation to aggregation sites probably need to be looked at over several spatial scales (Chap. 6). Do the currents at the aggregation site differ from areas just beyond the limits of the aggregation? What is the role of promontories in influencing current patterns. Do currents at sites favour offshore transport of eggs or their retention? Currents tend to run along reefs, unless there is a very pronounced projection out from the reef. In such cases, currents (if any are present) would tend to run along one side of the projection and then move offshore for a distance, producing a back eddy along the opposite side of the projection.

Measuring currents in channels where spawning might be occurring within the channel or at its mouth is also complicated (Fig. 9.3). If flow is strong enough, small eddies are produced on the edges of the channel and a modest indentation in the channel edge can produce a back eddy where flow is contrary to the general channel flow. This was seen for the blacktail snapper, *Lutjanus fulvus*, in Palau (Chap. 12). Depending on these small scale effects, the strategy for measuring currents at an aggregation site may have to be modified significantly.

9.6.13 Temperature

The use of satellite-derived sea surface temperatures to characterize thermal regimes at aggregation sites should be avoided unless there is no alternative. With present technology satellite sensors cannot measure the temperature over reefs, instead relying on nearby areas of open ocean, and surface temperatures may be higher than those just below the surface. Recording thermographs for underwater use are readily available, inexpensive and very robust. They should be carefully calibrated before and after deployments to ensure accuracy, particularly as the differences in temperature over seasons or between sites may be small. Sampling intervals should such that many measurements are taken per day, ideally every 30 min or less for spawning sites. The variation in temperature over time of days to weeks is also important as this variation provides a good measure of the amount of upwelling of deeper water over time to shallow reefs and may indicate seasonal upwelling potentially associated with increased phytoplankton production. If possible, vertical thermograph arrays also should be installed at shelf edge locations, as the structure of the water column may be important in influencing fish behaviour. The observation of Starr et al. (2007) that Nassau grouper descended for 2 months to depths averaging about 70 m, after spawning in shallower (20 m) water, would have been nicely complimented by having temperature information at these different depths.

9.6.14 Water Mass Properties

Shcherbina et al. (2008) deployed an autonomous underwater vehicle to examine bathymetric and water column properties around Glover's Reef, Belize, including a Nassau grouper aggregation site. While providing an abundance of intriguing data, the short sampling periods (5 days each in May and February) provide little insight into the seasonal oceanographic regimes that affect the aggregation site. This study would have been well complimented by long-term deployment of instruments (temperature at various depths, current meters) at sites around the atoll. Then the wealth of detailed data obtained from such a vehicle can be placed in a longer term perspective obtained from deployed instruments.

9.6.15 Drifters

Information on where eggs and larvae go after spawning is necessary to understand the importance of aggregations. The water into which gametes are released needs to be "tagged" in some manner, or the eggs themselves "tagged" (for those with planktonic eggs). Placing "drifters" into water containing eggs is one method of tracking propagules, however, there is no guarantee the drifter is actually staying with the desired parcel of water. Theoretically the gamete release from a spawn in a large transient aggregation could be tracked by drifter and the eggs and larvae subsequently sampled based on the location of the drifter. Real time verification of the coupling of drifter and propagules is limited by the inability to rapidly sort and identify larvae sampled near the drifter. At a later time after larval samples are processed there should be some ability to determine the connection between drifter track and larval presence.

The initial advection of eggs after spawning can be critical in determining where they will end up, either being incorporated into general oceanic circulation or retained locally by mechanisms associated with near-shore physical dynamics (see Chap. 6). Short term drifters (not satellite tracked) can be useful for tracking initial trajectories of eggs and larvae (see Fig. 5.16). Drifters can be "inert", their movement over time determined by following in a boat and logging positions with a hand held GPS, or "active", recording their positions in some manner. With a logging GPS installed on a marker float, recording its position over time, the drifter is recovered after some hours and data downloaded (Colin et al. 2003). Hatcher et al. (2004) used GPS drifters to track coral spawn. If left at sea for more than a few hours, it is often difficult to locate and recover the drifter to download data without some type of radio beacon. Some drifters use a combination GPS receiver/Family Radio Spectrum (FRS) radio (as a beacon) (Austin and Atkinson 2004) or cell phone to allow locating the drifter. Such a system may also allow some transfer of data electronically.

Satellite tracked drifters are the most capable for examining propagule dispersal over days to months providing location information nearly anywhere in the ocean and for time periods of months to years; but in many instances their capabilities far exceed the needs of a limited project. GPS/FRS drifters may cost only about US$200 (Austin and Atkinson 2004) while satellite tracked drifters cost a few thousand dollars, plus location services costs. For projects with limited needs and budgets (such as tracking egg movements for less than 24 h after spawning) simple systems may suffice, but for regional work satellite drifters are usually advisable.

9.6.16 Bathymetric Mapping

Combined with data from instrumentation of aggregation sites, detailed bathymetry of sites provides an ability to interpret the sometimes conflicting information regarding the possible benefits of sites in a physical and biological sense. The recording of single point position/depth data in the National Marine Electronic Association (NMEA) data from the GPS/depth sounder units has made it possible for researchers to easily map virtually any reef fish aggregation area in detail. Colin et al. (2003) and Kvernevik et al. (2002) described the basic system for single point bathymetric mapping of spawning aggregation sites. Heyman et al. (2007) used a lower frequency transducer (greater depth penetration) to map some aggregation sites on the Belize barrier reef and Kobara and Heyman (2008) did similar surveys in the Cayman Islands. Solid state NMEA data loggers (www.brookhouseonline.com) are simple, rugged and often allow longer data collection than portable computers. Single point data (latitude, longitude, depth) allows production of limited resolution maps, while side-scan and multi-beam sonar provide amazing resolution of large areas of bottom (Burton et al. 2005; Kadison et al. 2006; Nemeth et al. 2008). The congruence of GPS derived maps allows previous data that is GPS located to be applied to more detailed bathymetry at any time in the future. The location of the tiger grouper, *Mycteroperca tigris*, aggregation site reported by Sadovy et al. (1994) was determined by GPS receiver and this could later be plotted on the multi-beam sonar map of the Puerto Rico Virgin Island shelf showing it occurs on a broad promontory of the shelf edge east of Vieques, Puerto Rico (Fig. 9.12).

Multibeam sonar is presently the ultimate in method for producing detailed maps of the bottom for most spawning aggregation areas (Weaver et al. 2005; Nemeth et al. 2008; Kadison et al. 2006, 2009). Airborne LIDAR (Light Detection and Ranging) can map reef and aggregation areas in relatively clear water (Brock et al. 2004) but is limited in depth capabilities.

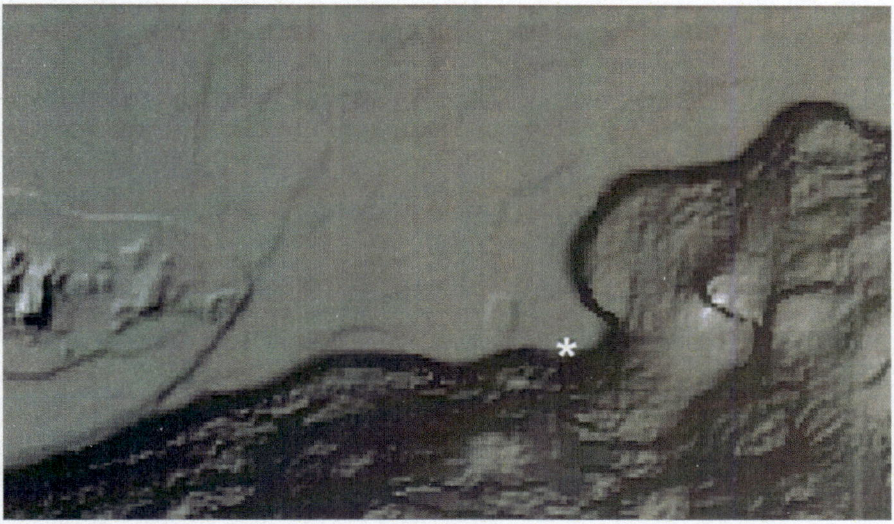

Fig. 9.12 The acquisition of detailed multibeam sonar images of marine substrates can provide a new perspective on the geomorphology of aggregation sites. The location of a spawning aggregation of tiger grouper, *Mycteroperca tigris*, is shown east of the island of Vieques in Puerto Rico. At the time this aggregation was studied the multibeam image was not available, but since the location of the aggregation was determined by GPS positioning, the location could later be plotted on the multibeam mapping, indicating the aggregation occurred near the end of projection of reef on the shelf edge, something which was not evident at the aggregation site (Base image courtesy Richard Nemeth)

9.6.17 Sociological Considerations

In some ways examining the relationship of humans to reef fish aggregations is relatively simple compared to dealing with what is happening in the ocean between biological entities and the physical environment. Examination of fisheries on aggregations is a common type of sociological consideration. Difficulties arise when fisheries are widely scattered, secretive (illegal?), destructive or information is considered proprietary. The Live Reef Fish Trade (LRFT) has been particularly difficult to gather meaningful data on, as it is lucrative for the top traders, often environmentally destructive and able to operate in remote areas because of the cruising range of the large vessels (Sadovy et al. 2003). Most participants do not want to draw attention to what they are doing. Aspects of this trade are considered in further detail in Chaps. 8, 10 and 11.

9.7 New Technology – Potential New Methods of Use in Studying Reef Fish Aggregations

Digital technology promises to provide new and added capabilities for data collection from aggregations. It is normally up to the researcher to comprehend the usefulness of these and figure out how to apply their capabilities to research work, as such devices are not designed specifically for the rather limited "fish spawning market". In the last two decades the advent of GPS positioning and digital video and still cameras has resulted in a major increase in information that can be gathered from aggregations. While new technological methods are appealing, great care needs to be exercised in deciding whether such methods are feasible in view of their generally greater costs and sometimes limited scope of measurement.

In diving technology increased ease and reduced costs of using mixed gas rebreathers, which produce no noise and few bubbles hold promise for less disturbance of aggregated fish while making observations. Given that many aggregations are located in shelf edge areas at depths of 25–40 m, the ability to control oxygen levels in breathing gas, either through use of enhanced oxygen mixtures (nitrox) or rebreathers, will increase limited bottom times for researchers.

The resolution and low light capabilities of digital still photography and videography allow new documentation of aggregations. The use of stationary autonomous monitoring stations equipped with cameras will allow new insights into fish behaviour. Stations which pan and tilt while taking photos could regularly cover the entire visible area around the station. Cameras could also be developed which sense motion, either through optical means or sonar, and photograph whatever is passing by. Deploying a series of these along migration pathways would provide remarkable information. Such could even be triggered by passage of an acoustically tagged fish, so that any other individuals accompanying it would also be documented.

Current meters and other electronic instruments will become less expensive and consequently sites can be instrumented with more sensors to allow finer resolution of physical factors. It would be amazing to have a large transient aggregation instrumented with an array of current meters to detect the movement of water throughout the site, rather than relying on a single meter at sites now.

The use of an AUV (autonomous underwater vehicle) to document an aggregation area at Glover's Reef (Shcherbina et al. 2008) opens a new avenue of investigation. Such vehicles are rapidly evolving and it would be instructive to deploy custom-designed AUVs at aggregation sites for documentation of sites in the absence of divers. Such a vehicle could have a hydroacoustic sonar system oriented horizontally to the vehicle which scans 180° ahead of the vehicle while simultaneously taking video or still images of the area as well as monitoring physical parameters in the water. Such a vehicle could be silent and slow-moving to avoid disturbing fishes. It could even have artificial intelligence to remain with an aggregation, detected by its sonar, and monitor it over a period of hours. Drawbacks to using AUVs for aggregation monitoring are the typical short durations of data gathering on dynamic water column properties and lack of complimentary measurements during and after deployments. AUVs should generally be used as part of a large programme with long-term monitoring objectives and capabilities.

Remote sensing is unlikely to become a significant method for discovery of aggregation sites given the limits of habitat visibility where most shelf edge aggregations occur. Satellite observations hold great promise for increased information on currents in the vicinity of aggregation sites and oceanic regions. High frequency radar coverage of some areas, such as the United States east coast, has greatly increased and with it currents can be monitored at potential aggregation sites in real time. With increasing knowledge of aggregation sites in areas such as the Florida Keys, such information would be valuable to incorporate in any considerations of larval transport.

Increased focus on genetics of fish populations, as well as reductions in cost for analyzing genetic samples, hold promise that the question of whether transient aggregations represent individual stocks that are largely self recruiting can be addressed. Genetic samples must still be obtained by sampling fishes, but opportunities to sample, even if they are stockpiled and stored for future use is a worthwhile activity, especially with threatened species.

Acknowledgements Numerous colleagues; particularly Michael Domeier, Terry Donaldson, Yvonne Sadovy, assisted with field work that allowed development and testing of methods described here. Support for field work was provided by the National Oceanic and Atmospheric Administration (USA) (for humphead wrasse), and other funders. My co-workers at the Coral Reef Research Foundation, Koror, Palau have greatly assisted this work, with particular support coming from Lori JB Colin. The Palau Conservation Society has worked with me to utilize and improve some of the techniques described, with special thanks to Asap Bukurrou and Scott Keifer. The Palau National Government and Koror State Government have graciously permitted this work to occur in their waters.

References

Aburto-Oropeza O, Hull PM (2008) A probable spawning aggregation of the leather bass *Dermatolepis dermatolepis* (Boulenger) in the Revillagigedo Archipelago, México. J Fish Biol 73(1):288–295

Appeldoorn RS, Hensley DA, Shapiro DY (1988) The use of various tracers in assessing the fate of pelagic eggs spawned by coral reef fishes. In: Proceedings of the sixth international coral reef symposium, Townsville, 1988

Appeldoorn RS, Hensley DA, Shapiro DY, Kioroglou S, Anderson BG (1994) Egg dispersal in a Caribbean coral reef fish. Thalassoma bifasciatum. II Dispersal off the reef platform. Bull Mar Sci 54:271–280

Aswani S, Hamilton RH (2004) Integrating indigenous ecological knowledge and customary sea tenure with marine and social science for conservation of bumphead parrotfish (*Bolbometopon muricatum*) in the Roviana Lagoon, Solomon Islands. Environ Conserv 31:1–15

Atkinson MJ (1987) Oceanography of Enewetak Atoll. In: Devaney DM, Reese E, Burch BL, Helfrich P (eds) The natural history of Enewetak Atoll, vol 1. U.S. Department of Energy, Marshall Island, DOE/EV/00703-T1

Austin J, Atkinson S (2004) The design and testing of small, low-cost GPS-tracked surface driters. Estuaries 27(6):1026–1029

Bell JD, Craik GJS, Pollard DA, Russell BC (1985) Estimating length frequency distributions of large fish underwater. Coral Reefs 4:41–44

Bolden SK (2000) Long-distance movement of a Nassau grouper (*Epinephelus striatus*) to a spawning aggregation in the central Bahamas. Fish Bull 98:642–645

Brock JC, Wright CW, Clayton TD, Nayegandhi A (2004) LIDAR optical rugosity of coral reefs in Biscayne National Park, Florida. Coral Reefs 23(1):48–59

Bull B, Doonan I, Hart A (2001) Diel variation in spawning orange roughy (*Hoplostethus atlanticus*) over a seamount feature on the Chatham Rise. N Z J Mar Freshw 35:43

Burton ML, Brennan KJ, Munoz RC, Parker ROJR (2005) Preliminary evidence of increased spawning aggregations of mutton snapper (*Lutjanus analis*) at Riley's Hump two years after establishment of the Tortugas south ecological reserve. Fish Bull 103:404–410

Campana SE (2001) Accuracy, precision and quality control in age determination, including a review of the use and abuse of age validation methods. J Fish Biol 59:197–242

Campana SE, Thorrold SR (2001) Otoliths, increments, and elements: keys to a comprehensive understanding of fish populations? Can J Fish Aquat Sci 58(1):30–38

Carter HJ (1994) Aspects of the ecology and reproduction of Nassau grouper (*Epinephelus striatus*) off the coast of Belize, Central America. Proc Gulf Caribb Fish Inst 1990:65–111

Choat JH, Davies CR, Ackerman JL, Mapstone BD (2006) Age structure and growth in a large teleost, *Cheilinus undulatus*, with a review of size distribution in labrid fishes. Mar Ecol Prog Ser 318:237–246

Colin PL (1992) Reproduction of the Nassau grouper, *Epinephelus striatus* and its relationship to environmental conditions. Environ Biol Fish 34:357–377

Colin PL (2010) Aggregation and spawning of the humphead wrasse, *Cheilinus undulatus* Rüppel (Pisces: Labridae): general aspects of spawning behaviour. J Fish Biol 76:987–1007

Colin PL, Clavijo IE (1988) Spawning activity of fishes producing pelagic eggs on a shelf edge coral reef, southwestern Puerto Rico. Bull Mar Sci 43:249–279

Colin PL, Shapiro DY, Weiler D (1987) Aspects of the reproduction of two groupers, *Epinephelus guttatus* and *E. striatus,* in the West Indies. Bull Mar Sci 40:220–230

Colin PL, Laroche WA, Koenig CC (1996) Development from egg to juvenile of the red grouper, *Epinephelus morio* (Pisces: Serranidae) in the laboratory. In: Arreguin-Sanchez F, Munro JL, Balgos MC, Pauly D (eds) Biology, fisheries and culture of tropical groupers and snappers. ICLARM conference proceedings 48, Makati, 1996

Colin PL, Sadovy YJ, Domeier ML (2003) Manual for the study and conservation of reef fish spawning aggregations. Society for the conservation of reef fish aggregations, special publication 1

Craig AK (1966) Geography of fishing in British Honduras and adjacent tropical waters. Louisiana state coast stud series no. 14, Louisiana State University Press, Belize

Domeier ML (2004) A potential larval recruitment pathway originating from a Florida marine protected area. Fish Oceanogr 13:287–294

Domeier ML (2005) Methods for the deployment and maintenance of an acoustic tag tracking array: an example from California's Channel Islands. Mar Technol Soc J 39(1):74–80

Ehrhardt NM, Deleveaux VKW (2007) The Bahamas' Nassau grouper (*Epinephelus striatus*) fishery – two assessment methods applied to a data-deficient coastal population. Fish Res 87:17–27

Erisman BE, Konotchick T, Blum S (2009) Observations of spawning in the leather bass, *Dermatolepis dermatolepis*, (Teleostei: Epiephelidae), at Cocos Island, Costa Rica. Environ Biol Fish 85(1):15–20

Gilmore RG, Jones RS (1992) Colour variation and associated behaviour in the Epinepheline groupers, *Mycteroperca microlepis* (Goode and Bean) and *M. phenax* Jordan & Swain. Bull Mar Sci 51:84–103

Hamilton RH (2005) Indigenous ecological knowledge (IEK) of the aggregating and noctural spawning behaviour of the longfin emperor, *Lethrinus erythropterus*. SPC Tradit Mar Resour Manag Bull 18:9–17

Hamner WM, Colin PL, Hamner PP (2007) Export-import dynamics of zooplankton on a coral reef in Palau. Mar Ecol Prog Ser 334:83–92

Hatcher GA, Reiss TE, Storlaszzi CD (2004) Application of GPS drifters to track Hawaiian coral spawning. Open-File Rep US Geol Surv 2004–1309. http://www.pubs.usgs.-gove/of/2004/1309/

Heyman WD, Kjerfve B (2008) Characterization of multi-species reef fish spawning aggregations at Gladden Spit, Belize. Bull Mar Sci 83(3):531–551

Heyman WD, Kjerfve B, Graham RT, Rhodes KL, Garbutt L (2005) Spawning aggregations of *Lutjanus cyanopterus* (Cuvier) on the Belize Barrier Reef over a 6 year period. J Fish Biol 67:83–101

Heyman WD, Ecochard JB, Biasi FB (2007) Low-cost bathymetric mapping for tropical marine conservation – a focus on reef fish spawning aggregation sites. Mar Geodesy 30(1):37–50

Johannes RE (1981) Words of the Lagoon. Fishing marine lore in the Palau district of Micronesia. University of California Press, Los Angeles

Johannes RE, Squire L, Granam T, Sadovy Y, Renguul H (1999) Spawning aggregations of groupers (Serranidae) in Palau, marine conservation research series publication no. 1, The Nature Conservancy

Johnston SV, Rosario A, Rivera JA, Timko MA, Nealson PA, Kumagai KK (2006) Hydroacoustic evaluation of spawning red hind (*Epinephelus guttatus*) aggregations along the coast of Puerto Rico in 2002 and 2003. Natl Mar Fish Serv Prof Pap 5:10–17

Kadison E, Nemeth RS, Herzlieb S, Blondeau J (2006) Temporal and spatial dynamics of *Lutjanus cyanopterus* and *L. jocu* (Pisces: Lutjanidae) spawning aggregations on a multi-species spawning site in the USVI. Rev Biol Trop 54(Suppl 3):69–78

Kadison E, Nemeth RS, Blondeau JE (2009) Assessment of an unprotected red hind (*Epinephelus guttatus*) spawning aggregation on the Saba Bank in the Netherland Antilles. Bull Mar Sci 85(1):101–118

Kiflawi M, Mazeroll AL (2006) Female leadership during migration and the potential sex-specific benefits of mass spawning in the brown surgeonfish (*Acanthurus nigrofuscus*). Environ Biol Fish 76:19–23

Kiflawi M, Mazeroll AI, Goulet D (1998) Does mass spawning enhance fertilization success in coral reef fish? A case study of the brown surgeonfish. Mar Ecol Prog Ser 172:107–114

Klimley AP, Brown ST (1983) Stereophotography for the field biologist: measurement of lengths and three-dimensional positions of free-swimming sharks. Mar Biol 74:175–185

Kobara S, Heyman WD (2007) Caribbean-wide geospatial analysis of the location of transient reef fish aggregation sites using remote sensing. Proc Gulf Caribb Fish Inst 59:463–465

Kobara S, Heyman WD (2008) Geomorphometric patterns of Nassau grouper (*Epinephelus striatus*) spawning aggregation sites in the Cayman Islands. Mar Geodesy 31:231–245

Koenig CC, Coleman FC (1998) Absolute abundance and survival of juvenile gags in sea grass beds of the northeastern Gulf of Mexico. Trans Am Fish Soc 127:44–55

Kulbicki M (1998) How the acquired behaviour of commercial reef fishes may influence the results from visual censuses. J Exp Mar Biol Ecol 222:11–30

Kvernevik T, Zambri M, Akhir M, Studholme J (2002) A low-cost procedure for automatic sea floor mapping, with particular reference to coral reef conservation in developing nations. Hydrobiologia 474(1):67–79

Lawson GL, Rose GA (2000) Small-scale temporal and spatial patterns of spawning in Atlantic cod (*Gadus morhua*) in coastal Newfoundland waters. Can J Fish Aquat Sci 5:1011–1024

Lindemann KC, Pugliese R, Waugh GT, Ault JS (2000) Developmental patterns within a multispecies reef fishery: management applications for essential fish habitats and protected areas. Bull Mar Sci 66(3):929–956

Lobel PS (1992) Sounds produced by spawning fishes. Environ Biol Fish 33:351–358

Luckhurst BE (1998) Site fidelity and return migration of tagged red hinds (*Epinephelus guttatus*) to a spawning aggregation site in Bermuda. Proc Gulf Caribb Fish Inst 50:750–763

Luckhurst BE, Hateley J, Trott T (2006) Estimation of the size of spawing aggregations of red hind (*Epinephelus guttatus*) using a tag recapture methodology in Bermuda. Proc Gulf Caribb Fish Inst 57:535–542

Mazeroll AI, Montgomery WL (1998) Daily migrations of a coral reef fish in the Red Sea (Gulf of Aqaba, Israel): initiation and orientation. Copeia 1998:893–905

Nemeth RS (2005) Population characteristics of a recovering US Virgin Islands red hind spawning aggregation following protection. Mar Ecol Prog Ser 286:81–97

Nemeth RS, Blondeau J, Herzlieb S, Kadison E (2007) Spatial and temporal patterns of movement and migration at spawning aggregations of red hind, *Epinephelus guttatus*, in the U.S. Virgin Islands. Environ Biol Fish 78:365–381

Nemeth RS, Kadison E, Blondeau JE, Idrisi N, Watlington R, Brown K, Smith T, Carr L (2008) Regional coupling of red hind spawning aggregations to oceanographic processes in the eastern

Caribbean. In: Grober-Dunsmore R, Keller BD (eds) Caribbean connectivity: implications for marine protected area management. Marine Sanctuary Conservation Service NMSP-08-07. U.S. Dept Commerce, NOAA, Office of National Marine Sanctuaries, Silver Spring, pp 170–183

Palau Conservation Society (2010) Enhanced monitoring of grouper spawning aggregation at Ebiil channel: final technical report, palau conservation society and society for the conservation of reef fish aggregations report, 31 pp

Patterson HE, Thorrold SR, Shenker JM (1999) Analysis of otolith chemistry in Nassau grouper (*Epinephelus striatus*) from the Bahamas and Belize using solution based ICP-MS. Coral Reefs 18:171–178

Patterson HE, Kingsford MJ, McCullock MT (2005) Resolution of the early life history of a reef fish using otolith chemistry. Coral Reefs 24(2):222–229

Petersen SW, Warner RR, Cohen S, Hess HC, Sewell AT (1992) Variable pelagic fertilization success: implications for mate choice and spatial patterns of mating. Ecology 73(2):391–401

Rand PS, Taylor JC, Eggleston DB (2006) A video method for quantifying size distribution, density and three dimensional spatial structure of reef fish spawning aggregations. Natl Mar Fish Serv Prof Pap 5:4–9

Randall JE (1963) Food habits of reef fishes of the West Indies. Stud Trop Oceanogr 5:665–847

Randall JE, Randall HA (1963) The spawning and early development of the Atlantic parrot fish, *Sparisoma rubripinne*, with notes on other scarid and labrid fishes. Zoologica 48:49–60

Rhodes KL, Sadovy Y (2002a) Reproduction in the camouflage grouper (Pisces: Serranidae) in Pohnpei, Federated States of Micronesia. Bull Mar Sci 70(3):851–869

Rhodes KL, Sadovy Y (2002b) Temporal and spatial trends in spawning aggregations of camouflage grouper, *Epinephelus polyphekadion*, in Pohnpei, Micronesia. Environ Biol Fish 63:27–39

Rochet M, Cadiou J, Frenkel VM (2006) Precision and accuracy of fish length measurements obtained with two visual underwater methods. Fish Bull 104:1–9

Sadovy YJ (1996) Reproduction in reef fishery species. In: Polunin NVC, Roberts CM (eds) Reef Fisheries, Chapman & Hall, UK, pp 15–59

Sadovy Y, Colin PL, Domeier ML (1994) Aggregation and spawning in the tiger grouper, *Mycteroperca tigris* (Pisces: Serranidae). Copeia 2:511–516

Sadovy YJ, Donaldson TJ, Graham TR, McGilvray F, Muldoon GJ, Phillips MJ, Rimmer MA, Smith A, Yeeting B (2003) The live reef food fish trade while stocks ast. Asian Development Bank, Manila

Sala E, Ballesteros E, Starr RM (2001) Rapid decline of Nassau grouper spawning aggregations in Belize: fishery management and conservation needs. Fish 26:23–30

Samoilys MA (1997) Periodicity of spawning aggregations of coral trout *Plectropomus leopardus* (Pisces: Serranidae) on the northern Great Barrier Reef. Mar Ecol Prog Ser 160:149–159

Shapiro DY, Sadovy Y, McGehee MA (1993) Size, composition and spatial structure of the annual spawning aggregation of the red hind, *Epinephelus guttatus* (Pisces: Serranidae). Copeia 1993(3):399–406

Shapiro DY, Garcia-Moliner G, Sadovy Y (1994) Social system of an inshore stock of red hind grouper *Epinephelus guttatus* (Pisces: Serranidae). Environ Biol Fish 41:415–422

Shcherbina AY, Gawarkiewicz GG, Linder GA, Thorrold SR (2008) Mapping bathymetric and hydrographic features of Glover's Reef, Belize, with a REMUS autonomous underwater vehicle. Limnol Oceanogr 53(5, part 2):2264–2272

Smith CL (1972) A spawning aggregation of Nassau grouper, *Epinephelus striatus* (Bloch). Trans Am Fish Soc 101:257–261

Starr RM, Sala E, Ballesteros E, Zabala M (2007) Spatial dynamics of the Nassau grouper *Epinephelus striatus* in a Caribbean atoll. Mar Ecol Prog Ser 343:239–249

Taylor JC, Eggleston DB, Rand PS (2006) Nassau grouper (*Epinephelus striatus*) spawning aggregations: hydroacoustic surveys and geostatistical analyses. Natl Mar Fish Serv Prof Pap 5:18–25

Warner RR (1988) Traditionality of mating-site preferences in a coral-reef fish. Nat Lond 335:719–721

Warner RR (1995) Large mating aggregations and daily long-distance spawning migrations in the bluehead wrasse, Thalassoma bifasciatum. Environ Biol Fish 44(4):337–345

Watanabe WO, Ellis SC, Ellis EP, Head WD, Kelly CD, Moriwake A, Lee CS, Bienfang PK (1995) Progress in controlled breeding of Nassau grouper *Epinephelus striatus*. Aqua 138:205–219

Weaver DC, Naar DR, Donahue BT (2005) Deepwater reef fishes and multibeam bathymetry of the Tortugas South Ecological Reserve, Florida Keys National Marine Sanctuary, Florida. Natl Mar Fish Serv Prof Pap 5:48–68

West G (1990) Methods of assessing ovarian development in fishes a review. Aust J Mar Freshw Res 41:199–222

Whaylen L, Pattengill-Semmens CV, Semmens BX, Bush PG, Boardman MR (2004) Observations of a Nassau grouper (*Epinephelus striatus*) spawning aggregation site in Little Cayman Island. Environ Biol Fish 70:305–313

Whaylen L, Bush P, Semmens BX, Johnson B, Luke KE, McCoy C, Heppell SA (2006) Aggregation dynamics and lessons learned from five years of monitoring at a Nassau grouper (*Epinephelus striatus*) spawning aggregation in Little Cayman, Cayman Islands. Proc Gulf Caribb Fish Inst 59:479–488

Whiteman EA, Jennings CA, Nemeth RS (2005) Sex structure and potential female fecundity in a *Epinephelus guttatus* spawning aggregation: applying ultrasonic imaging. J Fish Biol 66:983–995

Will TA, Reinert TR, Jennings CA (2002) Maturation and fecundity of a stock- enhanced population of striped bass in the Savanna River estuary, USA. J Fish Biol 60:532–544

Zeller DC (1997) Home range and activity patterns of the coral trout *Pletropomus leopardus* (Serranidae). Mar Ecol Prog Ser 154:65–77

Zeller DC (1998a) Ultrasonic telemetry: its application to coral reef fisheries research. Fish Bull 97:1058–1069

Zeller DC (1998b) Spawning aggregations: patterns of movement of coral trout *Plectropomus leopardus* (Serranidae) as determined by ultrasonic telemetry. Mar Ecol Prog Ser 162:253–263

Zeller DC, Russ G (1998) Marine reserves: patterns of adult movement of the coral trout (*Plectropomus leopardus* (Serranidae)). Can J Fish Aquat Sci 55:917–924

Chapter 10
The Role of Local Ecological Knowledge in the Conservation and Management of Reef Fish Spawning Aggregations

Richard Hamilton, Yvonne Sadovy de Mitcheson, and Alfonso Aguilar-Perera

Abstract Knowledge of the existence, location and timing of reef fish spawning aggregations is largely obtained from Local Ecological Knowledge in the fishing communities that exploit, or once exploited them. This information is typically collected by interviewing, followed, ideally, by validation by visiting and surveying reported aggregation sites. Conducting interviews is a relatively simple process that can be extremely productive but only if the interviewees are engaged and selected carefully (by gear, location, age, etc.), the interviewer is knowledgeable, prepared and gains the respect of the interviewee, and the various limitations of interviews as a source of information are clearly understood. Moreover, to ensure that information cannot potentially be misused and can be effectively applied to management and conservation, it is important that it is not only validated, and shared and communicated appropriately, but that it is integrated into the relevant scientific framework, and that confidentiality is respected as necessary. We review a range of studies from around the tropics based on the interview approach, evaluate its effectiveness against validated aggregations, and provide guidelines for what we believe to be good interview practices.

R. Hamilton (✉)
The Nature Conservancy, Indo-Pacific Resource Centre, 51 Edmondstone St, Brisbane, QLD 4101, Australia
e-mail: rhamilton@tnc.org

Y. Sadovy de Mitcheson
School of Biological Sciences, University of Hong Kong, Hong Kong, China
e-mail: yjsadovy@hku.hk

A. Aguilar-Perera
Departamento de Biología Marina, Universidad Autónoma de Yucatán, Km. 15.5, carretera Mérida-Xmatkuil, A.P. 4-116, Itzimná, C.P. 97315 Mérida, Yucatán, Mexico

Y. Sadovy de Mitcheson and P.L. Colin (eds.), *Reef Fish Spawning Aggregations: Biology, Research and Management*, Fish & Fisheries Series 35,
DOI 10.1007/978-94-007-1980-4_10, © Springer Science+Business Media B.V. 2012

10.1 Introduction

Local Ecological Knowledge (LEK), also frequently called traditional or indigenous ecological knowledge, refers to the cumulative knowledge of groups' or individuals' practices, experiences and beliefs about their natural environment. LEK contains empirical and conceptual aspects and is passed down over successive generations or intra-generationally (Berkes 1999; Hamilton 2005). LEK is a dynamic state of knowledge, kept alive in an oral form which makes it extremely fluid and flexible. It is constantly being tested and modified as individuals interact with their environment, adopt new technologies and make their own observations and refinements (Ruddle and Chesterfield 1977). Subsistence, artisanal, and commercial fishers often possess detailed LEK on the fisheries and environment upon which they depend (e.g. Johannes 1981; Neis et al. 1999a). As a general rule, those who spend more time on or in the water know the most, but much also depends on individuals' powers of observation and communication, curiosity about their environment and disposition for learning (Baird 2007).

The potential values of utilizing LEK for fisheries research and conservation are increasingly recognized, and there is a growing body of literature advocating its documentation and integration with more quantitative types of research and scientific methodology, for management planning, education and outreach, as well as for strengthening the case for conservation and management (Christie and White 1997; Johannes and Neis 2007; Sadovy de Mitcheson et al. 2008). Fishers can provide important information on such things as inter-annual, seasonal, lunar, diel, tide- and habitat-related differences in species behaviour, presence of concentrations of females with eggs, types and abundance of target species and their changes over time, and how these factors determine fishing strategies (Johannes et al. 2000). Fishers are often the first to discover the location of important habitats, such as nursery, feeding and spawning grounds of fishes on which scientists subsequently work (Johannes 1989; Hamilton and Walter 1999; Ames 2007). When time and care are taken to document LEK resource user's perceptions about their environment can also be understood, which can assist in directing future management efforts.

Local knowledge can be of great value for providing a historical perspective on the state of reef fish communities and is particularly important where there is no long-term monitoring or database, as for most tropical coastal fisheries. Specific examples range from declines in bonefish migrations in Kiribati disrupted by coastal construction to the rapid demise of bumphead parrotfish, *Bolbometopon muricatum*, and humphead, or Napoleon, wrasse, *Cheilinus undulatus*, populations in the Pacific (Johannes and Yeeting 2001; Sadovy et al. 2003; Dulvy and Polunin 2004). In the case of the Nassau grouper, *Epinephelus striatus*, in the Caribbean, informal discussions with local fishers identified a long since disappeared aggregation; this information was followed up by a questionnaire sent around the Caribbean soliciting local knowledge on the species, with results of the survey indicating similar losses and declines elsewhere and highlighting a problem with the species (Sadovy 1993; Sadovy and Eklund 1999). Suitably validated and treated within the appropriate bounds, LEK and historical anecdotal records can be useful proxies for biological and fishery information and can help to identify 'shifting baselines'; long-term and usually negative changes that are often not immediately or readily apparent to new generations of fishers or scientists working in a data-poor area (Ainsworth et al. 2008; Holm et al. 2001; Pauly 1995). A testament to the value of fishers LEK is its increasing application in marine research, species assessments and planning processes that are linked with conservation and management programmes (Sadovy and Cheung 2003; Aswani and Hamilton 2004; Aswani and Lauer 2006; Smith and Hamilton 2006; Stanley and Rice 2007; Sadovy de Mitcheson et al. 2008; Almany et al. 2010; Game et al. 2011).

One of the most widely applied uses of LEK in the marine setting is its role in the research of Fish Spawning Aggregations (FSAs). In many locations, fishers have known of FSAs for centuries or have experienced seasonal gluts in landings subsequently identified as FSAs (Johannes 1978; Colin, et al. 2003). In recognition of this, and because of the practical difficulties of discovering FSAs that typically form at highly localized areas for brief periods of time, biologists and coastal managers have long drawn on the local knowledge of fishers in the initial stages of their fieldwork (Johannes, et al. 1999; Robinson et al. 2004; Hamilton et al. 2005a, 2011; Sadovy de Mitcheson et al. 2008). Indeed, most aggregations of commercial fishes known today were initially identified from descriptions of fishers. The value of this knowledge is heavily dependent on the methodology used to collect, process and validate it and the way(s) in which it is integrated into science-based management.

Our objectives in this chapter are to evaluate the methods for collecting, assessing and applying LEK, using lessons from our own work and numerous other case studies, and to develop a robust methodology for the documentation of FSAs using LEK. We consider the collection and validation of information, and integration of LEK with scientific knowledge. We also examine pitfalls and identify caveats in the collection and use of LEK.

10.2 Documenting Local Ecological Knowledge

The methods used for documenting LEK derive from the social sciences and require some understanding of the language and cultural context in which the interactions take place (Briggs 1986; Mailhot 1993; Neis et al. 1999a). For instance, social scientists interested in fishers' knowledge typically use ethnographic methods such as *interviewing* and *participant observation*. Interview formats range from structured to unstructured, while participant observation requires that the researcher live in the local society among the resource users and is actively partaking in fishing or related activities. In Melanesia, these techniques have documented extremely detailed LEK e.g. (Hviding 1996; Aswani 1997; Foale 1998). Participant observation requires a large investment of time within a very narrow geographical area. Consequently most studies of LEK with a predetermined conservation or management focus rely predominantly on interviews. Other approaches to information collection are *analytical workshops*, where collective knowledge of participants is put into perspective, and *collaborative fieldwork*, where participants collaborate in scientific procedures (sampling, interpretation) (Fraser et al. 2006; Huntington 2000). Due to its widespread use, we focus in this chapter on interviews as a means of documenting LEK on fish spawning aggregations.

10.2.1 Interviews

Marine research and management interviews can range from informal exchanges to highly structured questionnaires (Briggs 1986; Huntington 1998, 2000; Gubrium and Holstein 2002). Some interviews may be analyzed by dedicated computer programmes [e.g. ANTHROPAC 4.92 (Borgatti 1996) ParFish www.fmsp.org.uk]. One commonly used interview method is the semi-structured form in which the interview flows relatively freely around a simple and predetermined core of questions. This allows the informant to introduce aspects of LEK with which they are particularly knowledgeable and may be unexpected or completely unknown to the interviewer. Interviews of single individuals are informative and detailed but biased to the particular point of view and experience of the informant. Group interviews may help the participants to encourage each other to provide information and assist in recall, but can be dominated by particular individuals (Huntington 1998).

Interviews can also be used as the basis for stock assessments in a way that not only collects information in unmonitored fisheries but can also engage stakeholders in management planning. Beyond spawning aggregations, aggregating species often need to be managed in the broader context of the whole fishery. Stock assessments, or at the least a better understanding of the overall fishery, may be needed. An example is ParFish, based on the logistical biomass growth model, in which interview data from fishers are used to construct 'priors' for the fishery model. Combined with other available information and opinions of fishers about other aspects of their fishery, a stock assessment can be conducted and management planning discussed.

Important advantages of the interviews are that they (1) are inexpensive, (2) are opportunities for establishing a rapport for increasing accuracy and honesty of answers, (3) give in-depth coverage of given topics (semi-structured interviews), (4) allow for data comparison for different geographical areas, (5) act as a starting point to establish a trust between informants and interviewer that can aid in building towards effective management, (6) allow for identification of misunderstandings regarding the marine environment, and (7) enable identification of fishery and outreach needs. Disadvantages include (1) the time required, (2) that responses are highly dependent on gear type used, (3) that some information may be difficult to analyze and compare or to put in a suitable scientific context, (4) that some recorded interviews may need translation if performed in the native language, potentially introducing error and reducing the ease of information exchange, (5) that interviewers need to be suitably skilled for semi-structured interviews, (6) that informants may be unwilling to share information or purposely provide misleading information, and (7) usually requires validation of information for management planning.

Important considerations in developing a scientifically meaningful interview methodology are sample size (how many interviews), how to standardize catch data to allow for comparisons over time and space, and how to capture the historic perspective in a meaningful way or such that it could be represented quantitatively. Care is needed in the design phase of the work to address such questions and use appropriate data collection techniques and analytical methods.

10.2.2 Documentation of LEK Requires Both Anthropological and Biological Skills

The seemingly simple process of asking a few questions about fish spawning aggregations must be carefully conducted and interviewers must be suitably trained and knowledgeable. If workers are poorly prepared and are unclear of their goals, they will not understand the social, cultural, economic or biological context of the fishery they are interested in. The result of poor preparation is that much time, energy, money and patience can be wasted. The frustration felt by interviewees who can recognize an interviewer who knows little of the topic at hand, could reduce their willingness to participate in the future, lead to misleading responses, cause them to disrespect the interviewer (Sect. 10.3), or lead to management inaction where action might be needed. It is also our experience that LEK surveys will only be done effectively if time and funding are set aside to do this type of ethnographic research and if interviewers are carefully selected and trained. The apparent simplicity of this (interview) approach is deceptive but interviewing well requires considerable preparation. As one of many examples, workers who did not know local species and understood very little about aggregations quickly ran into difficulty when they could not distinguish fish schools from spawning aggregations in the Philippines (YSM personal observation 2005).

10.2.3 Selection of Experts

Identifying who to interview in LEK surveys is critical (Davis and Wagner 2003). The gender, degree of involvement in a fishery and the type of fishing gear may substantially influence the distribution of LEK in a community (Sect. 10.3.1). At times the target group to interview may be obvious to researchers already familiar with an area. In the Solomon Islands for example, Hamilton (2003a) knew that night spear fishers were the only group that targeted bumphead parrotfish and were the obvious source of information. Similarly, in southern Manus, Papua New Guinea, preliminary interviews revealed that fishing at grouper spawning sites was traditionally restricted to one specific clan within a community, and that this clan has retained highly detailed LEK on these aggregations (Hamilton et al. 2005a).

Identification of expert fishers in a new area can be informal and opportunistic, or conducted in a structured manner (Fig. 10.1). Referrals from local associations,

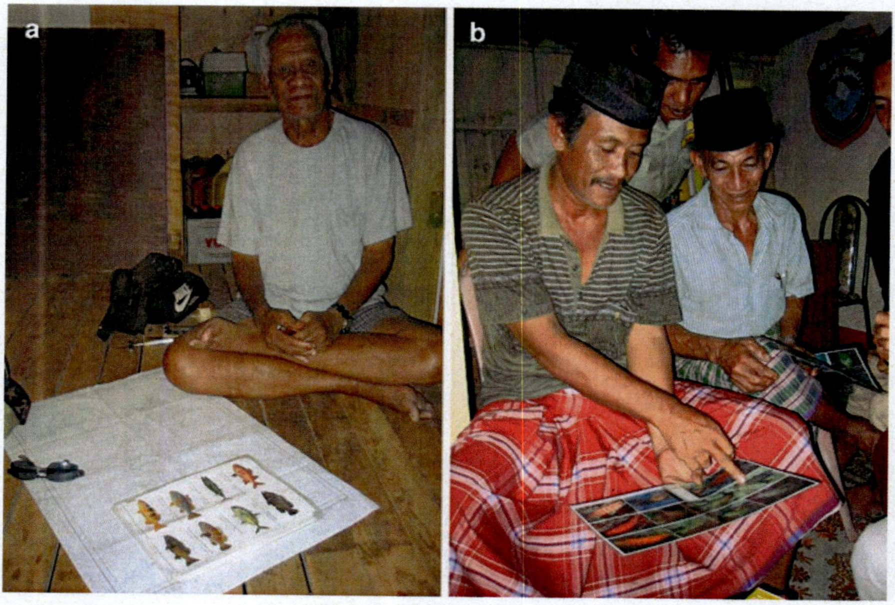

Fig. 10.1 (**a**) (*left*) Interview with patriarch fisher in Palau with local fish pictures and map (**b**) (*right*) Group interview with fishers in Indonesia with fish photos to aid identification (Photos: Yvonne Sadovy de Mitcheson)

from one or more community peers, or responses from a large sample of resource-users to select experts are useful (Ferguson and Messier 1997; Olsson and Folke 2001). 'Snowballing' is often used to identify additional experts whereby, at the end of an interview; fishers are asked to give the name(s) of other fishers experienced and knowledgeable in the area or species of interest (Neis et al. 1999b). Sadovy and Cheung (2003) had to interview everybody they could find with knowledge of the giant yellow croaker, *Bahaba taipingensis*, as few fishers of this species were still surviving, whereas, when looking at historical changes in a fishery, selection of patriarch or locally respected, often older, fishers is preferable (Johannes and Yeeting 2001; Alves et al. 2005; Gass and Willison 2005; Fraser et al. 2006). In Brazil, interviews revealed just a small number of "goliath grouper experts" (*Epinephelus itajara*), in each community, with referrals for interviews rapidly converging on just a few names (Gerhardinger et al. 2006).

10.3 The Data Collection Process

In this section, we examine methods for conducting and assessing interviews and other informal sources of information. Specifically, we examine how to select interviewees, some of the challenges of data gathering and interpretation, and other considerations such as confidentiality and data quality. Follow-up steps for management and conservation are considered while validation is addressed in Sect. 10.4.

10.3.1 LEK Is Not Distributed Equally

The depth and precision of LEK can vary enormously among fishers. Geography, gender, age, fishing method used, area fished, dependency on fishery resources and fishing traditions are all important factors. Knowledge is often proprietary and not always widely shared within a community; this alone produces a vast array of knowledge levels. Therefore, before gathering fishers' LEK it is important to learn as much as possible about the practices, species, trade and history of fisheries in the area, by reading and talking to people knowledgeable about the region or society. Such preparation allows interviewers to be better informed to select and, importantly, engage with appropriate interviewees, formulate appropriate questions and be able to supply answers to questions likely to arise.

10.3.2 Gear, Species, Location and Fish Handling Practices

The gear used, fishing locations, species of greatest interest, or the typical practice of cleaning fish will all determine interviewee experiences and knowledge, and influence questions and responses. For example, spearfishers may know about behaviours and colour changes of the fishes they see underwater, while inshore net fishers may know little of species taken offshore by hook and line but much about seasonally migrating species that follow the shoreline. In some communities, women may target fish inshore while men travel further offshore, so experiences concerning marine resources are often very different within a community, or even a household. In Brazil, long-line fishers were unaware of spawning aggregations of goliath grouper fished only by spearfishers from the same city; long-liners and spearfishers used different popular names for goliath grouper (Gerhardinger et al. 2006). If fishers do not clean their catch (removal of abdominal contents and gills) before sale, either because fish are sold 'whole' or because others in the community, such as a fisher's wife, do the cleaning, questioning fishers about ovary condition may be meaningless.

Different types of aggregating species may be easier than others to identify from interviews. Large *transient* aggregations of certain mullets (Mugilidae), rabbitfishes (Siganidae) and groupers (Serranidae), being brief and highly time- and location-specific, are more readily recognizable as spawning aggregations than are *resident* aggregators that gather in many smaller groupings throughout much the year at poorly defined sites (for definitions of aggregation type see Chap. 1). Surgeonfishes (Acanthuridae), for example, normally move around in schools but also form

Fig. 10.2 (**a**) Papua New Guinean fishers from New Ireland Province displaying, from *left* to *right*, a squaretail coral-, brown-marbled and camouflage grouper that were speared during the day from a known FSA (**b**) Solomon Island fisher displaying two bumphead parrotfish, *Bolbometopon muricatum*, speared at night while resting in large groups (Photos: Richard Hamilton)

resident aggregations on a regular, even daily basis, for much of the year and at specific sites; LEK does not tend to readily identify such species as aggregation spawners. Local ecological knowledge on valuable, iconic or distinctive species, such as the bumphead parrotfish or humphead wrasse may be easily remembered (Sadovy et al. 2003; Dulvy and Polunin 2004) (Fig. 10.2).

10.3.3 Individual Attributes of Interviewees

The depth of individual LEK depends on the extent of their reliance on marine resources, their age and length of time fishing and/or trading, as well as cultural, personal and economic considerations. Long-time traders may have a better perspective of overall changes in species composition, sizes or landings than individual fishers who only know their own fishing area. Full-time fishers who are dependent on marine resources in one area may be more aware of long-term changes in that area than those who move between fishing grounds, or are only partially or seasonally dependant on fishing. Again, careful selection of the local experts is critical to interview outcomes (Sect. 10.2.3).

Older fishers usually have a broader perspective on changes and problems in local fishing grounds and may be less inhibited than younger ones about speaking openly (Fig. 10.3). In Fiji many older fishers had reported substantial declines in bumphead parrotfish, humphead wrasse or large sweetlips (Haemulidae) (probably Giant sweetlips *Plectorhinchus obscurus)*, while younger men rarely reported seeing these species (Dulvy and Polunin 2004, YSM personal observation 2006).

Fig. 10.3 (**a**) Experienced Papua New Guinean fishers from Manus Province openly sharing their fishing experience and writing the local names of their reefs onto a satellite map (**b**) Women fishers often fish in different areas, usually more inshore, from men and can share these experiences in interviews, as in Fiji (Photos: (**a**) © Michael Berumen (**b**) Yvonne Sadovy de Mitcheson)

Conversely, in many places, older fishers may never have visited offshore aggregation sites because of lack of access prior to mechanization (outboard engines) of the fishery and a tendency for them to focus on inshore resources as they age. In areas where destructive fishing practices are used, fishers are reluctant to be interviewed; in Busuanga Island, for example, Philippines, many young men use cyanide as a fishing method to maximize their incomes but are reluctant to be interviewed because its use is illegal (YSM personal observation 2006).

10.3.4 Cultural and Social Contexts

Some cultures use or retain detailed information on resources. Others appear to have little awareness of particular species or trends in catches. Whether this is due to educational background, historic dependence on the sea, contemporary factors, cultural factors, beliefs or attitudes is not always apparent. Interviews across the Indian and Pacific Oceans in two studies show low awareness of spawning aggregations in India, Indonesia, Sri Lanka, Philippines and Thailand with much higher awareness in the Maldives and western Pacific islands (Sadovy de Mitcheson et al. 2008; Tamelander et al. 2008). In Indonesia and the Philippines many people were uprooted or have migrated to more promising fishing grounds and have little or no traditional knowledge of their present fishing areas (e.g. Sadovy and Liu 2004). It is also possible that local knowledge on FSAs may be absent because, due to heavy fishing, few aggregations remain. In many places such as parts of Southeast Asia, the fisheries are in such poor condition, with low catches predominantly of small fishes, that spawning aggregations of many species may no longer occur (Sadovy de Mitcheson et al. 2008). By contrast, many western Pacific coastal communities have long traditions of fishing on their customary reefs and fisheries remain in relatively good shape. Knowledge continues to be passed from one generation to the next; where communities rely heavily on marine resources the most detailed LEK is found (e.g. Johannes 1981).

10.4 LEK Can Provide Critical Information for Conservation and Management

LEK can assist marine conservation and management efforts by providing information on the physical and biological characteristics of aggregations, as well as insights into changes in fisheries over time. These attributes of LEK are elaborated below.

10.4.1 Physical and Biological Characteristics of Aggregations

Local fishers may know where and when some commercial species aggregate for spawning. If properly validated (Sect. 10.6) such knowledge may be relevant to conservation or management objectives. For example, LEK on seasons when FSAs form has helped to establish closed seasons for fisheries (e.g. in Palau, Johannes 1981; YSM personal observation 2010). LEK on seasonality is often incomplete (Sect. 10.4) and seasonality can vary even over small spatial scales in some species. In Melanesia, the peak spawning times of the brown-marbled grouper, *Epinephelus fuscoguttatus*, can be staggered by a month or more across sites separated by as little as tens of kilometres (Johannes and Lam 1999; Hamilton et al. 2011). While spawning seasons can be linked to the months in a western Gregorian calendar, in many locations this calendar is not used. For examples, Daw (2004) found that the Islamic calendar was excellent for identifying synchrony between FSAs and the phases of the moon (Daw 2004), but of limited use for describing seasonal trends as it shifts each year relative to the Gregorian calendar. Similarly, in the Pacific traditional lunar calendars identify the months in which aggregations occur, by associating them with the flowering of particular plants, as is the case of groupers in Fiji with the 'tavola' or tropical almond tree, *Terminalia catapa* (YSM personal observation 2005).

To identify an aggregation as reproductive through LEK, as opposed to groupings formed temporarily for another reason, the occurrence of both *spawning* and *aggregation* should be established using clear criteria and definitions (Domeier and Colin 1997, Chap. 1). It is critical that interviewers understand and consider both criteria during interviews. Care must also be taken with terminology; in some locations spawning aggregations may be referred to as a 'school' (Bahamas for Nassau grouper), so careful interpretation is needed. Moreover, aggregations of fish may be caught at certain times and places that are not reproductive, such as during feeding.

10.4.2 Historical Baseline on Changes in Aggregation Fisheries

LEK is used increasingly by scientists to assess fisheries that have limited or no data, or to reconstruct their history (Neis et al. 1999a). Johannes and Yeeting (2001) used information from villages in Tarawa, Kiribati, on declines in bonefish spawning migrations, and their possible causes (demise of customary marine tenure, causeways and gillnets blocking migration routes) that led to a community management plan. Interviews in many countries mirror biological studies indicating declines in large reef fish species, providing alternative sources of information for species assessments and management discussions, and promoting strong local buy-in (Sect. 10.6) (Sadovy et al. 2003; Aswani and Hamilton 2004; Sadovy de Mitcheson et al. 2008).

A few specific examples illustrate the value of such information as well as the challenges of using an approach that depends on experience and memory. In these examples, information from interviews could be quantified and usefully compared to indicate apparent long-term trends. In Palau, grouper aggregation sites yield mixed species landings (mainly squaretail coralgrouper *Plectropomus areolatus* and camouflage grouper, *E. polyphekadion*). Fishers from several communities switched

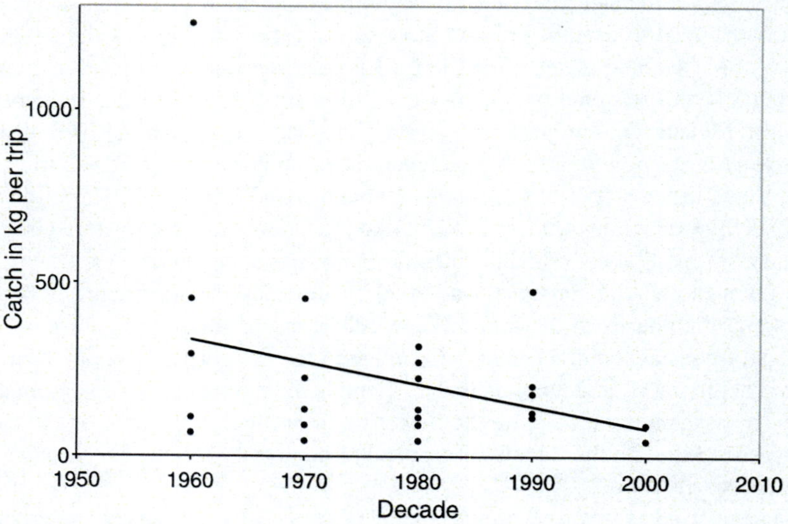

Fig. 10.4 Peak catches reported from multi-species spawning aggregations of groupers (*E. polyphekadion, E. fuscoguttatus* and *P. areolatus*) with trend line; catch data include all three species lumped together according to fisher interviews conducted in July 2003 in Palau (Adapted from Sadovy de Mitcheson et al. 2008)

among multiple locally known sites and reported landings reflecting the overall declines in catches perceived, rather than trends from a single aggregation or community (Fig. 10.4). In Fiji, for the camouflage grouper, it was most effective to direct questions at the scale of decades and to ask for recall of the highest catches encountered per trip (Fig. 10.5). The concept of average catch was either not understood or not easy to recall over the long term; maximum catch per boat in a day was the more memorable way in which landings were recalled. Consistent trends indicated from many independent fishers and from different communities allowed for acknowledgement and recognition of declines among stakeholders, and facilitated follow-up discussions and research planning. Clear and consistent trends, especially if expressed quantitatively, are also valuable for demonstrating interview outcomes persuasively to local government authorities, other interested stakeholders and the public.

10.4.3 Working with Local Partners and Information Exchange

Typically, interviews are conducted in partnership with local community and government officials, non-governmental organizations (NGO's), and educational institutions involved in conservation or management. Projects initiated or based

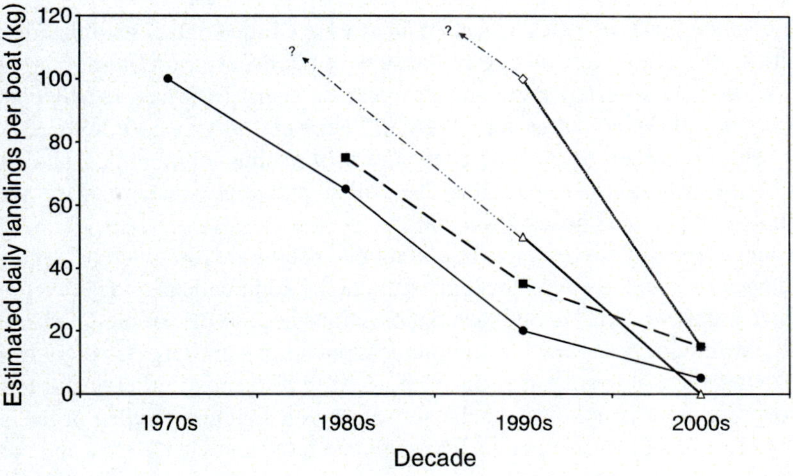

Fig. 10.5 Camouflage grouper, *Epinephelus polyphekadion*, peak landings from four different aggregations in Fiji and over at least four decades, according to fisher interviews conducted 2003–2005. Fishers were asked their highest recalled catches in each of the decades and the sites were validated as aggregation sites in 2005. *Dash/dot lines with '?'* point backwards to decades when fishers reported "hundreds" of kilograms per trip. Each *line* represents information from a separate aggregation site along contiguous coastline in northern Fiji (Sadovy 2006, YS, unpublished data)

outside of target countries should be closely aligned with the work of local partners to ensure that studies are appropriately designed and that significant outcomes are communicated and understood. Such steps build a platform for future management or conservation initiatives that generate understanding, interest and trust, identify educational and management needs, and avoid duplication of effort.

A big advantage of working with spawning aggregations is that they are easy to explain and understand. Everybody knows and appreciates the importance of reproduction, and most people are naturally interested to learn more about the resources on which they depend, especially when odd or interesting behaviours are concerned; some aggregations can be quite spectacular to see or fish. This provides excellent opportunities for two-way transfer of information; the interviewee responding to the questionnaire, and the interviewer able to discuss and explain the collective outcomes of interviews in light of experiences and relevant information from outside. This allows further opportunities for discussion after interviews and again highlights the need for well-prepared and knowledgeable interviewers who know something of fishery and marine resources of the area as well as the general biology and ecology of the species being reported (Sect. 10.2.2).

Local collaborators may or may not be knowledgeable about fishing communities in their own countries. In the Solomon Islands and Papua New Guinea many fisheries officers grew up in small fishing communities and have an excellent knowledge of and genuine interest in the fishery and fishing communities (RH personal observations 1996–2011). In some areas of Indonesia and parts of the

Pacific, by contrast, fisheries officers typically have little contact with fishing communities and their understanding of the state of the fishery can differ greatly from that of the fishers, often impeding progress on fisheries issues (YSM personal observation 2005). In the eastern Yucatan Peninsula of Mexico (Quintana Roo), most fisheries officers know little of the needs of fishing communities, and tend to treat fishers with distrust, regarding them all as potential poachers of lobster and conch (AA-P personal observation 2006).

After completing surveys, it is important that results are discussed in community meetings, presented to government departments or communicated via the radio and television. Further possible outreach opportunities need to be explored. Other interested and important parties for communications might be religious organizations and leaders, education departments, journalists, enforcement officers and tourism interests. Changes in policy and behaviour start with an understanding of the issues among users and the wider public. There remain large knowledge gaps and considerable misunderstanding in many countries in relation to the sea, in general, and the challenges faced by, and needs of, those who depend directly on its living resources. In many places, even fishery managers are little engaged in fisheries management and have insufficient training. It is worth highlighting here that in most countries fisheries issues, or at least small-scale coastal fisheries, are not taught in schools. Fish biology and ecology is not typically covered in school curricula so a general understanding of the sea is very limited in most communities.

10.5 LEK Can Be Incomplete, Inaccurate or Misleading

Interviews are a first step in a process enabling researchers to ask more specific questions and design focused studies and validation (Sect. 10.4) (Valbo-Jorgensen and Poulsen 2000). This section examines ways in which LEK falls short of providing all necessary information for management and conservation, and identifies what additional work is needed once interviews are completed.

10.5.1 Observation Versus Interpretation

Documentation of information and its interpretation are distinct activities. Divergences in interpreting the meaning or significance of an observation are common between the interviewer and interviewee (Ruddle et al. 1992). Learning how fishers or other interviewees view the sea and interpret the changes they witness, as well as identifying means to bridge information or perception gaps, is critical for planning educational materials and interventions. Therefore, interviews should include questions that explore the views, perspectives, and attitudes of interviewees and allow opportunity for them to ask questions. It is common, for example, for fishers to observe that fishes have declined in a given area over time, and then to interpret this as being due to the fish having moved elsewhere, gone deeper or changed behaviour, such as no longer biting hooks. The interpretation of the observation in this case is not that the fish are no longer in the sea (i.e. possibly overfished) but that they are still present in the area but can no longer be caught. Such a perspective would make the concept of spatial protection or effort reduction in an area unacceptable or even incomprehensible as a solution to the changes perceived in catches. Discussion of such perceptions opens up opportunities for additional possible explanations, including overfishing, or discussions of various options for management as well as the longer term implications of inaction.

Interviewers must exercise care in formulating unambiguous questions while anticipating possible confounding factors. For example, in asking about trends in catches over time, fishers may report no changes. The reality could indeed be no change, but the perception of the fisher could also be due to him/her spending longer at sea, or shifting between fishing grounds, behaviours that could mask overall declines in the fishery, to compensate for declining catch per unit effort. A progressive increase in the power of outboard engines, facilitated by well-intentioned government and international aid programmes has also led many fishers to move further offshore, often to the edges of shallow water coastal platforms, with an initial boost in landings, as largely unexploited fishing grounds were entered.

Even simple gear changes can markedly affect catch rates. Spearfishers in the Pacific boosted catch rates when new technologies as simple as prefabricated flippers (dive fins) dive masks, spearguns and underwater flashlights (which enabled night spearfishing) were introduced (Gillett and Moy 2006, YSM and RH personal observations 1996–2011). In many Pacific Island locations the relatively recent initiation of night spearfishing, especially when combined with compressed air diving, has resulted in substantially increased catches. This fishing method in particular is having profoundly negative impacts on populations of large species such as the bumphead parrotfish which sleeps on the reef at night and is easy to catch at this time (Hamilton 2003a). Spearfishing is considered a major problem when aggregations are targeted in Fiji and the Solomon Islands (Gillett and Moy 2006) and is also a considerable issue in Palau and Papua New Guinea (YSM and RH personal observations 2005). An interview with a Solomon Island spear fishermen demonstrates this:

> I remember a spear fishing trip in 1985, not long after I had learned to use fins, when we were asked to spear Topa (bumphead parrotfish) for an upcoming wedding… I speared 74 big Topa that night. I could have speared many more, but our canoe began to sink from the weight of all the Topa (Hamilton 2004, p 66).

In some instances, of course, LEK can be unreliable, and fishers' interpretations of their own observations incorrect, but interviewers should not be too swiftly judgemental (Johannes 1989). Tibby et al. (2007) reported on an assessment of a coastal dune lake in Australia in which local perceptions and accounts were inconsistent and did not always match historical evidence. Johannes (1981) recorded stories of octopi climbing trees to give birth to young, noting this to be biologically unlikely. Some initially unlikely reports, however, have turned out to be partially correct and highlight the need for interviewers to be open-minded. For example, Johannes and Neis (2007) reported how fishers informed marine biologists of an area off the coast of Belize where whale sharks aggregated and were swimming through milky clouds of water. The fishers assumed the whale sharks were spawning but their assumptions were wrong, as whale sharks are viviparous and do not release eggs. While their observations were accurate, the whale sharks were in fact feeding on spawn released by large aggregations of snappers (cubera snapper – *Lutjanus cyanopterus*, dog snapper – *L. jocu*) (Heyman et al. 2001). In Brazil, fishers reported that mullet, *Mugil platanus*, enter coastal rivers and estuaries to spawn during extensive spawning migrations, whereas scientists suggest that spawning occurs in the open ocean; although this has yet to be confirmed and the different instead accounts call for further work (Silvano and Valbo-Jorgensen 2008). Clearly, LEK is most useful in science and conservation if it is carefully collected, evaluated, cross-referenced and validated whenever possible (Johannes 1981; Ruddle et al. 1992; Usher 2000).

10.5.2 Only Knowing Part of the Picture

LEK in a local context is, by its nature, selective; it is a small part of a much bigger picture. Workers need to be just as clear about what they will not learn through interviews as about what they will (Sect. 10.3 and Table 10.1). One illustrative example is blacktail snapper, *Lutjanus fulvus*, in Palau. This is a small snapper evidently not known in aggregations or of particular interest to fishers, and yet predictably forms massive groups of ripe adults (Chap. 12.10). However, it has never been picked up in LEK despite numerous interviews in Palau over the last two decades or so (Johannes 1981; Sadovy 2007).

People also differ enormously in how observant they are. Spearfishers in Fiji, Papua New Guinea, Cook Islands, and the Solomon Islands, for example, provided impressive details on features such as colour changes, orientation relative to water currents and behaviours among aggregated groupers (Passfield 1996, YSM and RH personal observation 2005). These fishers could only have known such details from their own experiences thus providing strong evidence that spawning aggregations had been observed.

It is our experience that only the more obvious aggregations, or those of more valuable species, are clearly and consistently reported in interviews. In Fiji, for example, on a SCUBA dive to validate an aggregation site reported by many fishers to contain the camouflage and squaretail coralgroupers, another species, the speckled blue grouper, *Epinephelus cyanopodus*, was also found that no one had mentioned in interviews despite its presence in the fishery. We later discovered that only shallow-water spearfishers caught it and noted that our original interviews had not included these fishers (YSM personal observation 2006). In Palau, several groupers were reported to assemble in significant number in May and June in interviews (Johannes 1981), and these months were later included in a traditional seasonal closure 'bul' and in national laws. Aggregations also occur in July and August, as described by Johannes et al. (1999) based on field studies, and were reported in interviews to last from April to August (Sadovy 2007). While aggregations are also protected in July, August is still unprotected.

Likewise, interpretations of the condition of a resource can depend heavily on experiences based on gear types being used. In Brazil, those fishers using spears perceived the goliath grouper fishery to have declined; they attributed this to illegal fishing with underwater breathing apparatus, pollution and decreased availability of food resources, such as lobsters. Long-line fishers from Brazil, however, perceived abundance as being unchanged or even higher than before (Gerhardinger et al. 2006).

10.5.3 Knowledge Beyond Fishers

There are sources of knowledge, or means of corroborating interview-based information, other than fishers and fishery managers. Experienced and long-serving dive guides, local researchers and traders may have a wealth of knowledge on the status or history of a local fishery or on particular species. Historical documents can provide unique insights into past fisheries and, in some countries, trade, fish processing or market data, or even tax offices may yield useful information. For example, Sadovy and Cheung (2003) used LEK of older fishers and literature on swim bladders to reconstruct the history of the giant yellow croaker, also known as the Chinese bahaba, fishery. This large fish spawned in river mouths in southern China and was abundant in the mid twentieth century. Now it is on the verge of extinction after its spawning aggregations were heavily targeted for its valuable swim bladder. LEK regarding changes in its abundance was cross-checked by conducting reviews of the English and Chinese published and unpublished literature (e.g. Sadovy and Cheung 2003). In Cuba, an 1884 account on grouper migrations provides a fascinating insight into a fishery that has clearly changed profoundly in recent decades (Vilaro-Diaz 1884; Claro et al. 2009). In the Mexican Caribbean, early accounts by a geographer (Craig 1966) attracted the attention of researchers to study the Nassau grouper "corridas" (migrating runs) along the Mesoamerican Reef (Aguilar-Perera 1994). Traditional seasonal calendars in Fiji associate the times of the flowering of certain trees with increased abundance of certain grouper species (see above). Saenz-Arroyo et al. (2006) documented past marine fauna of the Gulf of California mainly through the use of accounts from early travellers, while Lajus et al. (2001) reconstructed Atlantic salmon fisheries in Russia mainly through the use of tax records.

10.5.4 Recall

Possible bias introduced from memory loss or distortion is a serious consideration with interviews that seek to document past conditions. The nature of the bias will depend on the questions, the importance of the species or issues to the interviewee and on how the information is to be used. The problem is likely to be greater the further back in time information is sought since memory fades with time (and age) (Neis et al. 1999a; Pascale et al. 2002). Memory aids that assist recall relative to special or historic events include major wars, an election, introduction of a new regulation, a marriage or birth, etc. (e.g. Ames 2007). Memory that is personal, long-lasting, and related to self-esteem can be positively biased (Thompson et al. 1996) such that maximum or 'best' catches,' largest' fish, for example, may be more memorable than 'average' fish size or low catches.

10.5.5 Exaggeration and Misleading Information

Unintentionally or otherwise, interviewees may exaggerate their responses or provide information that is misleading. In our experience, this is not a major problem. For certain sensitive issues, and if pressures on and interest in aggregations increase, it may become one. While difficult to detect, long-time experience with interviewing, internal consistency of answers (through questions that seek the same information in different ways, or across interviewees) and simple 'gut feeling' can help to identify many cases of misleading responses. Interviewees may exaggerate or mislead if they are trying to impress an interviewer, do not take the interview seriously, do not know answers to questions, are concerned about the use to which interview outcomes will be put, are in a hurry, tired or bored, as well as other possibilities (Silver and Campbell 2005, RH and YSM personal observations 2005). Problems often arise if interviews are being translated. YS had a case in Indonesia in which the translator decided, after the first few interviews, that he already knew the answers because he heard similar responses in earlier interviews. He then started to reply himself rather than to translate replies. Questionnaire design should include test questions for which the answer is known and photo-aids that include non-local species, as a "truth test".

10.5.6 Secretiveness

It is inevitable that interviewees might be secretive about an activity on which their livelihood depends, or with strangers. They may be competing with other fishers, not want to provide information indicating their income, feel threatened because they do not understand the purpose of the questions, or be fishing illegally. Collectively, we have experienced little indication of secretiveness, although there are exceptions and this is obviously difficult to judge. In most locations in Melanesia fishers willingly showed researchers the locations of spawning sites once the purpose of LEK surveys was explained. But in southern Choiseul, Solomon Islands, no one would discuss the locations of known leopard coralgrouper, *P. leopardus,* squaretail coralgrouper aggregation sites; these communities have a history of keeping the locations of these sites ‘private’ from other fishers, and will paddle off the site if they see another canoe approaching (RH personal observation 2004). In Mexico, fishers are often secretive about the locations of recently discovered sites. However, once the site becomes widely known, these fishers typically communicate. Currently, fishers from the Mexican Caribbean are collaborating with researchers to find ways to protect Nassau grouper spawning aggregations after learning from Honduran fishers about major declines in catches of this grouper (AA-P personal observation 2007). Where illegal fishing methods such as cyanide or dynamite are extensively used, as in parts of Indonesia or in the Philippines, it was hard to find cyanide fishers to agree to participate in interviews (YSM personal observation 2006).

Overall, where communities have effective local controls or where resources are not considered to be in short supply, we encountered little reluctance to answer general questions. However, questions that relate to money, especially when the interviewee is a trader or middleman, where catches are subject to taxation, or when government officers are present, may not be welcome. Government officials are sometimes wary or defensive, especially when they have little knowledge of the resources in question.

10.5.7 Basic Points to Remember When Interviewing

Basic points to remember when preparing for interviews and conducting interviews in the field are outlined in Appendix 10.1 and elaborated below.

10.5.8 Confidentiality

Most aggregation sites, revealed through fisher interviews, continue to be exploited but not managed. Although the interview process itself is often closely linked to ongoing local management initiatives and awareness-raising, this is by no means always the case. Management can take many years to implement, and even then enforcement may be weak or non-existent, hence release of detailed aggregation site information (including on maps and in final project reports) into the public domain is likely ill-advised, until a management framework is in effect. Moreover, the general question of the release of indigenous knowledge where there is no direct benefit to the source communities or consensus from the communities on the matter, and where this could possibly lead to increase in exploitation from outside pressures, is not only a practical but also an ethical issue.

The issues of confidentiality and accountability in marine studies in relation to sensitive areas or species have received little attention to date; certainly not to the same extent as in terrestrial situations where such concerns are often addressed. However, several fisheries agencies (for example the Great Barrier Reef Marine Park Authority in Australia and the Seychelles Fishing Authority) have a policy not to publicly release fish aggregation site locations (Jan Robinson, Martin Russell personal communication). In Palau and Fiji, local NGOs or authorities prefer not to release site data on aggregations for fear that local recreational and commercial fishers, many with powerful outboard boats and using SCUBA and spears, could quickly reach newly reported sites. Poaching is common in both countries, enforcement minimal, and fish stocks close to more highly populated areas are overfished leading to interest in expansion of fishing grounds (YSM personal observation 2005). For data-sharing purposes, confidentiality agreements could be developed.

10.6 Validating and Integrating LEK with Science to Support Management and Understanding

The major challenges in incorporating LEK into management agendas is assessing the reliability and accuracy of the information collected and using this information to gain support for management, if needed. This section addresses the issue of validation, communication and integration of interview-sourced information. Possible approaches range from visits to reported sites at the time(s) and place(s) indicated for spawning (although this does not exclude the possibility that spawning occurs at other times and places not recorded) to market visits, discussion with dive guides, fishery reports, cross-sector interviews (i.e. traders, wholesalers, etc.), underwater visual census (UVC) surveys, dedicated follow-up studies and assessment of consistency among or within communities of experiences reported.

To assess fisher's knowledge effectively, an interviewer must be certain of which species informants are talking about and where and when their observations were made. Having good fish reference materials greatly facilitates this process. Local fishing communities often have a highly developed folk taxonomy for food fishes and experienced fishers can quickly attribute local species names to photos of fish in a reference book. Photos may need to include both dead and live specimens, according to the likely experience of the interviewee. An experienced and prepared interviewer will already have a good knowledge of folk names but, nonetheless, should continue to cross-check local names with fish reference guides; local names can vary markedly even at small geographic scales and important species may have several different names that relate to different size classes of the same species. In the Yucatan Peninsula, for example, fishers from the Yucatan State call the black grouper, *Mycteroperca bonaci*, 'negrillo'; in contrast, those from Quintana Roo call it 'abadejo'. To fishers from the Yucatan, however, "abadejo" is the gag grouper, *M. microlepis* (AA-P personal observation 2006).

Maps, charts and satellite photos of reefs are useful for documenting fish migratory pathways, checking site names and identifying locations of FSA sites. To collect fine-scale information, interviewers can ask knowledgeable fishers to draw a sketch map. Spearfishers, in particular, can often provide a wealth of information on the physical features of a site, depth ranges of aggregating species, aggregation boundaries, water conditions, and variations in fish densities. However, not all fishers can read maps and spatial information should be carefully documented if aggregation sites cannot be visited.

Once validated, the importance of LEK cannot be overstated. Community support for conservation plans consistently emerges as one of the most important factors in maintaining long term effectiveness and support from local people (King and Faasili 1999; Johannes 2002; Aswani and Hamilton 2004). Moreover, community acknowledgement and understanding of key threats engender greater respect for regulations, as noted for marine protected areas (MPAs) in the Bahamas (Broad and Sanchirico 2008), and can garner government and community support for management, as in Pohnpei, Fiji and elsewhere.

10.6.1 Examination of LEK for Internal Consistency and Reliability

Internal consistency and reliability of LEK can be assessed by comparing within and between transcripts and through expert consensus. Carefully examining an interview transcript will help assess the likely accuracy of the information. If the information provided clearly contradicts itself, then its validity is thrown into question (Huntington 1998). When many fishers have been interviewed, multiple transcripts can be compared to identify components of LEK that are consistent and frequently raised, and highest reliability assigned to these aspects of LEK or to particularly reliable respondents (Neis et al. 1999b). A cautionary note is that not all 'experts' are created equal. In cases where a single individuals' LEK could potentially be highly relevant to management, extra effort is needed to cross-check with independent data sources or conduct independent research (Hamilton 2005).

Researchers can hold review sessions with a community after completion of interviews to discuss the LEK gathered and reach consensus. LEK documented during interviews is presented to a group of expert fishers for discussion and consensus-building as well as to raise further awareness about the purpose of the study, the implications of the results and to build support and explore possible responses (e.g. Huntington 1998).

10.6.2 Independent Validation of LEK

There is no substitute for independently validating LEK before incorporation into conservation and fisheries management (Usher 2000). Table 10.1 lists several attempts at validation to illustrate possible outcomes and lessons learned. Key points include that aggregations may last for more months than reported by LEK, with first and last months or the months with lower numbers of fish less obvious or possibly varying from year to year, and therefore often excluded from LEK (as in the case of August in Palau – see above). Highly variable responses tend to be found in areas with weak

Table 10.1 Outcomes of activities that sought to validate local ecological knowledge on aggregations

Location and species	Method(s) for collecting LEK	Assumption(s) of LEK tested using validation	Method(s) for validating LEK	Outcome	Source
Manus, Papua New Guinea, *Epinephelus fuscoguttatus* and *Plectropomus areolatus*	Semi-structured interviews	*E. fuscoguttatus* aggregate at a specific site in March, April, May and June. Aggregations of *P. areolatus* form at this site in every month of the year, peaking in the months that *E. fuscoguttatus* aggregate	UVC monitoring data collected at the aggregation site several days prior to the new moon from July 2005 to December 2007	Results confirm LEK: *E. fuscoguttatus* aggregate between March –August, and *P. areolatus* aggregate every month with highest densities typically in months when *E. fuscoguttatus* aggregate to spawn (Fig. 10.6)	Hamilton (2003b), Hamilton et al. (2004), and Manuai Matawai and RH, unpublished data
Manus, Papua New Guinea, *Epinephelus ongus*	Semi-structured interviews	*E. ongus* aggregates by the thousands to release eggs in the week leading up to and including the new moon in March, April, May and June. Aggregations disperse on or just after the new moon	Sampling fish from an aggregation site in June	Results support LEK. Gonads of 89 *E. ongus* captured at the FSA before and just after the new moon in June 2005 provided evidence that *E. ongus* aggregated for spawning at time and place indicated by fishers	Hamilton et al. (2004, 2005c)
Roviana Lagoon, Solomon Islands, groupers	Participant observation, semi-structured interviews	LEK details a spawning season for all groupers in Roviana Lagoon between October to January each year	UVC monitoring data collected monthly at one aggregation site in Roviana Lagoon several days prior to the new moon from April 2004 to June 2006	LEK on spawning season incomplete. UVC monitoring data show that *P. areolatus* aggregations of variable size occurred in virtually every month of the year, whereas aggregations of *E. fuscoguttatus* and *E. polyphekadion* occurred from December to April	Aswani (1997), Johannes and Lam (1999), Hamilton and Kama (2004), and Smith and Hamilton (2006)

Papua New Guinea and Solomon Islands, *Epinephelus polystigma*	Semi-structured interviews, participant observation	*E. polystigma* forms large nocturnal aggregations in shallow waters of river mouths and brackish mangrove regions throughout year, prior to new moon. Sites identified in several regions in Melanesia using maps	Comparison of LEK that was documented in different regions in Melanesia and sampling fish from one aggregation site in February	Results partially support LEK. Visited one site 3 days before new moon at night in 1 month (February) and found many small clusters of fish in very shallow water of a range of sizes, species identification confirmed by caught fish; 18 with ripe ovaries and running ripe testes	Johannes (2001) and Hamilton and Potuku (2007)
Kimbe Bay, Papua New Guinea, *E. fuscoguttatus and P. areolatus*	Semi-structured interviews	Fishers identified eight sites in Eastern Kimbe Bay where groupers aggregate to spawn. No annual season was identified	A single UVC survey was conducted at each site in the week leading up to the new moon in March 2005 and April 2006. Timing of these surveys was based on knowledge of peak aggregating period for these species in Manus Province	Results consistent with LEK. Snapshot surveys verified that five of these eight sites (63%) had *P. areolatus* and/or *E. fuscoguttatus* aggregations in the months surveyed. At verified sites groupers were present in relatively high densities (20–200 fish), groupers displayed aggressive behaviours consistent with spawning and gravid females were sighted. At remaining three sites no FSA were sighted on either occasion	Hamilton et al. (2005b) and Aitsi et al. (2006)

(continued)

Table 10.1 (continued)

Location and species	Method(s) for collecting LEK	Assumption(s) of LEK tested using validation	Method(s) for validating LEK	Outcome	Source
Solomon Islands. *Lethrinus erythropterus*	Semi-structured interview with one fisherman	Large aggregations (100s–1000s) of *L. erythropterus* form at a specific site in shallow water around the new moon for the purpose of spawning	Dived at the aggregation site on the new moon in 1 month (March) to make observations and sample fish	Results consistent with LEK. 300–500 *L. erythropterus* were aggregated in shallow water; 26 fish caught with ripe ovaries and running ripe testes	Hamilton (2005)
Kadavu Island, Fiji, *E. polyphekadion, E. fuscoguttatus* and *P. areolatus*	Semi-structured interviews in multiple communities in northern Kadavu	Aggregations of groupers occur at three reef passages between July -September with no obvious lunar cycle. Leaves of the local tree turn red/yellow and start to fall at time that fish appear at aggregations	Single 'snapshot' SCUBA surveys at two sites July followed by intensive diving (i.e. daily for 10 days and on both sides of indicated passage) at the site reported to be the most productive. Visited fish markets to identify and squeeze fish taken from Kadavu for eggs/sperm. Fish sampled on site	Where intensive diving was undertaken all three species aggregated in July and August, gravid females were sighted and chasing behaviour observed. Fish sold in markets had running sperm in early August and Tavola leaves started to go red and fall. Single dives conducted on other sites did not confirm fish aggregating but probably were conducted a little early according to men fishing at sites; this would be consistent with gonad inspection. Interestingly the dominance of different species varied with side of the passage monitored.	Sadovy de Mitcheson et al. (2008) and YSM and Aisake Batibasaga, unpublished data

Fiji, *E. polyphekadion* and *P. areolatus*	Semi-structured interviews in several communities along contiguous coastline	Spawning aggregations of groupers occur at four sites in northern Fiji during at least one of the months indicated	Dived at four reported aggregation sites on one of peak months indicated by interviews, July. Checked gonads of groupers being captured at this site by fishers	Confirmed presence of both species at three of four sites; consistent outcomes by four communities. Fourth site had no fish or boats despite excellent weather and local community reported it to be fished out. Found *E. cyanopterus* at one site that had not been otherwise reported – only noted by shallow water spearfishers. Traders at the main trading town of the region confirmed large numbers of the same three species during indicated months	Sadovy (2006) and YSM, personal observation
Palau, *E. polyphekadion*, *E. fuscoguttatus* and *P. areolatus*	Unstructured and semi structured interviews	Spawning months from April to July for *P. areolatus*. *E. fuscoguttatus* and *E. polyphekadion*	UVC surveys in every month of year at three sites during days leading up until full moon and until fish numbers declined after spawning. Market sampling of fish to examine gonad condition	Months reported correct but incomplete; August is also a month in which spawning aggregations form although can be smaller than other months. *Plectropomus areolatus* misidentified as *P. leopardus* in earlier study and at one site also was present in other months in small numbers	Johannes et al. (1999), Sadovy (2007), Johannes (1981), Patrick Colin, personal communication 2009, and Palau Conservation Society, unpublished data
Philippines Groupers	Semi-structured interviews	Some interviewees reported several small aggregation sites off NE Palawan of *Plectropomus* spp.	SCUBA dived at several sites at indicated times	Confirmed aggregation of fish for one of the indicated months. Gravid females sighted	YSM and Jose Ingles, unpublished collaboration with WWF Philippines

(continued)

Table 10.1 (continued)

Location and species	Method(s) for collecting LEK	Assumption(s) of LEK tested using validation	Method(s) for validating LEK	Outcome	Source
Seychelles	Semi- and unstructured interviews. 39 semi-structured over 10 months, catch, sizes, species, habitat, location, timing. Patriarch fishers targeted	Eight spawning sites for several species of grouper and a rabbitfish	Combinations of fish sampling for sizes and gonad condition, including GSI and histology, and dives at reported sites at indicated times and places	Validation of six out of eight reported sites and for all species reported except for *E. multinotatus* which may be overfished. Confirmed *P. punctatus, E. fuscoguttatus, E. polyphekadion* and *Siganus sutor.* Project raised awareness leading to further work and some protection	Robinson et al. (2004, 2007)
Puerto Rico, Caribbean	Interviews locally and questionnaires circulated regionally on status of past and present Nassau grouper aggregations and populations	That many of the Nassau grouper aggregations known historically have declined	Literature review, discussions with local biologists, visiting reported sites	Declines confirmed in many places quantitatively and qualitatively. Difficult logistically to reach many of the sites said to be existent. Still seeking funds to do so, e.g. Bahamas	Sadovy and Eklund (1999)

Mexico (eastern Yucatan Peninsula), Caribbean, *Epinephelus striatus*	Semi-structured interviews with commercial fishers.	A Nassau grouper spawning aggregation forms at a specific site around the full moon in December and January each year	UVC monitoring in December and January around the full moon period between 1988 and 1993, and examining gonads of fish captured at the site	Confirmed aggregation of 500–1000 ripe adults forms during the full moon period in the months of December and January	Aguilar-Perera (1994) and Aguilar-Perera and Aguilar-Dávila (1996)

Refer to Chap. 12 for individual species accounts on many of the species
LEK local ecological knowledge, *UVC* underwater visual census, *GSI* gonadosomatic index

aggregation signatures, which is common in particularly heavily exploited FSAs (e.g. Indonesia and Philippines – YSM personal observation 2006; Sadovy and Liu 2004). Validation efforts tend to focus initially on the reported sites and months. To establish full timing of aggregation, validation should ideally be conducted during all months of the year and for extended periods each month. Single "snapshot" visits to sites are of limited use for seasonal protection, although may effectively serve spatial protection initiatives by confirming the existence of an aggregation site. Interviews need to involve fishers, particularly those using different gears, to obtain a detailed profile of aggregating species and timing. CPUE information must be carefully collected and the alternative possible hypotheses for any trends explored. Fish should be sampled for reproductive condition whenever possible and this could be done in fish markets if the source of the fish is local (see GSI in Chap. 9).

10.6.3 Conducting Research to Validate LEK on FSA

Many aspects of LEK pertaining to aggregations lend themselves particularly well to being independently validated. Information on lunar and annual seasonality, FSA location, species composition and purpose for aggregating (i.e. reproduction or feeding) can be verified by UVC or by examining the gonad state of fish captured from a FSA (Chap. 1, Table 10.1).

10.6.4 Underwater Visual Censuses

Divers can often make direct observations on whether or not a FSA is present, the species aggregating and whether courtship or spawning is occurring. This process is relatively simple and effective if local knowledge on FSA seasonality is detailed and if the precise locations of FSA were previously documented with a GPS, or using a good quality map. Independent validation at a potential FSA site will often be done on SCUBA over days or a few weeks and typically includes activities such as estimating the numbers and species of fish present and collecting indirect or direct evidence of spawning (Chap. 1, Table 10.1). However, given the spatial and temporal variation of FSAs, several short or extended trips may be necessary or it will be difficult to determine what it means if no FSA is found. In this scenario, researchers cannot know whether local knowledge was incorrect, they simply missed the exact location, visited it during the wrong lunar or annual period or if FSA timing varies naturally.

Long-term UVC monitoring of a given FSA is typically conducted to determine whether conservation measures in place at a FSA are working, to fine-tune management or for research (Colin et al. 2003). While data collected from long-term monitoring programmes primarily serve to inform adaptive management, another important objective of monitoring is to validate LEK, especially in scenarios where management strategies have been developed based purely on LEK (Hamilton et al. 2005a). For example, year-round monthly UVC surveys from a FSA in Manus,

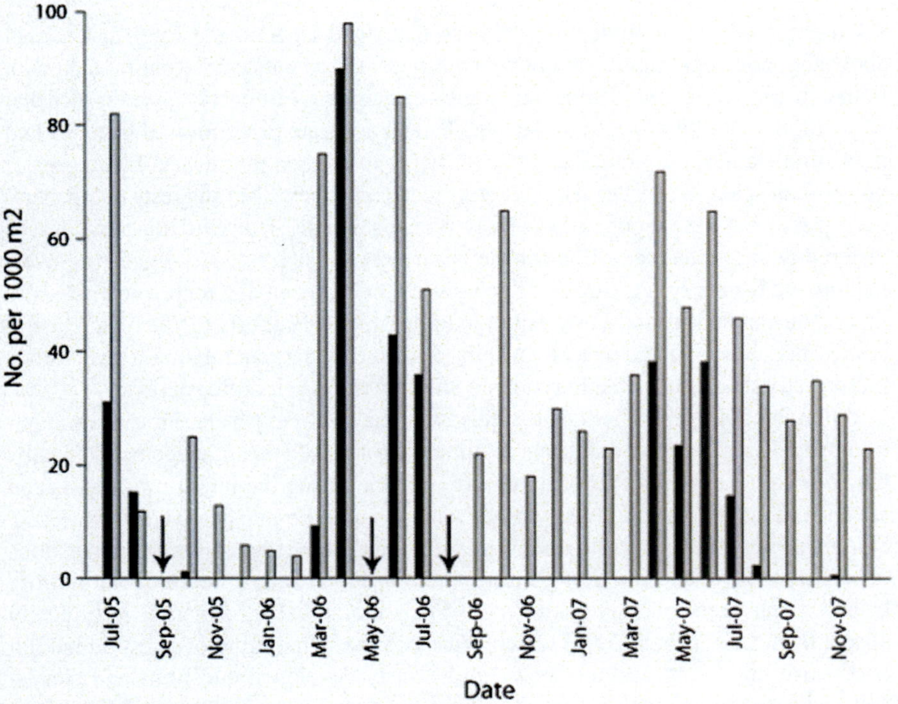

Fig. 10.6 Number of *Epinephelus fuscoguttatus* (*dark bars*; 20 m deep transect) and *Plectropomus areolatus* (*light bars*; 10 m deep transect) per 1,000 m^2 at a spawning aggregation site in Manus, Papua New Guinea, based on underwater visual census surveys conducted on one occasion several days prior to the new moon in each of the indicated months. *Arrows* indicate no data from months when surveys could not be conducted for logistical reasons (Manuai Matawai and RH unpublished data)

Papua New Guinea (Fig. 10.6) validated LEK pertaining to this FSA (Table 10.1) and informed adaptive management. In addition, follow-up projects can be developed that increase understanding and awareness of aggregations and their significance; examples include a simple tagging project in Fiji where fish were tagged and local fishers assisted by returning tagged fish they caught. Interest can be generated by such studies which become a focus for discussion and opportunity for outreach (YSM personal observation 2005).

10.7 Combining LEK and Science Assists Effective FSA Management

In this chapter, we attempt to show that if LEK is collected judiciously it can provide an excellent source of data on aggregating species, as well as the probable historical and current status of their fisheries, can create excellent opportunities for awareness and information exchange, and can assist in promoting and supporting management development. An interesting example of a direct comparison between scientific data (e.g. bottom trawl, fishery data) and LEK in the English Channel showed good overall agreement between fishers' statements and scientific data with both sources suggesting declines in some commercially important species (Rochet et al. 2008). The study noted that fishers had an accurate perception of changes and their time-frames, but not necessarily of their causes, and that fishers had a greater power than scientific survey data to detect recent changes. This suggests that fishers' perceptions have great potential as early warning signals. However, the two sources differed in suggested causation for the observed changes; science suggesting over-fishing and fishers suggesting that declines in cod had been due to movements away from their fishing ground. This example highlights the important role of science in addressing causation and of LEK in being able to capture general longer term trends more easily than is possible in typically short duration scientific surveys.

Although LEK on FSAs can be highly detailed, in our experience it is a combination of LEK and science that leads to effective forms of FSA management. Ideally, the two-way exchange of information that occurs during the interview process can advance management initiatives and develop a scientific perspective (Drew 2005). Scientists can play a valuable role by sharing their knowledge, answering questions fishers may have, and providing information on larval stages, recruitment and life history parameters of aggregating fish. Such biological information is typically absent from LEK (Foale 2006) which addresses the 'what' kind of question but not necessarily the 'why' and the 'how'. Judiciously developed questions and consistent and careful questioning can allow for a degree of quantification of LEK which can be very powerful when talking to politicians or others who might otherwise be unreceptive to what may sometimes be seen as anecdotal information. Scientists can also provide regional and global perspectives on FSAs -something even the most knowledgeable fishers typically lack- and one reason why workers must be familiar with aggregating species.

Often informing fishers of the critical biological role that FSAs play and the ease with which they can be destroyed is the only catalyst required to have communities initiate the process of managing their FSA (Hamilton et al. 2005a). Documenting LEK regarding the presence of FSAs, or changes in landings from FSAs, is often important for reinforcing what people have already experienced, and a better understanding of aggregations can help communities make sense of what they have learned from experience or tradition. Importantly, scientists should be open with their collaborators regarding data collected and analysed and not expect to withhold it until publication. Confirming that FSAs form does not necessarily mean that they all have to be protected. Moreover, it is not necessary to know the location of all aggregations if seasonal protection is likely to be more effective. Depending on the local context, seasonal rather than spatial protection may be more appropriate in many places, especially where enforcement of distant sites is highly unlikely; the best protected sites are those that nobody knows.

LEK is far more likely to be taken seriously by conservationists, scientists and managers if validated, and scientists are more likely to be taken seriously if what they say harmonizes with local understanding. LEK on FSA in Kimbe Bay, Papua New Guinea provides one such example. UVC 'snapshot' surveys conducted at FSAs in

Kimbe Bay independently confirmed that grouper aggregations formed at five out of eight sites previously identified through LEK. These verified sites were subsequently given a much higher ranking than non-verified sites when Marxan software was used to design a MPA network for Kimbe Bay by an NGO working in the area (Green et al. 2009). In Fiji, the participation during interview studies of the Research Division of the Fisheries Department and subsequent preliminary evaluation of several sites were important factors in the inclusion of grouper aggregation protection in both outreach work by the Division and in a draft revision of the national fishery ordinance. Management or conservation initiatives that are at least partly in line with traditional thinking have far more chance of success than approaches that are not familiar to, or consistent with, local beliefs and understanding.

Even communities might prefer to base their actions on more detailed studies. In 2004, traditional owners in Roviana Lagoon, Solomon Islands, expressed interest in placing a seasonal ban at one aggregation during the locally defined grouper spawning season (October to January) (Hamilton and Kama 2004). However, before placing a seasonal ban on this FSA they requested that an independent, long-term scientific monitoring be conducted. In June 2006, the results of 2 years of monthly monitoring were presented to the local community that claimed traditional ownership of this site (Table 10.1). On the basis of these scientific results, the community decided that a seasonal ban from October to January would not effectively conserve this grouper aggregation, and the community consequently voted to turn the aggregation site and the surrounding reef area into a community-based no-take MPA.

Attention attracted to aggregations can even lead to much broader initiatives, following validation. In the southern Philippines, the discovery of several small aggregation sites in TayTay Bay, northeastern Palawan, followed by on-site validations resulted not only in protection of the sites, but was also the catalyst for many communities to come together in the Bay to discuss marine protected area work in general, work that later helped to address concerns about the developing export trade in live fish, particularly the leopard coralgrouper (YSM personal observation 2006).

Finally, the case study from southern Manus (Table 10.1, Fig. 10.6) provides a contrasting example of where a community was so confident of their LEK on their FSA that they used it to immediately develop a community-based fisheries management strategy. In early 2004, the Pere community from Southern Manus placed a lunar-based ban on spearfishing and commercial fishing at the aggregation site in the 10 days leading up to and including the new moon in every month of the year. During this period only subsistence hook and line fishing was allowed at the aggregation site (Hamilton et al. 2004). In this particular case, long-term monitoring commenced only after management measures based on local knowledge were implemented, and long-term scientific monitoring simply confirmed what local fishers already knew regarding lunar and annual seasonality of the aggregations. This is not to say that monitoring did not serve an important role in adaptive management or broader conservation initiatives. In early 2007 when the Pere community saw monitoring data that indicated that the densities of groupers at their FSA had not improved, they voted to place a 1 year ban on all fishing at the aggregation site during lunar periods when aggregations formed. Initial efforts to protect and monitor this

FSA also led to increased awareness on a broad range of marine management issues in Pere, and in 2009 the Pere community and local government officials launched the Pere Environment and Conservation Area Management Plan (PECAMP 2009), which outlines management measures for three FSAs and many other important marine resources that occur within the traditional reef boundaries of Pere.

Acknowledgements We wish to express our sincere thanks to all of the fishers who have supported our work over the past two decades. Their local knowledge and the lessons they have taught us form the foundations of this chapter. We are grateful for the editing assistance provided by Rachel Wong and we would also like to thank Richard Walter, Kenneth Ruddle and Simon Foale for reviewing and providing helpful comments on an earlier version of this chapter. YSM and RH particularly acknowledge the early ground-breaking work of Bob Johannes in collecting LEK on spawning aggregations in the Pacific; both of us have learned a lot from our acquaintance and work with him and dedicate this chapter to him. Finally, we thank the David and Lucile Packard Foundation for the financial assistance they have provided for many of the studies that are reported herein.

Appendix 10.1 Basic Points to Remember When Preparing and Conducting Interviews and Applying Outcomes for Conservation, Management and Education

1. Preparing for a trip

- Clearly and concisely determine the intention/objective of the interview-based study.
- Obtain necessary permits and establish contacts or community permissions as well as understand local social protocols.
- Conduct meetings or give presentations to collaborators or government officials regarding purpose and relevance of work.
- Conduct background reading to prepare yourself about the fishery and fish species so that you can also provide information on the species to interviewees and assess information quality during interviews. Inform yourself not just of the target species but of aggregating species in general and experiences from overseas with aggregations so you can you provide examples and experiences from elsewhere.
- Learn the local names of the fish if necessary; they can change a lot, even between villages. Names can also refer to species groups and not just species. Note that different names might apply to different life history stages.
- Prepare cards with photos of fish from the area, both live and dead to accommodate different experiences of the species. Include photos from fish not in the area as a control.
- Purchase or prepare good marine maps of the area.
- Careful selection of interviewees, by gear, age group, area, peer review etc., is very important; stratify sampling if possible.
- Need to select appropriate vocabulary – be sure that key words or concepts are clearly conveyed in ways that will be unambiguously understood.

- In many settings providing items such as coffee, cigarettes, batteries or biscuits during the interview is culturally appropriate and helps to break down barriers between the interviewee and interviewer. However, determine whether or not it is appropriate to provide gifts to interviewees in each situation; differences in local practices can mean that giving rewards/incentive is sometimes insulting and sometimes expected. Rewards can show appreciation for time spent but should not be the incentive for the interview. Care is needed.
- Ensure proper dress codes – many communities are traditional and expect certain behaviour, especially by females.
- Carefully select interviewers; they should be knowledgeable about the resource, the fishery and the community, patient, open-minded and communicative. It is very important that interviewers are prepared to discuss their findings to communities, provide useful information to interviewees and gain their respect. Interviewers, ideally, should be knowledgeable on relevant issues internationally, including general matters of fishery management options.
- Consider filming or oral recording interviews, with permission, for later media or educational activities.

2. During a trip

- Make clear to the interviewee what the interview is about, why you are requesting it and what you will do with the information.
- Continually work to establish your own credibility through your knowledge – you will get respect and better responses. It would be useful to be introduced by credible people. Make clear that you respect the knowledge of interviewees
- Use open-ended and semi-structured questions during interviews and while participating in fishing, etc.
- Go fishing when possible with interviewee and inspect fish/catches when possible.
- Prepare a minimum subset of questions that are the most important to conduct: fishers might be tired and not have much time or patience.
- Be courteous and respectful and try to be engaging.
- Focus clearly on one species at a time and confirm species with photos. Always indicate in your notes what information applies to which species.
- Ask about opinions and likely causes of observed trends.
- Decide whether to conduct group or one-on-one interviews.
- Be open-minded and allow time for conversations to go off in multiple directions but also focus on the key questions you intend to cover – this is another reason the interviewer must have a sound knowledge of the subject. Don't dismiss information that sounds unlikely but follow up with further questions.
- Be patient and prepared to be flexible with your travel schedule – i.e. spend extra time in an area if it proves productive, or move on early if necessary and factor in delays.
- Repeat questions in different ways to check reliability of interviewees.

- Use every opportunity to exchange information and discuss interesting aspects of the life history of locally taken species. This is yet another reason for interviewers to be informed and prepared.
- Ask comparative questions, i.e. 'more or less fish than before?' and pick large time periods for temporal comparisons (such as decades). If you ask about proportions or percentages, make sure that this concept is understood.
- It may be necessary to talk about 'maximum' or 'best' catches since average or typical catches may not be well understood or not remembered. Try to quantify catches in kg or whatever is the local measure that is widely used (coolers, fish bundles, etc.), and catch rates in a consistent way. This will allow for a quantification and comparison of results.
- Don't just ask which species spawn and when; interviewees may have no idea about this even if they have seen spawning. Ask instead about the direct and indirect indicators of spawning such as seasonal highs in landings, eggs, concentrations, etc., good and bad seasons for catches. Ask about presence of eggs, moon phase, etc., behavioural or colour changes, etc. Adjust questions according to fishing method.
- Could work closely with local Government/NGOs who will later be involved in management while conducting interviews. Often it is better to not be accompanied by those who usually enforce or get taxes.
- It may be better to leave sensitive issues, like income, out of biological surveys.
- Make a note about possible reliability or otherwise of interviewee.
- Don't assume that everybody can easily read a map or have good recall or follows moon phases.
- Write down notes immediately; also allows for refining and going back to responses before leaving an area. This is especially important if recording interviews.
- Seek opinion about why changes occur if changes are noted.
- Be sensitive about difficulties that might arise due to gender of interviewer/ee.
- Seek to provide immediate feedback to the community and encourage discussion of interview outcomes before leaving the area.

3. Follow up to a trip

- Follow up with any promises made to communities/individuals. People often ask for photos, so be sure to make appropriate arrangements.
- Produce a report that is shared with communities and collaborators, and at least provide a preliminary report before leaving the country or very soon thereafter.
- Give presentations on outcomes of work and indicate how to apply the findings.
- If appropriate, talk about outcomes in various media.
- For non-nationals, be available even having left the country, for providing additional information.
- Follow up with educational materials if necessary.

- Be sure to reflect back to the communities visited, in the appropriate format, the outcomes of the interviews and the broader implications of the study findings. Identify possible studies that are needed to address original objectives or to address fisher concerns or questions identified during the interview process.
- Be sure to respect the information provided; this should not be released into the public domain unnecessarily, especially in the case of site-specific information, and not before the relevant conservation or management has been put in place. However, information will be needed for local planning.

References

Aguilar-Perera A (1994) Preliminary observations of the spawning aggregation of Nassau grouper, *Epinephelus striatus*, at Mahahual, Quintana Roo, Mexico. Proc Gulf Caribb Fish Inst 43:112–122

Aguilar-Perera A, Aguilar-Dávila W (1996) A spawning aggregation of Nassau grouper *Epinephelus striatus* (Pisces: Serranidae) in the Mexican Caribbean. Environ Biol Fish 45(4):351–361

Ainsworth CH, Pitcher TJ, Rotinsulu C (2008) Evidence of fishery depletions and shifting cognitive baselines in Eastern Indonesia. Biol Conserv 141(3):848–859

Aitsi JA, Sapul A, Hamilton R, Seeto S (2006) Reef fish spawning aggregations verification surveys, Kimbe Bay, West New Britain Province, Papua New Guinea, 23–28 Apr 2006. Final report. (restricted access version). Report prepared by the Pacific Island Countries Coastal Marine Program, The Nature Conservancy. TNC Pacific Island Countries Report No. 4/06

Almany GR, Hamilton RJ, Williamson DH, Evans RD, Jones GP, Matawai M, Potuku T, Rhodes KL, Russ GR, Sawynok B (2010) Getting communities involved in marine protected area research: two case studies from Papua New Guinea and Australia. Coral Reefs 29(3):567–576

Alves RRN, Nishida AK, Hernández MIM (2005) Environmental perception of gatherers of the crab 'caranguejo-uçá' (*Ucides cordatus*, Decapoda, Brachyura) affecting their collection attitudes. J Ethnobiol Ethnomed 1:10

Ames T (2007) Putting fishers' knowledge to work: reconstructing the Gulf of Maine cod spawning grounds on the basis of local ecological knowledge. In: Haggan N, Neis B, Baird IG (eds) Fishers' knowledge in fisheries science and management. UNESCO Publishing, Paris, pp 353–364

Aswani S (1997) Customary sea tenure and artisanal fishing in the Roviana and Vonavona Lagoons, Solomon Islands: the evolutionary ecology of marine resource utilization. PhD dissertation, University of Hawaii, Hawaii

Aswani S, Hamilton RJ (2004) Integrating indigenous ecological knowledge and customary sea tenure with marine and social science for conservation of bumphead parrotfish (*Bolbometopon muricatum*) in the Roviana Lagoon, Solomon Islands. Environ Conserv 31(1):69–83

Aswani S, Lauer M (2006) Benthic mapping using local aerial photo interpretation and resident taxa inventories for designing marine protected areas. Environ Conserv 33(3):263–273

Baird IG (2007) Local ecological knowledge and small-scale freshwater fisheries management in the Mekong River in Southern Laos. In: Haggan N, Neis B, Baird IG (eds) Fishers' knowledge in fisheries science and management. UNESCO Publishing, Paris

Berkes F (1999) Sacred ecology: traditional ecological knowledge and resource management. Taylor & Francis, Philadelphia

Borgatti SP (1996) ANTHROPAC 4.92. Analytic Technologies, Natick

Briggs CL (1986) Learning how to ask: a sociolinguistic appraisal of the role of the interview in social science research. Cambridge University Press, Cambridge, UK

Broad K, Sanchirico JN (2008) Local perspectives on marine reserve creation in the Bahamas. Ocean Coast Manag 51:763–771

Christie P, White AT (1997) Trends in development in coastal area management in tropical countries: from central to community orientation. Coast Manag 25:155–181

Claro R, Sadovy de Mitcheson Y, Lindeman KC, Garcia-Cagide AR (2009) Historical analysis of Cuban commercial fishing effort and the effects of management interventions on important reef fishes from 1960–2005. Fish Res 99(1):7–16

Colin PL, Sadovy YJ, Domeier ML (2003) Manual for the study and conservation of reef fish spawning aggregations. Society for the Conservation of Reef Fish Aggregations Special Publication No. 1 (Version 1.0), pp 1–98+iii

Craig AK (1966) Geography of fishing in British Honduras, vol 14, Coastal Studies Series. Louisiana State University Press, Baton Rouge

Davis AD, Wagner JR (2003) Who knows? On the importance of identifying "experts" when researching local ecological knowledge. Hum Ecol 31:463–489

Daw T (2004) Reef fish aggregations in Sabah, East Malaysia. Western Pacific fisher survey series: society for the conservation of reef fish aggregations, vol 5. (Confidential appendix). Society for the Conservation of Reef Fish Aggregations, 63 p, www.SCRFA.org

Domeier ML, Colin PL (1997) Tropical reef fish spawning aggregations: defined and reviewed. Bull Mar Sci 60:698–726

Drew JA (2005) Use of traditional ecological knowledge in marine conservation. Conserv Biol 19:1286–1293

Dulvy NK, Polunin NVC (2004) Using informal knowledge to infer human-induced rarity of a conspicuous reef fish. Anim Conserv 7:365–374

Ferguson MA, Messier F (1997) Collection and analysis of traditional ecological knowledge about a population of Arctic Tundra Caribou. Arctic 50:17–28

Foale SJ (1998) The role of customary marine tenure and local knowledge in fishery management at West Nggela, Solomon Islands. PhD dissertation, University of Melbourne, Melbourne

Foale SJ (2006) The intersection of scientific and indigenous ecological knowledge in coastal Melanesia: implications for contemporary marine resource management. Int Soc Sci J 58:129–137

Fraser DJ, Coon T, Prince MP, Dion R, Bernatchez L (2006) Integrating traditional and evolutionary knowledge in biodiversity conservation: a population level case study. Ecol Soc 11(2):4

Game E, Lipsett-Moore G, Hamilton R, Peterson N, Atu W, Kereseka J, Watts M, Possingham H (2011) Informed opportunism in practice; conservation planning in the Solomon Islands. Conserv Lett 4(1):38–46

Gass RE, Willison JH (2005) An assessment of the distribution of deep-sea corals in Atlantic Canada by using both scientific and local forms of knowledge. In: Freiwald A, Roberts JM (eds) Cold-water corals and ecosystems. Springer, Berlin/Heidelberg

Gerhardinger LC, Marenzi RC, Bertoncini AA, Medeiros RP, Hostim-Silva M (2006) Local ecological knowledge on the goliath grouper *Epinephelus itajara* (Teleostei: Serranidae) in southern Brazil. Neotrop Ichthyol 4(4):441–450

Gillett R, Moy W (2006) Spearfishing in the Pacific Islands. Current status and management issues. FAO/FishCode review no. 19. FAO, Rome

Green AL, Smith SE, Lipsett-Moore G, Groves CR, Peterson N, Sheppard S, Lokani P, Hamilton R, Almany J, Aitsi J, Bualia L (2009) Designing a resilient network of marine protected areas for Kimbe Bay, Papua New Guinea. Oryx 43:488–498

Gubrium JF, Holstein JA (2002) Handbook of interview research: context & method. Sage Publications, Thousand Oaks

Hamilton RJ (2003a) The role of indigenous knowledge in depleting a limited resource – a case study of the bumphead Parrotfish (*Bolbometopon muricatum*) artisanal fishery in Roviana Lagoon, Western Province, Solomon Islands. In: Putting fishers' knowledge to work conference proceedings, 27–30 Aug 2001, University of British Columbia, Canada. Fish Cent Res Rep 11(1):68–77

Hamilton RJ (2003b) A report on the current status of exploited reef fish aggregations in the Solomon Islands and Papua New Guinea – Choiseul, Ysabel, Bougainville and Manus Provinces. Western Pacific fisher survey series, vol 1. (Confidential appendix). Society for the Conservation of Reef Fish Aggregations, www.SCRFA.org

Hamilton RJ (2004) The demographics of bumphead Parrotfish (*Bolbometopon muricatum*) in lightly and heavily fished regions of the Western Solomon Islands. PhD dissertation, University of Otago, Dunedin

Hamilton RJ (2005) Indigenous ecological knowledge (IEK) of the aggregating and nocturnal spawning behaviour of the longfin emperor *Lethrinus erythropterus*. SPC Tradit Mar Resour Manag Knowl Bull 18:9–17

Hamilton RJ, Kama W (2004) Spawning aggregations of coral reef fish in Roviana Lagoon, Western Province, Solomon Islands: local knowledge field survey report. (Unrestricted access version). Report prepared for the Pacific Island Countries Coastal Marine Program, The Nature Conservancy, TNC Pacific Island Countries report no. 5/04

Hamilton RJ, Potuku T (2007) Local knowledge of white-dotted grouper (*Epinephelus polystigma*) aggregations in Melanesia. SPC Tradit Mar Resour Manag Knowl Inf Bull 22:3–14

Hamilton RJ, Walter R (1999) Indigenous ecological knowledge and its role in fisheries research design. A case study from Roviana Lagoon, Western Province, Solomon Islands. SPC Tradit Mar Resour Manag Knowl Bull 11:13–25

Hamilton RJ, Matawai M, Potuku T (2004) Spawning aggregations of coral reef fish in New Ireland and Manus Provinces, Papua New Guinea: local knowledge field survey report. (Unrestricted access version). Report prepared for the Pacific Island Countries Coastal Marine Program, The Nature Conservancy, TNC Pacific Island Countries report no. 4/04

Hamilton RJ, Matawai M, Potuku T, Kama W, Lahui P, Warku J, Smith AJ (2005a) Applying local knowledge and science to the management of grouper aggregation sites in Melanesia. SPC Live Reef Fish Inf Bull 14:7–19

Hamilton RJ, Lahui J, Warku J, Aitsi J, Sapul A, Seeto S (2005b) Local knowledge of reef fish spawning aggregations in Kimbe Bay, West New Britain Province, Papua New Guinea. (unrestricted access version). Report prepared for the Pacific Island Countries Coastal Marine Program, The Nature Conservancy, TNC Pacific Island Countries report no. 2/05

Hamilton RJ, Mous P, Smith A (2005c) Advanced workshop on monitoring reef fish spawning aggregations report, Pere, Manus Province, Papua New Guinea, 2–11 June 2005. Report prepared for the Pacific Island Countries Coastal Marine Program, The Nature Conservancy, TNC Pacific Island Countries report no. 6/05

Hamilton RJ, Potuku T, Montambault J (2011) Community based conservation results in the recovery of reef fish spawning aggregations in the Coral Triangle. Biol Conserv 144:1850–1858

Heyman WD, Graham RT, Kjerfve B, Johannes RE (2001) Whale sharks *Rhincodon typus* aggregate to feed on fish spawn in Belize. Mar Ecol Prog Ser 215:275–282

Holm P, Smith TD, Starkey DJ (2001) The exploited seas: new direction for marine environmental history. International Maritime Economic History Association, St. John's

Huntington HP (1998) Observations on the utility of the semi-directive interview for documenting traditional ecological knowledge. Arctic 51(3):237–242

Huntington HP (2000) Using traditional ecological knowledge in science: methods and applications. Ecol Appl 10:1270–1274

Hviding E (1996) Guardians of Marovo Lagoon, practice, place, and politics in maritime Melanesia. University of Hawaii Press, Honolulu

Johannes RE (1978) Reproductive strategies of coastal marine fishes in the tropics. Environ Biol Fish 3(1):65–84

Johannes RE (1981) Words of the lagoon: fishing and marine lore in the Palau District of Micronesia. University of California Press, Berkeley, p 245

Johannes RE (1989) Spawning aggregations of the grouper *Plectropomus areolatus* (Ruppell) in the Solomon Islands. In: Choat JH, Barnes DJ, Borowitzka MA, Coll JC, Davies PJ, Flood P,

Hatcher BG, Hopley D, Hutchings PA, Kinsey D et al (eds) Proceedings of the Sixth International Coral Reef Symposium, Townsville, Australia

Johannes RE (2001) A possible new candidate for grouper aquaculture. SPC Live Reef Fish Inf Bull 8:31–32

Johannes RE (2002) The renaissance of community-based marine resource management in Oceania. Annu Rev Ecol Syst 33:317–340

Johannes RE, Lam M (1999) The live reef food fish trade in the Solomon Islands. SPC Live Reef Fish Inf Bull 5:8–15

Johannes RE, Neis B (2007) The value of anecdote. In: Haggan N, Neis B, Baird IG (eds) Fishers' knowledge in fisheries science and management. UNESCO Publishing, Paris

Johannes RE, Yeeting B (2001) I-Kiribati knowledge and management of Tarawa's lagoon resources. Atoll Res Bull 489:1–25

Johannes RE, Squire L, Graham T, Sadovy Y, Renguul H (1999) Spawning aggregations of grouper (Serranidae) in Palau. The nature conservancy marine research series publication no. 1. The Nature Conservancy and the Forum Fisheries Agency

Johannes RE, Freemen MMR, Hamilton RJ (2000) Ignore fishers' knowledge and miss the boat. Fish Fish 1:257–271

King M, Faasili U (1999) Community-based management of subsistence fisheries in Samoa. Fish Manag Ecol 6:133–144

Lajus JU, Alekseeva YA, Davydov R, Dmitrieva Z, Kraikovski A, Lajus D, Lapin V, Mokievsky V, Yurchenko A, Alexandrov D (2001) Status and potential of historical and ecological studies of Russian fisheries in the White and Barents Seas: the case of the Atlantic salmon (Salmo salar). In: Poul Holm, Tim D. Smith, David J. Starkey (eds) The Exploited seas: new directions for marine environmental history. International Maritime Economic History Association, St. John's. Research in Maritime History No. 21

Mailhot J (1993) Traditional ecological knowledge; the diversity of knowledge systems and their study. Great Whale environmental assessment: background paper no. 4, Great Whale Public Review Support Office, Montreal, Quebec

Neis B, Schneider DC, Felt L, Haedrich RL, Fischer J, Hutchings JA (1999a) Fisheries assessment: what can be learned from interviewing resource users? Can J Fish Aquat Sci 56(10):1949–1963

Neis B, Felt LF, Haedrich RL, Schneider DC (1999b) An interdisciplinary method for collecting and integrating fisher's ecological knowledge into resource management. In: Newell D, Ommer RE (eds) Fishing places, fishing people: traditions and issues in Canadian small-scale fisheries. University of Toronto Press, Toronto, pp 217–238

Olsson P, Folke C (2001) Local ecological knowledge and institutional dynamics for ecosystem management: a study of Lake Racken watershed, Sweden. Ecosystems 4:85–104

Pascale P, Desgranges B, Benali K, Eustache F (2002) Episodic and semantic remote autobiographical memory in ageing. Memory 10(4):239–257

Passfield K (1996) Notes on grouper spawning aggregations in Tongareva, Cook Islands, June 1995. SPC Tradit Mar Resour Manag Knowl Bull 7:20

Pauly D (1995) Anecdotes and the shifting baseline syndrome of fisheries. Trends Ecol Evol 10(10):430

PECAMP (2009) Pere environment and conservation area management plan. Nali Sopat Penabu Local Level Government, Manus Province, Papua New Guinea, October 2009. http://conserveonline.org/workspaces/pacific.island.countries.publications/documents/pere-environment-and-conservation-area-management

Robinson J, Isidore M, Marguerite MA, Ohman MC, Payer RJ (2004) Spatial and temporal distribution of reef fish spawning aggregations in the Seychelles – an interview-based survey of artisanal fishers. West Indian Ocean J Mar Sci 3:63–69

Robinson J, Marguerite R, Payet M, Isidore M (2007) Investigation of the importance of reef fish spawning aggregations for the sustainable management of artisanal fisheries resources in Seychelles final MASMA Project report: March 2007, 62 pp

Rochet M, Prigent M, Bertrand JA, Carpentier A, Coppin F, Delpech J, Fontenelle G, Foucher E, Mahe´ K, Rostiaux E, Trenkel VM (2008) Ecosystem trends: evidence for agreement between fishers' perceptions and scientific information. ICES J Mar Sci 65(6):1057–1068

Ruddle K, Chesterfield RA (1977) Education for traditional food procurement in the Orinoco Delta. Ibero-Americana no. 53. University of California Press, Berkeley/Los Angeles

Ruddle K, Hviding E, Johannes RE (1992) Marine resources management in the context of customary tenure. Mar Resour Econ 7:249–273

Sadovy Y (1993) The Nassau grouper, endangered or just unlucky? Reef Encount 1993:10–12

Sadovy Y (2006) Validating camouflage grouper, *Epinephelus polyphekadion*, spawning aggregations: a preliminary study from Fiji fisher interviews. Secr Pac Commun Live Reef Fish Bull 16:26–28

Sadovy Y (2007) Report on current status and exploitation history of reef fish spawning aggregations in Palau. Western Pacific fishery survey series: society for the conservation of reef fish aggregations, vol 3. SCRFA and the Palau Conservation Society, pp 40, www.SCRFA.org

Sadovy Y, Cheung WL (2003) Near extinction of a highly fecund fish: the one that nearly got away. Fish Fish 4:86–89

Sadovy Y, Eklund AM (1999) Synopsis of biological information on *Epinephelus striatus* (Bloch, 1972), the Nassau grouper, and *E. itajara* (Lichtenstein, 1822) the jewfish. U. S Department of Commerce, NOAA Technical Report, NMFS 146. FAO Fisheries Synopsis 157

Sadovy Y, Liu M (2004) Report on current status and exploitation history of reef fish spawning aggregations in eastern Indonesia, Western Pacific fisher survey series: society for the conservation of reef aggregations, vol 6, www.SCRFA.org

Sadovy Y, Kulbicki M, Labrosse P, Letourneur Y, Lokani P, Donaldson TJ (2003) The humphead wrasse, *Cheilinus undulatus*: synopsis of a threatened and poorly known giant coral reef. Rev Fish Biol Fish 13(3):327–364

Sadovy de Mitcheson YS, Cornish A, Domeier M, Colin PL, Russell M, Lindeman KC (2008) A global baseline for spawning aggregations of reef fishes. Conserv Biol 22(5):1233–1244

Sáenz-Arroyo A, Roberts CM, Torre J, Carin˜o-Olvera M, Hawkins JP (2006) The value of evidence about past abundance: marine fauna of the Gulf of California through the eyes of 16th to 19th century travellers. Fish Fish 7:128–146

Silvano RAM, Valbo-Jorgensen J (2008) Beyond fishermen's tales: contributions of fishers' local ecological knowledge to fish ecology and fisheries management. Environ Dev Sustain 19(5):657–675

Silver JJ, Campbell LM (2005) Fisher participation in research: dilemmas with the use of fisher knowledge. Ocean Coast Manag 48:721–741

Smith AJ, Hamilton RJ (2006) Protecting and managing reef fish spawning aggregations in the Pacific: project final report. Report prepared by the Pacific Island Countries Coastal Marine Program, The Nature Conservancy. TNC Pacific Island Countries report no. 3/06

Stanley RD, Rice J (2007) Fishers' knowledge? Why not add their scientific skills while you're at it? In: Haggan N, Neis B, Baird IG (eds) Fishers' knowledge in fisheries science and management. UNESCO Publishing, Paris

Tamelander J, Sattar S, Campbell S, Hoon V, Arthur R, Patterson Edward JK, Satapoomin U, Chandi M, Rajasuriya A, Samoilys M (2008) Reef fish spawning aggregations in the Bay of Bengal: awareness and occurrence. In: Proceedings of the 11th international coral reef symposium, Ft. Lauderdale, 7–11 July 2008. Session number 22 Coral Reef Associated Fisheries, Submission ID: 314

Thompson CP, Skowronski JJ, Larsen SF, Betz A (1996) Autobiographical memory: remembering what and remembering when. L. Erlbaum Associates, Mahwah

Tibby J, Lane MB, Gell PA (2007) Local knowledge and environmental management: a cautionary tale from Lake Ainsworth, New South Wales, Australia. Environ Conserv 34(4):334–341

Usher PJ (2000) Traditional ecological knowledge in environmental assessment and management. Arctic 53(2):183–193

Valbo-Jorgensen J, Poulsen AF (2000) Using local knowledge as a research tool in the study of river fish biology: experiences from the Mekong. Environ Dev Sustain 2:253–376

Vilaro Diaz DJ (1884) Corrida y arribazon de algunos peces cubanos. Manuel Gomez de la Maza. La Habana

Chapter 11
Management of Spawning Aggregations

Martin W. Russell, Brian E. Luckhurst, and Kenyon C. Lindeman

Abstract This chapter examines the needs and tools for managing reef fish spawning aggregations. We present a global overview of the management of aggregations, and explore management options. We evaluate conventional fishery management and marine protected area options in relation to aggregation conservation, and examine examples of management successes and failures. Most management to date has been reactive, and there remains an overwhelming need for proactive management of aggregations. Long-term monitoring, appropriate fishery policy and extensive fisher and community consultation and outreach are key elements in instituting effective and adaptive management of spawning aggregations.

M.W. Russell (✉)
Society for the Conservation of Reef Fish Aggregations, c/o 215/1000 Ann Street,
Brisbane, Queensland 4006, Australia
e-mail: scrfa@hku.hk

B.E. Luckhurst
Marine Resources Division, Department of Environmental Protection,
Ministry of the Environment, Bermuda Government, c/o 2-4 Via della Chiesa,
Acqualoreto, Umbria 05020, Italy
e-mail: brian.luckhurst@gmail.com

K.C. Lindeman
Department of Marine & Environmental Systems, Florida Institute of Technology,
DMES, FIT, 150 W. University Blvd., Melbourne, FL 32937, USA
e-mail: lindeman@fit.edu

Y. Sadovy de Mitcheson and P.L. Colin (eds.), *Reef Fish Spawning Aggregations: Biology, Research and Management*, Fish & Fisheries Series 35,
DOI 10.1007/978-94-007-1980-4_11, © Springer Science+Business Media B.V. 2012

11.1 Historical Perspective

Over the past few centuries, spawning aggregations of reef fish species were minimally affected by fishing because of the limited technical capabilities and low numbers of fishers exploiting them. When working from small sail or rowing boats, and having few means of preserving the catch beyond drying or salting, people had relatively little incentive, or ability to take excessive amounts of spawning fish. Since aggregations persisted and, once found, could be repeatedly visited and noteworthy amounts of fish taken, a valuable body of traditional knowledge of spawning aggregations developed in some areas, such as in the western Pacific (Johannes 1978, Chap. 10). Throughout the tropics, aggregations were viewed as fortuitous fishing opportunities in certain seasons or for special cultural events, such as village celebrations. The first published report on reef fish spawning aggregations, of which we are aware, is from Cuba (Vilaró Díaz 1884). However, a wider public and scientific awareness of spawning aggregations did not develop in the Caribbean until the early 1970s, starting with a Nassau grouper (*Epinephelus striatus*) aggregation (Smith 1972). Declines in this species were already reported by the late 1970s (Olsen and Laplace 1979). In the Pacific, early reports of spawning aggregation fisheries included those of Johannes (1978, 1982). In the Indian Ocean, early reports of aggregation fishing for rabbitfish (*Siganus sutor*) were documented by Hornell (1927).

Only in the last few decades has it become apparent that fishing of aggregations and of many aggregating fishes in general, needs effective management to maintain ecologically important populations of (typically) large predatory fishes such as groupers and snappers. The first recorded management intervention in the Caribbean occurred in 1974 in Bermuda when a seasonal spawning area fishing closure (May to August) for red hind (*Epinephelus guttatus*) was implemented (Luckhurst 1996, 1998; Luckhurst and Trott 2009). In some Pacific Island countries traditional knowledge, marine tenure and customary management of aggregation fishing have long been a traditional part of the culture (Johannes 1982). However, in most countries little is known about aggregations outside of local communities, few are managed and scientific acknowledgement of their vulnerability has been slow to develop. In Southeast Asia and other locations where fishing effort is particularly high, spawning aggregations are rarely reported, despite the presence of species known to form them, and may well have already been extirpated (Sadovy de Mitcheson et al. 2008).

In the 1990s, management of aggregation-fishing began on a broader scale globally along with an increasing focus on ecosystem-based management (EBM). This was largely a reactive response to declines in overall fish catches and in landings from aggregations. Initiatives for protecting aggregations in the Caribbean and tropical western Atlantic, where more than half of all known aggregations have declined or are gone, have only come about in recent years (Luckhurst 2003, Society for the Conservation of Reef Fish Aggregations (SCRFA) database, www.SCRFA.org,

Fig. 11.1 (**a**) Average monthly (meses) catches (capturas) in tonnes of Nassau grouper, *Epinephelus striatus* in Cuba over 5-year intervals, 1962–2005. Collapse of the aggregation-fishery occurred between 1970 and 1980 (From Claro et al. 2009, with permission); (**b**) Nassau grouper aggregating (Photo: Enric Sala)

Sadovy de Mitcheson et al. 2008). In Cuba, for example, long term data sets show that targeting Nassau grouper aggregations during their aggregation season over a 50-year period resulted in a dramatic decrease in catches (Claro et al. 2009, Chap. 12.6, Fig. 11.1). Despite a growing focus on marine protected areas (MPAs) in Caribbean fishery management, relatively few aggregations are specifically incorporated early in the MPA planning process or subjected to conventional fisheries management approaches of input or output controls (Sect. 11.3.3, Table 11.1). A shift in management focus from species-specific management to ecosystem-based approaches to management since the 1990s has provided an important impetus for addressing protection of spawning aggregations and aggregation habitat.

11.2 The Need for Management

The growing human population, especially in coastal areas is spurring the global demand for fishes. There is a ready cash sale market for reef fishes, as well as developing export markets. Since many economically valuable species aggregate to spawn, the incentive to find and exploit spawning aggregations, which provide a predictable source of large numbers of marketable fish, has dramatically increased and, to date, is almost unchecked. Today, motorized vessels, electronic navigation, and fish-finding equipment allow for relatively easy access to aggregation sites that are increasingly being discovered or sought for commercial exploitation. Improved fishing gear, such as braided line and underwater breathing apparatus allow more efficient capture of fish than in earlier times. The advent of portable waterproof lights, in particular, has facilitated night-fishing causing further pressures on reef fishes, particularly in the Pacific region. Refrigeration and ice plants allow for storage of catch well beyond what can be immediately consumed or bartered, and typically are introduced without prior assessment of the resource base available. Expanding markets and technological developments enable rapid processing and sale of fish, including for export.

For many of the nearly 100 fish species reliably documented to spawn in aggregations (SCRFA database, www.SCRFA.org, Appendix), these gatherings may represent the major or sole opportunity to reproduce, because some species are not known to spawn outside of aggregations. Among the two general types of spawning aggregations, transient and resident (Chap. 1), the most intense commercial exploitation is targeted at transient aggregating species, because these species are usually large and of great commercial or food interest. Subsistence fishing occurs on both types of aggregations. Regardless of the type of aggregation, ensuring that fishes successfully spawn, complete their life cycle and produce the next generation is the foundation for a sustainable fishery and a key consideration for management and conservation. Moreover, aggregation conservation is essential for the maintenance of larval recruitment and population connectivity as part of larger scale ecosystem-based management.

There is a growing acknowledgement of the need to manage fishing on aggregations and fishes that aggregate to spawn because many species are also otherwise vulnerable, due to their biology, to overfishing. As one example, the outcome of a spawning aggregation workshop in the Caribbean in 2008 concluded: *It is unequivocal that exploited aggregations need either total protection from fishing or that fishing activity on them should be minimized through careful management. FSAs [fish spawning aggregation sites] can be viewed as bank accounts from which we can derive "interest", and it is important to recognize FSAs as critical areas for management. The challenge is how to achieve effective management. Major changes in perspectives are needed to create an awareness of why aggregations are important for sustaining the fisheries of many aggregating species and why they need management even when they appear to be capable of producing large catches each year. Management depends first of all on an appreciation of the vulnerability and role of aggregations and their value as sources of regeneration of fish stocks.* (Luckhurst et al. 2009). Indeed, the Food and Agriculture Organization's *Code of Conduct for Responsible Fisheries*, Sect. 6.8 (FAO 1995), explicitly recognizes the need for nursery and spawning areas to be protected from the impacts of human activities that threaten the viability of fishery resources.

In parts of the Indian and Pacific Oceans the switch from barter to a cash economy has stimulated commercial fishery development. In southeast Asia, growth in the international trade in live reef fishes for restaurants in Chinese communities is increasing pressure on reef fishes. This trade has expanded the number of countries from which it sources fish in an increasing radius from the main market and trade centre in Hong Kong, with source countries located well into the western and southern Pacific as well as parts of the Indian Ocean (Sadovy et al. 2003-Fig. 1.2). For live fish fishers and traders alike, aggregation-fishing is appealing because it can potentially supply large quantities of reef fish at large profit in a short time period. Several important live fish trade species, such as the camouflage grouper, *Epinephelus polyphekadion* and the squaretail coral grouper, *Plectropomus areolatus*, can be taken in large quantities when aggregating to spawn. Customary management measures such as marine tenure, may have reduced pressure on coastal fisheries in many parts of the Indo-Pacific in the past (Johannes 1982). However, because of technology advances and the move to a cash society, almost half of the known aggregations are either in decline or can no longer be found and may be functionally extinct (SCRFA database, www.SCRFA.org).

Fisheries management has had shifting objectives over time. It has moved from seeking to manage fish stocks for maximum sustainable yield, and avoiding growth overfishing for maximum yields, to the more conservative aim of ensuring sufficient spawning biomass. There is now a growing emphasis on ensuring sufficient reproductive capacity to ensure fish population persistence. The protection of spawning biomass is also a major consideration in the design of MPAs; as they are increasingly promoted as a tool for the management of reef-associated fisheries. This consideration is significant for spawning aggregation protection.

The importance of protecting aggregations cannot be over-emphasized, and there is a global need to recognize this as a matter of urgency given the already strong scientific basis for management of aggregations in general and the speed with which they can be depleted. Unfortunately, in many cases, aggregations are discovered and fished before any effort is made to gather appropriate data, resulting in an incomplete understanding by managers of the ecological and economic importance of aggregations. As a consequence, management actions are typically implemented too late to prevent overfishing or extirpation (Sadovy de Mitcheson et al. 2008). There is a need to promote the mainstreaming of aggregation fishing into management and conservation planning. The need for precautionary approaches is fundamental given incomplete knowledge and the risk of collapse through hyperstability (Table 11.1, Sect. 11.5). There are many historic examples of the time that it takes to fully protect aggregations once discovered, and in many instances, this protection comes too late.

The bottom line is that if an aggregation is known and fished, it is most likely already overfished and in need of management. Indeed, the best-protected aggregations are those that are not yet known. Moreover, management measures on fishing a species outside of aggregations may also be necessary depending on the type and intensity of the fishery, the species and on enforcement capabilities.

11.3 Management Options

The types of fisheries input and output management controls, and conservation options for management of aggregations vary widely. Management tools include species-specific protection (the sale, export or possession of a species may be prohibited seasonally or year-round), temporal and spatial protection (a general area or specific geographic location may be closed to fishing or entry, either seasonally or permanently), and fishery input and output restrictions such as limited entry to a fishery, catch quotas, fish size limits, or fishing gear limitations.

Because of the many information gaps regarding species-specific spawning behaviours and aggregation status, and concerns over declines in aggregating species, precautionary approaches are necessary to ensure sustainable management (NRC 1999). This is clearly articulated in the *FAO Code of Conduct for Responsible Fisheries* (FAO 1995), which states "*The absence of adequate scientific information should not be used as a reason for postponing or failing to take conservation and management measures… States should ensure that the level of fishing permitted is commensurate with the state of the fishery resource under exploitation*" using the 'best available scientific data'. Deciding upon and implementing an appropriate precautionary approach early enough to prevent declines is not straightforward given that immediate short term socio-economic considerations may influence decision makers to be less precautionary, or an aggregation fishery still appears to be in good condition (see 'hyperstability' below). As a result, management to date has largely been reactive (Sadovy de Mitcheson 2009). Clearly, the longer term benefits to the overall fishery of preserving high reproductive potential through maintaining aggregations need to be more widely appreciated. For some species, such as the Nassau grouper, it may well be that the best fishery production overall will occur if aggregations are fully and permanently protected and left to perform their reproductive function.

Single management measures alone are unlikely sufficient to manage aggregating species, especially when they are fished both during and outside of the aggregating time period, or if they are threatened. More emphasis needs to be placed on coupling management measures directed at protecting spawning aggregations with measures also directed at the non-aggregation component, i.e. the wider fishery. For many species, especially migratory species that move over large distances to and from transient aggregation sites as adults, multiple management approaches may be needed to minimize the risks of overfishing in all areas where the species of interest may be vulnerable to fishing. Multiple management measures should be complementary and may be needed if enforcement is a problem. For example, while a seasonal ban on the possession and sale of certain fish species may afford some protection, the addition of seasonal protection of their aggregation sites can further limit fishing mortality during the period of greatest vulnerability. A minimum fish size limit can also ensure that fish grow large enough to reproduce before recruiting to a fishery, irrespective of where and when the species is fished. There are several examples of multiple measures being applied to aggregating spawners (e.g. red hind: Luckhurst and Trott 2009; leopard coralgrouper, *Plectropomus leopardus*: Russell 2006; Nassau grouper: Bahamas, Cuba, Belize, Cayman Islands, Sadovy and Eklund 1999; Whaylen et al. 2004; Claro et al. 2009; Janet Gibson personal communication 2010, Table 11.1, Chaps. 12.3, 12.6, and 12.9).

Once management is in place, assessment of the effectiveness of management measures is essential to ensure sufficient protection and to demonstrate outcomes to the fishing and wider community. However, variable enforcement effectiveness and the confounding effects of having a range of management measures within a particular fishery can complicate assessments of management effectiveness. For example, a highly valued fishery for leopard coralgrouper in Australia has a combination of management measures including quotas, minimum size limits, MPAs and aggregation season protection. In 2008, after considerable debate including stakeholder workshops, the aggregation season protection was downgraded largely because of concerns in the fishing community about the impacts of the global economic crisis. This was a reactive management decision not based on biological demonstration of stock status or effectiveness of existing measures, but due to pressure placed on government to remove spawning season closures given the many other restrictive management arrangements in place. Moreover, there was little evidence from fishery-dependent data that aggregations were being targeted (Chap. 12.9). In the Bahamas, the Nassau grouper aggregation protection was likewise down-graded in response to similar concerns in 2008.

The most appropriate management for a particular location or species inevitably depends on local social and economic factors, as well as can the biology and conservation status of the target species, and is best made on a case-by-case basis. Protection may also need to be implemented both during and outside of aggregation periods. There are many potential mixes of management options, and all options require effective enforcement and monitoring on a long-term basis to allow for adaptive management, which is fundamental in any attempt to manage a complex system of biological, ecological, economic and social variables. It is important to ensure that management is planned for the long term. Many of the fish species vulnerable to overfishing are long-lived and slow to recover, requiring long-term management planning. For example, since protection in 2002 was established for Nassau grouper aggregations in the Cayman Islands there are only now indications of increasing numbers of fish according to diver surveys, mark-and-recapture data, and video laser observations (Scott Heppell personal observation 2010, Table 11.1). Given the long-lived nature of Nassau grouper, more time is likely to see larger benefits from protection.

Management can occur at the species or ecosystem level with special attention given to threatened species or habitats. Species-specific protection is useful when information on aggregation sites and timing are lacking, incomplete or liable to misuse by fishers if specifically identified. It is also particularly important in the case of vulnerable or threatened species. Controls to restrict the amount of fish being exported from a country may be particularly effective if there is particular concern for a species, if there are few export points to monitor, or few export companies involved. For the live reef fish food trade, placing the export control point at the main airports is more feasible in some countries than placing a quota on individual fishers because of the large number of fishing communities involved, and the inability to ensure accurate records of catches. For example, the international trade of humphead wrasse (*Cheilinus undulatus*) has been restricted through being listed on CITES Appendix II since 2004. This species is extensively caught throughout Indonesia, and after determining an export quota, Indonesia has specified air-only exports (i.e. none permitted by sea) to enable better export control (Sadovy de Mitcheson et al. 2010, Chap. 12.13). See also Section 8.7 and Table 8.2.

11.3.1 Temporal and Spatial Protection

Short-term site closures, sometimes linked to controls on sales, can be used to close an area for a specific time to avoid disturbing aggregated fish. Seasonal closures may be species-specific or cover a range of species. For example, on the Great Barrier Reef seasonal closures to fishing for all reef fishes occur each year during a few days around the October and November new moons, although the timing of the closures was primarily designed to manage aggregations of the most highly valued species in the assemblage, the leopard coralgrouper (Russell 2006, 2009, Chap. 12.9). Seasonal sales bans prohibit or control commercial trade during a specified period in relation to the spawning season. A sales ban attempts to control movement of a species at the market end of the supply chain, and may be relatively easier to enforce than at-sea patrols, because fish markets and restaurants can be monitored more readily and cost-effectively. Evidence of possession is usually unequivocal; a fisher or wholesaler either has or does not have the species of concern.

Seasonal closures of an aggregation site as a sole management measure can be implemented for the duration of the spawning season of the managed species, but this does not prevent fishing for the species outside the spawning aggregation area. This option relies on having information on both spawning season timing and spawning locations as well as some idea of the status of the population generally. Temporal protection does not require all spawning sites to be known. For example, spawning season closures on the Great Barrier Reef were determined without knowledge of where all aggregations occur, but with information on spawning timing of key species (Russell 2006, 2009). In the Bahamas during the 2008–2009 spawning season, one aggregation site for Nassau grouper (High Point) was closed to fishing during the spawning season (December to February), while possession and sale of Nassau grouper during this period was not permitted throughout the country. This measure reflects the fact that not all aggregation locations in the country are well-described (www.bahamas.gov.bs). Seasonal closures have been variously implemented but some are subject to renewal each year rather than being permanent measures. In 2009, in the southeastern United States, the South Atlantic Fishery Management Council established a temporal closure from January to April each year to protect spawning adults of grouper species (*Mycteroperca* spp. and *Epinephelus* spp.).

Permanent area closures can provide protection if aggregation sites are known and can encompass many aggregation sites if the areas are large enough. There is a higher overall fishery advantage of permanent spatial closures compared to short-term closures because the protected area can act as a source of recruits for a wide range of species not otherwise protected by seasonal or short-term area closures. Permanent MPAs can protect resident aggregations from fishing, but may not be as effective for transient aggregating species, because these fish may be vulnerable to fishing on the spawning migration routes outside the MPA (e.g. lane snapper, *Lutjanus synagris*, in Cuba, Claro and Lindeman 2003). Very large permanent closures encompassing not only aggregation sites but a large part of the fish stock can conserve populations and allow for spillover and recruitment subsidy, as well as habitat protection, but do not alone control the remaining fishery which can still become overexploited. Aggregation sites that are used for spawning by single or multiple species, many of them commercial, are known to occur in outer reef channels and along outer reefs that drop-off into oceanic waters (Heyman and Kjerfve 2008; Koenig et al. 2000; Sadovy de Mitcheson et al. 2008). These offshore habitats are not routinely included in protected area designations, or otherwise managed. In many areas, deep slope or drop off areas may be important for deepwater assemblages, may include migration pathways (Starr et al. 2007), and may represent the last refuge for species heavily fished in shallow water, so merit more attention from MPA planners. In 1987, the Cayman Islands' Department of the Environment began monitoring Nassau grouper populations because fishers reported smaller fish and lower catches. Monitoring at Little Cayman's west end began in 2002 when large numbers of Nassau groupers were harvested relative to the total numbers present. This aggregation site, along with seven others, was protected under legislation enacted in 2003 that prohibits fishing year-round within the protected areas and has been regularly monitored by a dedicated group of researchers. The protection is likely to be an important factor in stabilizing or even increasing numbers of fish at the site. Numbers now appear to be stable and there are indications of an increase (Whaylen et al. 2007).

The locations of some aggregations may shift short distances of up to several hundred metres between years (Kadison et al. 2007; Heyman and Kjerfve 2008), or the exact location and size of an aggregation may not be fully documented prior to introduction of site-based measures. Protected areas need to be of an adequate size to include a buffer area around the spawning area to allow for such variability and uncertainty (see also 'staging areas' Chap. 2). For example, in the Florida Keys there was insufficient spatial information on a black grouper (*Mycteroperca bonaci*) aggregation site when the boundaries for a no-take MPA were being set. Subsequent research determined that the aggregation occurred in deeper waters outside of the MPA boundary (Eklund et al. 2000), thus the original MPA afforded little protection to the black grouper aggregation. Mechanisms to adjust or add to MPA boundaries in such instances could be considered as part of adaptive management.

Since the best protection is when site locations are entirely unknown, it would be counterproductive to publicize discovery of little or formerly unknown sites in an area unless there is effective legislation already in place to ensure immediate protection from fishing on discovery, and confidence that enforcement would be successful.

11.3.2 Fishery Input and Output Controls

Conventional fisheries management tools such as limited entry, commercial catch quotas, recreational possession limits, fish size limits, and fishing gear controls are indirect controls for protecting fishes that aggregate to spawn. However, if used without some form of direct spatial, temporal or species protection aligned with spawning locations or timing, these measures alone are unlikely to be effective at preventing overfishing of aggregations (Chap. 8). Approaches to consider include regulating input and/or output controls for specific aggregations, regulating input and/or output controls for the overall fishery for the species, and a mixed strategy, which probably represents the strongest level of protection.

Limiting the number of fishing licenses or permits, and the amount of time fishers can fish will reduce the overall impact on aggregating species. However, an increase in fishing power or gear modification that often follows such controls can offset the management effect over time (Chap. 12.7).

Commercial catch quotas and recreational possession limits may be set at conservative levels fishery-wide, and if complied with are effective ways to ensure catch is limited to within a predetermined sustainable level (see Chap. 12.12). However, these limits can be difficult to enforce because of multiple catch landing locations, and ensuring catch records are accurate is typically a challenge. Setting a sustainable catch level, whether fishery-wide or specific to an aggregation, depends on having reliable fisheries data, which are not readily available for most fisheries. Where there is a significant recreational fishery, a possession ban for the managed species can be used in addition to a sales ban, which may accompany a seasonal ban at the time of aggregation. Another form of quota management is export control. Sales controls are most effective when there is sufficient enforcement capacity to monitor the markets. However, this becomes a challenge if there are numerous markets or landing points. Unless penalties are appropriate for breaching such bans, fishers may take the risk of marketing protected species, or fillet them to avoid identification because the potential financial gain is large in comparison to the penalties. One way to mitigate filleting is to implement a rule to keep skin attached to fish fillets to aid in species identification. In Belize (Janet Gibson personal communication 2010) and on the Great Barrier Reef, Australia (Martin Russell personal observation 2010), certain commercially caught reef fish must have a portion, or all, of the skin remaining on a fillet ('skin patch'). Recreational fishers are not permitted to fillet fish at sea.

Minimum and maximum fish size limits, although difficult to enforce especially in remote locations, can protect juvenile and large adult fish, contributing to addressing growth and recruitment overfishing. A minimum size limit set at a size large enough will allow a target fish species to spawn in at least one or more spawning seasons before recruiting to a fishery. A maximum size limit offers protection to large, highly fecund fish to remain in the population to help replenish the fish stock. However, a fisher, whether commercial, subsistence or artisanal, is unlikely to be willing to return an under-size fish to the water especially if it is a prime protein and economic source, and/or a fellow fisher may catch and retain that same fish. Managing for upper size limits to protect large females, or males of protogynous species, is challenging because of a general market preference for larger fish. Also, non-legal species taken from deeper waters are unlikely to survive release due to barotrauma; this is a substantial problem in many deeper water fisheries with fish size limits.

Fishing gear controls, such as limiting or prohibiting the use of SCUBA and hookah, traps, nets, spears, night fishing, poisons such as cyanide, and explosives can reduce overfishing of aggregations, among other objectives. For example, fish traps have been banned in Bermuda since 1990 to prevent overfishing of coral reef fishes, including aggregating species such as groupers and snappers (Luckhurst and Ward 1996). Gill and channel nets are prohibited to reduce pressure on migrations of reef and estuarine species in Cuba (Claro et al. 2009), and gill nets were prohibited on an aggregation site in Mexico (Aguilar-Perera 2004). Spear-fishing, especially at night with lights and using SCUBA, is now occurring in many Pacific countries and is considered a major threat to inshore fisheries (Gillett and Moy 2006).

When considering input and output controls, where the aggregation component of the fishery for a species is small compared to the non-aggregation component, management may be applied at the overall stock/fishery level. However, emphasis should be placed on minimizing the catch from aggregations as much as possible.

11.3.3 Which Management Option to Use?

Deciding the appropriate fisheries management or conservation approach will depend on the vulnerability of the species, how depleted its aggregations or stocks are and the potential risk of depletion. The biology and behaviour of target species and the type, economics and social context of the fishery are also key considerations. Management is essential and should be prioritized if the catchability of a species is greater during aggregation periods, and if aggregations are specifically targeted or are likely to be targeted by commercial fishing. Management measures required to rebuild aggregations will typically involve full protection using closures if a large component of the catch comes from aggregations.

The type of aggregation and the species involved, as well as the prevailing fishing pressure, will be major factors determining management choices. Transient aggregating species that form relatively few, large aggregations, as opposed to resident aggregating species, may need specific management such as spatial or seasonal closures, because they are potentially more vulnerable to overfishing. This is because a greater proportion of the adult fish population is predictably available to fishing in a defined area and over a relatively short time period, and there are likely to be fewer spawning sites compared to resident spawners. For example, for several species of grouper that aggregate together in Palau, there appears to be only about 10 known spawning sites in the country (Sadovy 2007), which could be protected. By contrast, resident aggregating species with numerous aggregation sites could be better managed by seasonal closures and/or other non-aggregation focused measures such as total allowable catches, species bans or broad spatial closures. Much depends on the available information regarding the timing and location of spawning, the number of aggregations, spawning behaviour and local enforcement capacity.

In the Caribbean, many examples of management of aggregating species have been introduced over the past two decades (Table 11.1). These examples demonstrate the highly variable combinations of management that can be adopted, with many involving multiple management measures and most focusing on transient aggregators. For example, in Bermuda, seasonal aggregation site closures were introduced, followed later by possession limits implemented during the spawning season to reduce the impact of potential poaching at aggregation sites during the closed season and to aid enforcement of the seasonal closure. For the Nassau grouper in Belize, aggregation protection proved a challenge and so a minimum size was also introduced to strengthen the management of reproductive capacity for this species (Janet Gibson personal communication 2010). In Cuba, more and more measures were progressively introduced as the fishery continued to decline (Claro et al. 2009).

To help determine appropriate management options, we provide general guidance that can be used for both commercial and subsistence fisheries, although each aggregation or aggregating species needs to be evaluated on a case-by-case basis as many different biological, ecological and socio-economic factors can vary considerably. For low level subsistence fisheries, precautionary measures are warranted, involving effort monitoring and possibly limiting effort because even subsistence fisheries can deplete an aggregation (Fig. 11.2).

If the aggregation sites or timings are known or partly known, the following are options for aggregation management that can be considered alone or in combination.

1. **Short-term site closure of aggregation site**. The aggregation site is closed for the duration of the spawning season, but fishing for this species is permitted outside the aggregation.
2. **Seasonal closure of fishery**. The fishery is closed for the spawning season with no catch permitted, and the aggregation site is not specifically protected.
3. **Short-term aggregation site and time closure for species**. Protect the aggregation site, and the species from being fished anywhere during its spawning season.
4. **Fully-protected area**. The managed species is protected from fishing year-round in a designated MPA, but may be fished outside of the MPA at any time. MPAs should be designed to include the entire aggregation sites and, if applicable, migration routes.

Fig. 11.2 Children (about 10 years old) holding up two of several humphead wrasse, *Cheilinus undulatus*, caught at a spawning aggregation in the Solomon Islands – possibly most of the aggregation was caught at the one time (Photo: © Michael Giningele)

Additional considerations include sales bans when there are large seasonal market gluts in catches that could depress prices and cause wastage during spawning periods (which could last several months) (Sect. 8.5.1, Fig. 11.3). Species that are threatened with extinction should be priorities for management action if aggregations are targeted. For size limits, skin patches can be required on fillets for ease of species identification (Sect. 11.3.2). Many years may be needed, especially in the case of long-lived species, before management yields notable increases in fish sizes or catches (Table 11.1).

11.3.4 Alternatives to Fishing Spawning Aggregations

Effective conservation of fished aggregations typically requires controls during aggregating periods and, in the short term, there may be direct and/or indirect economic and food source implications for fishers and their communities. Mitigation of negative effects is necessary, especially if it takes many years for the long-term objective of maintaining or restoring the reproductive function of aggregations and the benefits of higher catches associated with the non-aggregation season to be achieved. Management initiatives that stop aggregation fishing should, therefore, be implemented together with planning for alternative livelihoods, food sources, or with some form of financial incentive for those communities and fishers foregoing catch. Funded incentives and re-employment can reduce impacts on displaced fishers.

Fig. 11.3 Massive numbers of Gulf corvina, *Cynoscion othonopterus*, are caught at brief, peak, spawning periods in the northern Gulf of California, Mexico. Up to 1,000 tonnes of corvina may be caught and sold to buyers in 3–4 days, a glut that causes market prices to plummet. When the market is over-supplied, buyers cease purchasing and hundreds of tons of fish may be abandoned on the beach or dumped by fishers who cannot sell their catch (see also Fig. 8.13) (Brad Erisman, unpublished data 2010) (Photo: © Brad Erisman)

Below are examples illustrating approaches and challenges of addressing short-term problems following management.

A history of cooperation, collaboration and careful planning in Belize led to the introduction of alternative livelihoods following the closure of aggregations. Local fishers who once targeted aggregations have been trained as fly-fishing or SCUBA diving guides. Belize is now a major fly-fishing tourist destination, principally for permit (*Trachinotus falcatus*) (Graham and Castellanos 2005). Also, a complete transition from fishing to tourism occurred at Gladden Spit on the Belize barrier reef; where once large numbers of cubera snapper (*Lutjanus cyanopterus*) were caught during spawning, fishers now take tourist divers to witness the spawning event and the appearance of whale sharks (*Rhincodon typus*) feeding on the eggs released during spawning (Heyman et al. 2001). To reduce the possible negative impact of divers on the aggregations, there are limits on the numbers of divers, boats and on dive times.

An example of a fisher-to-tourism conversion exists in south-central Cuba, the Cayo Doce Leguas area of the Gardens of the Queen Archipelago. A small, floating hotel in a marine reserve caters to diving and catch-and-release fly-fishing. Many hotel employees were fishers displaced by the creation of the reserve over 10 years

Fig. 11.4 Spawning aggregations can be popular viewing opportunities for divers in some locations, such as for snapper in Palau (Photo: Mandy Etpison)

ago but now work as fishing guides, divemasters, cooks, and mechanics. There have been major reductions in fishing effort overall and on spawning aggregations, and substantial social and economic benefits have accrued to the displaced fishers (Giuseppe Omegna personal communication 2008, Pina Amargós 2008).

A few large aggregations have potential for diving tourism, but most are only of short-term interest each year for sport divers, and many are difficult to reach because they are off-shore and may occur in challenging diving conditions. For example, in Palau, tourist divers at aggregation sites typically spend only a few minutes observing aggregating groupers and snappers, because the fish are somewhat wary of divers (Patrick L Colin personal communication 2010). Although there has been some success in Belize (above), dive tourism on aggregations does not appear to have much potential as an alternative livelihood. The possible impact of divers on aggregated fish needs to be evaluated to determine if this is a beneficial alternative use since there are possible negative effects of divers on courtship and spawning (Heyman et al. 2010, Chap. 8 Fig. 11.4).

In Australia, the Great Barrier Reef Marine Park was re-zoned in 2004 by increasing the area closed to fishing from about 4% to 33% of the Park (Day et al. 2005). Although this was not a direct aggregation protection strategy, it did result in representative examples of most habitat types being protected from fishing, including six known aggregation sites. A range of measures, including financial incentives, fishing license buy-out, business exit and restructuring to alternative livelihoods, and employee and social-community assistance were administered by government to assist those adversely affected by the increase in the area closed to fishing. However, political positioning, industry lobbying, and inconsistencies in claims for assistance resulted in many of those affected by the increased protection being dissatisfied, and the cost to government was significantly more than originally estimated (Macintosh et al. 2010).

11.4 Management Successes and Failures

Most fishery departments were initially established to promote and develop fisheries. A major challenge to management is that government actions may simultaneously be taken to develop and preserve a fishery (Hilborn 2005). Government subsidies for the purchase or increase in size of boats and engines have allowed fishers to increase their fishing range without due consideration for the condition of, or impact on, the resource base at the same time that management is being considered. Combined with the increasing commercial demand for reef fishes, this has inevitably elevated pressure on aggregations. Introduction of ice plants in the Indo-Pacific as development initiatives, especially in remote locations, typically have not taken into account the underlying sustainability of the species likely to be affected. Development and management arms of government typically act independently in relation to fisheries.

Throughout the Caribbean, the status of over 70% of known exploited aggregations, assessed on the basis of both direct and indirect evidence of spawning, is unknown, while a further 20% are decreasing and very few show positive trends (database, www.SCRFA.org). Examples of successes and failures can be attributed to enforcement, political priorities, presence or absence of monitoring information and local interest. Several examples are illustrative (also refer to Cayman Islands example above).

An example of a management failure was the temporal fishing closure on a degraded Nassau grouper aggregation in Mahahual, Mexico. This closure was unsuccessful due to insufficient enforcement and that the aggregation no longer formed (Aguilar-Perera 2004, Alfonso Aguilar-Perera personal communication 2010). The potential for return of spawning fish has been further constrained by the construction of a major cruise ship terminal (about one million visitors per year) within 1 km of the aggregation site. This terminal not only changed some physical characteristics of the aggregation site but the busiest cruise ship months are during December and January, the peak spawning months (KCL personal communication 2010).

In Bermuda, a seasonal spawning closure was introduced in 1974 for red hind because fishers raised concerns to government about declines of two red hind aggregations due to perceived overfishing by commercial and recreational fishers. Fishers had already witnessed the overfishing of Nassau grouper spawning aggregations in Bermuda which led to their eventual extirpation, and did not want the same thing to happen to red hind. The initial steep decline in red hind landings (Fig. 11.5) is

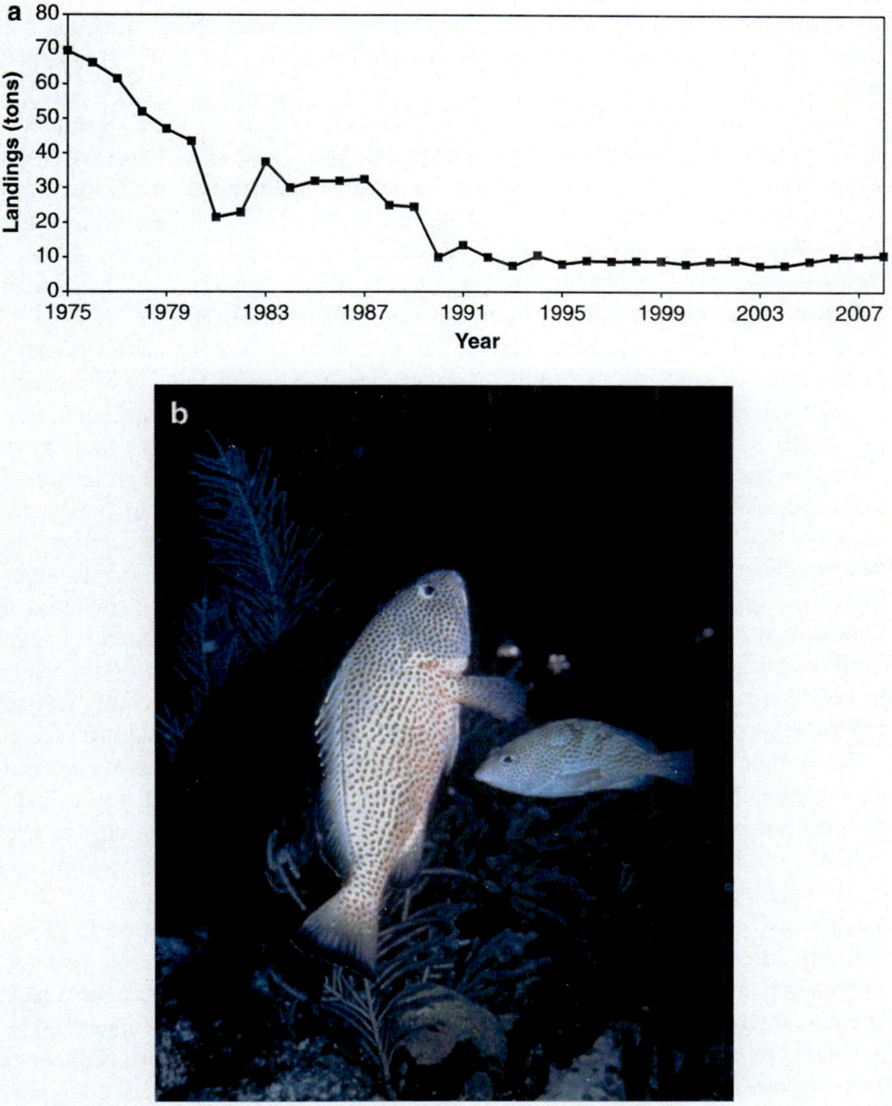

Fig. 11.5 (**a**) Reported landings of red hind, *Epinephelus guttatus* in Bermuda, 1975–2008 (From Bermuda commercial fishery statistical database). (**b**) Red hind at a spawning aggregation in Puerto Rico

thought to have been due to heavy fishing pressure on spawning aggregations. Seasonal site closures were in place from 1975, but compliance was poor due to ineffective enforcement during this early period. Landings continued to decline in the 1980s, but eventually stabilized after 1990 following a ban on fish traps, which restricted the fishery to line fishing only. Enforcement of the seasonal closures of spawning aggregations has become more effective over the past 10–15 years, thus the landings reflect active line fishing effort during this latter period, although fish numbers have not recovered to former levels (Luckhurst 1996, 1998; Luckhurst and Trott 2009).

In the British Virgin Islands, seasonal closures for the red hind have been ineffective due to lack of enforcement (Eristhee et al. 2006, Table 11.1). However, in the adjacent US Virgin Islands, red hind fish sizes and numbers dramatically increased in a closed area. This was due to adequate enforcement and the establishment of a comprehensive monitoring programme that enabled the Caribbean Fishery Management Council to better justify regulatory actions based on sound research information that demonstrated positive outcomes (Nemeth 2005).

In Belize, 11 Nassau grouper aggregation sites were declared as marine reserves permanently closed to fishing, while two other sites, now closed, initially had limited access by fishers under permit (Gibson et al. 2007, Janet Gibson personal communication 2010). As most of these sites are multi-species aggregation sites (Sala et al. 2001), the creation of reserves to protect Nassau grouper resulted in the protection of other species as well (Heyman and Kjerfve 2008). However, despite high expectations for the recovery of Nassau grouper at these sites, fish numbers continued to decline, likely due to illegal fishing at several sites (Gibson et al. 2007). Due to ongoing concern for the Nassau grouper, additional management measures including a minimum size limit were implemented at the national level in 2009 and have high local acceptance (Janet Gibson personal communication 2010). The progress made in Belize is largely due to a highly motivated working group that includes fishing groups, representatives from NGOs, government and the tertiary educational sector.

In the Indo-Pacific, of the few aggregations that are managed, the most common management tool is seasonal site protection, usually combined with a sales ban. A small number of sites are in MPAs. Although aggregation status in most of the region is unknown, where information is available over 30% of aggregations are decreasing, including in Fiji, Indonesia, Malaysia and Papua New Guinea (SCRFA database, www.SCRFA.org). In a relatively early management initiative in 1980, Pohnpei, Micronesia, introduced a sales ban from March to April each year for camouflage grouper. This proved to be largely ineffective most likely due to a lack of enforcement in the market (Rhodes and Sadovy 2002; Rhodes and Tupper 2007). In Palau, protection of grouper aggregations accompanied by a sales ban is likely to be a major reason that some aggregations still occur. Unfortunately, not all aggregating months are included in the sales ban, putting these aggregations at risk, and poaching, a major problem generally, continues (Sadovy 2007).

Notwithstanding the historic trend in delayed and reactive management, from the few examples above, illegal fishing and lack of enforcement are major problems. The potential for illegal fishing in any management decision needs to be identified. There is little advantage to implementing fish size limits or prohibiting the take of certain species if enforcement capacity is inadequate. Effective compliance with area closures depends on several factors including proximity of the closure to the coast, the size of the area, clarity in the boundaries of the protected areas, community acceptance of management measures and enforcement capacity. Disputes can easily arise between fishers and enforcers as to exactly where a protected area is located, particularly if the fishers do not use navigation equipment such as global positioning systems. Some MPA boundaries are marked by buoys, for example, in Palau, Fiji, Philippines, and Seychelles, but these buoys and moorings need regular maintenance and can be prone to theft. At-sea enforcement patrols and aerial surveillance are two key enforcement tools, but the latter is often not available in developing countries due to high cost. Vessel monitoring systems (VMS), whereby vessels can be followed because they each carry an identification device are expensive but practical options, e.g. in the Seychelles where VMS is used to monitor fishing activity at remote spawning sites (Jan Robinson personal communication 2010).

Increased compliance comes from building compliance-focused outreach into all local user group interactions from early stages. Outreach needs to penetrate throughout the different sectors of the fishing industry and communities involved in resource stewardship to gain acceptance and general understanding of the need for management. The goal is to have fishers recognize and agree that protection of the resource is necessary and that protection will benefit them in the longer term. To help achieve compliance, fishers need to be involved in management design and implementation, e.g. through community-based co-management supported by outreach. Positive dialogue between managers, scientists and fishers is an important element in achieving this goal. Enforcement must be effective, and meaningful sanctions must be imposed to discourage illegal fishing. Importantly, management agencies should provide regular feedback to fishers and communities about compliance and the impact of management on the status of the target fish population. Using data collected from monitoring programmes, management agencies must demonstrate the benefit of, and need for, the management to maintain support and interest by fishers and communities. It is fundamental that fishery and MPA managers work closely with fishers to establish trust and share information on aggregations.

11.5 The Challenge of Hyperstability

If only fishery-dependent data are available from an aggregation-fishery, decision-making may be compromised if continued high levels of catch per unit of effort (CPUE) mask actual declines in abundance of the fish population due to ‘hyperstability’ (Sadovy and Domeier 2005, Fig. 11.6). Because of hyperstability, fishery-independent data are particularly important for assessing aggregation status. A probable example of hyperstability is the Nassau grouper fishery in Cuba, where major fishing effort on the species was based on spawning aggregations that abruptly collapsed in the 1970s after providing high levels of landings from aggregations for many years (Claro et al. 2009, Table 11.1). Declines have continued and the fishery has not recovered; while multi-species reef fish landings in general remained reasonably stable overall, probably in response to increasing effort, Nassau grouper and several other aggregating species showed marked declines.

Another example of probable hyperstability is the Spanish mackerel (*Scomberomorus commerson*) fishery on the GBR. This line fishery has historically

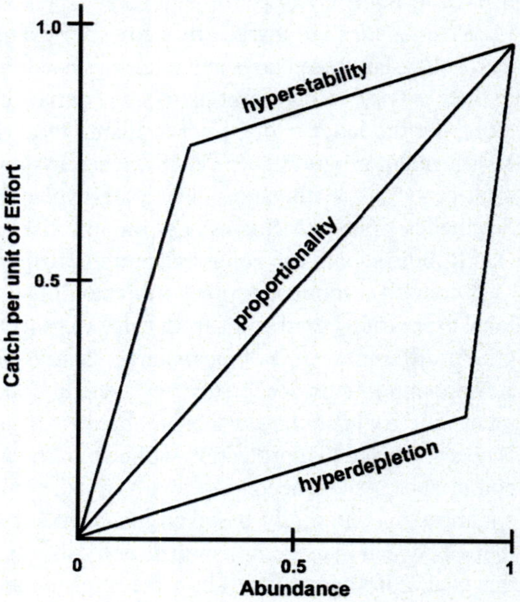

Fig. 11.6 Relationships between catch per unit of effort (CPUE) and abundance under hyperstability when fish or fisher behaviour results in elevated CPUE even as fish abundance declines until the stock suddenly collapses (from Sadovy and Domeier 2005 with permission)

concentrated on a peak aggregation time and location. Since 1988, about 65% of the annual catch was taken in October from a single spawning aggregation site. Although the fishery-dependent CPUE remains stable, fisher anecdotal information suggests that the aggregation used to be much larger and that there were more aggregations, which have now ceased to MWR exist or have diminished substantially, throughout the GBR prior to the 1980s (personal observation).

11.6 The Role of Non-governmental Interests and Bodies

Spawning aggregation management efforts have been associated with non-government organizations (NGOs), individuals, and communities since before the 1970s. However, it was not until the 1990s that a few international and local NGOs initiated stronger marine programmes. As foundation and donor support for NGO marine programmes grew, increased focus on marine protected areas and conservation developed among NGOs, with a relatively recent focus on fisheries sustainability.

People and organizations outside of government play important roles in capacity- and consensus-building, training, advice, auditing, and promoting statements of concern or calls for action. This is often a very effective way to transfer information and encourage and support community-based initiatives. However, NGOs can also be territorial, may have conflicting themes, and may exclude other organizations and individuals, rather than working together. Problems can arise if there are umbrella conservation or fishery agendas that are promoted irrespective of local capacities and needs, such as blanket promotion of MPAs in places where other measures might be more effective or appropriate. Poor data sharing among NGOs and between NGOs and government can also be a problem. The Belize aggregation working group mentioned above addressed this particular problem by having a password-protected database for sharing information. Capacity-building and education are important roles the NGOs can play to make management possible and more acceptable.

Positive examples of NGO involvement in spawning aggregation issues teach important lessons. The Tortugas Ecological Reserve, within the US Florida Keys National Marine Sanctuary, is a successful management initiative based on a collaborative approach, good compliance, substantial outreach, and scientific information (Cowie-Haskell and Delaney 2003). A coalition of national, state and local NGOs helped catalyze a government planning process to improve management of the Dry Tortugas during the late 1990s. This process culminated in the designation and implementation of two areas representing the Tortugas Ecological Reserve (Cowie-Haskell and Delaney 2003). NGOs played a significant role in the focus on snapper spawning aggregation protection, particularly at the Riley's Hump promontory (Tortugas South). By 2000, the reserve was supported by commercial fishers, in part due to NGO outreach. The support of respected, high-production fishers for a reserve to protect multi-species, multi-season aggregations provided credibility for agency and NGO efforts, as well as providing new fish spawning aggregation information (Lindeman et al. 2000). The Nassau grouper working group in Belize has made considerable progress towards managing this species through collaborations with a wide range of individuals and organizations, high motivation and long-term commitment.

The first international NGO entirely dedicated to fish spawning aggregation management and research is the Society for the Conservation of Reef Fish Aggregations (SCRFA), formed in 2000. One key initiative undertaken was the development of a global web-based database on known information about aggregations (Sadovy de Mitcheson et al. 2008). Other initiatives include research and training especially designed for studying and managing aggregation and outreach using widely disseminated materials in multiple languages and formats (www.SCRFA.org).

It is only recently that global recognition of the need for aggregation protection, both locally and globally, and the importance of managing human use of spawning aggregations has been formally recognized through various statements. In 2004, the Third IUCN World Conservation Congress recommended to "*Take action to protect reef fish spawning aggregations*", (http://cmsdata.iucn.org/downloads/wcc_res_rec_eng.pdf), and in 2006, the International Coral Reef Initiative (ICRI) made a statement on "*Coral reef fish spawning aggregations*" (http://www.icriforum.org/icri-documents/statements/icri-statement-coral-reef-fish-spawning-aggregations). Having such statements provides greater leverage for both government and NGOs to progress conservation initiatives for spawning aggregations. These international statements have given greater recognition to the importance of evaluating spawning aggregations as an essential part of fisheries and MPA management.

11.7 Future Efforts to Improve Management

It is essential to improve information on aggregations to implement and adapt management strategies. Catch data should be collected on both aggregation catches and landings of aggregating species, supplemented by fishery-independent data, such as underwater visual census at aggregation sites. There is justification for developing vulnerability indicator systems for data-poor contexts that would allow rapid evaluation of the need for management, direct or indirect, while more data are being collected (Johannes 1998). Work is also needed to increase overall understanding of the reproductive significance of aggregations for fisheries. However, in most parts of the world such information is not readily available. Moreover, there is a disparity of information among the Pacific, Indian and Atlantic Oceans, with the Caribbean being relatively information-rich. Public knowledge and understanding of aggregation management is essential. Education priorities should include demonstrating the need to protect aggregating fish, and providing relevant information on the performance of management. Tools should be developed to help ensure communities, fishers and the general public understand that the taking of spawning fish is in effect taking the next generation of fish and can negatively affect the fishery as a consequence. This is a simple but powerful concept for educational programmes.

There is a need for more emphasis on the political and social environments in which successful management occurs, and to engage the major resource users. Community-based management, where applicable, has the greatest chance of success if the values and traditions of the community are incorporated into the overall management scheme, and if communities are empowered with management authority and supported by government in matters of enforcement and capacity development. Researchers, managers and NGOs must convey a sense of urgency to act once aggregations are discovered and fished, and decision makers require a better knowledge base with respect to the economic and social consequences of a lack of action concerning the protection of aggregations. It is essential that management be implemented adaptively, with clear and transparent reasoning that is not overly influenced by politics or inappropriate agendas. Good science is essential for seeking the best possible management approach.

11.8 Conclusion

Managing spawning aggregations is a complex challenge that requires strong commitment, political will and an adaptive approach. Knowledge of the biology and behaviour of the target species is essential, particularly of the spatial and temporal dynamics of spawning. Years of directed research and monitoring may be needed to define the bio-physical system adequately, to account for inter-annual variability in the system and to enable the outcomes of management action to be assessed. Because of the extreme vulnerability of spawning aggregations to over-exploitation, management action is often desirable before the required information is available, and needs to be precautionary. Precautionary measures must be strong and proactive. There is already enough evidence for the need to conserve aggregations and the lack of more detailed information should not be used as an excuse for inaction. The examples of aggregation management and outcomes in the Caribbean (Table 11.1), demonstrate that the attempts made to mitigate overfishing of aggregations and aggregating fish have been less effective than anticipated because they were introduced too late and/or were not supported by good compliance or enforcement. Largely for this reason, management has become multi-layered, and recovery seems ever more remote as fisheries decline. However, examples of positive outcomes of management in the Caribbean show that aggregation recovery is possible.

The demise of spawning aggregations can occur very rapidly. Because recovery may not occur at all, or only very slowly, it is essential that managers use all the tools available to limit fishing mortality in aggregations, and on aggregating species, to within sustainable levels. Legislation that can rapidly be brought into effect once a new aggregation is found is particularly important as more aggregations are described and enter the wider literature. An essential element in achieving appropriate management is political will. Political will needs to be bolstered with sound information on aggregations and on aggregating species in general. The role of NGOs to gather information, provide outreach, and lobby for improved regulations or push for compliance with existing regulations is important for dealing with this complex challenge. It cannot be overstated that a well-planned fisher outreach and monitoring programme to assess the impact of management is essential to both evaluate the effects of management as well as to adapt to changes which may occur in the biological system. Consideration must be given to finding alternative food and income sources for those affected by management.

Ultimately, it is likely that aggregations will come to be considered as critically important sources of fisheries, and permanently closed to anything other than subsistence fishing to safeguard and maximize the reproductive capacity of a range of aggregating species.

Management effectiveness depends on knowledge of fish biology and spawning behaviour, monitoring, enforcement capacity, and social acceptance. In the tropics and sub-tropics, small-scale subsistence fisheries on aggregations with no external market pressures have and may continue to persist without management. However, given human population growth, the advent of commercial fishing, and increased pressure on subsistence fishers to generate cash, it is highly unlikely that aggregation fishing can be sustained under other circumstances. Moreover, there are far more examples of management failure than managing spawning aggregations successfully once they are discovered.

Acknowledgements We would like to thank Janet Gibson, Graciela Garcia-Moliner, Newton Eristhee, Rodolfo Claro, Michelle Meadows and Kim Iverson for assistance with Table 11.1. Rodolfo Claro and Richard Appeldoorn assisted with reviews of information. A Doherty Fellowship in the Department of Marine and Environmental Systems, Florida Institute of Technology, supported the work of Ken Lindeman. We also thank Jan Robinson, Seychelles Fishing Authority, for providing review comments, and Vicki Nelson, Madam En-dash Editing and Proofreading, for chapter edits.

Table 11.1 Examples of management actions to conserve spawning aggregations of reef fishes in the Caribbean and some of the outcomes

Country/ state	Species	Management measures	Year implemented	Results of measure	Comments	References
Bahamas	Nassau grouper	Andros (High Cay) Spatial/temporal closure (Dec–Feb)	1998	Landings still declining		Ray et al. (2000), www.breef.org
	Nassau grouper	Long Island Spatial/temporal closure (Dec–Feb)	1999	Landings still declining		Ray et al. (2000), BREEF (2002)
	Nassau grouper	Country-wide seasonal closure: no take or sale	2004–2010	Uncertain		www.breef.org
Belize	Cubera snapper	Some aggregation sites included in no-take MPAs	2003	Inadequate data to assess	Cubera snapper not targeted	Heyman et al. (2005)
	Mutton snapper	Gladden Spit Restricted access to site	2003	Fishing has caused declines	Decline in landings and CPUE	Graham et al. (2008)
	Nassau grouper	Permanent closure of 11 sites Permit access at 2 sites Closed season (Dec–March)	2003	Limited signs of recovery since site closures	Illegal fishing due to weak enforcement	Gibson et al. (2007) and Sala et al. (2001)
	Nassau grouper	Slot size limit (51–76 cm TL). Must be landed whole, with no filleting	2009	Too early to assess impact		Belize Fisheries Regulations 2009
Bermuda	Red hind	Spatial/temporal closure (May–Aug)	1974	After initial decline, landings stabilized, (Fig. 11.5) Mean size has increased markedly since closures	Site boundaries have changed over time, some monitoring since closure but not regularly	Luckhurst (1996, 1998) and Luckhurst and Trott (2009)

	Red hind	Seasonal bag limit (10 fish per boat per day)	1990	Reported landings declined during closure period	Unable to validate landings data	B. Luckhurst personal observation
	Black grouper	Spatial/temporal closure (May–Aug)	2005	Insufficient data to assess; heavily fished outside closure period	Assessment started in 2005, ongoing research	Luckhurst (2010)
	Black grouper	Spatial/temporal closure (May–Nov)	2008–2010	Aggregation period appears to be longer than first assumed	Extended closure period to allow further research and tagging	T. Trott (pers. comm., 2008) Luckhurst, unpublished data
British Virgin Islands	Red hind, white margate, black margate	Temporal closure (Jan–March). Take and marketing prohibited	2003	Enforced and markets monitored, but impact of management unclear		N. Eristhee personal communication
	Nassau grouper	Temporal closure (March–May). Take and marketing prohibited	2003	No longer targeted, by-catch only		N. Eristhee personal communication
Cayman Islands	Nassau grouper	Closed season (Nov–March); no trapping within 1 nm of sites during spawning season; no spearfishing at sites	2002	Ineffective, most sites disappeared since monitoring began in 1987	Declines in catch, size and CPUE	Bush et al. (2006)
	Nassau grouper	Total ban on fishing at all 6 spawning aggregation sites for 8 years	2003–2011	West End Little Cayman site rediscovered and heavily fished in 2001–2002		Whaylen et al. (2004) and Semmens et al. (2007)

(continued)

Table 11.1 (continued)

Country/ state	Species	Management measures	Year implemented	Results of measure	Comments	References
Cuba	Cubera snapper, grey snapper	Channel net bans in limited areas to protect pre-spawning migrations (July–Aug) Some effort reduction in MPAs	1996–2000	No recovery detected to date	Species are overfished. Data is pooled for both species	Claro and Lindeman (2003) and Claro et al. (2009)
	Lane snapper	Channel net bans to protect pre-spawning migrations Some effort reduction in MPAs	1978–2008	Partial recovery since 1995. Landings stabilizing at lower levels with variations related to fishing effort	Severely overfished in the 1970s	Claro et al. (2001), Claro and Lindeman (2003), and Claro et al. (2009)
	Mutton snapper	Channel net bans in limited areas to protect pre-spawning migrations Some effort reduction in MPAs	1996–2008	Landings stabilized at lower level	Commercial effort reduced since 1992, but other types of effort increased	Claro et al. (2001), Claro and Lindeman (2003), and Claro et al. (2009)
	Nassau grouper	Some effort reduction in MPAs	1992–2008	Limited data suggest some recovery in the north central zone	Information limited. Fishing effort is reduced and not stable	Claro et al. (2001), Claro and Lindeman (2003), and Claro et al. (2009)

Florida	Gag, black grouper, red grouper, yellowfin grouper, yellow-mouth grouper, red hind, rock hind, scamp, tiger grouper, coney, graysby	*State waters*: Atlantic and Florida Keys closed season (Jan–April). Gulf of Mexico closed season (Feb–March) *Federal waters*: Atlantic and Florida Keys closed season (Jan–April) and more than 10 MPAs prohibit take year-round. Gulf of Mexico closed season (Feb–March), and two MPAs closed (Nov–April), one area closed (Jan–April)	2010	Status is species dependent	State and two different federal fishery management councils have coordinated rules effectively across complicated geographies and jurisdictions	www.myfwc. com www.safmc.net/ www.gulfcouncil. org
	Nassau grouper, goliath grouper	All take prohibited in State and Federal waters	1992, 1990	Limited recovery for Nassau grouper. Some recovery for goliath grouper	Active aggregations currently unknown for Nassau grouper, known for goliath grouper	Sadovy and Eklund (1999); www. myfwc.com/
	Vermillion snapper	Fl State waters (Atlantic only) and Fl Federal waters (Atlantic only): closed season (Nov–March)	2009	Federal assessment pending		www.myfwc.com/ www.safmc.net/

(continued)

Table 11.1 (continued)

Country/ state	Species	Management measures	Year implemented	Results of measure	Comments	References
	Cubera snapper	Take prohibited at Tortugas South Ecological Reserve (Riley's Hump)	2001	Status uncertain, assumed to have improved	Deeper water aggregations in the area are not protected	Lindeman et al. (2000)
	Mutton snapper	Take prohibited at Tortugas South Ecological Reserve (Riley's Hump)	2001	Recently, increased number of fish at site observed by divers	Evidence of potential recovery	Lindeman et al. (2000), and Burton et al. (2005)
Puerto Rico	Red hind	Closure of 3 sites (Dec–Feb) in state waters on west coast Closure in Federal waters (Dec–Feb)	1995, 2004	Landings and CPUE on west coast increased from 1996–1999 Insufficient data to assess Federal closure	Possible positive effect but not conclusive. Ineffective enforcement	Matos-Caraballo (2002) and CFMC (2009)
	Red grouper, black grouper, tiger grouper, yellowfin grouper, yellow-edge grouper	Seasonal closure (Feb–April) in Federal waters	2006	Largely unknown	Before management, tiger grouper aggregation at Vieques Isl. showed large decline in landings and CPUE	Matos-Caraballo et al. (2006) and CFMC (2009)
	Mutton snapper	Temporal closure (April–May) in state waters; closed (April–June) in Federal waters	2004	Aggregations heavily fished.	Effect of measure largely unknown	CFMC (2009)

	Lane snapper	Seasonal closure. (April–June) in Federal waters	2006	Unknown		CFMC (2009)
	Blackfin snapper, vermillion snapper, black snapper, silk snapper	Seasonal closure (Oct–Dec) in Federal waters	2005	Unknown		CFMC (2009)
	Nassau grouper	No take in state and Federal waters	2004	Unknown		CFMC (2009)
U.S. Virgin Islands	Red hind	St. Croix (Lang Bank)	1993, 1999	Large increase in abundance following closure at St Croix		Beets and Friedlander (1998), Nemeth (2005), and CFMC (2009)
		Spatial/temporal closure (Dec–Feb) in Federal waters				
		St. Thomas (Hind Bank Reserve)				
		Spatial/temporal closure (Dec–Feb) in Federal waters				
		USVI Federal waters: closed (Dec–Feb)				
	Nassau grouper	No-take in state and Federal waters.	2006	2 sites disappeared at St. Thomas	Little evidence of aggregation recovery.	Olsen and LaPlace (1979) and Munro and Blok (2005)

(continued)

Table 11.1 (continued)

Country/state	Species	Management measures	Year implemented	Results of measure	Comments	References
	Yellowfin grouper, cubera snapper, dog snapper	St. Thomas (Grammanik Bank) Spatial/temporal closure (Feb–April) in Federal waters	2005	Estimated large decline in yellowfin grouper abundance; snappers unknown	Up to 50% decline of yellowfin grouper in 5 years; current status of snappers unknown.	Nemeth et al. (2006) and Kadison et al. (2007)
	Red grouper, black grouper, tiger grouper, yellowfin grouper, yellow-edge grouper	Seasonal closure (Feb–April) in USVI Federal waters	2006	Unknown		CFMC (2009)
	Blackfin snapper, vermillion snapper, black snapper, silk snapper	Seasonal closure (Oct–Dec) in USVI Federal waters	2005	Unknown		CFMC (2009)
	Mutton snapper, lane snapper	Seasonal closure (April–June) in Federal waters	2006	Unknown		CFMC (2009)

Data also available in the database of www.SCRFA.org

References

Aguilar-Perera A (2004) Detection of fishing effects on a Nassau grouper spawning aggregation from southern Quintana Roo, Mexico. Proc Gulf Caribb Fish Inst 55:543–556

Beets J, Friedlander A (1998) Evaluation of a conservation strategy: a spawning aggregation closure for red hind, *Epinephelus guttatus*, in the U.S. Virgin Islands. Environ Biol Fish 55:91–98

Belize Fisheries Regulations (2009) Statutory instrument – Nassau grouper and species protection, 2 p

BREEF (2002) Nassau grouper and queen conch in the Bahamas – status and management options, report by MacAlister Elliott and partners

Burton ML, Brennan KJ, Munoz RC, Parker RO Jr (2005) Preliminary evidence of increased spawning aggregations of mutton snapper (*Lutjanus analis*) at Riley's hump two years after the establishment of the Tortugas South Ecological Reserve. Fish Bull 103:404–406

Bush PG, Lane ED, Ebanks-Petrie G, Luke K, Johnson B, McCoy C, Bothwell J, Parsons E (2006) The Nassau grouper spawning aggregation fishery of the Cayman Islands – an historical and management perspective. Proc Gulf Caribb Fish Inst 57:515–524

Caribbean Fishery Management Council (CFMC) (2009) Quick reference to the fishing regulations history in the US Caribbean (prepared by Graciela García-Moliner)

Claro R, Lindeman KC (2003) Spawning aggregation sites of snapper and grouper species (Lutjanidae and Serranidae) on the insular shelf of Cuba. Gulf Caribb Res 14:91–106

Claro R, Baisre JA, Lindeman KC, García-Arteaga JP (2001) Cuban fisheries: historical trends and current status. In: Claro R, Lindeman KC, Parenti LR (eds) Ecology of the marine fishes of Cuba. Smithsonian Institution Press, Washington, DC

Claro R, Sadovy YS, Lindeman KC, García-Cagide A (2009) Historical analysis of commercial Cuban fishing effort and the effects of management interventions on important reef fishes from 1960–2005. Fish Res 99:7–16

Cowie-Haskell BD, Delaney JM (2003) Integrating science into the design of the tortugas ecological reserve. Mar Technol Soc J 37(1):68–79

Day J, Fernandes L, Jago B (2005) Planning for the future of the Great Barrier Reef. Qld Plan 45(3):12–14

Eklund AM, McClellan DB, Harper DE (2000) Black grouper aggregations in relation to protected areas within the Florida Keys National Marine Sanctuary. Bull Mar Sci 66:721–728

Eristhee N, Kadison E, Murray PA, Llewlyn A (2006) Preliminary investigations into the red hind fishery in the British Virgin Islands. Proc Gulf Caribb Fish Inst 57:373–384

FAO (Fisheries and Agricultural Organization of the United Nations) (1995) Code of conduct for responsible fisheries. FAO, Rome

Gibson J, Pott RF, Paz G, Majil I, Requena N (2007) Experiences of the Belize spawning aggregation working group. Proc Gulf Caribb Fish Inst 59:455–462

Gillett R, Moy W (2006) Spearfishing in the Pacific Islands: current status and management issues, vol 19, FAO/FishCode review. FAO, Rome

Graham RT, Carcamo R, Rhodes KL, Roberts CM, Requena N (2008) Historical and contemporary evidence of a mutton snapper (*Lutjanus analis* Cuvier 1828) spawning aggregation fishery in decline. Coral Reefs 27:311–319

Graham RT, Castellanos DW (2005) Courtship and spawning behaviours of carangid species in Belize. Fish Bull 103:426–432

Heyman WD, Kjerfve B (2008) Characterization of transient multi-species reef fish spawning aggregations at Gladden Spit, Belize. Bull Mar Sci 83:531–551

Heyman WD, Graham RT, Kjerfve B, Johannes RE (2001) Whale sharks *Rhincodon typus* aggregate to feed on fish spawn in Belize. Mar Ecol Prog Ser 215:275–282

Heyman WD, Kjerfve B, Graham RT, Rhodes KL, Garbutt L (2005) Spawning aggregations of *Lutjanus cyanopterus* (Cuvier) on the Belize Barrier Reef over a 6 year period. J Fish Biol 67:83–101

Heyman WD, Carr LM, Lobel PS (2010) Diver ecotourism and disturbance to reef fish spawning aggregations: it is better to be disturbed than to be dead. Mar Ecol Prog Ser 419:201–210

Hilborn R (2005) Faith-based fisheries. Fish 31(11):554–555

Hornell J (1927) Report on the fisheries and fish resources of the Seychelles Islands. H.M.S.O, London, 76 pp

Johannes RE (1978) Traditional marine conservation methods in Oceania and their demise. Annu Rev Ecol Syst 9:349–364

Johannes RE (1982) Traditional conservation methods and protected marine areas in Oceania. Ambio 11:258–261

Johannes RE (1998) The case for data-less marine resource management: examples from tropical nearshore finfisheries. Trends Ecol Evol 13:243–246

Kadison E, Nemeth RS, Herzlieb S, Blondeau J (2007) Temporal and spatial dynamics of *Lutjanus cyanopterus* (Pisces: Lutjanidae) and *L. jocu* spawning aggregations in the United States Virgin Islands. Rev Biol Trop 54(Suppl 3):69–78

Koenig CC, Coleman FC, Grimes CB, Fitzhugh GR, Scanlon KM, Gledhill CT, Grace M (2000) Protection of fish spawning habitat for the conservation of warm-temperate reef-fish fisheries of shelf-edge reefs of Florida. Bull Mar Sci 66:593–616

Lindeman KC, Pugliese R, Waugh GT, Ault JS (2000) Developmental patterns within a multispecies reef fishery: management applications for essential fish habitats and protected areas. Bull Mar Sci 66:929–956

Luckhurst BE (1996) Trends in commercial fishery landings of groupers and snappers in Bermuda from 1975 to 1992 and associated fishery management issues. In: Arreguin-Sanchez F, Munro JL, Balgos MC, Pauly D (eds) Biology, fisheries and culture of tropical groupers and snappers, ICLARM Conf Proc 48. ICLARM, Makati

Luckhurst BE (1998) Site fidelity and return migration of tagged red hinds (*Epinephelus gutttatus*) to a spawning aggregation site in Bermuda. Proc Gulf Caribb Fish Inst 50:750–763

Luckhurst BE (2003) Development of a Caribbean regional conservation strategy for reef fish spawning aggregations. Proc Gulf Caribb Fish Inst 54:668–679

Luckhurst BE (2010) Observations of a black grouper (*Mycteroperca bonaci*) spawning aggregation in Bermuda. Gulf Caribb Sci 22:1–8

Luckhurst BE, Trott TM (2009) Seasonally-closed spawning aggregation sites for Red Hind (*Epinephelus guttatus*): Bermuda's experience over 30 years (1974–2003). Proc Gulf Caribb Fish Inst 61:331–336

Luckhurst BE, Ward JA (1996) Analysis of trends in Bermuda's fishery statistical database from 1975 to 1990 with reference to fishery management measures implemented during this period. Proc Gulf Caribb Fish Inst 44:306–324

Luckhurst BE, Donaldson TJ, Sadovy de Mitcheson Y, Russell M (2009) Biology and management of spawning aggregations – lessons learnt and panel discussion: a half-day symposium sponsored by the Society for the Conservation of Reef Fish Aggregations (SCRFA). Proc Gulf Caribb Fish Inst 61:338–343

Macintosh A, Bonyhady T, Wilkinson D (2010) Dealing with interests displaced by marine protected areas: a case study on the Great Barrier Reef Marine Park Structural Adjustment Package. Ocean Coast Manag 53(9):581–588

Matos-Caraballo D (2002) Portrait of the commercial fishery of the Red hind *Epinephelus guttatus* in Puerto Rico during 1992–1999. Proc Gulf Caribb Fish Inst 53:446–459

Matos-Caraballo D, Posada JM, Luckhurst BE (2006) Fishery-dependent evaluation of a spawning aggregation of tiger grouper (*Mycteroperca tigris*) at Vieques Island, Puerto Rico. Bull Mar Sci 79:1–16

Munro JL, Blok L (2005) The status of stocks of groupers and hinds in the Northeastern Caribbean. Proc Gulf Caribb Fish Inst 56:283–294

National Research Council (NRC) (1999) Sustaining marine fisheries. National Academy Press, Washington, DC

Nemeth R (2005) Population characteristics of a recovering US Virgin Islands red hind spawning aggregation following protection. Mar Ecol Prog Ser 286:81–97

Nemeth R, Kadison E, Herzlieb S, Blondeau J, Whiteman EA (2006) Status of a yellowfin (*Mycteroperca venenosa*) grouper spawning aggregation in the US Virgin Islands with notes on other species. Proc Gulf Caribb Fish Inst 57:543–558

Olsen DA, LaPlace JA (1979) A study of a Virgin Islands grouper fishery based on a breeding aggregation. Proc Gulf Caribb Fish Inst 31:130–144

Pina Amargós F (2008) Efectividad de la Reserva Marina de Jardines de la Reina en la conservación de la ictiofauna. PhD dissertation. Universidad de la Habana, Centro de Investigaciones Marinas

Ray GC, McCormick-Ray MG, Layman CA, Silliman BR (2000) Investigations of Nassau grouper breeding aggregations at High Cay, Andros: Implications for a conservation strategy. Report to Department of Fisheries, Nassau, The Bahamas

Rhodes KL, Sadovy Y (2002) Temporal and spatial trends in spawning aggregations of camouflage grouper, *Epinephelus polyphekadion*, in Pohnpei, Micronesia. Environ Biol Fish 63:27–39

Rhodes KL, Tupper M (2007) A preliminary market-based analysis of the Pohnpei, Micronesia, grouper (Serranidae: Epinephelinae) fishery reveals unsustainable fishing practices. Coral Reefs 26:335–344

Russell M (2006) Leopard coral grouper (*Plectropomus leopardus*) management in the Great Barrier Reef Marine Park, Australia. SPC Live Reef Fish Inf Bull 16:10–12

Russell M (2009) Adaptive management of fish apawning aggregations on the Great Barrier Reef. Aust Proc Gulf Caribb Fish Inst 61:324

Sadovy Y (2007) Report on current status and exploitation history of reef fish spawning aggregations in Palau. Western Pacific Fishery Survey Series: Society for the Conservation of Reef Fish Aggregations, vol 3. SCRFA and the Palau Conservation Society, www.SCRFA.org

Sadovy Y, Domeier M (2005) Are aggregation-fisheries sustainable? Reef fish fisheries as a case study. Coral Reefs 24:254–262

Sadovy Y, Eklund AM (1999) Synopsis of biological data on the Nassau grouper, *Epinephelus striatus* (Bloch 1792) and the Jewfish, *E. itijara* (Lichtenstein 1822). NOAA Technical Report NMFS 146

Sadovy YJ, Donaldson TJ, Graham TR, McGilvray F, Muldoon GJ, Phillips MJ, Rimmer MA, Smith A, Yeeting B (2003) While stocks last: the live reef food fish trade. Manila: Asian Development Bank. http://www.adb.org/Documents/Books/Live_Reef_Food_Fish_Trade/default.asp

Sadovy de Mitcheson Y (2009) Biology and ecology considerations for the fishery manager. Chapter 2. In: Cochrane KL, Garcia SM (eds) A fishery manager's guidebook, 2nd edn. FAO/Blackwell Publishing, Oxford

Sadovy de Mitcheson Y, Cornish A, Domeier M, Colin PL, Russell M, Lindeman KC (2008) Reef fish spawning aggregations: a global baseline. Conserv Biol 22(5):1233–1244

Sadovy de Mitcheson Y, Liu M, Suharti S (2010) Gonadal development in a giant threatened reef fish, the humphead wrasse *Cheilinus undulatus*, and its relationship to international trade. J Fish Biol 77:706–718

Sala E, Ballesteros E, Starr RM (2001) Rapid decline of Nassau grouper spawning aggregations in Belize: fishery management and conservation needs. Fish 26(10):23–30

Semmens BX, Luke KE, Bush PG, Pattengill-Semmens CV, Johnson B, McCoy C, Heppell S (2007) Investigating the reproductive migration and spatial ecology of Nassau grouper (*Epinephelus striatus*) on Little Cayman Island using acoustic tags – an overview. Proc Gulf Caribb Fish Inst 58:199–206

Smith CL (1972) A spawning aggregation of Nassau grouper, *Epinephelus striatus* (Bloch). Trans Am Fish Soc 101:257–261

Starr RM, Sala E, Ballesteros E, Zabala M (2007) Spatial dynamics of the Nassau grouper *Epinephelus striatus* in a Caribbean atoll. Mar Ecol Prog Ser 343:239–249

Vilaró Díaz DJ (1884) Corrida y arribazón de algunos peces cubanos. Manuel Gómez de la Maza, La Habana, Cuba

Whaylen L, Pattengill-Semmens CV, Semmens B, Bush PG, Boardman MR (2004) Observations of a Nassau grouper, *Epinephelus striatus*, spawning aggregation site in Little Cayman, Cayman Islands, including multi-species spawning information. Environ Biol Fish 70:305–313

Whaylen L, Bush P, Johnson B, Luke K, McCroy C, Heppell S, Semmens B, Boardman MR (2007) Aggregation dynamics and lessons learned from five years of monitoring at a Nassau grouper (*Epinephelus striatus*) spawning aggregation in Little Cayman, Cayman Islands. BWI Proc Gulf Caribb Fish Inst 59:479–487

Chapter 12
Species Case Studies

12.1 Introduction

Reef fishes that aggregate to spawn, even those considered to be threatened or to have commercial importance, are typically little studied. Information on their biology and ecology is often widely scattered in the published and unpublished literature, and many species remain little understood. Yet, the foundation for both good species- and ecosystem-level management rests on an understanding of life history, including in relation to the aggregation-forming habit and the fisheries that species of interest form. The testing of hypotheses, such as the why and how of aggregation formation, also requires an understanding of biology. The wide taxonomic range of species that are confirmed as aggregation spawners (Appendix) is reflected in the diversity of aggregation strategies they exhibit and, along with the social context of their fishery, determines what conservation approaches are advisable, monitoring methods required and the importance of aggregation versus non-aggregation exploitation.

Case studies of the aggregation, reproduction as well as other basic life history attributes of the better-studied aggregating reef fishes are provided in this chapter with additional detailed information not included in the earlier chapters, as well as useful background material on the species in general. The 23 species highlighted, in addition to notes on a few others, range in size from small bluehead wrasse, *Thalassoma bifasciatum*, to large humphead wrasse, *Cheilinus undulatus*, and are variously found globally throughout the tropics and sub-tropics. Some species are exploited, some are not, while a few are considered to be threatened and in urgent need of management. Several studies cover previously unpublished accounts of individual research projects on a single or small group of fishes (longnose parrotfish, *Hipposcarus longiceps*; blacktail snapper, *Lutjanus fulvus*). Others attempt to summarize knowledge of a species for which there is already a relatively large body of work (red hind, *Epinephelus guttatus*; Nassau grouper, *E. striatus*; leopard coralgrouper, *Plectropomus leopardus*), although even in such

Y. Sadovy de Mitcheson and P.L. Colin (eds.), *Reef Fish Spawning Aggregations: Biology, Research and Management*, Fish & Fisheries Series 35,
DOI 10.1007/978-94-007-1980-4_12, © Springer Science+Business Media B.V. 2012

cases, we consider few species to be comprehensively studied. Many of the case studies include information obtained only recently or not published previously while some also provide useful profiles of management successes and challenges, lessons learned and data gaps. Some accounts were updated almost to the moment of going to press with this volume. Many of the case studies are referred to in other chapters.

A standard format was suggested to each of the contributing authors for the species accounts. However, since the level and type of information varies considerably among species, some accounts are longer, or somewhat differently organized, than others. We have attempted to provide interesting and informative photographs, nearly all previously unpublished, which illustrate the information included in the text. Knowledge of a general biological phenomenon, such as spawning aggregations, often starts with the study of individual species. As understanding grows and deepens it allows for the determination of general principles and the development and testing of theories regarding the evolution and benefits of the phenomenon. It is hoped the information on fishes presented here encourages and inspires others to continue gathering the basic information needed to advance the science of reef fish spawning aggregations.

12.2 Brown-Marbled Grouper – *Epinephelus fuscoguttatus*

Rachel Pears
Great Barrier Reef Marine Park Authority, P.O. Box 1379, Townsville, QLD 4810, Australia
e-mail: rachel.pears@gbrmpa.gov.au

12.2.1 General

Brown-marbled grouper (*Epinephelus fuscoguttatus*) is a large-bodied grouper (Serranidae), reaching approximately 1 m in length (Fig. 12.1a). This species is widespread throughout the Indo-Pacific, including the Red Sea, and is often confused with the similar-looking, co-occurring but smaller camouflage grouper (*E. polyphekadion*) (Heemstra and Randall 1993; Chap. 12.5; Craig et al. 2011). Brown-marbled grouper was listed as 'Near Threatened' on the (IUCN) Red List in 2004. Little is known about early life history in the wild, however some information is available from aquaculture (e.g. Lim et al. 1990; Lim 1993). Adults to use coral reef habitats, and the limited information on juveniles suggests they frequent seagrass beds (Sommer et al. 1996; Gell and Whittington 2002) and estuarine or low salinity areas (Sadovy 2000). Juveniles are collected from inshore reduced salinity waters for grow-out in cages prior to sale in the international live reef food fish trade (Sadovy et al. 2003a).

Fig. 12.1 Adult brown-marbled grouper (*Epinephelus fuscoguttatus*) (**a**) in typical colouration. (**b**) Male in courtship colouration only reported from spawning aggregation sites (Photos: Patrick L. Colin)

Biological information comes from a recent age-based study on Australia's Great Barrier Reef (GBR) (Pears et al. 2006, 2007), unless otherwise noted. Brown-marbled grouper are uncommon (less than 1 fish per 1,000 m^2), and relatively long-lived (lifespan can reach 42 years, Fig. 12.2). The length-weight relationship was estimated to be: W (g) = 1.16×10^{-5} FL (mm) 3.075 (n = 127, $r^2 = 0.98$).

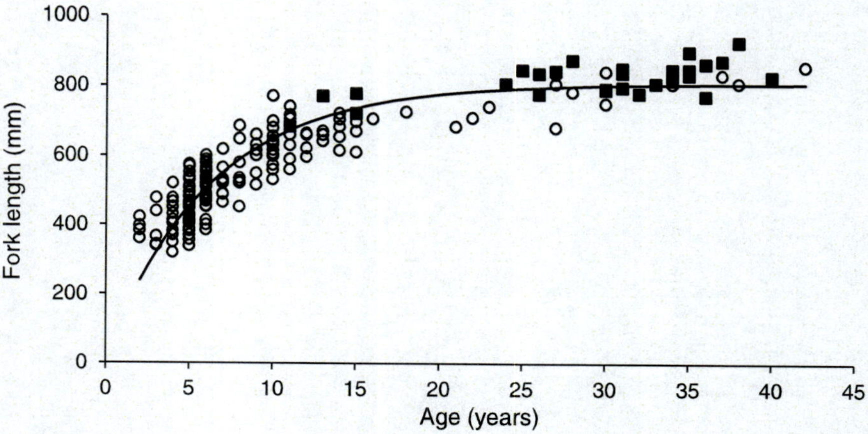

Fig. 12.2 Growth curve for brown-marbled grouper from the Great Barrier Reef: *closed squares* – males; *open circles* – females (Pears et al. (2006). With permission MEPS 1 February 2011)

The estimates of size and age at 50% sexual maturity in females during the spawning season (November to January on the GBR) are 57 cm total length and about 9 years. Large females attain 40+ years of age and make important reproductive contributions due to disproportionately high numbers of eggs. Males are restricted to size classes over about 70 cm FL (Johannes et al. 1999; Pears et al. 2006).

12.2.2 *Reproductive Biology*

Noteworthy aspects of the reproductive ecology of the species include late maturity, limited sexual activity among smaller mature-sized females during the spawning period, and the likelihood that some females do not undergo sex change. The sexual pattern for brown-marbled grouper appears to be monandric protogynous hermaphroditism in which juveniles mature and function as adult females before changing sex, although confirmation of sex change is still required.

The reported annual spawning periods for brown-marbled grouper are relatively narrow (e.g. 3–4 months), with seasonal timing and lunar periodicity varying widely across the Indo-Pacific (Johannes et al. 1999; Robinson et al. 2004; Pet et al. 2005; Pears et al. 2007; Robinson et al. 2008a, b, Fig. 8.3). Brown-marbled grouper has been reported to aggregate in relatively large numbers to spawn, with over 350 individuals being counted at spawning aggregation sites in Palau (Johannes et al. 1999) and a peak aggregation abundance of 1,050 individuals estimated in the Seychelles (Robinson et al. 2008b). Spawning aggregations are known from Australia, British Indian Ocean Territory, Federated States of Micronesia, Fiji, Indonesia, Kenya, Malaysia, New Caledonia, Palau, Seychelles and Solomon Islands (Cornish 2004; Hamilton et al. 2005; Pears et al. 2007; Robinson et al. 2008a, b;

Yvonne Sadovy de Mitcheson, Fiji, unpublished data, www.SCRFA.org). Spawning aggregation sites are reported to occur at reef channels (passes), outer reef slopes and drop offs (shelf edges). Aggregation sites of brown-marbled grouper are often in the same general area as those of camouflage grouper and other groupers, such as the squaretail coralgrouper (*Plectropomus areolatus*) or marbled coralgrouper (*P. punctatus*). Recent research in Pohnpei, Micronesia, has confirmed the use of reproductive migratory corridors and sex-specific variation in residency, with males arriving at the aggregation site earlier, frequenting the site for more months of the season and remaining present 50% longer than females (Kevin L. Rhodes, personal communication 2010). Males take on a distinctive courtship colouration (Fig. 12.1b).

12.2.3 Fishing History and Management

The history of fishing and management of the species varies widely. Management measures include spawning season sale bans, time/area closures, marine protected areas, size and catch limits, gear restrictions and limited entry. Two contrasting regions are presented, a well-managed fishery and a remote emerging fishery.

Relatively well managed fishery on the east coast of Queensland, Australia – Groupers are caught throughout the year as part of the multi-species Coral Reef Finfish Fishery by commercial fishers (using hook and line) and recreational fishers (using hook and line or spear). While the two primary target species in this fishery are the leopard coralgrouper (*P. leopardus*) and red throat emperor (*Lethrinus miniatus*), at least 40 species of groupers occur in the catches. Brown-marbled grouper (known in Australia as flowery rockcod) makes up about 1.2% of the total commercial catch by number of fish in fishing operations for processed reef fish (Pears et al. 2007). The estimated total recreational catch of reef fishes was 2,600 tonnes in 2005 (DEEDI 2010), a small proportion of which was brown-marbled grouper. Groupers over 6 kg are difficult to sell due to concerns about ciguatera poisoning. Large groupers such as brown-marbled grouper are highly appreciated by divers in the GBR tourism industry (Dean Miller, personal communication 2007).

A central Great Barrier Reef spawning aggregation site of this species and camouflage grouper has sometimes been targeted resulting in large catches, e.g. 1,500 kg per fishing boat over 2.5 days (Pears 2005; Pears et al. 2007). Numerous fishers were reported fishing this aggregation on occasions, e.g. six mother boats with about four fishing dories each. The current status of this aggregation is unknown. Any heavy fishing of spawning aggregations appears to be opportunistic on the GBR, and largely driven by market price. Australian fishers generally do not put other groupers in live holding tanks because they are considered to cause distress to leopard coralgrouper, the main target of the live fishery.

In 2003–2004, revised management arrangements for the fishery were introduced by the fisheries department in Queensland, Australia, in response to concerns about rapidly increasing fishing effort driven by the lucrative export market for live reef food fish (this is an international trade in live fish, mainly groupers with major demand centres in Hong Kong and mainland China, Sadovy et al. 2003a). Measures included commercial catch quotas (3,061 tonnes, of which 1,011 tonnes is for reef fish species excluding leopard coral grouper and red throat emperor), seasonal spawning closures, size limits and revised recreational bag limits. Between 2004 and 2008 the seasonal closure of the fishery was for 9 days around each new moon in October, November and December, chosen mainly to encompass the peak spawning period for leopard coral-grouper and some other reef fishes. In December 2008, the legislated 9-day closure was removed, and for 2009–2013, the seasonal closure period was reduced to 5 days around the new moons in October and November, in response to socio-economic concerns following the near-global financial crash of 2008. An exemption was also provided to allow eligible offshore charter operators to operate under permit during the closure periods.

The vast majority of the fishery operates within the GBR Marine Park and World Heritage Area, where, in 2004 protection of reef and shoal habitat was increased to 30% no-take areas, and all individuals of the genus *Epinephelus* over 100 cm in length are protected.

In December 2003, minimum and maximum size limits were introduced for brown-marbled grouper at 50 and 100 cm respectively. However, recent research indicated that the main reproductive individuals of the population were not well protected, since the fished component still included most of the active spawning stock and all of the males (Pears et al. 2006). To address this, the Queensland Government reduced the maximum size limit to 70 cm for this species in 2009. For brown-marbled grouper, only limited protection is likely from the seasonal closures since spawning events only partly coincide with the closure periods (Pears et al. 2007).

Remote artisanal fishery at Farquhar Atoll in the Seychelles, western Indian Ocean – The Seychelles contain several reported aggregation sites for brown-marbled grouper, including a site well-known to fishers at Farquhar Atoll shared with camouflage grouper. The history of this site is interesting because it still seems to be functional, despite a long history of targeted aggregation fishing and a season of heavy exploitation for the Hong Kong-centred live reef food fish trade.

Groupers are part of the demersal reef fish catch of the Seychelles artisanal fishery, which in 2006 caught 123 tonnes of grouper, mainly for domestic consumption. The Islands Development Company (IDC) manages the outer islands of the Seychelles, including Farquhar Atoll. IDC operates a commercial fishery for salted fish on Farquhar Atoll, and groupers, including brown-marbled grouper are targeted, with high catches (e.g. many tonnes of groupers) from periodic targeting of the aggregation site (Robinson et al. 2004; Aumeeruddy and Robinson 2006). There is also some subsistence fishing of grouper at Farquhar Atoll, including from the aggregation site, and commercial vessels from the main Seychelles islands are permitted to fish the outer reef, away from the aggregation site.

Groupers were also targeted at various island groups in the region during trials for a live reef fish fishery in 1998–1999, with approximately 26.4 tonnes exported from Farquhar Atoll (Aumeeruddy and Robinson 2006) with brown-marbled grouper an important part of the catch. Legislation was introduced in 2005 prohibiting the live reef food fish fishery in the Seychelles, since the fishery was not considered in the best interests of the country (Aumeeruddy and Robinson 2006). While heavy fishing likely reduced the population of brown-marbled grouper at Farquhar Atoll, occasional suspension of commercial fishing by IDC in the area for periods of up to three years, and the remoteness from the main populated islands, may have protected the spawning aggregation somewhat.

There are concerns throughout the Seychelles about the status of reef fish spawning aggregations targeted by traditional artisanal fishers (Robinson et al. 2004) and possibly in the live reef fish fishery trial at Cosmoledo. The Seychelles Fishing Authority has implemented a research programme for reef fish spawning aggregations, including for brown-marbled grouper, and is developing a management framework for grouper spawning aggregations, including proposed regulations for closed areas and seasons to protect key spawning sites and vulnerable species (SFA 2006). The spatial and temporal dynamics of spawning and aggregating behaviour were determined, with brown-marbled grouper aggregations lasting between 2 and 3 weeks in each of three consecutive months (November to February), males arriving at the aggregation before females, peak fish abundances a few days prior to the new moon, and the timing of spawning probably varying interannually within a few days leading up to and including the new moon (Robinson et al. 2008b). It also appears that large numbers of brown-marbled groupers remain on the spawning site between aggregating months. Grouper spawning aggregations are well-known to artisanal fishers in the Seychelles and are in need of precautionary management arrangements.

12.2.4 Lessons Learned, Challenges and Data Gaps

The degree of protection afforded to brown-marbled grouper spawning aggregations varies widely, and in many cases has been inadequate to prevent declines. For example, out of 34 records for the species in the global spawning aggregation database compiled by the Society for the Conservation of Reef Fish Aggregations (SCRFA), 11 spawning aggregations (32%) are decreasing and the status of 20 spawning aggregations (59%) are unknown (www.SCRFA.org May 2010). Within multi-species reef fisheries, late-maturing, aggregation-spawning groupers will be some of the more vulnerable species to overfishing during both aggregation and non-aggregation periods.

Brown-marbled grouper is vulnerable to overfishing as it is relatively uncommon and appears to use only a few spawning aggregation sites compared to species such as leopard coralgrouper or squaretail coralgrouper. Protection of these sites is a high priority, and should include migratory corridors and all potential spawning months to prevent the potential for sexual selection by the fishery. A better understanding of (1) spatial and temporal dynamics of additional spawning aggregation sites, (2) the main migration routes to aggregation sites and (3) the area from which a particular aggregation site draws individuals would help in determining the scales of potential fishing effects and in designing locally appropriate management strategies. Similarly, it will be important to determine biological characteristics, such as maturity and spawning periods, locally to ensure management arrangements are suitable. The biggest challenge is gaining ongoing local support for management measures to give adequate protection during aggregation and non-aggregation periods. Estimates of total fishing mortality and stock status, better species-level catch estimates and information on 'high-grading' practices (discarding to make room for more valuable or legal-sized fish) and post-release survival of discarded fish are also needed.

12.3 Red Hind – *Epinephelus guttatus*

Richard S. Nemeth
Centre for Marine and Environmental Studies, University of the Virgin Islands, St. Thomas, USVI 00802
e-mail: rnemeth@uvi.edu

12.3.1 General

Red hind (*Epinephelus guttatus*) are one of the more common groupers on coral reef habitats ranging from Brazil to Bermuda (Robins and Ray 1986). Red hind occupy shallow reefs from 3 to 50 m depth. Red hind reach maximum length and age at 50–55 cm (exceptionally to 72 cm) and 11, exceptionally to 22 years (Smith 1971; Olsen and LaPlace 1979; Thompson and Munro 1978; Randall 1983; Colin et al. 1987; Luckhurst et al. 1992; Sadovy et al. 1992). The home range of 22 fish (124–298 mm SL) ranged in size from 112 to 5,635 m^2 in shallow inshore areas of Puerto Rico over a 5-month period (Shapiro et al. 1994).

12.3.2 Reproductive Biology

Red hind are protogynous hermaphrodites and change sex from female to male at 32–41 cm total length (Sadovy et al. 1992; Nemeth 2005; Whiteman et al. 2005). Their spawning aggregations have been studied in a number of locations throughout the Caribbean including the US Virgin Islands (Olsen and LaPlace 1979; Beets and Friedlander 1999; Nemeth 2005; Whiteman et al. 2005; Nemeth et al. 2006), Puerto Rico (Colin et al. 1987; Sadovy et al. 1992, 1994; Shapiro et al. 1993),

Bermuda (Luckhurst 1998; Burnett-Herkes 1975) and Saba, Netherland Antilles (Kadison et al. 2009) and have been reported from the British Virgin Islands (Munro and Blok 2003; Eristhee et al. 2006) and Jamaica (Munro et al. 1973). Information on the spatial structure of red hind aggregations indicates that they consist of small haremic groups with one male defending three to five females and spawning occurs in pairs 1–2 m above the reef (Shapiro et al. 1993). Sadovy et al. (1994) suggest that red hints females are determinate spawners and spawn more than once during the course of the annual spawning season. Red hind ovaries can be 15% of body weight (gonado-somatic index or GSI) just before spawning (Kadison et al. 2009). Fecundity can range from ca. 100,000 eggs for a 31 cm female to ca. 1,500,000 eggs for a 41 cm female (Thompson and Munro 1978; Olsen and LaPlace 1979; Sadovy 1993a, b; Whiteman et al. 2005) (Fig. 12.3a, b). The spawning aggregation can consist of a few 100 to over 80,000 adults which occupy an area of 0.015–0.35 km^2 (Shapiro 1987; Shapiro et al. 1993; Nemeth 2005). Aggregation sites typically occur on the top of deep coral reef ridges which are located on or near the shelf edge (Colin et al. 1987; Beets and Friedlander 1999; Nemeth 2005). These locations may have specific oceanic conditions that facilitate larval retention on the island shelf (Nemeth et al. 2008; Cherubin et al. 2011).

In the Caribbean, red hind aggregations typically form the week before the full moon in the months of December, January and February (Shapiro et al. 1993; Beets and Friedlander 1999; Nemeth 2005). Depending upon the timing of the full moon, red hind in the Virgin Islands begin to arrive at spawning aggregation sites after mid December, show a distinctive peak in January about 20–40 day after the winter solstice and disperse by the end of February (Nemeth et al. 2007). Spawning typically occurs during periods of declining seawater temperature and slacking currents within a temperature range of 26–27.5°C and current speed of 2.5–3.5 cm s^{-1}. In more northern areas (i.e. Bermuda), the spawning season occurs May through July within the same temperature range but during periods of increasing water temperatures (Luckhurst 1998). At the aggregation sites, males defend territories and display a characteristic barred colouration in the postero-ventral area; the lips also become barred; females just before spawning exhibit highly distended abdomens (Fig. 12.4a, b).

Red hind migrate 5–33 km from home sites to spawning grounds at swim speeds of 3–25 km day^{-1} and encompass a catchment area of 90–500 km^2 (Sadovy et al. 1992; Luckhurst 1998; Nemeth et al. 2007). Males arrive several weeks early at spawning sites and stay for the duration of the spawning season whereas females may only spawn during 1 month. Gender-specific movements during the week of spawning cause daily fluctuations in female:male sex ratios at the aggregation site that range from more than 20:1 to less than 1:1. Whiteman et al. (2005) developed the use of ultrasound imaging to determine gender non-invasively. In between monthly spawning peaks, about 5–20% of the spawning population remain on the aggregation site comprising the largest males and females within the aggregation (Nemeth et al. 2007).

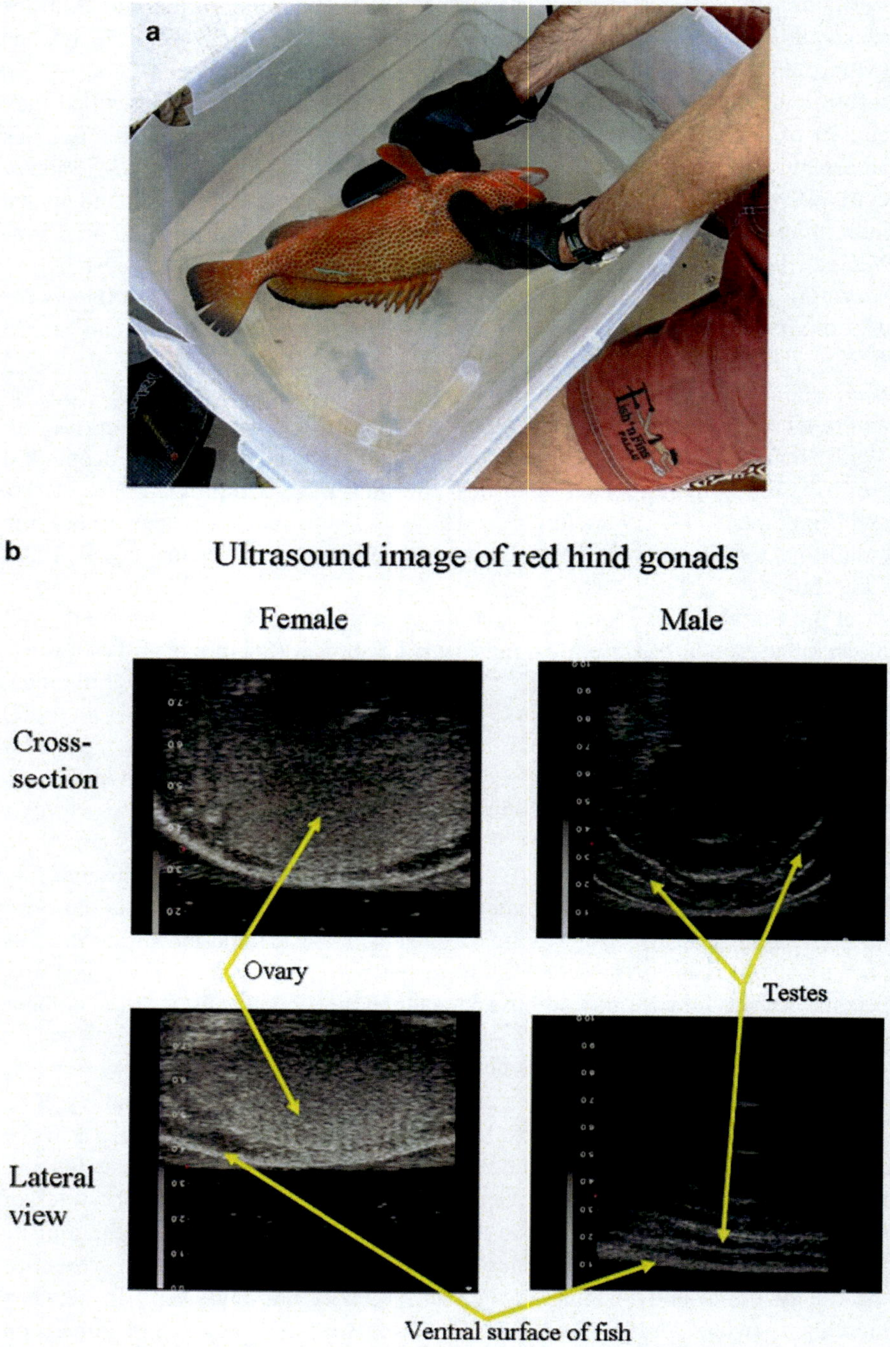

Fig. 12.3 (**a**) Tagged red hind being checked for gonad condition using ultrasound, and (**b**) cross-sectional and lateral views of red hind ovaries and testes as displayed by portable Teratech Terason 2000 ultrasound imaging system (Photos: Richard Nemeth)

Fig. 12.4 (**a**) Male red hind displaying at an aggregation site. Note the temporary barred patterns on the lips and postero-ventral areas seen at this time. (**b**) Female with swollen abdomen full of hydrated eggs just prior to spawning (Photos: Richard Nemeth, Charles Arneson)

12.3.3 Fishery and Management

Hinds and groupers (Serranidae) form a valuable component of reef fisheries throughout the Caribbean (Thompson and Munro 1978). The red hind, in particular, contributed 67–99% of the total fin fish catch in the US Virgin Islands (USVI) between 1987 and 1991 (Cummings et al. 1997) and over 30% of the grouper catch in Bermuda (Luckhurst 1996). Fishing of spawning aggregations can have a wide range of effects on target species and local fisheries. Most severe is the documented

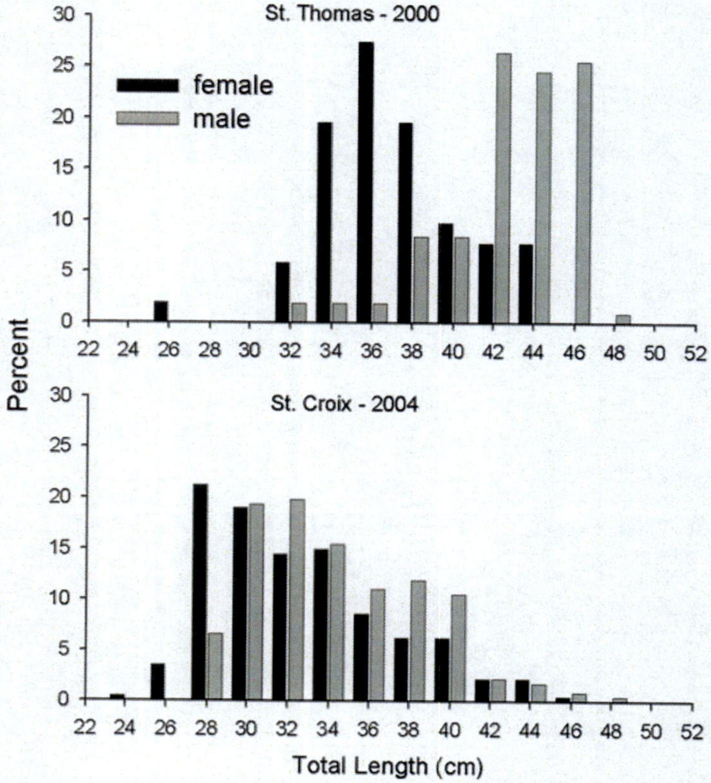

Fig. 12.5 Size frequency distribution of male and female red hind from spawning aggregations in St. Thomas (mean size: male = 41.8 cm, female = 36.2 cm) and St. Croix (mean size: male = 33.4 cm, female = 31.7 cm) following 10 years of seasonal closure. Note lack of bimodal size distribution in St. Croix spawning population

collapse of many spawning aggregations (Olsen and LaPlace 1979; Beets and Friedlander 1992; Colin 1992) and the resulting commercial extinction of several important species of groupers throughout the Caribbean (Luckhurst 1996; Huntsman et al. 1999; NMFS 1999). Unregulated fishing on aggregations may have contributed to a 65–95% decline of commercial grouper landings in Puerto Rico and Bermuda, respectively (Sadovy and Figuerola 1992; Sadovy et al. 1992; Luckhurst 1996).

Less dramatic, but equally important, are the subtle effects of fishing on spawning aggregations of protogynous species such as decreased average fish size, smaller size at sexual transformation and altered male:female sex ratio to an increased female bias (Coleman et al. 1996). For example, males of many grouper species, including red hind, remain higher in the water column defending spawning territories or displaying to potential mates (Colin et al. 1987; Gilmore and Jones 1992). Sex and/or size-selective fishing mortality at spawning aggregations will occur when these behaviourally more dominant males take baited hooks more frequently than females (Gilmore and Jones 1992; Levin and Grimes 2002). Reduced male abundance and mean size may result in sex change of females at a smaller size (Fig. 12.5) resulting in lower fecundity and reproductive output of the spawning population (Whiteman et al. 2005; Nemeth et al. 2006). For example, Whiteman et al. (2005) found that a 41 cm female has nearly 8 times greater fecundity than a 31 cm female (1,566,120 vs. 199,840 eggs).

In the Caribbean different management measures have been used to protect red hind spawning aggregations. These include permanent closures (St. Thomas, Bermuda), seasonal closures (St. Croix), and seasonal market closures (British Virgin Islands, western Puerto Rico). Not enough population information exists on the effects of these different management actions on sustaining red hind spawning aggregations and local populations. However, several studies in the USVI indicate that seasonal closures may improve population characteristics (length, sex ratio) of some aggregations but not others, whereas large permanent closures can facilitate rapid recovery of a spawning population and improve the catch of the local fishery (Nemeth 2005; Nemeth et al. 2006; Pickert et al. 2006).

12.4 Atlantic Goliath Grouper – *Epinephelus itajara*

Beatrice P. Ferreira, Mauricio Hostim-Silva, Athila A. Bertoncini, Felicia C. Coleman, and Christopher C. Koenig
Departamento de Oceanografia, Universidade Federal de Pernambuco, Av. Arquitetura s/n, Cidade Universitaria Recife, Pernambuco, Brazil, CEP 50740-550
e-mail: beatricepadovani@yahoo.com.br

Universidade Federal do Espírito Santo, DCAB, CEUNES, Rodovia BR 101 Norte, Km 60, Bairro Litorâneo s/n, São Mateus, ES, CEP 29.932-540
e-mail: mhostim@uol.com.br

Universidade Federal Fluminense, PPG- Biologia Marinha, ECOPESCA, Outeiro São João Batista s/n, Niterói, RJ, Brazil, CEP 24020-141 C.Postal 100.644
e-mail: athilapeixe@gmail.com

Department of Biological Science, Florida State University Coastal and Marine Laboratory, 3618 Coastal Highway, St. Teresa, FL 32358, USA
e-mail: coleman@bio.fsu.edu; koenig@bio.fsu.edu

12.4.1 General

Lichtenstein (1822) first described the goliath grouper as *Serranus itajara* (Family Serranidae) in a publication on the natural history of Brazil. In 1884, Jordan proposed its inclusion in the genus *Epinephelus* (Bloch 1793), sub-family Epinephelinae. Recently, Craig and Hastings (2007) suggested raising the sub-family status to family (Epinephelinae) and Craig et al. (2009), revealed through DNA sequencing that goliath grouper populations actually represent two distinct species: one in the eastern Pacific with the proposed new specific designation *E. quinquefasciatus* Bocourt and at least one in the Atlantic – currently subsumed under *E. itajara* (Lichtenstein).

The translation of its name to "Lord of the Rock" ("ita"= rock, "jara" =lord), provides a vivid description of this magnificent, top-level predator. Since its initial description, the English common name was shifted from giant grouper, to jewfish (Nelson et al. 2001), to goliath grouper by the Committee on Names of Fishes of the American Fisheries Society and the American Society of Ichthyologists and Herpetologists, and in the recent split by Craig et al. (2009) to the Atlantic goliath grouper and the Pacific goliath grouper. The giant grouper is now applied to the Indo-Pacific species, *E. lanceolatus*.

This profile addresses only the Atlantic species, which occurs in tropical waters of the western and eastern Atlantic Ocean. In the western Atlantic, it ranges from North Carolina to southern Brazil, including the Gulf of Mexico and the Caribbean Sea. In the eastern Atlantic, its distribution extends from Senegal to Congo, although it is rare in the Canary Islands.

12.4.2 Biology

The goliath grouper is one of the largest reef fishes on earth, and the largest grouper in the western Atlantic, reaching over 2 m in length and nearly 450 kg in weight (Heemstra and Randall 1993). The oldest individual known was 37 years old (female) and 26 years (male) (Bullock et al. 1992), but when that study was conducted, the population was already heavily overfished, so individuals may live even longer.

Despite their large size, goliath groupers feed at a rather low trophic level, preying largely on crustaceans (e.g. crabs, spiny lobsters and shrimp), slow-moving, typically bottom-associated fishes (e.g. stingrays, catfish, toadfish, burrfish and cowfish), but very few, if any, snappers or groupers. They may incidentally take octopus, and young sea turtles (Randall 1967). Goliath groupers ambush their prey, combining rapid lunges with suction feeding to engulf prey and swallow them whole. Their small sharp teeth prevent prey escape while their pharyngeal teeth serve to ratchet prey to the stomach.

These fish have ontogenetically distinct habitat preferences, including an association with mangrove coastlines as juveniles and with offshore reefs as adults (Koenig et al. 2007). Males and females form spawning aggregations, making spawning ascents from the reef while releasing gametes into the water column. The fertilized eggs then disperse in water currents, and are presumed to develop in a manner similar to that described for other *Epinephelus* species (see Colin et al. 1996; Sadovy and Eklund 1999), wherein fertilized eggs hatch within two days and develop into kite-shaped larvae with elongated second dorsal-fin and pelvic-fin spines. Pelagic larvae are transported from offshore sites to inshore mangrove nursery habitat over a 35–80 day period, after which they metamorphose into benthic juveniles at ~ 15 mm SL (Monica Lara, Jennifer Schull, David Jones and Robert Allman unpublished data). The benthic juveniles occupy mangrove leaf litter initially before taking up residence along mangrove shorelines for the duration of their extended (5–6 years) juvenile phase (Koenig et al. 2007; Lara et al. 2009). Rarely are they found in other habitats, although they occasionally appear in seagrass beds that are in close proximity to mangroves.

In southwest Florida juvenile goliath grouper migrate from mangrove nursery habitats as they mature (Bullock et al. 1992; Koenig et al. 2007) and reach sizes approaching 1,000 mm total length (TL), moving either to reefs in the immediate offshore area, north toward the Florida Panhandle, or south through the Florida Straits to the east coast of Florida (Koenig et al. 2007; CCK unpublished data). This pattern is borne out by tag returns and regional diving surveys of adult goliath grouper abundance that suggests density-dependent factors as a mechanism driving adults out from their centre of abundance. Adults re-enter mangrove habitats occasionally; this is one of few groupers known to venture into brackish waters as adults. In Brazil, large juveniles and adults have also been observed in Fernando de Noronha and Atol das Rocas, two offshore islands, where there are no mangrove formations (Ferreira et al. 2006). A shallow lagoon, at least in the case of Atol das Rocas, may be a nursery area.

Both juveniles (Koenig et al. 2007) and adults (Sadovy and Eklund 1999; Koenig et al. 2011) have limited home ranges, within which they exhibit some degree of territoriality, displaying to intruders with an open mouth, quivering body (Sadovy and Eklund 1999), and a booming sound generated by contraction of muscles attached to the swim bladder (Mann et al. 2009). The sound is apparently used for intra- and inter-specific communication. Tagging studies have shown that individuals can move over 150 km. In adults most of these movements appear related to travel to and from spawning sites or feeding sites. However, juveniles leaving the mangroves may also move hundreds of km, apparently in response to density-dependent factors on offshore reefs (Koenig et al. 2011; Pina-Amargós and González-Sansón 2009).

12.4.3 Reproduction and Aggregation

Smith (1965) classified all *Epinephelus* species as protogynous hermaphrodites. However, neither goliath grouper nor Nassau grouper (*E. striatus*) (Sadovy and Colin 1995) exhibit good evidence for functional hermaphroditism either in population or gonadal structure, based on criteria established by Sadovy and Shapiro (1987) and elaborated in Sadovy de Mitcheson and Liu (2008). Although bisexuals have been found in the goliath grouper population of the southeastern US (CCK and FCC unpublished data), the nature of this hermaphroditic pattern is unknown as neither the female – male sex ratio (ranging from 1.75:1 to 1:1) nor the size distribution is skewed. Indeed, small males occur in the population (Bullock and Smith 1991). Further, goliath grouper exhibit no definitive sexual dimorphism in body shape or colour (Bullock et al. 1992), though Colin (1994) noted a presumed spawning colour pattern on presumed males. Size and age of sexual maturation are similar between the sexes: male goliath grouper from the eastern Gulf of Mexico reach sexual maturity by the time they reach 1,155 mm TL, with 50% mature by the age

Fig. 12.6 Adult goliath grouper in a small spawning aggregation in Florida (Photo: © Walt Stearns)

of 5–6 years, and 100% mature by the age of 7 years. Females mature by the time they reach 1,225 mm TL and 6 years of age (Bullock et al. 1992).

The Atlantic goliath grouper is one of the few large groupers that form spawning aggregations on shallow (<50 m) offshore reefs (Coleman and Koenig 2003), especially those with high relief rocky ledges, or wrecks and artificial reefs with large holes and caves. Spawning aggregations can consist of up to 100 or more individuals and are spatially and temporally consistent – that is, spawning occurs at the same time from year to year, and fish exhibit strong inter-annual spawning site fidelity (Fig. 12.6).

Goliath grouper start forming spawning aggregations in the summer that persist through fall in the northern hemisphere, with perhaps some geographic variation in timing. Reproductive condition suggests a somewhat contracted spawning period in the northeastern Caribbean during July and August (Erdman 1976) and more protracted spawning in the Gulf of Mexico from June through December (Bullock et al. 1992), whereas acoustic recordings of spawning aggregations suggest that spawning occurs from early August to late October (Mann et al. 2009). The spawning period at any one site may persist for a few months each year and likely represents the total annual reproductive output (Sadovy and Eklund 1999).

There is some suggestion of lunar influence on the timing of spawning (Colin 1994; Hostim-Silva et al. 2005; Gerhardinger et al. 2006; Mann et al. 2009). While Colin (1994), Gerhardinger et al. (2006), and Hostim-Silva et al. (2005) assumed spawning periodicity was related to full moons, based on reproductive condition, Mann et al. (2009) suggest that spawning takes place on moonless nights, based on results from active and passive acoustic tracking [the species-specific low-frequency (50–100 Hz) pulsed sounds produced at night exhibit a lunar periodicity that is apparently associated with reproductive behaviour, Mann et al. 2008]. CKK and

James Locascio (unpublished data) verified this by collecting fertilized goliath grouper eggs (DNA verification by Matthew Craig, University of Puerto Rico) on two spawning aggregations and also collected eggs on all phases of the moon off central east Florida.

12.4.4 Conservation and Management

All of the information on the status of goliath grouper populations comes from a combination of local and scientific knowledge. It is primarily the former that revealed how scarce goliath grouper populations had become throughout their range, with many fishers noting that large aggregations and large individuals had all but disappeared from many sites, leaving only remnants of historical sites and only local knowledge of what had been present before (details below). This has gone on for so long that the historical baseline has shifted significantly as local knowledge erodes and older fishers are replaced by younger ones without similar experiences. This leaves photographic and newspaper records to provide the important historic perspective. McClenachan (2009) used these to great effect to show the dramatic declines occurring in the south Florida goliath grouper population since the early part of the twentieth century.

The decline of the Atlantic goliath grouper populations can be attributed to a combination of overfishing and loss of nursery habitat. Overfishing by commercial and recreational fishers occurred rapidly because the characteristics of these fish – their longevity, aggregating behaviour, and sedentary nature – combined with technological advances in positioning gear over the last 30 years – made them particularly vulnerable to fishing pressure (Ferreira and Maida 1995; Sadovy and Eklund 1999; Coleman et al. 2000). Many aggregations have evidently disappeared throughout the western Atlantic (Sadovy and Eklund 1999; Graham et al. 2008). In South Brazil, historical aggregations once having at least 60 individuals have dwindled to only a dozen fish, according to local knowledge (Gerhardinger et al. 2006). The coupled effects of overfishing and the loss of mangrove habitat extent and quality have probably accelerated population declines. Much of the habitat loss is due primarily to direct destruction from aquacultural, agricultural, and urban development, and from mosquito control impoundments, while water quality declined as a result of upland agricultural and urban runoff, and fresh water diversions (discussed in Koenig et al. 2007).

These population declines have resulted in fishery closures in both Brazil and the United States, and led the IUCN to list goliath grouper on the Red List as critically endangered, a category reserved for the most threatened and impacted species (IUCN 2010). Despite these closures, a significant bottleneck to complete recovery and sustainability continues without similar protection for mangrove habitat (Koenig et al. 2007).

There is a bright spot in this sea of bad news for goliath grouper. Indeed, fishing mortality in the United States has declined (by 50–90%) as a result of the fishery closure (Kingsley 2004; Porch et al. 2006) and the population in the southeastern

United States now shows clear signs of growth in both the juvenile (Koenig et al. 2007; Cass-Calay and Schmidt 2008) and adult populations (Kingsley 2004; Porch et al. 2006; Koenig et al. 2011). For instance, an aggregation off the southeast Florida coast that existed in the 1960s disappeared in the 1980s, to reappear in 2005 with about 65 individuals (CCK, personal observation). This is a good start, but it is insufficient to recover a population that is highly vulnerable to fishing. There remains considerable uncertainty about the effects of illegal fishing, catch-and-release mortality, and habitat loss associated with the US population.

Successes are even rarer when considered over the entire geographic range of this species. In both Belize and Brazil, for instance, juvenile populations have been heavily fished and showing severe signs of decline (FCC, CCK and Buddy Powell unpublished data; Ferreira et al. 2006). In spite of a 7-year moratorium on fishing the species in Brazil, there are no clear signs of recovery. The problem is intensified by the pressure exerted on juveniles, known as "meretes", which fishers do not recognize as having the same protected status as adults. Recent campaigns to raise awareness about the endangered condition and the protected status of the species are becoming more prevalent (www.merosdobrasil.org) and are hoped to have a positive effect.

There is pressure in the United States from some commercial and recreational fishermen to reopen the fishery for the goliath grouper, at least at a limited level. Uncertainty suggests proceeding in a precautionary manner. A short-term fishery benefit is easily exhausted in a vulnerable species. Further, we know little about the historical ecological role of goliath grouper as a dominant component of the reef communities, and the toll that fishing and habitat destruction may have taken (McClenachan 2009). Management practices should consider the potential for developing non-consumptive industries (keeping goliath grouper "on the fin"–alive) as well as considering the ecological contributions this species bestows on the marine communities within which it lives.

12.5 Camouflage Grouper – *Epinephelus polyphekadion*

Kevin L. Rhodes
College of Agriculture, Forestry and Natural Resource Management,
University of Hawaii at Hilo, 200 W. Kawili St., Hilo, HI, USA
e-mail: klrhodes_grouper@yahoo.com

12.5.1 General

The camouflage grouper, *Epinephelus polyphekadion* (=*microdon*) (Bleeker 1849), is a widely distributed Indo-Pacific serranid with life history characteristics typical of the sub-family Epinephelinae. The species ranges from the eastern coast of Africa through the Red Sea to the islands of the eastern Pacific, south to Lord Howe Island

and north to the Ryuku Islands of Japan (Heemstra and Randall 1993). Adult camouflage grouper are commonly associated with coral reefs in clear water and are solitary during non-reproductive periods. Detailed habitat use is unknown, but fisheries-dependent data suggest that adults prefer inshore reef areas, with the species apparently less common on outer reefs during non-reproductive periods (Rhodes and Tupper 2007). Like many other epinepheline groupers, the camouflage grouper is a top carnivore that preys on portunid crabs, other crustaceans, fishes, and molluscs (Parrish 1987; Michel Kulbicki, Yves Letourner and Pierre Labrosse unpublished data).

Camouflage grouper is reported to reach 90 cm total length (TL) (Heemstra and Randall 1986; Bouhlel 1988). The maximum age for the species is 31 years in both the Seychelles (Grandcourt 2005) and New Caledonia (Brett Taylor and KLR unpublished data), while in Pohnpei Rhodes et al. (2011) found the maximum age to be 22 years.

Several studies have investigated the early life-history of camouflage grouper under hatchery conditions (AQUACOP et al. 1989a; Al-Thobaity and James 1996; Tamaru et al. 1996; James et al. 1998). Fertilised eggs ranged from 0.73 to 0.85 mm in diameter and are transparent, positively buoyant (34–35 ppt) with a single oil droplet (AQUACOP et al. 1989a; Al-Thobaity and James 1996; Tamaru et al. 1996). Hatching occurred within 18–19 h at 28–30°C in Saudi Arabia and 20–21 h at 27°C in Palau. Newly hatched larvae ranged from 1.55 to 1.71 mm TL (James et al. 1998). No published reports of wild larvae are available. From published photographs, elongate fin spines and head spines were apparent, typical of serranid larvae studied to date (Al-Thobaity and James 1996). Larval transformation (post-flexion) occurred at 40 days [average size = 34.4 ± 4.9 mm TL at 45 days] and after 15 months juveniles reportedly grew to 800 g (0.28 ± 0.11 g $fish^{-1}$ day^{-1}) (James et al. 1998). In Palau, post-settlement camouflage grouper were observed in low branching corals with bushy macroalgae, massive coral heads and coral rubble. Post-settlement individuals appeared to become more cryptic as they grew and seemed to prefer more turbid waters in deeper rubble areas and coral heads (Tupper 2007). In Pohnpei, post-settlement individuals (~ 10 cm TL) were taken by hook and line from rubble patches in turbid waters near mangroves (KLR personal observation).

12.5.2 Reproduction and Aggregation

Sexual pattern in the camouflage grouper was initially described as protogynous, based on combined histochemical and histological techniques (AQUACOP et al. 1989b; Debas 1989; Duslé-Sicard et al. 1992) or as uncertain (Rhodes and Sadovy 2002a). More recently, Rhodes et al. (2011) used age and growth and histological gonad assessments to confirm functional gonochorism for the species, with the potential for sexual transition.

Although there is often substantial size overlap between sexes, adult males are sometimes larger than adult females (Rhodes and Sadovy 2002a; Robinson et al.

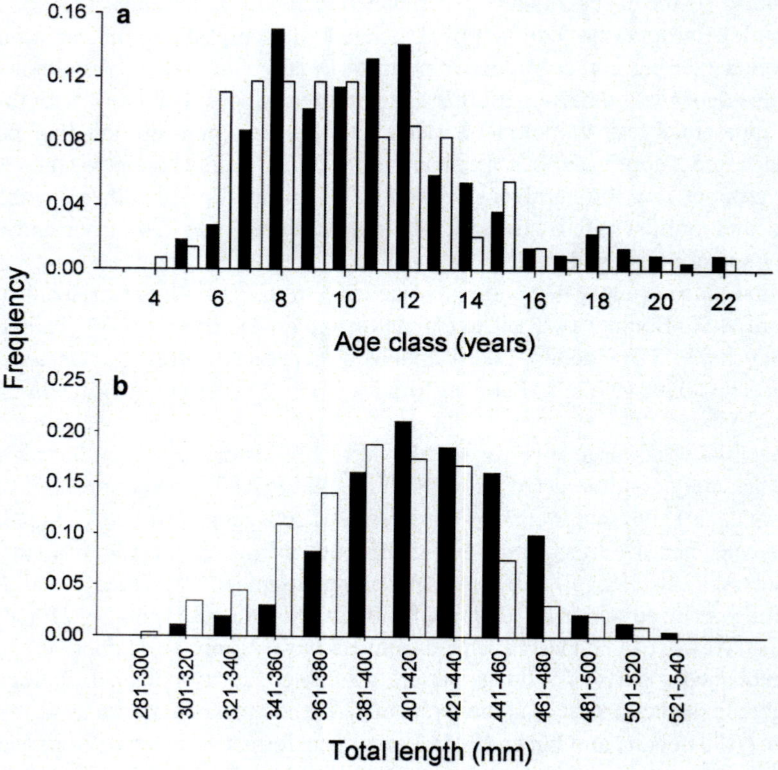

Fig. 12.7 Sex-specific (**a**) age-frequency and (**b**) length-frequency distributions taken from spawning aggregations in 1998 and 1999. *Black bars* represent males (n = 221 [**a**] and 360 [**b**]) and *white bars* represent females (n = 145 [**a**] and 292 [**b**]) (Rhodes et al. 2011)

2008a, b). Using sectioned sagittal otoliths from market and aggregation samples, Rhodes et al. (2011) found no significant differences between sexes for mean age or growth (Fig. 12.7). In Pohnpei, individuals enter the fishery at age 2 years, but are not found within spawning aggregations until 4 years. Growth (k) was estimated as 0.18 (Grandcourt 2005) in Seychelles and 0.25 $year^{-1}$ for Pohnpei (Rhodes et al. 2011), while natural mortality (M) was estimated at 0.13 and 0.144 $year^{-1}$, respectively. In Pohnpei, total mortality (Z) was 0.227 $year^{-1}$, based on the age-based catch curve, while the size (L_{50}) and age (t_{50}) at 50% female maturity was 327 mm TL and 5.4 years, respectively (Fig. 12.8).

Camouflage grouper aggregate to spawn and their aggregations (FSA) are reported from many areas within its range, including Micronesia (Rhodes and Sadovy 2002a), Melanesia (Hamilton et al. 2005), Palau (Johannes et al. 1999; Sadovy 2007; Chap. 12.8), Fiji (Yvonne Sadovy de Mitcheson personal observation 2009), FrenchPolynesia (Serge Planes personal communication 2010), Indonesia (Chris Rotinsulu, personal communication) and the Seychelles (Robinson et al. 2008a, b). Abundance within FSAs may number in the hundreds to many thousands

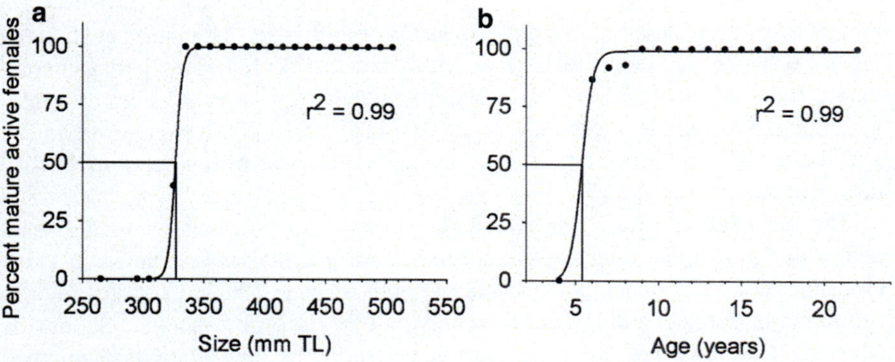

Fig. 12.8 Estimates of female maturity based on percentage of mature active females within the spawning aggregation during the spawning season. *Boxes* indicate 50% maturity estimates for (**a**) size (L_{50}) and (**b**) age (t_{50}). Sample sizes: (**a**) n = 128, (**b**) n = 125 (Rhodes et al. 2011)

Fig. 12.9 Very large spawning aggregation of *Epinephelus polyphekadion* in Polynesia (Photo: © Paul and Paveena McKenzie www.wildencounters.net)

of individuals (Fig. 12.9). Estimates of over 10,000 individuals have been reported from a single spawning aggregation in Pohnpei from underwater census surveys (Rhodes 1999). More recent estimates at the same site found only a few hundred individuals following intense fishing starting in 1999 (Rhodes et al. 2011), which probably also contributed to the recently observed size and age truncation within the population. Estimates from camouflage grouper FSAs in Palau and FrenchPolynesia ranged from a few hundred to many thousands of individuals (Johannes et al. 1999; Palau Conservation Society 2010; Paul McKenzie and Serge Planes, personal communication), while approximately 1,900 individuals were reported from an aggregation in the Seychelles (Robinson et al. 2008a, b), about 200 from one site in Fiji and about 500 at a second site in Fiji (Yvonne Sadovy de Mitcheson unpublished data; Chap. 12.8, Fig. 12.15).

The seasonal and lunar timing of aggregation formation and spawning varies across the geographic range of the species, from just before new moon in Palau (Johannes et al. 1999) and the Seychelles (Robinson et al. 2008a, b), to just prior to full moon in Pohnpei and Fiji (Rhodes and Sadovy 2002a; Fiji-Yvonne Sadovy de Mitcheson unpublished data). In most areas surveyed, aggregations form over 2–3 months, with peaks in abundance during a single month. Variation occurs in arrival times between males and females, with males commonly present at the spawning site for longer periods and arriving prior to females (Rhodes and Sadovy 2002a; Robinson et al. 2008a, b; Chap. 8, Fig. 8.10).

Some evidence of reproductive migratory corridors has been recorded from acoustic telemetry studies. Small groups of adult camouflage grouper were observed moving along common routes toward the aggregation site prior to reproductive periods in Pohnpei from up to 8 km away from aggregation sites. In Fiji, one tagged fish was recaptured 15 km away from an aggregation site (Yvonne Sadovy de Mitcheson unpublished data). Spawning site fidelity has been demonstrated from both conventional (spaghetti) (Johannes et al. 1999; Robinson et al. 2008a, b) and acoustic tagging studies (KLR unpublished data), with inter-annual site fidelity recorded. In the Seychelles, two tagged individuals returned over two consecutive reproductive months and at least one individual showed inter-annual site fidelity 352 day after initial tagging (Robinson et al. 2008a, b). In Pohnpei, one acoustically tagged female returned to the aggregation 1 year after tagging and was present for 4 days. In Palau, multi-year tagging studies suggested that (1) individuals may return to spawning sites over multiple (2–3) months within the spawning season, (2) at least some individuals return to the same site inter-annually, (3) individuals may travel at much as 5–10 km from spawning sites between reproductive and non-reproductive periods and, (4) individuals appear not to utilize more than one spawning site during reproduction, even though a number of aggregations may form in relatively close proximity (Johannes et al. 1999).

Camouflage grouper spawning aggregations are typically found on seaward-facing portions of reefs or channels at 4–40 m depths (e.g. Rhodes and Sadovy 2002b; Robinson et al. 2008a, b). During aggregation periods, territorial disputes are frequent among males. Disputes typically involve two individuals facing each other in a cheek-to-cheek position, with occasional bouts of pushing by one individual at the other's vent or abdomen (Chap. 9, Fig. 9.6). Snout-to-snout snapping and sidling is also common within aggregations. Colour changes involve blanching to a pale grey or whitish color with a prominent black saddle, a blackish 'slash' through the eyes and darkened nostrils (Rhodes and Sadovy 2002a; Robinson et al. 2008a, b). Examination of oocyte development in Pohnpei suggests spawning occurs nocturnally between dusk and dawn (Rhodes and Sadovy 2002a, b). In Palau, observations were made at between 0530 and 0600 of pairs and small groups of six to eight groupers that moved about 2 m above the substrate and spawned together at the seaward entrance to an outer reef channel; the outgoing current had started to pick up and spawning was seen every 10 min (Jim Forrest and Jeanette Denby, personal observation 2011).

After spawning, individuals disperse from the aggregation site for several days prior to re-forming the aggregation in the subsequent spawning month. In Pohnpei, the spawning season coincides with periods of stable water temperatures between 28°C and 29°C. In Palau, the aggregation season occurs when water temperatures are between about 28.5°C and 29.5°C (Chap. 5, Fig. 5.17) and in Fiji at about 24°C (annual range at aggregation site is approximately 24–28.5°C) (Yvonne Sadovy de Mitcheson unpublished data).

For studies of adult female camouflage grouper from Pohnpei, fecundity was estimated at 1,350 oocytes g^{-1} body weight for individuals examined between 22.8 and 43.0 cm SL (403–2,618 g body weight) (Rhodes and Sadovy 2002a). Histological assessments of reproductively active female camouflage grouper showed synchronous development of eggs prior to spawning, with all or nearly all ripe oocytes spawned (Rhodes and Sadovy 2002a). Repeat spawning by individual females within a reproductive month was confirmed (Rhodes and Sadovy 2002b).

12.5.3 Fisheries, Management and Conservation

The camouflage grouper is highly prized throughout much of its range and often contributes substantially to local catches (Rhodes and Tupper 2007; Rhodes et al. 2008; Robinson et al. 2008a, b; Yvonne Sadovy de Mitcheson, personal observation). The species is taken with a wide variety of gears, but is most targeted by spear or hook and line, and often at spawning aggregations or along reproductive migratory pathways (e.g. Johannes et al. 1999; Robinson et al. 2008a, b; Rhodes et al. 2011). Camouflage grouper are sought after by local subsistence and commercial fishers, as well as by foreign commercial fishing entities that include the Southeast Asia-based live reef food fish trade, centred in Hong Kong, to which the species contributes substantially (Sadovy et al. 2003a, b). Aggregation-fishing has led to a reduction of spawning aggregation abundance in many places and the likely loss of spawning aggregations in some areas (Sadovy and Domeier 2005; Rhodes et al. 2011). Impacts to the age and size distribution of camouflage grouper were shown in Pohnpei following only 1 week of intense aggregation fishing (Rhodes et al. 2011). In some areas, camouflage grouper are found to be ciguatoxic (Myers 1999; Michel Kulbicki, personal communication). The species can be hatchery-reared but this does not occur commercially.

Currently, camouflage grouper is listed as near threatened on the IUCN Red List (Russell et al. 2006). Management specifically directed at the species was introduced in Pohnpei and Palau focusing on spawning sites and during reproductive periods. Australia has general restrictions on the amount of catch of groupers (e.g. Northern Territory) and species-specific limits on both the minimum (50 cm TL) and maximum sizes (70 cm TL) that can be taken (i.e. Queensland). Some additional protection may be provided by no-take zoning restrictions within certain sections the Great Barrier Reef Marine Park. Bans or restrictions on the live reef fish trade are in place in some areas of its distribution (e.g. Solomon Islands, Palau).

For directed management of camouflage grouper, sales (Pohnpei, Palau) and catch bans (Palau only) are in place during all or most of the reproductive season. No-take (marine) protected areas (MPA) specific to spawning aggregation sites are in place in both Palau and Pohnpei. In Palau, fishing is not permitted during the reproductive season and some aggregation sites are protected from fishing as a result of either national or state legislation, or due to traditional fishing bans (*bul*) (Johannes et al. 1999). In Pohnpei, two large spawning aggregations are protected as no-take zones. Catch and sales bans during reproductive periods and seasonal (reproductive) or permanent no-take MPAs centered on aggregation sites can be particularly effective management tools for addressing overfishing of the species. Small-scale MPAs will likely be less effective than larger-scale no-take zones that incorporate migratory corridors and home range habitats. Where practical, management should also limit the catch of juveniles by setting size restrictions. Exports, including those to the live reef fish trade, are generally discouraged unless information on sustainable yield is available and strong controls are in place to limit or restrict the targeting of reproductively active fish.

Although camouflage grouper are important components of subsistence and commercial fisheries, many details of its life history are still lacking for management, such as habitat use that includes home range area, habitat shifts during development (ontogeny) and patterns of movement during reproductive life history. Seasonal catch and sales bans, while useful in protecting target species, such as camouflage grouper, can shift fishing pressure to other commercially important species (Rhodes et al. 2008). Therefore, such measures should be instituted under a comprehensive management strategy that considers all vulnerable species and habitats.

Ecosystem-based management strategies are encouraged, since these encompass a more holistic view of individual species needs together with reef community and ecosystem function. For aggregating grouper, such as camouflage grouper, monitoring and enforcement challenges are often daunting owing to the dynamic nature of populations during reproduction, oftentimes wide distribution of spawning sites within management jurisdictions and typically substantial resource requirements for monitoring and enforcement, especially for those sites far from shore. Some challenges can be overcome by focusing on sites of highest vulnerability, utilizing traditional knowledge (once validated) of reproductive periods to establish fishing restrictions and identifying practical points of enforcement and monitoring interdiction, such as at fish markets or landing sites. Precautionary, adaptive, traditional and co-management approaches offer a wider range of options and may provide greater potential for success through increased monitoring and enforcement capacity and from greater flexibility in responding to changes in fish population or ecosystem status.

12.6 Nassau Grouper – *Epinephelus striatus*

Yvonne Sadovy de Mitcheson, Scott A. Heppell, and Patrick L. Colin
School of Biological Sciences, The University of Hong Kong,
Hong Kong, China
e-mail: yjsadovy@hku.hk

Department of Fisheries and Wildlife, Oregon State University, Corvallis, OR, USA
e-mail: scott.heppell@oregonstate.edu

Coral Reef Research Foundation, P.O. Box 1765, Koror, Palau 96940
e-mail: crrf@palaunet.com

12.6.1 Background

The Nassau grouper, *Epinephelus striatus* (Family Serranidae) ranges from Bermuda and Florida (USA), throughout The Bahamas and Caribbean Sea. The previous report of *Epinephelus striatus* from the Brazilian coast south of the equator is unsubstantiated (Heemstra and Randall 1993; Craig et al. 2011). While not among the largest of all groupers, the species is sizeable, reaching about 25 kg and 120 cm total length (TL) (Sadovy and Eklund 1999). The Nassau grouper was once very important commercially in the insular Caribbean. Because of severe overfishing throughout its range, as described below, there are few places where commercial catch of Nassau grouper is still viable. The species is listed as endangered on the IUCN Red List (Albins et al. 2009).

12.6.2 General Biology and Early Life History

The Nassau grouper occurs in clear waters around high relief coral reefs and over rocky substrate down to at least 130 m. Adults feed primarily on fishes and crustaceans, with piscivory becoming more important as they grow (Randall 1965). The species reaches at least 29 years (Bush et al. 2006) with von Bertalanffy growth parameters of $L_{inf}=97.4$ cm standard length (SL); K 0.063–0.185; log ungutted weight (g) = −4.67 + 3.03 log SL (mm); TL (mm) = 2.24 + 1.11SL (mm) (population parameters are reviewed in Sadovy and Eklund 1999).

Genetic data indicate high gene flow among samples taken in Florida, Cuba, Belize and The Bahamas (Sedberry et al. 1996), thus it is presumed that Nassau grouper maintains a single, genetically homogeneous population. A mitochondrial DNA analysis of samples collected recently from within the Cayman Islands (Alexis Jackson unpublished data) also failed to demonstrate anything other than a homogeneous stock structure. However, these studies are not conclusive and await the application of more sensitive genetic methods.

The Nassau grouper is known to spawn in large aggregations but its social behaviour outside of aggregations is little studied due to a lack of areas in its range with significant populations. Investigations around Lee Stocking Island in the Exuma Chain of The Bahamas found that the seaward area of the island, bordering on deep Exuma Sound, had a relatively large undisturbed population (Colin 1992). A number of specific individuals at this location were regularly surveyed, each fish identifiable by individual variations in the markings on the body. Most individuals moved over several hundred m^2 and many had overlapping home ranges with a social hierarchy. When a dominant fish encountered a smaller conspecific the latter would quickly adopt the "submissive" bicolour pattern seen among aggregated fish as the two fish approached each other (Colin 1992). Since Nassau grouper have no external sex differences, it is impossible to ascribe a sex to the individuals observed. During the spawning season the largest fish within the study area disappeared and was not seen for 4 weeks. Another, smaller individual had apparently taken over as the dominant individual during the absence of the larger fish and did not allow the latter to regain its position. There ensued for several weeks a continuing struggle between these two fish manifested by aggressive encounters whenever they met, and both fish displayed tears and scars from their encounters (PLC, personal observation). Similar observations of change to a submissive colour pattern when transiting another Nassau's territory were made off Little Cayman Island (SAH and Brice Semmens, personal observation).

The relationship of spawning and larval development to the physical environment is detailed in Chaps. 5 and 7. Pelagic eggs are produced in the aggregation and after a pelagic larval (Chap. 7, Fig. 7.5) period of about 41–43 days (range 37–45) settlement occurs (Colin et al. 1997). Most information on Nassau grouper larval recruitment comes from Lee Stocking Island in The Bahamas during the period 1989–1991 (Colin et al. 1997; Shenker et al. 1993). Colin et al. (1997) deployed moored nets in a tidal channel in early 1989 and 1990 and successfully captured numerous recruiting Nassau grouper on the new moons of February 1989 and 1990. While Shenker et al. (1993) found an association of recruitment through tidal channels with a winter storm event, other years (Colin et al. 1997) did not demonstrate the same relationship of atmospheric conditions with recruitment. Daily otolith increments back-dated to full moon spawning in December and January, which is consistent with the known aggregating season for the species in The Bahamas. Recent satellite-based current drifter tracking in the Cayman Islands indicates that not only are there eddies that form over the aggregation in the days immediately after spawning that can potentially retain larvae close to where they are produced, but that long-term patterns indicate a high likelihood of pre-settlement fish being returned over the ~40 day pelagic larval duration (Fig. 12.10) (Heppell et al. 2008). This suggests that local spawning and its timing may be critical features associated with the long-term sustainability of local populations.

The habitats into which newly settled Nassau grouper recruit have received some attention, but because of the cryptic nature of small grouper recruits, it is risky to make too many assumptions based solely on habitats that can be readily sampled for recruitment versus less accessible ones. Eggleston (1995) sampled habitats on the Bahama

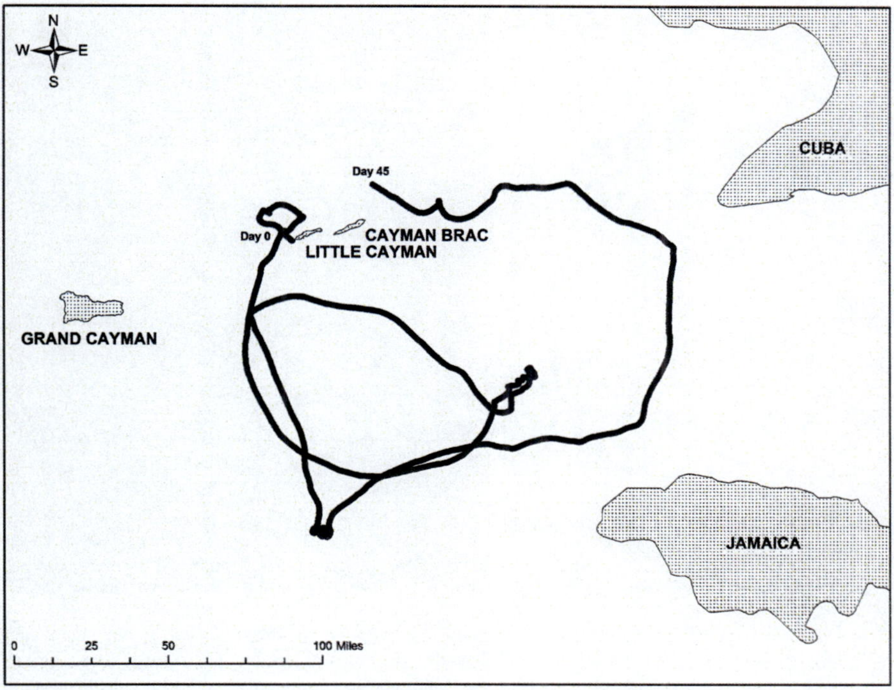

Fig. 12.10 Path of a Pacific Gyre (trademark) Surface Velocity Profile drifter released from the west end of Little Cayman Island ("Day 0") and tracked for a period of 45 days ("Day 45"). The drogue was centered at a depth of 10 meters (SAH unpubl. data)

banks and concluded that coral-algal areas are the preferred habitat for settling Nassau grouper. In the same area offshore, Colin et al. (1997) documented the occurrence of newly settled Nassau grouper juveniles in rubble mounds by sand tilefish (*Malacanthus plumieri*). Settlement inshore in macroalgae was reported by Dahlgren and Eggleston (2001). Aguilar-Perera (2006) reported young Nassau grouper (60–80 mm TL) associated with seagrass patches, around an old tyre and in a dissolution hole in shallow bedrock in Mexico. Claydon et al. (2009) demonstrated use of discarded conch shells in shallow sand areas surrounding the Turks and Caicos Islands. There is currently no evidence to indicate which settlement habitat might be preferred, and only with comprehensive and destructive sampling methods such as rotenone could we determine whether the species settles into more cryptic reef habitats.

12.6.3 Reproductive Biology

Details of reproduction and aggregation activity are known from several areas of the greater Caribbean (Colin 1992; Sadovy and Colin 1995; Whaylen et al. 2004; 2007; Sala et al. 2001; Aguilar-Perera 2006; Starr et al. 2007). The Nassau grouper has a

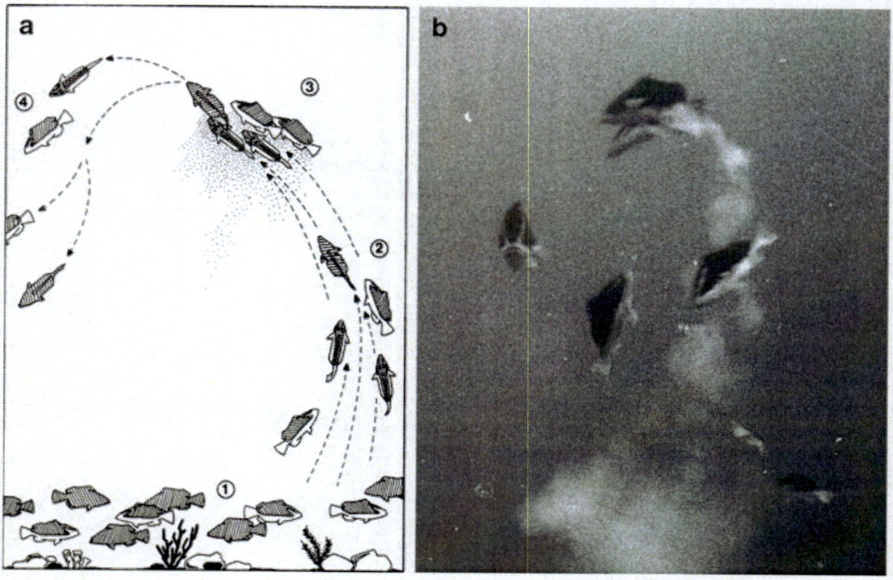

Fig. 12.11 (**a**) Spawning in the Nassau grouper, *Epinephelus striatus*: (*1*) fish move into water column; (*2*) small group rises led by dark female and bicolour males; (*3*) release of gametes; (*4*) return to substrate. (**b**) Group-spawn in the Bahamas (Reproduced from Sadovy (1996). With permission from Chapman & Hall publishers. Photo: Patrick L. Colin)

complex sexual and social system; outside of the spawning season, adult Nassau grouper are relatively sedentary and solitary. During spawning, Nassau grouper aggregate in large groups with reproduction occurring within the large aggregations by the formation of small clusters of fish comprised of one or possibly several, dark-coloured females followed by many bicoloured fish, probably males (see Frontispiece, Chap. 7 Fig. 7.3, Fig. 12.11). The group rises up into the water column, releases large clouds of egg and sperm and rapidly returns to the substrate. Spawning occurs during a narrow time window at dusk (Colin 1992).

The Nassau grouper is somewhat unusual amongst groupers in being a gonochoristic species that spawns in large (as much as tens of thousands of fish) aggregations. While it is possible that large females might be able to change sex to males and that the species can, therefore, exhibit diandric protogyny, few large fish evidently remain due to overfishing. Sexual maturity for both sexes is reached between 40 and 45 cm TL and at about 4 years of age (Sadovy and Colin 1995; Sadovy and Eklund 1999). Reproductively active males in aggregations have some of the largest testes relative to body size of any fish known, with the gonadosomatic index (GSI) approaching 10% and thus matching that observed for ovarian weight in females. High male GSI and lack of protogyny, combined with androgen profiles that indicate low levels of aggressiveness (SAH unpublished data) and behavioural observations that confirm this (Colin 1992), collectively, strongly suggest sperm competition as the reproductive strategy adopted by this species.

The Nassau grouper was the first species scientifically documented to spawn in spectacularly abundant aggregations (Smith 1972). Although the species has never been observed to spawn outside of aggregations, this possibility cannot be excluded (Sadovy and Colin 1995). Although the occurrence of aggregations had long been known (e.g. Vilaro Diaz 1884; Sadovy 1993b), until publication of Smith's (1972) remarkable underwater photos and description of a spawning aggregation in The Bahamas, estimated at between 30,000 and 100,000 fishes, the number of fish that gathered was not quantified. The smallest aggregation of Nassau grouper observed to spawn numbered about 20 fish (Colin 1992) and only a single spawning, with a few individuals participating, was seen in these fish. The group was observed for some time without other evidence of spawning. It may be that the presence of large numbers of fish (or greater than some threshold number) in an aggregation may be important for inducing spawning amongst clusters of fish from within the aggregation (Sect. 8.4.2). The Grouper Moon project, run by the Reef Environmental Education Foundation (REEF), has also documented (Brice Semmens, personal observation) that aggregation dynamics change for depleted aggregations. Fish at a small aggregation on Cayman Brac (~1,500 fish) aggregated for a longer period of time than those at a larger aggregation on an adjacent island, and larger fish aggregate for longer than smaller fish. This means that fish in aggregations that are being depleted potentially become more vulnerable to fishing because they are present for a longer period of time, and the largest, most reproductively valuable fish are the most susceptible.

Aggregations occur at a predictable time and place each year around the time of the full moon, usually between December and March, although in Bermuda aggregation-spawning occurred in the northern summer period (Sadovy and Eklund 1999). Spawning in the aggregation has been documented to occur within 20 min of sunset in The Bahamas (Colin 1992) but as early as an hour before sunset to 20 min after sunset in the Cayman Islands (Whaylen et al. 2007). Individual fish are known to travel as far as 220 km to spawn (Bolden 2000), migrating in groups to aggregation sites which tend to be located close to deep waters on outer reefs (Colin 1992; Carter et al. 1994) (Fig. 12.12). Not only are spawning aggregations stable in location over many years, but individual animals are known to repeatedly use the same aggregation: uniquely patterned individual animals ("natural tags") were observed returning to the same aggregation over multiple years (Whaylen et al. 2007) and by acoustic telemetry (Semmens et al. 2006, 2009). The acoustic telemetry work demonstrates that Nassau grouper not only return to their specific aggregation between years (Semmens et al. 2009), but that individuals can return to the same aggregation to spawn in different months within a year (Semmens et al. 2006).

12.6.4 Fishery and Commercial Importance

The Nassau grouper once formed what were almost certainly the largest aggregations of any western Atlantic grouper, and perhaps any grouper in the world. Aggregation fisheries formed the basis of large and important seasonal fisheries,

Fig. 12.12 (**a**) Nassau grouper aggregated during the day when fish tend to stay close to the substrate, as in the photo (Belize). (**b**) Later in the afternoon and as spawning approaches the fish concentrate into 'balls' and move up into the water column and towards a drop-off to deeper water (Cayman Islands) (Photos: © Enric Sala © Scott Heppell)

particularly in insular and offshore areas of the Caribbean (Sadovy and Eklund 1999). However, due to over-fishing and lack of fishery management over the last 4 decades, the species is widely considered to be commercially extinct. Historically the biggest fisheries were probably in The Bahamas, Cuba and Belize because of their naturally large reef areas. In The Bahamas, one of the few countries where a Nassau grouper fishery is still active, the species once accounted for about 50% of the total annual finfish landings in the country, but the Nassau grouper has declined substantially (Sadovy 1997; Sadovy and Eklund 1999; Ehrhardt and Deleveaux 2007; Sadovy de Mitcheson et al. 2008). In Cuba the Nassau grouper fishery has evidently collapsed, relative to its former abundance (Claro et al. 2009) and in Belize landings are a tiny proportion of previous landings (Sala et al. 2001; Janet Gibson, personal communication 2007)

Only where the species is managed, or not subjected to significant commercial fisheries (as in the Turks and Caicos Islands and Cayman Islands and possibly in parts of The Bahamas) do viable fisheries apparently persist. Even in locations where they are currently managed, however, such as the Cayman Islands, there are no total quota controls, aggregation protection is not necessarily permanent, and there have been substantial declines in catch, catch per unit effort and mean body size at aggregations (Bush et al. 2006). In The Bahamas, although there is still a viable fishery, reported annual landings have dropped at least threefold over approximately 15 years (most recent data suggest about 400 tonnes-Bahamas Department of Marine Resources) and the status of most aggregations is unknown. Throughout its entire range, more than half of all known aggregation sites ($n \approx 100$ in total) are believed to be extirpated (Sadovy and Eklund 1999; Sadovy de Mitcheson et al. 2008). In the Turks and Caicos, although there is relatively little focus on the Nassau grouper at present, there is considerable consumer interest (expected to grow along with tourism) in the species and a significant proportion of fish in one recent study is already taken below sexual maturation. Therefore pressure on the Nassau grouper in this country is expected to grow, and the species is not yet managed (Landsman et al. 2009).

Some aggregations fished to extirpation have not recovered in decades (Beets and Friedlander 1992), and it is possible that aggregations persist when younger fish learn migration routes by following older fish to traditional spawning sites. When the older fish are fished out, this "traditional knowledge" may be lost and the aggregation may be incredibly slow to reform. No aggregation, once extirpated, has ever been observed to recover although there are recent indications of a possible increase in numbers at one aggregation site in the US Virgin Islands (Rick Nemeth, personal observation). If true, this means, among other things, that restocking is unlikely to be an appropriate recovery strategy for the species.

The Nassau grouper is recognized for its economic importance for dive tourism and encounters with such large groupers are very appealing to divers (Rudd and Tupper 2002). Sala et al. (2001) estimated that tourist diving on grouper spawning aggregations would produce 20 times the income produced from fishing it in Belize. While it is suggested that the impacts of divers on spawning might be acceptably low, a more detailed analysis of Nassau grouper is recommended before the impacts of divers can be understood (Heyman et al. 2010). Certainly large schools of fish just prior to spawning are influenced by divers in the water and slowly start to move away from them (YS, personal observation 2006).

Nassau grouper are fished commercially and recreationally by handline, longline, fish traps, spear guns, and gillnets. Spawning aggregations are heavily targeted, often yield most of the annual catches and are mainly exploited by handlines or fish traps, although gillnets are also used in Mexico IUCN red list assessment (Albins et al. 2003). Fishery-dependent data in the United States showed marked declines and severe fluctuations in catch rates through the 1980s resulting in a ban on retention of Nassau grouper in 1990. The species is variously protected in Belize, Bahamas, Bermuda, Cayman Islands, Cuba, Puerto Rico, United States Virgin Islands, and Mexico. In spite of such protection enforcement is wanting in most locations (Sadovy and Eklund 1999; Sala et al. 2001; Whaylen et al. 2004; Aguilar-Perera and Aguilar-Dávila 2007; Sadovy de Mitcheson et al. 2008; Aguilar-Perera et al. 2009; Claro et al. 2009; www.SCRFA.org database) (Table 12.1). There is a need for both national and regional management measures for recovery of this threatened species (García-Moliner and Sadovy 2008; Albins et al. 2009).

Table 12.1 Status of documented Nassau grouper spawning aggregations

Country	Number of aggregations known	Number no longer known to form	Status of known extant aggregations		
			Declining	Unknown	Healthy
Bahamas	18		3	15	
Bermuda	2	2			
Belize	10	2	4	4	
Cayman Islands	6	3	2	Possibly two more	1
Cuba	33	2	11	20	
Dominican Republic	1	1			
Honduras	3		1	2	
Mexico	6	1	1	4 (more reported)	
Puerto Rico (USA)	2	2			
Turks and Caicos (GB)	1			1	
US Virgin Islands *(British Virgin Islands aggregation(s) status unknown)*	2	2	One possibly undergoing recovery		
Total	84	15	22	46	1

Source: Society for the Conservation of Reef Fish Aggregations (SCRFA) Global Database (www.SCRFA.org)

Note: The SCRFA database includes only verified Spawning Aggregation sites. Many more have been mentioned in the literature, but their existence historically or persistence could not be verified by in-country experts. Therefore the compilation here represents a minimum reliable dataset on the past and present distribution of Nassau grouper spawning aggregations. Nassau grouper countries in which spawning aggregations have not been recorded or confirmed are: Anguilla (*GB*), Antigua and Barbuda, Aruba, Barbados, Colombia, Costa Rica, Dominica, Grenada, Guadeloupe (*FR*), Haiti, Guyana, French Guiana, Guatemala, Jamaica, Martinique, Montserrat, Netherlands Antilles (Curaçao), Nicaragua, Panama, Saint Kitts and Nevis, Saint Lucia, Saint Vincent and the Grenadines, Suriname, Trinidad and Tobago, USA, and Venezuela (Craig Dahlgren unpublished)

12.6.5 *Conservation, Management and Monitoring*

Sadovy and Eklund (1999) and García-Moliner and Sadovy (2008) concluded that the Nassau grouper was in urgent need of management throughout much of its geographic range, at both local and regional levels. Noting that standard fishery management measures, such as quotas and size limits are rarely in use for any species throughout the Caribbean, are typically not effective for larger reef species in multi-species fisheries, and that enforcement of regulations in tropical reef fish fisheries is typically weak, they suggested that the greatest protection for such species would likely be afforded by reduction or elimination of fishing effort or other controls on effort and gear, especially on spawning aggregations, along with properly sited marine reserves to protect important habitats for juveniles and adult fishes. This assessment was echoed by the American Fisheries Society (Coleman et al. 2000), which, in addition, advocated that no-take zones be accompanied by a reduction in total allowable catch to reduce the shift of fishing effort from these areas to adjacent ones.

Especially worrisome for the recovery of the species is the fact that (1) the number of aggregations has declined (Sadovy de Mitcheson et al. 2008), (2) traditional spawning aggregations appear not to return, such as in the US Virgin Islands (Beets and Friedlander 1992), although there are indications of increased numbers at one site (Rick Nemeth, personal communication 2007), and (3) Nassau grouper numbers in extant spawning aggregations have also declined (Sadovy and Eklund 1999). In Belize, for example, one of the last viable Nassau grouper spawning aggregations decreased from 15,000 to fewer than 3,000 individuals (a decline of over 80%) in the 25 years leading up to 2001 (Sala et al. 2001).

To address the problem of overfishing of Nassau grouper, at least six measures should be considered: (1) prevent over-fishing of spawning aggregations – overfishing is a major factor in decline of the species; the most practical measure to reduce overexploitation is to not fish aggregations; (2) protect sub-adult fish from capture – a minimum size limit for Nassau grouper should be established at or above the size of sexual maturation. This measure is in place in a number of countries (Bahamas, Cuba, Belize, etc.) although enforcement is typically lacking; (3) protect remnant populations from fishing – this is necessary in locations where Nassau grouper populations are reduced to the point where there is no evidence of reproductive activity, or where there may be refuges for spawning stock, or important life history areas, (4) improve compliance with fishery restrictions – enforcement of fishery closures and other existing controls on capture of Nassau grouper is needed. This includes addressing problems of illegal, unmonitored and unregulated trade in The Bahamas, and elsewhere; (5) implement regional management strategies – given the possibly panmictic nature of the species and its extensive seasonal migrations, regional management is needed. The SPAW (Specially Protected Areas and Wildlife) protocol might be applicable; (6) Decide on the best use of Nassau within countries – decisions may be needed in countries with remaining viable populations of Nassau grouper as to whether this species should be retained for local trade and food security, be part of a significant export trade, or conserved for tourism and reef health. It is very likely that the most productive management approach in any situation is to protect spawning aggregations for reproduction, without allowing any fishing; the eggs and larvae produced can maintain the fishery at other times of the year.

Finally, an ecosystem approach, based on expanding the network of marine protected areas (MPAs) throughout the Caribbean and addressing key aspects of the species and the assemblage of which it forms a part will be key to its long-term persistence as a viable fishery, as a dive-tourism draw, and as a member of the reef ecosystem in which it lives. For example, since Nassau grouper are often captured as part of a multi-species snapper/grouper fishery, and since returning undersize fish is often not acceptable to fishers, no-take zones prevent individuals from being taken as unwanted bycatch, as well as provide a haven from spearfishing. Given that Nassau grouper make ontogenetic habitat shifts from nearshore to deeper reef areas,

MPA design would need to take into account these ontogenetic movements, the importance of non-reproductive residential areas, spawning aggregation sites and the migration routes to and from those sites.

Monitoring of Nassau populations and their aggregations is crucial. Where they are monitored, continued monitoring should be required. Where Nassau grouper populations are not monitored, monitoring plans should be initiated. Monitoring metrics should include an assessment of total population size if possible, but aggregation size, spawning frequency, recruitment of new adults to the spawning stock, overall harvest rates, and mortality rate estimates should be included at a minimum. A Caribbean-wide genetic assessment should be conducted to assess overall population structure and connectivity.

Given the difficulties of monitoring the numbers of Nassau groupers in large spawning aggregations, especially in three dimensional balls up in the water column just prior to spawning, hydroacoustic surveys are occasionally used to document fish numbers. Such techniques hold great promise, but are only meaningful where careful ground-truthing and independent evaluation (usually by divers) are carried out concurrent with hydroacoustic work (Taylor et al. 2006). Acoustic surveys that lack controls, and fail to consider the known parameters of spawning aggregations have not yet proven reliable and may significantly overestimate the numbers of fish present (Ehrhardt and Deleveaux 2007). Given the multi-species nature of nearly all Nassau grouper aggregations (Whaylen et al. 2004), the presence of many other (non-aggregating) reef fishes in their normal populations, and other unknowns, considerable caution is required in interpreting hydroacoustic results. They cannot, at this stage of understanding, be used for assessing fish numbers for stock assessments.

In-water assessments, while challenging, may prove to be the most feasible and reliable method for assessing aggregation size. Visual assessment can have broad confidence intervals, but may be used for tracking of trends. Various methods have been suggested for counting fish numbers just prior to spawning when in a large ball up in the water column, a particularly challenging task (Fig. 12.10b) (Sala et al. 2001; Whaylen et al. 2007). Mark-recapture techniques can allow for additional estimates of population size through simple Lincoln-Peterson estimates or more complex techniques. Acoustic telemetry (Starr et al. 2007, Semmens et al. 2006, 2009) can be used to evaluate spawning aggregation visit frequency by individual fish, and possibly also mortality rates. The REEF database can be used to track general trends, through underwater visual census surveys, in Nassau grouper abundance in non-aggregation times (e.g. Stallings 2009) on islands where dive tourism is popular. SAH and coworkers (unpublished data) have developed a technique to use size distribution data from video-based laser length measurements as a means to quantify population recovery in a protected aggregation and identify new recruitment to the adult spawning population.

Although the species can be spawned and raised in hatcheries (e.g. Watanabe et al. 1995), the use of such cultured fish to restore populations is currently unrealistic due to limited understanding of survival rates on release, the factors involved in spawning aggregation formation and expense. In an experimental release of hatchery-reared Nassau grouper, the number of tagged fish declined over a 9-month period after release in the US Virgin Islands. Based on the study and its review of available information, Nassau grouper stock enhancement was concluded only to be an option when simpler management methods for attaining wild stock recovery have failed (Roberts et al. 1995).

12.6.6 Gaps in Knowledge

Despite the fact that the Nassau grouper is one of the best-studied large-bodied reef fish in the Caribbean, we still have substantial knowledge gaps that need to be filled to improve management and ensure the long term persistence of this species. In particular, a better understanding of population structure, the catchment areas for aggregations (i.e. how far adults travel to reach specific aggregation sites), site fidelity and minimum viable spawning aggregation size is needed on the adult spawning side. Traditional measures of population productivity need to be determined in order to determine what reasonable exploitation levels might be. Such determinants of productivity would of course need to be coupled with effective management measures to sustain viable populations and rebuild depleted ones. Larval dispersal distances, source-sink dynamics, and the frequency of strong year classes are important aspects of this work.

12.7 Gag Grouper – *Mycteroperca microlepis*

Christopher C. Koenig and Felicia C. Coleman
Department of Biological Science, Florida State University Coastal and Marine Laboratory, 3618 Coastal Highway, St. Teresa, FL, 32358, USA
e-mail: koenig@bio.fsu.edu; coleman@bio.fsu.edu

12.7.1 General

The gag, *Mycteroperca microlepis* (Goode and Bean 1879) (Family Serranidae), is a large, relatively slender-bodied grouper with dark vermiculate markings on the body. Geographic range extends throughout the warm-temperate and sub-tropical waters of the western Atlantic. They occur primarily from North Carolina to the Yucatan Peninsula, Mexico, and are very abundant on the West Florida Shelf. This species is rare in Bermuda, recorded only once in Cuba, and reported from eastern Brazil (Heemstra and Randall 1993; Heemstra et al. 2002). Males are sometimes called "blackbellies" or "charcoal bellies" because of the presence of dark blotchy colorations on the belly and ventral sides below the pectoral fins (Heemstra et al. 2002) (Fig. 12.14).

Fig. 12.13 Juvenile gag grouper (Photo: © Don De Maria)

12.7.2 Biology

The life-cycle of the gag is relatively complex, exhibiting ontogenetically distinct habitat preferences. After a pelagic larval stage of 40–60 days, a post-larval stage recruits through inlets, harbours and saltmarsh creeks from March through June to settle in estuaries and other shallow coastal areas for 4–6 months, and an adult stage associated with offshore reefs of varying depths (Keener et al. 1988). Juveniles prefer natural or artificial structure, including seagrass beds, oyster reefs, wrecks, pilings and dredged canals as habitat. Off South Carolina oyster reefs seem to be the favoured juvenile habitat (Keener et al. 1988; Mullaney 1994; Mullaney and Gale 1996) while in Gulf of Mexico seagrass beds are preferred (Koenig and Coleman 1998; Renan et al. 2006) (Fig. 12.13). They typically remain in these habitats until the first cold fronts of the autumn (October), then egress to offshore reefs. In offshore areas gag occupy natural and artificial reefs, including wrecks, hard bottom, live bottom, shelf-edge scarps, ledges, sponge/coral habitat and other habitats providing vertical relief. Small juveniles feed on crustaceans; shifting to piscivory just prior to moving offshore (Mullaney 1994; Koenig and Coleman 1998). Adults are top-level predators (Sedberry 1988; Coleman et al. 1996; McGovern et al. 1998a) feeding heavily on schooling fishes such as round scad *Decapterus punctatus*, scaled sardines, *Harengula jaguana*, and tomtates, *Haemulon aurolineatum* (Weaver 1996; Lindberg et al. 2002). Gag have a maximum adult total length (TL) of about 1.3 m with a maximum weight of 39 kg. They are long-lived (up to 35 years) and slow-growing.

Fig. 12.14 Adult gag grouper showing male black belly colouration and bearing a tag (Photo: © Chris K. Koenig)

12.7.3 Reproduction and Aggregation

Females start maturing at around age 3 or 4 years with 100% reaching maturity by age 6 (Murdy et al. 1997; Heemstra et al. 2002). They are protogynous hermaphrodites, changing sex from female to male over the course of their life-time (Coleman et al. 1996; Koenig et al. 1996). Collins et al. (1998) found that males in the Gulf of Mexico range in size from about 886–1,290 mm TL, and in age from about 7–22 years old. Sex change is believed to occur only in a social context when fish are together in groups (CCK unpublished data). The primary time that males and females co-occur is during the northern hemisphere winter spawning season on offshore reefs where they aggregate at depths exceeding 40 m. They appear to reside separately at other times of the year. Indeed, females form pre-spawning aggregations in relatively shallow water (20 m) before moving to shelf-edge reefs (50–100 m) for spawning (Coleman et al. 1996; Koenig et al. 1996; McGovern et al. 1998b; Sedberry et al. 2006). Outside of the spawning season, males remain on spawning sites while females move into shallower water. So the window of opportunity to initiate sex change through social mediation appears to be limited in time and place to the aggregations.

12.7.4 Fishery and Population Status

Gag is extremely important economically throughout its range both to commercial and recreational fisheries. It is the second most common grouper species in commercial catches of the United States (exceeded by red grouper, *Epinephelus morio*) and the most important grouper in recreational catches. Gag commercially landed in the 1980s and early 1990s were often misreported as black grouper (*Mycteroperca bonaci*). The two species are in large part separated geographically, with gag being replaced along the west Florida shelf to the south by black grouper, and the misreports are attributed to both true misidentifications and to attempts to provide a more palatable market name for gag. Regardless of the reason, misreporting was less prevalent in the southeastern US Atlantic coast than in the Gulf of Mexico, where black grouper was more abundant.

The gag is managed in the United States by the Gulf of Mexico Fishery Management Council and the South Atlantic Fishery Management Council in their respective areas. Gulf of Mexico gag have been included in the reef fish fishery management plan since 1984, but were essentially unregulated until 1990 when a minimum size limit (MSL) of 51 cm TL, and a recreational aggregate daily grouper bag limit of five fish were imposed (Reef Fish Management Plan, Amendment 1). In 2000, the MSL was increased for both commercial and recreational sectors (to 61 and 56 cm TL, respectively) and a 1 month closure (15 February through 15 March, peak spawning period for gag) was introduced for the commercial fishery (although most of the fish are caught by the recreational sector). Two marine reserves (Madison-Swanson Marine Reserve and the Steamboat Lumps Marine Reserve) were established in the northeastern Gulf of Mexico to investigate the efficacy of using spatial closures as fishery management tools. The MSL for the recreational fishery increased 2.54 cm each year until 2002 when it reached 61 cm TL.

Current management is primarily through a license limitation, quota with size limits for both commercial and recreational fisheries, bag limits for recreational fisheries, limited gear restrictions (including elimination of traps, restricted areas for use of long-line gear required use of circle hooks for the hook and line fishery), and some area and seasonal closures. The primary area closures include the two Gulf of Mexico areas indicated above, and the *Oculina* Banks Marine Reserve along the US western Atlantic coast. Gag aggregations occur too deep (50–120 m depth) for any prospect of tourist diving operations, limiting non-consumptive use.

Gag are clearly vulnerable to intense fishing pressure because of their life-history traits, their habitat preferences, and their aggregating behaviour (McGovern et al. 1998a; Coleman et al. 2000; Harris and Collins 2000; Musick et al. 2000). The fact that gag associate with reefs means that they are relatively easy to catch year-round, especially with the advent of sophisticated positioning gear. The fact that they form spawning aggregations that are consistent in time and space means that they are easy and routine targets for fishers (a practice that started in the 1970s). Fish caught at the depths where larger gag occur suffer significant mortality after release. Indeed, at times, the mortality from catch-and-release in the recreational fishery for gag has exceeded the landed catch, even when based on a relatively conservative release mortality rate estimate of 20%. This problem is not limited to the recreational fishery as during seasonal closures of the gag fishery, gag are still present in the commercial bycatch when other reef fishes are targeted.

Overfishing has the propensity to skew size and age ratios affecting reproductive capacity of the population. It can also alter normal sex ratios, as has been clearly seen for gag, with the percentage of males declining by over an order of magnitude between the 1970s and 1990s in both the Gulf of Mexico and the Atlantic populations (Coleman et al. 1996; McGovern et al. 1998a; Chap. 8, Fig. 8.12). Declining male sex ratios on spawning sites occur due to male removal as males are present on aggregation sites year-round (Coleman et al. 2010). The decline in the proportion of males may be related to more aggressive feeding by males on spawning aggregations, causing them to be caught preferentially to females (Gilmore and Jones 1992). An alternative explanation is that males are more vulnerable to capture after the spawning season when females are dispersed and fishermen continue to fish spawning sites. Evidence for this hypothesis comes from the data of Collins et al. (1998) which show that the catch of males is highest within several months after the spawning season. If the second hypothesis explains most of the loss of males, then year-round closures of the spawning sites would be necessary to ensure the historical sex ratio in the gag population. The removal of males and presumptive (transitional) males through fishing could feed back to disrupt the social system responsible for sex change, reducing the social interactions that occur during aggregations triggering the production of transitionals, the percentage of which is highest during the post-aggregation period. The potential fishery consequences of male loss are: (1) reduced spawning opportunities for females, and (2) lowered fertilization rates through sperm limitation.

The overwhelming evidence for gag stocks, and for many other species, has been that: (1) loss of the largest reproductive individuals because they are targeted by fisheries and the largest fish tend to occur on spawning sites; (2) recruitment is now highly variable, and (3) there has been a decline in the number of historical aggregation sites. Once sites are fished out, they tend not to recover or reappear. Individually, any of these effects can have devastating repercussions on a population and they can interact to put groupers at particularly high risk of economic extinction and raise the spectre of becoming evolutionarily threatened.

The gag population was considered stable in 1994 (Schirripa and Goodyear 1994), "experiencing overfishing" in 1997 (Schirripa and Legault 1997), then both "overfished" and "experiencing overfishing" in 2001 (Turner et al. 2001) when increased discard mortality (undersize fish that are released, but die due to barotraumas or hooking mortality), attributed to increased size limits (particularly in the recreational fishery), represented 40–50% of the entire number of fish killed in recent years. Despite the "overfished" status, landings by both commercial and recreational sectors increased sharply. The gag populations are now considered to be "experiencing overfishing" and "approaching an overfished condition" along the Atlantic coast of the SE US and Gulf of Mexico gag are currently 'overfished' and 'undergoing overfishing' (SEDAR 2009). GMFMC must develop a plan to end overfishing and rebuild the stock by August of 2011. Currently, gag is included on the International Union for the Conservation of Nature and Natural Resources (IUCN) List of Threatened Species as a species of least concern (IUCN 2008).

Coastal development and fishery impacts adversely affect species in coastal nursery habitats, from salt marshes and seagrass beds to oyster reefs, important to juvenile gag. Destruction and pollution of juvenile habitat is of significant concern for its influence on year-class strength (Coleman et al. 1999, 2000). There is considerable concern now that red tide events along the west Florida shelf during the mid-2000s have had a significant negative effect on year class strength that has reverberated through the fishery. The red tide event of 2005 in the eastern Gulf of Mexico is thought to have added 18% mortality to fishing and natural mortality (GMFMC 2009).

12.7.5 Conservation Action and Recommendations

In the mid 1990s, Coleman et al. 1996 reported on the disturbing decline in the proportion of male gag at the same time the National Marine Fisheries Service reported significant declines in gag stocks, suggesting that the species was approaching an "overfished" condition (Schirripa and Legault 1997). This set in motion a series of events for the Gulf of Mexico Fishery Management Council that resulted in the establishment of two marine reserves, Madison Swanson Marine Reserve and Steamboat Lumps Marine Reserve, the first ones in the Gulf of Mexico (GMFMC 1999). The research conducted in these reserves has relied heavily on the input and participation of commercial and recreational fishers, resulting in significant acceptance by those involved. Results had initially been promising, indicating that males are remaining on site long after the spawning season, and that the catch per unit area of male gag within reserve boundaries is much higher than that in reference sites outside of the reserves (CCK and FCC unpublished data). This case is significant because (1) it was the first time that behavioural aspects of exploited stocks were considered, particularly the social structure that influences sex change, and (2) involvement of commercial fishers has led to significant acceptance, even by those strongly opposed to their formation (Neils 2003). While the positive effects of reserves appeared in a relatively short period of time and were believed to be due to strong enforcement, subsequently enforcement has been uneven and the disappearance of positive results largely attributed to lack of enforcement after Hurricanes Ivan in 2004, as well as Hurricanes Katrina, and Rita in 2005, when enforcement agents were otherwise occupied.

To foster conservation of the gag, there is a need to initiate broad studies of the behaviour, genetics and physiology of sex reversal, maturation and migration, timing and control of sex reversal, formation of shallow pre-spawning aggregations, and movement of fish to spawning sites for gag. While Koenig and Coleman (1998) started and continue to monitor an index of abundance for juvenile gag in the early 1990s and from these data, have been able to identify strong year classes. There is a need to continue and expand the geographic coverage to evaluate recruitment success and predict trends in adult population size. More detailed information, rather than just landings data, is needed regarding the status of the stock based on the percentage catch of (1) mature fish caught; (2) individuals with optimum length; and (3) large spawning fish in the catch, based on the work of Froese (2004). Management plans need to be based on the outcomes of such alternative stock assessments that will ensure adequate sex ratios for successful spawning and fertilization of eggs. The habitats used by juvenile gag need to be more carefully mapped and their loss reduced by working with developers, local governments, the US Army Corps of Engineers and other partners to deter development in sensitive areas. Given the prospects for continued development, there is a pressing need to designate and implement marine reserves that protect all sensitive life stages, including prespawning habitat, spawning habitat, shallow seagrass (early juveniles) and nearshore reef (late juveniles) nursery habitat.

12.8 Squaretail Coralgrouper – *Plectropomus areolatus*

Kevin L. Rhodes and Yvonne Sadovy de Mitcheson
College of Agriculture, Forestry and Natural Resource Management,
University of Hawaii at Hilo, 200 W. Kawili St., Hilo, HI, USA
e-mail: klrhodes_grouper@yahoo.com

School of Biological Sciences, The University of Hong Kong, Hong Kong, China
e-mail: yjsadovy@hku.hk

12.8.1 General

The squaretail coralgrouper, *Plectropomus areolatus* (Rüppell 1830), is a widely distributed and commercially important medium-sized grouper that occurs in lagoon and seaward coral-rich areas of the tropical Indo-Pacific at a depth of 5 to more than 50 m (Heemstra and Randall 1993; Hutchinson and Rhodes 2010). The squaretail coralgrouper is distributed from the Red Sea to the Phoenix Islands and Samoa, north to the Ryukyu Islands and south to Australia, with populations in the Maldives and Chagos Island. Its distribution is disjunct at the coast of Southeast Asia (Myanmar and Thailand). Squaretail coralgrouper feeds exclusively on fishes. Tupper (2007) found early juveniles almost exclusively in coral rubble habitats on the slopes of tidal channels, at a narrow depth range of 5–7 m. In Manus, Papua New Guinea, early juveniles were found in rubble habitat in back reef sandy bottom 'pools' surrounded by low-growing coral (Rick Hamilton, personal communication). The maximum reported size for squaretail coralgrouper is 73.0 cm total length

(TL) (Heemstra and Randall 1993). The species is reported to be ciguatoxic in some areas (Wong et al. 2005; Oshiro et al. 2010).

Age assessments from sectioned sagittal otoliths showed maturity for females and males at 2 and 3 years, respectively, with both sexes represented in the oldest age class of 12 years (KLR unpublished data). In the eastern Torres Straits of northern Australia, Williams et al. (2008) found a maximum age of 14 years old. In the eastern Torres Strait, total mortality (*Z*) for squaretail coralgrouper was estimated at 0.40 $year^{-1}$ and the growth coefficient (*K*) was 0.09 $year^{-1}$ (Williams et al. 2008).

12.8.2 Reproduction and Aggregation

The sexual pattern of the squaretail coralgrouper is unknown. Size of sexual maturation is approximately 35–40 cm FL and the species spawns in aggregations ranging in abundance from 10s to 1,000s of individuals. Squaretail coralgrouper aggregations overlap somewhat spatially and temporally with those of camouflage grouper (*Epinephelus polyphekadion*) and brown-marbled grouper (*Epinephelus fuscoguttatus*) but where depths of greater than 12–15 m are found, they are typically in shallower areas than the other two species (Sadovy 2005). In many parts of its range, squaretail coralgrouper spawning aggregations form over 4–6 months (e.g. Johannes et al. 1999; Rhodes and Tupper 2008; Williams et al. 2008), while in Papua New Guinea and parts of Indonesia, aggregations can form monthly (Hamilton et al. 2004; Pet et al. 2005) (Fig. 12.15). In Papua New Guinea, monthly spawning throughout the year was confirmed from back-calculation of otoliths (Rick Hamilton, personal communication). The lunar timing of aggregation formation and spawning varies regionally, with new moon spawning in Palau (Palau Conservation Society 2010) Indonesia (Pet et al. 2005) and Melanesia (Hamilton et al. 2005), but at full moon in Pohnpei (Hutchinson and Rhodes 2010). Aggregations in Pohnpei coincide with seasonally low water temperatures (range=28.2–29.4°C), but occur during seasonal upper-level temperature extremes in the eastern Torres Strait (29–30°C) (Pitcher et al. 2004).

Individuals of both sexes appear to follow migratory corridors to reach aggregation sites, the males moving individually or in small groups, the females in larger groups (Rhodes and Tupper 2008). At the spawning site, individuals stay approximately 10 days (Rhodes and Tupper 2008). Throughout the aggregation period, males establish and aggressively defend territories until spawning (Fig. 12.16). When sex ratios are highly male-biased, possibly as a result of overfishing, females are relentlessly chased by males (Johannes et al. 1999). Six colour morphs are known, some sex-specific (Johannes et al. 1999). Observations of pair-spawning 1 h prior to early morning high tide were made in Indonesia (Pet et al. 2005). After the aggregation period individuals return to home range areas that in Pohnpei measured 0.048 ± 0.018 km^2 S.E. (Hutchinson and Rhodes 2010). Acoustic tagging results in Pohnpei suggested relatively small catchment areas (200–300 km^2) for the species (Rhodes and Tupper 2008).

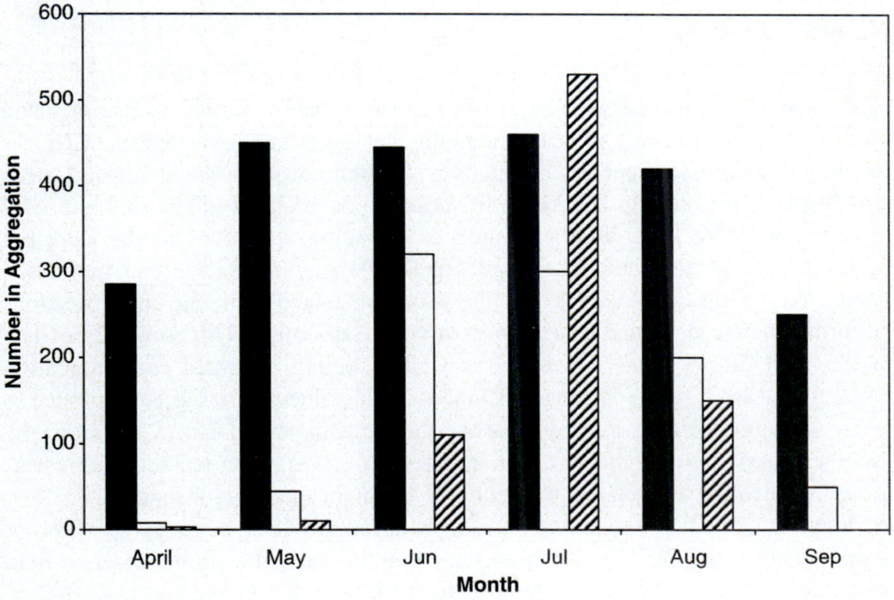

Fig. 12.15 In Palau at one multi-species aggregation site underwater surveys recorded elevated numbers of *Plectropomus areolatus* (*black bar*), *Epinephelus fuscoguttatus* (*white bar*) and *E. polyphekadion* (*dashed bar*) between April and September 2009 with *P. areolatus* attaining about 460 individuals and showing elevated numbers for more months than the other two species; aggregations do not form in other months (Source: Palau Conservation Society 2010)

Fig. 12.16 Photo: Male *Plectropomus areolatus* patrolling a territory in a spawning aggregation. The split tail may be the result of aggressive male-male interactions. Photo: © Steve Lindfield

12.8.3 Fishery

The squaretail coralgrouper is intensively fished throughout much of its range and taken with a wide variety of gears, including net, spear and hook and line (Rhodes et al. 2008). In multi-species aggregations, it often occurs in the shallowest areas and hence is particularly accessible to spearfishers, amongst others (YS, personal observation 2006) As a direct consequence of fishing, squaretail coralgrouper has undergone localized population-level declines in Palau, Pohnpei, Kiribati, Indonesia, Papua New Guinea, Maldives and the Solomon Islands, among other locations according to anecdotal and fishery observations (Hamilton and Matawai 2006; Chan et al. 2008; Rhodes and Tupper 2008; Being Yeeting, personal communication 2007, www.SCRFA.org database). Much of this decline appears to be attributed to spawning aggregation fishing by local and foreign commercial fishers, including the Southeast Asia-based trade that supplies live reef fish to Asian markets and restaurants in which the squaretail coralgrouper is a popular species (Johannes et al. 1999; Sadovy et al. 2003a, b; Hamilton and Matawai 2006; Wilson et al. 2010). A recent assessment found that 70% of a squaretail coralgrouper aggregation was removed in just 5 days as a result of aggregation-fishing for the trade (Wilson et al. 2010). In Pohnpei, individuals were fully recruited into the fishery at age 4 years. The species is not commercially produced in hatcheries.

12.8.4 Management and Conservation

Similar to other aggregating Indo-Pacific groupers, there is very little management specific to squaretail coralgrouper. In Palau, a sales ban is in place during most of the species' reproductive season, with the exception of August. In Pohnpei, the sales ban covers only two months of the 5-month spawning season. In both locations, many ripe fish are taken outside the ban period when aggregations occur. (Palau Conservation Society 2010). Palau also has a catch ban to coincide with the sales ban, again with the exception of August, and a year-round export ban on live fish. Both countries have marine protected areas (MPA) protecting spawning aggregation sites. In Palau, spatial protection is in place at nine known spawning aggregations as a result of either national and state legislation or traditional fishing bans (*bul*) as either permanent or seasonal MPAs (Johannes et al. 1999). Among commercial catches in Pohnpei, 76% of squaretail grouper females in a recent study were below the reported 50% size of sexual maturity, while only 3% of the total landings were mature squaretail coralgrouper males (Rhodes and Tupper 2007), suggesting that size-at-catch restrictions could benefit this population. As a consequence of dramatic, fisheries-induced reductions in population size at a number of locations, the IUCN Red List Authority has designated squaretail coralgrouper as vulnerable (Chan et al. 2008).

12.8.5 Challenges to Monitoring and Management

Squaretail coralgrouper are important components of subsistence and commercial fisheries, and proper management and conservation are needed for regional economic and food security. Like many other aggregating groupers, the squaretail coralgrouper is highly vulnerable to overfishing especially because of the apparent ease in locating and overexploiting spawning aggregations. The species is challenging to monitor because it can aggregate in many months of the year and numbers can vary substantially both between months and during a single lunar phase (Johannes et al. 1999; Pet et al. 2005; Rhodes and Tupper 2008; KLR unpublished data; Chap. 9).

As with many remaining transient aggregating species, spawning sites are often remote from human communities and widely dispersed, thereby creating challenges for monitoring fishing activities and enforcing site-based fishing regulations. Nonetheless, a combination of aggregation-based MPAs and seasonal catch restrictions, when properly monitored and enforced, can be valuable management tools. The existence of reproductive migratory corridors for squaretail coralgrouper in Pohnpei shows the need for their inclusion into aggregation-based MPAs (Rhodes and Tupper 2008). Larger scale MPAs around spawning sites may also provide protection for a greater portion of the reproductive population given the relatively small catchment area indicated in Pohnpei (Hutchinson and Rhodes 2010). Where markets are centralized or landing sites are known, seasonal sales and catch bans around spawning times can be effective in protecting reproductive individuals.

Where feasible, maximum allowable size-at-catch regulations should be implemented to protect larger, more fecund, females and males. Where information and human resources are limited, managers may want to focus on sites of highest vulnerability and utilize traditional knowledge of reproductive seasonality, subsequently verified, to help design seasonal fishing restrictions. The identification of practical points of enforcement and monitoring interdiction, such as at fish markets or landing sites, can assist in reducing monitoring and enforcement costs. A combination of precautionary, adaptive, traditional and co-management approaches offers a range of options, which should be reinforced by regular monitoring of the fishery and enforcement.

12.9 Leopard Coralgrouper – *Plectropomus leopardus*

Melita A. Samoilys
CORDIO (Coastal Oceans Research and Development – Indian Ocean)
East Africa, Mombasa, Kenya
e-mail: melita.samoilys@gmail.com

12.9.1 General

The leopard coralgrouper, *Plectropomus leopardus* (Lacepède 1802), also known in English as coral trout and leopard grouper, is distributed in the western Pacific from southern Japan to southern Queensland, west to Lombok, Indonesia and Western Australia, and east to Caroline Islands and Fiji. It occurs at depths of 3–100 m (Heemstra and Randall 1993).

The leopard coralgrouper is a medium-sized diurnally active grouper easily observed underwater, therefore vulnerable to spear fishing but also amenable to accurate visual census surveys (Samoilys and Carlos 2000). Biological parameters for the leopard coralgrouper, listed in Table 12.2, are largely from studies conducted on the Great Barrier Reef (GBR), Australia. The leopard coralgrouper grows rapidly in the first 2–3 years of life, and matures relatively early for a grouper (Table 12.2). The juvenile diet consists largely of benthic crustaceans, but this shifts just prior to maturity to a piscivorous diet (St John 1999). Recruitment of juveniles appears to be largely driven by current patterns and geomorphology (Doherty et al. 1994). Leopard coralgrouper larvae have been reared experimentally from eggs and sperm sourced from spawning aggregations on the GBR (Rimmer et al. 1994), Japan, and elsewhere in Southeast Asia, with commercial scale production recently reported from Hainan Island, Mainland China; no data are available on production and there are problems in ensuring the red body colouration in cultured fish (Yvonne Sadovy de Mitcheson, personal communication).

12.9.2 Reproductive Biology

The leopard coralgrouper is a protogynous hermaphrodite, maturing as female first at 2–3 years changing to male at around 7 years. Females are batch spawners, on average spawning 28 times per year (Table 12.2), with a high annual fecundity compared with other serranids (Sadovy 1996). From studies on the northern GBR the leopard coralgrouper is known to form well-defined, temporally predictable spawning aggregations at the same sites over decades (Samoilys 1997; Samoilys et al. 2001). Aggregations form for a 5 day period around the new moon over 3 months, September to November. Males display distinctive spawning colouration (Fig. 12.17) and control territories of 50–250m^2 where they pair-spawn. A small group of moderate sized individuals at the periphery of the aggregation appeared to be males but were not observed to court. However in one pair-spawning a single smaller male streaked into the egg cloud and released sperm. Pair-spawning in aggregations is confined to a 30 min period at sunset. However, based on tagging studies and estimates of leopard coralgrouper densities, only around 14% of individuals appear to participate in aggregation spawning (Samoilys 2000). Further, at least 50% of all spawnings may occur outside aggregations because a female spawns on average 28 times a year, yet aggregations are only available for 15 days

Table 12.2 Biological parameters for leopard coralgrouper, *Plectropomus leopardus*, from the GBR, unless otherwise noted

Parameter	Value	Source
Maximum size (females)	63.0 cm FL	Ferreira (1995)
Maximum size (males)	66.0 cm FL	Williams et al. (2008)[a]
Longevity – maximum age	14 years	Ferreira and Russ (1994)
	17 years	Williams et al. (2008)[a]
	19 years	Loubens (1980)[b]
Asymptotic Length, L_∞ (von Bertalanffy model)	52.2 cm SL	Ferreira and Russ (1994)
	74.6 cm FL	Williams et al. (2008)[a]
Growth coefficient, K (von Bertalanffy model)	0.35	Ferreira and Russ (1994)
	0.07	Williams et al. (2008)[a]
Total mortality, Z	0.44	Williams et al. (2008)[a]
Natural mortality rate, M	0.147	Russ et al. (1998)
Age at first maturation (50%)	2–3 years	Ferreira (1995)
Size at first maturation (50%)	21–25 cm FL (1992–1993);	Ferreira (1995)
	32–36 cm FL (1990–1992)	Samoilys (2000)
Mean age at sex change, A_{50}	7.2 years	Williams et al. (2008)[a]
Sex change range	3–12 years	Ferreira (1995)
Mean size at sex change, L_{50}	46–50 cm FL	Samoilys (2000)
Mean size at sex change, L_{50}	50.2 cm FL	Williams et al. (2008)[a]
Sex change size range	30–55 cm TL	Ferreira (1995)
Operational sex ratio (F:M)	Large variation with location and exposure to fishing – ranging from 0.9 to 5.5	Ferreira (1995) Adams et al. (2000) Samoilys (2000)
Mean annual fecundity:	4.4 million eggs (hydrated) per female	Samoilys (2000)
Mean batch fecundity:	157,702 eggs per batch spawn	Samoilys (2000)
Annual spawning frequency:	28 batch spawns per year	Samoilys (2000)
Average population densities	4.8/1,000 m^2 (GBR average)	Ayling et al. (2000)
	5.3/1,000 m^2 (max = 23/1,000 m^2)	Zeller and Russ (2000)
	9.3/ 1,000 m^2 (protected, inshore)	Evans and Russ (2004)
		Evans and Russ (2004)
	6.3/1000 m^2 (un-protected, inshore)	Nardi et al. (2004)[c]
	28–60/1,000 m^2 (protected)	Nardi et al. (2004)[c]
	8/1,000 m^2 (un-protected)	
Pelagic larval duration:	25.2 days	Doherty et al. (1994)

FL fork length, *SL* standard length
[a]Torres Straits, Australia, which lies between Queensland and Papua New Guinea
[b]New Caledonia
[c]Houtman Albrolhos Islands, Western Australia

per year. Further, around 50% of females (based on presence of hydrated oocytes) are believed to spawn during the first quarter lunar phase when aggregations are not known to occur.

The mating patterns, fast growth and early maturation of the leopard coralgrouper suggest a relatively complex and flexible reproductive strategy compared with

Fig. 12.17 Large male (65 cm FL) *Plectropomus leopardus* displaying courting and spawning colouration at Lizard Island Site BR 1, 1730 h 7 September 2010 (Video credit Alex Vail, Lizard Island Research Station)

other groupers which may be the reason that *P. leopardus* is a relatively abundant serranid on the GBR and perhaps more resistant to fishing pressure.

12.9.3 Commercial Fisheries

The members of the genus *Plectropomus* ('coral trout' or coralgrouper) are major commercial finfishes taken in Australia, predominantly on the GBR. In commercial catches they comprise three dominant species (leopard coralgrouper – *P. leopardus*, blacksaddle coral grouper – *P. laevis* and spotted coralgrouper – *P. maculatus)*, and four lesser species. Commercial log books do not distinguish among the *Plectropomus* species, but independent research shows that the leopard coralgrouper comprises at least 80% of the *Plectropomus* commercial catch (Samoilys et al. 2002a; Mapstone et al. 2004). It is also the most abundant *Plectropomus* species on the GBR (Ayling and Ayling 1992).

Commercial fishers in Australia use simple hook and line, generally with one hook, and frozen pilchard as bait, fishing from small tender boats to a mother vessel. Recreational fishers use handlines, rods and spearguns. Mother vessels have large freezer and live holding tank capacity and may stay at sea for 3–4 weeks. Artisanal fisheries in Fiji and New Caledonia take leopard coralgrouper by hook and line and spear gun, and incidentally by trap and net. Leopard coralgrouper is sometimes taken using cyanide in the Philippines and Indonesia.

Commercial log book data since 1989 show increases in catch and effort of *Plectropomus* spp. on the GBR, followed by declines (Table 12.3, Fig. 12.18) in

Table 12.3 Commercial *Plectropomus* spp. catch data from the GBR. (Sources: Samoilys et al. 2002a, b; Mapstone et al. 2004, DPI&F 2009)

Year	Catch (tonnes)	Live fish value (US$/kg; converted from Aus$ in April 2011)	Effort (boats)	Effort (days)
1989	925	–	410	–
1996	1,672	47	714	–
2000	1,528	61	666	–
2001	1,540–2,420	–	700	29,054
2007	1,114	34–42	–	12,000

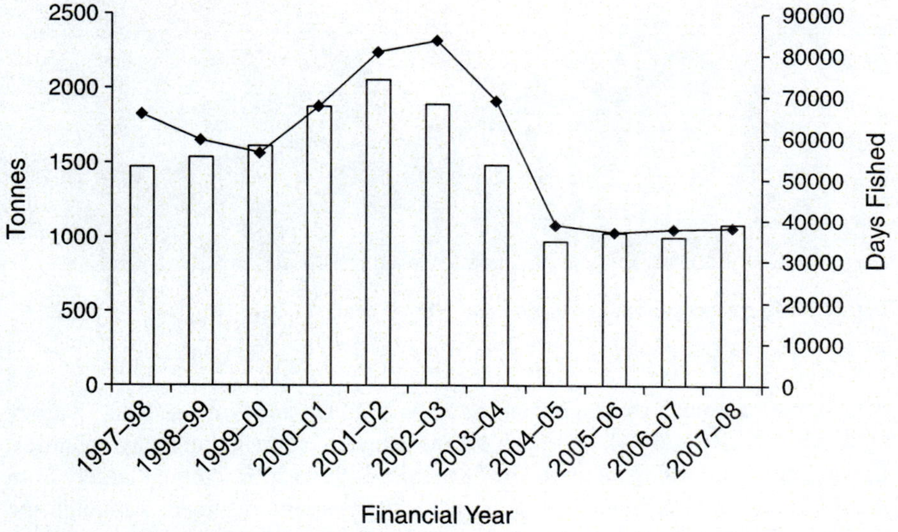

Fig. 12.18 Total commercial catch and effort of coral trout species on the Great Barrier Reef, Australia, 1989–2007. Most of this catch is made up of leopard coralgrouper (Source: DPI&F CFISH Database, 2007, State of Queensland database 2009). A catch quota was introduced in 2004. *Bars* indicate weight in tonnes and *line* indicates number of dory (tender) days fished

response to a total allowable catch (TAC) introduced in 2004 (see below). In 2000 the commercial harvest had a wholesale value of US$ 16.3 million. The leopard coralgrouper is also taken on the GBR by private recreational fishers and by commercial recreational fishing charters. The leopard coralgrouper is also caught commercially and recreationally from the Houtman Albrolhos islands in Western Australia (WA). The WA fishery is managed through minimum size limits, recreational bag limits and area closures and is not considered overexploited (Nardi et al. 2004).

In the mid 1990s the trade in live leopard coralgrouper, primarily to Hong Kong, grew in Australia. The portion of live fish in the commercial catch increased to around 25% initially and by 2007 to 80%. By 1998 live coral trout fetched up to eight times the price of whole or filleted coral trout (Samoilys et al. 2002a; DPI&F 2009).

The export of live leopard coralgrouper is a significant commercial fishery in the Asia-Pacific region, with fish taken primarily from Indonesia and the Philippines,

Fig. 12.19 Large leopard coralgrouper on sale in Fiji (Photo: © Randy Thaman)

with limited amounts from the Pacific Ocean (Sadovy and Vincent 2002; Sadovy et al. 2003; SCRFA 2004), and fresh (dead) fish are also traded in many countries. Imports into Hong Kong increased by 58% from 1999 to 2002, coming largely from Australia and the Philippines as other countries' exports declined. Although the majority of leopard coralgrouper in Hong Kong's live fish markets are above the size at first reproduction, this is misleading regarding size of capture because many were wild-caught as juveniles and grown out in captivity to attain market-preferred sizes, then exported, particularly from the Philippines (Sadovy et al. 2003a).

Artisanal fisheries for leopard coralgrouper exist in the Pacific Islands, including Fiji, Papua New Guinea and New Caledonia (Samoilys et al. 1995) (Fig. 12.19). There is no strong evidence of targeted fishing of spawning aggregations of leopard coralgrouper on the GBR though this has not been well studied (Turnbull and Samoilys 1997). However, the potential for increased catchability of leopard coralgrouper while migrating to aggregations has been demonstrated through simulation (Fulton et al. 1999). Further, the collapse, recovery and further collapse of one spawning aggregation in the Cairns region of the GBR was attributed to commercial fishers targeting this site (Samoilys et al. 2001), and these concerns have led to spawning closures (see below). It is likely that easily located aggregation sites are vulnerable to targeted fishing but the relatively small aggregations, probable >1 aggregation per reef, small migration distances and non-aggregation spawning make the species less vulnerable than many other aggregating groupers. No intentional targeting of spawning aggregations is reported from Torres Strait (Williams et al. 2008).

Spawning aggregations in this species have been analysed with respect to timing, size, location and distance travelled to sites (Samoilys 1997; Zeller 1998). A summary of these features illustrates the differences between large aggregate-spawning groupers in the tropical Atlantic as opposed to the western Pacific. The sizes of aggregations of the leopard coralgrouper were relatively small with a maximum of 128 fish (Samoilys 1997) but with much lower numbers in many aggregations. Zeller (1998) recorded a mean distance of 912 m for movement between *P. leopardus* home ranges and aggregation sites recorded at Lizard Island. This study concluded that most instances of aggregate spawning occurred on the reef on which the individual was resident. Both authors expressed concern as to the longer term continuity of these aggregations in the face of fishing pressure which prompts questions with respect to their small size. Are small aggregations in this species a consequence of fishing pressure? Recent observations by Alex Vail (Lizard Island Research Station) on spawning by *P. leopardus* at a site recorded by Zeller (1998) (Site BR1) helped clarify these issues.

During the 15 years elapsed since Zeller's observation (Zeller 1998), a similar sized aggregation was recorded at the same site spawning at dusk during the new moon phase. Further south, an aggregation on Scott Reef off Cairns has persisted for 20 years (1989–2009) with an average of 109 (range: 47–345) fish at the site which has not declined (Samoilys, Roelofs, Russell and Squire unpublished data). Detailed population abundance surveys and tagging of *P. leopardus* in a 6.5km^2 area around a neighbouring aggregation site at Elford Reef show that the site draws fish from a catchment area of 1.5km^2 with an estimated population of 3,169 adult fish, moving up to 1,200m to the site. Tagging, visual observations and histology suggest a maximum of only 600 of these 3,000 adult fish use the aggregation site to spawn (Samoilys 2000). It is concluded that numerous small spawning aggregations of a resident nature are a consistent feature of the reproductive biology of this abundant western Pacific serranid, but aggregation-spawning is only one mechanism, with a large number of individuals (>80%) spawning in pairs or small groups..

Small aggregations of the species have also been reported through fisher interviews in the Solomon Islands, Papua New Guinea, Fiji, Malaysia, Philippines, Indonesia and Fiji; very large aggregations of this species have nowhere been reported (www.SCRFA.org-country reports; Yvonne Sadovy unpublished data).

12.9.4 Conservation and Management Action and Concerns

Conservation and management action is largely limited to Australia where the commercial Queensland fishery is limited entry, licensed, with gear restrictions, quotas and a minimum size limit of 38 cm total length (TL), managed through the Department of Primary Industries and Fisheries (DPI&F) regulations (*Fisheries (Coral Reef Fin Fish Fishery) Management Plan 2003)*. The size limit prevents fishing of the first three age classes, representing both the fastest growth period as well as when females start reproducing (Ferreira 1995), generally considered a sound management approach. However this does not factor in the relatively larger contribution of big females to reproductive output due to higher fecundity. Recreational fishers are limited to 7 leopard coralgrouper per person. Charter fishing operators are regulated by permits.

The Queensland reef fishery operates within the GBR Marine Park, an area of 344,400 km^2, established in 1975 and managed by the Great Barrier Reef Marine Park Authority (GBRMPA), a Commonwealth Statutory Authority. The GBRMPA has responsibility 'to provide for the long term protection and conservation of the environment, biodiversity and heritage values …', and explicitly does not have authority for fisheries management, but has area closures where fishing is prohibited. Studies on the impacts of these closures on leopard coralgrouper reveal complex results, but generally higher densities are recorded on unfished reefs (e.g. Evans and Russ 2004), and females are larger, older and more abundant on closed reefs suggesting reef closures are an effective strategy (Vincent and Sadovy 1998; Adams et al. 2000).

Despite extensive fisheries and protected area controls (including no-fishing areas, size limits, partial aggregation closures and quotas), significant declines in leopard coralgrouper abundance in the central and southern sections of the GBR have been reported (Ayling et al. 2000). This observation, the 50% increase in number of active commercial boats in the fishery, the decline in catch rate since the early 1990s and GBRMPA's recognition that their zoning was inadequate and not representative, led to substantial reviews and revision of legislation in the early 2000s. In 2004 DPI&F introduced three 9-day closures (around the new moon) to protect spawning aggregations of key target species, particularly leopard coralgrouper aggregations and migrations (Turnbull and Samoilys 1997). In addition, DPI&F introduced a TAC for coral trout of 1,288 mt reflected in the recent drop in catch (Table 12.3). Meanwhile the GBRMPA completed a massive re-zoning of the Park in 2004, resulting in 33% of the Park being closed to fishing (Fernandes et al. 2005).

Recent studies on the effects of these closures on leopard coralgrouper show that after 1.5–2 years of protection, the density of coral trout increased by a factor of 1.7 in areas that had been closed to fishing in 2004, while density and biomass decreased slightly in the areas that remained open to fishing (Russ et al. 2008; Evans et al. 2006). A review and interview-based surveys with the fishers regarding the spawning closures by Queensland Fisheries in 2008, resulted in a change, effective 2009–2013, to two 5-days closures around the new moon in October–November, largely driven by commercial fishers, concerns over loss in revenue and their lack of conviction that closures were necessary biologically. These shorter closures will still provide good protection of leopard coralgrouper spawning aggregations which only last for 5 days, and spawning is concentrated in Oct–Nov.

The Houtman Albrolhos Islands in WA were declared a Fish Habitat Area (FHA) by the Department of Fisheries with 17% of the FHA closed to fishing in 1994. After 8 years of closure the leopard coralgrouper increased three to sevenfold attributed to reduced fishing mortality and their relatively small home ranges (Nardi et al. 2004). No information is reported on leopard coralgrouper spawning aggregations in WA.

Historic data on leopard coralgrouper from the GBR before the substantial commercial fishery operated are not available and therefore it is difficult to assess whether biological population parameters obtained from exploited populations reflect "normal" populations. The difference in densities of leopard coralgrouper on reefs closed to fishing between WA and the GBR (Table 12.2) is revealing and may indicate severely depleted populations on the GBR.

The species is considered to be near threatened on the IUCN Red List based on declining populations in different countries (www.iucnredlist.org).

12.9.5 Lessons Learned, Specific Management Challenges

Despite a significant amount of research and management of leopard coralgrouper on Australia's GBR there remain concerns and difficulties:

- As one of a group of coral trout species, some of which are hard to distinguish, disaggregating accurate information on leopard coralgrouper is difficult.
- The live fish trade represents a market demand for small ("plate-size" = 25–40 cm TL) fish increasing fishing pressure on fish entering the fishery (38 cm TL legal size), and fishing inshore in shallow reef areas (<20 m) where fishers perceive young fish occur.
- Understanding the impacts of the GBR spawning closures on leopard coralgrouper is difficult because no monitoring system was established when the closures were introduced, and only two spawning aggregations have been monitored for any length of time on the entire GBR.
- Only four reefs have been specifically protected for their leopard coralgrouper spawning sites (Russell and Pears 2007).

The status of leopard coralgrouper populations in Asian and Pacific Island countries, the extent to which leopard coralgrouper contributes to the live fish export fishery, and whether aggregation-fishing is involved in this fishery are all largely unknown. The species is very highly valued at the retail end of the trade so there is high interest in sourcing wild fish. This fishery is typified by declining catches of high value groupers in some areas of Indonesia, Malaysia, Philippines and Vietnam, though imports to Hong Kong appear to be stable (Sadovy et al. 2003). The Philippines continues to be a major source of leopard coralgrouper for the Hong Kong–based trade in live reef fish. Fish are either caught and exported or caught, if below the preferred marketable size, and placed in cages to grow to a more profitable size. This has led to a heavy fishery in juvenile leopard coralgrouper which has had severe negative effects in Palawan where they are trying to bring the export of the species under control (e.g. Padilla et al. 2003; Yvonne Sadovy de Mitcheson, personal communication 2010).

Management and conservation of leopard coralgrouper should employ a range of measures to limit total catch and protect spawning stock biomass, including size limits, effort controls, area controls and seasonal spawning closures. Spawning closures alone are not likely to be adequate because aggregation spawning may only partially contribute to the total reproductive output of the species (at least on the GBR).

For better management information is needed on a number of issues.

- Biological and ecological studies of leopard coralgrouper from across its range, particularly from the Indian Ocean, Philippines, Fiji, Japan and Indonesia to evaluate the parameters already determined from eastern Australia (Table 12.2).
- Verification of whether leopard coralgrouper aggregations are targeted for the live fish trade in Philippines, Indonesia and Fiji.
- Data on species-specific abundance, catch and effort in the artisanal fisheries of the Pacific Islands to properly assess the population status of this species.
- Knowledge of reproductive parameters and spawning behaviour of populations outside Queensland, particularly where fishing pressure is light, to properly assess the impacts of fishing leopard coralgrouper on the GBR.
- As the most heavily fished reef predator on the GBR, further studies on the impacts of lowered population densities of leopard coralgrouper on reef communities could provide important information for understanding ecosystem effects of fishing. Studies should include Western Australia to examine the high population densities reported there.
- A thorough understanding of the spawning dynamics of leopard coralgrouper on the GBR such as: what factors control whether females use aggregations or not; does size and age of females affect aggregation behaviour; what is the impact of fishing pressure on these factors and are there regional differences?
- Monitoring of aggregations and of the impacts of spawning closures as part of fisheries management.

12.10 *Lutjanus fulvus* – Blacktail Snapper with Notes on Other Species of Atlantic and Indo-Pacific Snappers (Lutjanidae)

Yvonne Sadovy de Mitcheson, Patrick L. Colin, and Jiro Sakaue
School of Biological Sciences, The University of Hong Kong, Hong Kong, China
e-mail: yjsadovy@hku.hk

Coral Reef Research Foundation, PO Box 1765, Koror, Palau 96940
e-mail: crrf@palaunet.com

Southern Marine Laboratory, P.O. Box 1598, Koror, Republic of Palau 96940
e-mail: bigrun2006@hotmail.co.jp

12.10.1 General

The blacktail snapper, *Lutjanus fulvus* (Forsskal), is a common Indo-west Pacific reef fish occurring from the Red Sea to southeastern Oceania and ranges north to south from Japan to New South Wales in Australia. It was successfully introduced

Fig. 12.20 (**a**) Group of blackfin snapper, *Lutjanus fulvus*, from a spawning aggregation on the side of a deep tidal channel in Palau. (**b**) A small portion of a spawning aggregation of *Lutjanus fulvus* on the side of a barrier reef channel in Palau. The fish are mixed males and females, averaging about 19 cm standard length, with their abdomens visibly swollen with gonads during the late afternoon. It is not known exactly how and when the fish spawn (Photos: Patrick L. Colin)

to Hawaii, but is not common there (Randall 2007). It reaches about 34 cm standard length (SL) and is found in lagoons, protected reefs, mangroves and river mouths to depths to about 75 m (Allen and Talbot 1985; Myers 1999) (Fig. 12.20a, see also Chap. 9, Figs. 9.7 and 9.8). It is principally a nocturnal feeder on benthic crustaceans and fishes. Its yellowish body, yellow anal fin, dark caudal fin and large eye make it easily identified (Allen and Talbot 1985). Due to its relatively small size, it is of minor fishing importance throughout its range, with only limited artesanal or commercial catch according to available information (e.g. Sadovy 2007). Little has been published regarding its spawning biology, although in Tahiti it was reported to spawn around the full moon (Randall and Brock 1960) this section also includes notes on some other lutjanid species.

12.10.2 Reproduction and Aggregation

At the time of the full moon in January 2007 the blacktail snapper was found in Palau to form a large spawning aggregation numbering tens of thousands of fish on the side of a deep tidal channel (Fig. 12.20). There is no traditional knowledge recorded regarding such aggregations of this fish (Johannes 1981 and later surveys) but almost certainly the aggregation would have been observed by the numerous spearfishermen of Palau (see below). In many other areas of the Pacific, however, the aggregations would be easy targets for fishing with explosives and might become a target in Palau if larger fish species decline in the reef fishery.

Spawning has not been observed despite attempts from late afternoon until a half hour after sunset. Spawning may occur at dawn, at night, or at some other time of day. The fish are disturbed by observers, constantly moving away, perhaps preventing easy observations of spawning. The one known aggregation in Palau occurred on the slope of a barrier reef channel, but located well inside the channel over 1 km from the mouth of the channel. The fish ranged from the shallow reef flat along the channel edge (1 m deep at low tide) down the 30–45° slope to around 20 m depth. The aggregating fish occurred as 3–4 discrete "schools" along about 150 m of channel edge. At peak numbers schools were estimated at 10,000–20,000 fish, based on the estimated density within small areas and the overall areas covered by the schools. They are present at the site during the entire day, starting at dawn, but may leave the area around sunset (see Fig. 9.8). The site was visited for surveys during the full moon period for 10 months of 2007 (2 months missed due to weather). The aggregation was present all 10 months and it seems likely it occurs every month of the year. The aggregation duration most months was from the day before the full moon to 3 days after, a total of 5 days with large numbers of fish present. In a few months, the aggregation was not present on the day before the full moon, and in one case it did not appear until the day after the full moon.

While the presence of tens of thousands of snappers in a limited area where normally there might be only a handful fish implied that a spawning aggregation was likely occurring, without a direct indication of spawning it was not correct to consider this group of fish as a spawning aggregation. At least one of the three direct criteria accepted by SCRFA as direct evidence of a spawning aggregation; observations of spawning, hydrated oocytes and post-ovulatory follicles, needed to be met (Colin et al. 2003; Chap. 1). In this case, specimens from the aggregation were

Fig. 12.21 Female *Lutjanus fulvus* (20.0 cm SL, 292 g weight) collected 5 March 2007 from Palau aggregation site. The ovaries (48.3 g weight, GSI 16.5%) with hydrated eggs are shown relative to their position in the body cavity (Photo: Patrick L. Colin)

speared; both sexes were found to be very ripe with hydrated oocytes in females collected during afternoons, indicating imminent spawning (Fig. 12.21). The fish also became visibly swollen in the afternoons when aggregation was occurring.

Females collected from the aggregation site (n=26) ranged between 17 and 22 cm SL (mean 19.1) with weights of 157 and 390 g. Females with hydrated eggs (assumed to be just prior to spawning) had Gonadosomatic Index values (GSI is gonad weight/body weight × 100) of 11–16.5%. Ripe, but not hydrated females, from the aggregation had GSI values around 2–6% (Fig. 12.22a). An increase in GSI for females during the day was seen (Fig. 12.22b) prior to a presumed sunset or evening spawning. Males were similar in size, ranging from 17.5 to 21.5 cm SL (mean 18.6), weights of 141–276 g and GSI values of 1–3% (Fig. 12.23). There are no external sex differences.

Why the one section of channel edge is used as an aggregation site is unclear. The location and limits of the site have been consistent throughout the period of observations of over 3 years. The location is nearly 1.5 km inside the channel mouth and occurs only on one side of the channel. The entire channel length on both sides has been surveyed during the time aggregation was occurring without any other major aggregations being found. One area had a few schools numbering at most a few hundred fish, but these comprised only 1% or less of the estimated size of the larger aggregation. The area where the aggregation occurs becomes quite turbulent when tidal currents are strong and actually has a reverse eddy along much of its length with the current along the channel slope running opposite that found in the general channel. Whether the fish are keying on this current or some other factor is under study.

There is, a present, no known fishing pressure on the aggregation. The area where it occurs is a nominal protected area; however, the area is subject to regular subsistence and limited commercial line and spear-fishing. Although the fish are not

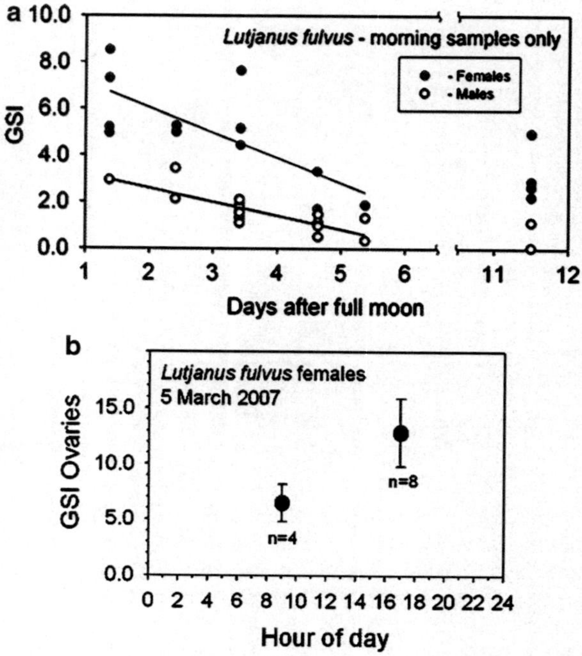

Fig. 12.22 (**a**) Gonado somatic Index (GSI) for *Lutjanus fulvus* according to days after the full moon, January–March 2007. Only fish collected during the morning were used for this graph to avoid ovaries that are large due to hydration. Fish collected at 11.5 days were taken at the aggregation site, but fish were not aggregated. (**b**) Ovaries of *L. fulvus* collected 1 day after the full moon from the spawning aggregation show a large increase in GSI between the morning (11 a.m.) and afternoon (5 p.m.) due to hydration of eggs prior to spawning

Fig. 12.23 School of *Lutjanus bohar* in the area of a spawning aggregation at the south end of Peleliu Island, Palau (Photo: © Nial McCarthy)

present at the aggregation for about 85–90% of each lunar month, it is unknown how far the snappers migrate or what routes are taken.

Based on the oceanographic and physical parameters of the one known aggregation, it will be interesting to see whether other aggregations have similar requirements. The one known aggregation described here contradicts some of the widely held tenets of spawning aggregation occurrence. It is found along the side of a channel a considerable distance from the channel mouth, the currents at the site (a reverse eddy contrary to the general channel current flow) tend to imply eggs are not going to be easily dispersed into the open ocean from the aggregation site and beyond this possible effect of current conditions there is nothing distinctive about the aggregation area. Moreover, there is no obviously distinct seasonality in the formation of aggregations at this site.

12.10.3 Fishery and Traditional Knowledge

The lack of recorded traditional ecological knowledge (TEK) is interesting in that aggregation occurs in a location that is shallow and often in protected waters (strong westerly storms being the exception). It has undoubtedly been seen by local fishermen, yet no one has reported this information. This could be due to one of several possibilities. First, perhaps no fisheries person has asked the right fisherman (i.e. spearfisherman) about aggregations in the area where this occurs (see Sect. 10.2.3). Perhaps, due to the lack of exploitation or because the fish are relatively small and not of high commercial value, it was not considered as anything worthy of mention even if somebody asked. The most likely explanation is a combination of these factors which should elicit caution in thinking that interviews of fishermen by outsiders ever truly gathers all TEK available about fish and fisheries. It points to the limits of recorded TEK and the oftentimes disparity between what might be of interest to a biologist and a subsistence fisherman. There is no management of the species or aggregation site.

12.10.4 Aggregation and Spawning Among Other Snappers (Lutjanidae)

The reproduction of snappers is starting to become better known with several Atlantic species the most-studied to date. These include the cubera snapper, *Lutjanus cyanopterus*, and the mutton snapper, *L. analis*. Aggregation-spawning has been documented for some western Atlantic species (Wicklund 1969; Heyman et al. 2005; Kadison et al. 2006) with a number of other species having probable to possible records (see Appendix) that are being validated, and will hopefully be published in the next few years. For the Indo-west Pacific there were no previously published reports of verified spawning aggregations of the family. However in Palau, schools of some snappers previously assumed to be spawning aggregations

(Johannes 1978, 1981) have now been verified as such. Transient aggregations are confirmed for the twin-spot snapper, *Lutjanus bohar* (known as "red bass" in Australia), discussed subsequently, and the blue-lined sea bream, *Symphorichthys spilurus* (Chap. 12.11). Suzuki and Hioki (1979) reported group-spawning by 10 or more captive common blue-stripe snapper *Lutjanus kasmira* in an aquarium during evening hours. Sala et al. (2003) described spawning in species from the eastern Pacific (e.g. *Lutjanus argentiventris* and *L. novemfasciatus*) and some may have true spawning aggregations. All lutjanid species that have been observed to spawn to date do so in large sub-groups that break out of the main aggregation i.e. group-spawning. Dog snapper *Lutjanus jocu* also spawns in aggregations (see Fig. 9.11). Some of the larger snappers appear to be quite long-lived, making them potentially easy to overfish. If they are also fished on their spawning aggregations, then they are likely to need management to ensure that the fishery continues in the long term.

12.10.4.1 Twin-Spot Snapper – *Lutjanus bohar*

The twin-spot snapper occurs in the Indo-Pacific from east Africa to the Marquesas and Line Islands, north to the Ryukyu Islands, and south to Australia. It mainly lives around oceanic islands and is found in groups of up to thousands of fish (Fig. 12.24). It is an important commercial fish, as well as a popular game fish, taken by a wide range of fishing gears. There is little fishery information on the twin-spot snapper but it is known to be fished on its aggregations in some locations and has shown declines in some fisheries. Typically marketed dead, the species is also found in the Hong Kong-based live reef food fish trade, it is avoided in some areas because it can be ciguatoxic (Gillespie et al. 1986; Wright et al. 1986; R Hamilton, personal communication 2003; Sadovy et al. 2003a; Marriott and Mapstone 2006; Sadovy 2007).

The species is large and long-lived, attaining 56 years of age and up to 90 cm total length (TL). The species is probably gonochoristic (Allen and Talbot 1985; www.fishbase.org; Marriott and Mapstone 2006; Mariott et al. 2007) with females maturing at about 43 cm fork length (FL) (about 9 years).

In Palau, the species has long been reported to form aggregations (Johannes 1981; Sadovy 2007) and spawning was recently recorded from Palau just after dawn (Fig. 12.24). These aggregations are massive and, as a spectacle of nature, rival those of the cubera snapper found in Belize (Heyman et al. 2005). As with some other species of fishes spawning in Palau, the released eggs and sperm are often targeted by black snapper, *Macolor niger*, only seconds after release. They are reported to form probable spawning aggregations and may spawn every month in Papua New Guinea and the Solomon Islands during full moon and the third lunar quarter in groups of several hundreds of fish (R Hamilton, personal communication 2003). Fish in Palau are known to aggregate and spawn at the southern end of Peleliu Island, the southernmost extension of the shallow reef around the main Palau group. The aggregations are closely associated with those of the

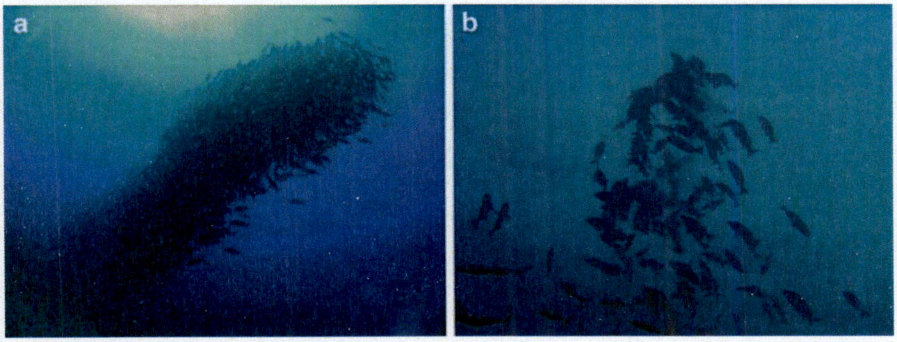

Fig. 12.24 (**a**) Spawning aggregation of twin-spot snapper angled up off the bottom at Peleliu Island, Palau at 10.30 am after spawning has occurred. (**b**) Spawning of twin spot snapper with large groups rising from the main aggregation to release gametes near the surface at dawn (Photos: © Jiro Sakaue)

blue-lined sea bream and these groups are receiving detailed attention. In fisher interviews in Palau, the only snapper consistently identified to aggregate with eggs was *Lutjanus bohar* (*L. gibbus,* humpback red snapper, was also mentioned by one fisher). The time when this species had eggs was reported to vary, from several days after or before full moon to the new moon phase for many months of the year; the months of February to August being identified specifically. Groups of fish with eggs were particularly located outside channel mouths and outer reef slope. At these times, thousands of kg of fish could be caught; catches were reported to have been high in the 1980s, but to have since declined (Sadovy 2007). In Johannes (1981) *L. bohar* species was reported to gather with eggs at the full moon from April to July, possibly in more months, at the outer reef slope.

The formation of large aggregations and group-spawning behaviour was observed for the twin-spot snapper in Palau where, in at least one location, it shares the same spawning site with the blue-lined sea bream, although the duration of the annual spawning season appears to be longer in the twin-spot snapper (December–June), when compared to the blue-lined sea bream (February–May). Aggregation size is probably no more than 10,000 fish and the number of fish changes in different months. The twin-spot snapper begins to aggregate 5–6 days before the full moon and spawning finishes at the full moon or the following day. Spawning occurs for about 3–4 days in any one lunar cycle in the twin-spot snapper, which is shorter than for the blue-lined sea bream (6 or more days). While spawning occurs during an easterly current at the tip of a protruding reef, the fish will rest in a sheltered area, where there is no current, 200 m away from the spawning site when not spawning. At spawning, which occurs at sunrise, a small group of fish move upwards into the water, with a female leading two or more males. Spawning does not stop, even if a large predator or diver approaches (JS, personal observation, unpublished data).

12.10.4.2 Mutton Snapper – *Lutjanus analis*

The mutton snapper is distributed in the western Atlantic, where it occurs on the continental shelf and clear water around islands, in bays, estuaries, mangrove and reef areas. The species forms small schools during the day which disband during the night, and juveniles are found in both brackish and marine waters. The species can attain 94 cm TL, 40 years of age and at least 15 kg. Sexual maturation is reached at about 40 cm FL in males and about 46 cm FL in females and the species is probably gonochoristic (Allen and Talbot 1985; Bortone and Williams 1986; Huntsman 1996; Burton 2002; Cummings 2007; Murray and Bester 2009; www.fishbase.org).

The mutton snapper spawns in large transient aggregations offshore consistently at the same sites year after year on outer reef slopes, reef promontories and drop-offs. Aggregations last for several weeks in each of several months a year and have been recorded from April to September during the full moon or third lunar quarter period (Burton 2002; Claro and Lindeman 2003; Burton et al. 2005; Graham et al. 2008).

The mutton snapper, sometimes marketed as 'red snapper', is fished commercially and recreationally with a wide range of fishing gears. Its spawning aggregations have been targeted by commercial fishers, including with the use of set nets in some places. Aggregations contribute around 50% of the total annual catch of the species in some areas. Annual landings of the species have declined substantially in many countries, probably largely due to aggregation-fishing but also because juveniles are often taken in catches because of the large size of sexual maturation of the species relative to most other species in the multi-species fisheries of the region. In Cuba, maximum catches were obtained in May and June, during the peak spawning season, and declines in landings or catch per unit of effort, and sometimes in average length taken, have been noted in Cuba, Belize, Jamaica, Puerto Rico, and Florida (Thompson and Munro 1983, Claro et al. 2001; Cummings 2007; Graham et al. 2008; Claro et al. 2009).

The species is little managed. Even where there are some management measures, such as in Cuba and Florida, either they are not enforced or aggregation protection has resulted in a shift of effort from spawning to non-spawning seasons. In one protected reserve fish numbers in the aggregation increased after a number of years of protection (Burton 2002; Burton et al. 2005). The species can be produced in hatcheries but there is no commercial production by culture (Watanabe et al. 1998; Benetti et al. 2002). The species was listed as vulnerable on the IUCN Red List due to overfishing and lack of management and the species status is being updated (www.iucnredlist.org).

12.10.4.3 Cubera Snapper – *Lutjanus cyanopterus*

The cubera snapper is the largest lutjanid in the western Atlantic and is distributed north to Nova Scotia, Canada, Bermuda, and south to the Brazilian coast of Santa Catarina State. It reaches a maximum of 160 cm TL and 57 kg, is probably gonochoristic and reaches sexual maturity at about 65 cm TL. It is reef-associated with

Fig. 12.25 Cubera snapper spawning in Belize (**a**) males and females releasing eggs and sperm. (**b**) Fish returning to the substrate after spawning (Photos: © Doug Perrine/SeaPics.com)

young fish typically found in inshore mangrove areas and seagrass beds, sometimes in estuaries and freshwater canals (Lindeman and DeMaria 2005). It spawns in aggregations gathering in large numbers mainly from April to July at consistent locations and times around the full moon and spawning at around sunset (Heyman et al. 2005; Kadison et al. 2006; Fig. 12.25). Aggregations occur on outer reef slopes, reef promontories sandy drop-off areas of the shelf edge and deep reefs and fish gather in their hundreds to many thousands and the snapper is known to spawn in the same areas and times as the dog snapper, *Lutjanus jocu*, and the grey snapper, *Lutjanus griseus* (Allen and Talbot 1985; Huntsman 1996; Domeier and Colin 1997; Claro and Lindeman 2003; Heyman et al. 2005; Murray and Bester 2009). Whale shark, *Rhincodon typus*, are known to feed on the spawn of the cubera snapper (Heyman et al. 2001; Chap. 1, Fig. 1.3a).

The cubera snapper is a popular commercial fish and is also taken recreationally, although it is sometimes ciguatoxic. It is fished by line and set nets on its aggregations in many places, and in some locations most of the annual catch is from the spawning aggregations. In Belize, aggregations are also fished at night. Marked declines in catches have been noted in some areas, including Puerto Rico, Cuba and Belize. The species receives little management although a site in Belize where the aggregation is visited by whale shark receives some protection for its ecotourism value (Heyman et al. 2001). Since juveniles occupy inshore areas, the condition of mangroves and seagrass beds is important to the species. The species was listed as vulnerable on the IUCN Red List due to overfishing, especially on its aggregations, and lack of effective management and the species status is being updated (Lindemann et al. 2000; Government of Belize 2003; Claro and Lindeman 2003; Heyman et al. 2001, 2005; Lindeman and DeMaria 2005; Claro et al. 2009; www.iucnredlist.org).

12.11 The Blue-Lined Sea Bream – *Symphorichthys spilurus*

Jiro Sakaue, Hiroshi Akino, and Hitoshi Ida
Southern Marine Laboratory, P.O. Box 1598, Koror, Republic of Palau 96940
e-mail: bigrun2006@hotmail.co.jp

Day Dream Palau, P.O. Box 10046 Abby's Bld, 1st Floor, Malakal Koror, Republic of Palau 96940
e-mail: akino@daydream-dive.com

Kitasato University School of Marine Lifesciences, Sanriku-cho, Ofunato-shi, Iwate-prf., 022-0101, Japan
e-mail: palau-sakana@ab.cyberhome.ne.jp

12.11.1 General

The blue-lined sea bream, *Symphorichthys spilurus* (Günther 1874), also known as the threadfin bream or sailfin snapper, is a species of the snapper family, Lutjanidae, distributed in the subtropics and tropics of the western Pacific. The species is dramatically distinctive in having long filamentous rays on the dorsal and anal fins, and occurs in shallow waters over sandy substrates around coral reefs. It is typically seen by divers as single individuals in inshore lagoon and shallow reef areas but is known to undergo seasonal migrations away from resident reefs to form large spawning aggregations at specific times and places each year (Myers 1999). In Palau these aggregations have been known by divers for decades. The species reaches 60 cm total length (TL) and feeds on fishes, molluscs and sand-dwelling crustaceans. It is of minor commercial importance in fisheries and for the aquarium trade.

12.11.2 Reproductive Biology

The blue-lined sea bream is little studied although its reproductive biology has been examined in Palau. The species aggregates in large numbers to spawn during a limited reproductive season each year (Fig. 12.26). These aggregations are a spectacle well-known to divers for decades, and occur in at least three areas, one each to the north, west and south of Palau. Based on the timing of aggregation formation, there appear to be two reproductive periods, February to May and September to November. The southern aggregation occurs at Peleliu Island and was studied from 2006 to 2009 by diving. Underwater photos were used to assist estimating fish numbers and to record colouration and behaviour.

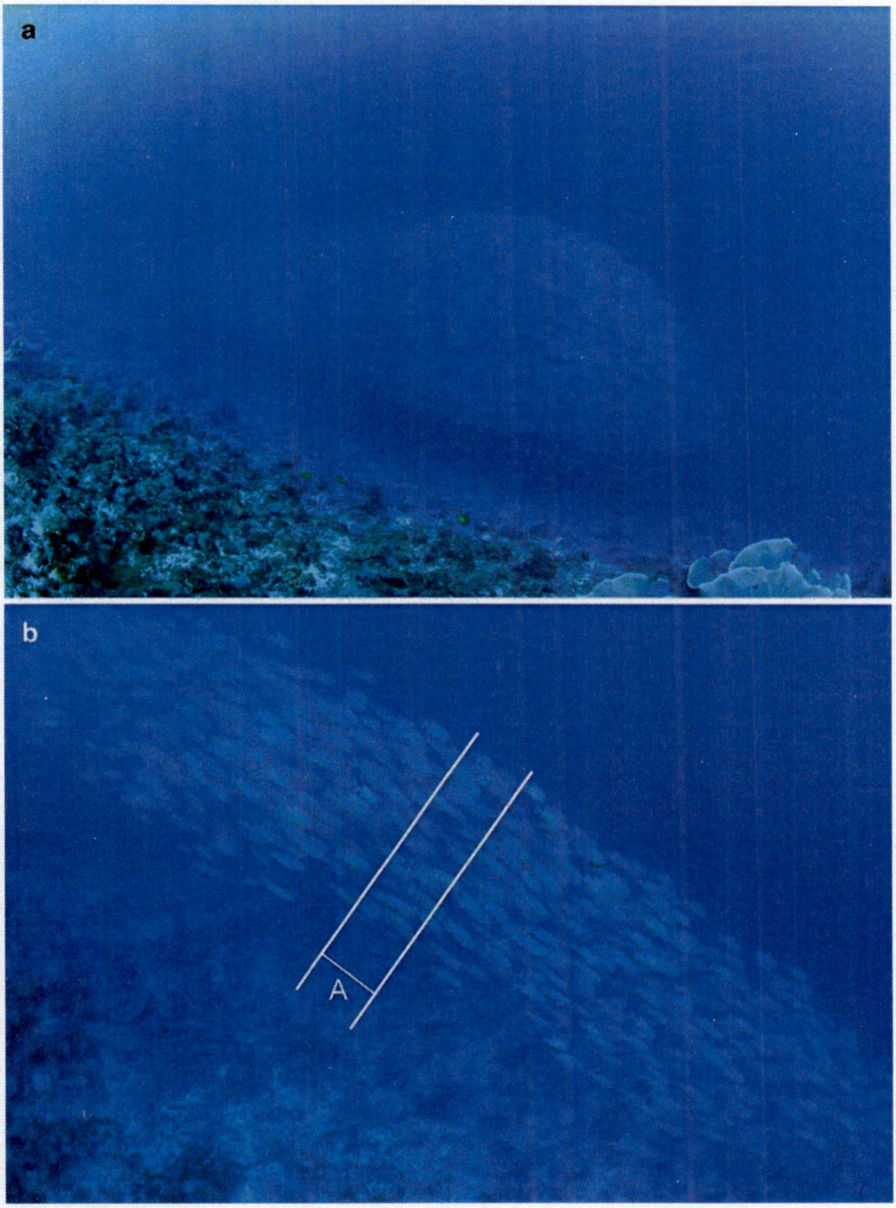

Fig. 12.26 (**a**) Large moving aggregation of *Symphorichthys spilurus* and (**b**) illustration of method used to estimate fish numbers by using photos and a focal area (**A**). See text for details (Photos: Jiro Sakaue)

The species aggregated in huge numbers at the study site in densely packed groups of numbers estimated at many tens of thousands possibly exceeding 50,000 fish. These numbers were estimated by photographing schools and counting numbers of fish at a focal location, then adjusting for the density of the school

(10 replicate counts of school depth), which varies over time, the speed of movement and the time taken for the entire school takes to pass the focal location (Fig. 12.26b). These groupings form twice a year and, at other times, the site does not contain the species. Water temperatures during aggregation are generally 28.0°–29.5°C (Patrick Colin, personal communication, 2008). All three sites appear to have distinctive promontory-like features and are located on outer reefs. Aggregation has a lunar pattern, starting after the full moon, with fish numbers peaking at around the third quarter and disappearing by the following new moon. During the aggregating season, the species makes diel migrations of about 2 km moving from sheltered areas in daylight hours to shallow areas at night, with a minor migration to the deeper spawning site at dawn. Interestingly, the stomachs of fish during the aggregation period were found to be empty and they would not take bait at this time.

Spawning occurs in small sub-groups that form from within the large aggregations which appear to move up and down in depth in response to prevailing water movements (Fig. 12.27a). Fish stay near the substrate and are more spread out over a large area in strong currents and become more concentrated and three-dimensional as fish move up in the water column when currents are less strong. The sub-groups are believed to be composed of a single female and several males. Males take on a temporary and distinctive colouration during courtship with the blue lines and the inner side of the dorsal and caudal fins becoming darker with the body colour in general intensifying. As spawning approaches, the dark coloured males of a single subgroup approach females, which are lighter coloured and have clearly swollen abdomens, to chase and nuzzle them (Fig. 12.27b). Several males may simultaneously engage in this courtship behaviour with a single female. The female then begins to rush into the upper layers of the aggregation followed by several males and she leads the male group at high speed and with rapidly changing direction. The spawning occurs in a rapid burst of swimming either vertically or horizontally that culminates in a release of gametes at its apex and the rapid return of the fish to the substrate and the aggregation.

The eggs are pelagic, 0.80 ± 0.01 mm in diameter, with a single oil globule, and hatch about 24 h after fertilization into larvae of 2.3 mm TL that appear to be typical of lutjanid larvae.

12.11.3 Management and Conservation

In Palau the species is not common in the local markets and was not reported as aggregating recently by fishers interviewed on species exploited from spawning aggregations (Sadovy 2007). However, fishers reported being aware of schools of this species in the 1980s and commented that these could be fished out from areas by spearfishing (Johannes 1981). Since the species does not appear to be common on Palau reefs, each spawning aggregation may draw fish in from a wide catchment area, making them susceptible to overfishing (Johannes 1981). The blue-lined seabream is listed as an important marine resource for a number of western Pacific countries (e.g. Nichols 1991; Hamilton et al. 2004).

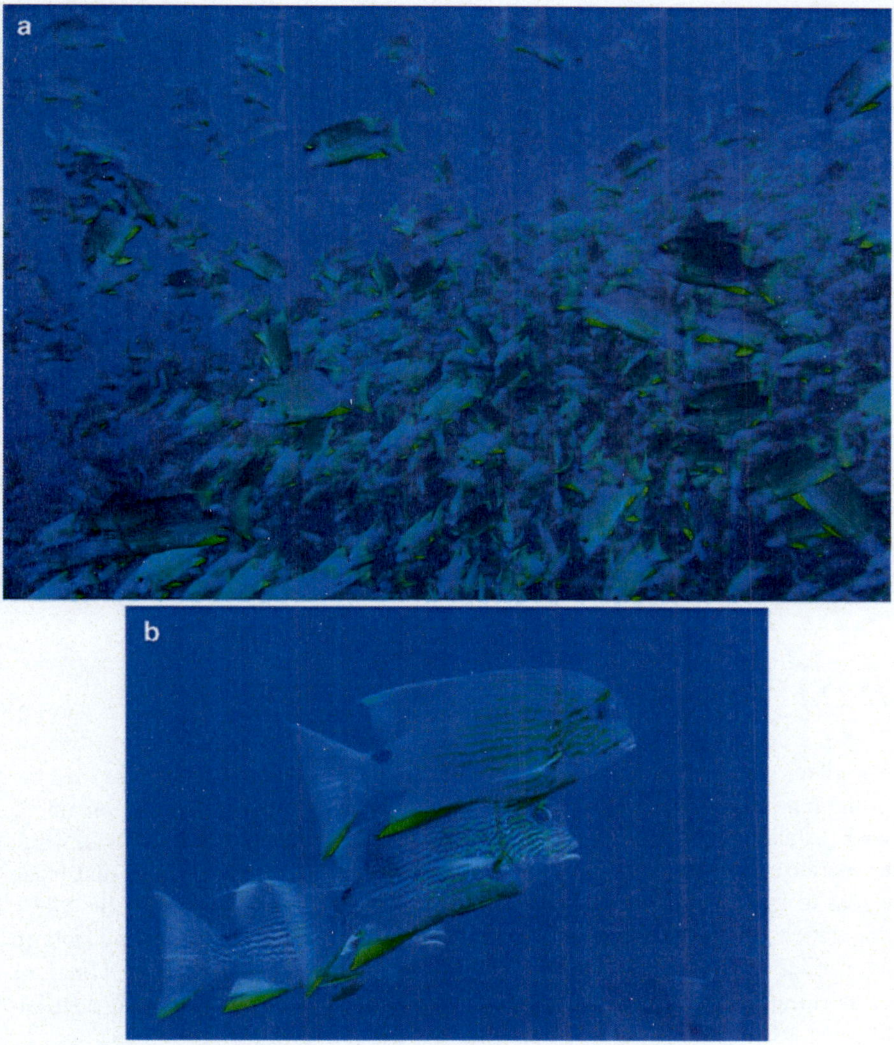

Fig. 12.27 (**a**) Large stationary aggregation of *Symphorichthys spilurus* and (**b**) subgroup of female (lighter and leading) and two males (Photos: Jiro Sakaue)

Reports (non-validated by field observations) from fisher interviews at two sites in Papua New Guinea (Hamilton et al. 2004) suggested similar movements to and from aggregation areas as reported by the present study. Day time aggregation sites were located at a passage wall and large rocky outcrop with caves while night time sites were on a seaward facing reef slope. Aggregations lasted about 3–5 days and were reported to form monthly in one location and for only 3 months at the other. Observations were provided by spearfishers who claimed that their flashlights and spearing disturbed the fish 'floating' up in the water column. Fish aggregations numbered 100–200 and daytime aggregations had been known for decades, while the night time aggregation at one site was discovered by a spearfisher in 2003 (Hamilton et al. 2004); fishers can remove up to 60 fish from an aggregation in less than 2 h. The species is used for subsistence and artisanal purposes and catches appear to be stable.

The species has not been assessed for its conservation status. While it is typically not managed, at one site in PNG a ban was placed in May 2004 on day and night-time spearfishing at this site in the 10 days leading up to and including the new moon. This lunar ban is in force for every month of the year. Only subsistence hook and line fishing is allowed here during this lunar period (Hamilton et al. 2004).

12.12 Silver Seabream ('Snapper') – *Pagrus auratus*

Gary Jackson
Department of Fisheries, Western Australian Fisheries & Marine Research Laboratories, P.O. Box 20, North Beach, Western Australia, 6920 Australia
e-mail: Gary.jackson@fish.wa.gov.au

12.12.1 General

The silver seabream or 'snapper', *Pagrus auratus,* is a large and long-lived member of the family Sparidae (seabreams and porgies, referenced as 'silver seabream' in www.fishbase.org) (Fig. 12.28). The species is widely distributed throughout warm temperate and subtropical waters of the western Indo-Pacific from China and Japan across to India and down to Indonesia, Australia and New Zealand (Paulin 1990). Populations in the northern and southern hemispheres are independent and isolated but similar enough to be considered the same species (www.fishbase.org). Snapper support important commercial and recreational fisheries throughout their distribution and are of some interest to the aquaculture sector (Kailola et al. 1993).

Juveniles typically inhabit sheltered marine waters such as bays and inlets, often over mud and seagrass (Kailola et al. 1993). Adults may be associated with reefs but are also found over mud and sand out to depths of more than 300 m (Kailola et al. 1993). Snapper can attain 130+ cm total length (TL) and 20+ kg in weight (Gomon et al. 2008). Fish commonly reach 40+ cm TL and mature at 20–30 cm TL (www.fishbase.org). The species is an opportunistic demersal predator with a broad-ranging diet that can include crustaceans, bivalves, cephalopods, marine worms, starfish, sea urchins and fishes (Coutin et al. 2003). Snapper can live to 40+ years in Australia (Norriss and Crisafulli 2010) and 50–60 years in New Zealand (Francis et al. 1992).

Snapper in Shark Bay, Western Australia, have been extensively studied since the 1980s, greatly enhancing our understanding of the species biology and management of its exploitation (Moran et al. 2005; Jackson 2007). Stock structure in the region is complex. Genetics, tagging, otolith chemistry and other techniques have shown little

Fig. 12.28 A large snapper (approx. 60 cm TL) from the Eastern Gulf, Shark Bay, Western Australia (Photo: © Department of Fisheries, Western Australia)

or no mixing of juvenile, sub-adult and adult snapper between oceanic waters outside Shark Bay and the inner gulfs, or between the separate inner gulf areas (Jackson 2007) (Fig. 12.29). Consequently, four separate stocks are recognised: an 'oceanic' stock that supports a major commercial line-based fishery outside the Bay and separate 'Eastern Gulf', 'Denham Sound' and 'Freycinet Estuary' stocks that support an important recreational boat-based fishery in the gulfs (Stephenson and Jackson 2005). Research on snapper in these inner gulfs has been particularly instructive because it has revealed considerable variation in life history characteristics within a relatively small geographic area and it has been able to track how the separate stocks have responded to management intervention over the last 15 or so years (Jackson 2007).

12.12.2 Reproductive Biology

Snapper are functional gonochorists with prematurational sex change (Francis and Pankhurst 1988). Spawning in the Shark Bay region occurs from April–October (Wakefield 2006; Jackson et al. 2010). Peak spawning in the inner gulfs occurs around May–July in the Eastern Gulf and Denham Sound and around August-October in the Freycinet Estuary (Table 12.4). Females reach 50% maturity at 35–42 cm fork length (FL) and 3–5 years of age, and males at 24–33 cm FL and 2–3 years of age (Table 12.4). Snapper are batch spawners that display asynchronous gonadal development and indeterminate fecundity (Mackie et al. 2009).

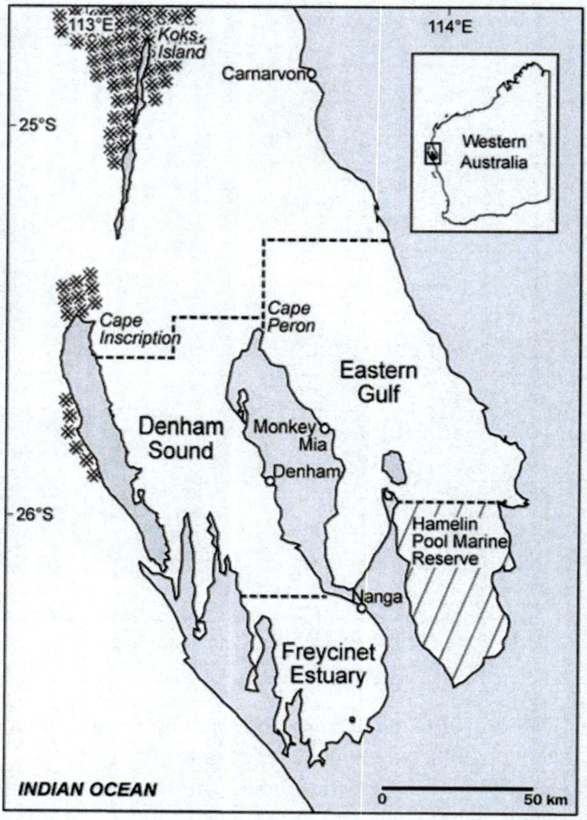

Fig. 12.29 The main recreational fishing areas in the inner gulfs of Shark Bay, Western Australia. The boundaries between Denham Sound, Freycinet Estuary and Eastern Gulf approximate the distributions of three separate inner gulf snapper stocks. Hamelin Pool is a marine nature reserve and closed to all fishing. The main commercial fishing grounds for the oceanic snapper stock are shown by the starred symbol

Spawning is highly synchronised and occurs at the same key spawning locations each year. During the spawning season, spawning mostly occurs over 3–4 days around the new, and to a lesser extent, full moon, mainly between the early afternoon and evening (Jackson et al. in press). Batch fecundity is size-related; a 19-cm-FL female produces ~1,500 hydrated oocytes at each spawning compared with ~650,000 hydrated oocytes by a 71-cm-FL female (Mackie et al. 2009).

12.12.3 *Fishery and Management*

Commercial line-fishing in oceanic waters and recreational boat-based fishing in the gulfs have both developed due to the higher catch rates/larger catches made possible through the targeting of spawning aggregations during austral winter. Commercial snapper fishing in oceanic waters dates back to 1908 (Cooper 1997), and in the inner gulfs to the 1920s–1930s (Edwards 2000). Fishing effort in the gulfs increased through to the 1950s then steadily declined in the 1960s–1970s (GJ unpublished data) as boats shifted to the main commercial grounds in oceanic waters (Fig. 12.29) in pursuit of higher catch rates and in response to management action (Moran et al. 2005). Commercial snapper fishing of the oceanic stock came under formal management in 1987 with the creation of the Shark Bay Snapper Managed Fishery that has been managed with individual quotas since 2000 and a Total Annual Commercial Catch (TACC) since 2003 (Jackson and Lai 2008). Commercial catches of snapper in the inner gulfs nowadays are only small (<2 tonnes per year) and limited to bycatch taken by local beach seine fishing operations (Jackson et al. 2008).

Recreational fishing in the inner gulfs was observed to steadily increase between the late 1970s and 1990s (Jackson et al. 2003). Between the early 1980s and early 1990s, road access into Shark Bay (Denham) improved and recreational snapper fishers became more efficient with advances in fishing technology (e.g. Global Positioning System and colour fish sounders) and increased knowledge of the main aggregations. By the mid-1990s, there was concern that recreational fishing of the winter spawning aggregations in the Eastern Gulf in particular had reached unsustainable levels (Marshall and Moore 2000; Anderson 2004). In response to the community concern and to obtain much needed information for management, research commenced in 1997 that included daily egg production method (DEPM) surveys to estimate snapper biomass (Jackson et al. in press) and recreational fishing surveys to estimate recreational catches (Sumner et al. 2002).

Based on research results that indicated that all three snapper stocks had been over-exploited, stricter management was progressively introduced from 1997 onwards, to reduce recreational snapper catches in the winter months. These measures included a 5-year moratorium on snapper fishing in the Eastern Gulf (June 1998–March 2003), a 6-week spawning closure in the Freycinet Estuary (from 2000 onwards) and introduction of a Total Allowable Catch (TAC) for each stock for the first time in 2003 (Stephenson and Jackson 2005; Jackson 2007) (Table 12.4). Different combinations of management measures have been used since 2003 (Table 12.4) including a novel management quota-tag system in the Freycinet Estuary (Jackson et al. 2005; Mitchell et al. 2008). Recreational fishers in this area are only permitted to retain a snapper when in possession of a quota-tag (Fig. 12.30). A limited number (n = 1040 per year, Table 12.4) of these quota-tags are only available via a lottery-based allocation process each year (before the season starts) where fishers are limited to only 2 tags per person. During fishing, once a snapper is landed and is to be retained, a tag is inserted through the mouth and secured using a tamper-proof barrel-style locking mechanism prior to landing (Fig. 12.30).

The management measures in place between 2003 and 2007 were highly effective and limited snapper catches to 13–28% of TAC in the Eastern Gulf, to 32–48% of TAC in Freycinet Estuary and to 37–102% of TAC in Denham Sound (GJ unpublished data). The extensive management interventions since 1997 have been very successful in recovering local snapper stocks; breeding stocks (spawning biomass) in the Eastern Gulf and Denham Sound have rebuilt to above the management target levels and the breeding stock in Freycinet Estuary is recovering (Jackson et al. 2008).

Table 12.4 Summary of biological and management characteristics for snapper stocks in the inner gulfs of Shark Bay, Western Australia

	Denham Sound	Eastern Gulf	Freycinet Estuary
Reproduction:			
Peak spawning	May–July	May–July	August–October
Female L_{50} (cm)	40	35	42
Female L_{max} (cm)	75	75	79
Female A_{50} (years)	5.5	3.2	4.5
Female A_{max} (years)	19	17	29
Male L_{50} (cm)	28	24	33
Male L_{max} (cm)	73	60	79
Male A_{50} (years)	2.7	1.6	2.7
Male A_{max} (years)	17	15	31
Sex ratio	~1:1	~1:1	~1:1
Fecundity	$F = 9.44 \times 10^{-5} FL^{3.36}$	$F = 9.44 \times 10^{-5} FL^{3.36}$	$F = 9.44 \times 10^{-5} FL^{3.36}$
Growth:			
Females			
K (year^{-1})	0.14	0.18	0.17
L_{∞} (cm)	76	76	77
Males			
K (year^{-1})	0.18	0.17	0.17
L_{∞} (cm)	66	75	77
Stock status (last assessment in 2008):	$B = \sim 42\%$	$B = \sim 45\%$	$B = \sim 25\%$
Current management:	TAC 15 tonnes, bag limit (1 fish) & slot limit (>50<70 cm TL)	TAC 15 tonnes, bag limit (1 fish) & slot limit (>50<70 cm TL) Plus 3-month spawning closure (May–July)	TAC 5 tonnes, slot limit (>50<70 cm TL) Plus 6-week spawning closure plus management tags (1040 available for recreational fishers via lottery each year)

All lengths are fork length. F batch fecundity. Growth parameters are based on von Bertalanffy model. B spawning biomass relative to unexploited level. TAC Total Allowable Catch (information from Jackson 2007; Jackson et al. 2010)

12.12.4 Lessons Learnt in Managing Snapper During Spawning Season and Data Gaps

Size and bag limits do not limit recreational fishing effort and on their own have only very limited effect on reducing the overall level of catch or fishing mortality. The size and bag limits that have been used since 1998 have resulted in very large numbers of snapper being released each year (~40,000 in all three areas combined,

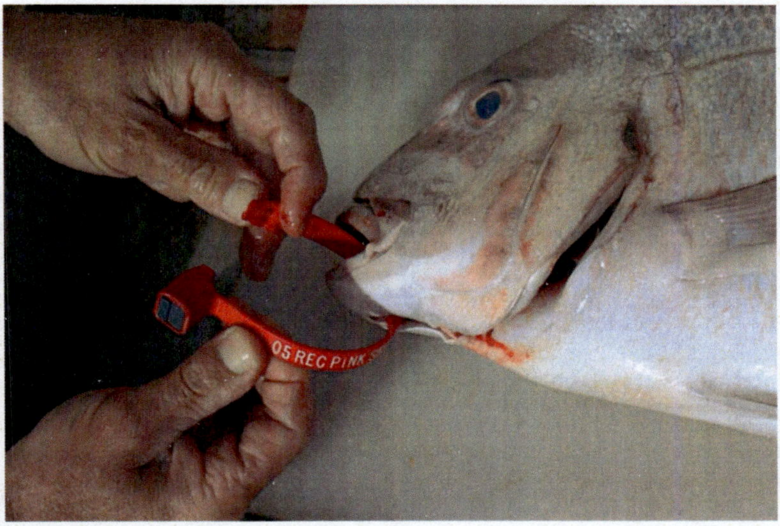

Fig. 12.30 Management-tags used to limit the recreational catch of snapper to the TAC in the Freycinet Estuary, Shark Bay, since 2003. Tags are inserted through lower jaw and secured with tamper-proof lock (Reproduced from Mitchell et al. 2008)

GJ unpublished data). However, snapper are considered to be more robust than many other species and in the shallow waters of the inner gulfs (mostly < 20 m), discard mortality due to barotrauma is not considered to be significant (Lenanton et al. 2009). Commercial trawl fisheries for prawns and scallops operate in limited areas of Denham Sound where numbers of age (years) 0+ and 1+ snapper (10–15 cm TL) may be taken as bycatch. Research has shown that the prawn fishery reduces juvenile snapper survival from 8% to 6% per year. This equates to a loss in annual yield similar in magnitude to that resulting from the return of undersized fish (discard mortality) in the recreational fishery (Moran and Kangas 2003).

Spatial closures can be effective when the spawning aggregations are within a well-defined and manageable area, as they are off Monkey Mia in the Eastern Gulf. A similar approach was discussed in relation to the management of snapper in Denham Sound (in 2002) where the spawning grounds cover an extensive area but the local community was not prepared to support the large spatial closure required. Spatial closures also present challenges to authorities as they need to be effectively marked on-the-water and regularly monitored by compliance officers.

Temporal closures to protect spawning aggregations can be effective in some cases. Their effectiveness depends on their timing and duration that in turn require a sound knowledge of the reproductive biology of the target species. Because snapper in the inner gulfs spawn over 3-4 months and are aggregated throughout this period, spawning closures need to be extensive to be effective. The success of management in the Eastern Gulf during 2003–2005 was mostly due to the 4-month duration of the spawning closure; catches were 13-23% of the TAC during this period but rose to 25-28% of the TAC in 2006-2007 following a reduction in the closure to 3 months duration. In contrast, the 6-week spawning closure in the Freycinet Estuary did not reduce the recreational snapper catch to the management target when it was first introduced in 2000 (Sumner and Malseed 2002). The closure saw a subsequent seasonal shift in recreational fishing effort in 2001 and 2002 as fishers visited the area at alternative times (July or October) to avoid the closure, and the snapper catch remained at a similar level as a result (Jackson et al. 2005).

The estimates of recreational catch from boat ramp surveys indicate that management quota-tags have been very successful in limiting the recreational catch in the Freycinet Estuary since 2003 (Jackson et al. 2008). The quota-tag system works to limit participation, with the total number of tags limited, and consequently exploitation rates are reduced. The success of the tags is related to the limited number of access points that assists with compliance efforts, low levels of release mortality due to the shallowness of waters in the Estuary, and as fishing is targeted at spawning aggregations, bycatch of other species is minimal. The tags are subject to high levels of compliance and acceptance with fishers 7–8 years after their introduction.

Elsewhere in Western Australia, large snapper migrate into the marine embayments of Cockburn Sound and Warnbro Sound, near Perth (32° S, 116° E), to form spawning aggregations between September and January each year (Wakefield 2006). These aggregations were historically targeted by commercial and recreational fishers but have been protected by seasonal closures each year since 2000. Both areas are now closed to all snapper fishing between October and January each year.

In South Australia, commercial and recreational fishers have historically targeted spawning aggregations in areas such as northern Spencer Gulf and Gulf St Vincent (34–35°S, 137–138°E) between November and January. Spawning season closures have been used since 2000, with all of November now closed to snapper fishing statewide (Anthony Fowler personal communication).

Adult snapper migrate into Port Phillip Bay near Melbourne, Victoria (38° S, 145° E), during October-December to spawn (Coutin et al. 2003). While both recreational and commercial fishers have historically targeted the annual spawning-related movements of snapper around the Bay, no specific management measures are in place to manage the spawning aggregations. Similarly, no dedicated measures are currently used to manage snapper during the spawning season in New South Wales and Queensland.

12.13 Humphead Wrasse – *Cheilinus undulatus*

Patrick L. Colin and Yvonne Sadovy de Mitcheson
Coral Reef Research Foundation, PO Box 1765, Koror, Palau 96940
e-mail: crrf@palaunet.com

School of Biological Sciences, The University of Hong Kong, Hong Kong, China
e-mail: yjsadovy@hku.hk

Fig. 12.31 The humphead wrasse, *Cheilinus undulatus*, is the world's largest wrasse and an iconic species on coral reefs visited by tourist divers. (**a**) Adult males can be over 1.5 m total length and have the characteristic bulbous head; adult females are smaller, up to about 1 m TL, (**b**) small individuals do not have a bump on the head and have somewhat different colouration (Photos: Patrick L. Colin)

12.13.1 General

The humphead wrasse, *Cheilinus undulatus* Rüppell, is the biggest labrid in the world and one of the largest bony fishes found on coral reefs (Fig. 12.31). A review Sadovy et al. (2003b) reports a maximum reported total length (TL) of over 200 cm (perhaps as much as 250 cm) with weights of 191 kg. However, lengths greater than about 150 cm are only rarely observed (Choat et al. 2006). It is a member of the charismatic reef megafauna, which includes sharks, turtles and large reef fishes. Large males with their notably bulbous heads are high on the list of nearly every tourist diver as a "must see" (Fig. 12.31a) and this feature (as well as large size and similar name) occasionally causes it to be confused with the bumphead parrotfish, *Bolbometopon muricatum* (where both sexes have bulbous heads) by uninformed divers. The prominent cephalic hump occurs in both sexes and increases significantly with body size (Fig. 12.31a). Since females do not appear to grow bigger than about 100 cm TL, the largest fish are males and these have the most developed hump (Liu and Sadovy de Mitcheson 2011). Smaller fish have different colouration and detail (Figs. 12.31a, b, and 12.32a). The species is also known by other English common names, such as Napoleon fish and Maori wrasse. Sadovy et al. (2003b) has a listing of 24 non-English common names for the species.

12.13.2 Distribution and Feeding

The humphead wrasse occurs throughout much of the tropical Indo-west Pacific, from the Red Sea and east Africa in the west, but is absent from some marginal areas of the central Pacific, such as Hawaii, Rapa and Pitcairn.

Many aspects of its biology were summarized by Sadovy et al. (2003b). It is largely a carnivore, searching the reef and often turning over rocks (Fig. 12.32b), looking for molluscs and fishes (Randall et al. 1978). The fish has powerful pharyngeal dentition capable of crushing most molluscs and echinoderms. It is sometimes shadowed by other reef predators, probably seeking opportunistic feeding as the wrasse turns over rocks. It is one of the few predators on the crown-of-thorns starfish, *Acanthaster planci*, and a decrease in the population of the fish may have been a factor in population explosions of the coral-eating sea star. In unfished or lightly fished areas its densities range from 2 to 27, rarely more than 10 fish, per 10,000 m^2. Where fished, densities drop tenfold or more to <1 fish per 10,000 m^2 (Sadovy et al. 2003b).

12.13.3 Early Life History and Juvenile Life

The planktonic eggs are 0.66–0.67 mm in diameter, spherical with an oil droplet 0.14 mm in diameter and no pigment. This egg size is towards the lower size limit of labrid eggs, and is similar to other Indo-west Pacific labrids that are up to three orders of magnitude smaller in body weight. This small egg/large fish relationship is also found in the world's largest scarid, the bumphead parrotfish, which has a similarly small egg (0.65 mm diameter) (Colin and Bell 1991). Humphead wrasse have spawned in hatcheries, but the larvae have proven difficult to rear due mainly to the small mouth and, therefore, small food requirements. The few larvae that have been reared to the juvenile stage took about 25 days post hatch to reach 2 cm in length (Slamet and Hutapea 2005). At 6 months of age, fish are only 5–6 cm TL. At present it appears unlikely that hatchery production is going to provide a ready supply of young to form the basis of an aquaculture programme or for release into the wild.

Little is understood about settlement or dispersal although the species is known to spawn in aggregations of up to several hundreds of fish. In Palau, current following drifters launched in areas with spawned eggs were transported a short distance

Fig. 12.32 (a) Juvenile humphead wrasse differ from larger individual in being more black and white. They are shy. The eye slashes, however, are distinct and found in all sizes of fish (see also Figs. 12.31a, b, and 12.36). (b) Large male humphead wrasse feeding on a shelf edge reef in Palau. The humphead wrasse looks under rocks and overhangs, occasionally turning over rocks with its jaws. A twinspot snapper (*Lutjanus bohar*), a carangid and white tip reef sharks shadow the wrasse, ready to grab any creature disturbed by its activities (Photos: Patrick L. Colin, Mandy T. Etpison)

off the reef on the falling tide after spawning, then moved parallel to the reef with prevailing currents for several hours. Where aggregations have been eliminated, such as the Layang Layang site in Malaysia (below), the impact on recruitment and population maintenance is unknown, but it seems difficult for a population to be reestablished, particularly if nearby reefs were similarly overfished.

Little is known of the habitats and life history of juveniles. The review by Sadovy et al. (2003b) reported juveniles above about 3 cm (Fig. 12.32a) to occur in coral-rich lagoon areas, thickets of branching *Acropora* corals, in sea grass beds, murky outer river areas with patch reefs, shallow sandy areas adjacent to coral reef lagoons and in shore mangrove and seagrass areas. Dorenbosch et al. (2006) reported juveniles from 2.5 to 27.5 cm TL (but most commonly 5–7.5 cm TL) in seagrass habitats, while on coral reefs juveniles as small as 10 cm TL were rarely seen. Tupper (2007) considered coral-algal flats in Palau to be essential habitat for juveniles; however they also occur in a variety of other reef habitats. Due to their cryptic nature they are often difficult to see and techniques, such as small rotenone stations (Smith-Vaniz et al. 2006), would be needed to identify all types of habitats where they occur. There is a general ontogenetic shift from more inshore habitats to the offshore reefs and deep reef slopes where adults normally live (Sadovy et al. 2003b).

One acoustic tagging study in New Caledonia, using a single small probable female (45 cm TL), found the fish remained near its tagging point for about 25 days, then disappeared (Chateau and Wantiez 2007). Since individuals can be recognized by markings on head and body (e.g. Fig. 12.31a), if the fish can be approached closely enough to photographed, then a collection of identification photos can be prepared.

12.13.4 Reproduction, Growth and Sexual Pattern

The humphead wrasse can live several decades and is a protogynous hermaphrodite. Choat et al. (2006) has provided a demographic picture for the fish (Fig. 12.33) on the Great Barrier Reef (GBR), using sagittal otoliths for ageing, where juveniles take approximately 5 years to reach sexual maturity at about 35 cm standard length (SL). At approximately 9 years and 65 cm SL, or above, some females change sex to males. Remaining females continued to grow but did not exceed about 90 cm, or 32 years. Males grow faster, reaching lengths of 1.5 m but were not found to reach more than about 25 years in age. Specimens examined from Indonesia, GBR and Papua New Guinea, field observations in Palau and Malaysia, and minimum size for females to spawn in captivity in Indonesia indicate that first female sexual maturation typically occurs between about 40 and 50 cm TL (Sadovy et al. 2003b; Colin 2010; Sadovy de Mitcheson et al. 2010; Steve Oakley unpublished data). Being protogynous, most males develop by the sex change of functional females, although a few males appear to develop directly from the juvenile stage (Sadovy et al. 2003b).

Humphead wrasse spawn in aggregations which have been confirmed in the South China Sea off Sabah, Malaysia, the Great Barrier Reef and Palau (detailed in Colin 2010). It is a resident aggregator, somewhat surprising since most other large reef fishes that aggregate, such as groupers, have transient aggregations. When ready to spawn, males display themselves above the reef in a courtship posture with the anal fin pointed, the caudal fin scissored closed and the dorsal fin close to the body (Fig. 12.34). The face becomes blue in males and the head markings become

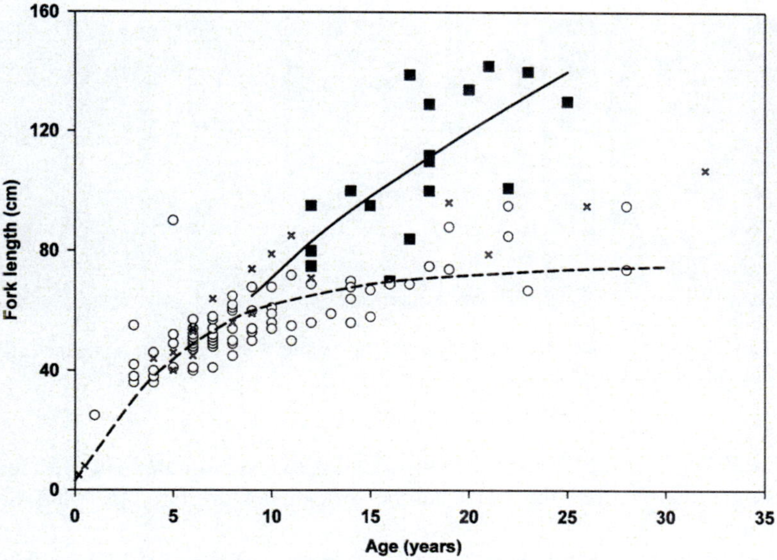

Fig. 12.33 Size at age (as determined by sagittal otoliths) plot displayed by sex with fitted von Bertalanffy growth function curves (sizes at t_0 constrained to 0). Females indicated by *circles*, males by *squares* with *crosses* for unsexed fish (Redrawn from Choat et al. 2006). With permission Marine Ecology Progress Series)

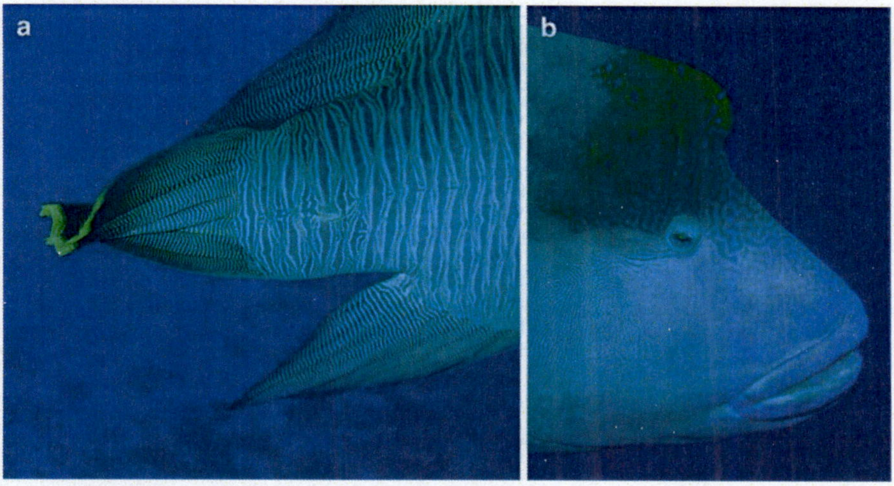

Fig. 12.34 When swimming above the reef in courtship posture, male humphead wrasse fold the caudal fin so the rays are scissored over one another (**a**) and the head becomes bluish with the marking around the eye and head much less distinct compared to (**b**) males during non-courting times (see Figs. 12.31a and 12.32b for comparison) (Photos: Patrick L. Colin)

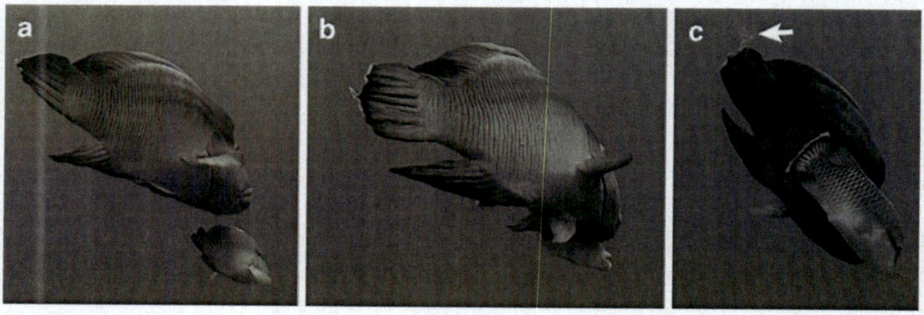

Fig. 12.35 Spawning sequence of *C. undulatus*. The smaller fish is the female and the egg/sperm cloud is *arrowed* (Photos: Mandy T. Etpison)

faint. In Palau Colin (2010) found up to 15 males and an estimated 100–150 females in a single aggregation (the exact number of females present was impossible to determine).

When ready to spawn the females rise up in the water column where they join with the males in a relatively sedate gamete release (Fig. 12.35) after which the female returns to the reef (see also Colin 2010). The male remains in the water column to spawn with additional females. In all areas spawning occurred during daylight a few hours (2–2.5 h in Palau) after high tide near the surface and the eggs were carried some distance off the reef by currents.

At Layang Layang there was only one aggregation for the entire 11 km long atoll with an outer reef area of about 12 km^2. Fish migrated to the western end of the elongate atoll aggregating on the projection of reef at the end (Nicholas Pilcher, personal communication 2006) and as of 2006 this aggregation was likely fished out by Live Reef Fish Trade (LRFT) boats. The LRFT is a large international trade, centred in Hong Kong and China for which the humphead wrasse is heavily sought after as one of the most economically valuable species.

In Palau there appear to be numerous aggregations at intervals along the outer slope on the barrier reef. There have also been observations of spawning at some promontory areas, such as Blue Corner and New Drop Off. Nothing is known about reproduction at inshore reefs, such as in the Rock Islands, but mature fish do occur there. It seems likely they actually spawn in these inshore areas (fringing reefs around Rock Islands) at presently unknown locations, as it would be very difficult to migrate the distance to the shelf edge on a regular basis (Colin 2010).

12.13.5 Commercial and Other Use

The species has high commercial and cultural value (Sadovy et al. 2003a). In parts of the Pacific it is considered a delicacy, saved for special occasions as a food item or its consumption limited to high status individuals in the community. It is not commonly available, and brings a high price dead in local markets. It also occupies

Fig. 12.36 This juvenile or small female humphead wrasse, in an aquarium at a Hong Kong restaurant, is soon destined for a dinner table. The typical sizes of fish on sale are between about 30 and 50 cm TL, below or at the size of sexual maturation (Sadovy et al. 2003a, b) (Photo: © Stanley Shea)

a special place in the Live Reef Fish Trade, being highly sought after and commanding an extravagant price (sometimes more than US$120 per kg) at the restaurant table in Southeast Asia (Sadovy et al. 2003a, b; YSM, personal observation) (Fig. 12.36). This status has exposed the species to extreme overfishing in many locations. Adults have virtually disappeared from some areas, with large fish only seen in areas that are effectively protected or that are otherwise inaccessible to fishing. Extensive regions where the species is still fished for export, such as eastern Indonesia, have very low numbers remaining according to underwater visual census (Sadovy et al. 2007). Indonesia is the major exporting country for the species although the Philippines provides many fish to international trade, albeit illegally. Sadovy et al. (2003a, b) showed a relationship between fishing pressure and fish densities at 24 localities, while individuals for the LRFT are often taken using destructive fishing practices, such as cyanide, which is commonly used to catch juveniles, for collection. In many areas, the species is taken by spear and is particularly easy to catch at night from its sleeping holes. The great majority of individuals in the live trade are typically 30–40 cm TL or less such that the fishery is largely one of juveniles; many smaller fish are taken and grown out in captivity until they attain the typical size range (Sadovy de Mitcheson et al. 2010).

Fig. 12.37 (**a**) Humphead wrasse is one of the fishes that all tourist divers want to see on reefs. Such "charismatic megafauna" is extremely valuable for tourism, but often such fishes become acclimatized to divers and are easy targets for spear or line fishing. (**b**) This humphead wrasse, known in Palau as "Abu", was a loved resident of the world-famous Blue Corner dive site in Palau and when he finally disappeared, after surviving numerous spear wounds and fish hooks, was memorialized with an obituary in the local newspaper (Photo: © Mandy T. Etpison, 2011)

The species is highly sought-after by divers. Individual large fish, found in the same area for years, are extremely valuable to local dive operators. The loss of a single large male can affect the local diving tourism industry (Fig. 12.37).

12.13.6 Management and Conservation

To stem the demise of this species it was listed on CITES Appendix II and as endangered on the IUCN Red List, both in 2004 (www.iucnredlist.org; www.humpheadwrasse.info). Some countries, such as Australia, Palau, and Fiji, have extended complete protection to the species, although there is continual pressure to reopen the species to fishing. Sadovy (2005) reported on the status of the fish in various areas of Indonesia, as well as attempting to quantify the international trade, in an effort to improve management for sustainable trade as mandated by CITES listing in Appendix II.

Growing knowledge of the biology of the humphead wrasse certainly indicates that a single management strategy is not suitable for all "large" species of reef fishes. It is a resident aggregator, spawning in a manner that is similar to various other wrasses while many other "large" fishes have transient aggregations.

The Layang Layang spawning aggregation described above disappeared after LRFT vessels caught several metric tonnes of them at the atoll around 2005–2006 (Nicholas Pilcher, personal communication 2006). In October 2006 only 2 individuals were seen in 2 weeks of survey work at Layang Layang, but a LRFT vessel was still working that atoll, and others nearby, for other species. They certainly would have taken any humphead wrasse found however few fish were present. Because of its luxury seafood status, consumers continue to seek it even when prices increase as supplies decrease (YSM personal observation 2006).

For effective management more information is needed but a number of protective measures are applicable. Since the humphead wrasse spawns at widely dispersed sites in relatively small groups over much, if not all, of the year, closure of specific sites, unless they are the only natural or the last remaining aggregation site, would not be effective to protect the local population. Sites outside the closure area would receive even more pressure and most could be overfished. There could be a seasonal closure (or opening), but this would make little biological sense unless there was a specific seasonality of aggregation and spawning. Ideally, large marine protected areas could encompass a number of aggregation sites to ensure persistence of spawning stock and activity. Management anywhere the species is exploited beyond subsistence levels will almost certainly require size limits (upper and lower) to ensure persistence of males and that juveniles survive to reproductive age. Annual quotas could be applied (for methodology see Sadovy et al. 2007), however, monitoring and management for effective implementation of such regulations is unlikely to be financially viable for such a limited fishery.

Since the species is now partially protected due to the Appendix II CITES listing which requires a sustainable management plan to be in place before exports are permitted, efforts should be focused on stopping illegal trade, which is considerable (YSM unpublished data; www.humpheadwrasse.info CITES page). There is also a need to document recovery of populations in areas where fishing has stopped. Other threats that remain include spearfishing at night, when fish are in their sleeping holes, particularly using SCUBA gear. Many countries do not allow the species to be exported and Malaysia and Indonesia have recently introduced annual export quotas of zero and 5,400 fish (for 2010; down from 8,000 for the previous few years). However, a recent review of the realities of sustainably managing this species, taking into account the costs of monitoring and enforcement concludes that any exports are unlikely to be sustainable suggesting that the species is not suited to the international export trade, in general, given its biology, the management necessary and the problems with successful enforcement of regulations (Gillett 2010).

12.14 Bluehead Wrasse – *Thalassoma bifasciatum*

Robert R. Warner
University of California Santa Barbara, Santa Barbara, CA, USA
e-mail: warner@lifesci.ucsb.edu

12.14.1 General

Bluehead wrasse (*Thalassoma bifasciatum* [Bloch, 1791]) is member of the wrasse (Labridae) family. Bluehead wrasse live in relatively shallow water over coral reefs and other hard-bottom substrates throughout the tropical

Fig. 12.38 Terminal male phase (*lower left*) and initial phase (*upper right*) which can be exhibited by both females and initial phase males (Photo: © Doug Perrine/SeaPics.com)

Western Atlantic, from Bermuda, Florida (USA), southeastern Gulf of Mexico and throughout the Caribbean Sea to northern South America. Maximum recorded size is about 25 cm total length (TL).

After a pelagic larval phase of about 50 days, bluehead wrasse settle out onto reefs and metamorphose into small (<10 mm standard length, or SL) juveniles. The fishes mature at about 35 mm SL and 3–4 months of age. All small adult fish are in the initial colour phase (IP), some of which are females and some males (Fig. 12.38). If an individual lives long enough, it transforms into the distinctive terminal colour phase (TP) at about 75 mm SL. Maximum longevity is about 3 years (Warner 1984a, 2002).

12.14.2 Reproduction

All TP individuals are males, and thus there are two sources for these individuals: they may result from simple colour change from an IP male, or they may be the result of both colour and sex change of an IP female. Thus the basic polymorphism in this species is not between males and females, but between protogynous hermaphrodites (operating either as females or secondary males) and non-sex-changing (primary) males. Smaller TP males range over the reef as bachelors or "floaters", but larger individuals maintain permanent territories around mating sites.

While local populations are genetically mixed due to extensive dispersal in the pelagic larval stage, after settlement juveniles and adults do not leave the reef on which they find themselves. Mating takes place year-round at specific sites on the reef, and on any particular day all mating occurs in a predictable 2-hour mid-afternoon period. Individual females mate nearly every day, and these females tend to be highly faithful to particular mating sites (Warner 1986).

Because the eggs produced by females are pelagic and there is no parental care, the eggs themselves immediately become part of the plankton. Mating and egg release occur at the down-current ends of reefs, and individual females must migrate from the up-current feeding areas to the down-current spawning sites. In many cases, this is only a short distance and takes just a few minutes; however, certain reef configurations result in daily spawning migrations that can comprise more than 10% of the daily activity (Warner 1995).

Down-current mating sites consist of upward projections from the edge of a reef, and serve to minimize the distance a female must travel up into open water to release her pelagic eggs near the surface (Warner 1988). Tagging studies have shown that once a mating site is chosen by a female, she migrates to that particular site, migrating to spawn at it by the same specific route on a regular basis (Warner 1986, 1995). Wholesale removals and replacements of populations demonstrated that mating site location itself is determined by tradition, involving social learning passed on through females; the same mating sites are used over many successive generations based on sites selected by females (Warner 1988). Successive whole-population replacements showed that females are in fact capable of mating site resource assessment, but under normal circumstances they avoid individual assessment and only copy other females (Warner 1990a). Females do not engage in intensive mate assessment or searches for mates. Experiments that shifted males from one mating site to another indicate that sites, rather than males, are the objects of female choice (Warner 1987). Finally, by replacing only males or females on a reef it was shown that females determine mating site location without reference to male behaviour (Warner 1990b). Thus female mating behaviour appears to operate without much reference to males (Warner and Schultz 1992).

The mating system is polymorphic. Large males can control all the mating sites on small reefs when fish densities are low. However, on larger reefs with many fish the most successful mating sites are economically undefendable by large territorial males (Warner and Hoffman 1980a), and most of the matings go to IP males in group-spawning aggregations (Warner and Hoffman 1980b; Warner 1984b). The group spawning aggregations on large reefs with many blueheads can range up to many thousands of individuals, taking place every day of the year (Warner 1995) (Fig. 12.39). Spawning aggregations are not the only means of reproduction in this species, and individual females can opt to group-spawn on 1 day and pair-spawn with a large territorial male on another. Aggregations are only found on larger reefs where large numbers of fishes exist.

12.14.3 Commercial and Research Value

The bluehead wrasse is not fished for food, other than incidentally, although it is taken occasionally for the marine aquarium trade. The daily spawning aggregations can be easily observed and have been the object of sport diving.

Fig. 12.39 A group spawning by females with initial phase males (Photo: © Doug Perrine/SeaPics.com)

Because the bluehead wrasse mates daily in large aggregations, is not heavily exploited and is easily manipulated, we can answer some questions about spawning aggregations that are very difficult to answer in larger, less frequently mating fishes. For the bluehead wrasse, at least, we know that the sites of spawning aggregations are determined by females, and that information on the location of spawning sites is passed from older females to younger through social learning (or copying). If females are induced to change spawning sites, males simply follow the female lead.

Spawning aggregations of this species are exceedingly easy to observe and manipulate. However, the lessons learned from this small, abundant prey fish may not apply to the larger commercially important predators that form spawning aggregations.

12.15 Bumphead Parrotfish – *Bolbometopon muricatum*

R.J. Hamilton and J.H. Choat
The Nature Conservancy, Indo-Pacific Resource Centre, 51 Edmondstone Street, South Brisbane, QLD 4101, Australia
e-mail: rhamilton@tnc.org

School of Marine and Tropical Biology, James Cook University, Townsville, QLD 4811, Australia
e-mail: john.choat@jcu.edu.au

12.15.1 General

Bolbometopon muricatum (Valenciennes 1839), the largest of the scarine labrids (Choat et al. 2006), is an excavating parrotfish (Scaridae). The genus is monotypic comprises a distinct lineage within the group and is notable for its massive jaws and associated musculature (Bellwood 1994). The evolutionary history of the scarine labrids identifies them as a geographically widespread but relatively recent group, confined mainly to coral reefs. The lineage containing *Bolbometopon* diverged approximately 13 my bp (Alfaro et al. 2009). Although the parrotfishes have a number of distinctive morphological and nutritional traits recent phylogenetic analyses place them within the broader grouping of the Family Labridae (Westneat and Alfaro 2004). We now know that large wrasses such as *Cheilinus undulatus* (Chap. 12.13) are more closely related to large parrotfishes such as *Bolbometopon* than they are to many of the smaller wrasses that inhabit coral reefs. *B. muricatum* is widespread on Indo-Pacific coral reefs with a distribution covering 37.6×10^6 km^2 and extending from the Red Sea through the Indian Ocean and Indo-Australian archipelago to the central and southern Pacific excluding the Hawaiian Islands and the Marquesas. There is also some doubt as to whether this species occurs in the Society Islands (Rob Myers, personal communication in 2009).

B. muricatum is readily recognized by its large size with mature males often exceeding 1 m in total length. The largest recorded specimen is from the Solomon Islands; (1,390 mm, fork length, weight 52 kg). *B. muricatum* is a major contributor to the production of sediment on reefs and individuals can remove an average of 5.7 tonnes of carbonate material from reef surfaces per year of which up to 50% may be in the form of living coral (Bellwood et al. 2003). In areas of high abundance schools of this species may remove ~280 tonnes of carbonate material per hectare per year. The main nutritional activity is scavenging protein by consuming sessile animals (including coral), detritus and elements of the epilithic algal complex on shallow reef surfaces which is processed in the pharyngeal mill before digestion.

Adult *B. muricatum* consistently form schools for feeding and reproductive purposes (Hamilton et al. 2008), and is inactive nocturnally (a labrid trait) sleeping in schools in shallow water. The primary adult habitats are reef fronts and crests usually on exposed reefs. It is commonly observed feeding on reef surfaces and live coral between 5 and 15 m although feeding schools will move into shallow water (1–2 m) with the rising tide. Although the lower depth range is in the vicinity of 40 m the feeding activities are largely restricted to shallow reef fronts and crests. School size varies from approximately 8–10 individuals up to large groups of 100–150 adults (Dulvy and Polunin 2004); in an unfished population on the Great Barrier Reef (GBR), Australia, schools ranged from 1 to 79 individuals with a mean of 36. Mean abundance on outer reefs of the GBR was 30 individuals per hectare. Solomon Islands and Papua New Guinea (PNG) populations are subject to fishing including night spearing. Here densities varied from 0 per ha in the Autonomous Region of Bougainville to 14 per ha in the Western Province of Solomon Islands, with other fished areas ranging from 0.3 to 4.2 per hectare. Numbers in Fiji and

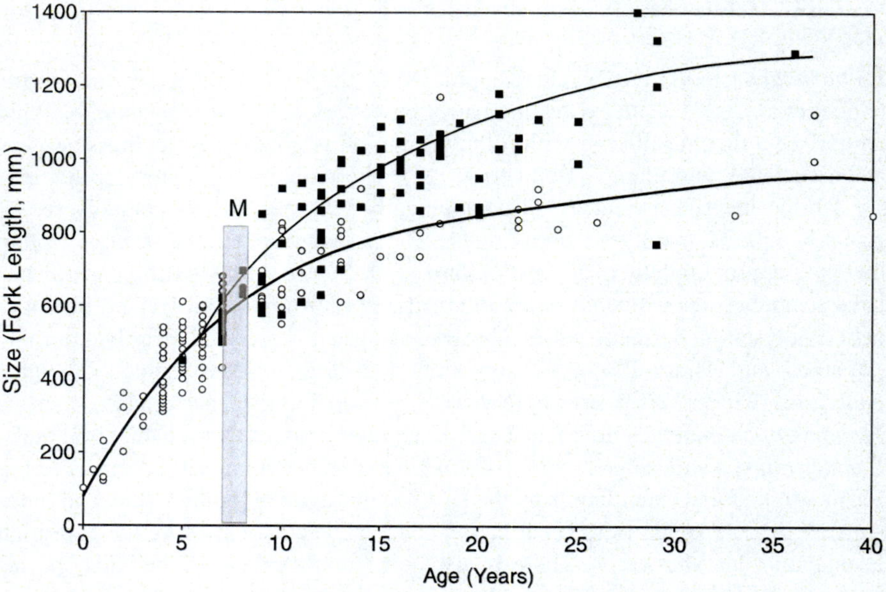

Fig. 12.40 Composite size-at-age plot by sexual identity for populations of *Bolbometopon muricatum*. Individuals < 300 mm FL were collected from lagoon habitats Roviana, Solomon Islands and lagoon habitats Cocos-Keeling Island. Mature individuals were collected from the Solomon Islands and the Great Barrier Reef. Males denoted by squares, females by circles. **M** Age at sexual maturity ±95% CI (From Hamilton et al. 2008)

Samoa are very low and substantially less than the fished areas of the Solomons and PNG. In demographic terms this species is one of the longer lived parrotfishes achieving a maximum age of 40 years, and it shares a number of features with other large members of the family. These include an indeterminate growth curve and thus a relatively low growth coefficient (k) value, and a significantly higher growth rate in males although females achieve greater maximum ages. Female sexual maturation in Solomon Islands populations occurs at 7.5 ± 0.8 years (Fig. 12.40).

Size frequency estimates of *B. muricatum* populations on reef fronts which are the preferred adult habitat are usually skewed towards the upper size ranges. Juveniles and newly recruited individuals occur in sheltered lagoonal habitats and inshore reefs with progressive colonization of more exposed habitats with increasing size (Aswani and Hamilton 2004). The dependency on a sheltered lagoonal environment is seen through comparisons of regional abundance patterns. Christmas Island, an isolated Indian Ocean island with a fringing reef habitat and no local fishery for this species lacks a lagoonal habitat. This area has a very low abundance of *B. muricatum*, only 3% of that recorded from an unfished reef system in the same regional area (Rowley Shoals;16 per hectare) with a large area of juvenile habitat in the form of lagoonal systems. There is also increasing evidence that this species is more abundant on reefs associated with large islands and continental land masses as opposed to ocean reefs surrounded by deep water.

12.15.2 Reproduction

Spawning occurs in aggregations on reef fronts and passes although there is no evidence that schools merge to achieve a larger mass of reproductive individuals (Gladstone 1986; Hamilton et al. 2008). Observations to date suggest most spawning occurs early in the day. Direct observation of reproductive activity by Purwanto (Conservation Coordinator with the joint TNC-WWF Program Wakatobi) suggests pair-spawning within schools. As with many large parrotfishes there is no permanent pattern of sexual dichromatism although males adopt temporary colour signals associated with spawning (Fig. 12.41). Histological evidence shows that the dominant sexual pattern of *B. muricatum* is gonochorism with high incidences of anatomical but non-functional hermaphroditism. *B. muricatum* also differs from most other parrotfishes in that all males pass through an immature female phase with the developing gonad being morphologically female (Sadovy de Mitcheson and Liu 2008). Adult testes retain the ex-ovarian lumen and peripheral sperm sinuses in the gonad wall. It is presently unclear if sex reversal occurs at any point in the geographical range of this widespread species (Hamilton et al. 2008).

12.15.3 Fisheries and Conservation

Coral reef fisheries target this species over its entire range, the exceptions being marine reserves and areas remote from human habitation. Although spawning occurs in groups there is little evidence that reproductive aggregations are the primary target for fishers. Fishing practices tend to be ocean-specific. Western Pacific fishers usually employ nocturnal diving to target schools sleeping in shallow water on reef fronts and passes. Some harvesting, especially in Indonesia, is based on daylight spearing with sale in local markets (Fig. 12.42). At Indian Ocean sites such as the Cocos-Keeling Islands, using large nets to surround schools feeding in shallow water is the method of choice. Some netting occurs in Melanesia but this is relatively uncommon and represents a more traditional approach before the availability of masks and underwater illumination. By focusing on schooling behaviour as well as the shallow foraging and sleeping depth range all methods can result in large catches of adults over short time periods. Nocturnal spearing on SCUBA allows for the greatest catches as an entire school may be harvested in a single fishing episode.

The species is marked fresh and chilled. It does not occur in the live reef fish trade, possibly a reflection of the extreme difficulty in maintaining this species in captivity (Hamilton 2004). *B. muricatum* is represented in local markets at a large number of locations over the geographic range including Jedda (Red Sea), Indonesia and New Caledonia. Given the large size and ease of return from netting and nocturnal spearing it is anticipated that harvests of relatively unexploited populations will increase. The status of the fishery in the western Pacific is instructive. In Papua New Guinea and the Solomon Islands *B. muricatum* is heavily exploited by night spear

Fig. 12.41 Spawning sequence in the bumphead parrotfish, note the temporary colouration of the larger male seen at this time. Clockwise from *top left*: male and female come close and start to align themselves up in the water column; after sperm and egg release the fish return quickly to the substrate (Photos: © Mandy T. Etpison)

fishers once cash markets for this species develop. In many locations stocks are now greatly depleted but *B. muricatum* continues to be harvested opportunistically by night spear fishers who are primarily spearing smaller reef fishes and harvesting sea cucumbers. In some very remote areas that lack well-developed fish markets, *B. muricatum* aggregations continue to be targeted for subsistence purposes when large amounts of food are required for cultural events such as feasts.

Fig. 12.42 (**a**) Spearfisher in Solomon Islands; the bumphead parrotfish is readily taken in large numbers by spear and (**b**) sold in local markets fresh or chilled (Fiji) (Photos: © Richard Hamilton, © Randy Thaman)

In many locations in PNG and the Solomon Islands *B. muricatum* are no longer consumed for subsistence purposes, but targeted to supply domestic markets, such as provincial and national fish markets, restaurants, and mining and logging camps. It is now generally recognized that this species is in need of protection. There are numerous anecdotal reports indicating major reductions in the abundances and catch rates of *B. muricatum* shortly after the introduction of night spearfishing (e.g. Dulvy and Polunin 2004; Hamilton 2004; Sadovy 2007; Hamilton et al. 2010). *B. muricatum* is presently listed as vulnerable following an IUCN 2010 Red List assessment. In addition the species is actively being considered for listing as endangered or threatened under the US Endangered Species Act through the National Marine Fisheries Service NOAA.

In the western Pacific nocturnal spearfishing is increasingly practiced with at least 100 species targeted as it provides a greater CPUE than diurnal fishing (Gillett and Moy 2006). Even though depleted, *B. muricatum* is a bonus for the nocturnal fisher so reduction in numbers will not curtail harvesting of this species. In the Indian Ocean and Red Sea nocturnal spearfishing is less likely to be practiced although large numbers may be captured by netting schools in shallow water. Although this species is not important in the international live reef fish trade, local fishing is sufficient to deplete it across whole archipelagos and reef systems in the Pacific and Indian Oceans (Dulvy and Polunin 2004). Clearly there is no basis for spawning closures as a management option for this species and it is unlikely that any system of bag or size limits will be useful in management. The most effective management option would be a species-level restriction on sales in local markets. If the legislative machinery is available then establishment of no-take reserves is the most appropriate option for longer-term protection. However given the size and foraging range this would require a minimal reserve area ~ 6 km^2.

Donaldson and Dulvy (2004) note that large size and known longevity indicate low replacement rates and high vulnerability to fishing. However this is not the problem with respect to *B. muricatum*. Indeed there are many coral reef species with longer life spans, slower growth rates and very low recruitment rates that are not in such a vulnerable position as this species. The recent assessment of extinction probabilities in mammals (Davidson et al. 2009) shows that there are unique pathways to extinction for species with different lifestyles and combinations of traits. These traits included density, group size, body mass and habitat mode. In the case of *B. muricatum* specific behavioural and foraging aspects of its biology and not life history parameters underlie its vulnerability. Consistent and predictable schooling behaviour associated with all activities (foraging, spawning, sleeping), the cohesiveness of schools and their restriction to shallow water makes this conspicuous species highly vulnerable to nocturnal and diurnal spearing and netting of foraging groups. In addition the well established negative relationship between body size and abundance means that this species will be relatively rare over its considerable geographic range.

12.16 Longnose Parrotfish – *Hipposcarus longiceps*

Patrick L. Colin
Coral Reef Research Foundation, PO Box 1765, Koror, Palau 96940
e-mail: crrf@palaunet.com

12.16.1 General

The longnose parrotfish *Hipposcarus longiceps* is a common medium-sized parrotfish (maximum 60 cm SL) found from the Line and Tuamotu Islands throughout the tropical western Pacific and as far east as a few locations in the eastern Indian Ocean. It is replaced by Indian Ocean longnose parrotfish, *Hipposcarus harid*, in the remainder of the Indian Ocean. The adult initial phase males and females are distinctive with a grey body and yellow caudal fin. The terminal phase males are generally light blue or green with a dark or light band in the central area of the body. The juveniles are quite different in colour having a single orange stripe.

This parrotfish occurs widely in reef environments, being found on inshore reefs as well as offshore clear water reefs. They are herbivores, grazing rocky substrata around reefs. They are superficially similar to the redfin parrotfish, *Sparisoma rubripinne*, of the western Atlantic both in colouration and behaviour. Little work has been conducted on this species, most of it in Palau.

12.16.2 Aggregation and Spawning Behaviour

There has been very little published on the aggregation or courtship of the longnose parrotfish. The closely related Indian Ocean longnose parrotfish has received more attention (see below), but is still very poorly known. For Palau, Johannes (1981) recorded that a local knowledgeable fisherman reported longnose parrotfish (as *S. harid*) to frequently aggregate at the extremities of underwater promontories on the outer reef slope to spawn. Johannes (1981) also reports that in Pohnpei longnose parrotfish (again as *S. harid*) spawns around the new moon in March, April and May.

Previously the only observations of spawning by longnose parrotfish were those of Colin and Bell (1991) who reported pair-spawning in a tidal channel in April, May and September at Enewetak Atoll. They did not observe aggregation or group-spawning. Longnose parrotfish generally spawned in the morning at Enewetak following high tide, however one pair-spawning was also seen on an oceanside reef just after mid-day. There is also a record of aggregation spawning by longnose parrotfish from Canton Island (Stone 2004), in a popular magazine article, which reported aggregation spawning at Canton Island early in the morning by a group of perhaps 5,000 individuals starting some time after 6:30 AM and continued for an hour. Date and lunar phase were not specified, although spawning was said to occur on an outgoing tide.

New observations were made at Blue Corner, a promontory area of the western barrier reef of Palau (Fig. 12.43), which is a popular tourist diving site due to the regular presence of many sharks, other large fishes and turtles. Longnose parrotfish are often present at the site, but the numbers and occurrence of aggregation and spawning are very incompletely documented. Observations of spawning were first brought to the attention of the author by Jim Forrest and Jeanette Denby of the yacht

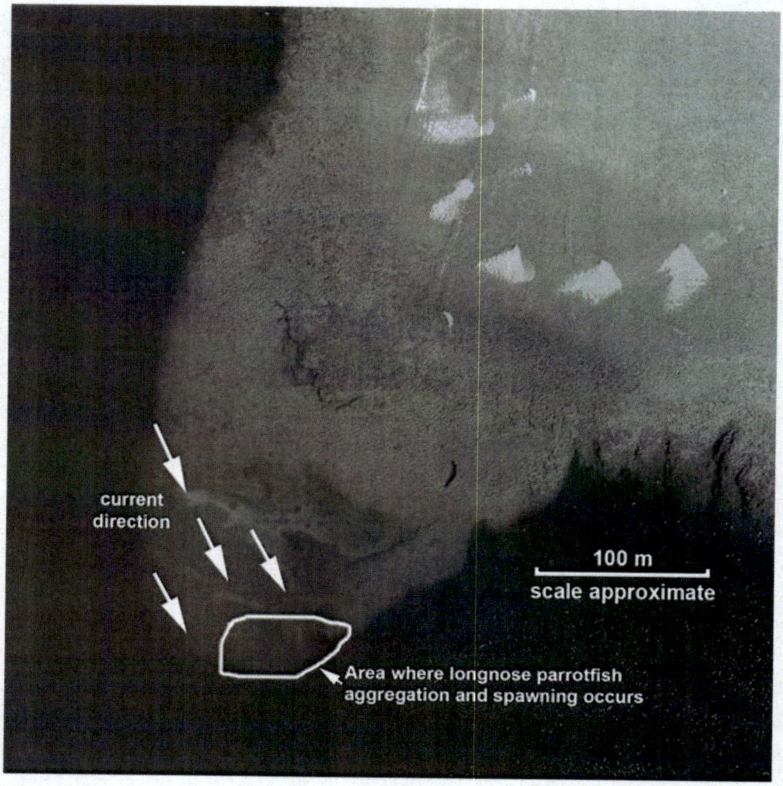

Fig. 12.43 Aerial photograph of Blue Corner, Palau barrier reef showing the location of the aggregation site of longnose parrotfish relative to the reef and current (Photo: Patrick L. Colin)

"Dancer" who were diving regularly at the site. Mandy Etpison also documented the spawning at this site. A large school of a few thousand longnose parrotfish was seen at Blue Corner in the aggregation area in 11 April 2009, 2 days after the full moon. Aggregation and spawning has been seen during April, May and December 2010, occurring 3 days after the new moon in April and 6 days after the full moon at other times. The lunar cycling is not fully documented.

All aggregation fish had a grey body with yellow tail. Schooling fish on the reef swam very actively, often changing direction as a school. Terminal phase males actively courted both before and at the time aggregation spawning occurred. Courting males swam rapidly around a possible temporary territory using pectoral sculling, and had the caudal fin slightly compressed and the mouth held open (Fig. 12.44). The white to brown bar on the body becomes apparent when males were courting females. The colour could be changed within seconds to the opposite, but the role of the two different patterns is not yet clear. The terminal male also had a pink anterior portion to the pectoral fin, which is not normally seen. Sharks and other fishes are often swimming amongst them with no reaction from the parrotfish or predation attempts by sharks.

Fig. 12.44 (**a**) Courting terminal male longnose parrotfish undergo colour changes, swim with the mouth open and the caudal fin compressed slightly, sculling with the pectoral fins. (**b**) At other times the male can have the bar on the body dark, perhaps when it is not patrolling its territory (Photo: Patrick L. Colin)

The aggregation in which spawning was seen numbered about 200–300 fish, which either stayed together in a single group or occasionally broke into two separate groups located a short distance apart. The fish swam relatively close to the substrate, and gradually rose up from the bottom, constantly changing direction as a whole. Some individuals at the top of the group swam excitedly and more fish join them, culminated in a spawning rush. In some cases at least 70 fish are seen in photos of a single group rising in a spawning rush (Fig. 12.45). It appears that not all fish release at the top of the rush, however, subgroups within the ascending fish may release a separate batch of eggs and sperm close to the initial release (Fig. 12.45d). In another spawn there were about 30 fish in the starting group, but several more rushed in from open-water directly towards the rising lead female as the group ascended. The rush is quick, lasting only a few seconds, as has been seen in most parrotfishes. It appears that a single, probably female leads the ascent (Fig. 12.45a) with a number of probable males clustering around her as they rise (Fig. 12.45b). At the peak the lead female spurts forward, probably releasing eggs at that point, and then turns above and back towards the group. Multiple males also appear to release sperm and other groups within the ascending fish may also release gametes at slightly different locations (Fig. 12.45). In photographs of two spawns the lead

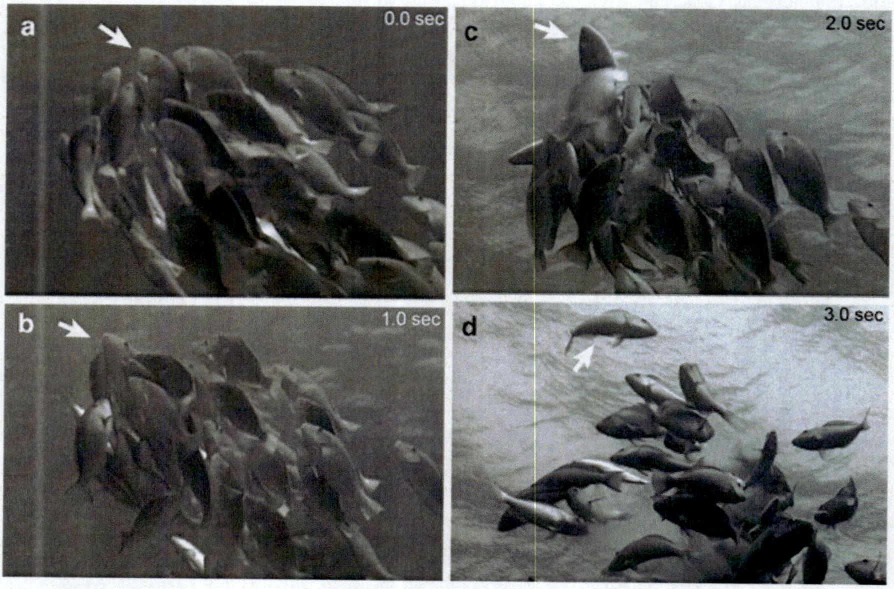

Fig. 12.45 Spawning by longnose parrotfish, Blue Corner, Palau. (**a**) A single female (*white arrow* in all photos) leads the spawning ascent, several probable males follow. (**b**) As spawning begins the presumed female breaks forward from the pack. (**c**) Lead female exits the group, while presumed males just behind her start releasing sperm. (**d**) Additional females probably release their eggs and in only a few seconds the group has broken apart. Elapsed time is shown in each frame (Photo: Patrick L. Colin)

Fig. 12.46 Predation by black snapper, *Macolor niger*, on eggs of longnose parrotfish after spawning. The predators anticipate the group spawning and move in within a few seconds to inhale water and eggs by buccal pumping (Photo: Patrick L. Colin)

female was clearly seen and after the lateral break with egg release the female headed directly down and away from the group. After release of the gametes, the group descended down quickly and then swam horizontally just above the bottom.

The longnose parrotfish generally leaves a highly visible gamete cloud, and at times there is immediate predation on newly released spawns by black snapper, *Macolor niger*. If the parrotfish release their gametes close to the bottom, nearly every spawn is preyed upon by one to several black snappers using the buccal pumping method of ingesting water with eggs (Fig. 12.46). This method was first documented by Colin and Bell (1991) and has been seen in many locations to prey of eggs of numerous spawning species. The gill rakers of black snappers can filter the eggs from the water. However, not all spawns are preyed upon, particularly if the longnose parrotfish release their gametes high (4–10 m) above the bottom. In such cases, the black snapper which may be visible in the area, ignore the spawn, perhaps because they would not be able to reach the area in time to exploit the highly concentrated eggs before dispersal occurs. The egg of the longnose parrotfish is spindle-shaped, 1.53–1.60 mm long by 0.54–0.58 mm wide, with relatively blunt ends (compared to the more acuminate eggs of some species of *Scarus*), no oil droplet and very clear (Colin and Bell 1991; PLC unpublished data).

12.16.3 Important Lessons from This Species

It is important to note that despite its common presence on many reefs, and value as a food fish, almost nothing has previously been published on the reproduction of this species. The group spawning behaviour was unreported. In Palau these observations were made at a dive site called "Blue Corner" which is visited each day by at least 100 SCUBA divers led by professional dive guides. When the author visited this site, tourist divers were taking photos of the aggregation and spawning. Despite this activity, no observer had reported or documented this spawning in any form that would reach the general knowledge base. Such information could come from a popular article in a dive magazine, natural history magazine, submission of information to a database (e.g. SCRFA database www.SCRFA.org), or posting of video or photographs of aggregation and spawning on line (Youtube, other websites).

12.16.4 Comparison with the Indian Ocean Longnose Parrotfish, Hipposcarus harid

Gladstone (1996) reported Indian Ocean longnoses near Farasan Island, Red Sea (Saudi Arabia) to aggregate in March or April, with aggregations occurring in only 1 month but differing between years. The fish were captured when they swam onto the reef flat by groups of men who used monofilament gill nets. He examined 10 females 15–25 cm length and while all had large, ripe ovaries, spawning was not observed. The schools were not easily disturbed and Gladstone (1996) suggested that this might be an example of the spawning stupor, where "this apparent ease with which the longnose parrotfish were approached and captured is often observed among schools of food fishes that are preparing to spawn" (Johannes 1981).

12.17 Striped Parrotfish – *Scarus iserti*, and Bullethead Parrotfish – *Chlorurus sordidus* and Notes on Other Small Parrotfishes (Scaridae)

Patrick L. Colin
Coral Reef Research Foundation, PO Box 1765, Koror, Palau 96940
e-mail: crrf@palaunet.com

12.17.1 General

The parrotfishes (Scaridae) are abundant on reefs throughout the tropics. They are one of the dominant groups of herbivores on reefs, converting benthic plant production into fish biomass and then into vast amounts of spawning products which may be exported from the reef as a result. All species produce planktonic eggs through both pair- and group-spawning. Only a limited number of species are known to aggregate for spawning, and all those are resident aggregations (Appendix). The numbers of species with aggregating behaviour will surely rise as increased attention leads to new records. Those small scarids which aggregate provide interesting comparisons between family members of similar size as well as with the larger parrotfishes. Some of the larger species, such as the bumphead parrotfish, *Bolbometopon muricatum*, and long-nose parrotfish, *Hipposcarus longiceps*, are being considered in separate case studies (Chaps. 12.15 and 12.16). The striped parrotfish, *Scarus iserti*, in the western Atlantic and the bullethead parrotfish, *Chlorurus sordidus*, in the Indo-west Pacific are excellent examples of small scarids and are probably the best known in their respective regions. The females and initial phase males are similar in size (Fig. 12.47a, b), and both species have colourful terminal phase males (Fig. 12.47c).

12.17.2 Reproduction and Aggregation

Small species are found in both the Indo-west Pacific and the Atlantic and have similar patterns of both pair- and group-(aggregation) spawning. Scarids have both adult "initial phase" fish, which can be female or male, and "terminal phase" males. Initial phase fish generally group-spawn (and most should meet the definition of a spawning aggregation) while terminal phase males pair-spawn with initial phase

Fig. 12.47 (**a**) The initial phase of the striped parrotfish, *Scarus iserti*, has broad dark stripes or an overall grey colouration. (**b**) The initial phase of the bullethead parrotfish, *Chlorurus sordidus*, has a dark body, often with a dark blotch on the side. (**c**) the terminal phase male of the bullethead parrotfish has bright green and yellow colouration, which becomes more intense when it is courting, as seen here (Photo: Patrick L. Colin)

females. The "small" parrotfishes can be considered those with a maximum standard length (SL) of 20–40 cm (with some mature as small as 15 cm SL), whereas the largest scarids reach from 60 to over 120 cm SL.

The reefs of the western Atlantic with the striped parrotfish do not have strong tidal currents and spawning seems to occur only during late afternoon. For bullethead parrotfish, aggregation and spawning has been found in areas where tidal currents exist in channels and along reefs and strongly influence timing of spawning.

The first records of spawning by striped parrotfish (as *S. croicensis*) were reported by Randall and Randall (1963), Ogden and Buckman (1973) and Barlow (1975).

Colin (1978) documented differences in spawning activity between winter and summer periods in Jamaica, as well as the afternoon spawning of the species. Pair-spawning occurs from early to late in the day. Striped parrotfish can occur in schools during its morning foraging activities, using the large numbers of fish combined with smaller numbers of some other species to overwhelm herbivorous territorial damselfishes that maintain algal "farms" (Robertson et al. 1976). The damselfish can only drive off a few fish, while the remainder graze aggressively at the exposed food patch. This behaviour, known as "mobbing", is also seen in the IWP greenthroat parrotfish, *Scarus prasiognathos*, which works with surgeonfishes such as *Acanthurus blochii*, to the same end.

In the striped parrotfish, it appears the same schools that overwhelm the damselfishes remain cohesive and form a spawning aggregation each afternoon. In late afternoon striped parrotfish aggregated over a reef pinnacle (the highest point among nearby reef fingers) on the edge of a drop off. They engaged in group-spawning (Fig. 12.48) for about an hour starting about 90 min before sunset (Colin 1978). A single terminal phase male was also at the site, trying to get females interested in spawning with him (largely in vain). The striped parrotfish spawning aggregation investigated in 1975 (Colin 1978) was still present in 1988 (Colin 1996), but its present status is not known. In Puerto Rico Colin and Clavijo (1988) had an aggregation of about 100 fish in a study area on a shelf edge coral reef. The fish spawned as groups emerging from the aggregation in late afternoon, actively during the winter but at a lower level during summer. Their aggregation site was close to, but slightly different than that of two species of *Acanthurus*. Again pair-spawning with a terminal male also occurred in the area at the same time.

Bullethead parrotfish normally aggregate and spawn on a tidal schedule, typically just after high tide where there is a distinct tidal signal (Yogo et al. 1980; Moyer 1989; Kuwamura et al. 2009). It often has the highest number of observed spawns of any reef fish (Sancho et al. 2000a, b; Kuwamura et al. (2009). In some areas the local currents, generated by winds or wave pumping, may be out of phase with the tidal amplitude, and the fishes seem to use the currents rather than absolute tide level to cue spawning (Sancho et al. 2000a). In most locations the species spawns when high tide is during morning periods (Hamner et al. 2007; Kuwamura et al. 2009; PLC unpublished data).

At Johnston Atoll, where currents in channels did not often correspond to tide levels, Sancho et al. (2000a) reported that bullethead parrotfish avoided spawning on inflowing currents (ocean to lagoon) at a tidal channel, delaying spawning until currents reversed to outflowing. If currents reversed again, flowing into the lagoon, the fish interrupted their ongoing spawning. Sancho et al. (2000a) also found some differences in the spawning behaviour with currents in two different study sites at Johnston Atoll. This indicates that a specific tidal phase is not used as a fixed environmental cue across the geographic range of this species, but that fish are responding to directional water flows. Meyer et al. (2010) observed a school of roughly 800 bullethead parrotfish at the surface over 20–40 m depths in mid-day in the NW Hawaiian islands, but based on the known spawning behaviour of this fish, this appears not to be a spawning aggregation and is most likely a migration between feeding areas. The suggestion of this schooling as an example of a "spawning

Fig. 12.48 Striped parrotfish, *Scarus iserti*, spawning activity, showing group-spawning of clusters of fish within larger aggregations. Discovery Bay, Jamaica (Photo: Patrick L. Colin)

stupor" (*sensu* Johannes 1981) indicates how confused the interpretation of this supposed phenomenon has become (see Chap. 5).

During morning hours bullethead parrotfish eggs, spawned mostly in the hour after high tide, dominated the zooplankton being exported from coral reef in Palau (Hamner et al. 2007) and during afternoon high tides there was little or no spawning by scarids and the fish eggs streaming off the reef were dominated by those of surgeonfishes. Sancho et al. (2000a) reported afternoon spawning (peak abundance and activity 1,300–1,700 h) to be most common in bullethead parrotfish, although it was not observed spawning at dusk.

12.17.3 *Other Small Species of Scarids*

In the Indo-west Pacific small scarids known to aggregate include Bleeker's parrotfish – *Chlorurus bleekeri*, and greenthroat parrotfish – *Scarus prasignathos*, and others reported need further confirmation (surf parrotfish – *S. rivulatus*, Myers 1999). While longnose parrotfish, *Hipposcarus longiceps* (formerly known as "*H. harid*"), is at the upper limit of "small" scarids, and its Indian Ocean replacement has been previously reported to group-spawn (Gladstone 1996), nearly all observations have been of pair-spawning (Chap. 12.16). The behaviour of this species closely resembles its western Atlantic counterpart the redfin parrotfish, *S. rubripinne*, even regarding similarities in colour patterns of initial and terminal phase fish between the two species. Some other IWP genera, *Calotomus* and *Leptoscarus* are reported to spawn in groups of individuals (Robertson et al. 1982), but whether these represent aggregations migrating to a spawning area is unknown. The seagrass parrotfish, *Leptoscarus vaigiensis*, is a gonochorist (no sex change), the first parrotfish reported to be so (Robertson et al. 1982)

The occurrence of group- and pair-spawning in the same area is common in many species, with terminal phase males often courting females that are also involved in group-spawning activities. It appears there is a "critical mass" factor in determining whether group spawning will occur based on the numbers of initial phase fish present: Bleeker's parrotfish pair-spawn hundreds of times on a study reef in Palau, but never group-spawned. About 100 m down the reef a group of about 50 initial phase Bleeker's parrotfish group-spawn. The example is good evidence of the caution needed when making generalities about spawning in a particular species, particularly based on experience at one small site. Colin and Bell (1991) examined spawning by scarids and labrids within an area of reef channel at Enewetak Atoll and although they found 13 species of scarids spawning in this area, none would be considered to aggregate. The small bullethead parrotfish, *C. sordidus*, which aggregates in other areas of Enewetak, was not common in their study area, hence no aggregations were seen there.

Since small scarids in the IWP seem to spawn with a tidal timing, the situation with bullethead parrotfish is interesting. It appears in several locations that spawning occurs only when tides are in the morning; at other tidal phases they either do not spawn or spawn at some time that has not been noted. In Palau, the aggregation of bullethead parrotfish is replaced by surgeonfishes (striped bristletooth – *Ctenochaetus striatus* and brown surgeonfish – *Acanthurus nigrofuscus*) when high tides are in the afternoon. Most scarids have unusual and distinctive spindle-shaped pelagic eggs about 1.2–2.2 mm in length and only 0.4–0.6 mm wide. The eggs of some have bent pointed ends (Colin and Bell 1991). Only a small number of species have more typical spherical eggs (e.g. bicolor parrotfish – *Cetoscarus bicolor* and bumphead parrotfish – *Bolbometopon muricatum*). One possible explanation for this egg type is that the shape may have less resistance to vertical movement in the water column due to buoyancy and this shape might either help to keep them closer to the surface or increase their ascent rate away from benthic predators.

For some of the larger parrotfishes, such as bicolor parrotfish, it seems unlikely they have group-spawning, as they are usually not common on the reef and pair-spawning is well known, while no indications (courtship, etc.) of group-spawning have been seen. Other larger scarids, such as bumphead parrotfish appear to spawn as pairs (and potentially small groups) out of schools which may or may not be true spawning aggregations (Chap. 1).

12.17.4 Conservation and Fisheries Importance

These fishes are not always thought of as commercially important, but in areas where reef fish resources are highly exploited, they are a major part of the subsistence and commercial catch. As such, aggregation sites potentially represent an area where fishes could be captured by fish trap, nets or spearfishing in some abundance.

These fishes are important for maintaining the health of coral reef systems, as they and other herbivores can limit the growth of benthic algae which can threaten coral populations, particularly during recovery phases after a bleaching event or storm disturbance. In areas where herbivores have been overharvested, algae quickly take over much of the rocky bottom where they were previously held in check. They are also important ecologically in that they convert a large amount of primary and secondary production into pelagic gametes that are exported from the reef (Hamner et al. 2007).

12.18 Bigeye Trevally – *Caranx sexfasciatus* with Notes on Other Jacks (Carangidae)

Patrick L. Colin
Coral Reef Research Foundation, PO Box 1765, Koror, Palau 96940
e-mail: crrf@palaunet.com

12.18.1 General

The bigeye trevally, *Caranx sexfasciatus* Quoy and Gaimard, is a common, medium-sized carangid reaching about 85 cm fork length (FL) found on outer reefs from the Red Sea to Central America. It often occurs in schools of many hundreds to thousands of individuals swimming in tight whorls which produce amazing images for underwater photographers (Fig. 12.49). The schools might seem to represent spawning aggregations, however, this appears not to be the case, and this case study illustrates the difficulties in deciding what is and isn't a spawning aggregation.

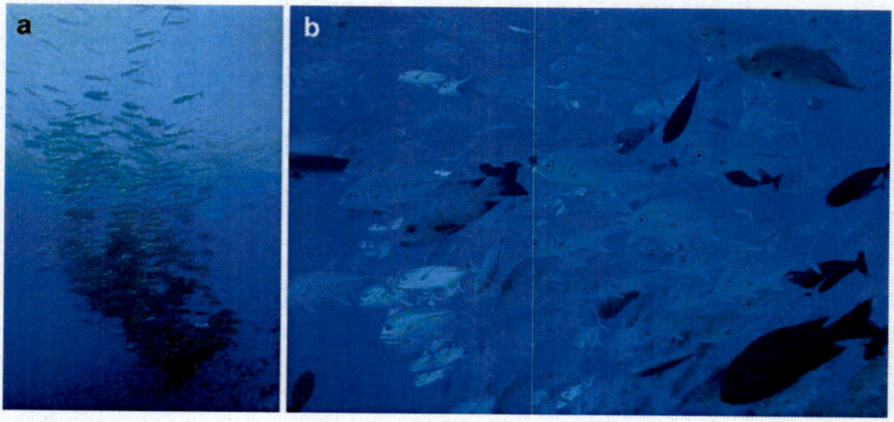

Fig. 12.49 (**a**) The large numbers of bigeye trevally, *Caranx sexfasciatus*, found in schools make it seem they might be spawning aggregations, but when the fish spawn they separate from the school into pairs. (**b**) Schools of bigeye trevally can have other fishes mixed in, such as these *Macolor niger* snappers seen here at Blue Corner in Palau (Photos: Patrick L. Colin)

The Carangidae are a family of strong swimming schooling open-water carnivores, with muscular bodies and highly forked caudal fins with heavy supporting elements (scutes, etc.) at its base. There are about 150 species, of which several dozen occur in most coral reef areas. The species in the genus *Caranx* range in size from a few 10s of cm to confirmed lengths of over 1.7 m length, with weights up to 80 k for the giant trevally (*Caranx ignobilis*). They are highly prized sport fishes, the giant trevally a legendary fighting fish. There are only a few published reports of spawning by family members, and thus only a modest amount is known about spawning in the family. One question is whether carangids truly have spawning aggregations, or do they have what is known as "simple migratory spawning"?

12.18.2 Reproduction and Aggregation

Carangids are gonochorists and the sexes are not easily distinguished. For the species included in the Appendix the observations of spawning are not questioned, but whether the species engage in true aggregation or simple migratory spawning is not positively known. Hence they are listed in Appendix as "probable" aggregation spawners. Domeier and Colin (1997), as well as Domeier (Chap. 1), defines simple migratory spawning as "migration and spawning of pairs or small groups of fishes from a non-spawning area to a spawning area". A true spawning aggregation is that of "a unique phenomenon of behavioural ecology where an entire sub-population of individuals halt their normal routine, migrate, gather and spawn" (Chap. 1). Since many species of carangids normally school in large numbers, the density of fish that might occur when spawning is taking place may not be higher than normal, just the location where this occurs might (or might not) be different from the normal range. Whether carangid spawning is considered "aggregation" or "simple migratory" the fish are effectively releasing their gametes within a restricted area at specific times, so the oceanographic mechanisms that affect them would be similar or identical to those for aggregation spawners. Whether carangid spawning is more like transient or resident aggregations is also a question. Data are scarce, but some species may well act as resident aggregators, while other more resemble transient.

Hamilton and Walter (1999) reported on traditional knowledge of movements of carangids in the Roviana lagoon, Solomon Islands. Some species were reported to move back and forth through channels on the tides. Bigeye trevallies were found to have their movements influenced by lunar phase in the Honiavasa Passage and were present most often during the full moon period. In Palau there is little traditional knowledge of aggregation of bigeye trevally (e.g. Johannes 1981 does not list it).

The only published report on a spawning aggregation of bigeye trevally is that of Sala et al. (2003) from the Gulf of California in the eastern Pacific. They reported bigeye trevally to spawn there around the full moon of July–September at an offshore islet with about 500–1,500 fish usually present. The school was absent at other times of year. The fish spawned during the day as pairs at 10–20 m over a 30 m deep bottom out of the larger group. The aggregation area as a whole covered about 10,000 m^2.

This same behaviour is seen in the western Pacific locations. In Palau, a school of several hundred fish is often present at Blue Corner and the fish stay near or over the drop-off. In the afternoon they remain in the same area, but pairs break away and swim slightly inshore over a 20 m deep bottom during courtship (Fig. 12.50). The males become quite dark, almost black, while the female retains normal colour, and the male stations himself slightly below and occasionally slightly behind the female. If another male tries to approach the female, the intruder is kept away by the male blocking any attempt of its rival to approach from behind the pair. Figure 12.50a shows how widely dispersed the pairs are during courtship, each pair a unit from the school which still retains its integrity a short distance away, and the pairs swim slowly into the current, maintaining their position over the bottom.

Elsewhere, bigeye trevallies are often seen in schools with pairs subsequently separating themselves out. Near Port Moresby in Papua New Guinea, a swirling school of bigeye trevally on the "horseshoe reef" barrier reef switched their location to the opposite side of the reef as the current changed (PLC unpublished data).

12.18.3 Other Species of Carangids

Sala et al. (2003) also reported aggregation spawning in amberjack, *Seriola lalandi*. They observed a school with about 80 fish for 3 days starting 3 days before the April full moon, and while they did not observe spawning, females had hydrated eggs and the density of fish was much greater than normal. By SCRFA criteria, this is considered a valid spawning aggregation record (Chap. 1)

Fig. 12.50 (**a**) Male (*dark*) – female pairs of bigeye trevally, *Caranx sexfasciatus*, preparing to spawn at Blue Corner, Palau are widely dispersed and do not approach one another. The remaining school maintains its integrity a short distance away. (**b**) Close up of single pair of bigeye trevally, with the male dark and female silvery, photographed 2 February 2009, 16:23 h, 10 m depth over 20 m deep bottom (Photos: Patrick L. Colin)

Von Westernhagen (1974) reported giant trevally to spawn as pairs from a school located on a shoal 35–45 m deep in a channel between two islands in the Philippines. At the edges of the school, three to four males could be seen vigorously pursuing females and eventually one male remained with a single female and the pair sank to the bottom, swam in tight circles and released gametes. In Palau schools of giant trevally are often found in high current areas, particularly at the southern end of Peleliu, and some evidence of courtship by pairs has been seen at times. Males, like other carangids, appear to become dark during courtship (Fig. 12.51).

For those carangids in which spawning has been observed, all known spawning occurs as pairs or small groups which separate from a larger school or aggregation. Males become dark, often black for silvery species, but females of some species

Fig. 12.51 Male giant trevally, *Caranx ignobilis*, in courtship colouration? Photographed at Peleliu, Palau (Photo: Patrick L. Colin)

seem to exhibit some dark pigmentation during courtship (Fig. 12.52). These colours are under nervous control and can be quickly replaced by the typical silvery colouration. When fish separate as pairs from the larger group, they undergo a lengthy period of swimming with the presumed male behind and slightly below the female. At this time, if other fish (presumably males attempting to intrude) approach the pair, the presumed male will turn and chase the approaching fish, or still swimming, attempt to prevent the intruding fish from coming near the female.

A variety of reports of carangid spawning aggregations, not verified, originated from Palau. Johannes (1981) and Myers (1999) reported *Caranx melampygus* to aggregate as a 1,000 or more at the south tip of Peleliu to spawn in April on the new moon and for golden trevally *Gnathanodon speciosus* indicated they aggregate in shallow water to spawn around full moon from November to May.

For western Atlantic species Graham and Castellano (2005) reported permit, *Trachinotus falcatus*, in Belize to spawn on two occasions (April and August) 7 and 9 days after full moon. Subgroups of 5–9 fish from a larger school of up to 300 fish ascended above the group and released gametes with a larger, presumed female, leading smaller fish. The yellow jack, *Carangoides bartholomaei*, was seen to spawn once in April (9 days after full moon) 1 h before sunset, with groups rising from a larger school at 40–45 m to spawn at about 35 m depth. Also in Belize, Heyman and Kjerfve (2008) reported observing spawning in six species of carangids, while a seventh species was included based on courtship and reports of fishers. They considered carangids to be "semi-pelagic spawners", based on the location (high above reef bottoms on the shelf edge) where they aggregated and spawned. Observations of spawning in horse-eye jack, *Caranx latus*, were reported for April, July and August (later indicated as only July and August) with subgroups of 15–20 fish swimming away from a larger school of about 150–700 individuals and ascending in a vertical twisting rush to release gametes 1–2 m below the surface. For crevalle jack, *Caranx hippos*, behaviour was similar, although fish in the

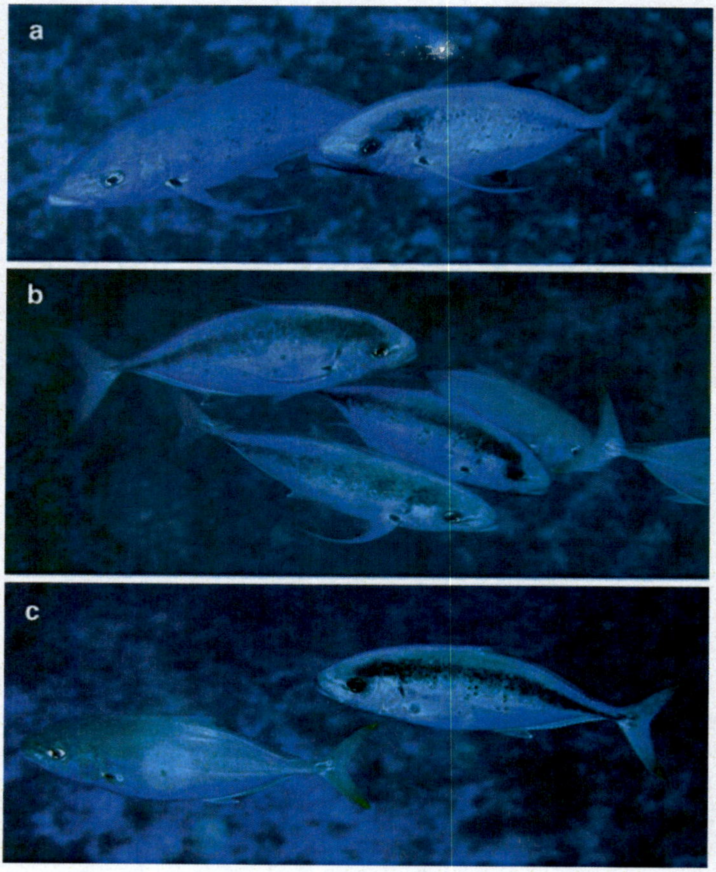

Fig. 12.52 The blue trevally, *Carangoides ferdau*, was seen courting and spawning at Chuuk Atoll, Micronesia above one of the WWII wrecks in the lagoon. (**a**) The probable males have dark areas, typical of courting male carangids. (**b**) The male with the dark dorsal stripe was attempting to prevent three other males from approaching the female, whose caudal fin can be seen on the right. (**c**) Typical relationship between female (*silvery*) and male (*dark*) during courtship (Photos: Patrick L. Colin)

aggregation assumed courtship colouration (rather than when swimming as pairs), with fish courting as pairs and then spawning as groups from the larger school.

12.18.4 Commercial Importance

Carangids are of major importance for subsistence, sport and commercial fisheries. Given that they often occur in schools, there does not seem to be much directed exploitation of schools of spawning fishes. There is a need for more information about this family, and even a limited amount of field work at the right times and locations could yield significant new knowledge.

12.19 The Moorish Idol – *Zanclus cornutus*

Patrick L. Colin, Mandy T. Etpison, and Paul Collins
Coral Reef Research Foundation, PO Box 1765, Koror, Palau 96940
email: crrf@palaunet.com

NECO Marine, Koror, Palau
e-mail: etpison@palaunet.com

Department of Science, University of Glamorgan, Treforest Campus, Pontypridd, Wales, UK
e-mail: paul@abolutediving.uk

12.19.1 General

The Moorish idol, *Zanclus cornutus*, (the only member of the monotypic family Zanclidae) is one of the iconic fishes of the Indo-Pacific, yet its biology and reproduction are poorly known. It has an extremely broad distribution from the east coast of Africa and the Persian Gulf across the Indian and Pacific Oceans to the western shores of the Americas. It is related to the surgeonfishes, sharing a somewhat similar larval type (Johnson and Washington 1987). Its maximum reported total length is about 23 cm, but at least in western Pacific areas most fish are much smaller. The species is widespread where there is hard substrate and down to 180 m. It usually occurs in schools that range from a few fish to over 100 animals. It feeds mainly on sponges. It is a popular aquarium fish but is difficult to maintain (Myers 1999). Despite the distinctive nature of this species very little work has been conducted on it.

12.19.2 Reproduction and Aggregation

For many years schools of hundreds of Moorish idols have been observed at numerous locations on the western barrier reef of Palau from December to March, with peaks in January–February (Fig. 12.53). These have been reported as probable spawning aggregations (e.g. Etpison 1997, 2004), but had not previously been verified under SCRFA criteria as spawning aggregations (Chap. 1).

A few reports of pair-spawning by Moorish idol exist. Colin and Bell (1991) saw three pair spawnings (no aggregation) on the edge of a tidal channel at Enewetak Atoll just after high tide near dawn. Sancho et al. (2000a, b) listed 5 spawns (both pair and group) by Moorish idol at Johnston Atoll without further detail. A few pair-spawns, outside of aggregations, have been observed by the senior author on the Palau barrier reef about 2 h after high tide in mid-day time periods.

In Palau, while knowledge is still imperfect, aggregations of Moorish idols usually occur around the first and last quarter moons with peak numbers of fish on the day before, the day of, and the day after the quarter moon. At Blue Corner and other sites, several species of surgeonfishes also form what are believed to be spawning

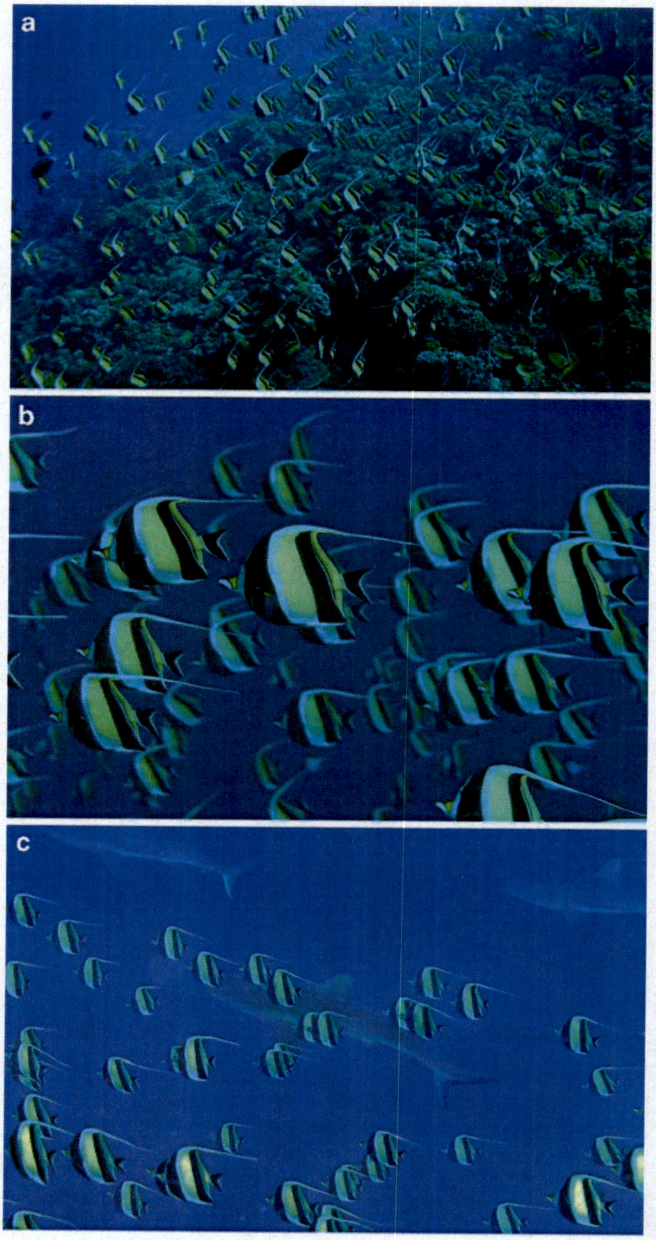

Fig. 12.53 (**a**) Moorish idols a in spawning aggregation swimming just above the reef. (**b**) When aggregations of Moorish idols swim in one direction, they move in unison. (**c**) Spawning aggregations occur amongst populations of many predators, include grey reef sharks (*Carcharhinus amblyrhinchus*), seen here in the background (Photos: Patrick L. Colin)

aggregations (not verified) in the same general area. Moorish idol aggregations occur at several sites along the western barrier reef in Palau; mostly at spots used for dive tourism (which is presumably why the observations were made). These aggregations probably occur throughout the Palau barrier reef system and likely number a hundred or more. At Blue Corner, the best-known site, schools of about 500 to a few thousand fish mill about variously over a distance of approximately 200 m. When the fish move horizontally they do so in unison as a school, all oriented in the same direction (Fig. 12.53), an impressive sight.

In areas such as "Blue Corner" and Saes Corner a variety of predators have been observed "hunting" both aggregated Moorish idols and surgeonfishes (Fig. 12.36). These predators include twin-spot snapper (*Lutjanus bohar*), giant trevally (*Caranx ignobilis*), humphead wrasse (*Cheilinus undulatus*), grey reef sharks (*Carcharhinus amblyrhynchus*) and white tip reef sharks (*Triaenodon obesus*) (Fig. 12.53c, see also Fig. 12.32). Often multi-species groups of predators pursue the school of moorish idols and surgeonfishes, trying to separate individuals which are then attacked and eaten. Sometimes the pursued fish will take shelter under a rock and the predators will then focus their attention on that site, trying to scare or drive the single unfortunate fish out where it can be taken easily. Groups of sharks sometimes drive the aggregations of Moorish idols as a whole away from the reef, and force the school towards the surface, where they attack the Moorish idols, taking many individuals. At Blue Corner several dive boats have observed schools of Moorish idols at the surface being "herded" and eaten by grey reef sharks with their dorsal fins breaking the surface and Moorish idols jumping on the surface of the water trying to avoid the predators. Whether these were fish that had left the shelter of the reef to spawn is not known, but possible. This may occur at or near the time of spawning, but is incompletely documented as the behaviour is both difficult and dangerous for divers to observe in the water. Moorish idols are small compared to the sharks attacking them, and it is somewhat surprising that the grey reef sharks are so vigorous in pursuing such small prey.

Overall the temporary presence of the schools of Moorish idols and surgeonfishes in an area where they normally do not occur in abundance and in areas that lack large amounts of shelter probably provides an opportunity for predation potentially associated with spawning aggregations, but not targeted specifically on courting or spawning fishes.

A possible spawning movement was observed on 4 Feb 2009, the day after the first quarter moon. The Blue Corner aggregation was watched from about 9 am until noon. The fish moved back and forth along the seaward drop-off within 1–3 m of the bottom (Fig. 12.53) between about 6 and 20 m depth. However, just after 11 am and without any interaction with any predators, the entire group of about 430 fish swam into open water up and away from the reef, reaching near the surface some distance from the drop off (Fig. 12.54). It is presumed the fish were preparing to spawn, however, once near the surface a dive boat ran through the middle of the school, temporarily dispersing them, and then strong currents made it risky to continue to remain with the group, so actual spawning was not observed.

Fig. 12.54 A group of about 430 Moorish idols (the number counted from a photographic print) swimming off the reef and towards the surface for spawning (Photo: © Mandy T. Etpison)

Spawning was seen on 16 Feb 2011, 2 days before full moon, when an aggregation in excess of 100 individuals was observed at 4 pm on a rising tide at a reef point called Siaes Corner. The aggregation was videotaped for approximately 40 min moving from one side of the reef point corner, and as described earlier, the fish were "herded" (forced by the sharks to move back and forth around this reef corner at depths of 11–20 m) and occasionally attacked by approximately 80 grey reef sharks. During this time small groups of 10–30 Moorish idols broke off many times from the main group and proceeded toward the surface in a spiral motion to spawn within 6 m of the surface. After eggs and sperm were released the spawning groups returned immediately to the main group. The grey reef sharks were observed on several occasions to attack individuals that had broken off from the main group to spawn (in addition to general attacks on the group) and were successful in capturing a few of these individuals.

Confirmation of spawning condition was achieved by inspection of ripe gametes in 15 February 2010 (1 day after new moon) eight reproductively active females 96–114 mm SL were speared in late afternoon on the outer reef north of Ulong Channel, Palau, outside of any aggregation. The fish were in a small school, but not in a large aggregation such as is seen at Blue Corner and Siaes Corner. Their gonad weights were 1.16–3.36% of body weight with hydration evident in oocytes in mid-afternoon (Fig. 12.55). Two larger females, 122 and 125 mm SL, taken at the same time had no visible gonad development and possibly might have already spawned some days before.

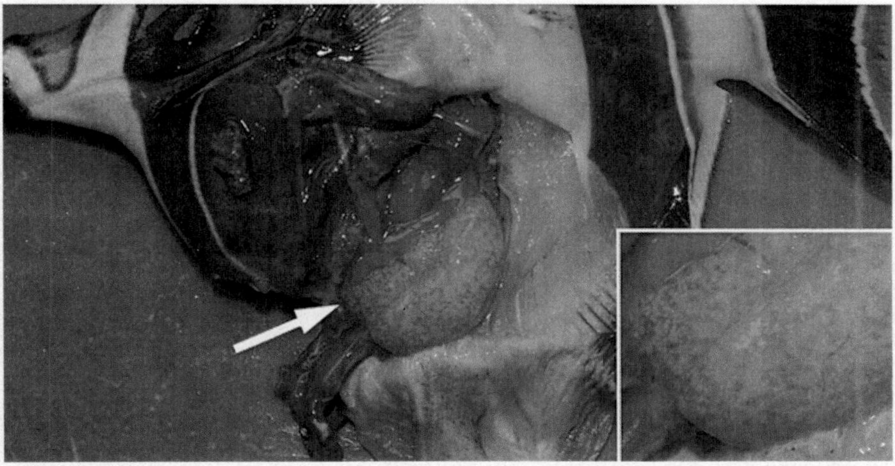

Fig. 12.55 Ripe female Moorish idol, *Zanclus cornutus*, showing the relative size of the ovary (*white arrow*). The insert shows a close-up of the ovary with clear hydrated oocytes easily visible, indicating this female was almost ready to spawn (Photo: Patrick L. Colin)

12.19.3 Data Gaps

There is still much to be learned about spawning of Moorish idols, but the present information provides a basis for further study. The criteria of SCRFA for occurrence of spawning aggregations have been met, including the criteria of a four times higher density and actual spawning observations. The females with hydrated eggs (a criterion that indicates imminent spawning) were collected from isolated non-aggregating, individuals and the lunar phase of their capture differs somewhat from that of aggregation and spawning observations. These females could have engaged in pair-spawning either within or outside of aggregations.

The Moorish idol spawning aggregations documented here are quite different from those known for other species. By present definitions these would certainly be considered as transient aggregations (TA) with strong seasonal and lunar periodicity. But contrary to most TA species the Moorish idol is small and common; the opposite of a goliath grouper or cubera snapper.

12.20 Ocean Surgeonfish – *Acanthurus bahianus* and Blue Tang – *Acanthurus coeruleus* in Puerto Rico

Patrick L. Colin and Ileana J. Clavijo
Coral Reef Research Foundation, P.O. Box 1765, Koror, Palau 96940
e-mail: crrf@palaunet.com

Department of Biology and Marine Biology, University of North Carolina, Wilmington, NC, USA
e-mail: clavijo@uncw.edu

12.20.1 General

While the surgeonfishes (Acanthuridae) are found worldwide in tropic and warm temperate waters, only two species which occur in the tropical Western Atlantic; *Acanthurus bahianus* (ocean surgeon) and *Acanthurus coeruleus* (blue tang) are considered here (Fig. 12.56). A third species, *Acanthurus chirurgus* (doctorfish), is not covered in detail.

Surgeonfishes are not important commercial food fishes in the western Atlantic, but are commonly taken in trap fisheries, mostly for subsistence use. Wide variation in growth rates and maximal sizes across the geographic range of the species has been reported (Robertson et al. 2005a). Densities of fish on different reef systems range from 1 to 16 (averaging 5) fish per 100 m^2 (Robertson et al. 2005b).

Surgeonfishes are herbivores whose constant grazing on the reef helps keep populations of algae in check and leaves areas clear for invertebrate larvae to settle. Despite their importance in reef fish communities, surprisingly little has been written regarding the reproduction, particularly their spawning biology, of Atlantic species. Their biology was last reviewed by Reeson (1983) and information on spawning aggregations had not been published at that time. Other aspects of their life histories have received some attention (Robertson 1988a, b; Robertson et al. 2005a, b; Rocha et al. 2002).

Reeson (1983) reported ocean surgeonfish to first mature at about 92 mm total length (TL) while most fishes probably mature at about 126–134 mm TL and the largest fish was about 200 mm TL (sexes not reported). Blue tang matured starting at 109 mm TL, with most fishes mature at 120–130 mm TL and the largest fish were 220–240 mm TL. Robertson (1988a, b) believed that Atlantic surgeonfishes in Panama reach sexual maturity about 2 years after settlement and can live up to about 10 years. Robertson et al. (2005a) found ocean surgeonfish to have the highest recorded growth rate for a surgeonfish, and corroborated a 10 year age for large fish from Belize. They also showed that fish in other locations may achieve considerably longer lives, as much as 30 years in Bermuda.

12.20.2 Aggregation and Spawning

Resident spawning aggregations were first discovered, by chance, by the authors in December 1976 in Puerto Rico and have now been reported (Colin and Clavijo 1988; Colin 1996; Deloach 1999) for all three species; ocean surgeonfish (Puerto Rico), blue tang (Puerto Rico, The Bahamas, Belize) and doctorfish (Belize). Pair-spawning is also known in all three species, often occurring at the same sites and times as aggregation spawning.

After the initial discovery of the Puerto Rican aggregations, we undertook a study from late 1977 through 1980 to document the locations, timing and general

Fig. 12.56 Colour patterns (differential shading) in courting Atlantic surgeonfishes: (**a**) pair-spawning male ocean surgeonfish, (**b**) pair-spawning male blue tang, (**c**) group spawning blue tang (males and females?) (Photo: Patrick L. Colin)

biology of ocean surgeonfish and blue tang. They were found to spawn in both aggregations and as pairs. Although ripe doctorfish were found near the study area and some gonad data obtained, they were never seen to spawn. Colin and Clavijo (1988) and Colin (1985) provided some information on these aggregations, but did not include details regarding gonad indices, size frequency, and population density, which are now included here. Colin (1996) reported on the occurrence of these aggregations 12 years after their original discovery and described a small aggregation of blue tang in The Bahamas which disappeared shortly after its discovery. In 2011 the site was visited again and while fishes were still aggregating and spawning in the same general area, numbers were less than previously.

Location of aggregations: The aggregations occurred in a discrete area of the outer reef along the SW Puerto Rican shelf edge. The reef was mapped in some detail using the tape and compass method (see Chap. 9) and the locations of the aggregations plotted on this base map (Fig. 12.57). The site was revisited in 1988 (Colin 1996) and again in 2011. While their locations did not appear to have shifted between the 1970s and 1988, in 2011 both aggregations were found to have moved slightly further east (but less than 50 m distance) and a bit more towards the drop off edge. The consistency of these aggregation locations over 34 years is indicative of the stability of sites for a surgeonfish with resident aggregation.

The aggregation site and sizes of fish aggregating: The locations of the aggregation site in Puerto Rico are shown in Fig. 12.57. In the late 1970s the aggregation areas for ocean surgeonfish and blue tang overlapped, the smaller blue tang site lying totally inside the area used by *A. bahianus*. Currents at the site were mild and did not have any particular correlation with spawning occurrence. All aggregations of ocean surgeonfish and the largest aggregations of blue tang occurred during the coldest portion of the year, when water temperatures are about 24–25.5°C. The water is clearer during the northern winter period also, with visibilities occasionally of 30 m, but generally 20–25 m. The aggregations sites utilized by the Puerto Rico surgeonfishes remained exactly the same for at least 12 years (Colin 1996) and varied only slightly over 34 years.

Fishes from the spawning aggregations were speared and data on standard length, sex and gonad weight were taken. Ocean surgeonfish females from the Puerto Rico aggregation ranged in size from 115 to 150 mm standard length (SL), while males were 115–205 mm SL (Fig. 12.58). There was a significant size difference in males and females from the aggregation at the $p<0.05$ level. Blue tang from the aggregation ranged in size from 105 to 180 mm SL for females and 110 to 180 mm SL for males. There was no difference in size of males and females for blue tang.

Seasonal, lunar and daily spawning and migration patterns: Data were obtained on the presence/absence and, if present, the size of the aggregation and whether spawning was occurring for both species on 139 days during the main spawning season (December to March), as well as on an additional 40 days in the "off season" of April to November. Observations were much more difficult during the latter period as the water is much less clear (visibility as little as 9–12 m) making finding the aggregation and observing spawning difficult.

During the main spawning period (December to March) the ocean surgeonfish was aggregated and spawning at all lunar phases (Figs. 12.59 and 12.60) with low

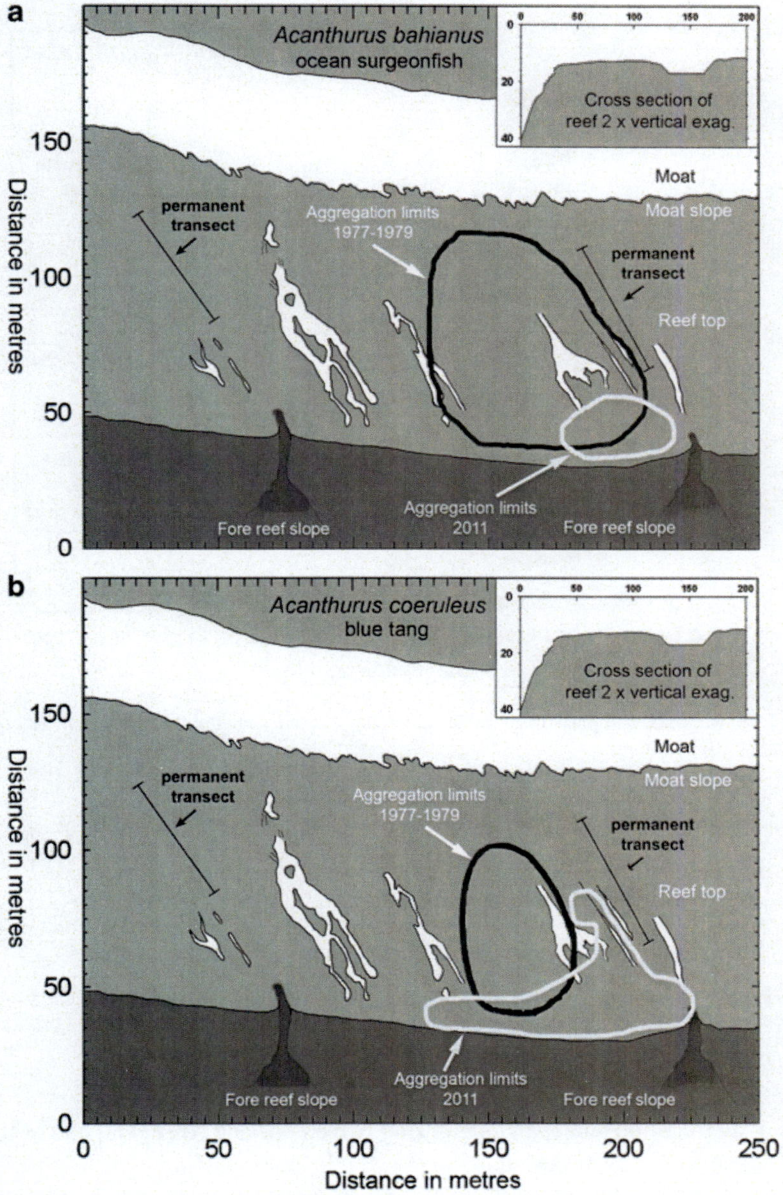

Fig. 12.57 Comparison of spawning aggregation areas of surgeonfishes on a shelf edge reef off southwestern Puerto Rico between 1977–1979 and 2011. (**a**) Aggregation area of ocean surgeonfishes decreased considerably between the 1970s and 2011 and the numbers of fish decreased to only about 5–10% of the roughly 20,000 seen in the 1970s. (**b**) Aggregation area of blue tang remained similar in size, but changed shape and shifted somewhat in location. This map is based on Colin and Clavijo (1988, Fig. 6)

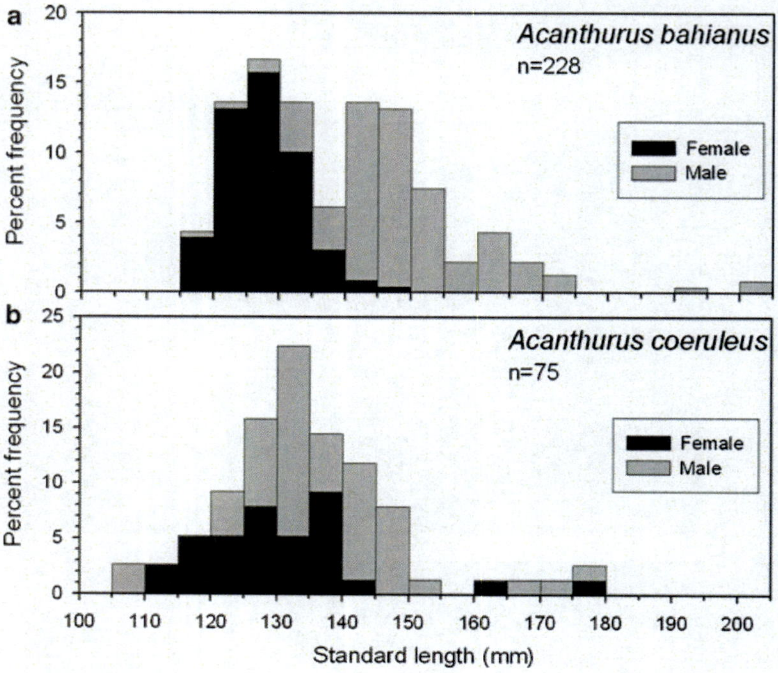

Fig. 12.58 (**a**) Size comparison of male and female from spawning aggregation. (**a**) ocean surgeonfish (**b**) blue tang

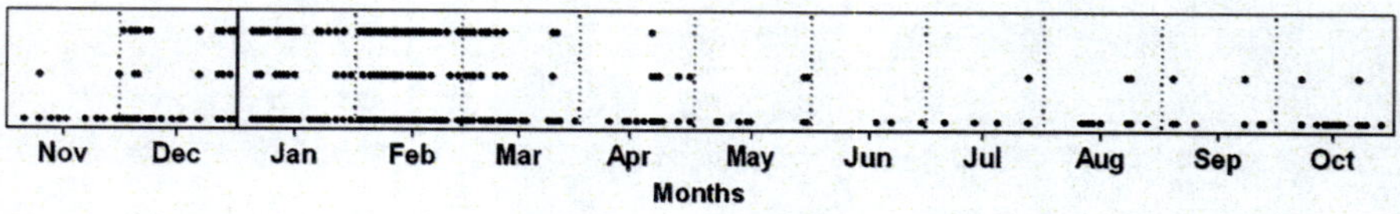

Fig. 12.59 The *black blocks* show the days of confirmed observation of aggregation and spawning occurrence in ocean surgeonfish (*upper*) and blue tang (*middle*) against days when observations were made that attempted to document aggregation occurrence (*lower*)

levels of aggregation and spawning during November and April. Outside of those times no group spawnings, and only a single pair-spawning in August, was observed. For blue tang aggregation and group-spawning were seen in all months except June and November, although the winter period (December–March) certainly had larger aggregations and more spawning. Blue tang, however, had a very clear lunar periodicity of aggregation spawning during the winter period, with spawning occurring from just before the full moon until a few days after the new moon (Fig. 12.60).

We believe that most ocean surgeonfish migrate to the spawning area each afternoon, because the densities of fish found in the study area during morning periods

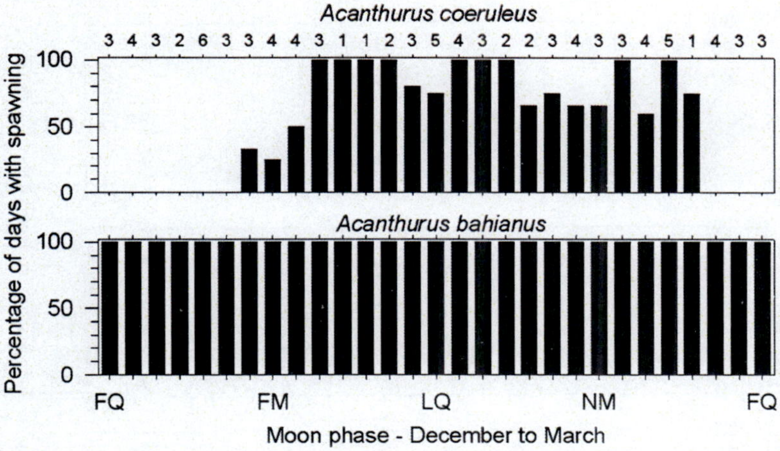

Fig. 12.60 These histograms show the occurrence of spawning on different lunar phases during the winter period (December to March) for blue tang (*upper*) and ocean surgeonfish (*lower*). The number of days when observation were made on a given day of the lunar calendar during those months are shown along the upper axis of the upper histogram. Blue tang showed lunar periodicity and ocean surgeonfish spawned every day of the lunar month

were not high enough to form aggregations of the size seen. We found a density of about 4 adult ocean surgeonfish per 100 m^2 of bottom during the non-spawning period (June-August). With an estimated 20,000 individuals in the spawning aggregation, a semi-circular area (since the spawning site is at the shelf edge and surgeonfishes do not penetrate deep water in abundance) 500–600 m in radius from the aggregation site would contain that number of fish. Although never quantified, it appeared there were discrete groups of surgeonfishes ("trains" in the sense of Myrberg et al. 1988) coming to the aggregation sites along specific paths across the sandy basin ("the moat") shoreward of the outer reef.

The daily patterns of spawning rushes of ocean surgeonfish in the aggregation were documented during the peak spawning season in February; blue tang were too shy to be counted and fled an approaching diver. Typically spawning consists of flurries of spawning rushes for 2–5 min followed by short quiescent periods of only 1–2 min (Fig. 12.61). When 5 days data were combined, the overall pattern to spawning activity is apparent with spawning starting about 90 min before, peaking 70–50 min before and ceasing about 30 min before sunset (Fig. 12.62). In the limited field of view from which spawning rushes were documented for 5 days during peak activity in February, the number of rushes seen per afternoon was 198–228. As this view encompassed only a portion of the aggregation area, only a fraction of the total numbers of spawnings were recorded.

Gonad indices of aggregated fishes: There is a clear correlation between increased gonad indices in both male and female ocean surgeonfish and the season for

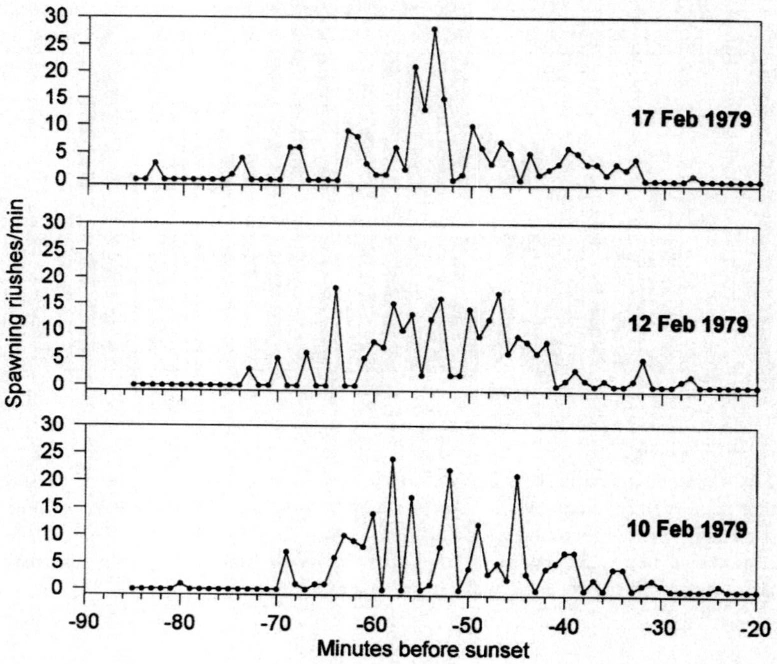

Fig. 12.61 The spawning of groups of ocean surgeonfish from the aggregation is episodic with bursts of spawning a few minutes long followed by quiescent period of 1–3 min. Spawning activity initiates somewhat gradually and tapers off about 30–40 prior to sunset

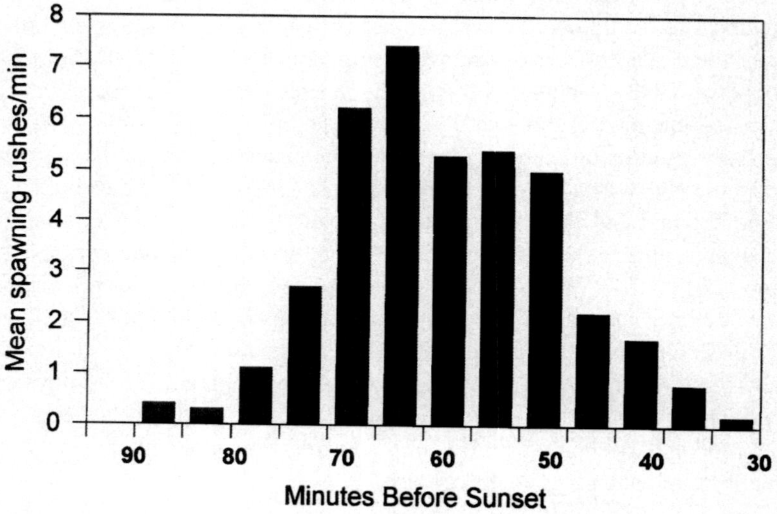

Fig. 12.62 The spawning occurrence of ocean surgeonfish observed over five successive afternoons relative to time before sunset shows that spawning lasts about 60 min overall, peaking about 1 h before sunset

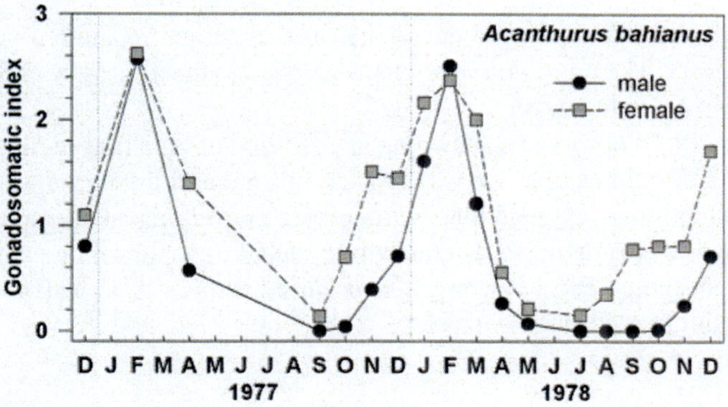

Fig. 12.63 The gonadosomatic indices (GSIs) of ocean surgeonfish for males and females are similar; most other fishes generally have female GSI higher than males during spawning periods unless there is sperm competition

aggregation spawning (Fig. 12.63). Gonad indices were high from November to March, with February the highest, while they were low from April until October. The range of gonad indices of individuals in any given month can be quite large. Male gonad indices are often higher than females, with some males having testes reaching 4–7% of body weight. There was no clear positive correlation between gonad indices and body weight for either sex during the aggregation season.

The only potentially comparable work on surgeonfish gonad development is that of Reeson (1983) who examined qualitative gonad stage for surgeonfish from both coastal reefs off Kingston, Jamaica and offshore banks south of Jamaica. Both blue tang and ocean surgeonfish from mainland Jamaica near Kingston had less gonad development than fishes collected from oceanic banks south of Jamaica. Ocean surgeonfish from near Kingston showed a pattern similar to that for Puerto Rico with gonads large from January to May, and a peak in April. On the offshore banks, levels of ripe gonads remained high throughout most of the year.

Duality of spawning – group and pair-spawning behaviour: Both ocean surgeonfish and blue tang pair spawn, in addition to their aggregation-spawning. Apparent large males hold and defend benthic territories in and around the aggregation site and are among the largest individuals found in the study area. When courting and spawning, their colour patterns were quite different from most other fish; the male ocean surgeonfish takes on a white underbelly (Fig. 12.56a) and blue tang a white mark on the upper side of the body (Fig. 12.56b). Ocean surgeonfish pair-spawned during and after the aggregation spawning period (Colin and Clavijo 1988), up until sunset, while blue tang pair spawned after group-spawning had ceased.

Notes on spawning aggregations of blue tang in The Bahamas and Belize: An aggregation of a few hundred blue tang at Lee Stocking Island, Bahamas, occurred between May and October 1989 (Colin 1996). The aggregation was discovered by following migrating fish to an area about 20 m deep more than 1 km inside the insular shelf edge where they group spawned. After November this aggregation disappeared, and it was not present at the same time and location a year after its initial discovery. The migration of adults, which led to the discovery of the aggregation site, also ceased to occur.

Deloach (1999) observed a spawning aggregation of blue tang at Turneffe Atoll, Belize, for 3 days in March, 7 days after the full moon. Large groups swam along the drop off prior to aggregation, with several converging at the spawning site. During group-spawning many fish had white caudal fins. Spawning continued until 20 min before sunset. Pair-spawning also occurred at this site, as well as off Bimini, Bahamas from May through August.

12.20.3 *Conservation of Surgeonfish Aggregations*

Given the vast lengths of shelf edge reefs, there must be many surgeonfish aggregations along them. Since the time and seasonality of aggregations are now known, searching in late afternoon during winter seasons may make it relatively easy to locate more aggregations. As specific aggregation sites are discovered, they might make excellent tourist diving sites and be subject to special protection.

The long-term existence of surgeonfish (and other resident aggregation species) aggregations is potentially an important resource for monitoring general health of reef fish populations. Dispersed fishes gather a specific locations and the stability of their numbers can be an important measure of general reef fish populations. Revisiting known aggregation sites allows the assessment of changes over time, limited by the methods used for estimating numbers of fishes involved.

Obviously, some spawning aggregations can cease to exist "naturally" and there may be a lower threshold of population when they can naturally cease. The apparent disappearance of the blue tang aggregation at Lee Stocking Island, in the absence of any known exploitation, is interesting, and may have been of a marginal "critical mass" for an aggregation to form when first noted.

Acknowledgements Work in Puerto Rico in 1977–1979 was supported by a grant from the National Geographic Society. The return visits to the site in 2011 were supported by the Caribbean Coral Reef Institute through award number NA09NOS4260243 from NOAA/CSCOR, Richard Appeldoorn Principal Investigator and by the Society for the Conservation of Reef Fish Aggregations.

12.21 Striped Bristletooth – *Ctenochaetus striatus* and Brown Surgeonfish *Acanthurus nigrofuscus* with Notes on Other Indo-West Pacific Surgeonfishes (Acanthuridae)

Patrick L. Colin
Coral Reef Research Foundation, PO Box 1765, Koror, Palau 96940
e-mail: crrf@palaunet.com

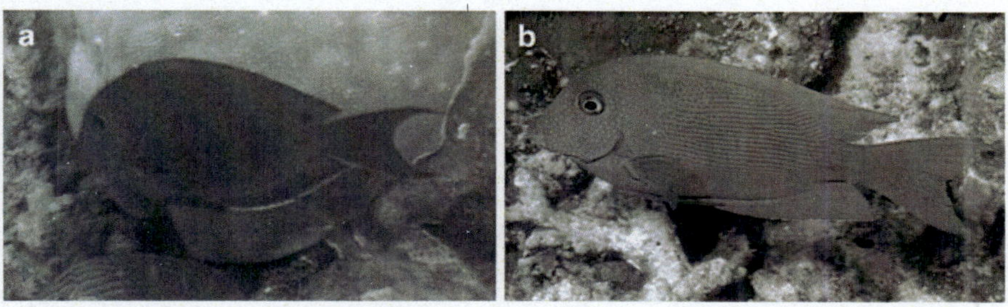

Fig. 12.64 (**a**) Brown surgeonfish, *Acanthurus nigrofuscus*, (**b**) Striped bristletooth, *Ctenochaetus striatus* (Photos: Patrick L. Colin)

12.21.1 General

In the Indo-west Pacific (IWP) two surgeonfishes, the striped bristletooth (*Ctenochaetus striatus*) and the brown surgeonfish (*Acanthurus nigrofuscus*), are excellent examples of fishes with resident aggregations that engage in short migrations and often spawn daily. Perhaps more is known about their aggregation and spawning than for any other surgeonfish. The brown surgeonfish is common in most areas of the IWP and the striped bristletooth is found widely in the Indo-Pacific, excluding the Hawaiian, Marquesan and Easter islands (Fig. 12.64).

The surgeonfishes (Acanthuridae), with about 80 species in 6 genera (*Acanthurus*, *Ctenochaetus*, *Naso*, *Paracanthurus*, *Prionurus*, *Zebrasoma*), occur worldwide in tropic and warm temperate waters, being one of the dominant families on coral reefs (Randall 2002). Most of the species are found in the Indo-Pacific, with the few western Atlantic *Acanthurus* covered separately (Chap. 12.20). They derive their common name from the sharp spines found at the base of the caudal fin, which serve both defensive and offensive purposes. The family includes species which feed on benthic algae, floating algae and zooplankton. Surgeonfishes are widely harvested for subsistence use in the Indo-west Pacific, but are not generally considered of major commercial importance. In some areas a few species, such as the bluespine unicornfish, *Naso unicornis*, are so highly prized and are specially sought by spearfishers for commercial sale and subsistence use. The eggs of acanthurids are spherical, or nearly so, and relatively small, 0.58–0.70 mm in diameter. Surgeonfishes are characterized by long larval life, with the acronurus larvae which reaches a large size before recruiting. Late-stage larvae are strong swimmers (Leis and Carson-Ewart 2000; Fisher 2005) capable of sustained swimming speeds greater than most ambient currents for long durations. Stobutzki and Bellwood (1997) reported them to be able to swim 95 km in 8 days in a laboratory flume.

Like most reef fish larvae, little is known of the distribution of surgeonfish larvae at sea. The acronurus larvae, however, is easy to identify, so the advantage of being able to easily distinguish surgeonfishes in larval samples may foster increased knowledge of the family. Genetic studies have supported the retention or return of acanthurid larvae to source regions rather than broad dispersal. Planes et al. (1996)

and Planes and Fauvelot (2002) reported strong genetic stratification of ocean populations of convict surgeonfish across the Pacific, indicating limited dispersal and self-recruitment in this species.

12.21.2 Reproductive Biology and Aggregation

Many species of surgeonfishes are known to aggregate to spawn in most areas where they occur, and aggregation-spawning is certainly a worldwide attribute of this family. Records of aggregation-spawning are widespread geographically, and where there is sufficient information, most species exhibit consistent patterns of aggregation and spawning throughout their ranges. No sex change is reported among any surgeonfish.

***Acanthurus nigrofuscus* – brown surgeonfish.** The spawning of the brown surgeonfish has been examined in the Red Sea, the western Indian Ocean, Great Barrier Reef, American Samoa, Palau and other locations (Johannes 1981; Robertson 1983; Fishelson et al. 1987; Myrberg et al. 1988; Craig 1998; PLC unpublished data). Robertson (1983) provided the first information on aggregation and spawning from three locations; Aldabra Atoll, Great Barrier Reef and Palau. In Palau, peak aggregation and spawning occurred on days just after mid- to late afternoon high tides (co-occurring with the striped bristletooth) while no spawning was seen on days with early morning high tides. Fish migrated a short distance from feeding areas, in 'trains' only a few metres wide. Pair spawning also occurred in feeding territories of males and females, however some males which appeared to set up temporary territories near aggregation areas were not observed to pair-spawn.

Spectacular spawning aggregations of many thousands of brown surgeonfish off Eilat, Israel in the northern Red Sea were initially documented by Myrberg et al. (1988), Fishelson et al. (1987) and Kiflawi et al. (1998). There two spawning aggregations of brown surgeonfish formed each afternoon during summer months (June–August). The two aggregations were separated by about 1450 m along a fringing reef slope. The fish in each aggregation spawn as a large group just before sunset after migrating as much as 1.5 km from feeding grounds. Fishelson et al. (1987) found differences in the life history patterns of two populations of brown surgeonfish living close together. One population had to migrate daily between feeding and resting grounds, while the second was resident in a single area which provided both requirements. Tidal patterns, local currents and daily temperature variation showed no relationship with spawning. Further work on the feeding ecology of the population has been published by Fishelson et al. (1987) and Montgomery et al. (1989). Fishelson et al. (1985) documented an unusual fat body external to the abdominal cavity in several surgeonfishes, including brown surgeonfish, which appears to serve as an energy storage site to fuel gonadal function during the spawning season. Mazeroll and Montgomery (1995, 1998) and Kiflawi and Mazeroll (2006) described the migrations of brown surgeonfish to and from spawning aggregations (for more detail see Chaps. 2, 5 and 9 this volume).

While most records of aggregation have been from fore-reef environments, Robertson (1983) reported one aggregation occurring at a channel mouth at Aldabra

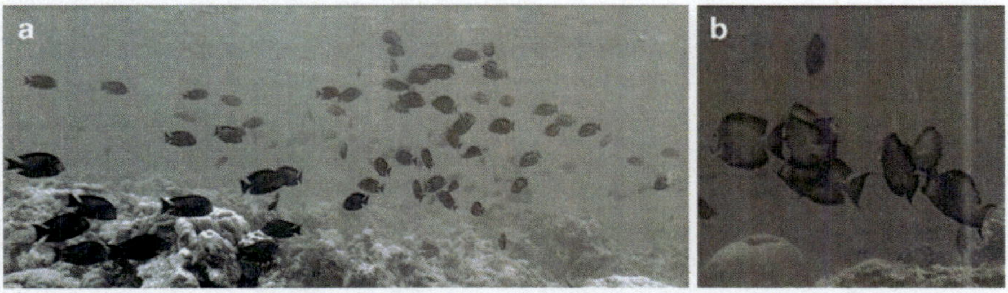

Fig. 12.65 (**a**) Striped bristletooth *Ctenochaetus striatus* aggregation along the side of the West Channel, Palau. (**b**) A group of striped bristletooth surgeonfish preparing to spawn. The *colour pattern* shown was reported to Robertson (1983) to represent group spawning males, but the sex is not confirmed for these fishes (Photos: Patrick L. Colin)

Atoll, Indian Ocean with fish migrating at least 300 m to that site. Along with its great geographic distribution, the species spawns at a wide range of temperatures. In the Red Sea spawning was limited to summer months, with temperatures of 24–25°C, and only 18.5–20°C during winter (Myrberg et al. 1988). In Palau spawning occurs year-round over the annual temperature range of 28.0–29.5°C (PLC unpublished data).

***Ctenochaetus striatus* – striped bristletooth.** Many aggregations of striped bristletooth can occur along IWP reef edges, where the fish spawn in response to tides and currents (Fig. 12.65). Robertson (1983) first reported aggregation and spawning in the striped bristletooth from Palau, Aldabra Atoll and Great Barrier Reef. The pattern of its spawning is quite similar to that of the brown surgeonfish, with peak spawning occurring after mid to late afternoon high tides (Robertson 1983) and occurs in the same areas in both Palau and Lizard Island. The maximum size and weight of striped bristletooth from Palau and Aldabra were similar (183 mm SL, 178 g versus 161 mm SL, 205 g) unlike the brown surgeonfish. Despite the occurrence of *C. striatus* at Eilat, no record of it spawning with *A. nigrofuscus* has been published.

In Palau striped bristletooth form resident aggregations which spawn just after high tide, as the current starts to move off the reef toward the open sea (Robertson 1983; PLC unpublished data). Brown surgeonfish aggregations mix in among the striped bristletooth and spawn nearly simultaneously. These surgeonfish probably migrate only a short distance, at most a few 100 m, to the spawning site as they are quite abundant and the numbers that occur would not require a large area to assemble. In Indonesia, groups were found to aggregate at intervals along a barrier reef and to spawn as the current changed direction with the tide (PLC unpublished data). Local current regimes can sometimes outweigh general reef currents in determining when spawning occurs. At the West Channel in Palau, groups of striped bristletooth initiated spawning along the channel edge while the general tide was still rising when winds drove slight currents from the shallow reef into the deeper channel (Fig. 12.65).

There are also many known occurrences of large episodic recruitments when thousands of small surgeonfishes, such as striped bristletooth (Fig. 12.66), appear on reefs almost overnight (Sancho et al. 1997; PLC unpublished data). Why such

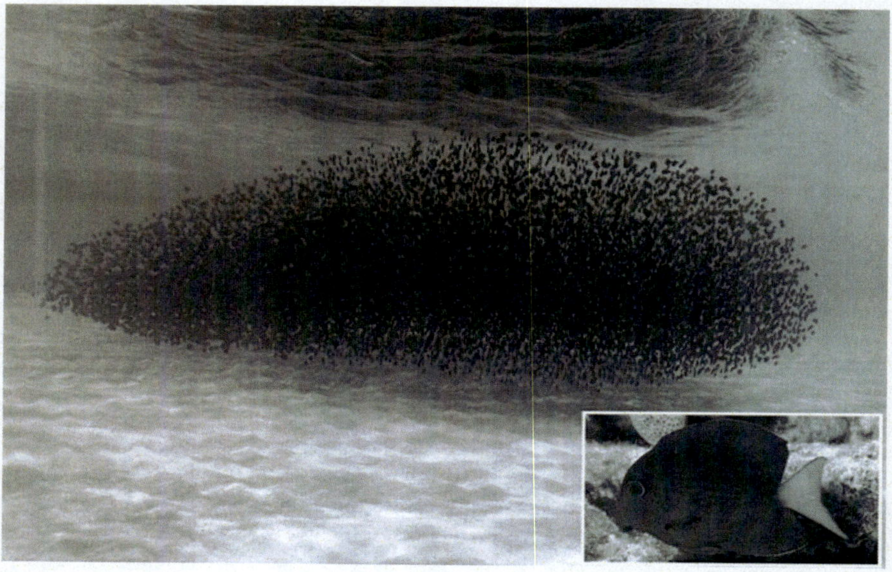

Fig. 12.66 Massive settlement of *Ctenochaetus* sp. (probably striped bristletooth) in Palau in spring 2009 over a sandy bottom in a lagoon area (photo: Mandy Etpison). Insert – juvenile striped bristletooth showing pale caudal fin (Photo: Patrick L. Colin)

events occur given the fairly regular and long durations (weeks to month) spawning of many species, is an interesting subject. Perhaps a given cohort is lucky in encountering ideal conditions for survival, while others are doomed to not succeed?

Robertson (1983) reported three possible patterns of diel spawning in surgeonfishes; early morning (blue-banded surgeonfish – *Acanthurus lineatus*, white-spotted surgeonfishes – *A. guttatus*), peak in late afternoon (brown surgeonfish – *A. nigrofuscus*, striped bristletooth – *C. striatus*) and dusk spawners (Atlantic *Acanthurus*, brushtail tang – *Zebrasoma scopas* in some areas). The last pattern may be most common where tidal range and currents are minimal. Among those species of surgeonfishes known to have spawning aggregations, most also engage in pair-spawning, often at the same times and places as aggregations. In Palau these dual modes of spawning produced high numbers of eggs, principally from aggregations of brown surgeonfish and striped bristletooth, (Hamner et al. 2007) along with smaller outputs of eggs from pair spawning.

12.21.3 Aggregation Occurrence Across the Family

Among the six genera of surgeonfishes, only *Paracanthurus* and *Prionurus* lack records of aggregation. *Acanthurus* has the largest number of species and the most spawning aggregation records. Where surgeonfish aggregations have been followed for some years, the locations of aggregations have remained consistent over that

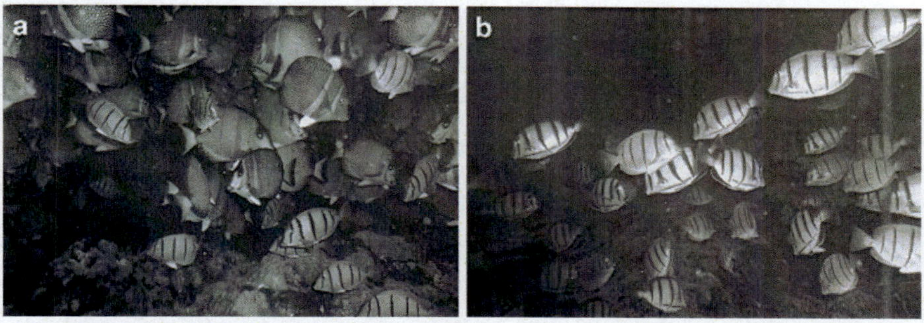

Fig. 12.67 (**a, b**) Aggregations of white-spotted surgeonfish, *Acanthurus guttatus*, and convict surgeonfish, *Acanthurus triostegus*, occur in reef channels in American Samoa (Photos: © Peter C. Craig)

time period (Colin 1996; Mazeroll and Montgomery 1998; Robertson 1983). Sancho et al. (2000a, b) found the presence of caudal spines does not deter the low level of predation on spawning adult surgeonfishes; levels of successful predation on blue-lined surgeonfish, *Acanthurus nigroris*, at Johnston Island were similar to that on parrotfishes, which lack the caudal spines. In addition to brown surgeonfish, information is available for the following species:

***Acanthurus triostegus* – Convict surgeonfish (manini)**. The exceptionally common convict surgeonfish is found on reef flats throughout the Indo-west Pacific region and was the first species of *Acanthurus* to have its spawning documented. Randall (1961a) reported on both pair- and group-spawning in Hawaii, and later Polynesia (Randall 1961b). Johannes (1981) reported aggregations to occur at the outer end of a channel through the barrier reef in Palau between the "fourth and tenth of the lunar month" between May and August, and perhaps to a lesser extent in other months. Robertson (1983) observed group- (and pair-) spawning on ebb tides in Aldabra Atoll, Palau and the Great Barrier Reef and illustrated colour patterns. Craig (1998) found convict surgeonfish spawning in reef channels during early morning in American Samoa, where it co-occurs with aggregations of white-spotted surgeonfish, *Acanthurus guttatus* (Fig. 12.67).

***Acanthurus lineatus* – blue-banded surgeonfish.** This is extremely common on shallow outer reefs. Johannes (1981) observed early morning aggregation spawning in April at the south end of Peleliu, Palau while Robertson (1983) also documents its early morning spawning elsewhere in Palau on shallow fore reefs. Robertson (1983) also saw a few group-spawns and one pair-spawn. The species has never been seen to spawn at other that in the morning, either as pairs or groups.

***Acanthurus nigroris* – blue-lined surgeonfish**. During extensive observations at Johnston Island, Sancho et al. (2000a) recorded more total spawns (all group spawns) by this species than any other, except bullethead parrotfish, *Chlorurus sordidus*, (even pair spawning species). Blue-lined surgeonfish spawned only for a portion of the lunar month, plus the channel where they were studied had unidirectional

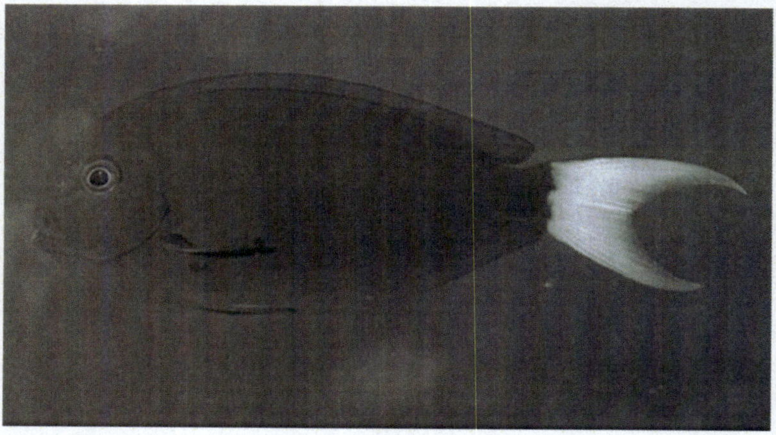

Fig. 12.68 The Thompson's surgeonfish, *Acanthurus thompsoni*, spawns in small groups along the outer reef faces in clear water areas (Photo: Patrick L. Colin)

outflowing currents during the period of observations. Based on more limited observations, Lobel (1978) earlier had reported group spawning in Hawaii.

***Acanthurus guttatus* – white-spotted surgeonfish**. Another very common *Acanthurus* that has received little attention is the white-spotted surgeonfish. Craig (1998) documented early morning spawning by *A. guttatus* in reef channels in American Samoa, where it co-occurs with aggregations of convict surgeonfish (Fig. 12.68), but no other records of spawning were found.

***Acanthurus thompsoni* – Thompson's surgeonfish.** The small outer reef Thompson's surgeonfish has been seen to group-spawn at Peleliu in Palau (Fig. 12.68) on a few occasions. Only small aggregations have been seen.

***Acanthurus blochii* – ringtail surgeonfish.** The ringtail surgeonfish migrates nearly daily along the fore-reef in Palau to an aggregation site near the side of a channel. It is one of the "medium-sized" scarids and acanthurids to make this daily migration, while smaller species in both families move only a short distance to the reef front to spawn. "Trains" of ringtail surgeonfish move to and from the aggregation site, in combination with greenthroat parrotfish, *Scarus prasignathos*.

***Acanthurus achilles*: Achilles tang.** Sancho et al. (2000a, b) reported 9 group-spawnings by Achilles tang, *Acanthurus achilles*, at a channel site at Johnston Island, while blue-lined surgeonfish, *A. nigroris*, engaged in over 3,000 spawnings at the site over the course of their multi-year study.

Ctenochaetus spp. – The only species of *Ctenochaetus*, other than the striped bristletooth, reported to aggregate is *Ctenochaetus strigosus*, the goldring bristletooth. Lobel (1978) observed group spawning by goldring bristletooth in Hawaii, while Sancho et al. (2000a, b) recorded several hundred group spawns at Johnston Island in a reef channel.

Fig. 12.69 Schools of orangespine unicornfish, *Naso literatus*, and blackstreak surgeonfish, *Acanthurus nigricauda*, can occur on outer reefs of Palau, such at that seen here at Blue Corner. These may possibly represent spawning aggregations, but is this as yet unproven (Photo: © Mandy T. Etpison)

Unicornfish (*Naso* spp.) – Whether members of the unicornfishes, genus *Naso*, which school above coral reefs, have true spawning aggregations is uncertain. The genus includes zooplankton feeders as well as species that feed on drifting algae and benthic materials. Pair spawning has been reported in several instances, but not confirmed group-spawning. They typically swim high in the water column, are often difficult to approach and, being favourites of spearfishermen in some areas, often wary.

Large schools of orangespine unicornfish, *Naso literatus*, are potentially spawning aggregations and are a well-known phenomenon on many outer reef slopes of Palau during spring (Etpison 2004). Schools of many hundreds to thousands appear on the reef slopes (Fig. 12.69), often mixed with a few other types of acanthurids, such as the blackstreak surgeonfish, *Acanthurus nigricauda*, which may also potentially be aggregating. Here they are harassed and occasionally attacked by sharks and other predators during the day, a spectacle often viewed by tourist divers. Johannes et al. (1999) reported orangespine unicornfish to aggregate without specific information and at present these must be considered unconfirmed records. Large schools occur during the spring along outer reef faces, such as Blue Corner, and may be spawning aggregations (Fig. 12.70). The late larval stage of the subfamily with *Naso* has some differences from the acronurus of most other surgeonfishes and is identified by a different name, the "keris".

Much of the information concerning species of unicornfishes is anecdotal or poorly documented. Johannes (1981) reported fishermen indicating that bluespine unicornfish, *Naso unicornis*, in Palau move along outer reef slopes and have great

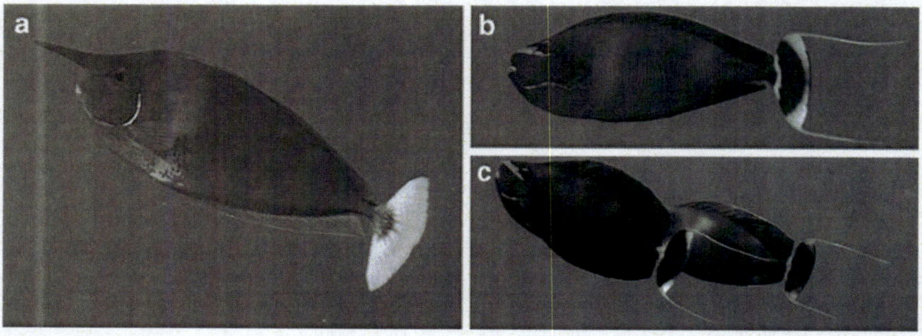

Fig. 12.70 Male courtship colouration and behaviour of *Naso* species: (**a**) spotted unicornfish, *Naso brevirostris* (**b**) bignose unicornfish, *Naso vlamingii* male in courtship colors; (**c**) Two male bignose unicornfish, *Naso vlamingii*, engaged in a "territorial" dispute (Photos: Patrick L. Colin)

seasonal variation in numbers with different timing in different areas. Locally called "chum", the bluespine unicornfish is one of the most prized fishes in Palau and fishers have paid considerable attention to its habits. Based on gonads fishermen believed fish to spawn around both new and full moons. In more recent interviews in Palau, fishers noted the species to be found in large groupings when it has eggs and when it can be caught in large numbers, 250 kg or more in a single fishing trip. Many were once taken using the leaf sweep, a method little used today having been replaced with gillnets. Fish are found with eggs in many months and, although accounts between fishers varied, months often identified were February and August, at both full and new moon times (Sadovy 2007). Rhodes (2003) similarly reported the slender unicornfish, *Naso lopezi*, to group-spawn and again this needs confirmation. Finally, based on information from fishers Johannes (1981) reported whitemargin unicornfish, *N. annulatus*, to live and spawn along the inner barrier reef edge and slope near Ngeremlengui, Palau and to spawn on new and full moons in May. Courtship and territorial defense is often seen among species of *Naso* on outer reef slopes in Palau (Fig. 12.70).

Tang (*Zebrasoma* spp.) – There are a few records of group-spawning by yellow tang, *Zebrasoma flavescens*, in Hawaii (Lobel 1978). Sancho et al. (2000a, b) recorded over 900 group-spawns (and no pair-spawns) in a channel at Johnston Island, so this seems likely a confirmed aggregating species. For brushtail tang, *Zebrasoma scopas*, Randall (1961a, b) reported it to group-spawn although in most areas pair-spawning seems most common (PLC unpublished data). For sailfin tang, *Z. velliferum*, Robertson (1983) recorded one group-spawn among numerous other pair-spawns and it is unknown how common possible aggregative spawning might be in this species. In the latter two species there may be a critical mass effect, where group-spawning occurs only when populations reach a certain level (Chap. 12.14).

***Paracanthurus* and *Prionurus* –** Robertson (1983) reported an aggregation of 25–30 palette surgeonfish, *Paracanthurus hepatus*, at Escape Reef, Great Barrier Reef that would break up into sub-groups of a male and several females. Pair-spawnings were seen, but no group-spawning. For species of *Prionurus*, there are no known group- or pair-spawning observations in the technical literature. The genus has several species and is found in cooler areas, being most common in semi-tropical to temperate waters; they occur in some tropical areas, such as Bali, where there are regular cold water upwellings.

12.21.4 *Fisheries and Ecological Importance*

Surgeonfish are of considerable local importance in subsistence fisheries, and some species come under intense fishing pressure. In Palau many fishers have commented on the disappearance of bluespine unicornfish *N. unicornis*, from some areas and the generally smaller size of those taken today. When Palauans from the most populated region of Koror visit remote reef sites, such as Helen Reef where fishing pressure is low, they are impressed with the size and abundance of "chum" found. The larger *Acanthurus*, as well as *Ctenochaetus* are readily taken by local spearfishers, but it is not known whether spawning aggregations are particularly targeted for these fishes. Robertson (1983) reported brown surgeonfish from Palau, where they are subject to spearfishing, about half the size and weight of fishes from remote Aldabra Atoll, Indian Ocean. A number of colourful species, such as palette surgeonfish and yellow tang are popular aquarium fishes.

Acanthurids are dominant herbivores in many areas, turning benthic primary production into planktonic gametes which are exported. Hamner et al. (2007) found large amounts of fish eggs, a significant portion of which were produced by aggregation spawning striped bristletooth and brown surgeonfish pulsing off of a reef in Palau on falling tides. Some of these eggs returned near the reef, while others were transported along the reef or, in a few cases, taken further offshore, on the next rising tide.

12.22 Dusky Rabbitfish, *Siganus fuscescens*, in Palau

Ann Hillmann Kitalong
The Environment, Inc., P.O. Box 1696, Koror, Palau 96940
e-mail: kitalong@palaunet.com

12.22.1 *General*

The dusky rabbitfish, *Siganus fuscescens* (Houttuyn 1782), is known as *meas* in Palau (Fig. 12.71) and is widely distributed in the Indo-west Pacific from the Andaman Islands to Vanuatu, Japan, Korea, Australia, New Caledonia, Palau, Yap,

Fig. 12.71 Small group of *Siganus fuscescens* which has just arrived at the spawning area (Photo: Patrick L. Colin)

Woliae and possibly Pohnpei. It is a member of the family Siganidae, which contains 2 genera and 25 species, 12 of which are found in Palau (Myers 1999). Palauan fish were previously confused with *S. canaliculatus*, an Indian Ocean and Southeast Asian species, (Woodland 1990). The family is known for their venomous dorsal, anal pelvic spines which can inflict a painful wound, and as an important food fish in some areas. Species in the family for which there is information have adhesive demersal eggs which are shed by the females and externally fertilized by males.

In Palau the dusky rabbitfish (Siganidae) is an important grazer inhabiting lagoons, coastal reefs, bays and shallow seagrass flats. In the latter habitat adults and juveniles feed diurnally on the epiphytes of seagrasses and often school with juvenile parrotfishes, surgeon fishes, and goatfishes. Juvenile dusky rabbitfish (2.5–10 cm) are found year round with greater numbers during January and May in some areas. Newly settled juveniles form large schools on seagrass beds and school sizes decrease with age. Subadults and adults roam the seagrass flats during high tide and retreat to deep lagoon waters as the tide ebbs.

12.22.2 Reproduction and Aggregation

The species is known for forming large "pre-spawning aggregations" (PSA), in which fish ready to spawn congregate as groups in areas inshore or inside shallow reefs and then migrate *en masse* across the reef in late afternoon just prior to spawning which occurs later in the afternoon or early evening on the outer reef face. At this time tides are usually at or near low water.

Rabbitfish spawn year round with the main spawning season in Palau between February and June, and occasionally a secondary peak between September and November; detailed studies conducted over various time periods provide finer temporal scale information on fish densities and biomass. The smallest sexually mature males were measured at 12.5 cm SL (standard length) for males and 15 cm SL for females (AHK unpublished data 2004). Gonads from a confiscated catch (taken during closed spawning season) represented 11% of the total body weight for 18 females and 8% for 33 males. The sizes of fish caught from PSAs in 1990–1991 (mean fork length-FL = 18.6 cm, sd = 2.1 cm, n = 526) were similar to those caught in 2005 (mean FL = 18.6, sd = 1.9 cm, n = 82), but in 2005 there were many fewer adult fish greater than FL 23.0 cm, implying loss of the largest size category of spawning fish.

Spawning of dusky rabbitfish has not been witnessed in the wild, but spawning activity was observed in the tanks in the early morning. Males actively court females who eventually go to the bottom of sandy areas and release a batch of eggs which the male fertilizes (Thomas Taro, personal communication 2008). Johannes (1981) described from fisher interviews how, just prior to spawning, fish tend to mill in slow tight circles and form pairs. Fishers stated that rabbitfish migrate from adjacent areas to spawn in Airai Bay in southeastern Babeldaob, the main island of Palau and it is thought that there are resident populations as well as migratory populations within the bay (Clarence Kitalong, personal communication 2008). Pre-spawners are also known to release gametes in the nets or into the boats after capture.

In aquaculture studies in Palau females released about 500,000 adhesive eggs at a time (Drew 1973) that are denser than seawater and settle to the substrate where they are scattered on the bottom (De los Santos et al. 2008). Broodstock spawned year round (De los Santos et al. 2008) with egg production on the 4th–8th day after full moon (De los Santos et al. 2008). New vitellogenic oocytes appear on day 2 after the first spawning and are fully mature by day 30. When a greater percentage of the most advanced oocytes fully develop, they form a batch and separate from the adjacent group of smaller pre-vitellogenic oocytes, indicating that the dusky rabbitfish is a multiple spawner with an ovary belonging to the group-synchronous type of oocyte development. Batch fecundity ranged from 0.52 to 2.56 million eggs. Fecundity (F) increased exponentially ($F = .0536854\,L^{5.07292}$) with fork length (Hoque et al. 1999). One female in captivity spawned repeatedly in successive months (Hasse et al. 1977). Eggs develop within 32 h, phytoplankton is found in larval guts within a day of hatching. The planktonic stage is 21–23 days and newly metamorphosed fish average 23 mm in standard length.

12.22.3 *Commercial/Subsistence Importance*

Dusky rabbitfish have long supported subsistence and, more recently, modest commercial fisheries. While data are limited, annual commercial production for dusky rabbitfish from 1976 to 2006 indicate 5-year cycles, with peak catches during 1980, 1985 and 1990 and the highest reported landings of over 30 metric tonnes in 1990

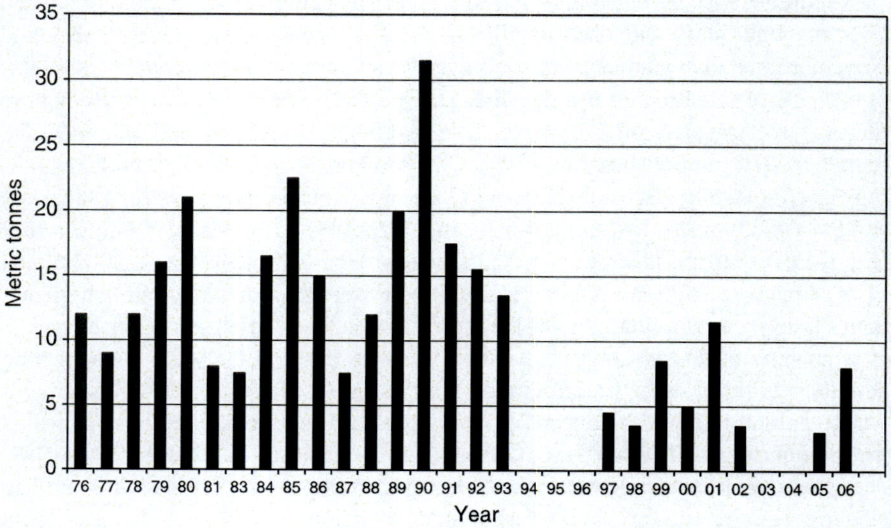

Fig. 12.72 Reported commercial capture of *Siganus fuscescens* in Palau from 1976 to 2006. (Data Source: Bureau of Marine Resources 1992)

(Fig. 12.72). This cyclic pattern may reflect maximum sustainable yield had been reached in peak years, causing declines in subsequent years, with climate cycles recruitment variability or socio-economic factors contributing to the decreases, or to the patterns in general. Since 1994, peak catch years have been equivalent to the lowest of the last three decades. During 1990 and 1991 length converted catch curves were used to derive exploitation rates suggesting that over a decade earlier the fish stocks of dusky rabbitfish were already optimally exploited (E_{opt} =0.507)(Kitalong and Dalzell 1991). There has been a generalized impression that siganid "runs" are greatly depleted and calls for conservation are being made of fishers. A recent newspaper editorial in Palau entitled 'We seldom see them anymore' bemoaned the loss of large annual catches of rabbitfishes and several other schooling fishes (Palau Horizon, page 4, April 17, 2009).

Much of the fishery on dusky rabbitfish has focused on capture of fish moving to spawning sites from inshore areas. The timing of migrations is well known, starting 3–6 days after the new moon and lasting for a few days from February to June each year (Fig. 12.73). Dusky rabbitfish are caught during the day with throw-nets (still the most common method), throw spears, surround nets and spear guns. Night fishing with spearguns and flashlights is common today. Most fishing activity observed by the authors occurred from February-April and in July. During 2004 and 2006, more commercial landings were reported before and after the closed season during February and June. However, commercial landings (which should have been zero) were still reported throughout the March-May closed season. Interviews with fishers indicated that they had experienced reductions in both sizes of fish and numbers caught during their fishing careers (Sadovy 2007).

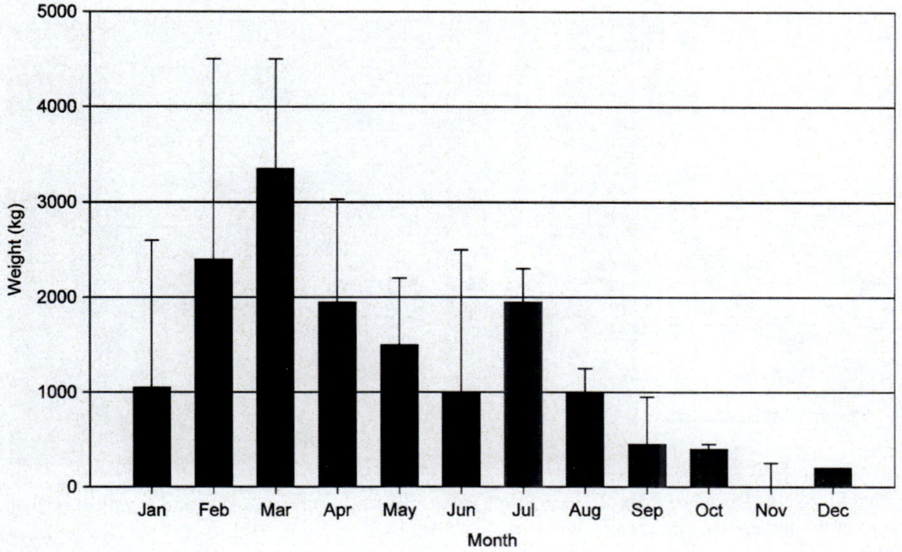

Fig. 12.73 Mean monthly landings (sd = 1) of *Siganus fuscescens* from 1983 to 1992 (Data Source: Bureau of Marine Resources, Republic of Palau 1992)

In Palau at least 38 inshore areas were once known (McVey 1972) where fish gathered 3–7 days after the new moon reportedly year round as dense schools of gravid fish prior to migrating across shallow reefs to reef front spawning locations. The present status of most of these sites is unknown. Airai Bay, on southeastern Babeldaob, is known as one of the largest dusky rabbitfish PSA areas in Palau. Such aggregations had 50–1,000 fish in past years, although much lower numbers are seen now. During 1990–1992 Kitalong and Oiterong (1992) observed up to 8 PSA schools in Airai Bay with 100–500 fish each, as well as other schools with as few as 15 fish. In Airai Bay PSAs begin forming in the southeast and southwest areas, then move along and across the reef flats and channels to the outer reef where fish are known to release their eggs and sperm in shallow breaking surf at the reef edge. Fishermen standing in shallow water with cast nets position themselves along the migration path in the late afternoon, mostly along two natural sandy depressions along the eastern reef where the fish concentrate as they head to the outer reef. They throw their nets to capture fish as they pass over the shallow (0.5 m or less depth) areas.

12.22.4 Conservation and Management Concerns and Action

In the early 1990s fishers became concerned about the decline in the number and sizes of PSAs. In 1994 legislation ("Marine Protection Act") banned fishing for dusky rabbitfish from March 1 to May 31 during the supposed peak spawning season. For the years of 1994–1996 there were no reported commercial landings,

Fig. 12.74 (**a**) Location map of Airai State, Palau. (**b**) Location (*asterisks*) and numbers of fish within the largest pre-spawning aggregations (*PSA*) observed from 2003 to 2006. *Boxes* indicate numbers of fish within a given PSA with position. Small groups (less than 100 fish) of PSA are shown as *black dots* and clustered along the two monitoring transects at Site 1 and Site 2 (QuickBird Image courtesy Palau Automated Land and Resources Information Systems (PALARIS), Ministry of Resources and Development)

followed by relatively low production in subsequent years (Fig. 12.72). The decline after 1994 may be due to one or some cumulative results of fewer fish, fewer fishers, the impact of the national seasonal closure, limitations on exports, pollution, increased boating activity impacting habitats and the 1998–1999 ENSO event. In 2006 the Marine Protection Act was modified to shift the closed season to February 1 to March 31, reducing the closure from 3 to 2 months, in order to allow fishing during April and May to increase landings of the fish. Overall the data have become less reliable due to several problems, with recent years showing relatively large landings during the closed season, an obvious flaunting of the closed season regulation (Fig. 12.75).

The locations, sizes and numbers of PSAs were quantified in Airai Bay from 2004 to 2006. At least two sites were surveyed at least twice a month with more intensive monitoring (5 days/month) during the peak spring season. Two known sites (sites 1 and 2 with similar numbers and densities of PSAs) were selected for study in 2004 based upon an earlier study (Kitalong and Oiterong 1992) with additional sites based upon traditional knowledge of fishers. Rabbitfish were counted along the natural contour of either the seagrass beds or the coral reef at each site by snorkelers (Fig. 12.74b). During 2004–2006 only two schools of more than 100 fish were seen with an overall decline in the number and sizes of PSAs compared to the early 1990s. Fishermen, however, provided contradictory information, reporting to have counted 23 spawning aggregations ranging from 80 to 1,000 fish per aggregation during 3 h in April 2006, in Airai Bay, indicating the tenuous nature of siganid abundance data. Stationary monitoring over a longer period of time at a given location may be an important alternative method of assessing aggregations at specific locations.

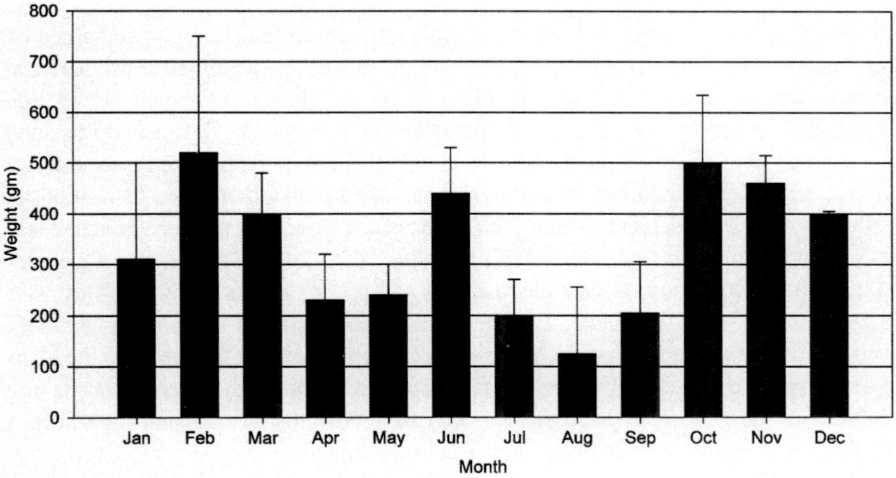

Fig. 12.75 Mean total reported commercial landings (sd = 1) during 2004 and 2006. Closed season is March to May (Data Source: Bureau of Marine Resources Database)

12.22.5 *Traditional Management and Its Challenges*

Prior to the 1994 Marine Protection Act, a traditional "bul" (a traditional management tool to restrict harvest of a resource that is considered low in abundance) or law was set in 1991 by the chiefs of Airai to restrict fishing for rabbitfish during the time of the spring spawning event. Management was less complicated when groups of men fished together using canoes and rafts. Today the number of fishers has dropped so fewer community members are at sea on a daily basis to safeguard their resources. Faster boats are in use by fishers from other areas which can quickly come and go without being detected, making traditional law enforcement more difficult.

Government enforcement of closed seasons and protected areas is limited by resources for enforcement and the relatively large areas to protect. It had been hoped that the presence of National Government Fish and Wildlife officers would deter illegal fishing during the closed season; however during each month of closed season officers still discovered occurrences of illegal fishing. In 2005 one two-time violator of the law was fined and his net and catch confiscated. The Chief of Fish and Wildlife, Kammen Chin, has stressed the importance of perseverance and consistency in terms of enforcement, highlighting the need for cooperation and collaboration between the communities, State, and national entities.

In 2006 the National seasonal closure period of March 1 to May 31 was shortened to February 1 to March 31 using the logic that February is considered the start of the peak season and the closure should be at the beginning of the season, not its end. Recommendations to extend the closed season from February to May, rather than to shorten it, were rejected. Complicating enforcement of the limited closure period, rabbitfish PSAs and spawning sites are open to fishing for other species during the closed season. Boat inspections and close up observations are difficult and not welcomed by fishers when focusing their fishing efforts on other species (Kammen Chin, personal communication 2006). Enforcement efforts during the closures at known aggregation sites have not proven cost-effective for the Division of Fish and Wildlife and State and community conservation officers. The 2006 modification of the Marine Protection Act allows seizure of banned fish on land or at the markets rather than at sea. Inspections at fish markets during February and March have proven more cost and time effective, but have done little to stop subsistence fishing (which does not go to market) on the sites during the closure. Four village meetings, as well as one meeting with the Governor and Legislatures of Airai State, were held in 2006 to exchange information and concerns about the aggregations of *S. fuscescens*. Several chiefs commented that the mangroves have expanded and were displacing seagrass habitats that were good fishing habitats in the past. They requested a study to address factors causing the seaward expansion of mangroves including increased sedimentation (Golbuu et al. 2003).

Aquaculture has been suggested as a means to rejuvenate natural populations. Rabbitfish have proven amenable to culture in tanks and there is considerable interest in using wild-caught juveniles as the basis of 'grow-out' commercial aquaculture of dusky rabbitfish. Current obstacles for fish farmers are the lack of consistent supply of juveniles, predation by birds and lack of protection of farms from poachers. Moreover, aquaculture will not solve overfishing of wild populations unless the fishermen turn from capture fisheries for the species to their culture and no additional fishing pressure is introduced.

Acknowledgements Clarence Liz Bausouch and Mcknight assisted with all the field monitoring for this paper. Clarence Kitalong, Elizabeth Matthews, Bausouch Ngiramur, and McKnight McArthur contributed to the overall work. Support was provided by the National Oceanographic and Atmospheric Agency (NOAA) Habitat Conservation Program, Palau Marine Resources Pacific Consortium, members Ngaremeliwei of Airai, Palau Conservation Society, Coral Reef Research Foundation, Palau Land and Resource Information Systems (PALARIS), and the Bureau of Marine Resource. We thank all the fisherfolk who shared their traditional ecological knowledge about dusky rabbitfish, allowed us to observe their fishing activities and patiently answered our many questions.

12.23 Yellowmargin Triggerfish – *Pseudobalistes flavimarginatus*, with Notes on Other Triggerfishes

Patrick L. Colin
Coral Reef Research Foundation, PO Box 1765, Koror, Palau 96940
e-mail: crrf@palaunet.com

12.23.1 General and Reproduction

The yellowmargin triggerfish, *Pseudobalistes flavimarginatus*, is found from the Red Sea to the Tuamotus in the central Pacific. The family of the triggerfishes (Balistidae) has about 40 species in 11 genera distributed throughout tropical and

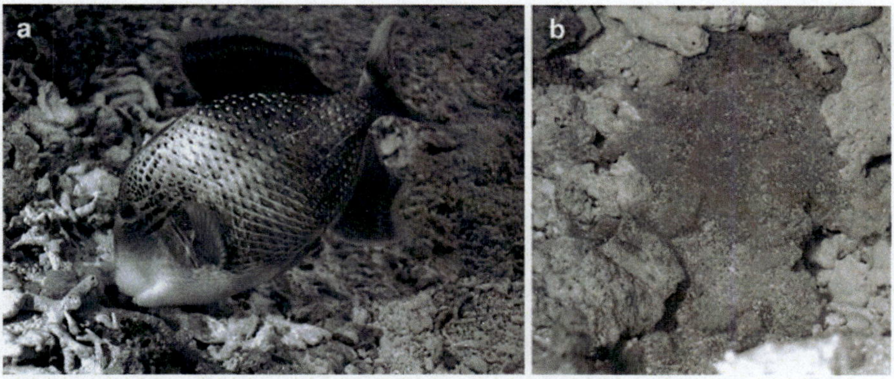

Fig. 12.76 (**a**) Yellowmargin triggerfish, *Pseudobalistes flavimarginatus*, tending nest. (**b**) Egg mass of yellowmargin triggerfish at base of nest pit amongst rubble pieces (Photo: Patrick L. Colin)

most sub-tropical waters where they are common members of reef associated communities. For some species it is known they have adhesive and negatively buoyant eggs, generally deposited as a mass in the bottom of depressions ("nests") in sandy or rubble bottoms (Fig. 12.76).

The yellowmargin triggerfish appears typical for balistids with concentrations of fish and nests occurring at specific locations and times, meeting the basic definition of a spawning aggregation. They differ from most other aggregating species in their benthic eggs (rather than planktonic). Like other balistids, yellowmargin triggerfish eggs have a short incubation period and their eggs hatch in less than 1 day into larvae which resemble those of reef fishes with pelagic eggs. It seems likely eggs hatch near simultaneously, making the demersal aggregation sites near point-sources of larvae. Other groups with true demersal eggs, such as the damselfishes, have the eggs attached individually to the bottom with an incubation lasting up to several days.

At Ulong Channel (Ngerumkaol) in Palau, a grouper aggregation site, yellowmargin triggerfish nested in the sandy areas of the channel adjacent to hard bottom areas where groupers gathered. While surveys on successive days showed changes in nest distribution (Fig. 12.77) and nest sites were abandoned between full and new moon periods, nest presence was consistent in the same areas year to year. Lobel and Johannes (1980), Johannes (1981) and Myers (1999) reported them to nest in sand-bottomed channels and shallow barrier reef cuts 3 days before to 1 day after both new and full moons from April through August, and November and December. It is likely spawning occurs year-round. Nests are depressions up to 2 m wide and 0.7 m deep (Fig. 12.78) with rubble at their bottoms. Eggs are deposited at the base in a spongy fist-sized cluster weighted down with a piece of rubble to be negatively buoyant. One cluster examined by Lobel and Johannes (1980) had about 430,000 small eggs, only 0.55 mm diameter. In many triggerfishes the male is believed to excavate the nest pits, while both males and females may guard and aerate the eggs by blowing water on them (Fig. 12.78a). Guarding fishes can be surprisingly aggressive, rushing at any intruders, including human divers who occasionally are bitten

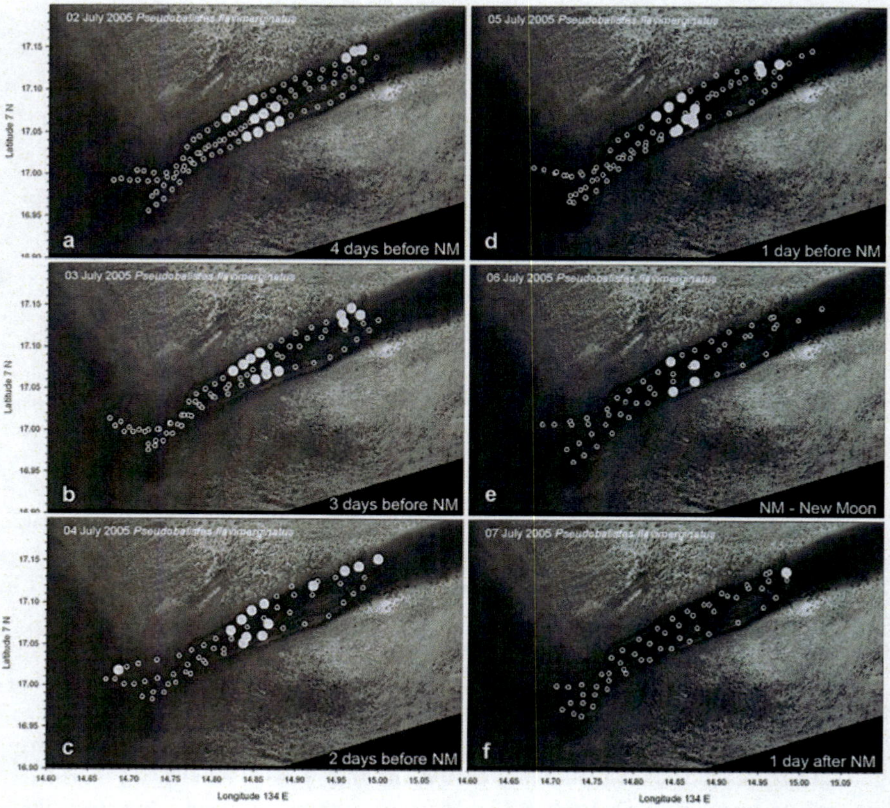

Fig. 12.77 (**a–f**) Distribution of yellowmargin triggerfish nests at Ulong Channel, 2–7 July 2005 (6 July – new moon). The white circles show locations where nests were encountered while *open circles* indicate areas where no nest occurred

(Millington and Randall 1990). Predators will eat the eggs if the guarding fish is driven away.

Gladstone (1994) detailed a lek-like mating system in yellowmargin triggerfish on the Great Barrier Reef. Large numbers of males migrate to traditional mating grounds where they establish territories (and dig nests?). Females arrive several days later and select a male for spawning. Mating was semi-lunar; eggs were present in nests several days before both new and full moon days with high tide near sunset. Mating lasted 2–3 days, starting 2–5 days before the new and full moons. Spawning (not observed) may occur in the early morning and up to three egg masses, probably from separate females, have been observed in one egg chamber. Females concurrently tend nests (aerating eggs) within the territory of one male. Eggs are gone the morning after spawning and may well hatch in early evening after spawning. Fish abandon the spawning ground after the last eggs have hatched and males return 9–13 days later to begin the new spawning cycle.

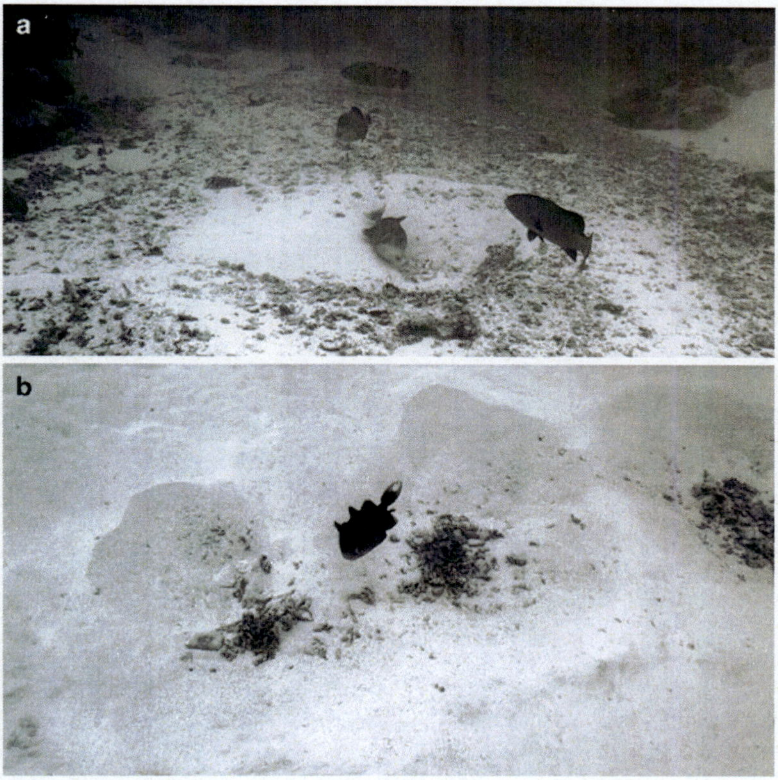

Fig. 12.78 (**a**) Nests of yellowmargin triggerfish at Ulong Channel on rubble bottom 8–12 m deep where aggregated groupers transit. The groupers are attacked if they venture too close to nests. (**b**) Multiple triggerfish nests converging on a sandy bottom. Yellowmargin triggerfish guarding and aerating the eggs (Photos: Patrick L. Colin)

12.23.2 *Other Species of Balistids*

The characteristics of nest-digging and guarding of eggs by balistids allows detection of aggregation and spawning for periods of a day or more, whereas the evidence of other fishes aggregating and spawning is often gone in a few minutes to a few hours, simplifying monitoring of aggregation occurrence.

The reproduction of the large titan triggerfish, *Balistoides viridescens*, is largely unreported, but appears similar to that of the yellowmargin triggerfish (Fig. 12.79), including attacks on divers approaching guarded nests. For other genera, there are spotty reports of spawning. In the Indo-west Pacific Kuwamura (1997) found both sexes of Picassofish, *Rhinecanthus aculeatus*, maintained long-term territories; some for up to 8 years, thus this species may not be considered an aggregator (no increase in density of adults). Male territories overlapped 2–3 female territories with only

Fig. 12.79 **(a–b)** The titan triggerfish, *Balistoides viridescens*, is the largest triggerfish and a common inhabitant of Indo-west Pacific reefs. It digs nests in areas similar to those used by yellowmargin triggerfish and guards its egg mass, indicated by *dark arrow*, with vigour (Photo: Patrick L. Colin)

females caring for eggs. Pair-spawning occurred for about 1 week around new and full moon, with some females spawning as many as three times during that time. Spawning took place around sunrise with eggs hatching just after sunset on the same day. For *Xanthichthys* Kawase (2003) reported females to aerate eggs by blowing water at them, and both sexes guarded the eggs. Kawase and Nakazono (1994) as well as Ishihara and Kuwamura (1996) reported on spawning in halfmoon triggerfish, *Sufflamen chrysopterus*. Sahayak (2005) reported on reproductive biology of bridle triggerfish, *Sufflamen fraenatus*, but did not have any observations of spawning behaviour. The clown triggerfish, *Balistoides conspicillum*, appears to spawn as individuals, as isolated fish (males?) have been encountered guarding nests.

For western Atlantic species even less is known. Nellis (1980) briefly noted that ocean triggerfish, *Canthidermis sufflamen*, constructed benthic nests at 15–20 m depth on the edge of reef and sand during August–September 1973 in St. Croix, USVI. It appears nothing has been published on the spawning of the queen triggerfish, *Balistes vetula*, the largest balistid in the Atlantic Ocean (Robertson 1988a, b).

For the circumtropical black durgeon, *Melichthys niger*, which can be exceptionally common on outer reef faces, apparently nothing has been published. In some areas, such as Palau, there are millions of juveniles and small adults on the outer reef faces, and their ecological importance is certainly very high (Fig. 12.80). Vast shoals of them can be found in the water column over reef dropoffs, and they are a major predator on newly spawned fish eggs. How there can be virtually nothing known of the reproduction of such a ubiquitous species is surprising, but also an excellent opportunity to advance knowledge.

Finescale triggerfish, *Balistes polylepis*, aggregate to spawn and nest from May to September in the southern Gulf of California, Mexico (Sanchez-Velasco et al. 2009;

Fig. 12.80 Schools of *Melichthys niger* above shelf edge reef at Peleliu Island, Palau (Photo: Patrick L. Colin)

Brad Erisman, personal observation). Commercial fishers target these aggregations using nighttime hookah fishing and gillnets (Chap. 8, Fig. 8.5).

Triggerfishes generally are of minor commercial importance. Many are good to eat, but ciguatera poisoning is a concern in some areas. They can be ecologically important, since many are very common and some eat algae and zooplankton, as well as preying on fish eggs.

References

Adams S, Mapstone BD, Russ GR, Davies C (2000) Geographic variation in the sex ratio, sex specific size, and age structure of *Plectropomus leopardus* (Serranidae) between reefs open and closed to fishing on the Great Barrier Reef. Can J Fish Aquat Sci 57:1448–1458

Aguilar-Perera A (2006) Disappearance of a Nassau grouper spawning aggregation off the southern Mexican Caribbean coast. Mar Ecol Prog Ser 327:289–296

Aguilar-Perera A, Aguilar-Dávila W (2007) A spawning aggregation of Nassau grouper *Epinephelus striatus* (Pisces: Serranidae) in the Mexican Caribbean. Environ Biol Fish 45:351–361

Aguilar-Perera A, González-Salas C, Villegas-Hernández H (2009) Fishing, management, and conservation of the Nassau Grouper, *Epinephelus striatus*, in the Mexican Caribbean. Gulf Caribb Fish Inst 61:313–319

Albins MA, Hixon MA, Sadovy Y (2009) Threatened fishes of the world: *Epinephelus striatus* (Bloch, 1792) (Serranidae). Environ Biol Fish 86(2)

Alfaro ME, Brock CD, Banbury BL, Wainwright PC (2009) Does evolutionary innovation in pharyngeal jaws lead to rapid lineage diversification in labrid fishes? BMC Evol Biol 9:255

Allen GR, Talbot FH (1985) Review of the snappers of the genus *Lutjanus* (Pisces: Lutjanidae) from the Indo-Pacific, with the description of a new species. Indo Pac Fish 11:1–87

Al-Thobaity S, James CM (1996) Developments in grouper culture in Saudi Arabia. INFOFISH Int 1(96):22–29

Anderson C (2004) Preserving the patch – plotting a future for Shark Bay snapper. Western Fisheries, September 2004

AQUACOP, Fuchs J, Debas L (1989a) Spawning of broodstock of the grouper *Epinephelus microdon* in French Polynesia. Natural cycle characteristics and first results in captivity. J World Aquaculture Soc 19(1):14A

AQUACOP, Debas L, Fostier A, Fuchs J, Weppe M, Nedelec G, Benett A, Cauty C, Jalabert B (1989b) The sexuality of cultured hermaphroditic fish species: analysis of morphological and endocrinological features in a protogynous hermaphrodite, *Epinephelus microdon,* as a basis for further research to control reproduction in the grouper. In: Proceedings of the 9th symposium advances in tropical aquaculture, Tahiti, pp 543–557

Aswani S, Hamilton RJ (2004) Integrating indigenous ecological knowledge and customary sea tenure with marine and social science for conservation of bumphead parrotfish (*Bolbometopon muricatum*) in the Roviana Lagoon, Solomon Islands. Environ Conserv 31:69–83

Aumeeruddy R, Robinson J (2006) Closure of the live reef food fish fishery in Seychelles. SPC Live Reef Fish Inf Bull 16:3–9

Ayling AM, Ayling AL (1992) Abundance, distribution and length frequencies of a group of large piscivorous fishes, *Plectropomus* spp. (Pisces: Serranidae) on the Great Barrier Reef. Unpublished Report to GBRMPA

Ayling AM, Samoilys MA, Ryan D (2000) Trends in common coral trout populations on the Great Barrier Reef. A Report to Queensland Fisheries Management Authority. Inf. Ser. Dep. Prim. Ind. (Queensl.); QDPI, Brisbane

Barlow GW (1975) On the sociobiology of four Puerto Rican parrotfishes (Scaridae). Mar Biol 33:281–293

Beets J, Friedlander A (1992) Stock analysis and management strategies for red hind, *Epinephelus guttatus*, in the U.S. Virgin Islands. Proc Gulf Caribb Fish Inst 42:67–79

Beets J, Friedlander A (1999) Evaluation of a conservation strategy: a spawning aggregation closure for red hind, *Epinephelus guttatus*, in the U.S. Virgin Islands. Environ Biol Fish 55:91–98

Bellwood DR (1994) A phylogenetic study of the parrotfishes family Scaridae (Pisces: Labroidei), with a revision of genera. Rec Austr Mus Supp 20:1–86

Bellwood DR, Hoey AS, Choat JH (2003) Limited functional redundancy in high diversity systems: resilience and ecosystem function on coral reefs. Ecol Lett 6:281–285

Benetti DD et al (2002) Growth, survival, and feed conversion rates of hatchery-reared mutton snapper *Lutjanus analis* cultured in floating net cages. J World Aquaculture Soc 33(3):349–357

Bolden SK (2000) Long-distance movement of a Nassau grouper (*Epinephelus striatus*) to a spawning aggregation in the central Bahamas. Fish Bull 98:642–645

Bortone SA, Williams JL (1986) Species profiles: life histories and environmental requirements of coastal fishes and invertebrates (South Florida): gray, lane, mutton and yellowtail snappers. U.S. Fish and Wildlife Service, Biological Report 82.

Bouhlel M (1988) Poissons de Djibouti. RDI International, Inc., Placerville

Buslé-Sicard S, Debas L, Fourcault B, Fuchs J (1992) Ultrastructural study of sex inversion in a protogynous hermaphrodite, *Epinephelus microdon* (Teleostei, Serranidae). Reprod Nutr Dev 32:393–406

Bullock LH, Smith GB (1991) Seabasses (Pisces: Serranidae), Memoirs of the Hourglass Cruises VIII (II) Florida Marine Research Institute. Department of Natural Resources, St. Petersburg, Florida

Bullock LH, Murphy MD, Godcharles MR, Mitchell ME (1992) Age, growth and reproduction of jewfish, *Epinephelus itajara*, in the eastern Gulf of Mexico. Fish Bull 90:243–249

Bureau of Marine Resources (1992) Annual report. Ministry of Resources and Development. Republic of Palau, 92 p

Burnett-Herkes J (1975) Contribution to the biology of the red hind, *Epinephelus guttatus*, a commercially important serranid fish from the tropical western Atlantic. Ph.D. dissertation, University of Miami, Coral Gables, Florida 154 pp

Burton ML (2002) Age, growth and mortality of mutton snapper, *Lutjanus analis*, from the east coast of Florida, with a brief discussion of management implications. Fish Res 59:31–41

Burton ML, Brennan KJ, Munoz RC, Parker ROJR (2005) Preliminary evidence of increased spawning aggregations of mutton snapper (*Lutjanus analis*) at Riley's Hump two years after establishment of the Tortugas south ecological reserve. Fish Bull 103:404–410

Bush PG, Lane ED, Ebanks-Petrie GC, Luke K, Johnson B, McCoy C, Bothwell J, Parsons E (2006) The Nassau grouper spawning aggregation fishery of the Cayman Islands – an historical and management perspective. Proc Gulf Caribb Fish Inst 57:515–524

Carter JG, Marrow J, Pryor V (1994) Aspects of the ecology and reproduction of Nassau grouper, *Epinephelus striatus*, off the coast of Belize, Central America. Proc Gulf Caribb Fish Inst 43:65–111

Cass-Calay SL, Schmidt TW (2008) Monitoring changes in the catch rates and abundance of juvenile goliath grouper using the ENP creel survey, 1973–2006. Endanger Species Res. doi:10.3354/esr.00139

Chan T, Sadovy Y, Yeeting B (2008) *Plectropomus areolatus*. In: IUCN 2010. IUCN Red List of Threatened Species. Version 2010.1. http://www.iucnredlist.org

Chateau O, Wantiez L (2007) Site fidelity and activity patterns of a humphead wrasse, *Cheilinus undulatus* (Labridae), as determined by acoustic telemetry. Environ Biol Fish 80:503–508

Cherubin, L, Nemeth RS, Idrisi N. (2011) Flow and transport characteristics at a spawning aggregation site in St Thomas (US Virgin Island). Ecological Modelling doi:10.1016/j.ecolmodel.2011.05.031

Choat JH, Davies CR, Ackerman JL, Mapstone BD (2006) Age structure and growth in a large teleost, *Cheilinus undulatus*, with a review of size distribution in labrid fishes. Mar Ecol Prog Ser 318:237–246

Claro R, Lindeman KC (2003) Spawning aggregation sites of snapper and grouper species (Lutjandidae and Serranidae) on the insular shelf of Cuba. Gulf Caribb Res 14:91–106

Claro R, Baisre JA, Lindeman KC, García-Arteaga JP (2001) Cuban fisheries: historical trends and current status. In: Claro R, Lindeman KC, Parenti LR (eds) Ecology of the marine fishes of Cuba. Smithsonian Institution Press, Washington, DC

Claro R, Sadovy de Mitcheson Y, Lindeman KC, Garcia-Cagide AR (2009) Historical analysis of Cuban commercial fishing effort and the effects of management interventions on important reef fishes from 1960–2005. Fish Res 99(1):7–16

Claydon JAB, Jacob SE, Wagner CW, Ryan SK (2009) Juvenile habitat for Nassau grouper and spiny lobster is enhanced by discarded Queen conch shells. Proc Gulf Caribb Fish Inst 61:173–176

Coleman FC, Koenig CC (2003) Population density, demographics, and predation effects of adult goliath grouper. National Oceanic and Atmospheric Administration, Final Report MARFIN Grant Number 0548-0175. National Oceanic and Atmospheric Administration Silver Spring, MD, USA

Coleman FC, Koenig CC, Collins LA (1996) Reproductive styles of shallow-water groupers (Pisces: Serranidae) in the eastern Gulf of Mexico and the consequences of fishing spawning aggregations. Environ Biol Fish 47:129–141

Coleman FC, Koenig CC, Eklund AM, Grimes CB (1999) Management and conservation of temperate reef fishes in the grouper-snapper complex of the southeastern United States. Amer Fish Soc Symp 23:233–242

Coleman FC, Koenig CC, Huntsman GR, Musick JA, Eklund AM, McGovern JC, Chapman RW, Sedberry GR, Grimes CB (2000) Long-lived reef fishes: the grouper-snapper complex. AFS Policy Statement. Fisheries 25(3):14–20

Coleman FC, Scanlon KM, Koenig CC (2010) Groupers on the edge: shelf-edge spawning habitat in and around marine reserves of the northeastern Gulf of Mexico. Professional Geographer 63(4):1–19

Colin PL (1978) Daily and summer-winter variation in mass spawning of the striped parrotfish, *Scarus croicensis*. Fish Bull 76(1):117–124

Colin PL (1985) Spawning of western Atlantic surgeonfishes. Natl Geogr Res Rep 18:243–250

Colin PL (1992) Reproduction of the Nassau grouper, *Epinephelus striatus*, (Pisces: Serranidae) and its relationship to environmental conditions. Environ Biol Fish 34:357–377

Colin PL (1994) Preliminary investigations of reproductive activity of the jewfish, *Epinephelus itajara*. Proc Gulf Caribb Fish Inst 43:138–147

Colin PL (1996) Longevity of some coral reef fish spawning aggregations. Copeia 1996(1):189–192

Colin PL (2010) Aggregation and spawning of the humphead wrasse, *Cheilinus undulatus* (Pisces: Labridae): general aspects of spawning behaviour. J Fish Biol 76:987–1007

Colin PL, Bell LJ (1991) Aspects of the spawning of labrid and scarid fishes (Pisces: Labroidei) at Eniwetak Atoll, Marshall Islands with notes on other families. Environ Biol Fish 31:229–260

Colin PL, Clavijo IE (1988) Spawning activity of fishes producing pelagic eggs on a shelf edge coral reef, southwestern Puerto Rico. Bull Mar Sci 43(2):249–279

Colin PL, Shapiro DY, Weiler D (1987) Aspects of the reproduction of two groupers, *Epinephelus guttatus* and *E. striatus*, in the West Indies. Bull Mar Sci 40:220–230

Colin PL, Koenig CC, LaRoche WA (1996) Development from egg to juvenile of the red grouper (*Epinephelus morio*) (Pisces: Serranidae) in the laboratory. In: Arreguin-Sanchez F, Munro JL, Balgos MC, Pauly D (eds) Biology, fisheries, and culture of tropical groupers and snappers, ICLARM Conference Proceedings 48. ICLARM, Manila, pp 399–414

Colin PL, Laroche WA, Brothers EB (1997) Ingress and settlement in the Nassau grouper, *Epinephelus striatus* (Pisces: Serranidae), with relationship to spawning occurrence. Bull Mar Sci 60(3):656–667

Colin PL, Sadovy YJ, Domeier ML (2003) Manual for the study and conservation of reef fish spawning aggregations, Special Publication No. 1 (Version 1.0). Society for the Conservation of Reef Fish Aggregations. (www.SCRFA.org)

Collins LA, Johnson AG, Koenig CC, Baker MS Jr (1998) Reproductive patterns, sex ratio, and fecundity in gag *Mycteroperca microlepis* (Serranidae), a protogynous grouper from the northeast Gulf of Mexico. Fish Bull 96:415–427

Cooper R (1997) Shark Bay legends. L.J. Cogan, Geraldton

Cornish A (2004) *Epinephelus fuscoguttatus*. In: IUCN 2007. 2007 IUCN red list of threatened species. www.iucnredlist.org

Coutin PC, Cashmore S, Sivakumuran KP (2003) Assessment of the snapper fishery in Victoria. Fisheries Research and Development Corporation Final Report, Project 97/128. Department of Primary Industries, Victoria

Craig PC (1998) Temporal spawning patterns of several species of surgeonfishes and wrasses in American Samoa. Pac Sci 52:35–39

Craig MT, Hastings PA (2007) A molecular phylogeny of the groupers of the subfamily Epinephelinae (Serranidae) with a revised classification of the Epinephelini. Ichthyol Res 17:1–17

Craig MT, Graham RT, Torres RA, Hyde JR, Freitas MO, Ferreira BP, Hostim-Silva M, Gerhardinger LC, Bertoncini AA, Robertson DR (2009) How many species of goliath grouper are there? Cryptic genetic divergence in a threatened marine fish and the resurrection of a geopolitical species. Endanger Species Res 7:167–174

Craig MT, Sadovy de Mitcheson YJ, Heemstra PC (2011) Groupers of the World: A Field and Market Guide. Grahamstown: NISC (Pty) Ltd. 424 pp

Cummings N (2007) Important aspects of the life history of mutton snapper, *Lutjanus analis*, with emphasis on populations in the Caribbean. SEDAR-14 Assessment Workshop Report No 2

Cummings NJ, Parrack ML, Zweifel JW (1997) The status of red hind and coney in the U.S. Virgin Islands between 1974 and 1992. Proc Gulf Caribb Fish Inst 49:354–397

Dahlgren CP, Eggleston DB (2001) Spatio-temporal variability in abundance, size and microhabitat associations of early juvenile Nassau grouper *Epinephelus striatus* in an off-reef nursery system. Mar Ecol Prog Ser 217:145–156

Davidson AD, Hamilton MJ, Alison G, Boyer AG, Brown JH, Ceballosa G (2009) Multiple ecological pathways to extinction in mammals. Proc Natl Acad Sci USA 106:10702–10705

De los Santos MA, Uehara D, Dwight I (2008) Hatchery production of dusky rabbitfish (meas) *Siganus fuscescens* fingerlings. Extension Bulletin No. 7. Cooperative Research and Extension Palau Community College

Debas L (1989) Étude, biométrique, histologique at endocrinologique de la sexualité du mérou *Epinephelus microdon* dans le milieu naturel et en élevage: caractérisation de l'hermaphroditisme protérogyne, description du phénomène d'inversion et mise en évidence du phénomenè de réversion. Thèse doctoral University d'Marseille, France

DEEDI (2010) Annual status report 2009 Coral Reef Fin Fish Fishery. Department of Employment, Economic Development and Innovation, Brisbane

Deloach N (1999) Reef fish behavior: Florida, Caribbean. New World Publications, Jacksonville

Doherty PJ, Fowler AJ, Samoilys MA, Harris DA (1994) Monitoring the replenishment of coral trout (Pisces:Serranidae) populations. Bull Mar Sci 54(1):343–355

Domeier ML, Colin PL (1997) Tropical reef fish spawning aggregations: defined and reviewed. Bull Mar Sci 60:698–726

Donaldson TJ, Dulvy NK (2004) Threatened fishes of the world: *Bolbometopon muricatum* (Valenciennes 1840) (Scaridae). Environ Biol Fish 70:373

Dorenbosch M, Grol MGG, Nagelkerken I, van der Velde G (2006) Seagrass beds and mangroves as potential nurseries for the threatened Indo-Pacific humphead wrasse, *Cheilinus undulatus* and Caribbean rainbow parrotfish, *Scarus guacamaia*. Biol Conserv 129:277–282

DPI&F (2009) Annual status report: coral reef fin fish fishery

Drew A (1973) The comparative ecology of *Siganus lineatus* and *Siganus canaliculatus*: two fishes of the Palau Islands. MSc thesis, Northern Illinois University, Dekalb

Dulvy NK, Polunin NVC (2004) Using informal knowledge to infer human-induced rarity of a conspicuous reef fish. Anim Conserv 7:365–374

Edwards H (2000) Shark Bay through four centuries 1616 to 2000. Shire of Shark Bay, Denham

Eggleston DB (1995) Recruitment in Nassau grouper *Epinephelus striatus*: post-settlement abundance, microhabitat features, and ontogenetic habitat shifts. Mar Ecol Prog Ser 124(1–3):9–22

Ehrhardt NM, Deleveaux VKW (2007) The Bahamas' Nassau grouper (*Epinephelus striatus*) fishery - two assessment methods applied to a data-deficient coastal population. Fish Res 87:17–27

Erdman DS (1976) Spawning patterns of fishes from the northeastern Caribbean, vol VIII, no 2. Departamento de Agricultura, San Juan, Puerto Rico

Eristhee N, Kadison E, Murray PA, Llewelyn A (2006) Preliminary investigations into the red hind fishery in the British Virgin Islands. Proc Gulf Caribb Fish Inst 57:373–384

Etpison MT (1997) Palau: portrait of paradise, 3rd edn. NECO Marine, Koror

Etpison MT (2004) Palau - natural history. Tkel Corp, Koror

Evans RD, Russ GR (2004) Larger biomass of targeted reef fish in no-take marine reserves on the Great Barrier Reef, Australia. Aquat Conserv Mar Freshw Ecosyst 14:505–519

Evans RD, Williamson DH, Sweatman H, Russ G, Emslie M, Cheal A, Miller I (2006) Surveys of the effects of rezoning of the GBR Marine Park in 2004 on some fish species – preliminary findings. Marine and Tropical Sciences Research Facility, Townsville

Fernandes L, Day J, Lewis A, Slegers S, Kerrigan B, Breen D, Cameron D, Jago B, Hall J, Lowe D, Innes J, Tanzer J, Chadwick V, Thompson L, Gorman K, Simmons M, Barnett B, Sampson K, De'ath G, Mapstone B, Marsh H, Possingham H, Ball I, Ward T, Dobbs K, Aumend J, Slater D, Stapleton K (2005) Establishing representative no-take areas in the Great Barrier Reef: large-scale implementation of theory on marine protected areas. Conserv Biol 19(6):1733–1744

Ferreira BP (1995) Reproduction of the common coral trout *Plectropomus leopardus* from the central and northern Great Barrier Reef. Bull Mar Sci 56(2):653–669

Ferreira BP, Maida M (1995) Projeto Mero: apresentacão de resultados preliminares. Bol Tecn Cient CEPENE. Tamandaré 3(1):201–210

Ferreira BP, Russ GR (1994) Age validation and estimation of growth rate of the coral trout, *Plectropomus leopardus*, (Lacepede 1802) from Lizard Island, northern Great Barrier Reef. Fish Bull US 92:46–57

Ferreira BP, Maida M, Castro CB, Pires DO, Prates APL, Marx D (2006) The status of coral reefs in Brazil. In: Proceedings of 10th international coral reef symposium, Okinawa, pp 1011–1015

Fishelson L, Montgomery WL, Myrberg AA Jr (1985) A new fat body associated with the gonad of surgeonfishes (Acanthuridae: Teleostei). Mar Biol 86:109–112

Fishelson L, Montgomery WL, Myrberg AA Jr (1987) Biology of surgeonfish *Acanthurus nigrofuscus* with emphasis on changeover in diet and annual gonadal cycles. Mar Ecol Prog Ser 39:37–47

Fisher R (2005) Swimming speeds of larval coral reef fishes: impacts on self-recruitment and dispersal. Mar Ecol Prog Ser 285:223–232

Francis MP, Pankhurst NW (1988) Juvenile sex inversion in the New Zealand snapper *Chrysophrys auratus* (Sparidae). Aust J Mar Freshw Res 39:625–631

Francis RICC, Paul LJ, Mulligan KP (1992) Ageing of adult snapper (*Pagrus auratus*) from otolith annual ring counts: validation by tagging and oxytetracycline injection. Aust J Mar Freshw Res 43:1069–1089

Frias-Torres S (2006) Habitat use of juvenile goliath grouper *Epinephelus itajara* in the Florida Keys, USA. Endanger Species Res 2:1–6

Froese R (2004) Keep it simple: three indicators to deal with overfishing. Fish Fish 5:86–91

Fulton E, Kault D, Mapstone B, Sheaves M (1999) Spawning season influences on commercial catch rates: computer simulations and *Plectropomus leopardus*, a case in point. Can J Fish Aquat Sci 56:1096–1108

García-Moliner G, Sadovy Y (2007) Nassau grouper. In: Proceedings of the 16th Annual Gulf and Caribbean Fisheries Institute, pp 596–602

García-Moliner G, Sadovy Y (2008) The case for regional management of the Nassau grouper, *Epinephelus striatus*. Proc Gulf Caribb Fish Inst 60:596–602

Gell FR, Whittington MW (2002) Diversity of fishes in seagrass beds in the Quirimba Archipelago, northern Mozambique. Mar Freshw Res 53:115–121

Gerhardinger LC, Marenzi RC, Bertoncini AA, Medeiros RP, Hostim-Silva M (2006) Local ecological knowledge on the goliath grouper *Epinephelus itajara* (Teleostei: Serranidae) in Southern Brazil. Neotrop Ichthyol 4(4):441–450

Gillespie C, Lewis RJ, Pearn JH, Bourke ATC (1986) Ciguatera in Australia: occurrence, clinical features, pathophysiology and management. Med J Aust 145:584–590

Gillett R (2010) Monitoring and management of the humphead wrasse, *Cheilinus undulatus*, FAO Fisheries and Aquaculture Circular No. 1048. FAO, Rome

Gillett R, Moy W (2006) Spearfishing in the Pacific Islands – current status and management issues. Secretariat of the Pacific Community/Food and Agriculture Organization of the United Nations, Noumea/Rome. Gillett, Preston and Associates Inc.

Gilmore RG, Jones RS (1992) Color variation and associated behavior in the Epinepheline groupers, *Mycteroperca microlepis* (Goode and Bam) and *M. phenax* (Jordan and Swain). Bull Mar Sci 51:84–103

Gladstone W (1986) Spawning behaviour of bumphead parrotfish *Bolbometopon muricatum* at Yonge Reef, Great Barrier Reef. Jpn J Ichthyol 33:326–328

Gladstone W (1994) Lek-like spawning, parental care and mating periodicity of the trigger fish *Pseudobalistes flavimarginatus* (Balistidae). Environ Biol Fish 39:249–257

Gladstone W (1996) Unique annual aggregation of longnose parrotfish (*Hipposcarus harid*) at Farasan Island (Saudi Arabia, Red Sea). Copeia 1996(2):483–485

GMFMC (1999) Regulatory Amendment to the Reef Fish Fishery Management Plan to Set 1999 Gag/Black Grouper Management Measures (revised). August 1999 Gulf of Mexico Fishery Management Coucil, Tampa, Florida

GMFMC (2009) Scoping Document for Reef Fish Amendment 32: to set 2011 red grouper and gag annual catch limits and establish a gag rebuilding plan. Gulf of Mexico Fishery Management Council, 2203 North Lois Avenue, Suite 1100, Tampa, Florida 33607

Golbuu Y, Victor S, Wolanksi E, Richmond RH (2003) Trapping of fine sediment in an enclosed bay, Palau, Micronesia. Estuar Coast Shelf Sci 56:1–9

Gomon M, Bray D, Kuiter R (2008) Fishes of Australia's southern coast. Museum Victoria, Melbourne

Government of Belize (2003) Statutory Instrument No. 161 of 2003. Fisheries (Spawning Aggregation Site Reserves) Order, 2003

Graham RT, Castellanos DW (2005) Courtship and spawning behaviors of carangid species in Belize. Fish Bull 103:426–432

Graham RT, Carcamo R, Rhodes KL, Roberts CM, Requena N (2008) Historical and contemporary evidence of a mutton snapper (*Lutjanus analis* Cuvier, 1828) spawning aggregation fishery in decline. Coral Reefs 27:311–319

Grandcourt E (2005) Demographic characteristics of selected epinepheline groupers (Family Serranidae; Subfamily Epinephelinae) from Aldabra Atoll, Seychelles. Atoll Res Bull 539: 199–216

Hamilton R (2003) A report on the current status of exploited reef fish aggregations in the Solomon Islands and Papua New Guinea – Choiseul, Ysabel, Bouganville and Manus Provinces, Western Pacific Fisher Survey Series: Society for the Conservation of Reef Fish Aggregations, vol 1. (confidential appendix), p 52

Hamilton RJ (2004) The demographics of bumphead parrotfish (*Bolbometopon muricatum*) in lightly and heavily fished regions of the Western Solomon Islands. Ph.D. thesis, University of Otago, Dunedin, p 273

Hamilton RJ, Matawai M (2006) Live reef food fish trade causes rapid declines in abundance of squaretail coral grouper (*Plectropomus areolatus*) at a spawning aggregation site in Manus, Papua New Guinea. SPC Live Reef Fish Inf Bull 16:13–18

Hamilton R, Walter R (1999) Indigenous ecological knowledge and its role in fisheries research design: a case study from Roviana Lagoon, Western Province, Solomon Islands. SPC Tradit Mar Resour Manag Knowl Inf Bull 11:13–25

Hamilton RJ, Matawai M, Potuku, T (2004) Spawning aggregations of coral reef fish in New Ireland and Manus Provinces, Papua New Guinea: local knowledge field survey report, (Unrestricted access version). TNC Pacific Island Countries Report No. 4/04. Report prepared for the Pacific Island Countries Coastal Marine Program, The Nature Conservancy. http://conserveonline.org/search?q=hamilton+%2B+local+knowledge&btnG=&site

Hamilton RJ, Matawai M, Potiki T, Kama W, Lahui P, Warku J, Smith AJ (2005) Applying local knowledge and science to the management of grouper aggregation sites in Melanesia. SPC Live Reef Fish Inf Bull 14:7–19

Hamilton RJ, Adams S, Choat JH (2008) Sexual development and reproductive demography of the green humphead parrotfish (*Bolbometopon muricatum*) in the Solomon Islands. Coral Reefs 27:153–163

Hamilton R, Paiva F, Aitsi J, Potuku P, Siota C, Lokani P (2010) Northern Bougainville Marine Resource Assessment, Autonomous Region of Bougainville. Technical report of survey conducted from the 1st–25th November 2008. A report by the Marine Program of the Asia Pacific Conservation Region, The Nature Conservancy. 2/10.

Hamner WM, Colin PL, Hamner PP (2007) Export-import dynamics of zooplankton on a coral reef in Palau. Mar Ecol Prog Ser 334:83–92

Harris PJ, Collins MR (2000) Age, growth and age at maturity of gag, *Mycteroperca microlepis*, from the southeastern United States during 1994-1995. Bull Mar Sci 66:105–117

Hasse JJ, Madraisau BB, McVey JP (1977) Some aspects of the life history of *Siganus canaliculatus* (Park) (Pisces: Siganidae) in Palau. Micronesica 13:297–312

Heemstra PL, Randall JE (1986) Serranidae. In: Smith MM, Heemstra PC (eds) Smith's Sea Fishes. Springer, Berlin

Heemstra PC, Randall JE (1993) Groupers of the World (family Serranidae, subfamily Epinephelinae). An annotated and illustrated catalogue of the grouper, rockcod, hind, coral grouper and lyretail species known to date. FAO Fisheries Synopsis No. 125, Vol 16. FAO, Rome, p 382

Heemstra PC, Anderson WDJ, Lobel PS (2002) Serranidae. In: Carpenter KE (ed) The living marine resources of the western central Atlantic. Bony fishes part 1 (Ascipenseridae to Grammatidae), FAO Species Identification Guide for Fishery Purposes. American Society of Ichthyologists and Herpetologist Special Publication 5. FAO, Rome

Heppell, SA, Semmens BX, Pattengill-Semmens CV, Bush PG, Johnson BC, McCoy CM, Paris C, Gibb J, Heppell SS (2008) Tracking potential larval dispersal patterns from Nassau grouper aggregation sites: Evidence for local retention and the Importance of Place. Proc Gulf Caribb Fish 61:abstract

Heyman WD, Kjerfve B (2008) Characterization of multi-species reef fish spawning aggregations at Gladden Spit, Belize. Bull Mar Sci 83(3):531–551

Heyman WD, Graham RT, Kjerfve B, Johannes RE (2001) Whale sharks *Rhincodon typus* aggregate to feed on fish spawn in Belize. Mar Ecol Prog Ser 215:275–282

Heyman WD, Kjerfve B, Graham RT, Rhodes KL, Garbutt L (2005) Spawning aggregations of Lutjanus cyanopterus (Cuvier) on the Belize Barrier Reef over a 6 year period. J Fish Biol 67:83–101

Heyman WD, Carr LM, Lobel PS (2010) Diver ecotourism and disturbance to reef fish spawning aggregations: it is better to be disturbed than to be dead. Mar Ecol Prog Ser 419:201–210

Hoque MM, Takemura A, Matsuyama M, Matsuura S, Takano K (1999) Lunar spawning in *Siganus canaliculatus*. J Fish Biol 55(6):1213–1222

Hostim-Silva M, Bertoncini AA, Gerhardinger LC, Machado LF (2005) The "Lord of the Rock's" conservation program in Brazil: the need for a new perception of marine fishes. Coral Reefs 24:74

Huntsman G (1996) *Lutjanus analis*. IUCN 2009. IUCN red list of threatened species. http://www.iucnredlist.org

Huntsman GR, Potts J, Mays RW, Vaughan DS (1999) Groupers (Serranidae, Epinephelinae): endangered apex predators of reef communities. In: Musick JA (ed) Life in the slow lane ecology and conservation of long-lived marine animals. American Fisheries Society, Bethesda, pp 217–232

Hutchinson N, Rhodes KL (2010) Home range estimates for squaretail coralgrouper, *Plectropomus areolatus* (Ruppell 1830). Coral Reefs 29:511–519

Ishihara M, Kuwamura T (1996) Bigamy or monogamy with maternal egg care in the triggerfish, *Sufflamen chrysopterus*. Ichthyol Res 43:307–313

IUCN (2004) IUCN Red List of Threatened Species. www.iucnredlist.org. IUCN-The World Union, Gland/Cambridge

IUCN (2008) The IUCN Red List of Threatened Species. May 21, 2009. http://www.iucnredlist.org/details/14050

IUCN (2010) IUCN red list of threatened species. Version 2010.1. www.iucnredlist.org

Jackson G (2007) Fisheries biology and management of pink snapper, *Pagrus auratus,* in the inner gulfs of Shark Bay, Western Australia. Ph.D. thesis, Murdoch University, Perth

Jackson G, Lai E (2008) Gascoyne Demersal Scalefish Fishery status report. In: Fletcher WJ, Santoro K (eds) State of the Fisheries Report 2007/08. Department of Fisheries, Western Australia

Jackson G, Lenanton R, Sumner N et al (2003) Research and management of snapper, *Pagrus auratus*, stocks in the inner gulfs of Shark Bay, Western Australia. In: Coleman A (ed) Proceedings of the Third World Recreational Fishing Conference. Fisheries Report no. 67, Northern Territory Department of Business, Industry and Research and Development, Darwin

Jackson, G, Sumner N, Cribb A et al (2005) Comparing conventional social-based and alternative output-based management models for recreational fisheries using Shark Bay pink snapper as case study. Fisheries Research and Development Corporation Final Report, Project 2003/066. Department of Fisheries, Western Australia

Jackson G, Norriss J, Lai E (2008) Inner Shark Bay Scalefish Fishery status report. In: Fletcher WJ, Santoro K (eds) State of the Fisheries Report 2007/08. Department of Fisheries, Western Australia

Jackson G, Norriss JV, Mackie MC et al (2010) Spatial variation in life history characteristics of snapper (*Pagrus auratus*) within Shark Bay, Western Australia. NZ J Mar Freshw Res 44:1–15

Jackson G, Cheng YW, Wakefield CB (in press) An evaluation of the daily egg production method to estimate spawning biomass of snapper (*Pagrus auratus*) stocks in inner Shark Bay, Western Australia, following more than a decade of surveys 1997–2007. Fish Res

James CM, Al-Thobaiti SA, Rasem BM, Carlos MH (1998) Comparative growth of brown-marbled grouper *Epinephelus fuscoguttatus* (Forsskål) and camouflage grouper *E. polyphekadion* (Bleeker) under hatchery and growout conditions. Asian Fish Sci 11:133–147

Johannes RE (1978) Reproductive strategies of coastal marine fishes in the tropics. Environ Biol Fish 3:65–84

Johannes RE (1981) Words of the lagoon: fishing and marine lore in the Palau District of Micronesia. University of California Press, Berkeley/Los Angeles

Johannes RE, Squire L, Graham T, Sadovy Y, Renguul H (1999) Spawning aggregations of groupers (Serranidae) in Palau. The Nature Conservancy Marine Research Series Publication No. 1, August 1999. The Nature Conservancy report, available online at http://www.conserveonline.org/.../per%20aggregations%20report%20Aug99.pdf

Johnson GD, Washington BB (1987) Larvae of the moorish idol, *Zanclus cornutus*, with comparisons with other larval acanthuroids. Bull Mar Sci 40(3):494–511

Kadison E, Nemeth RS, Herzlieb S, Blondeau J (2006) Temporal and spatial dynamics of *Lutjanus cyanopterus* and *L. jocu* (Pisces: Lutjanidae) spawning aggregations on a multi-species spawning site in the USVI. Rev Biol Trop 54:69–78

Kadison E, Nemeth RS, Blondeau JE (2009) Assessment of an unprotected red hind (*Epinephelus guttatus*) spawning aggregation on Saba Bank in the Netherlands Antilles. Bull Mar Sci 85:101–118

Kailola PJ, Williams MJ, Stewart PC et al (1993) Australian fisheries resources. Bureau of Resource Sciences, Canberra

Kawase H (2003) Spawning behavior and biparental egg care of the crosshatch triggerfish, *Xanthichthys mento* (Balistidae). Environ Biol Fish 66:211–219

Kawase H, Nakazono A (1994) Reproductive behaviour of the flagtail triggerfish *Sufflamen chrysopterus.* In: Proceedings of the Seventh International Coral Reef Symposium, Guam, vol 2, pp 905–907

Keener P, Johnson GD, Stender BW, Brothers EB, Beatty HR (1988) Ingress of postlarval gag, *Mycteroperca microlepis*, through a South Carolina barrier island inlet. Bull Mar Sci 42:376–396

Kiflawi M, Mazeroll AL (2006) Female leadership during migration and the potential for sex-specific benefits of mass spawning in the brown surgeonfish (*Acanthurus nigrofuscus*). Environ Biol Fish 76:19–23

Kiflawi M, Mazeroll AL, Goulet D (1998) Does mass spawning enhance fertilization in the coral reef fish. Mar Ecol Prog Ser 172:107–114

Kingsley MCS (ed) (2004) The goliath grouper in southern Florida: assessment review and advisory report. Report prepared for the South Atlantic Fishery Management Council, the Gulf of Mexico Fishery Management Council, and the National Marine Fisheries Service. Southeast Data and Assessment Review

Kitalong A, Dalzell P (1991) A preliminary assessment of the status of inshore coral reef fish stocks in Palau. The South Pacific Commission, Noumea, New Caledonia

Kitalong A, Oiterong E (1992) Quantification of spawning aggregations of *Siganus canaliculatus*) in Airai Bay. 1990–1992. Marine Resources Division (Palau), Technical Report 92.5

Koenig CC, Coleman FC (1998) Absolute abundance and survival of juvenile gags in sea grass beds of the northeastern Gulf of Mexico. Trans Am Fish Soc 127:44–55

Koenig CC, Coleman FC, Collins LA, Sadovy Y, Colin PL (1996) Reproduction of gag (*Mycteroperca microlepis*) (Pisces:Serranidae) in the eastern Gulf of Mexico and the consequences of fishing spawning aggregations. In: ICLARM Conference 48. ICLARM Conference Proceeding, Manila, pp 307–323

Koenig CC, Coleman FC, Eklund AM, Schull J, Ueland J (2007) Mangroves as essential nursery habitat for goliath grouper (*Epinephelus itajara*). Bull Mar Sci 80(3):567–586

Koenig CC, Coleman FC, Kingon K (2011) Pattern of recovery of the goliath grouper *Epinephelus itajara* population of the southeastern US. Bull Mar Sci 87: doi:10.5343/bms.2010.1056

Kuwamura T (1997) Evolution of female egg care in haremic triggerfish, *Rhinecanthus aculeatus*. Ethology 103(12):1015–1023

Kuwamura T, Sagawa T, Suzuki S (2009) Interspecific variation in spawning time and male mating tactics of the parrotfishes on a fringing coral reef at Iriomote Island, Okinawa. Ichthyol Res 56(4):354–362

Landsman SJ, Jadot C, Ashley M, Claydon JAB (2009) Investigation of the Nassau grouper (*Epinephelus striatus*) fishery in the Turks and Caicos Islands: implications for conservation and management. Proc Gulf Caribb Fish 61:82–89

Lara MR, Schull J, Jones DL, Allman R (2009) Early life history stages of goliath grouper, *Epinephelus itajara (Pisces: Epinephelidae)* from ten thousand Islands, Florida. Endanger Species Res 7:221–228

Leis JM, Carson-Ewart BM (eds) (2000) The larvae of Indo-Pacific coastal fishes, vol 2, Fauna Malesiana Handbook. Brill, Leiden/Boston/Koln

Lenanton R, StJohn J, Wise B et al (2009) Maximising survival of released undersized west coast reef fish, Fisheries Research Report No 191. Department of Fisheries, Western Australia

Levin PS, Grimes CB (2002) Reef fish ecology and grouper conservation and management. In: Sale PF (ed) Coral reef fishes dynamics and diversity in a complex ecosystem. Academic, London, pp 377–389

Lim LC (1993) Lariculture of the greasy grouper *Epinephelus tauvina* F. and the brown-marbled grouper *E. fuscoguttatus* F. in Singapore. J World Aquaculture Soci 24(2):262–274

Lim LC, Chao TM, Khoo LT (1990) Observations on the breeding of brown-marbled grouper, *Epinephelus fuscoguttatus* (Forskal). Singap J Prim Ind 18(2):66–84

Lindberg WJ, Mason D, Murie DJ (2002) Habitat-mediated predator-prey interactions: implications for sustainable production of gag grouper in the eastern Gulf of Mexico. Final Project Report, Florida Sea Grant Program (project no. R/LRB-49)

Lindeman KC, DeMaria D (2005) Juveniles of the Caribbean's largest coral reef snapper do not use reefs. Coral Reefs 24:359

Lindemann KC, Pugliese R, Waugh GT, Ault JS (2000) Developmental patterns within a multispecies reef fishery: management applications for essential fish habitats and protected areas. Bull Mar Sci 66(3):929–956

Liu M, Sadovy de Mitcheson Y (2011) Forehead morphology of the humphead wrasse *Cheilinus undulatus* (Perciformes: Labridae) in relation to body size. Copeia 2:315–318

Lobel PS (1978) Diel, lunar, and seasonal periodicity in the reproductive behavior of the pomacanthid fish, *Centropyge potteri*, and some other reef fishes in Hawaii. Pac Sci 32(2):193–206

Lobel PS, Johannes RE (1980) Nesting, eggs and larvae of triggerfishes. Environ Biol Fish 5:251–252

Loubens G (1980) Biologie de quelques especes de poissons du lagon Neo-Caledonien. III. Croissance. Cahiers de l'Indo-pacifique 2:101–153

Luckhurst BE (1996) Trends in commercial fishery landings of groupers and snappers in Bermuda from 1975 to 1992 and associated fishery management issues. In: Arreguin-Sanchez F, Munro JL, Balgos MC, Pauly D (eds) Biology, fisheries and culture of tropical groupers and snappers, vol 48. ICLARM, Manila, pp 286–297

Luckhurst BE (1998) Site fidelity and return migration of tagged red hinds (*Epinephelus guttatus*) to a spawning aggregation site in Bermuda. Proc Gulf Caribb Fish Inst 50:750–763

Luckhurst BE, Barnes J, Sadovy Y (1992) Record of an unusually large red hind, *Epinephelus guttatus* (Serranidae), from Bermuda with comments on its age. Bull Mar Sci 51:267–270

Mackie M, Jackson G, Tapp N et al (2009) Macroscopic and microscopic description of snapper (*Pagrus auratus*) gonads from Shark Bay, Western Australia, Fisheries Research Report No 184. Department of Fisheries, Western Australia

Mann D, Locascio J, Schärer M, Koenig C, Nemeth M, Nelson M, Coleman F, Appeldoorn R (2008) Passive acoustic detection of grouper sound production. J Acoust Soc Am 123(5):3098 (abstract)

Mann DA, Locascio JV, Coleman FC, Koenig CC (2009) Goliath grouper (*Epinephelus itajara*) sound production and movement patterns on aggregation sites. Endanger Species Res 7:229–236

Mapstone BD, Davies CR, Little LR, Punt AE, Smith ADM, Pantus F, Lou DC, Williams AJ, Jones A, Ayling AM, Russ GR, McDonald AD (2004) The effects of line fishing on the Great Barrier Reef and evaluations of alternative potential management strategies, CRC Reef Research Centre Technical Report No. 54. CRC Reef Research Centre, Townsville

Mariott RJ, Mapstone BD, Begg GA (2007) Age-specific demographic parameters, and their implications for management of the red bass, *Lutjanus bohar* (Forsskal 1775): a large, long-lived reef fish. Fish Res 83(2–3):204–215

Marriott RJ, Mapstone BD (2006) Geographic influences on and the accuracy and precision of age estimates for the red bass, *Lutjanus bohar* (Forsskal 1775): a large tropical reef fish. Fish Res 80(2–3):322–328

Marshall DG, Moore SA (2000) Tragedy of the commons and the neglect of science: planning and management in the Shark Bay World Heritage Area. Environ Plan Law J 17:126–137

Mazeroll AI, Montgomery WL (1995) Structure and organization of local migrations in brown surgeonfish (*Acanthurus nigrofuscus*). Ethology 99:89–106

Mazeroll AI, Montgomery WL (1998) Daily migrations of a coral reef fish in the Red Sea (Gulf of Aqaba, Israel): initiation and orientation. Copeia 1998(4):893–905

McClenachan L (2009) Historical declines in south Florida goliath grouper populations. Endanger Species Res 7:175–181

McGovern JC, Wyanski DM, Pashuk O, Manooch CSI, Sedberry GR (1998a) Changes in the sex ratio and size at maturity of gag *Mycteroperca microlepis*, from the Atlantic coast of the south-eastern United States during 1976-1995. Fish Bull 96:797–807

McGovern JC, Sedberry GR, Harris PJ (1998b) Status of stocks of reef fishes in the South Atlantic Bight. 1983-1996. Proc Gulf Caribb Fish Inst 50:871–895

McVey JP (1972) Observations on the early-stage formation of the rabbitfish *Siganus fuscescens* (=*S. canaliculatus*) at Palau Mariculut Demonstation Centre. South Pacific Island Fish Newsletter No. 6 (September):11-12

Meyer CG, Clark CE, Dale JJ (2010) Unusual surface schooling behavior by bullethead parrotfish (*Chlorurus sordidus*). Coral Reefs 29(4):881

Millington JT, Randall JE (1990) Triggerfish bite – a little-known marine hazard. J Wilderness Med 1:79–85

Mitchell RWD, Baba O, Jackson G et al (2008) Comparing management of recreational *Pagrus* fisheries in Shark Bay (Australia) and Sagami Bay (Japan): conventional catch controls versus stock enhancement. Mar Pol 32:27–37

Montgomery WL, Myrberg AA Jr, Fishelson L (1989) Feeding ecology of surgeonfishes (Acanthuridae) in the northern Red Sea, with particular reference to *Acanthurus nigrofuscus* (Forsskal). J Exp Mar Biol Ecol 132:179–207

Moran M, Kangas M (2003) The effects of the trawl fishery on the stock of pink snapper, *Pagrus auratus*, in Denham Sound, Shark Bay. Fisheries Research Bulletin No 31. Department of Fisheries, Western Australia

Moran M, Stephenson P, Gaughan D et al (2005) Minimising the cost of future stock monitoring, and assessment of the potential for increased yields, from the oceanic snapper, *Pagrus auratus*, stock of Shark Bay. Fisheries and Research Development Corporation Final Report, Project 2000/138. Department of Fisheries, Western Australia

Moyer JT (1989) Reef channels as spawning sites for fishes on the Shiraho Coral Reef, Ishigaki Island, Japan. Jpn J Ichthyol 36:371–375

Mullaney MJ (1994) Ontogenetic shifts in diet of gag, *Mycteroperca microlepis* (Goode and Bean), (Pisces: Serranidae). Proc Gulf Caribb Fish Inst 43:432–445

Mullaney MD Jr, Gale LD (1996) Ecomorphological relationships in ontogeny: anatomy and diet in gag, *Mycteroperca microlepis* (Pisces: Serranidae). Copeia 1996(1):167–180

Munro JL, Blok L (2003) The status of stocks of groupers and hinds in the Northeastern Caribbean. Proc Gulf Caribb Fish Inst 56:283–294

Munro JL, Gout VC, Thompson R, Reeson P (1973) The spawning season of Caribbean reef fishes. J Fish Biol 5:69–84

Murdy EO, Birdsong RS, Musick JA (1997) Fishes of the Chesapeake Bay. Smithsonian Institution Press, Washington, DC

Murray R, Bester C (2009) Cubera snapper biological profile. http://www.flmnh.ufl.edu/fish/Gallery/Descript/CuberaSnapper/CuberaSnapper.html

Musick JA, Harbin MM, Berkeley SA, Burgess GJ, Eklund AM, Findley L, Gilmore RG, Golden JT, Ha DS, Huntsman GR, McGovern JC, Parker SJ, Poss SG, Sala E, Schmidt TW, Sedberry GR, Weeks H, Wright SG (2000) Marine, estuarine, and diadromous fish stocks at risk of extinction in North America (Exclusive of Pacific Salmonids). Fisheries 25:6–30

Myers RF (1999) Micronesian Reef Fishes: A Comprehensive Guide to the Coral Reef Fishes of Micronesia, 3rd edition. Coral Graphics, Guam

Myrberg AA Jr, Montgomery WL, Fishelson L (1988) The reproductive behavior of *Acanthurus nigrofuscus* (Forsskal) and other surgeonfishes (Fam. Acanthuridae) off Eilat, Israel (Gulf of Aqaba, Red Sea). Ethology 79:31–61

Nardi K, Jones GP, Moran MJ, Chen YW (2004) Contrasting effects of marine protected areas on the abundance of two exploited reef fishes at the sub-tropical Houtman Abrolhos Islands, Western Australia. Environ Conserv 31(2):160–168

Neils P (2003) Tasting the No-Take tonic. Florida State University Research in Review, Tallahassee, pp 18–37

Nellis DW (1980) Reproduction in the ocean triggerfish *Canthidermis sufflamen*. Caribb J Sci 16(1–4):167

Nelson JS, Crossman EJ, Espinosa-Perez H, Findley LT, Gilbert CR, Lea RN, Williams JD (2001) Recommended change in the common name for a marine fish: goliath grouper to replace jewfish (*Epinephelus itajara*). Fisheries? Am Fish Soc 26(5)):31–34

Nemeth RS (2005) Population characteristics of a recovering US Virgin Islands red hind spawning aggregation following protection. Mar Ecol Prog Ser 286:81–97

Nemeth RS, Blondeau J, Herzlieb S, Kadison E (2007) Spatial and temporal patterns of movement and migration at spawning aggregations of red hind, *Epinephelus guttatus*, in the U.S.Virgin Islands. Environ Biol Fish 78:365–381

Nemeth RS, Herzlieb S, Blondeau J (2006) Comparison of two seasonal closures for protecting red hind spawning aggregations in the US Virgin Islands. In: Proceedings of the 10th International Coral Reef Conference, Okinawa, 4:1306–1313

Nemeth RS, Kadison E, Blondeau JE, Idrisi N, Watlington R, Brown K, Smith T, Carr L (2008) Regional coupling of red hind spawning aggregations to oceanographic processes in the Eastern Caribbean, NOAA Special Publications. NOAA, Silver Spring, pp 170–183

Nichols PV (1991) Republic of Palau, Marine Resources Profiles. Fisheries Development Section, Forum Fisheries Agency, FFA Report Number 91/59

NMFS (1999) Our living oceans. Report on the status of U.S. living marine resources, 1999. Report No. NMFS-F/SPO-4. Department of Commerce, Washington, DC

Norriss JV, Crisafulli B (2010) Longevity in Australian snapper *Pagrus auratus* (Sparidae). J R Soc WA 93:129–132

Ogden JC, Buckman NS (1973) Movements, foraging groups, and diurnal migrations of the striped parrotfish *Scarus croicensis* Bloch (Scaridae). Ecology 54:589–596

Olsen DA, LaPlace JA (1979) A study of Virgin Islands grouper fishery based on a breeding aggregation. Proc Gulf Caribb Fish Inst 31:130–144

Oshiro N, Yogi K, Asato S, Sasaki T, Tamanaha K, Hirama M, Yasumoto T, Inafuku Y (2010) Ciguatera incidence and fish toxicity in Okinawa, Japan. Toxicon 56(5):656–661

Oxenford HA, Fanning P, Cowen RK (2008) Spatial distribution of surgeonfish (Acanthuridae) pelagic larvae in the eastern Caribbean. In: Grober-Dunsmore R, Keller BD (eds) Caribbean connectivity: implications for marine protected area management, Marine Sanctuary Conservation Series ONMS-08-07. U.S. Dept. Commerce, NOAA, Silver Spring, MD, USA

Padilla JE, Mamauag S, Braganza G, Bruca lN, Yu D, Morales A (2003) Sustainability assessment of the live reef-fish for food industry in Palawan Philippines. Southeast Asia Policy Programme WWF-Philippines, Quezon City

Palau Conservation Society (2010) Enhanced monitoring of grouper spawning aggregation at Ebiil channel. Final Technical Report, June 2010. Palau Conservation Society and the Society for the Conservation of Reef Fish Aggregations (www.SCRFA.org)

Parrish JD (1987) The trophic biology of snappers and groupers. In: Polovina JJ, Ralston S (eds) Tropical snappers and groupers: biology and fisheries management. Westview Press, Boulder

Paulin CD (1990) *Pagrus auratus*, a new combination for the species known as snapper in Australasian waters (Pisces: Sparidae). NZ J Mar Freshw Res 24:259–265

Pears RJ (2005) Biology and management of the Flowery Cod and Camouflage Cod: how similar are they? In: Tobin R et al (eds) Fishing for more: a student-stakeholder workshop on the biology, ecology, sociology and economics of fisheries, CRC Reef Technical Report 59. CRC Reef Research Centre, Townsville

Pears RJ, Choat JH, Mapstone BD, Begg GA (2006) Demography of a large grouper, *Epinephelus fuscoguttatus*, from Australia's Great Barrier Reef: implications for fishery management. Mar Ecol Prog Ser 307:259–272

Pears RJ, Choat JH, Mapstone BD, Begg GA (2007) Reproductive biology of a large, aggregation-spawning serranid, *Epinephelus fuscoguttatus* (Forsskal): management implications. J Fish Biol 71:795–817

Pet JS, Mous PJ, Muljadi AH, Sadovy YJ, Squire L (2005) Aggregations of *Plectropomus areolatus* and *Epinephelus fuscoguttatus* (groupers, Serranidae) in the Komodo National Park, Indonesia: monitoring and implications for management. Environ Biol Fish 74:209–218

Pickert P, Kelly T, Nemeth RS, Kadison E (2006) Seas of Change: Spawning aggregations of the Virgin Islands. In: Pickert P, Kelly T (eds). DVD by Friday's Films, San Francisco, CA 28 min

Pina-Amargós F, González-Sansón G (2009) Movement patterns of goliath grouper *Epinephelus itajara* around southeast Cuba: implications for conservation. Endanger Species Res 7:243–247

Pitcher R, Condie S, Ellis N, McLeod I, Haywood M, Gordon S, Skewes T, Dunn J, Dennis D, Cotterell L, Austin M, Venables B, Taranto T (2004) Torres Strait seabed & water-column data collation, biophysical modeling and characterization. Final Report to the National Oceans Office, CSIRO Marine Research, 117 p

Planes S, Favelot C (2002) Isolation by distance and vicariance drive genetic structure of a coral reef fish in the Pacific Ocean. Evolution 52(2):378–399

Planes S, Bonhomme F, Galzin R (1996) A genetic metapopulation model for reef fishes in oceanic islands: the case of the surgeonfish, *Acanthurus triostegus*. J Evol Biol 9:103–117

Porch CE, Eklund AM, Scott GP (2006) A catch-free stock assessment model with application to goliath grouper (*Epinephelus itajara*) off southern Florida. Fish Bull 104:89–101

Randall JE (1961a) A contribution to the biology of the convict surgeonfish of the Hawaiian Islands, *Acanthurus triostegus sandvicensis*. Pac Sci 15:215–272

Randall JE (1961b) Observations on the spawning of surgeonfishes (Acanthuridae) in the Society Islands. Copeia 1961:237–238

Randall JE (1965) Food habits of the Nassau grouper (*Epinephelus striatus*). Assoc Isl Mar Lab Caribb 6:13–16

Randall JE (1967) Food habits of reef fishes of the West Indies. Stud Trop Oceanogr Miami 5:665–847

Randall JE (1983) Caribbean reef fishes, second edition, TFH Publications, Neptune City, NJ

Randall JE (2002) Surgeonfishes of the world. Mutual Publishing, Honolulu

Randall JE (2007) Reef and shore fishes of the Hawaiian Islands, University of Hawaii Sea Grant Program, Honolulu

Randall JE, Brock VE (1960) Observations on the ecology of epinepheline and lutjanid fishes of the Society Islands, with emphasis on food habits. Trans Am Fish Soc 89(1):9–16

Randall JE, Randall HA (1963) The spawning and early development of the Atlantic parrot fish, *Sparisoma rubripinne*, with notes on other scarid and labrid fishes. Zoologica: NY Zool Soc 48:49–60

Randall JE, Head SM, Sanders APL (1978) Food habits of the giant humphead wrasse, *Cheilinus undulatus* (Labridae). Environ Biol Fish 3:235–238

Reeson PH (1983) The biology, ecology and bionomics of the surgeonfishes, Acanthuridae. In: Munro JL (ed) Caribbean coral reef fishery resources. ICLARM, Manila

Renan X, Brule T, Lecomte-Finiger R (2006) First evidence of a nursery habitat for juvenile gag in the southern Gulf of Mexico. Trans Am Fish Soc 135:595–603

Rhodes KL (1999) Grouper aggregation protection in proactive Pohnpei. SPC Live Reef Fish Inf Bull 6:14–16

Rhodes KL (2003) SCRFA spawning aggregation survey: federated States of Micronesia, Western Pacific fisher survey sSeries, vol 2. Society for the Conservation of Reef Fish Aggregations.

Rhodes KL, Sadovy Y (2002a) Reproduction in the camouflage grouper, *Epinephelus polyphekadion* (Pisces: Serranidae), in Pohnpei, Federated States of Micronesia. Bull Mar Sci 70(3):851–869

Rhodes KL, Sadovy Y (2002b) Temporal and spatial trends in spawning aggregations of camouflage grouper, *Epinephelus polyphekadion* (Bleeker 1849) in Pohnpei, Micronesia. Environ Biol Fish 63:27–39

Rhodes KL, Tupper MH (2007) A preliminary market-based analysis of the Pohnpei, Micronesia, grouper (Serranidae; Epinephelinae) fishery reveals unsustainable fishing practices. Coral Reefs 26:335–344

Rhodes KL, Tupper MH (2008) The vulnerability of reproductively active squaretail coralgrouper (*Plectropomus areolatus*) to fishing, based on acoustic and conventional tagging surveys in Pohnpei, Micronesia. Fish Bull 106:194–203

Rhodes KL, Tupper MH, Wichilmel CB (2008) Characterization and management of the commercial sector of the Pohnpei coral reef fishery, Micronesia. Coral Reefs 27:443–454

Rhodes KL, Taylor BM, McIlwain J (2011) Demographic profile of a spawning aggregation of camouflage grouper, *Epinephelus polyphekadion*. Mar Ecol Prog Ser 421:183–198

Rimmer MA, Garrett RN, Samoilys MA (1994) In vitro fertilization of eggs of coral trout, *Plectropomus leopardus* (Serranidae), collected from an aggregation site on the Great Barrier Reef, Australia. Bull Mar Sci 54(1):356–358

Roberts CM, Quinn N, Tucker JW, Woodward PN (1995) Introduction of hatchery-reared Nassau grouper to a coral reef environment. N Am J Fish Manag 15:159–164

Robertson DR (1983) On the spawning behavior and spawning cycles of eight surgeonfishes (Acanthuridae) from the Indo-Pacific. Environ Biol Fish 9:192–223

Robertson DR (1988a) Abundances of surgeonfishes on patch-reefs in Caribbean Panama: due to settlement or post-settlement events. Mar Biol 97:495–501

Robertson DR (1988b) Extreme variation in settlement of the Caribbean triggerfish, *Balistes vetula*. Copeia 1988(3):698–703

Robertson DR, Sweatman HP, Flectcher EA, Cleland MG (1976) Schooling as a mechanism for circumventing the territoriality of competitors. Ecology 57(6):1208–1220

Robertson DR, Reinboth R, Bruce RW (1982) Gonochorism, protogynous sex-change and spawning in three sparisomatinine parrofishes from the western Indian Ocean. Bull Mar Sci 32(4):868–879

Robertson DR, Ackerman JL, Choat JH, Posada JM, Pitt J (2005a) Ocean surgeonfish *Acanthurus bahianus*. I. The geography of demography. Mar Ecol Prog Ser 295:229–244

Robertson DR, Choat JH, Posada JM, Pitt J, Ackerman JL (2005b) Ocean surgeonfish *Acanthurus bahianus*. II. Fishing effects on longevity, size and abundance? Mar Ecol Prog Ser 295:245–256

Robins CR, Ray GC (1986) A field guide to Atlantic coast fishes. Houghton Mifflin Company, Boston

Robinson J, Isidore M, Marguerite MA, Öhman MC, Payet RJ (2004) Spatial and temporal distribution of reef fish spawning aggregations in the Seychelles - an interview-based survey of artisanal fishers. West Indian Ocean J Mar Sci 3:63–69

Robinson J, Samoilys M, Kimani P (2008a) Reef fish spawning aggregations in the Western Indian Ocean: current knowledge and implications for management. In: Obura DO, Tamelander J, Linden O (eds) Ten years after bleaching – facing the consequences of climate change in the Indian Ocean, CORDIO Status Report 2008. Coastal Oceans Research and Development in the Indian Ocean/Sida-SAREC, Mombasa, pp 263–276. http//:www.cordioea.org

Robinson J, Aumerruddy R, Jörgensen TL, Öhman M (2008b) Dynamics of camouflage (*Epinephelus polyphekadion*) and brown marbled grouper (*Epinephelus fuscoguttatus*) spawning aggregations at a remote reef site, Seychelles. Bull Mar Sci 83(2):415–431

Rocha LA, Bass AL, Robertson DR, Bowen BW (2002) Adult habitat preferences, larval dispersal, and the comparative phylogeography of three Atlantic surgeonfishes (Teleostei: Acanthuridae). Mol Ecol 11:243–252

Rudd MA, Tupper MH (2002) The impact of Nassau grouper size and abundance on scuba diver site selection and MPA economics. Coast Manag 30(2):133–151

Russ GR, Lou DC, Higgs J, Ferreira BP (1998) Mortality rate of a cohort of the coral trout, *Plectropomus leopardus*, in zones of the Great Barrier Reef Marine Park closed to fishing. Mar Freshw Res 49:507–511

Russ G, Cheal AJ, Dolman AM, Emslie MJ, Evans RD, Miller I, Sweatman H, Williamson DH (2008) Rapid increase in fish numbers follows creation of world's largest marine reserve network. Curr Biol 18(12):R514

Russell M, Pears RJ (eds) (2007) Workshop summary: management and science of fish spawning aggregations in the Great Barrier Reef Marine Park, 12–13 July 2007. Museum of Tropical Queensland, Townsville, Australia. GBRMPA, Townsville

Russell B, Situ A, Cornish A (2006) *Epinephelus polyphekadion*. In: IUCN 2010. IUCN Red List of Threatened Species. Version 2010.4 www.iucnredlist.org

Sadovy Y (1993a) Biology and Fishery of the red hind in Puerto Rico and the United States Virgin Islands. Caribbean Fishery Management Report. Caribbean Fishery Management Council, San Juan, Puerto Rico

Sadovy Y (1993b) The Nassau grouper, endangered or just unlucky? Reef Encount 13:10–12

Sadovy YJ (1996) Reproduction of reef fishery species. In: Polunin NVC, Roberts CM (eds) Management of reef fisheries. Chapman & Hall, London

Sadovy Y (1997) The case of the disappearing grouper: *Epinephelus striatus* (Pisces: Serranidae). J Fish Biol 46:961–976

Sadovy Y (2000) Regional survey for fry/fingerling supply and current practices for grouper mariculture: evaluating current status and long-term prospects for grouper mariculture in Southeast Asia. Final report to the Collaborative APEC Grouper Research and Development Network (FWG 01/99)

Sadovy Y (2005) Troubled times for trysting trio: three aggregating groupers in the live reef food-fish trade. SPC Live Reef Fish Inf Bull 14:3–6

Sadovy, Y (2007). Report on current status and exploitation history of reef fish spawning aggregations in Palau. Western Pacific fishery survey series: Society for the Conservation of Reef Fish Aggregations, vol 3. SCRFA and the Palau Conservation Society, pp 40

Sadovy Y, Shapiro DY (1987) Criteria for the Diagnosis of Hermaphroditism in Fishes. *Copeia* 1987(1):136–156

Sadovy de Mitcheson Y, Liu M (2008) Functional hermaphroditism in teleosts. Fish Fish 9:1–43

Sadovy de Mitcheson Y, Cornish A, Domeier M, Colin P, Russell M, Lindeman K (2008) A global baseline for spawning aggregations of reef fishes. Conserv Biol 22(5):1233–1244

Sadovy de Mitcheson Y, Liu M, Suharti S (2010) Gonadal development in a giant threatened reef fish, the humphead wrasse *Cheilinus undulatus*, and its relationship to international trade. J Fish Biol 77:706–718

Sadovy Y, Colin PL (1995) Sexual development and sexuality in the Nassau grouper, *Epinephelus striatus* (Bloch) (Pisces: Serranidae). J Fish Biol 46:961–976

Sadovy Y, Domeier M (2005) Are aggregation-fisheries sustainable? Reef fish fisheries as a case study. Coral Reefs 24:254–262

Sadovy Y, Eklund AM (1999) Synopsis of biological data on the Nassau grouper, *Epinephelus striatus* (Bloch 1792) and the jewfish *E. itajara* (Lichtenstein 1822). NOAA Technical Report NMFS 146, U.S. Dept. of Commerce, Seattle, Washington

Sadovy Y, Figuerola M (1992) The status of red hind fishery in Puerto Rico and St. Thomas, as determined by yield-per-recruit analysis. Proc Gulf Caribb Fish Inst 42:23–38

Sadovy YJ, Vincent ACJ (2002) Ecological issues and the trades in live reef fishes. In: Sale PF (ed) Coral reef fishes dynamics and diversity in a complex ecosystem. Academic, San Diego

Sadovy Y, Figuerola M, Roman A (1992) Age and growth of red hind *Epinephelus guttatus* in Puerto Rico and St. Thomas. Fish Bull 90:516–528

Sadovy Y, Rosario A, Roman A (1994) Reproduction in an aggregating grouper, the red hind, *Epinephelus guttatus*. Environ Biol Fish 41:269–286

Sadovy Y, Kulbicki M, Labrosse P, Letourneur U, Lokani P, Donaldson TJ (2003a) The humphead wrasse, *Cheilinus undulatus*: synopsis of a threatened and poorly known giant coral reef fish. Rev Fish Biol Fish 13:327–364

Sadovy YJ, Donaldson TJ, Graham TR, McGilvray F, Muldoon GJ, Phillips MJ, Rimmer MA, Smith A, Yeeting B (2003b) While stocks last: the live reef food fish trade. Asian Development Bank, Manila, 147pp

Sadovy Y, Punt AE, Cheung W, Vasconcellos M, Suharti S (2007) Stock assessment approach for the Napoleon fish, *Cheilinus undulatus*, in Indonesia: a tool for quota-setting for data-poor fisheries under CITES Appendix II Non-Detriment Finding requirements, FAO Fisheries Circular No. 1023. FAO, Rome, 71pp

Sahayak S (2005) Reproductive biology of the masked triggerfish *Sufflamen fraenatus*. J Mar Biol Assoc India 47(1):70–76

Sala E, Ballesteros E, Starr RM (2001) Rapid decline of Nassau grouper spawning aggregations in Belize: fishery management and conservation needs. Fisheries 26(10):2–30

Sala E, Aburto-Oropeza O, Paredes G, Thompson G (2003) Spawning aggregations and reproductive behavior of reef fishes in the Gulf of California. Bull Mar Sci 72:103–121

Samoilys MA (1997) Periodicity of spawning aggregations of coral trout, *Plectropomus leopardus* (Pisces: Serranidae) on the northern Great Barrier Reef. Mar Ecol Prog Ser 160:149–159

Samoilys MA (2000) Reproductive dynamics of an exploited serranid on the Great Barrier Reef. Ph.D. thesis, James Cook University, Townsville

Samoilys MA, Carlos G (2000) Determining methods of underwater visual census for estimating the abundance of coral reef fishes. Environ Biol Fish 57:289–304

Samoilys M, Fuentes H, Tuwai I, Tikomainiusiladi B, Leqata J, Oreihaka E, Mobiha A, Potuku T, Die D, Connell S, Lincoln Smith M, Wilson M, Watson R (1995) Application of underwater visual census to assessing coral reef fish stocks in the tropical Pacific. Report prepared for the Australian Centre for International Agricultural Research (ACIAR). ACIAR Project Number 9304, Final Report, September 1995

Samoilys MA, Squire LC, Roelofs A (2001) Long term monitoring of coral trout spawning aggregations on the Great Barrier Reef: implications for fisheries management. In: Abstract - Indo Pacific Fish Conference, Durban, South Africa, May 2001

Samoilys M, Slade SJ, Williams LE (2002a) Coral trout. In: Williams LE (ed) Queensland's fisheries resources: current conditions and recent trends 1988–2000, Q102012. Department of Primary Industries Queensland, Brisbane

Samoilys M, Williams L, Slade S (2002b) Coral reef line fishery. In: Williams LE (ed) Queenland's fisheries resources – current condition and recent trends 1988-2000. Queensland Department of Primary Industries, Brisbane

Sanchez-Velasco L, Lavın MF, Peguero-Icaza M et al (2009) Seasonal changes in larval fish assemblages in a semi-enclosed sea (Gulf of California). Cont Shelf Res 29:1697–1710

Sancho G, Ma D, Lobel PS (1997) Behavioural observations of an upcurrent reef colonization event by larval surgeonfish *Ctenochaetus strigosus* (Acanthuridae). Mar Ecol Prog Ser 153:311–315

Sancho G, Solow AR, Lobel PS (2000a) Environmental influences on the diel timing of spawning in coral reef fishes. Mar Ecol Prog Ser 206:193–212

Sancho G, Petersen CW, Lobel PS (2000b) Predator-prey relations at a spawning aggregation site of coral reef fishes. Mar Ecol Prog Ser 203:275–288

Schirripa MJ, Goodyear CP (1994) Status of gag stocks of the Gulf of Mexico. Assessment 1.0. CRD-93/94-61. National Marine Fisheries Service, Southeast Fisheries Science Center, Miami Laboratory, Miami, Florida

Schirripa MJ, Legault CM (1997) Status of the gag stocks of the Gulf of Mexico. Assessment 2.0. Stock Assessment, 1 October 1997 Southeast Fisheries Science Center, Sustainable Fisheries Division, Miami, Florida

SCRFA (2004) Fisher interviews (www.SCRFA.org)

SEDAR (2009) Report of assessment workshop, stock assessment of gag in the Gulf of Mexico: SEDAR update assessment. SEDAR (http://www.sefsc.noaa.gov/sedar/), Charleston, South Carolina

Sedberry GR (1988) Food and feeding of the black sea bass, *Centropristis striata*, in live bottom habitats in the South Atlantic Bight. J Elisha Mitchell Sci Soc 104:35–50

Sedberry GR, Stevenson DE, Chapman RW (1996) Stock identification in potentially threatened species of grouper (Teleostei: Serranidae: Epinephelinae) in Atlantic and Caribbean waters. Final report to the National Marine Fisheries Service, 51p

Sedberry GR, Pashuk O, Wyanski DM, Stephen JA, Weinbach P (2006) Spawning locations for Atlantic reef fishes off the southeastern US. Proc Gulf Caribb Fish Inst 57:463–514

Semmens BX, Bush PG, Heppell SA, Johnson BC, McCoy CM, Luke SKE, Pattengill-Semmens CV, Heppell SS, Whaylen L (2006) Defining the biogeography of a spawning aggregation to inform marine reserve planning and evaluation in the Cayman Islands. Proc Gulf Caribb Fish 59:638 (Abstract)

Semmens BX, Bush P, Heppell SA, Johnson B, McCoy CM, Pattengill-Semmens C, Heppell SS (2009) The spatial ecology of a remnant Nassau grouper stock on Cayman Brac, Cayman Islands. Proc Gulf Caribb Fish 61:337 (Abstract)

SFA (2006) Annual Report 2006. Seychelles Fishing Authority, Mahe

Shapiro DY (1987) Reproduction in groupers. In: Polovina JJ, Ralston S (eds) Tropical snappers and groupers: biology and fisheries management. Westview Press, Boulder, pp 295–327

Shapiro DY, Sadovy Y, McGehee MA (1993) Size, composition, and spatial structure of the annual spawning aggregation of the red hind, *Epinephelus guttatus* (Pisces: Serranidae). Copeia 2:399–406

Shapiro DY, Garcia-Moliner G, Sadovy Y (1994) Social system of an inshore population of the red hind grouper, *Epinephelus guttatus* (Pisces: Serranidae). Environ Biol Fish 41:415–422

Shenker JM, Maddox ED, Wishinski E, Pearl A, Thorrold SR, Smith N (1993) Onshore transport of settlement-stage Nassau grouper *Epinephelus striatus* and other fishes in Exuma Sound, Bahamas. Mar Ecol Prog Ser 98(1–2):31–43

Slamet B, Hutapea JH (2005) First successful hatchery production of Napoleon wrasse at Gondol Research Institute for Mariculture, Bali. SPC Live Reef Fish Inf Bull 13:43–44

Sluka R, Chiappone M, Sullivan KM, Wright R (1996) Habitat and life in the Exuma Cays, the Bahamas: the status of groupers and coral reefs in the Northern Cays. The Nature Conservancy, Coral Gables

Smith CL (1965) The patterns of sexuality and the classification of serranid fishes. Am Mus Novit 2207:1–20

Smith CL (1971) A revision of the American groupers: *Epinephelus* and allied genera. Bull Am Mus Nat Hist 146:69–241

Smith CL (1972) A spawning aggregation of Nassau grouper, *Epinephelus striatus* (Bloch). Trans Am Fish Soc 101:257–261

Smith-Vaniz WF, Jelks HL, Rocha LA (2006) Relevance of cryptic fishes in biodiversity assessments: a case study at Buck Island Reef National Monument, St. Croix. Bull Mar Sci 79(1):17–48

Sommer C, Schneider W, Poutiers JM (1996) FAO species identification guide for fishery purposes. The living marine resources of Somalia. FAO, Rome

Sponaugle S, Cowen RK (1996) Larval supply and patterns of recruitment for two Caribbean reef fishes, *Stegastes partitus* and *Acanthurus bahianus*. Mar Freshw Res 47:433–447

St John J (1999) Ontogenetic changes in the diet of the coral trout reef grouper *Plectropomus leopardus* (Serranidae): patterns in taxa, size and habitat of prey. Mar Ecol Prog Ser 180:233–246

Stallings CD (2009) Fishery-independent data reveal negative effect of human population density on Caribbean predatory fish communities. PLoS One 4(5):1–9

Starr RM, Sala E, Ballesteros E, Zabala M (2007) Spatial dynamics of the Nassau grouper *Epinephelus striatus* in a Caribbean atoll. Mar Ecol Prog Ser 343:239–249

Stephenson P, Jackson G (2005) Managing depleted snapper stocks in inner Shark Bay, Western Australia. In: Kruse GH, Gallucci VF, Hay DE, Perry RI, Peterman RM, Shirley TC, Spencer PD, Wilson B, Woodby D (eds) Assessment and management of new and developing fisheries in data-limited situations. Alaska Sea Grant College Program, Anchorage

Stobutski IC, Bellwood DR (1997) An analysis of the sustained swimming abilities of pre- and post-settlement coral reef fishes. J Exp Mar Biol Ecol 175:275–286

Stone G (2004) Phoenix islands: a coral reef wilderness revealed. National Geographic Magazine February2004; vol205, p 48 onwards

Sumner NR, Malseed BE (2002) Quantification of changes in recreational catch and effort on inner Shark Bay snapper species following implementation of responsive management changes, Fisheries Research and Development Corporation Final Report, Project 2000/139. Department of Fisheries, Western Australia

Sumner NR, Williamson PC, Malseed BE (2002) A 12-month survey of recreational fishing in the Gascoyne bioregion of Western Australia during 1998-99, Fisheries Research Report No 139. Department of Fisheries, Western Australia

Suzuki K, Hioki S (1979) Spawning behavior, eggs, and larvae of the lutjanid fish, *Lutjanus kasmira*, in an aquarium. Jpn J Ichthyol 26(2):161–166

Tamaru CS, Carlstrom-Trick C, Fitzgerald WJ, Ako H (1996) Induced final maturation and spawning of the camouflaged grouper *Epinephelus microdon* captured from spawning aggregations in the Republic of Palau, Micronesia. J World Aquaculture Soc 27:363–372

Taylor JC, Eggleston DB, Rand PS (2006) Nassau grouper (*Epinephelus striatus*) spawning aggregations: hydroacoustic surveys and geostatistical analyses. NMFS Prof Pap 5:18–25

Thompson R, Munro JL (1978) Aspects of the biology and ecology of Caribbean reef fishes: Serranidae (hinds and groupers). J Fish Biol 12:115–146

Thompson R, Munro JL (1983) The biology, ecology and bionomics of the snappers, Lutjanidae. In: Munro JL (ed) Caribbean coral reef fishery resources, ICLARM studies and reviews 7. International Center for Living Aquatic Resources Management, Manila

Tupper M (2007) Identification of nursery habitats for commercially valuable humphead wrasse *Cheilinus undulatus* and large groupers (Pisces: Serranidae) in Palau. Mar Ecol Prog Ser 332:189–199

Turnbull CT, Samoilys MA (1997) Effectiveness of spawning closures in managing the line fishery on the Great Barrier Reef. Report to the Reef Fish Management Advisory Committee of the Queensland Fisheries Management Authority

Turner SC, Porch CE, Heinemann D, Scott GP, Ortiz M (2001) Status of gag in the Gulf of Mexico, Assessment 3.0. Sustainable Fisheries Division Contribution SFD 01/02 - 134, August 2001. Southeast Fisheries Science Center, Miami, Florida

Vilaro Diaz DJ (1884) Corrida y arribazon de algunos peces cubanos. Manuel Gomez de la Maza, La Habana

Vincent ACJ, Sadovy Y (1998) Reproductive ecology in the conservation and management of fishes. In: Caro T (ed) Behavioural ecology and conservation biology. Oxford University Press, New York

Von Westernhagen H (1974) Observation on the natural spawning of *Alectis indicus* (Ruppell) and *Caranx ignobilis* (Forsk) (Carangidae). J Fish Biol 6:513–516

Wakefield CB (2006) Latitudinal and temporal comparisons of the reproductive biology and growth of snapper, *Pagrus auratus*, in Western Australia. Ph.D. thesis, Murdoch University, Perth

Warner RR (1984a) Deferred reproduction as a response to sexual selection in a coral reef fish: a test of the life historical consequences. Evolution 38:148–162

Warner RR (1984b) Mating behaviour and hermaphroditism in coral reef fishes. Am Sci 72:128–136

Warner RR (1986) The environmental correlates of female infidelity in a coral reef fish. In: Uyeno T, Arai R, Taniuchi T, Matsuura K (eds) Indo-Pacific fish biology: Proceedings of the second international conference on Indo-Pacific fishes. Ichthyological Society of Japan, Tokyo

Warner RR (1987) Female choice of sites versus males in a coral reef fish, *Thalassoma bifasciatum*. Anim Behav 35:1470–1478

Warner RR (1988) Traditionality of mating-site preferences in a coral-reef fish. Nature 335:719–721

Warner RR (1990a) Resource assessment vs. traditionality in mating site determination. Am Nat 135:205–217

Warner RR (1990b) Male vs. female influences on mating site determination. Anim Behav 39:540–548

Warner RR (1995) Large mating aggregations and daily long-distance spawning migrations in the bluehead wrasse, *Thalassoma bifasciatum*. Environ Biol Fish 44:337–345

Warner RR (2002) Synthesis: environment, mating systems, and life-history allocations in the bluehead wrasse. In: Dugatkin L (ed) Model systems in behavioral ecology. Princeton University Press, Princeton

Warner RR, Hoffman SG (1980a) Local population size as a determinant of mating system and sexual composition in two tropical marine fishes (*Thalassoma* spp.). Evolution 34:508–518

Warner RR, Hoffman SG (1980b) Population density and the economics of territorial defense in a coral reef fish. Ecology 61:772–780

Warner RR, Schultz ET (1992) Sexual selection and male characteristics in the bluehead wrasse, *Thalassoma bifasciatum*: mating site acquisition, mating site defense, and female choice. Evolution 46:1421–1442

Watanabe WO, Ellis SC, Ellis EP, Head WD, Kelley CD, Moriwake A, Lee C-S, Bienfang PK (1995) Progress in controlled breeding of Nassau grouper (*Epinephelus striatus*) broodstock by hormone induction. Aquaculture 138:205–219

Watanabe W, Ellis EP, Ellis SC, Chaves J, Manfredi C (1998) Artificial propagation of mutton snapper, *Lutjanus analis*, a new candidate marine fish species for aquaculture. J World Aquaculture Soc 29:176–187

Weaver DC (1996) Feeding ecology and ecomorphology of three sea basses (Pisces:Serranidae) in the northeastern Gulf of Mexico. Thesis (M.S.), University of Florida, Gainesville

Westneat MW, Alfaro ME (2004) Phylogenetic relationships and evolutionary history of the reef fish family Labridae. Mol Phylogenet Evol 36:370–390

Whaylen L, Pattengill-Semmens CV, Semmens BX, Bush PG, Boardman MR (2004) Observations of a Nassau grouper (*Epinephelus striatus*) spawning aggregation site in Little Cayman Island. Environ Biol Fish 70:305–313

Whaylen L, Bush P, Johnson B, Luke K, McCroy C, Heppell S, Semmens B, Boardman MR (2007) Aggregation dynamics and lessons learned from five years of monitoring at a Nassau grouper (*Epinephelus striatus*) spawning aggregation in Little Cayman, Cayman Islands, BWI. Proc Gulf Caribb Fish Inst 59:413–421

Whiteman EA, Jennings CA, Nemeth RS (2005) Sex structure and potential female fecundity in a red hind (*Epinephelus guttatus*) spawning aggregation: applying ultrasonic imaging. J Fish Biol 66:983–995

Wicklund R (1969) Observations on spawning of lane snapper. Underw Nat 6:40

Williams AJ, Currey LM, Begg GA, Murchie CD, Ballagh AC (2008) Population biology of coral trout species in the eastern Torres Strait: implications for fishery management. Cont Shelf Res 28:2129–2142

Wilson J, Rhodes KL, Rotinsulu C (2010) Aggregation fishing and local management within a marine protected area in Indonesia. SPC Live Reef Fish Inf Bull 19:7–13

Wong CK, Hung P, Lee KLH, Kam KM (2005) Study of an outbreak of ciguatera fish poisoning in Hong Kong. Toxicon 46(5):563–571

Woodland DJ (1990) Revision of the fish family Siganidae with descriptions of two new species and comments on distribution and biology. Indo Pac Fish 19:1–136

Wright A, Dalzell PJ, Richards AH (1986) Some aspects of the biology of the red bass, *Lutjanus bohar* (Forsskal), from the Tigak Islands, Papua New Guinea. J Fish Biol 28:533–544

Yogo Y, Nakazono A, Tsukahara H (1980) Ecological studies on the spawning of the parrotfish, *Scarus sordidus* Forsskal. Sci Bull Fac Agric Kyushu Univ 34:105–114

Zeller DC (1998) Spawning aggregations: Patterns of movement of the coral grouper *Plectropomus leopardus* (Serranidae) as determined by ultrasonic telemetry. Mar Ecol Prog Ser 162: 253–263

Zeller DC, Russ GR (2000) Population estimates and size structure of *Plectropomus leopardus* (Pisces:Serranidae) in relation to no-fishing zones: mark-release-resighting and underwater visual census. Mar Freshw Res 51:221–228

Chapter 13
Conclusion

Yvonne Sadovy de Mitcheson and Patrick L. Colin

13.1 Introduction

This book draws together the many aspects of reef fish spawning aggregations, as biological ecological and oceanographic phenomena intimately linked with abiotic conditions and exposed to increasing levels of exploitation that sometimes threaten them. A better understanding and appreciation of why, how, when and where they form is a major step towards ensuring their continued existence.

Hypotheses proposed to address the possible adaptive significance of aggregation formation can be broken down into two main themes: those that address benefits to the larvae and those with benefits to the adults. While these hypotheses are still largely untested, and are discussed in more detail below, it is reasonable to assume that the selective pressures that led to the development of aggregative spawning resulted in a reproductive strategy that presents an increased level of reproductive success for each individual, compared with non-aggregating species. Indeed, this reproductive strategy may well account for the fact that many of the most productive tropical, and indeed temperate, fish species, that are the basis of major global fisheries, exhibit this habit. It is also noteworthy that many major fishery declines involve aggregating species. It is surprising, therefore, that spawning aggregations have not attracted much greater management attention despite the fact that the majority are thought to be undergoing declines due to fishing and many no longer form. Aggregation-fishing appears to be the major threat to several important

Y. Sadovy de Mitcheson (✉)
Division of Ecology & Biodiversity, University of Hong Kong, Pok Fu Lam Road, Hong Kong, China
e-mail: yjsadovy@hku.hk

P.L. Colin
Coral Reef Research Foundation, P.O. Box 1765, Palau 96940, Koror
email: crrf@palaunet.com

Y. Sadovy de Mitcheson and P.L. Colin (eds.), *Reef Fish Spawning Aggregations: Biology, Research and Management*, Fish & Fisheries Series 35,
DOI 10.1007/978-94-007-1980-4_13, © Springer Science+Business Media B.V. 2012

species of groupers (Serranidae) and croakers (Sciaenidae), both families for which International Union for Conservation of Nature (IUCN) Red Listing has now been conducted for all species, something that should set off alarm bells with conservationists and fishery managers worldwide (www.iucnredlist.org; Craig et al. 2011).

Although many papers have been published on spawning aggregations, vexing and fundamental questions remain, among them: (1) why spawn in an aggregation, (2) how do spawning aggregations originally form where and when they do, (3) from how big a geographic, i.e. catchment, area do adults come to the spawning site, (4) how far do the offspring disperse from the aggregation site and (5) what determines the number of individuals gathering at different sites? There are serious fisheries and economic issues that could draw immediate benefit from the answers to these and other (see below) questions that would facilitate management and strengthen the case and urgent need for long-term preservation.

Despite the large number of reef-associated species (>100 Appendix) known to aggregate to spawn, an appreciation of the wider implications of the aggregating habit for coastal fisheries and the marine ecosystem, beyond their immediate biological role in reproduction, has been slow in materializing. The phylogenetic diversity, different aggregation types, high numbers (from hundreds to tens of thousands) of fish that briefly assemble at certain sites, as well as sometimes extensive (ten to several hundred kilometres) migrations collectively involve huge biomass movements of different trophic levels across broad expanses of reef. Initially described in a fisheries context in the 1800s by Vilaro Diaz (1884), serious scientific study of an aggregation was not conducted until the 1960s (Randall and Randall 1963), on a parrotfish species. The first descriptive studies of aggregations of a large commercial species, the Nassau grouper, *Epinephelus striatus*, were undertaken in the 1970s (Smith 1972; Olsen and LaPlace 1979). The widespread occurrence of fish spawning aggregations became apparent in the late 1970s and 1980s, with publications covering single geographic regions (e.g. Johannes 1981). The threats of fishing to aggregations was first recognised for the Nassau grouper (Sadovy 1993), while the ecosystem role of aggregations was considered more recently with recognition of the apparent food value of eggs produced by massive snapper aggregations during whale shark migrations (Heyman et al. 2001).

As long-term indicators of ecological stability, spawning aggregations are perhaps the most definitive iconic event available to marine scientists and managers. For some specific locations there are records of aggregations stretching back 40–50 years, with anecdotal and oral accounts from fishing communities for much longer periods. Since many aggregations involve large exploited reef fishes and have been extirpated or greatly reduced by exploitation (Sadovy et al. 2008), this fact alone should have set off alarm bells throughout the management community several decades ago. While there is a growing realization that aggregations matter, and, in many cases are critical to maintenance or recovery of fisheries resources, we could find few fishery management success stories and few aggregations yet effectively incorporated into marine protected areas (MPAs).

A compelling observation that strongly highlights the need for action is that, where aggregating species are found and exploited, and especially in the case of transient spawners, aggregations are less likely to be present where fishing pressure is high. For example, in the Philippines and Indonesia, both intensively fished and with very little fishery management, few aggregations have been reported relative to the number of fisher interviews conducted (SCRFA country reports: www.SCRFA.org). On the other hand, in the western and central Pacific, where fishing pressure is increasing but is overall considerably more moderate, there are many reports of aggregations, particularly in the lesser fished areas (Sadovy de Mitcheson et al. 2008). We also know that management can restore depleted aggregations. Such observations suggest that (1) healthy aggregations are associated with healthy fisheries, (2) it is important and possible to proactively manage remaining aggregations and (3) aggregations might prove to be useful indicators of fishery condition.

As scientists we take the long view, a necessity when so many fish populations and reefs in general are deteriorating (e.g. Wilkinson 2008). Fish populations can eventually recover from overfishing provided the right management environment is available and the species is not critically depleted. However, it is possible that biological or physiological factors may slow or even stall recovery at low population sizes, as has evidently been the case for Atlantic cod (Rowe and Hutchings 2003), signalling that we cannot wait until the last minute (i.e. numbers of animals have been much reduced) before we act, and still assume recovery is inevitable. In extreme cases, this may mean total closure of a fishery. Such a drastic measure may have recreational and commercial fishers up in arms, and will almost certainly be politically challenging, especially if there are no alternative livelihood options, yet may be the only way to sustain the fishery. As in all such matters, political will, backed by public support and built on a foundation of good science are critically important for success (Sale et al. 2005). It was with an eye firmly on the need for good science and understanding that this book was conceived.

13.2 What the Chapters Say

We summarize the major findings presented in the chapters of this book, highlighting when the authors have arrived at general principles previously unrecognized, or provide a new level of understanding of some aspect of aggregations. We then identify key research needs, discuss promising areas to explore, and reflect on specific management and conservation implications. For literary citations, please refer to original chapters.

A first important step is to clearly define what we mean by the term 'spawning aggregation'. To be useful, the definition has a clear ecological, biological and practical significance. The tone for this book is set by Michael Domeier in Chap. 1 in which the definition is clarified as a "unique phenomenon of behavioural ecology where an entire sub-population of individuals halt their normal routine, migrate, gather and spawn", with more specific guidance given below. The need for a straightforward definition applicable to nearly all situations is considerable, as confusion about what is and is not a spawning aggregation has at times prevented a realistic consideration of economic and management implications, or can be misleading when testing hypotheses on why when and where aggregations form. For example, an area where sharks congregate to release their young may be vulnerable to fishing, and possibly in need of management, but would be an inappropriate location to test hypotheses on why some bony fishes (teleosts) release pelagic eggs in spawning aggregations. Like any definition, however, there are challenges and few definitions are perfect. For example, for species that live and also spawn in schools, such as many jacks and trevallys (Carangidae) and the bumphead parrotfish (*Bolbometopon muricatum*), how is the definition best applied to meet its objectives?

Definitions aid communication and understanding, and their careful crafting and consistent application are important. Throughout this book we tried to clarify terms referred to in various ways in the literature, e.g. 'promontory', 'group'-spawning, 'spawning aggregation', etc. In the case of 'aggregations', for example, to verify that a given grouping of fish is a spawning aggregation, we specify that one or more of the following conditions must be met as *direct* indicators of spawning; (1) undisputed spawning observations, (2) females found with hydrated eggs, or (3) presence of post-ovulatory follicles in the ovaries of aggregating females. *Indirect* indicators are also identified in Chap. 1. The suggestion that a 'grouping' represents a minimum four-fold or more increase in numbers at a spawning site is somewhat arbitrary but intended to distinguish it from background levels of abundance in the non-spawning season. Most aggregating fishes have considerably higher relative densities when aggregated.

Two basic types of spawning aggregation, termed 'resident' (RA) and 'transient' (TA) appear to be useful concepts, distinguishing whether fishes remain "within or nearby" (resident) or "well outside" (transient) their home range during aggregation. However, as more species are studied in more detail, it appears that the scope of aggregation type is likely to be more of a continuum, with clear distinctions at the extremes but with no absolute dividing line between TA and RA. Examples include the leopard coralgrouper, *P. leopardus* and the blackfin snapper, *Lutjanus fulvus* (see case studies for these species in Chap. 12.10 for details).

In attempting to understand the wider ecological role of aggregating species, Rick Nemeth addresses the part they may play in ecosystem function in Chap. 2. He develops the concept of a 'functional migration area' that encompasses a mosaic of habitats through which a species migrates, including the specific spatial components of catchment area, staging area, courtship arena, and inclusive of the spawning aggregation site. The flux in fish biomass and the energy transfer between feeding grounds and spawning sites is a hitherto overlooked ecological and ecosystem component of adult connectivity. Information on predator-prey dynamics indicates that some fishes feed during migration and at spawning sites, while certain piscivores and egg predators feed on aggregated fishes or their spawning products. Mapping fish movements and documenting the physical and ecological components of functional migration areas are important steps in the management of multi-species tropical fisheries and shallow marine ecosystems.

Phillip Molloy, Isabel Côté and John Reynolds take a broad behavioural ecology perspective of aggregation-spawning in Chap. 3 in which they ponder the important and intriguing question of 'why spawn in aggregations'. They examine the theoretical frameworks, costs and benefits of fish spawning aggregations as well as those of a wide range of other, non-fish, group-breeding taxa and ask what insights can be gained from examining the habit in other taxa? In non-fish species, aggregating for reproduction evolved due to limitation of suitable space or habitat thereby forcing animals to occupy higher densities than normal at mating sites. Alternatively there may be active choice by individuals for proximity to conspecifics because of benefits to breeding adults or their young.

Social reproduction benefits are in two general categories: enhanced reproduction and reduced predation. Reduced predation within a breeding group can result from dilution effects (the likelihood of any individual being the target of a predator decreases as group size increases), earlier detection of predators (increasing probability of escape) or deterring/confusing predators through cooperative efforts among the group. In addition, the synchrony of reproductive activities in a group may temporarily satiate predators, leading to lower predation *per capita* when offspring are vulnerable. For fishes, at least, as covered in other chapters, there is little indication that predation, either on spawning adults or on eggs after release, has a major role in structuring aggregations, despite the likely benefit of short-term transport of eggs away from benthic predators.

From what is known about a few well-studied non-fish species, including the variation in social systems, travel costs, the nature of the aggregations themselves, and the phylogenetic diversity of aggregation-spawning fishes, there is unlikely to be a single evolutionary explanation for aggregation spawning. The lack of studies with experimental manipulations precludes the evaluation of some hypotheses. Moreover, although predictions from competing hypotheses are not fundamentally impossible to test, they are harder to tackle in some taxa than in others. Studies of bird breeding strategies, for example, have developed much farther than for fishes because birds are often easier to mark for individual recognition, the parentage of the young is easier to determine, they often lend themselves better to experimentation, and unexploited populations are more widely accessible for study. Although in other ways fishes lend themselves well to comparative study, it is curious that, unlike many other taxa, the possible importance of specific habitats for marine fish aggregations has received little attention.

In Chap. 4, with a focus on fish taxa, Howard Choat examines the biological features associated with aggregation spawning. He finds that maximum body size, trophic ecology, and anatomy are more predictive of aggregation versus non-aggregation habits than are other life history features. As long as the different aggregation types (RA and TA) share basic properties of size, nutritional ecology and anatomy, he found that they manifested similar spawning behaviours regardless of whether fish species are protogynous or gonochoristic, short- or long-lived or have slow or fast population turnover rates. Neither larval retention nor broad dispersal are seen as critical elements in the evolution of spawning aggregations. Choat hypothesizes that differences in aggregate spawning patterns and their underlying processes will occur in the Pacific and Atlantic oceans, a reflection of the different histories, oceanic environments and habitat structures of these two ocean basins.

Differences in fish species, reflecting natural abundance levels and associated with habitat structure, may also be a factor determining the sizes and even types of aggregation. It is clear that one major gap in our understanding of the aggregation habit is how it is related to fish abundance and/or densities and what implications changes in these might have in the case of exploited populations.

Hypotheses regarding the "where and when" of aggregations, covered by Pat Colin in Chap. 5, generally focus on whether aggregations produce dispersal or retention of propagules. In the past, the most common hypothesis was that reef fish populations were selected to spread their progeny across areas as wide as possible to ensure that at least some of the population finds conditions suitable for survival. The idea was that aggregations form in areas that promote the offshore dispersal of eggs and larvae into oceanic waters for wide dispersal or perhaps for return by meso-scale eddies after lengthy periods at sea. More recently, focus shifted to ideas of larval populations being retained near their sources with aggregation sites and times possibly playing a role in retention of eggs and larvae, rather than dispersal. While such hypotheses need testing, it is quite likely that populations exhibit characteristics of both strategies and are quite sensitive to locality differences.

A review of the diversity of locations, timing and conditions between TA and RA spawning aggregations, as well as comparisons with non-aggregation spawning sites, provides insight into factors likely to be influencing aggregation-spawning at both local and regional levels. Nearly all TA species either spawn on the reef drop-off or in areas closely connected (channels) with outer reef areas. RA species also spawn in such areas, but additionally utilize more inshore areas. Different species at multi-species TA sites do not usually spawn at exactly the same times but all spawning is completed within a typically restricted spawning season. This suggests that, within a region, spawning may be structured around local abiotic regimes with good evidence that temperature is a major shaping factor. Such physical conditions could be cues for adults in the area or may be important for the success of the larvae, particularly for feeding, a variation of the famous match/mismatch hypothesis (Sinclair 1988). At a wider spatial scale, differences between the western Atlantic and the Indo-Pacific regions may also influence aggregation patterns. The Atlantic is characterized by modest tidal amplitudes (1 m or less), predominance of insular shelves without barrier reefs, presence of few atolls and weak tidal currents on reefs. The Indo-Pacific has generally higher tidal amplitudes (up to several metres), abundant barrier reefs with channels, many atolls, and strong tidal currents on reefs (due to geomorphology and tidal amplitude). Such physical differences and how these also might affect fish densities and distributions is a little-considered but potentially very interesting area for comparative study.

An understanding of why and how fish spawning aggregations form and persist may emerge by examining commonalities in the location and timing of aggregation and spawning (Chap. 5) combined with what is known of the specific oceanography of aggregation sites covered in Chap. 6 by Bill Hamner and John Largier. Transient and resident aggregations with long-term persistence at specific sites are found throughout the world. The former tend to occur on outer reefs (with or without promontories), in channels, and at the ends of islands and atolls. The latter occur across a wider range of habitats including broad insular shelves and inner reefs. The multi-species use of both types of aggregation sites is common, suggesting, perhaps, something special about the sites themselves. There is intriguing evidence that migration routes to and from aggregation sites are learned from older fishes, with shelf edges or other prominent reef features used as a "highway" for long-distance movements. This suggests that not only should workers pay attention to the possible presence of juveniles at aggregation sites but that such highways may also need to be managed; certainly some are known to fishers and are actively exploited.

Although very little is known of the specific oceanography associated with aggregation spawning, Chaps. 6 and 7 highlight the various methods available and identify questions to be explored. Initial modelling of egg/larval dispersal during the days prior to larval initiation of feeding, for example, indicates that the larval density starting from a concentrated near-point source of spawning aggregations soon approaches that of larvae from fishes with more widespread spawning. This suggests that a point source of large numbers of eggs or larvae has little to do with the evolution of aggregation-spawning. Moreover, it seems unlikely that food density becomes more critical for the survival of aggregation-produced larvae than for other fish larvae and hence that there is no apparent disadvantage to higher larval numbers. Although it is highly unlikely that coherent clouds of larvae stay together for extended periods of time, some physical mechanisms may concentrate larvae when many are in the water at the same time and in the same general area. Several aggregating species are reported to settle in particularly large cohorts (e.g. Nassau grouper) or form the basis for capture-based aquaculture (e.g. certain mullets, rabbitfish, bonefish and groupers) whereby settling fish can be caught in enormous numbers in specific areas and in short time frames. These concentrations have enabled the establishment of fisheries that catch large numbers of very young animals and then grow them out in captivity until they reach market size (Lovatelli and Holthus 2008).

The factors that determine the timing of spawning, at each of annual, diel, and lunar scales, is of much interest and was variously addressed in several chapters. Tropical fish species in general have extended spawning seasons, probably because of the relative lack of seasonal changes compared to temperate regions. However, aggregating species, and most transient aggregators in particular, are interesting in that their spawning patterns tend to be very limited seasonally unlike other tropical species; indeed such species may only spawn for a few hours over a few days in a couple of months each year. What determines this annual seasonality is not known, but temperature may be one factor (as suggested by data on the Nassau grouper and camouflage grouper, *Epinephelus polyphekadion*) and the availability of larval food, another. Regarding the lunar cycle, most TA species have a strong lunar component to the timing of spawning which can be full or new moon related. Some species, like the camouflage grouper, might even cue to a different lunar phases in different areas. Lunar synchronicity could be determined by optimum times for hatching and/or larval settlement, or it could simply be a cue for coordinating reproductively ripe adults. RA species tend to show more of a tidal component to the timing of spawning than TA species. As for the diel cycle, present knowledge indicates that most TA species spawn at dusk, during the night or at dawn, while RA species spawn during the day, often in relation to tidal cycles. Despite the trends indicated, the adaptive significance of timing of spawning is not known.

Turning to the biological and socio-economic implications of fishing on spawning aggregations and aggregating species, Chap. 8 by Yvonne Sadovy de Mitcheson and Brad Erisman explores the extent to which they can be harvested sustainably beyond low-level subsistence fishing and looks with abundant eggs ready to hatch; sources of the next generation at attitudes towards their management. In all documented cases, unmanaged exploitation has resulted in population depletions. While it is clear that aggregation fisheries are important and widespread globally, it is also evident that they are typically poorly documented or managed. This is partly due to the fact that large catches give an 'illusion of plenty', belying the need for any intervention and thwarting proactive management planning. Interestingly, while the idea of protecting females that are clearly with abundant eggs close to hatching is widely accepted in the case of egg-bearing ("berried") lobsters and crabs, similar thinking has never been applied to female fish that are clearly full of eggs. Likewise, while few people would deny that turtle nesting colonies need protection and that nesting seabird colonies should be left in peace, few fish spawning aggregations receive similar care. Rather, they remain targets for fishing rather than for protection.

From a biological perspective, there is a need to incorporate elements of fish mating systems into fishery models to generate different scenarios against which to assess observations, make predictions or conduct comparative assessments. As aggregations get smaller from overfishing, we need to consider whether there are threshold densities below which recovery is likely to be much compromised because of behavioural components, such as the Allee Effect. For species that change sex, are the opportunities provided by aggregations to assess numbers of adult males and females important for cuing sex change events and hence determining operational sex ratios?

Economic considerations should address long-term objectives as well as more immediate short-term pressures. For example, gluts of fish produced by harvesting TAs can be wasteful and lead to reduced unit prices as well as contribute to overfishing of the species. It may, in the long term, be far more beneficial for the fishery if fishing is restricted to the non-spawning season, producing stable catches throughout the year, better prices and a more productive fishery because spawning is allowed to take place undisturbed. In the development of fisheries resources, it is not uncommon for the status of the resource, and hence its capacity to withstand development, to be ignored. The widespread introduction of ice plants in the Pacific to encourage and facilitate marketing from more remote locations, for example, took no account of the underlying resource base that would be affected.

Turning now to the important task of improving information on aggregations and monitoring those that are or could be managed, in Chap. 9 Pat Colin covers the methods available to quantify and document the various aspects of spawning aggregations, highlighting those methods that are most effective under different circumstances, as well as various shortcomings. This chapter provides considerable updates on the earlier SCRFA Methods Manual (Colin et al. 2003). Fortunately, methods are becoming more standardized, allowing for better comparisons among species and localities and over time. The application of new technology is particularly promising for answering some of the more challenging questions about aggregation dynamics, but fundamental questions persist. For example, the meaningful counting of fish in aggregations of thousands to tens of thousands, when fish mass in the water column in three-dimensional 'balls', when numbers change daily, and when fish are moving around calls for innovative approaches and careful planning. Maybe techniques applied to the quantification of flocks of birds and the use of video and other technology could help to improve accuracy. New 3D image technology holds promise for better quantification of numbers and sizes of fish in aggregations. It has become very clear that the monitoring of fish numbers in aggregations under anything more than the most benign situations, with low numbers of fish in a small shallow area, requires a high degree of expertise and should not be conducted without careful planning.

Studies of the physical characteristics of aggregation and spawning sites, as well as the biology of species, are needed to place the aggregation within physical and ecological contexts. Bathymetric surveys and other mapping of aggregation sites, along with the use of "control sites" without aggregations, are important for determining what characterizes used sites. Similarly, physical measurements (temperature, currents, etc.) need to be continued outside of the times and locations of aggregation and spawning, and ideally year-round, to be able to detect conditions that might be specifically associated with locations and seasons of spawning.

In most places we learn of aggregations through local ecological knowledge (LEK), usually from fishing communities, while LEK is an important driver of protective actions in some parts of the world, especially in some Pacific Island countries. In Chap. 10 Rick Hamilton, Yvonne Sadovy de Mitcheson and Alfonso Aguilar-Perera examine the role of LEK in documenting and managing aggregations. Many that are revealed through LEK have persisted for decades at subsistence fishing levels, only to become reduced after the addition of commercial pressure. Carefully collected LEK provides highly useful information on historical and current status of aggregation fisheries and provides valuable perspectives on attitudes and understanding of local marine resources. However, it is important to collect such information carefully and systematically and to validate it independently, whenever possible, before integration into management planning.

Although LEK can be very detailed, it is the combination of LEK and science that is most likely to lead to effective fish spawning aggregation management. The interview process for LEK can provide excellent opportunities for awareness and information exchange between fishers and researchers, but interviewers need to be knowledgeable and well-trained. Scientists can play a valuable role by sharing their knowledge and answering questions fishers may have. Scientists can also provide regional and global perspectives on fish spawning aggregations (FSAs), something even the most knowledgeable fishers typically lack; many local fishery departments and educational bodies likewise have little knowledge of this aspect of the marine ecosystem and local fisheries. Often, informing fishers of the critical biological role that FSAs play and the ease with which they have been destroyed in other places is the catalyst for communities to initiate the process of managing their own aggregations.

The challenge of management is tackled by Martin Russell, Brian Luckhurst and Ken Lindeman in Chap. 11. Management is particularly difficult since most aggregations, if fished, are already likely to be highly impacted, usually declining, in need of recovery and subject to political considerations, as they are often major components of annual catches of fishers. On the other hand, if aggregations are not fished, there is unlikely to be management in place. Prompt management action is often needed even when the desired information is not available. We understand enough about aggregating species and their exploitation to make meaningful decisions and take a precautionary approach to their management. There is, however, no 'one-size-fits-all' for managing spawning aggregations. Political will is essential for driving desired management outcomes supported with good scientific data. It can be prompted by pressure from informed non-governmental organizations (NGOs) and a public who can bring considerable weight to bear on governments to act responsibly and assist in raising understanding and awareness of the issues. Enforcement will continue to be one of the weakest links in the chain due to limited funds and manpower, and innovative solutions are badly needed because poaching is a major problem in most countries. A well-planned monitoring programme is essential to evaluate the effects of management and for adaptive management.

Finally, we included a chapter on species or family-level summaries. Aware of the growing interest in aggregating species on the one hand and the often-poor access to species-specific information, we invited experts to compile overviews of the biology, fishery and other relevant information from the published and unpublished literature for 23 species, and their lesser studied relatives, accounts. The accounts cover most of the diversity of aggregating species but differ in their focus according to work that has been done. It is striking to see how little information is available for many species that form important components of reef-associated fisheries. This chapter highlights the need for further general study on many species and the need for study in multiple locations.

13.3 Need for Scientific and Socioeconomic Understanding of Aggregations and Aggregating Species in Relation to Management

There is an ongoing need to expand both geographic and species coverage of aggregating species, improve basic scientific understanding, test hypotheses, and better appreciate the human components of associated fisheries, including how best to manage and monitor in the long term. While there is considerable interest within the scientific and conservation communities in spawning aggregations, there persists an overall lack of appreciation of the importance of accurate scientific and fishery information for setting up credible management or conservation initiatives and assessing the effectiveness of protective action. It is critically important for long term success in management to demonstrate outcomes to the fishing community, general public and politicians. Interest in the issues and confidence in progress must be maintained for long-term success. There are likely to be pressures to open previously closed aggregations to fishing for short-term economic boosts (as happened briefly in the

Bahamas after the economic turndown of 2008), or when some recovery is apparent. Ultimately it seems likely that aggregations should be permanently closed to commercial fishing so as to persist as sources of productive fisheries (through the eggs produced) rather than being targets of fishing. The wider issue of the economic benefits of healthy aggregations for the non-aggregation fishery, therefore, should become the major consideration in fisheries assessment and planning, rather than exploitation of aggregations themselves.

One major reason for regularly surveying aggregations is to determine with confidence whether aggregation sizes have changed or remain stable as a result of management. The very real challenge of collecting rigorous information from aggregations is widely unappreciated; it is costly and requires well-trained, committed and informed workers to regularly and consistently conduct underwater surveys on aggregations or to meaningfully sample associated fisheries. Often diving must be done under difficult physical conditions. Large concentrated groups of fish that briefly form at deep sites and move around are difficult to count with precision/accuracy and such work can be frustrating, and even dangerous. Moreover, it must be done over certain periods of the year to be consistent and provide comparable data on fish numbers and horizontal extent of aggregations. For transient aggregations, for example, several days may need to be monitored because fish numbers change daily and there may be some variability in the day of the lunar month that peak numbers appear in relation to lunar cycle. As we learn more about how to work with aggregations and aggregating species, the methodology is improving and more workers are gaining the necessary experience and training for conducting surveys and research. Nonetheless, such monitoring will continue to be a somewhat specialized activity and frequent updating and retraining should be part of any serious long-term monitoring programme to ensure consistency and rigour. While local community involvement is often a critically important aspect of such programmes and should continue to be so, meaningful underwater monitoring is typically not streamlined enough to exclude the involvement of expert oversight.

To illustrate some of the challenges, two specific examples of monitoring of transient aggregating species will serve. In Palau, the Ebiil spawning site has been studied on and off since the mid-1990s, most intensively in the 2000s. However, for many of these years, the full extent of the aggregation site was not known because the area was not surveyed or marked out. This resulted for many years in the exclusion of a deeper zone where the third of the three species initially reported, the camouflage grouper, was found to spawn. The earlier exclusion of this deeper area from regular monitoring led to a false conclusion that it had been entirely lost from the site. Following an intensive season and mapping of the site with expert input, a robust monitoring protocol was introduced and documented to allow for future replication (Palau Conservation Society 2010). In Belize, many of the Nassau grouper aggregation sites have been surveyed since 2003. However, dive teams change over time, retraining in the original methodology is not routinely conducted, details of sampling not fully documented (such as areas surveyed and how counts were conducted), and the horizontal extent of some of the surveyed sites are not known. This situation has led to a review of the procedures since the maximum counts recorded were, in some cases, the maximum number of fish actually seen rather than the peak numbers at the site for the year as determined by a clear and consistent counting protocol. The reality is that the Nassau grouper is exceptionally difficult to survey because the aggregation changes from a two-dimensional to a three-dimensional form according to time of day and fish numbers. Therefore, estimating numbers is a challenge akin to counting birds in large moving flocks and needs very careful planning.

Overall, there is a surprising lack of new information being generated for those species already known to aggregate, new localities of aggregations or of additional species suspected to have spawning aggregations (see sections B and C in Appendix). Aggregations reported in fisher interviews should be validated to confirm timing, species and locations. Nonetheless, the locations of aggregations of commercially important species, unless already widely known, should generally be treated as privileged data for use by managers and other concerned parties; this respects the users of the site and avoids the possibility of revealing the location that could result in additional fishing pressure. Detailed information on aggregations assists in management and conservation planning and enables enforcement to be finely tuned to the times of peak numbers, saving on manpower and other costs.

While observational work, such as underwater surveys, is important, there is also a need for specimen and fishery-based research to determine catch rates, sexual state, fish sizes, and fisher perspectives, etc. Much can be achieved based on very simple and inexpensive methodologies and will be very useful if samples are representative and large enough to have scientific validity. Moreover, much can be done non-destructively, a major concern for protected or threatened species. As one example, while sagittal otoliths are typically used to age tropical fishes, work on the threatened goliath grouper has shown that clipped dorsal fins can be used for ageing, obviating the need to kill fish. For reproductive work on spawning seasonality and timing, simple GSI methods can be applied (Chap. 9), while live fish can be cannulated for more precise information on timing of reproduction in any one lunar cycle. Simple training can enable workers to obtain scientifically robust information easily and inexpensively from individual fish obtained from fishers or markets.

In areas where fisheries are based on aggregations or on fishes migrating to aggregation sites, efforts should be made to work with fishers to gather data. Positive interactions with fishers, including buying specimens from them at premium prices often at the fishing site or paying to gather data (measure fish, determine sex, weigh gonads), take gonads for histology, etc. can allow for messages of conservation and information on species to be exchanged. Often seeing that the fish they are catching are the egg-bearing females, or discussion of decreasing fish sizes and numbers of catches over time, may have an eye-opening effect on fishers, which can be amplified by the simple presence of scientists trying to ascertain how to maintain fish populations for the eventual benefit of the fishery. This is particularly true if older individuals who might be concerned about what they are leaving behind and who have seen big changes in their lifetimes are involved. These fishers can be a particularly important source of understanding and inspiration for younger fishers with less experience.

Given how little is understood of aggregating species, hopes should not be pinned on radical or alternative solutions to overfishing, such as restocking or mariculture (i.e. farming of marine species). Both depend on proper management being in place and that usually involves some control on fishing effort. Moreover, restocking may not be effective for the recovery of much depleted aggregating species if information on aggregation location is traditionally transmitted. Moreover, while mariculture is an additional means of production, it is not a solution to overfishing unless directly linked i.e. unless 1 tonne produced by mariculture reduces fishing pressure by 1 tonne. To our knowledge this does not occur in practice.

13.4 Key Scientific Data Gaps for Aggregating Species

A number of fundamental gaps in knowledge impede our ability to effectively manage aggregation-based fisheries and to protect threatened and endangered species that form aggregations. From the work presented in this book, we identify 13 key areas that call for greater attention and suggest possible means and approaches for addressing them.

- Hypothesis testing on why aggregation sites and times are important. Such questions have bearing on their long-term persistence, the need to manage them, the risks of losing them, etc.
- There remains a need for empirical data on the effects of reduced aggregation sizes on reproductive activity to determine whether minimum threshold sizes exist below which spawning ceases in small aggregations. Related to this is the question of whether extirpated aggregations can reform after protection. Also relevant is a better understanding of how population abundance and density levels affect aggregation formation and mating behaviour.
- There is a dearth of information on the biological and socioeconomic effects of different management policies on aggregations and the fisheries of aggregating species at both aggregating and non-aggregating times.
- More effort is needed to directly survey and monitor known aggregations, since fisheries data may not reflect the status of aggregations (i.e. due to hyperstability). New technologies to survey aggregations at locations where diver surveys may or may not be possible need careful ground-truthing. Hydroacoustic surveys are perhaps the best example of the need for independent ground-truthing. New video and photographic techniques, with sophisticated image-processing, hold much promise for quantifying aggregations. Methods for counting fish in large three-dimensional groupings, such as some snappers and the Nassau grouper, need to be developed.
- Justifications for the management or protection of aggregation sites are needed, based on economic analyses of aggregation fisheries, alternative use activities, assessments of the importance of their fishery and ecosystem value, and consideration of aggregation sites for inclusion as important biodiversity areas, ecosystem-based management, and ecotourism, etc. The long-term economic benefits of the complete protection of aggregations, as sources of young for exploitation in the non-reproductive season, need serious consideration.

- The feasibility of using aggregations as' indicators' of fishery condition should be explored. This would highlight the importance of aggregations for fisheries and in their biological and ecosystem roles.
- The benefits of and need for managing and protecting aggregations should be communicated to stakeholder groups, policy makers, and the general public so that consensus support and compliance for policies is obtained. Indeed, more attention to education on the marine environment and marine resources is needed at all levels, through to tertiary sectors.
- General discovery phase work would expand both the species known to have spawning aggregations and improve understanding of the broad geographic extent of aggregations for single species, as well as possible intraspecific variations associated with habitat, fish densities or other factors. It is important to apply rigorous protocols for surveys and to carefully document methods used sufficiently rigorously to allow for replication.
- Validation of existing traditional ecological knowledge and fisher information is also needed in many areas.
- Detailed data on the environmental conditions at aggregation sites would allow for geographic comparisons of conditions across geographic areas and identify correlations of abiotic factors with aggregation and spawning. Instrumentation of sites is important, even if this involves nothing more than recording thermographs. Mapping of sites allows for long term monitoring of status and movement.
- Knowledge of the dynamics of the early life history of planktonic larvae of reef fishes would aid understanding of the effects of temporally and spatially concentrated spawning on recruitment and determination of population parameters that relate to the survival of larvae, dispersal distances after spawning and the ocean environments where larvae grow prior to settlement.
- The physiology associated with aggregation spawning, such as studies on the hypothesis of possible maternal benefits from aggregation could provide insights into why it occurs. It should be straightforward to develop testable hypotheses regarding egg quality, hormone levels in adults, and other aspects that could be comparable among transient, resident and non-aggregating species.
- Determination of catchment areas and genetic population structuring would be valuable for examining adult connectivity and spatial scales for conservation and management initiatives. The determination of dispersal catchments of propagules from spawning sites provides the counterpoint perspective on population structure and connectivity, and disparities between migration catchments and dispersal catchments will have great importance in shaping genetic population structure.

To examine some of these questions, smaller fishes, particularly those with resident aggregations that are unexploited or important in subsistence fisheries, provide many opportunities to study the biology and oceanography of aggregations, but such studies are rarely done. Instead, most effort is going to larger, commercially important species, which are certainly important, but often much rarer and more difficult to work with. Work on both types is needed.

To move forward in our understanding of fish aggregations, it would be helpful to study more model systems where individuals can be tracked and their reproductive success output recorded. For example, studies of groupers have been very informative about travel costs and individual variation in behaviour, and studies of Atlantic cod, *Gadus morhua*, in large tanks have been helpful for measuring individual differences in behaviour and reproductive success.

Some questions can be addressed using a comparative approach by focusing on better-documented species exposed to different levels of fishing pressure and policies in different areas. A good example would be the Nassau grouper for which many aggregations are heavily fished (many extirpated), but some are relatively intact. Comparing the timing, density, and behaviour of fish in exploited versus protected or unexploited spawning aggregations would be instructive, as would examining changes in spawning behaviour when traditional aggregations become depleted or recover from over-fishing, or when there are marked changes in population abundances or densities. Such studies would help elucidate the possible costs and benefits of spawning in aggregations, and the implications of severely reduced aggregation sizes. They might also shed light on the specific management needs and effectiveness of management actions taken to protect species.

Quality of data and analysis is variable and available from many different sources. Study of aggregating species is reported in a wide range of outputs, ranging from publications by non-governmental organizations, governments, university theses, intergovernmental organizations, consultancies, private companies, and individuals, to biologists and others in the peer-reviewed scientific and other literature. While much of this information is solid, the quality of non peer-reviewed 'grey' material is variable, may not be subject to quality-control, and should be treated carefully (Corlett 2011). Whenever possible, original literature, rather than secondary citations, should be referred to. On the other hand, we do not need to limit our information sources to the classic channels. Suitably validated, as for example for fisher interviews, much can be learned from reports by recreational divers, dive guides and others who spend a lot of time in the water.

13.5 Protection and Management of Aggregations

The negative impacts of fishing appear to be substantially greater on transient than on resident aggregating species and Chapter one pointed out differences between resident and transient aggregations that have important conservation and management implications. The record shows that transient aggregators tend to need more management attention than resident aggregators, for obvious reasons. Moreover, when considering the establishment of MPAs to protect spawning populations, multiple widespread resident aggregations can be protected by setting aside some percentage of an overall area, whereas MPAs to protect transient aggregations must be either very large or very well-placed, because the latter are relatively fewer and sparser. Catchment areas become a critical factor in transient aggregators if the scale of management efforts is to be appropriately determined, and other links may be important, such as to possible nursery areas or migration routes. Implementation of MPAs without any other management measure is rarely likely to be sufficient due to their typically small sizes and because they do not address fishing effort, the key cause of overfishing. Once designated, MPAs need enforcement, aggregation sites need to be monitored to evaluate management effectiveness and fishing activity may well need to be controlled on the species at non-aggregation times and places. While we cannot know the location of all spawning aggregations, and indeed are better off not knowing them all, it is becoming clear that many of the more vulnerable transient aggregations occur in outer reef shelf/drop-off habitats which merit far greater consideration for management and incorporation into MPAs (Sadovy de Mitcheson et al. 2008) than they have attracted to date. Conversely, confirming that FSAs form does not necessarily mean that they all have to be protected, and it is not necessary to know the location of all aggregations in an area if seasonal protection is the most effective approach. FSA management needs to be done on a case-by-case basis and requires broad cross-sector consultations and discussion.

Where sufficiently detailed studies of fisheries management have been carried out, the factors important for success are evident. They include enforcement, consultation, perception, monitoring, and planning that takes into account the realities of the biology of target species. For example, for long-lived species a decade or more might be needed before benefits become apparent. In the US Virgin Islands, it is clear from regular monitoring that sizes and numbers of red hind are increasing after more than a decade of protection. However not all fishers are convinced that management has been successful or that they have benefited and there is a need to understand why. In Florida, goliath grouper numbers, at least among young fish, is increasing following what appears to be effective protection, although time is needed to assess the impact on the population as a whole. In Belize, a strong cross-sectoral working group on grouper spawning aggregations successfully rallied broad support and raised considerable interest within the fishing community and other public sectors for Nassau grouper protection.

A major ongoing challenge is with enforcement, and failure in this area not only undermines management but can also lead to successively more measures being needed as the fishery continues to decline. For example in Cuba recognition of overfishing led to an increasing set of management measures being implemented as the fishery declined and recovery was not forthcoming (Claro et al. 2009). On the other hand, multiple measures can make management outcomes difficult to evaluate, and in Australia led to a weakening of aggregation protection for the leopard coralgrouper because other methods (licensing and minimum sizes) were thought, without scientific evidence, to be sufficient. The establishment of regulations or designation of MPAs are often seen as ends in themselves and yet it is only when these are properly effected that improvements can occur. Enforcement is not only expensive but must be a long-term commitment, given the prevalence of poaching, whether by government or by NGOs, to promote it. In Belize, for example, some NGOs directly assist the government financially to ensure that patrols of Nassau grouper aggregations can continue. Other approaches can aid enforcement such as the requirement of keeping skin flaps on fillets for species identification (Belize), a vessel monitoring system (Australia), sales bans combined with a public campaign to discourage people eating protected species (Bahamas) or stricter penalties and a fully engaged judiciary that commits to follow through on violations of the law. Detailed information on aggregation timing may assist in focusing enforcement needs over very short periods.

Another major challenge is to establish long-term monitoring programmes to demonstrate the outcomes of management, and adjust measures as necessary. This requires robust sampling protocols that consistently produce meaningful numbers and long-term commitment to fund these activities as well as to ensure that information gets back to stakeholders. Monitoring that involves teams of workers means a presence on the aggregation site, is excellent for training and raising awareness and provides information that can be discussed and used for management planning. Long-term monitoring does not have to be annually based but, done regularly and expertly, serves to increase the understanding of the effects of management and maintain interest.

The bringing together of science and LEK is critically important for placing observations in a broad context and for framing management in a way that is both acceptable and meaningful. Where LEK is well-developed and an important cultural driver, documenting LEK regarding the presence of FSAs or changes in landings from FSAs is important for reinforcing what people have already experienced. A biological understanding of aggregations can help communities make sense of what they have directly witnessed, or learned from tradition. LEK is far more likely to be taken seriously by conservationists, scientists and managers, and be useful for management if validated (for example for specific timing of spawning), and scientists are more likely to be taken seriously if what they say harmonizes with local understanding. Attention attracted by the plight of aggregations can even lead to much broader initiatives and act as a catalyst for communities to come together to discuss marine management and conservation issues in general. Locally focused films about aggregations have attracted attention in many countries and are a valuable educational tool.

Looking ahead, management success brings with it new challenges. When management leads to better catches or more fish being seen by divers on aggregations, this will almost inevitably lead to pressure to reopen closed aggregation-fisheries. This is an area that needs considerable forward planning and discussion to establish an overall philosophy with respect to key reproductive gatherings, taking account of the species through to the ecosystem perspective. Given increasing pressure on fisheries resources and the critical role that aggregations play in maintaining vulnerable fished species, it is likely that the greatest benefit would be to close them to fishing altogether (as long as there is a non-aggregation fishery involved). We believe, based on the weight of evidence to date, that, unless aggregations are fished for limited subsistence use only, they are unlikely to be sustained without considerable management intervention.

Clearly, we need to shift our perspective, recognizing that the greatest value of aggregations is in the young they produce; the 'interest' derived from the spawning adults, the 'capital'. Protected aggregations may mean much more productive fisheries overall and year-round. Education, especially to young fishers, conservation workers and budding fishery staff, is needed to communicate the notions of 'shifting baseline' (Pauly 1995), 'hyperstability', 'ecosystem-based management' and 'adaptive management'. Maybe it is time for a major paradigm shift, from seeing spawning aggregations as special opportunities for fishing, to understanding them as particularly important times at which fish need to be protected, ensure that there will continue to be future generations as well as guarantee opportunities in future to witness and study these marvellous biological events.

References

Claro R, Sadovy de Mitcheson Y, Lindeman K, Garcia-Cagide C, Garcia-Cagide AR (2009) Historical analysis of Cuban commercial fishing effort and the effects of management interventions on important reef fishes from 1960–2005. Fish Res 99(1):7–16

Colin PL, Sadovy YJ, Domeier ML (2003) Manual for the study and conservation of reef fish spawning aggregations. Society for the Conservation of Reef Fish Aggregations. Special publication No. 1 (Version 1.0), pp 1–98, www.SCRFA.org

Corlett RT (2011) Trouble with the gray literature. Biotropica 43(1):3–5

Craig MT, Sadovy de Mitcheson YJ, Heemstra PC (2011) Groupers of the world: a field and market guide. Grahamstown: NISC (Pty) Ltd, 424 pp

Heyman WD, Graham RT, Kjerfve B, Johannes RE (2001) Whale sharks *Rhincodon typus* aggregate to feed on fish spawn in Belize. Mar Ecol Prog Ser 215:275–282

Johannes RE (1981) Words of the lagoon fishing marine lore in the Palau district of Micronesia. University of California Press, Los Angeles

Lovatelli A, Holthus PF (2008) Capture-based aquaculture. Global review. FAO Fisheries technical paper. No. 508. FAO, Rome, pp 5–39. (Also available at: http://www.fao.org/docrep/011/i0254e/i0254e00.htm)

Olsen DA, LaPlace JA (1979) A study of Virgin Islands grouper fishery based on a breeding aggregation. Proc Gulf Caribb Fish Inst 31:130–144

Palau Conservation Society (2010) Enhanced monitoring of grouper spawning aggregation at Ebiil channel: final technical report, Palau Conservation Society and Society for the Conservation of Reef Fish aggregations (www.SCRFA.org)

Pauly D (1995) Anecdotes and the shifting baseline syndrome of fisheries. Trends Ecol Evol 10(10):430

Randall JE, Randall HA (1963) The spawning and early development of the Atlantic parrot fish, *Sparisoma rubripinne*, with notes on other scarid and labrid fishes. Zoologica 48:49–60

Rowe S, Hutchings JA (2003) Mating systems and the conservation of commercially exploited marine fish. Trends Ecol Evol 18:567–572

Sadovy Y (1993) The Nassau grouper, endangered or just unlucky? Reef Encounter 13:10–12

Sadovy de Mitcheson Y, Cornish A, Domeier M, Colin P, Russell M, Lindeman K (2008) A global baseline for spawning aggregations of reef fishes. Conserv Biol 22(5):1233–1244

Sale PF, Cowen RK, Danilowicz BS, Jones GP, Kritzer JP, Lindeman KC, Planes S, Polunin NVC, Russ GR, Sadovy YJ, Steneck RS (2005) Critical science gaps impede use of no-take fishery reserves. Trends Ecol Evol 20(2):74–80

Sinclair M (1988) Marine populations Washington Sea Grant. University of Washington Press, Seattle, Washington, USA

Smith CL (1972) A spawning aggregation of Nassau grouper, *Epinephelus striatus* (Bloch). Trans Am Fish Soc 101:257–261

Vilaro Diaz DJ (1884) Corrida y arribazon de algunos peces cubanos. Manuel Gomez de la Maza, La Habana

Wilkinson C (2008) Status of coral reefs of the world: 2008. Global Coral Reef Monitoring Network, Townsville

Abbreviations and Acronyms

AUV	Autonomous Underwater Vehicle
CBA	Capture-based aquaculture
CBL	Coastal boundary layer
CITES	Convention on International Trade in Endangered Species
CPUE	Catch Per Unit of Effort
EBM	Ecosystem-based management
EP	Eastern Pacific
FAO	Food and Agriculture Organisation (United Nations)
FL	Fork length
FRS	Family Radio Spectrum
FSA	Fish spawning aggregation
GBR	Great Barrier Reef Australia
GIS	Geographic information system
GPS	Global Positioning System
GSI	Gonado-somatic Index
ITQ	Individual Transferable Quota
IUCN	International Union for Conservation of Nature
IUU	Illegal unreported and unregulated
IWP	Indo-west Pacific
LEK	Local Ecological Knowledge
LIDAR	Light Detection and Ranging
LRFT	Live Reef Fish Trade
MPA	Marine Protected Area
NGO	Non-Governmental Organization
NMEA	National Marine Electronic Association
NOAA	National Oceanographic and Atmospheric Administration
RA	Resident Aggregation
ROV	Remote Operating Vehicle
SCRFA	Society for the Conservation of Reef Fish Aggregations

Y. Sadovy de Mitcheson and P.L. Colin (eds.), *Reef Fish Spawning Aggregations: Biology, Research and Management*, Fish & Fisheries Series 35,
DOI 10.1007/978-94-007-1980-4, © Springer Science+Business Media B.V. 2012

SCUBA	Self-Contained Underwater Breathing Apparatus
SL	Standard length
SST	Sea Surface Temperature
TA	Transient Aggregation
TAC	Total Allowable Catch
TEK	Traditional Ecological Knowledge
TL	Total length
TWA	Tropical west Atlantic
UVC	Underwater Visual Census
VMS	Vessel Monitoring System

Glossary

Advection in the ocean this is the physical transport of water and contained materials or properties which is primarily horizontal; generally considered to move "place to place", rather than up and down

Allee effect an effect of population density on population growth, by which there is a fall in reproductive rate at very low population densities and a positive relationship between population density and the reproduction and survival of individuals (Allee 1931)

Aquaculture FAO defines aquaculture as "the farming of aquatic organisms, including fish, molluscs, crustaceans and aquatic plants. Farming implies some form of intervention in the rearing process to enhance production, such as regular stocking, feeding, protection from predators, etc. Farming also implies individual or corporate ownership of the stock being cultivated. For statistical purposes, aquatic organisms which are harvested by an individual or corporate body which has owned them throughout their rearing period contribute to aquaculture, while aquatic organisms which are exploitable by the public as a common property resource, with or without appropriate licences, are the harvest of fisheries." *Capture-based aquaculture* (CBA) is also a significant supplier of aquacultured fish that has a heavy dependence on wild-sourced organisms. Marine aquaculture is known as *mariculture*.

Amplitude, tidal the distance between high and low tides

Barotrauma refers to an overinflated swim bladder which occurs as a result of fish being brought up from depth. Typically, a fish's air bladder inflates and deflates to enable the fish to suspend at a particular depth: if a fish is brought up from deep water, there is insufficient time for the air bladder to naturally get rid of air and so the sudden release of pressure experienced in shallower waters results in an overinflated air bladder that presses on vital internal organs, and the fish's stomach may protrude from its gullet. Serious damage is typically caused often leading to death unless the pressure is rapidly, artificially, released

Benthic associated with living on or near the sea-bottom, irrespective of the depth of the sea

Y. Sadovy de Mitcheson and P.L. Colin (eds.), *Reef Fish Spawning Aggregations: Biology, Research and Management*, Fish & Fisheries Series 35,
DOI 10.1007/978-94-007-1980-4, © Springer Science+Business Media B.V. 2012

Bipartite a two-part life cycle typical of many fish species whereby early life is planktonic, followed by settlement into the juvenile/adult habitat when a transformation (body form, feeding, behaviour) takes place from the pelagic larval from to the juvenile/adult form

Capital breeding reliance on stored energy for oocyte production due to the need to store resources to cover the costs of transport associated with migratory episodes and the restriction of reproduction to a limited number of episodes during the year (Warner 1995)

Catchability the extent to which a stock is susceptible to fishing; quantitatively, the proportion of the stock removed by a defined unit of fishing effort. Catchability is usually an inverse function of stock biomass

Catchment area At the largest spatial scale the catchment area encompasses the sum of home ranges and migration routes of a local spawning population that uses a specific aggregation site during the annual reproductive cycle. It refers to the sum of individuals that use the aggregation site under consideration and does not refer to individuals or species. The *Functional Migration Area* differs from the Catchment Area in that the FMA is a subset of the Catchment area but encompasses the greater majority of the spawning population and the specific habitats they utilize during migration

Capture-based aquaculture (CBA) is the practice of collecting "seeds" (see below) from the wild from early life history stages to adults and subsequent growing-out them in captivity to marketable size, using *aquaculture* techniques. This definition can clearly distinguish CBA from HBA (hatchery-based aquaculture), which is a practice of producing and using "seeds" from hatcheries through manipulation of adult maturation and reproduction and larval and juvenile rearing. "Seeds" are the aquatic organisms used to farm (i.e. grow-out) in captivity for varying times; these organisms can be captured and collected from the wild (e.g. for CBA) or hatched in hatcheries (e.g. HBA). These organisms cover a wide range of life history stages, from larvae to juveniles to adults, defined on the basis of morphology, including size, and sexual maturation stage (Lovatelli and Holthus 2008)

Ciguatera a condition brought about by eating fish that contain ciguatoxin and associated with gastrointestinal and neurological problems

CITES Convention on International Trade in Endangered Species of Wild Fauna and Flora which regulates the international trade to ensure that it is sustainable and does not threaten traded species with extinction (www.CITES.org)

Cohort a group of fish all of the same age belonging to the same *stock*

Connectivity the quality or condition of being connected. In a fish and fisheries context this refers to "ecological connectivity", where populations are able to sustain themselves through dispersive *propagules* or other means across the range of the population and "genetic connectivity", where a population retains its genetic integrity through dispersal of genetic material from physically isolated populations at suitable intervals

Convection in the oceans this is the physical transport of water or air and contained materials or properties which is primarily vertical, largely due to differences in density

Core the definition of core spawning area depends upon the mode of reproduction of individual species. For species that establish spawning territories and pair-spawn at aggregation sites the core area is defined as the area which consistently has the highest densities of aggregating. In large aggregations multiple core spawning sites may exist. For those species that spawn in large groups, the core area is the location where the majority of spawning adults ascend into the water column and spawn and this core area can change within the courtship arena even within a single evening (Chap. 2)

Courtship arena this is where males and females begin to interact during the specific reproductive period or lunar phase associated with spawning

Demersal dwelling at or near the bottom of a body of water

Diffusion the dispersal of particles or propagules by random movements of those particles or *propagules*. Can be either physically or biologically driven

Dispersal the outward movement of items away from a source

Dispersion the act of dispersing

Ecosystem-based management (EBM) an environmental management approach that recognizes the full array of interactions within an ecosystem, including humans, rather than considering single issues, species, or ecosystem services in isolation

Eddy a circular current of water or air moving contrary to the direction of and often along the edge of a larger current

Entrainment to establish and carry an object within a moving volume of water

Far field in oceanography the distance more than 1–2 wave lengths distant from a point or object

Fishing effort the total fishing gear in use for a specified period of time: when two or more kinds of gear are used they must be adjusted to some standard type for comparison

Fork length refers to the length of a fish measured from the tip of the snout to the centre of the caudal fin if this is of forked form

Functional migration area (FMA) this includes migration pathways, spatial and temporal habitat use during the spawning season, all intra- and interspecific interactions and predator-prey dynamics that occur within the *catchment area* during the spawning season from the moment the adults depart their home ranges until the time they return The FMA also takes into account the transfer of energy resulting from feeding, defaecation and release of propagules at the spawning site (Nemeth 2009)

Gamete reproductive cell, sperm in males and oocytes (eggs) in females

Gonadosomatic index (GSI) gonado-somatic index is a way to measure the relative size of the gonad to the somatic (body) weight. Usually GSI is depicted by the equation: GSI = 100 × gonad weight over body weight (minus gonad weight). It is a useful and relatively easy measure of assessing reproductive seasonality because GSI peaks just prior to spawning

Gonochorism separation of the sexes in different individuals, as opposed to *hermaphroditism*

Group-spawning spawning that occurs in groups of individuals that break out periodically from an aggregation containing much larger numbers of individuals.

Typically group spawns appear to comprise a single, possibly a few, females, and a larger number of following males with the production of pelagic eggs. *Sperm competition* is likely to be involved and group-spawning males typically have higher ripe *GSI* than pair-spawning males

Gyre a ring-like system of ocean currents normally rotating in an anti-cyclonic (clockwise) direction in the northern hemisphere

Haremic spawning a mating system that consists of a male that spawns exclusively with a group of females that he defends directly from other males or which utilize territory, such as feeding territory, that he defends. Spawning is in pairs

Hermaphroditism refers to individuals who have both unequivocal ovarian tissue and testicular elements that are reproductively active either simultaneously or sequentially. This is considered to be functional hermaphroditism. Sequential hermaphroditism can take two forms; *protandry* in which an organism is born as a male, and then changes sex to a female, and *protogyny* in which the organism starts as a female, and then changes sex to a male. In protogynous species, there are sometimes two male forms; an initial phase male and a terminal phase male. *Initial phase* males tend to resemble females and spawn in groups with other females and are not territorial. *Terminal phase* males are territorial and have a distinctively bright colouration. Individuals are born as males or females, but if they are born males, they are not born as terminal phase males. Females and initial phase males can become terminal phase males. Usually, the most dominant female or initial phase male replaces a terminal phase male by sex change, when those males die or abandon the group

Hyperstability a phenomenon in which an observed index of stock abundance (e.g. *catch per unit of effort*, or CPUE) remains stable while the abundance (population size) of the stock in question is actually declining (Fig. 8.6)

Income breeding reliance on concurrent intake: whereby daily reproduction is directly subsidized by continued feeding as would be expected for many resident spawning species (Warner 1995)

Indicators of spawning (direct and indirect) three criteria can verify directly that fish are gathering for the purpose of spawning; (1) undisputed spawning observations, (2) females with hydrated eggs and (3) presence of post-ovulatory follicles in the ovaries of aggregating females. A fourth means of directly documenting the presence of a spawning aggregation is added in Chap. 1: (4) identification of very early stage eggs and larvae, collected for example by plankton tows, that can be positively associated with the aggregating species. Indirect signs can be used for documenting new aggregations for species already known to form spawning aggregations. Indirect signs can include behaviours or colour patterns, if these are demonstrably known to be associated only with spawning, as well as *gonadosomatic index* (see GSI) data or the presence of swollen abdomens (indicating the presence of hydrated eggs) in a large percentage of the aggregated individuals. In the absence of witnessing the spawning event, it is not realistically possible to gather enough information to document spawning without sampling ovaries or larvae (Colin et al. 2003, Chap. 1)

Initial phase see *Hermaphroditism*

IUCN International Union for the Conservation of Nature – the largest global conservation non-governmental organization, which among other things, produces the Red List of threatened species (www.iucnredlist.org)

Kernel, dispersal a dispersal curve (a probability density function) of *propagules* formed from a spawning event at an earlier time

Lagrangian examination of a flow field where the track of a specific parcel or particle is observed as it moves through space and time. A current following drifter is an example of Lagrangian tracking

Live reef fish trade the (largely) international trade in live reef fish for the restaurant and live fish markets. The trade tends to focus on high value, luxury food fish, mainly groupers but including some snappers and wrasses. The trade is largely to satisfy the Chinese consumer market and involves a relatively low diversity of species, a few of which can be raised by mariculture. Wild fish supply a large portion of the trade (Sadovy et al. 2003)

Local ecological knowledge (LEK) refers to the cumulative knowledge of groups' or individuals' practices, experiences and beliefs about their natural environment. LEK contains empirical and conceptual aspects and is passed down over successive generations or intra-generationally and is a dynamic state of knowledge, kept alive in an oral form which makes it extremely fluid and flexible

Mariculture see *aquaculture*

Marine tenure locally specified entitlements to marine territories and resources claimed and exercised by the 'guardians' of those territories and resources

Mass spawning spawning (i.e. gamete release) by multiple males and females as a single mass

Mesoscale "medium-scale" oceanographic features, usually ascribed to rotating eddies that are on the order of 50–200 km across. Depending on their direction of rotation and hemisphere of occurrence, may be cold core or warm core

Metamorphosis abrupt physical change in the transformation from the larval to the adult condition

Near field in oceanography a distance usually 1 wave length (not exactly specified) distant from a point or object

Nekton ocean life which is capable of swimming sufficiently fast to make progress against ocean currents

Overfishing (biological) when *fishing effort* exceeds the ability of the targeted population to either replace itself (recruitment overfishing) or to produce maximum yield (growth overfishing)

Pair-spawning spawning that typically occurs between a single male and a single female; can produce either *pelagic* or *demersal* eggs. Occasionally the pair may be joined by a *sneak* spawner

Pelagic the zone of open water not close to the bottom, including areas both above the continental (and insular) shelves and deep ocean

Pelagic juvenile a developing life history stage, usually applied to fishes, which lives in open water (not benthic) but has acquired the morphology and many characteristics of the juvenile stage, but not colouration and other characters that are acquired once the stage has taken up residence on the bottom

Planktonic living in mid-water and not swimming sufficiently strongly to be able to make progress against ocean currents, essentially drifting wherever currents might carry them

Precautionary principle states that if an action or policy has a suspected risk of causing harm to the public or to the environment, in the absence of scientific consensus that the action or policy is harmful, the burden of proof that it is *not* harmful falls on those taking the action

Promontory literally a point of land projecting out into the water, described in this volume as a projection of reef or land whose limbs form an angle of less than 90°.

Propagule a life history stage of an organism that can effectively give rise to a new individual or individuals. Normally used for fishes in reference to early life history stages, including eggs and/or larvae

Protandry see *hermaphroditism*

Protogyny see *hermaphroditism*

Recruitment There are two widely applied uses of this term. (1) most commonly it refers to the amount of fish added to the exploitable stock each year due to growth and/or migration into the fishing area. For example, the number of fish that grow to become vulnerable to the fishing gear in 1 year would be the recruitment to the fishable population that year. This term is also used in referring to the number of fish from a year class reaching a certain age. For example, all fish reaching their second year would be age 2 recruits. (2) in fish ecology the term is sometimes used interchangeably with *settlement*

Red List see *IUCN*

Resident aggregation (RA) this type of spawning aggregation draws individuals from a relatively small and local area to an aggregation site, compared to a transient aggregation, following migration of a few hours or less and often lying within the home range of the participating individuals. RAs usually (1) occur at a specific time of day over numerous days, (2) last only a few hours or less, (3) occur daily over an often lengthy reproductive period of the year, and (4) can occur year round. A single day of spawning for an individual participating in a resident spawning aggregation represents a small fraction of that individual's annual reproductive effort

Restocking the translocation of adults, or release of juveniles, to create protected breeding populations that naturally rebuild stocks in the fishery. Restocking should be used only as a "last resort", such as in the case of threatened species, and not as a substitute for a precautionary management. However, restocking fish in marine waters has not yet led to documented long-term recovery of reproductive populations

Retention the ability to remain in an area, through some sort of biological or physical mechanism

Settlement the transition of a life history stage from the planktonic to benthic environment, as in marine fishes, usually at the end of the larval stage

Sexual selection a type of natural selection in which the sexes acquire distinct forms either because the members of one sex choose mates with particular features or because in the competition for mates among the members of one sex only those with certain traits succeed

Shear dispersion the differential movement of water or particles along a shear line.

Shifting baseline a term used to describe the way significant changes to a system are measured against previous baselines, which themselves may represent significant changes from the original state of the system (Pauly 1995)

Sneaking the situation where a male does not take part in courtship activities, but ries to surreptitiously release sperm at the site of *gamete* release after other fish have spawned, probably in hope of fertilizing some of the eggs. Similar to *streaking*

Spawning aggregation is a repeated concentration of conspecific marine animals, gathered for the purpose of spawning, that is predictable in time and space. The density/number of individuals participating in a spawning aggregation is at least four times that found outside the aggregation. The spawning aggregation results in a mass point source of offspring. Sometimes referred to as *FSA*

Sperm competition Sperm competition is competition between sperm of two or more males for the fertilization of an ovum. Species that have sperm competition are typically characterized by large male testes (and high reproductive *GSI*)

Sperm limitation a situation in which there may be insufficient males (sperm) to fertilize all available viable eggs (oocytes); this might be brought about by heavy fishing selectivity for males which great imbalances the reproductive sex ratio

Staging area an area where migration pathways begin to converge and certain aggregating fish species rest, feed or visit cleaning stations during the spawning season and in the vicinity of the spawning site

Standard length (SL) refers to the length of a fish measured from the tip of the snout to the posterior end of the last vertebra or to the posterior end of the mid-lateral portion of the hypural plate; SL excludes the length of the caudal fin

Stock a part of a fish population usually with a particular migration pattern, specific spawning grounds, and subject to a distinct fishery. A fish stock may be treated as a total or a spawning stock. Total stock refers to both juveniles and adults, either in numbers or by weight, while spawning stock refers to the numbers or weight of individuals which are old enough to reproduce. There may or may not be a specific genetic structure associated with a stock; extrinsic factors (immigration and emigration) are considered to be insignificant

Streaking similar to *sneaking*, but an individual rushes upward near simultaneously with a spawning rush by a pair or more of fish and releases gametes, usually sperm, at the site of gamete release by other fish. Often the streaking individual is attacked by spawning males

Sustainable fishery one in which the harvesting of the target species is conducted in such a way, and at a rate, that it does not threaten the long-term health of the stock (by exceeding the natural ability of the fished population to replace itself) and the ecosystem on which it depends, or does not inhibit recovery of the stock or the ecosystem if it has previously been reduced below appropriate levels

Swath width strictly speaking this refers to the strip of the Earth's surface from which data are collected by a satellite. The longitudinal extent of the swath is defined by the motion of the satellite with respect to the surface, whereas the swath width is measured perpendicularly to the longitudinal extent of the swath. In relation to underwater visual censuses, the longitudinal component is the transect while the total width surveyed about the transect is the swath.

Terminal phase see *Hermaphroditism*

Threatened species species assessed to be at risk of extinction if current practices, whatever they should be, continue and if they are not suitably managed for *sustainability*. Various criteria are used to assess conservation status, including level of threat to fish species at both national and international levels. The most widely used set of criteria and categories to assess conservations status are those of *IUCN* used in the *Red List*

Tidal jet an area of advecting water near the surface of lower density than underlying water that is driven by tidal currents, usually emanating from a channel or other narrow feature

Total allowable catch (TAC) this is the total regulated catch from a stock in a given time period, usually a year

Total length (TL) refers to the length from the tip of the snout to the tip of the longer lobe of the caudal fin, usually measured with the lobes compressed along the mid-line. It is a straight-line measure, not measured over the curve of the body

Traditional ecological knowledge (TEK) see LEK

Transect a linear survey which crosses an area where data are gathered on what is present along the transect. A transect may incorporate a width (*swath*) either side of the line where data are gathered, or be values present only along the line (line intercept)

Transient aggregation (TA) this type of spawning aggregation draws individuals from a relatively large area compared to a resident aggregation. Individuals must travel days or weeks to reach the aggregation site. TAs often (1) occur during a very specific portion of 1 to a few months of the year; (2) persist for a period of days or at most a few weeks and (3) do not occur year round. A single transient spawning aggregation may represent the total reproductive effort for participating individuals

Vorticity The rate of rotation of a fluid, such as water, and its measurement

References

Allee WC (1931) Animal aggregations: a study in general sociology. University of Chicago Press, Chicago

Colin PL, Sadovy YJ, Domeier ML (2003) Manual for the study and conservation of reef fish aggregations. Soc Conserv Reef Fish Aggreg Spec Publ 1:1–98, www.SCRFA.org

Lovatelli A, Holthus PF (eds) (2008) Capture-based aquaculture. Global overview. FAO fisheries technical paper. No. 508. FAO, Rome, 298p

Nemeth RS (2009) Dynamics of reef fish and decapod crustacean spawning aggregations: underlying mechanisms, habitat linkages and trophic interactions. In: Nagelkerken I (ed) Ecological interactions among tropical coastal ecosystems. Springer, Netherlands

Pauly D (1995) Anecdotes and the shifting baseline syndrome of fisheries. Trends Ecol Evol 10:430

Sadovy YJ, Donaldson TJ, Graham TR, McGilvray F, Muldoon GJ, Phillips MJ, Rimmer MA, Smith A, Yeeting B (2003) The live reef food fish trade while stocks last. Asian Development Bank, Manila

Warner RR (1995) Large mating aggregations and daily long-distance spawning migrations in the bluehead wrasse *Thalassoma bifasciatum*. Environ Biol Fish 44:337–345

Appendix: Species That Form Spawning Aggregations

Section A

Positive Records – Spawning aggregation first confirmed by at least one of three direct methods (positive spawning observations, hydrated eggs in females, post-ovulatory follicles in ovaries) as well as a minimum of 4 × increase (or more) in numbers.

Family	Location	Type	Pair Sp?	First record/other info
Genus and species				
Acanthuridae (planktonic)				
Acanthurus achilles	P	R	–	Sancho et al. (2000a)
Acanthurus bahianus	A	R	Y	Colin and Clavijo (1988)c
Acanthurus blochii	P	R	–	unpub (PLC)a
Acanthurus coeruleus	A	R	Y	Colin and Clavijo (1988)c
Acanthurus guttatus	P	R	Y	Craig (1998)
Acanthurus lineatus	P	R	–	Robertson (1983)a
Acanthurus nigrofuscus	P	R	–	Robertson (1983)c
Acanthurus nigroris	P	R	–	Sancho et al. (2000a)
Acanthurus thompsoni	P	R	–	Patrick L. Colin c
Acanthurus triostegus	P	R	–	Randall (1961a)a
Ctenochaetus striatus	P	R	Y	Robertson (1983)a
Ctenochaetus strigosus	P	R	–	Sancho et al. (2000a)
Zebrasoma flavescens	P	R	Y	Sancho et al. (2000b)
Zebrasoma scopas	P	R	Y	Randall (1961a, b)
Balistidae (demersal, parent care)				
Balistes vetula	A	R	–	unpub (RN)
Canthidermis sufflamen	A	R	–	Whaylen et al. (2004)
Pseudobalistes flavimarginatus	P	R	–	Lobel and Johannes (1980)c
Balistes polylepis	P	T	–	Sanchez-Velasco et al. (2009)
Caesionidae (planktonic)				
Caesio teres	P	R	–	Bell and Colin (1986)
Pterocaesio diagramma	P	U	–	Thresher (1984)

(continued)

Y. Sadovy de Mitcheson and P.L. Colin (eds.), *Reef Fish Spawning Aggregations: Biology, Research and Management*, Fish & Fisheries Series 35,
DOI 10.1007/978-94-007-1980-4, © Springer Science+Business Media B.V. 2012

(continued)

Family	Location	Type	Pair Sp?	First record/other info
Carangidae (planktonic)b				
See Heading Section B. below				
Labridae (planktonic)				
Cheilinus undulatus	P	R	–	Colin (2010)c
Clepticus parrae	A	R	–	Colin and Clavijo (1988)a
Halichoeres bivittatus	A	R	–	Warner and Robertson (1978)
Halichoeres tenuispinis	P	R	–	Nakazono (1979)
Stethojulius interrupta terina	P	R	–	Nakazono (1979)
Thalassoma amblycephalum	P	R	–	personal observation (PLC)
Thalassoma bifasciatum	A	R	Y	Randall and Randall (1963)c
Thalassoma cupido	P	R	–	Moyer (1974)
Thalassoma hardwickii	P	R	–	Craig (1998)a
Thalassoma lucasanum	P	R	–	Hobson (1965)a
Thalassoma quinquevittatum	P	R	Y	Craig (1998)a
Lethrinidae (planktonic)				
Lethrinus atkinsoni	P	U	–	Ebisawa (1999)
Lethrinus erythropterus	P	T	–	Hamilton (2005)
Lethrinus nebulosus	P	U	–	Ebisawa (1990)
Lutjanidae (planktonic)				
Lutjanus analis	A	T	–	Domeier et al. (1996)a
Lutjanus apodus	A	T	–	unpub (RN)
Lutjanus argentiventris	P	T	–	Sala et al. (2003)
Lutjanus bohar	P	T	–	Hamilton (2003a)
Lutjanus campechanus	A	T	–	Lindeman et al. (2000)
Lutjanus cyanopterus	A	T	–	Heyman et al. (2005)
Lutjanus fulvus	P	T	–	Patrick L. Colin, c
Lutjanus jocu	A	T	–	Domeier et al. (1996)a
Lutjanus novemfasciatus	P	U	–	Sala et al. (2003)
Lutjanus synagris	A	T	–	Wicklund (1969)
Lutjanus vitta	P	U	–	Hamilton (2003)
Symphorichthys spilurus	P	T	–	Sakue et al., c
Mugilidae (planktonic)				
Crenimugil crenilabris	P	U	–	Helfrich and Allen (1975)
Mugil cephalus	A	U?	–	Sadovy (2004)
Mullidae (planktonic)				
Pseudupeneus maculatus	A	T?	Y	Colin and Clavijo (1978)
Pomacentridae (demersal, parental care)				
Abudefduf saxatilis	A	T	Y	Cummings (1968)a
Abudefduf spp.	I	–	–	Fishelson (1970)

(continued)

(continued)

Family	Location	Type	Pair Sp?	First record/other info
Chromis multilineata	A	–	Y	Myrberg et al. (1967)
Chromis hypsilepis	P	–	–	Gladstone (2007)
Chromis atropectoralis?	P	–	Y	unpub (PLC)
Scaridae (planktonic)				
Bolbometopon muricatum	P	R	–	Gladstone (1986)c
Chlorurus sordidus	P	R	Y	unpub (PLC)c
Hipposcarus longiceps	P	R	Y	Patrick L. Colin, c
Scarus bleekeri	P	R	Y	unpub (PLC)
Scarus iserti	A	R	Y	Randall and Randall (1963)c
Scarus prasignathos	P	R	–	unpub (PLC)
Scarus psittacus	P	R?	Y	Sancho et al. (2000b)
Sparisoma rubripinne	A	R	Y	Randall and Randall (1963)
Sparisoma viride	?	?	?	van Rooij et al. (1996)
Serranidae (planktonic)				
Dermatolepis dermatolepis	EP	U	–	Erisman et al. (2009)
Epinephelus adscensionis	A	T	–	unpub (RSN, PLC, YSM)
Epinephelus fuscoguttatus	P	T	–	Johannes et al. (1999)c
Epinephelus marginatus	Med	R?	Y	Zabala et al. (1997)
Epinephelus guttatus	A	T	–	Colin et al. (1987)c
Epinephelus polyphekadion	P	T	–	Johannes et al. (1999)c
Epinephelus quinquefasciatus				Matthew Craig, personal communication
Epinephelus striatus	A	T	–	Smith (1972)c
Mycteroperca bonaci	A	T	–	Eklund et al. (2000)
Mycteroperca jordani	EP	T	–	Saenz-Arroyo et al. (2005)
Mycteroperca microlepis	A	T	–	Coleman et al. (1996)a, c
Mycteroperca phenax	A	T	–	Coleman et al. (1996)
Mycteroperca prionura	EP	T	–	Sala et al. (2003)
Mycteroperca rosacea	EP	T	–	Erisman et al. (2007)
Mycteroperca tigris	A	T	–	Sadovy et al. (1994)
Mycteroperca venenosa	A	T	–	Starr et al. (2007)?
Plectropomus areolatus	P	T	–	Johannes (1988)c
Plectropomus leopardus	P	T/R?	–	Samoilys and Squire (1994)c
Plectropomus punctatus	I	U	–	Robinson et al. (2008)
Siganidae (demersal, no care)				
Siganus fuscescens	P	T	–	Johannes (1981)c
Siganus sutor	I	U	–	Robinson et al. (2004)
Zanclidae				
Zanclus cornutus	P	T	Y	Colin et al., c

Section B

Probable spawning aggregation – determined by indirect methods (fisheries data, observation of groupings underwater).

Family	Location	Type	Pair?	First record
Genus and species				
Carangidae (planktonic)b				
Caranx bartholomaei	A	U	–	Graham and Castellanos (2005)b
Caranx hippos	A	U	–	Heyman and Kjerfve (2008)b
Caranx ignobilis	P	T?		von Westernhagen (1974)
Caranx latus	A	U	–	Heyman and Kjerfve (2008)b
Caranx ruber	A	U	–	Heyman and Kjerfve (2008)b
Caranx sexfasciatus	P	U	–	Sala et al. (2003)b, c
Decapterus macarellus	A	U	–	Whaylen et al. (2004)a
Seriola lalandi	EP	U	–	Sala et al. (2003)b
Trachinotus falcatus	A	U	–	Graham and Castellanos (2005)b
Centropomidae (planktonic)				
Lates calcarifer	P	T	–	Moore and Reynolds (1982)
Labridae (planktonic)				
Lachnolaimus maximus	A	R	–	Colin (1982)a
Lutjanidae (planktonic)				
Lutjanus argentimaculatus	P	U	–	Johannes (1981)a
Lutjanus argentiventris	P	U	–	Sala et al. (2003)
Lutjanus griseus	A	T	–	Domeier et al. (1996)a
Lutjanus rivulatus	P	U	–	Hamilton (2003)
Muraenidae (planktonic)				
Gymnothorax herrei	P	U	–	Ferraris (1985)
Serranidae (planktonic)				
Epinephelus itajara	A	T	–	Bullock et al. (1992)a
Plectropomus laevis	P	T	–	unpub (PLC, YSM)
Siganidae				
Siganus guttatus	P	U	–	Daw (2004)
Siganus puellus	P	U	–	Moyer unpublished MSa
Siganus randalli	P	U	–	Rhodes (2003)
Siganus spinus	P	U	–	Rhodes (2003)
Siganus vermiculatus	P	U	–	Rhodes (2003)
Siganus virgatus	P	T	–	Moyer unpublished MS
Siganus vulpinus	P	T	–	Moyer unpublished MS

Section C

Possible spawning aggregation – confirmation needed (interview or other anecdotal records mostly).

Family	Location	Type	Pair?	First record
Genus and species				
Acanthuridae (planktonic)				
Acanthurus mata	P	U	–	Johannes (1981)
Naso literatus	P	U	–	Johannes et al. (1999)a
Naso lopezi	P	U	–	Rhodes (2003)
Naso unicornis	P	U	–	Johannes et al. (1999)
Carangidae (planktonic)				
Caranx tille	P	U	–	Hamilton (2003)?
Elopidae (planktonic)				
Megalops atlanticus	A	U	–	Crabtree (1995)
Gerridae (planktonic)				
Gerres sp.	P	U	–	Johannes and Yeeting (2001)
Kyphosidae (planktonic)				
Kyphosus bigibbus	P	U	–	Rhodes (2003)
Kyphosus cinerascens	P	U	–	Rhodes (2003)
Kyphosus vaigensis	P	U	–	Rhodes (2003)
Scaridae				
Chlorurus frontalis	P	R	Y	Rhodes (2003)
Chlorurus microrhinos	P	R	Y	Rhodes (2003)
Labridae (planktonic)				
Choerodon anchorago	P	U	–	Johannes (1981)
Thalassoma duperrey	P	U	Y	Sancho et al. (2000b)
Lethrinidae (planktonic)				
Lethrinus harak	P	U	–	Rhodes (2003)
Lethrinus miniata	P	U	–	Sadovy (2007)
Lethrinus olivaceus	P	U	–	Rhodes (2003)
Lethrinus xanthochilus	P	U	–	Rhodes (2003)
Monotaxis grandoculis	P	U	–	Rhodes (2003)
Lutjanidae (planktonic)				
Lutjanus decussatus	P	U		Nanami et al. (2010)
Lutjanus gibbus	P	U	–	Johannes (1981)a
Macolor niger	P	U	–	unpub (PLC)
Ocyurus chrysurus	P	U	–	Garcia-Cagide et al. (2001)
Mugilidae (planktonic)				
Chelon macrolepis	P	U	–	Johannes and Yeeting (2001)
Liza vaigiensis	P	U	–	Johannes (1981)a
Neomyxus leuciscus	P	U	–	Rhodes (2003)
Valamugil seheli	P	U	–	Johannes and Yeeting (2001)

(continued)

(continued)

Family	Location	Type	Pair?	First record
Genus and species				
Mullidae (planktonic)				
Mulloides flavolineatus	P	U	–	Johannes (1981)
Mulloides vanicolensis	P	U	–	Rhodes (2003)
Scombridae (planktonic)				
Acanthocyubium solandri	API	U	–	Johannes (1981)
Grammatorcynus bicarinatus	P	U	–	Johannes (1981)
Rastrelliger kanagurta	P	U	–	SCRFA?
Scomberomorus commerson	P	U	–	Johannes (1981)
Serranidae (planktonic)				
Cephalopholis argus	P	U	–	Hamilton (2003)a
Cephalopholis miniata	P	U	–	Hamilton (2003)
Cephalopholis sexmaculata	P	U	–	Hamilton (2003)
Cephalopholis sonnerati	P	U	–	Hamilton (2003)
Cephalopholis urodeta	P	U	–	Hamilton (2003)
Epinephelus lanceolatus	P	U	–	Sadovy de Mitcheson and Liu (2008)
Epinephelus maculatus	P	U	–	Rhodes (2003)
Epinephelus malabaricus	P	U	–	SCRFA (2004)
Epinephelus merra	P	U	–	Hamilton (2003)a
Epinephelus rivulatus	P	U	–	Mackie (2000)
Epinephelus tauvina	P	U	–	Johannes (1981)
Epinephelus coioides	P	U	–	Hamilton (2003)
Epinephelus corallicola	P	U	–	Hamilton (2003)
Epinephelus cyanopodus	P	U	–	Rhodes (2003)
Epinephelus multinotatus	P	U	–	Hamilton (2003)
Epinephelus ongus	P	U	–	Hamilton (2003)
Epinephelus spilotoceps	P	U	–	Hamilton (2003)
Epinephelus trimaculatus	P	U	–	Hamilton (2003)
Plectropomus maculatus	I	U	–	Hamilton (2003)
Plectropomus oligacanthus	P	U	–	Hamilton (2003)
Paranthias colonus	EP	U	–	Sala et al. (2003)
Paranthias furcifer	A	U	–	Posada (1996)
Sphyraenidae (planktonic)				
Sphyraena barracuda	AIP	U	–	Hamilton (2003)
Sphyraena genie	P	U	–	Hamilton (2003)

Records indicated as "unpub" are confirmed records by experienced observers, authors in this volume. Often times there is additional information to support the observations, such as photographs or video of the spawning, with gamete cloud visible, collection of eggs, collection females with hydrated eggs and others

See also species case studies in Chapter 12. ? denotes unclear if first record

Key:

Type: *T* transient, *R* resident, *D* demersal, *DP* demersal parental care, *U* uncertain

Location: *A* Atlantic, *P* Pacific, *EP* Eastern Pacific, *I* Indian

a Additional supporting evidence is available

b While spawning observations are not questioned, whether the species engage in true aggregation or simple migratory spawning is not positively known, hence these are listed as "probable" aggregation spawners

c case study this volume as part of Chap. 12

References

Nakazono A (1979) Studies on the sex reversal and spawning behaviour of five species of Japanese labrid fishes. Rep Rish Res Lab Kyushu Univ 4:64

Bell LJ, Colin PL (1986) Mass spawning of *Caesio teres* (Pisces: Caesionidae) at Enewetak Atoll, Marshall Islands. Environ Biol Fish 15:69–74

Bullock LH, Murphy MD, Godcharles MF, Mitchell ME (1992) Age, growth and reproduction of jewfish *Epinephelus itajara* in the eastern Gulf of Mexico. Fish Bull 90:243–249

Coleman FC, Koenig CC, Collins LA (1996) Reproductive styles of shallow-water groupers (Pisces: Serranidae) in the eastern Gulf of Mexico and the consequences of fishing spawning aggregations. Environ Biol Fish 47:129–141

Colin PL (1982) Spawning and larval development of the hogfish, *Lachnolaimus maximus* (Pisces: Labridae). Fish Bull 80(4):853–862

Colin PL (2010) Aggregation and spawning of the humphead wrasse, *Cheilinus undulatus* (Pisces: Labridae): general Aspects of spawning behaviour. J Fish Biol 76:987–1007

Colin PL, Clavijo IE (1978) Mass spawning by the spotted goatfish, *Pseudupeneus maculatus* (Bloch) (Pisces:Mullidae). Bull Mar Sci 28:780–782

Colin PL, Clavijo IE (1988) Spawning activity of fishes producing pelagic eggs on a shelf edge coral reef, southwestern Puerto Rico. Bull Mar Sci 43:249–279

Colin PL, Shapiro DY, Weiler D (1987) Aspects of the reproduction of two groupers, *Epinephelus guttatus* and *E. striatus,* in the West Indies. Bull Mar Sci 40:220–230

Crabtree RE (1995) Relationship between lunar phase and spawning activity of tarpon, *Megalops atlanticus*, with notes on the distribution of larvae. Bull Mar Sci 56:895–898

Craig PC (1998) Temporal spawning patterns of several species of surgeonfishes and wrasses in American Samoa. Pac Sci 52:35–39

Cummings WC (1968) Reproductive habits of the sergeant major, *Abudefduf saxatilis*, (Pisces: Pomacentridae), with comparative notes on four other damselfishes in the Bahama Islands. PhD Dissertation, University of Miami, FL

Domeier ML, Koenig C, Coleman F (1996) Reproductive biology of the gray snapper (*Lutjanus griseus*), with notes on spawning for other western Atlantic snappers (Lutjanidae). In: Arreguin-Sanchez F, Munro JL, Balgos MC, Pauly D (eds) Biology, fisheries and culture of tropical groupers and snappers. ICLARM conference proceedings, vol 48. ICLARM, Makati City, pp 449–201

Ebisawa A (1990) Reproductive biology of *Lethrinus nebulosus* (Pisces: Lethrinidae) around the Okinawa waters. Nipp Suisan Gakkai 56:1941–1954

Ebisawa A (1999) Reproductive and sexual characteristics in the Pacific yellowtail emperor, *Lethrinus atkinsoni*, in waters off the Ryukyu Islands. Ichthyol Res 46:341–358

Eklund AM, McClellan DB, Harper DE (2000) Black grouper aggregations in relation to protected areas within the Florida Keys National Marine Sanctuary. Bull Mar Sci 66:721–728

Erisman BE, Buckhorn ML, Hastings PA (2007) Spawning patterns in the leopard grouper, *Mycteroperca rosacea*, in comparison with other aggregating groupers. Mar Biol 151:1849–1861

Erisman BE, Konotchick T, Blum S (2009) Observations of spawning in the leather bass, *Dermatolepis dermatolepis*, (Teleostei: Epinephelidae), at Cocos Island, Costa Rica. Environ Biol Fish 85:15–20

Ferraris CJ Jr (1985) Redescription and spawning behavior of the muraenid eel *Gymnothorax herrei*. Copeia 1985(2):518–520

Fishelson L (1970) Behaviour and ecology of a population of *Abudefduf saxatilis* (Pomacentridae, Teleostei). Anim Behav 18:225–237

Garcia-Cagide A, Claro R, Koshelev BV (2001) Reproductive patterns of fishes of the Cuban shelf. In: Claro R, Lindeman KC, Parenti LR (eds) Ecology of the marine fishes of Cuba. Smithsonian Institution Press, Washington, DC, pp 73–114

Gladstone W (1986) Spawning behavior of the bumphead parrotfish *Bolbometapon muricatum* at Yonge Reef, Great Barrier Reef. Jpn J Ichthyol 33:326–328

Gladstone W (2007) Temporal patterns of spawning and hatching in a spawning aggregation of the temperate reef fish *Chromis hypsilepis* (Pomacentridae). Mar Biol 151:1143–1152

Graham RT, Castellanos DW (2005) Courtship and spawning behaviors of carangid species in Belize. Fish Bull 103:426–432

Hamilton RJ (2003) A report on the current status of exploited reef fish aggregations in the Solomon Islands and Papua New Guinea – Choiseul, Ysabel, Bougainville and Manus Provinces. In: Western Pacific Fisher Survey Series, vol 1. Society for the Conservation of Reef Fish Aggregations (www.SCRFA.org)

Hamilton RJ (2005) Indigenous Ecological Knowledge (IEK) of the aggregating and nocturnal spawning behaviour of the longfin emperor *Lethrinus erythropterus*. SPC Tradit Mar Resour Manag Knowl Bull 18:9–17

Helfrich P, Allen PM (1975) Observations on the spawning of mullet, *Crenimugil crenilabis* (Forskal) at Enewetak, Marshall Islands. Micronesica 11:219–225

Heyman WD, Kjerfve B (2008) Characterization of multi-species reef fish spawning aggregations at Gladden Spit. Belize Bull Mar Sci 83(3):531–551

Heyman WD, Kjerfve B, Graham RT, Rhodes KL, Garbutt L (2005) Spawning aggregations of *Lutjanus cyanopterus* (Cuvier) on the Belize Barrier Reef over a 6 year period. J Fish Biol 67:83–101

Hobson ES (1965) Diurnal-nocturnal activity of some inshore fishes in the Gulf of California. Copeia 1965(3):291–302

Johannes RE (1981) Words of the Lagoon. Fishing marine lore in the Palau district of Micronesia. University of California Press, Los Angeles

Johannes RE (1988) Spawning aggregation of the grouper, *Plectropomus areolatus* (Ruppell) in the Solomon Islands. Proceedings of 6th International Coral Reef Symposium, Townsville, Australia, vol 2, pp 751–755

Johannes RE, Yeeting B (2001) I-Kiribati knowledge and management of Tarawa's lagoon resources. Atoll Res Bull 489:1–25

Johannes RE, Squire L, Graham T, Sadovy Y, Renguul H (1999) Spawning aggregations of groupers (Serranidae) in Palau (unpublished report), The Nature Conservancy Marine Research Series, Publ. No. 1, 144 pp

Lindeman KC, Pugliese R, Waugh GT, Ault JS (2000) Developmental patterns within a multispecies reef fishery: management applications for essential fish habitats and protected areas. Bull Mar Sci 66:929–956

Lobel PS, Johannes RE (1980) Nesting, eggs, and larvae of triggerfishes (Balistidae). Environ Biol Fish 5:251–252

Mackie M (2000) Reproductive biology of the half moon grouper, *Epinephelus rivulatus*, at Ningaloo reef, Western Australia. Environ Biol Fish 57:363–376

Moore R, Reynolds LF (1982) Migration patterns of barramundi *Lates calcarifer* in Papua New Guinea. Aust J Mar Freshw Res 33:671–682

Moyer JT (1974) Notes on the reproductive behavior of the wrasse *Thalassoma cupido*. Jpn J Ichthyo 21(1):34–36

Moyer JT (date uncertain) Siganids, pomacanthids, and the question of monogamy. Unpublished manuscript in the collection of P.L. Colin

Myrberg AA, Montgomery WL, Fishelson L (1988) The reproductive behavior of *Acanthurus nigrofuscus* (Forskal) and other surgeonfishes (Fam. Acanthuridae) off Eilat, Israel (Gulf of Aqaba, Red Sea). Ethology 79:31–61

Nanami A, Okuzawa K, Yamada H, Suzuki N, Aonuma Y (2010) Reproductive activity in the checkered snapper, *Lutjanus decussatus*, off Ishigaki Island. Okinawa Icthyol Res. doi:DOI: 10.1007/s10228-010-0155-5

Posada JM (1996) Life-history, reproductive biology and sexual pattern of the Creole Fish, *Paranthias furcifer* (Valenciennes, 1828) (Pisces: Serranidae). PhD Department of Marine Sciences, University of Puerto Rico, Mayaguez, Puerto Rico

Randall JE (1961a) A contribution to the biology of the convict surgeonfish of the Hawaiian Islands, *Acanthurus triostegus sandvicensis*. Pac Sci 15:215–272

Randall JE (1961b) Observations on the spawning of surgeonfishes (Acanthuridae) in the Society Islands. Copeia 1961:237–238

Randall JE, Randall HA (1963) The spawning and early development of the Atlantic parrot fish, *Sparisoma rubripinne*, with notes on other scarid and labrid fishes. Zool N Y Zool Soc 48:49–60

Rhodes KL (2003) SCRFA spawning aggregation survey: federated States of Micronesia. In: Western Pacific fisher survey series, vol 2. Society for the Conservation of Reef Fish Aggregations. (www.SCRFA.org)

Robertson DR (1983) On the spawning behavior and spawning cycles of eight surgeonfishes (Acanthuridae) from the Indo-Pacific. Environ Biol Fish 9:192–223

Robinson J, Isidore M, Marguerite MA, Ohman MC, Payer RJ (2004) Spatial and temporal distribution of reef fish spawning aggregations in the Seychelles – an interview-based survey of artisanal fishers. West Indian Ocean J Mar Sci 3:63–69

Robinson J, Aumeeruddy R, Jörgensen TL, Öhman MC (2008) Dynamics of camouflage (*Epinephelus polyphekadion*) and brown marbled grouper (*Epinephelus fuscoguttatus*) spawning aggregations at a remote reef site, Seychelles. Bull Mar Sci 83:415–431

Sadovy Y (2004) A report on the current status and history of exploited reef fish aggregations in Fiji. In: Western Pacific fisher survey series, vol 4. Society for the Conservation of Reef Fish Aggregations. (www.SCRFA.org)

Sadovy Y (2007) Report on current status and exploitation history of reef fish spawning aggregations in Palau. In: Western Pacific fishery survey series. Society for the Conservation of Reef Fish Aggregations, vol. 3. SCRFA and the Palau Conservation Society, 40 pp. (www.SCRFA.org)

Sadovy de Mitcheson Y, Liu M (2008) Functional hermaphroditism in teleosts. Fish Fish 9:1–43

Sadovy Y, Colin PL, Domeier ML (1994) Aggregation and spawning in the tiger grouper, *Mycteroperca tigris* (Pisces: Serranidae). Copeia 2:511–516

Saenz-Arroyo A, Roberts CM, Torre J et al (2005) Using fishers' anecdotes, naturalists' observations and grey literature to reassess marine species at risk: the case of the Gulf Grouper in the Gulf of California. Mex Fish Fish 6:121–133

Sala E, Aburto-Oropeza O, Paredes G, Thompson G (2003) Spawning aggregations and reproductive behavior of reef fishes in the Gulf of California. Bull Mar Sci 72:103–121

Samoilys MA, Squire LC (1994) Preliminary observations on the spawning behavior of coral trout, *Plectropomus leopardus* (Pisces: Serranidae), on the Great Barrier Reef. Bull Mar Sci 54(1):332–342

Sanchez-Velasco L, Lavın MF, Peguero-Icaza M (2009) Seasonal changes in larval fish assemblages in a semi-enclosed sea (Gulf of California). Cont Shelf Res 29:1697–1710

Sancho G, Petersen CW, Lobel PS (2000a) Predator-prey relations at a spawning aggregation site of coral reef fishes. Mar Ecol Prog Ser 203:275–288

Sancho G, Solow AR, Lobel PS (2000b) Environmental influences on the diet timing of spawning in coral reef fishes. Mar Ecol Prog Ser 206:193–212

SCRFA (no date) on-line database of reef fish spawning aggregations. Society for the Conservation of Reef Fish Aggregations. (www.SCRFA.org)

Smith CL (1972) A spawning aggregation of Nassau grouper, *Epinephelus striatus* (Bloch). Trans Am Fish Soc 101:257–261

Starr RM, Sala E, Ballesteros E, Zabala M (2007) Spatial dynamics of the Nassau grouper *Epinephelus striatus* in a Caribbean atoll. Mar Ecol Prog Ser 343:239–249

Thresher RE (1984) Reproduction in reef fishes. TFH Publication, Neptune City, USA

van Rooij J, Kroon F, Videler J (1996) The social and mating system of the herbivorous reef fish *Sparisoma viride*: one-male versus multi-male groups. Environ Biol Fish 47:353–378

von Westernhagen H (1974) Observation on the natural spawning of *Alectis indicus* (Ruppell) and *Caranx ignobilis* (Forsk) (Carangidae). J Fish Biol 6:513–516

Warner RR, Robertson DR (1978) Sexual patterns in the labroid fishes of the western Caribbean, I: the wrasses (Labridae). Smithson Cont Zool 254:1–27

Whaylen L, Pattengill-Semmens CV, Semmens BX, Bush PG, Boardman MR (2004) Observations of a Nassau grouper (*Epinephelus striatus*) spawning aggregation site in Little Cayman Island. Environ Biol Fish 70:305–313

Wicklund R (1969) Observations on spawning of lane snapper. Underw Nat 62:40

Zabala M, Garcia-Rubies A, Louisy P, Sala E (1997) Spawning behaviour of the Mediterranean dusky grouper *Epinephelus marginatus* (Lowe, 1834) (Pisces, Serranidae) in the Medes Islands Marine Reserve (NW Mediterranean, Spain). Scientia Marina 61(1):65–77

Index

A

Acanthaster planci. *See* Crown-of-thorns-seastar
Acanthuridae 2, 22, 29, 36, 40, 43, 44, 71, 87, 93, 98, 100, 337, 518, 526–535, 595, 599
Acanthurus achilles. *See* Achilles tang
Acanthurus bahianus. *See* Ocean surgeonfish
Acanthurus blochii. *See* Ringtail surgeonfish
Acanthurus chirurgus. *See* Doctorfish
Acanthurus coeruleus. *See* Blue tang
Acanthurus guttatus. *See* White-spotted surgeonfish
Acanthurus lineatus. *See* Blue-banded surgeonfish
Acanthurus nigricauda. *See* Blackstreak surgeonfish
Acanthurus nigrofuscus. *See* Brown surgeonfish
Acanthurus nigroris. *See* Blue-lined surgeonfish
Acanthurus thompsoni. *See* Thompson's surgeonfish, 532
Acanthurus triostegus. *See* Convict surgeonfish
Achilles tang, 532
Acoustic tracking, 25, 420
Acronurus, 205, 208, 210, 527, 533
Advection, 85, 88, 105–107, 138, 143, 152, 161, 163, 164, 166, 168–170, 180, 181, 183, 186, 204, 318, 321, 587
Aggregation
  costs and benefits, 60–62, 65–67, 70, 268, 322, 325, 378, 386, 389, 542
  definition, non-fish species, 2, 4, 38, 97, 104, 200, 213–214
  management, 2, 22, 74, 110, 118, 225, 286, 331, 371, 405, 567
  multi-species, 7, 33, 47, 147–148, 251, 252, 342, 388, 409, 411, 436–438, 447, 448, 466, 515, 570, 572, 573
  pre-spawning, 31, 236, 240, 241, 396, 441, 444, 536, 540
  recovery, 14, 74, 217, 226, 228, 237, 241, 253, 256, 260, 261, 263, 264, 275, 299, 388, 393, 394, 396–399, 421, 422, 435–438, 507, 568, 569, 574, 576, 579, 582
  resident, 7, 8, 18. 23, 25, 28, 31, 33, 35, 38, 86–89, 91, 92, 95–110, 119, 121, 132, 134, 148, 160, 170–177, 185–187, 203, 206, 209, 210. 226, 252, 260. 275, 317, 318, 337, 338, 374, 379, 381, 382, 455, 468, 482, 486, 502, 509, 518, 520, 526–529, 570, 572. 580, 592
  site, 3, 22, 58, 89, 118, 160, 202, 226, 286, 335, 372, 407, 568
  temporal variation, 10, 162, 358
  transient, 7–9, 18, 23, 24, 26, 29, 31, 33, 35, 36, 38, 39, 42, 48, 86–89, 91, 92, 96–100, 102–110, 119, 121, 131, 135, 154, 166, 167, 170–172, 175–181, 186, 199, 200, 202, 206, 218, 226, 252, 258, 260, 266, 274–276. 304, 317, 318. 321, 324, 325, 337, 374, 376, 379, 381, 382, 449, 464, 466, 482, 486, 509, 517, 568, 570, 572, 573, 577. 580–582, 594
  validation, 12, 16, 295, 315, 333, 334, 337, 344, 350–358, 361, 580
  working group (Belize). 388, 391, 582
Albula vulpes. *See* Bonefish
Albulidae, 205, 240
Aldabra atoll, 528, 529, 531, 535
Allee effect, 77, 248, 256, 259–264, 574, 587
Amberjack, 509

Y. Sadovy de Mitcheson and P.L. Colin (eds.), *Reef Fish Spawning Aggregations: Biology, Research and Management*, Fish & Fisheries Series 35,
DOI 10.1007/978-94-007-1980-4, © Springer Science+Business Media B.V. 2012

American Samoa, 528, 531, 532
Angelfishes, 6, 12, 65, 70, 92, 97
Appendicularian, 206
Aquaculture. *See* Mariculture
Artificial fertilization, 289
Ascent rates, eggs, 147, 318
Astractoscion nobilis. *See* White seabass
Atlantic cod, 69, 76, 77, 250, 258, 259, 262, 263, 271, 295, 569, 581
Atlantic Ocean, 108, 392, 418, 546, 571
Atlantic wreckfish, 67, 266
Australia, 8, 16, 28, 90, 151, 174, 178, 197, 212, 234, 235, 242, 244, 254, 259, 266, 270, 345, 349, 377, 380, 385, 406–409, 427, 445, 446, 450–455, 457, 458, 464, 472–474, 476, 478, 486, 490, 491, 535

B

Bahaba taipingensis. *See* Giant yellow croaker
Bahamas
  36, 91, 119, 121, 123–125, 129, 131, 132, 134, 136, 137, 140, 141, 144, 147, 177, 207, 211, 212, 218, 228, 270, 300, 315, 319, 341, 351, 356, 377, 378, 394, 429, 430, 432–437, 518, 520, 525, 526, 577, 583
  Exumas, 125, 136, 211, 300, 430
Balistes polylepis. *See* Finescale triggerfish
Balistidae, 31, 36, 43, 87, 91, 93, 98, 120, 195, 542, 595
Balistoides viridescens. *See* Titan triggerfish
Barotrauma, 255, 381, 443, 477, 587
Barred sand bass, 243
Bathymetry, 120, 178, 288, 322
Belize
  Gladden Spit, 149, 384, 394
  Glovers Reef, 27, 30, 134
Benthic, 31, 36–38, 46, 74, 97, 99, 102, 104, 105, 120, 133, 142, 147, 150, 160, 170, 175, 181, 192, 193, 195, 196, 205, 206, 208, 216, 217, 296, 418, 450, 459, 502, 506, 507, 525, 527, 533, 535, 543, 546, 571
Bermuda, 119, 139, 143, 145, 183, 208, 229, 372, 381, 382, 386, 387, 394, 412, 413, 415–417, 429, 433, 435, 436, 439, 466, 488, 518
Bigeye, Hawaiian, 213
Bigeyes, 143, 209, 213, 507–510
Bigeye trevally
  age and growth, 511
  aggregations, 143, 507–509
  length–weight, 507, 508, 511
  management, 510, 511
  recruitment, 508–509
  reproductive biology, 508–509
Bigmouth mackerel, 42
Biogeography, 89–91
Black durgeon, 546
Black grouper, 147, 350, 379, 395, 397, 398, 400, 442
Black marlin, 11, 15, 16
Black rockfish, 199
Black saddle coralgrouper, 99, 100, 148, 452, 598
Black snapper, 150, 399, 400, 464, 500, 501
Blackstreak surgeonfish, 533
Blacktail snapper
  aggregations, 240, 346, 405, 458–469
  habitats, 259, 264, 266, 467
  length–weight, 461
  management, 240, 405, 463, 464, 466–468
  reproductive biology, 465–467
Bleeker's parrotfish, 506, 597
Blob sculpin, 64, 73
Blue-banded surgeonfish, 40, 100, 530, 531, 595
Bluefin trevally, 40, 42, 71, 511
Bluehead wrasse
  age and growth, 488
  aggregations, 14, 61, 69, 74, 75, 92, 119, 152, 168, 169, 199, 203, 204, 206, 213, 214, 261–263, 292, 405, 489, 490
  aggregation sites, 30, 168, 261, 292
  currents, 203, 489
  eggs and hatching, 489
  habitats, 487–488
  larval development, 488
  range, 213, 405, 488, 489
  recruitment, 14, 213, 262
  reproductive biology, 488–489
  tagging, 489
Blue-lined sea bream, 119, 596
  aggregations, 137, 143, 464, 465, 468–472
  conservation, 470–472
  eggs and hatching, 470
  fisheries, 251, 468
  length–weight, 468
  management, 470–472
  reproductive biology, 137, 468–470
  spawning behaviour, 468, 470, 473, 474, 476–478
Blue-lined surgeonfish, 531, 532
Bluespine unicornfish, 242, 527, 533–535
Blue tang, 64, 95, 300, 517–523, 525, 526
Blue trevally, 512

Body size, 39, 67, 88, 97, 102, 108, 109, 250, 252, 254, 257–259, 261, 292, 293, 432, 435, 480, 496, 571
Bolbometopon muricatum. See Bumphead parrotfish
Bonefish, 205, 240–242, 332, 341
Brazil, 119, 336, 337, 346, 347, 412, 417–419, 421, 422, 429, 439, 466
Brotulid, 9
Brown-marbled grouper
  age and growth, 407, 408
  aggregations, 31, 34, 92, 137, 141, 146, 147, 232, 244, 247, 407–412
  aquaculture, 406
  conservation, 411
  eggs and hatching, 408
  fisheries, 409–411
  larval development, 408
  length–weight, 406–408, 410
  management, 409–411
  reproductive biology, 408–409
Brown puller damselfish, 9, 73, 74
Brown surgeonfish, 29, 42, 61, 68, 69, 95, 132, 148, 151, 170, 209, 263, 319, 506, 526–535
Brushtail tang, 530, 534
Buccal pumping, 150, 500, 501
Bul, 346, 428, 448, 541
Bullethead parrotfish, 42, 132, 143, 148, 170, 502–507, 531
Bumphead parrotfish
  age and growth 491
  aggregations 67, 99, 103, 143, 332, 337, 338, 480, 493, 502
  conservation 493–496
  fisheries 493–496
  length–weight 491, 492
  management 332, 337, 338
  range 491–493, 496
  reproductive biology 492–493
Buoyancy, 143, 147, 166, 192, 195, 204, 288, 298, 317, 318, 506
Butterflyfishes, 92, 97

C
Caesionidae, 36, 38, 43, 44, 87, 93, 98, 595
Caesio teres. See Yellowback fusilier
California sheephead, 213
Calotomus spp, 96, 506
Camouflage grouper
  age and growth, 423, 424
  aggregations, 11, 31, 137, 148, 375, 409, 411, 423
  conservation, 148, 338, 341–343, 375, 426–428
  eggs and hatching, 139, 423, 573
  fecundity, 427
  fisheries, 409, 410, 423, 427–428
  length–weight, 423, 424, 427
  management, 427–428
  reproductive biology, 31, 423
  tagging, 426
  temperature, 446
Canthidermis sufflamen. See Ocean triggerfish
Capacity
  building, 390, 391
  management, 110, 249, 272, 375, 380, 382, 388, 390–393, 428, 574
Capital breeding, 103, 106, 110, 588
Capture-based aquaculture (CBA), 209, 226, 240, 247, 573, 587, 588
Carangidae, 2, 22, 36, 40, 43, 65, 92, 94, 98, 210, 251, 507–512, 570, 596, 598, 599
Carangoides bartholomaei. See Yellow jack
Carangoides ferdau. See Blue trevally
Caranx hippos. See Crevalle jack
Caranx ignobilis. See Giant trevally
Caranx latus. See Horse-eye jack
Caranx melampygus. See Bluefin trevally
Caranx sexfasciatus. See Bigeye trevally
Caribbean (Sea), 22, 23, 26, 29, 37, 39, 40, 43, 46, 50, 90, 91, 96, 108, 109, 119, 145, 177, 213, 241, 242, 271, 302, 303, 332, 347, 349, 372–374, 382, 386, 388, 392–394, 412, 413, 415–418, 420, 429, 431, 434, 436–439, 488, 526
Catchability, 4, 248, 270, 273, 381, 454, 588
Catchment area, 7, 8, 21, 24–28, 36, 47, 48, 97, 98, 109, 413, 449, 455, 470, 568, 570, 588
Catch per unit of effort (CPUE), 15, 16, 229, 231, 233, 235, 236, 248, 250, 254, 265, 270, 272, 345, 358, 389, 390, 394, 395, 398, 431, 435, 496
Cayman Islands, 27, 36, 46, 125, 127–129, 131, 177, 217, 254, 315, 322, 377, 379, 386, 395, 429, 430, 433–436
Chaetodontidae, 92, 93, 98
Chanos chanos. See Milkfish
Charismatic mega-fauna, 479, 486
Cheilinus undulatus. See Humphead wrasse (Napoleon fish)
Chinese bahaba, 244, 254, 347

Chlorophyll maximum, 183, 184
Chlorurus bleekeri. See Bleeker's parrotfish
Chlorurus sordidus. See Bullethead parrotfish
Chromis hypsilepis. See Brown puller
Ciguatera, 409, 547
CITES. *See* Convention on International Trade in Endangered Species
Clupei pallasi. See Pacific herring
Coastal boundary layer, 171–175, 177, 178, 180, 181, 183, 185
Cod, Atlantic, 69, 76, 77, 250, 258, 259, 262, 263, 271, 295, 569, 581
Code of Conduct (FAO), 250, 375, 376, 587
Cohort, 106, 154, 163, 191, 192, 196, 197, 199, 209, 215, 530, 588
Common blue-stripe snapper, 464
Competition
  male-male, 60, 76
  sperm, 60, 61, 67, 68, 275, 432, 525, 593
Coney, 23, 147, 212, 230, 300, 397
Confidentiality, 331, 337, 349–350
Connectivity
  adult, 22, 33, 48, 49, 287, 438, 570, 580
  larval, 14, 22, 48, 49, 272, 374
Conservation
  extirpation, 47, 76, 148, 375
  IUCN, 250, 251, 253, 391, 421, 427, 444, 448, 457, 486, 590
  Red List, 240, 251, 253, 421, 427, 448, 457, 486, 592
  status, 228, 230, 232, 234, 236, 238, 249, 251, 253, 254, 377, 472
  threatened species, 325, 377, 378, 444, 578, 593
Convention on International Trade in Endangered Species (CITES), 238, 245, 378, 486, 487, 588
Convergence, 48, 185
Convict surgeonfish, 3, 15, 208, 210, 456, 528, 531, 532
Corals, stony, 191, 213, 214
Courtship
  arena, 21, 26–33, 48, 49, 570, 589
  behaviour, 28, 260, 262, 263, 273, 298, 470, 497–501, 509, 511, 534
  colour, 28, 30, 65, 263, 470, 506, 509
  display, 28, 262, 482
CPUE. *See* Catch per unit of effort
Crabs
  coconut, 214
  hermit, 38, 214
  terrestrial, 191, 214
Crenimugil crenilabis. See Fringelip mullet
Crevalle jack, 511
Critical period, 139, 197, 198
Crown-of-thorns-seastar, 480
Ctenochaetus striatus. See Striped bristletooth
Cuba, 37, 91, 177, 228, 251, 252, 254, 347, 372, 373, 377, 379, 381, 382, 384, 389, 396, 429, 434–437, 439, 466, 467, 482
Cubera snapper
  age and growth, 15, 166, 200
  aggregations, 15, 31, 37, 119, 131, 141, 145, 151, 152, 166, 200, 251, 273, 346, 384, 394, 463–468, 517
  habitats, 31, 141, 251
  management, 251, 273, 346, 384, 464, 467
Cupid wrasse, 42
Current
  direction, 73, 120, 135–137, 166, 168, 173, 174, 177, 178, 181, 185, 295, 296, 498, 529
  drifters, 128, 129, 137, 138, 171, 172, 174, 175, 177–179, 193, 290, 317, 321–322, 480
  meter, 296, 319, 320, 324
  quiet, 136, 152
  speed, 73, 117, 135, 136, 152, 166, 183, 295–297, 413
Cynoscion othonopterus. See Gulf corvina

D

Daisy parrotfish, 96
Damselfishes, 504
Day length, 23, 142
Demersal eggs, 1, 9, 11, 91, 118, 154, 192, 195, 199, 214, 218, 536, 543
Demography, 97, 99–105
Density
  dependence, 14, 262, 419
  population, 75, 520, 579
Depth sounders, 290, 322
Dermatolepis dermatolepis. *See* leather bass
Diffusion, 138, 159, 161, 163, 164, 166, 169, 170, 186, 589
Digital photography, 298
Disease, 72
Dispersal
  eggs, 14, 105, 106, 162, 163, 167, 191, 195, 200, 209, 257, 316, 322, 480, 489, 501, 572, 573, 580
  kernel, 161, 162, 591
  larval, 22, 88, 167, 257, 439, 573
  offshore, 14, 138, 152, 153, 572
Diurnal spawning, 496
Diving tourism, 269, 303, 385, 486
Doctorfish, 518, 519

Dog snapper
aggregations, 31, 71, 143, 312, 314, 464, 467
reproductive biology, 408–409
Dory snapper, 102
Drift cards, 291, 317
Drifter vial, 128
Dusky grouper, 65
Dusky rabbitfish
age and growth, 536, 537
aggregations, 536–537, 542
aquaculture, 537, 542
conservation, 538–542
eggs and hatching, 536, 537, 539
fisheries, 538, 541, 542
habitats, 540
larval development, 537
length–weight, 537, 538
management, 539–541
reproductive biology, 537
Dusky spinefoot, 92
Dye release, 289, 291

E

Early life history, 117, 141, 152, 154, 160–170, 172, 191–218, 317, 406, 423, 429–431, 480–482, 580
Ecology
behavioural, 1, 3, 58, 64, 508, 569, 570
nutritional, 85, 86, 88, 97, 99, 103, 104, 107, 108, 571
Economy
analysis, 249, 250
cash, 264, 374, 375
socioeconomic, 47, 225, 264
Ecosystem, 21–50, 161, 272–274, 276, 293, 377, 405, 428, 437, 458, 568, 570, 579, 580, 583
Ecosystem-based management, 273, 372, 374, 428, 579, 584, 589
Eddy(ies), 128, 164, 166, 177–183, 187, 320, 461, 463, 589
tidal, 159, 177, 178, 187
topographic, 172, 176–181, 185
Education, 225, 226, 240, 269–271, 274–275, 304, 332, 344, 362–365, 391, 392, 433, 580, 583
Eggs
ascent rate, 147, 318, 506
demersal, 1, 9–11, 91, 118, 154, 192, 195, 199, 214, 218, 536, 543
hydrated, 11, 12, 16, 87, 199, 289, 292–294, 415, 451, 461, 509, 517, 570, 595, 600
parental care, 9, 91, 489
pelagic demersal, 1, 9, 10, 91–94, 118, 192, 195, 214, 216, 218, 543
planktonic, 11, 118, 147, 154, 163, 191, 194, 289, 321, 480, 502
Elasmobranch, 4, 5, 15
Embayments, 170, 172–175, 187, 478
Emperors, 36, 92, 97, 207, 209, 267
Endangered species, 245, 496, 579
Energetic costs, 62, 72
Energy storage, 103, 528
Enewetak Atoll, 73, 497, 506
Enforcement, 18, 59, 232, 253, 275, 276, 344, 349, 350, 360, 376, 377, 380, 382, 386–389, 392–394, 398, 428, 435–437, 444, 449, 487, 541, 542, 576, 578, 582, 583
Epinephelus cyanopodus. *See* Speckled blue grouper
Epinephelus fuscoguttatus. *See* Brown-marbled grouper
Epinephelus guttatus. *See* Red hind
Epinephelus itajara. *See* Goliath grouper
Epinephelus marginatus. *See* Dusky grouper
Epinephelus multinotatus. *See* White-blotched grouper
Epinephelus ongus. *See* White-streaked grouper
Epinephelus polyphekadion. *See* Camouflage grouper
Epinephelus polystigma. *See* White-dotted grouper
Epinephelus quinquefasciatus. *See* Pacific Goliath grouper
Epinephelus striatus. *See* Nassau grouper
Equinox, 145
Exploitation, 4, 14, 50, 58, 118, 226, 240–248, 250, 254, 255, 260, 264, 267, 269, 273–275, 287, 349, 374, 376, 405, 410, 437, 439, 463, 472, 478, 512, 526, 538, 567, 568, 574, 576, 577, 579

F

Fantail filefish, 213
FAO. *See* Food and Agriculture Organisation (United Nations)
Fecundity
annual, 292, 450, 451
batch, 259, 292, 317, 451, 474, 476, 537
body shape, 292

Fertilization, 118, 140, 141, 164, 165, 167, 196, 200, 204, 214, 256, 260–262, 288, 289, 291, 316, 317, 445, 470, 593
  success rate, 164, 200, 445
Fidelity
  home range, 73, 299
  spawning site, 47, 263, 299, 304, 420, 426, 439
Fiji, 91, 141, 148, 227, 231, 241, 247, 250, 266, 338, 339, 341–343, 345–347, 350, 351, 354, 355, 359, 361, 388, 389, 408, 409, 424, 426, 427, 450, 452, 454, 455, 458, 486, 491, 495
Finescale triggerfish, 246, 546
Fishery closures
  spatial, 376, 378–380, 382, 394, 395, 399, 400, 442, 477
  temporal, 376, 378–380, 386, 394, 395, 398–400, 477
Fishing
  commercial, 22, 49, 215, 225, 228–247, 251, 253, 255, 257, 261, 262, 264–268, 272, 274, 287, 302, 332, 361, 375, 380–382, 386, 387, 393, 409–411, 421, 422, 427–429, 433–436, 442, 443, 448, 449, 452–456, 460, 461, 468, 472–475, 477, 484–486, 489–490, 507, 512, 537–539, 577
  cyanide, 244, 340, 349, 381, 452, 485
  effect of, 248–264, 268, 269, 271, 272, 287, 416, 421, 422, 458
  effort, 25, 49, 226, 243, 248, 257, 270, 273, 303, 372, 382, 385, 388, 389, 396, 410, 436, 437, 453, 458, 475, 476, 478, 542, 579, 582, 588, 589, 591
  explosive, 244, 381, 460
  global, 86, 90, 226, 567, 574
  hookah, 245, 246, 255, 381, 547
  hook and line, 229, 231–235, 237, 242, 244, 245, 296, 337, 361, 409, 423, 427, 442, 448, 452, 472
  licence, 380, 385, 442, 455
  longline, 16, 228, 229, 237, 435
  net, 168, 171, 194, 204, 206–208, 211, 212, 218, 228, 233–239, 241, 244, 245, 247, 255, 265, 291, 316, 317, 337, 381, 396, 430, 448, 452, 466, 467, 493, 501, 507, 537–539, 541
  poison, 381, 409, 547
  recreational, 16, 215, 225, 226, 237, 240–243, 255, 257, 264, 268, 269, 380, 386, 409, 421, 422, 435, 442, 443, 453, 466, 472–478
  speargun, 228–230, 233, 234, 242, 244, 345, 452, 538
  subsistence, 225, 233, 240–242, 267, 276, 332, 361, 374, 381, 382, 393, 410, 427, 428, 449, 461, 463, 472, 507, 512, 518, 535, 537–539, 542, 574, 575, 580, 583
  traditional, 240–242, 268, 340, 372, 428, 448, 463
  trawling, 229, 230, 239, 244, 255, 270, 360, 477
Fish market, 49, 118, 142, 232, 234, 237, 240, 244, 255, 264, 266–269, 273, 293, 302, 324, 347, 350, 354, 355, 358, 374, 375, 378, 380, 381, 383, 384, 388, 393, 395, 409, 417, 428, 448, 449, 454, 457, 470, 484, 493–496, 542, 578, 591
Fitness, 59, 64, 66, 72, 77, 199, 200, 259
Five-lined snapper, 99
Flathead grey mullet, 237, 596
Flexion, 195, 196, 198
Fluorescein, 291, 316
Flushing, 172, 173, 175
Food and Agriculture Organisation (United Nations) (FAO), 250, 375, 376, 587
Food webs, 22–24, 37, 38, 47, 50, 273
Forage, 97
France, 65
Freckled goatfish, 208
French Polynesia, 3, 424, 425
Friction, 167, 181, 182
Fringelip mullet, 151
Fronts, oceanic, 192
Functional migration area (FMA), 24–28, 37–48, 570, 589
Fusiliers, 36, 38, 65, 147, 201

G

Gadus morhua. See Atlantic cod
Gag
  age and growth, 441
  aggregations, 64, 68, 212, 230, 255, 259, 261, 263, 267, 397, 441
  conservation, 230, 397, 444–445
  fisheries, 230, 442–444
  habitats, 255, 440, 444, 445
  larval development, 440
  length–weight, 440
  male colouration, 439, 441
  management, 230, 397, 442
  recruitment, 212, 262, 440
  reproductive biology, 259, 261, 262, 441

Gamete cloud, 42, 46, 164, 166–170, 186, 201–203, 291, 317, 501, 600
Genetic techniques, 295
Geographic information systems (GIS), 254
Geomorphology, 35, 37, 119–133, 152, 154, 161, 288, 303, 323, 450, 572
Giant seabass, 237
Giant sweetlips, 338
Giant trevally, 508, 510, 511, 515
Giant yellow croaker (Chinese bahaba), 239, 244, 254, 336, 347
GIS. *See* Geographic information systems
Global Positioning System (GPS), 120, 146, 254, 289, 290, 293–295, 300, 305, 306, 311–313, 321–324, 358
Glovers Reef, 27, 30, 324
Glut, 264–268, 275, 333, 383, 384, 574
Gnathanodon speciosus. See Golden trevally
Gobies, 60
Gobiidae, 60, 93, 98
Golden trevally, 511
Goldline spinefoot, 59, 598
Goldring bristletooth, 532
Goliath grouper, 598
  age and growth, 418–420, 578
  aggregations, 145, 148, 229, 230, 262, 337, 397, 418–422, 517, 578, 582
  aquaculture, 421
  conservation, 229, 230, 397, 421–422
  fisheries, 229, 230, 347, 422
  genetics, 417
  habitats, 418, 419, 421
  larval development, 208, 418
  length–weight, 418, 419
  management, 229, 230, 397, 421–422, 578, 582
  reproductive biology, 419–421
Gonad, 6, 68, 100, 103, 218, 226, 227, 288, 289, 292–294, 300, 352, 354–358, 414, 423, 459, 461, 493, 516, 519, 520, 523–525, 534, 537, 578, 589
Gonadosomatic index (GSI), 11, 68, 69, 103, 107, 293, 356–358, 413, 432, 461, 462, 525, 578, 589, 590
Gonochorism, 104, 423, 493, 589
GPS. *See* Global Positioning System
Graysby, 23, 230, 397
Grey snapper, 251, 396, 467
Greenthroat parrotfish, 134, 504, 506, 532
Grey mackerel, 235
Grouper
  black, 147, 230, 350, 379, 395, 397, 398, 400, 442
  black-saddle coralgrouper, 99, 148, 452
  blue speckled, 72
  brown-marbled, 14, 28, 31, 34, 38, 92, 137, 141, 146, 148, 232, 233, 242, 244, 247, 338, 406–412, 446
  camouflage, 3, 14, 28, 29, 31, 34, 38, 92, 137, 139, 141, 143, 146, 148, 231, 232, 242, 247, 257, 258, 263, 305, 338, 341–343, 346, 375, 388, 406, 409, 410, 422–428, 446, 573, 577
  dusky, 65
  gag, 64, 68, 212, 230, 255, 259, 261–263, 267, 350, 397, 439–445
  Goliath, 142, 145, 148, 208, 229, 230, 262, 336, 337, 347, 397, 417–422, 517, 578, 582
  graysby, 23, 230, 397
  gulf, 233, 243
  leopard, 67, 76, 200, 234, 245, 256, 259, 262, 263, 450
  leopard coralgrouper, 5, 8, 28, 31, 33, 42, 99, 109, 110, 137, 141, 143, 145, 197, 234, 252, 259, 270, 348, 361, 377, 378, 405, 409–411, 449–458, 570, 582
  Nassau, 13, 14, 22, 23, 27, 29–31, 36, 39, 42, 46, 49, 58, 67, 71–73, 75, 76, 96, 97, 104, 110, 119, 121, 123, 125, 127–129, 131, 132, 134–137, 139–144, 146, 147, 150, 152, 162, 166, 168, 169, 177, 193, 196, 200, 202, 207, 208, 211, 215, 217, 218, 228, 229, 241, 251, 252, 254, 257–260, 263, 267, 269, 271, 274, 299, 304, 312, 315–319, 321, 332, 341, 347, 349, 356, 357, 372, 373, 376–379, 382, 386, 388, 389, 391, 394–397, 399, 405, 419, 429–439, 568, 573, 577–579, 581, 582
  Pacific Goliath, 230, 418
  red grouper, 75, 198, 230, 252, 397, 398, 400, 442
  red hind, 25, 27, 29, 31, 33, 36, 38, 42, 46, 64, 67, 74, 75, 97, 128, 130, 131, 135, 136, 144, 146, 148, 152, 177, 230, 231, 258, 259, 272, 300, 302, 304, 312, 315, 372, 377, 386–388, 394, 395, 397–399, 405, 412–417, 582
  squaretail coral-grouper, 31, 33, 34, 36, 38, 39, 92, 99, 137, 141, 142, 146, 148, 208, 211, 232, 233, 235, 241, 247, 252, 259, 338, 341, 346, 348, 375, 409, 411, 445–449
  tiger, 23, 31, 66, 75, 230, 231, 259, 263, 322, 323, 397, 398, 400

Grouper (*cont.*)
white-blotched, 356, 600
white-dotted, 353
white-streaked, 352, 600
yellowfin, 29, 31, 36, 38, 42, 103, 146, 230, 296, 397, 398, 400
Grunts, 92, 215, 251
Gulf corvina, 238, 254, 264, 265, 267, 384
Gulf grouper, 233, 243
Gut contents, 38, 206, 289, 293
Gyre, 170, 175, 182, 431, 590

H

Habitat linkages, 23, 33–37
Haemulidae, 92, 94, 98, 215, 251, 338
Harem, 23, 67, 143
Hatching times, 9, 140, 159, 160, 170, 195, 197, 423, 573
Hawaii, 213, 422, 445, 459, 480, 491, 504, 527, 531, 532, 534
Herbivore, 8, 48, 97, 152, 187, 497, 502, 507, 518, 535
Hermaphroditism
protandry, 100, 590, 592
protogyny, 100, 104, 408, 412, 419, 441, 450, 482, 488, 590, 592
Hipposcarus harid. See Hipposcarus longiceps
Hipposcarus longiceps. See Longnose parrotfish
Hogfish, 72, 108, 199, 598
Holocene transgression, 119
Home range, 7, 8, 22–28, 33, 37, 47, 48, 64, 73, 134, 171, 176, 299, 304, 412, 419, 428, 430, 446, 455, 456, 570, 588, 589, 592
Horizontal mixing, 166, 167
Horse-eye jack, 511
Humphead wrasse (Napoleon fish)
age and growth, 99, 199, 480, 482–484
aggregations, 8, 89, 103, 127, 142, 147, 166, 199, 273, 302, 338, 378, 383, 480–482, 484, 486, 487, 515
aquaculture, 480
conservation, 273, 486–487
eggs and hatching, 199, 480
fisheries, 485
length–weight, 479–480, 482
management, 378, 486–487
range, 134, 485
reproductive biology, 88, 99, 143, 482–484
Hydration, 103, 198, 462, 516
Hydroacoustic surveys, 16, 272, 285, 289, 295, 315, 324, 438, 579
Hydrodynamics, 105, 107, 130, 289
Hyperstability, 15, 33, 235, 248, 257, 272, 276, 375, 376, 389–390, 579, 584, 590
Hyplostethus atlanticus. See Orange roughy

I

Income breeding, 106, 590
Indian Ocean, 16, 90, 95, 119, 213, 242, 372, 375, 408, 410, 449, 458, 491–493, 496, 497, 501–502, 506, 528, 529, 535, 536
Indicator, 6, 11, 12, 15–17, 132, 254, 270, 274, 364, 392, 568–570, 580, 590
Individual transferable quota (ITQ), 266, 268
Indonesia, 90, 269, 336, 340, 343, 348, 349, 358, 378, 388, 408, 424, 446, 448, 450, 452, 453, 455, 457, 458, 472, 482, 485–487, 493, 529, 569
Indo-Pacific, 23, 39, 50, 72, 73, 94–96, 106, 108–110, 209, 247, 252, 375, 386, 388, 406, 408, 418, 422, 445, 448, 458–468, 472, 490, 491, 513, 527, 572
Indo-West Pacific (IWP), 16, 86, 90, 96, 108–110, 117, 119, 123, 125, 136–139, 143, 146, 148, 152, 194, 203, 208, 212, 458, 463, 480, 502, 504, 506, 526–535, 545, 546
Initial phase, 67–69, 75, 488, 490, 497, 502, 503, 506, 590
International Union for the Conservation of Nature (IUCN), 239, 250, 251, 253, 391, 406, 421, 427, 429, 435, 444, 448, 457, 466, 468, 486, 496, 568, 590, 592, 593
Interviews
questions, 12, 118, 241, 289, 299, 332, 334, 335, 337, 340–349, 351, 360, 362–365, 455, 463
techniques, 302, 333–335
Isopod, parasitic, 46, 72
Israel, 528
ITQ. *See* Individual transferable quota

J

Jacks, 2, 22, 36, 46, 65, 92, 97, 210, 251, 507–512, 570
Jamaica, 91, 413, 436, 466, 504, 505, 525
Jet, tidal, 88, 136–138, 174–176, 593
Jobfish, 42, 71
Johnston Atoll, 42, 71, 183, 504, 513

K

Kelp bass, 76, 243
Kiribati, 235, 241, 242, 332, 341, 448

L

Labridae, 36, 37, 40, 43, 44, 67, 87, 93, 98, 100, 170, 194, 487, 491, 596, 598, 599
Lachnolaimus maximus. See Hogfish
Ladyfishes, 22, 36
Landsat, 124, 132, 133, 303
Lane snapper, 251, 379, 396, 399, 400, 596
Langmuir circulation, 181, 185
Large yellow croaker, 15, 239, 244, 254, 258
Larimichthyes crocea. See Large yellow croaker
Larval
- advection, 88, 105, 107, 161, 170–181
- density, 184, 194, 197, 573
- development, 88, 91, 128, 141, 153, 192, 430
- dispersal, 22, 88, 167, 257, 439, 573
- duration, 106, 430, 451
- morphology, 192
- recruitment, 374, 430
- retention, 14, 49, 73, 88, 105–107, 172, 413, 571
- spines, 195, 204, 216, 423

Lateral trapping, 170, 173–175, 186, 187
Latitude, 10, 139, 140, 142, 215, 217, 290, 322
Leather bass, 296, 303, 597
Legislation, 231, 236, 241, 379, 380, 393, 411, 428, 448, 456, 539
Lek, 31, 60, 62, 67, 75, 77, 91, 544
Leopard coralgrouper
- age and growth, 99, 450, 451
- aggregations, 5, 8, 28, 31, 33, 109, 137, 141, 197, 234, 252, 259, 270, 348, 377, 409, 411, 450, 454–458, 570, 582
- catch and effort data, 452, 453, 458
- conservation, 455–457
- eggs and hatching, 42, 450
- fisheries, 452–456
- habitats, 28, 31, 109, 456
- larval development, 450
- length–weight, 453, 455, 457
- management, 455–458
- range, 28, 455–458, 570
- recruitment, 145, 450
- reproductive biology, 450–452
- spawning colouration, 28, 450, 452

Leopard grouper, 76, 200, 234, 245, 256, 259, 262, 263, 450
Leptocephalus, 205
Lethrinidae, 36, 87, 92, 93, 98, 207, 267, 596, 599
Lethrinus erythropterus. See Longfin emperor
Lethrinus olivaceus. See Longface emperor
Life history, 13, 86–88, 96, 99–105, 118, 119, 141, 144, 145, 152– 154, 160–170, 172, 191–218, 227, 249, 251, 271, 287, 288, 317, 360, 362, 364, 405, 406, 422, 423, 428–431, 437, 442, 473, 480–482, 496, 528, 571, 580, 588, 591, 592
Light measurement, 319
Lionfish, 23, 50
Livelihood
- alternative, 383–385, 569
- guide, 384
- tourism, 384, 385

Live reef (foof) fish trade (LR[F]FT), 110, 232, 247, 266, 323, 406, 410, 427, 428, 464, 484–486, 493, 496, 585, 591
Local ecological knowledge (LEK), 118, 299, 331–365, 575, 583, 585, 591, 594
Longevity, 87, 88, 135, 227, 249, 250, 292, 421, 451, 488, 496
Longface emperor, 242, 599
Longfin emperor, 143, 354, 596
Longitude, 290, 322
Longnose parrotfish
- age and growth, 497, 501
- aggregations, 63, 150, 203, 497–501
- eggs and hatching, 63, 150, 165, 499–501

Lunar
- cycle, 146, 177, 187, 354, 465, 498, 523, 573, 577, 578
- month, 142, 146, 178, 463, 523, 531, 577
- periodicity, 139, 293, 408, 420, 517, 522, 523
- phase, full moon, 26, 144, 145, 302, 509
- phase, new moon, 144, 145, 573
- phase, semi-lunar, 26, 118, 136, 144, 145, 172, 177, 187, 302, 312, 449, 452, 497, 509, 517, 520, 523, 573, 589

Lutjanidae, 22, 36, 40, 43, 44, 87, 93, 98, 100, 108, 171, 198, 231, 458–468, 596, 598, 599
Lutjanus analis. See Mutton snapper
Lutjanus argentimaculatus. See Mangrove red snapper
Lutjanus argentiventris. See Yellow snapper
Lutjanus bohar. See Twinspot snapper
Lutjanus cyanopterus. See Cubera snapper
Lutjanus fulviflamma. See Dory snapper
Lutjanus fulvus. See Blacktail snapper
Lutjanus griseus. See Grey snapper
Lutjanus jocu. See Dog snapper

Lutjanus kasmira. See Common blue-stripe snapper
Lutjanus novemfasciatus. See Pacific dog snapper
Lutjanus quinquelineatus. See Five-lined snapper
Lutjanus synagris. See Lane snapper

M
Macolor niger. See Black snapper
Malaysia, 388, 408, 455, 457, 481, 482, 487
Males
  primary, 100, 104, 488
  secondary, 32, 488
  terminal phase, 68, 69, 75, 96, 497, 498, 502–504, 506, 590
Management
  adaptive, 358, 359, 361, 377, 380, 576, 584
  community-based, 361, 389, 390, 392
  failure, 386–389, 393
  precautionary, 411, 422, 592
  quota, 266, 380, 442, 475, 478
  sales ban, 230, 380, 383, 388, 428, 448
  size limits, 376, 377, 380, 383, 388, 394, 436, 442, 455, 457, 487, 496
  spatial, 22, 23, 26, 47–49, 110, 341, 345, 376, 378–380, 382, 392, 395, 399, 411, 442, 448, 580
  success, 266, 386–389, 391, 392, 406, 568, 583
  temporal, 26, 378–380, 386, 392, 395, 399, 411, 477
Mangrove red snapper, 13, 598
Mapping, 46, 49, 120, 148, 288–290, 299, 322–323, 570, 575, 577, 580
Marbled spinefoot, 92
Mariculture, 236, 578, 579, 591
Marine Protected Area (MPA)
  no-take, 361, 379, 394, 428, 437
  permanent, 379, 428, 448
  seasonal, 8, 18, 240, 361, 379, 428, 448, 487
Marine tenure, 341, 372, 375, 591
Mark-recapture studies, 25
Mass balance, 163
Match/mismatch hypothesis, 88, 197, 215, 572
Mate choice, 60, 61, 65–67, 70, 76, 199, 257, 259, 260, 262
Maternal benefits, 580
Mating success, 64
Megalops atlanticus. See Tarpon
Melanesia, 146, 333, 341, 348, 353, 424, 446, 493
Mesoamerican reef, 27, 347
Mesoscale eddy, 182
Mexico, 46, 119, 131, 227–230, 233, 234, 237, 238, 244–246, 254–256, 259, 261, 264, 267, 344, 349, 357, 381, 384, 386, 397, 418–420, 431, 435, 436, 439–444, 488, 546
Migrations
  distance, 23–28, 38, 49, 62, 121, 123, 132, 134, 135, 168, 288, 299, 304, 318, 319, 454
  pathways, 22–26, 29, 37–39, 48, 49, 287, 289, 291, 319, 324, 379, 589, 593
  pre-spawning, 240, 241, 396
  route, 24, 27–29, 35, 38, 73, 264, 341, 379, 382, 412, 435, 438, 573, 582, 588
  spawning, 24, 33–38, 48, 49, 92, 96, 97, 227, 236, 240–242, 304, 341, 346, 379, 396, 489
  vertical, 30, 160, 215
Migratory corridors, 409, 411, 426, 428, 446, 449
Milkfish, 236, 255
Mixing, oceanographic, 119, 138
Monitor, 47, 118, 119, 248, 270–272, 276, 286, 304, 324, 361, 378, 380, 389, 444, 449, 576, 579
Moon. *See* Lunar
Moorish idol
  age and growth, 513
  aggregations, 209, 213, 513–517
  length–weight, 513, 516
  recruitment, 210
  reproductive biology, 513–517
Moratorium, 253, 422, 474
Morphology
  functional, 100, 102
  lateral compression, 102
Mortality
  fishing, 32, 49, 72, 76, 377, 393, 412, 416, 421, 443, 456, 476
  juvenile, 87
  natural, 181, 250, 251, 292, 424, 444, 451
Mugil cephalus. See Flathead grey mullet
Mugilidae, 2, 36, 40, 87, 237, 240, 337, 596, 599
Mugil platanus. See Mullet
Mullet, 2, 36, 151, 227, 237, 240, 241, 244, 247, 251, 264, 337, 346, 573

Mullidae, 36, 43, 44, 87, 241, 596, 600
Multi-beam sonar, 133, 322
Mutton hamlet, 208
Mutton snapper
  age and growth, 466
  aggregations, 231, 251, 302, 394, 466
  conservation, 394, 466
  fisheries, 231, 394, 466
  management, 394, 466
  reproductive biology, 466
Mycteroperca bonaci. See Black grouper
Mycteroperca jordani. See Gulf grouper
Mycteroperca microlepis. See Gag
Mycteroperca phenax. See Scamp
Mycteroperca rosacea. See Leopard grouper
Mycteroperca tigris. See Tiger grouper
Mycteroperca venenosa. See Yellowfin grouper

N
Napoleon fish. *See* Humphead wrasse
Narrow-barred Spanish mackerel, 389, 600
Naso annulatus. See Whitemargin unicornfish
Naso literatus. See Orangespine unicornfish
Naso lopezi. See Slender unicornfish
Naso spp. *See* Unicornfish
Naso unicornis. See Bluespine unicornfish
Nassau grouper
  age and growth, 135
  aggregations, 14, 22, 30, 36, 37, 46, 50, 58, 71, 73, 96, 119, 121, 123, 125, 127–129, 131, 132, 135–137, 141, 147, 152, 166, 177, 212, 215, 217, 251, 252, 254, 257, 260, 271, 274, 299, 304, 312, 315–317, 319, 321, 341, 349, 357, 372, 373, 376–378, 382, 386, 388, 389, 430, 432, 433, 435–439, 568, 577, 581, 582
  aquaculture, 573
  behaviour, 119, 251, 430
  conservation, 215, 228–229
  fisheries, 215, 228–239, 254, 271
  habitats, 30, 141, 251, 430–431
  length–weight, 429
  management, 228–239, 251, 267, 382, 386, 388, 436, 439
  spawning sites, 31, 36, 42, 58, 129, 162, 177
  range, 119, 140, 299, 304, 391, 429, 430, 436
  recruitment, 430, 438
  reproductive biology, 104, 376
  spawning, 14, 29, 31, 37, 42, 46, 58, 67, 72, 73, 75, 76, 110, 119, 121, 134, 136, 139, 141, 142, 144, 152, 162, 166, 169, 177, 202, 212, 215, 217, 251, 317, 321, 341, 349, 357, 372, 376, 389, 432, 435–438, 582
Nekton, 160, 193
Nesting, 59, 61, 257, 574
Nets, sampling
  channel, 207
  crest, 207 (*see also* Fishing)
New Caledonia, 408, 423, 451, 452, 454, 482, 493, 535
NGOs. *See* Non-governmental organizations (NGOs)
Nitrox, 324
Nocturnal
  fishing, 143, 496
  spawning, 96
Non-governmental organizations (NGOs), 342, 350, 361, 364, 388, 390–393, 576, 582
Nutrient, 37, 106, 110

O
Ocean surgeonfish, 95, 517–526, 595
Ocean triggerfish, 31
Ocyurus chrysurus. See Yellowtail snapper
Oil globule, 195, 199, 316, 470
Oocytes, 107, 292, 427, 451, 460, 461, 474, 516, 517, 537
Oophagous, 42
Orange roughy, 10–11, 244, 258, 270, 295, 599
Orangespine unicornfish, 533
Otoliths, 145, 194, 195, 205, 206, 208, 211, 212, 289, 293, 296, 424, 430, 446, 472, 482, 483, 578
  chemistry, 296, 472
  increment, 145, 194, 208, 212, 430
Ovary
  GSI, 103
  hydration, 103, 462
  post-ovulatory follicle, 12
  pre-vitellogenic eggs, 537
Oxygen, 142, 288, 295, 324

P
Pacific dog snapper, 143, 464, 596
Pacific goliath grouper, 230, 418
Pacific herring, 245
Pacific Ocean
  Central, 90
  east, 513

Pagrus auratus. See Silver seabream
Palau, 34, 38, 39, 63, 89, 92, 121, 125–127, 132, 134, 137, 138, 140–144, 146, 148, 150, 151, 170–173, 175, 176, 183–186, 203, 210–214, 217, 232, 236, 241, 244, 247, 259, 302, 306, 308, 310, 314, 317, 320, 336, 341, 342, 345, 346, 350, 351, 355, 382, 385, 388, 389, 408, 423–428, 446–448, 459–465, 468, 470, 480–482, 484, 486, 497, 498, 500, 501, 505, 506, 508–511, 513, 515, 516, 528–543, 546, 547, 577
Palau Conservation Society, 310–312, 325, 356, 426, 446–448, 542, 577
Palette surgeonfish, 534, 535
Panama, 213, 436, 518
Papua New Guinea, 90, 162, 241, 247, 335, 338, 339, 343, 345, 346, 352, 353, 359, 360, 388, 445, 446, 448, 451, 454, 455, 464, 471, 482, 491, 493, 509
Paracanthurus hepatus. See Palette surgeonfish
Paralabrax clathratus. See Kelp bass
Paralabrax nebulifer. See Barred sand bass
Parrotfish, 23, 31, 32, 36–38, 63, 65, 67, 71, 73, 75, 76, 92, 97, 99, 102–104, 134, 147, 170, 171, 185, 194, 198, 203, 204, 267, 317, 491–493, 497–499, 501–503, 506, 507, 531
Permit, 12, 64, 67, 160, 242, 247, 266, 362, 380, 384, 388, 394, 410, 456, 511
Philippines, 91, 236, 255, 335, 340, 349, 355, 358, 361, 389, 452, 454, 455, 457, 458, 485, 510, 569
Photography, aggregations, 305–306
Pinguipedidae, 92, 93, 98
Plankton, 11, 88, 150, 160, 168, 171, 173–175, 181, 187, 193, 194, 196, 204, 208, 210, 219, 317, 489
Plectorhinchus obscurus. See Giant sweetlips
Plectropomus areolatus. See Squaretail coralgrouper
Plectropomus laevis. See Black saddle coralgrouper
Plectropomus leopardus. See leopard coralgrouper
Poaching, 232, 253, 268, 275, 350, 382, 388, 576, 582
Pohnpei, 29, 121, 241, 258, 267, 351, 388, 409, 423–428, 446, 448, 449, 497, 536
Polyprion americanus. See Atlantic wreckfish
Pomacanthidae, 12, 44, 65, 92, 93, 98
Pomacentridae, 44, 60, 87, 93, 147, 195, 596
Porgies, 22, 36, 141, 472
Post-ovulatory follicle. *See* ovary
Predation
  adult, 39, 71–72
  on aggregated fishes, 104
  egg, 42–46, 63, 117, 136
  on eggs, 71–72, 571
  piscivory, 39–42
  on spawning fishes, 500, 531, 571
Pre-spawning aggregation. *See* Aggregation
Pre-vitellogenic eggs. *See* Ovary
Prionurus spp., 535
Propagules, 24, 88, 105, 106, 118, 128, 136, 152–153, 160, 161, 163, 166, 170, 172, , 187, 193, 198, 200, 209, 217, 319, 321, 572, 580
Protogyny, 100, 432, 590
Pseudobalistes flavimarginatus. See Yellowmargin triggerfish
Pseudocaranx dentex. See White trevally
Psychrolutes phrictus. See Blob sculpin
Public perspectives, 240
Puerto Rico, 46, 121, 132, 133, 148, 168, 228, 231, 241, 254, 298, 315, 322, 323, 356, 398, 412, 416, 417, 421, 435, 436, 466, 467, 504, 517–526

Q
Quadrats, survey, 312

R
Rabbitfish, 2, 9, 91, 92, 97, 120, 147, 195, 209, 213, 236, 241, 244, 247, 337, 356, 372, 540–542
Rebreather, diving, 324
Recruitment
  mass, 5, 136, 210, 212, 213
  pathways, 163
  pulses, 209–211, 213
  self-, 47, 106, 107, 136, 216, 528
Redfin parrotfish, 40, 95, 122, 148, 497, 506, 597
Red hind, 597
  age and growth, 412
  aggregations, 25, 31, 36, 38, 42, 46, 67, 130, 131, 136, 177, 258, 259, 272, 300, 302, 304, 312, 315, 386, 387, 413, 415–417
  habitats, 74, 412
  management, 231, 372, 394, 417
  reproductive biology, 412–415

Red List (IUCN), 239, 251, 253, 406, 421, 427, 429, 435, 448, 457, 466, 468, 486, 496, 590, 593
Red Sea, 90, 119, 132, 185, 292, 406, 422, 445, 458, 480, 491, 493, 496, 501, 507, 528, 529, 542
Reef
  barrier, 127, 128, 131, 136–138, 143, 148, 152, 162, 170–176, 186, 203, 211, 212, 306, 459, 460, 484, 497, 509, 513, 515, 529, 531, 534, 543, 572
  bend, 123, 124
  corner, 122, 124, 126, 130, 182, 203, 515, 516, 533
  drop-off areas, 121, 141, 269, 467
  pinnacle, 504
  projection, 122–128, 131, 172, 177, 187, 320, 323, 484, 489
  promontory, 121, 122, 124–126, 128, 177, 322, 470, 484
Remote sensing, 132, 295, 303, 324
Reproductive
  biomass, 226
  output, 39, 59, 72, 102, 103, 107, 226, 257, 258, 261–263, 271, 417, 420, 456, 457
  season, 11, 23, 24, 48, 49, 64, 65, 211, 254, 428, 448, 449, 468, 579
  success, 13, 48, 60, 64, 70, 76, 77, 87, 119, 120, 154, 247, 292, 567, 581
Resident aggregation, definition, 7–8, 18
Restocking, 14, 264, 435, 578, 579, 592
Retention, 14, 30, 49, 73, 88, 105–107, 128, 136, 153, 169, 170, 172–174, 177, 183, 187, 192, 200, 215, 216, 299, 320, 413, 435, 527
Retention zone, 170, 172, 173
Rhincodon typus. *See* Whale shark
Rhythym
  circadian, 10
  infradium, 10
Riley's Hump, Florida, 128, 130, 217, 391, 398
Ringtail surgeonfish, 134, 532, 595
Rivulated parrotfish, 96
Roe, 237, 241, 244, 246, 264, 276

S

Saba Bank, 128, 130, 177
Sailfin snapper. *See* Blue-lined sea bream
Sailfin tang, 534
Salinity, 10, 119, 142, 196, 215, 218, 288, 295, 296, 318, 319, 406
Samoa, 445, 492, 528, 531, 532
Sandperches, 92
Satellite images, 132, 133, 178, 300, 303
Saudi Arabia, 423, 501
Scamp, 103, 230, 255, 259, 263, 397, 597
Scaridae, 22, 36, 40, 43, 44, 65, 92, 194, 267, 491, 502–507, 597, 599
Scarus iserti. *See* Striped parrotfish
Scarus prasignathos. *See* Greenthroat parrotfish
Scarus rivulatus. *See* Rivulated parrotfish
Schooling, 8, 15, 65, 151, 161, 204, 260, 440, 493, 496, 498, 504, 508, 538
Scomberomorus commerson. *See* Narrow-barred Spanish mackerel
Scomberomorus semifasciatus. *See* Grey mackerel
Scombridae, 36, 41, 42, 45, 235, 600
SCRFA. *See* Society for the Conservation of Reef Fish Aggregations
Seabass, 36, 237, 238, 272
Sea bream, 119, 137, 143, 206, 242, 250–253, 262, 266, 464, 465, 468–478
Self-recruitment, 47, 136, 528
Semicossyphus pulcher. *See* California sheephead
Seriola lalandi. *See* Amberjack
Serranidae, 22, 36, 40, 41, 43, 44, 87, 92, 93, 100, 108, 171, 197, 240, 245, 337, 406, 415, 417, 429, 439, 568, 597, 598, 600
Sex
  change, 69, 75, 76, 110, 199, 259, 263, 288, 408, 413, 416, 432, 441, 443, 444, 451, 473, 482, 488, 574, 590
  ratio, 69, 70, 230, 231, 248, 257–259, 261, 292, 413, 416, 417, 419, 443, 445, 446, 451, 476, 574
Sexual
  maturation, 227, 249, 251, 293, 419, 435, 437, 446, 466, 482, 485, 492
  selection, 75–76, 199, 256, 257, 260, 411, 592
Seychelles, 28, 29, 247, 349, 356, 389, 408, 410, 411, 423, 424, 426
Shark
  bull, 15, 16
  lemon, 42
  nurse, 5, 65
  scalloped hammerhead, 65
  tiger, 65
  whale, 15–17, 42, 46, 48, 71, 150, 346, 384, 467, 568
Shark Bay, Western Australia, 266, 472–474, 476
Shifting baseline, 270, 333, 592
Shoemaker spinefoot, 356, 372, 597

Siganidae, 2, 36, 87, 91, 93, 98, 120, 195, 236, 240, 337, 536, 597, 598
Siganus argenteus. *See* Streamlined spinefoot
Siganus fuscescens. *See* Dusky rabbitfish
Siganus guttatus. *See* Goldline spinefoot
Siganus luridus. *See* Dusky spinefoot
Siganus rivulatus. *See* Marbled spinefoot
Siganus sutor. *See* Shoemaker spinefoot
Silver seabream
  age and growth, 473, 476
  aggregations, 242, 266
  aquaculture, 472
  fisheries, 242, 266, 473
  genetics, 472
  length–weight, 472
  management, 242, 266, 472–477
  reproductive biology, 474
Simple migratory spawning, 3, 4, 8, 205, 508, 600
Site fidelity, 47, 263, 299, 304, 420, 426, 439
Size
  average, 109, 208, 230, 234, 423
  body, 39, 67, 88, 97, 102, 104, 108–110, 250, 252, 254, 257–259, 261, 292, 293, 432, 435, 480, 496, 571
  dimorphism, 75, 419
  frequency, 260, 261, 288, 316, 416, 492, 520
  maximum, 27, 99, 100, 104, 108, 381, 410, 428, 451, 529
Slender unicornfish, 534, 599
Snapper
  black, 44, 63, 150, 399, 400, 464, 500, 501, 508, 599
  blacktail, 240, 306, 308, 320, 346, 405, 458–468, 596
  common bluestripe, 464
  cubera, 15, 31, 37, 119, 141, 143, 145, 147, 150–152, 162, 166, 168–170, 200, 202, 251, 273, 317, 346, 384, 394, 396, 398, 400, 463, 464, 466–467, 517
  dog, 31, 43, 71, 312, 314, 346, 400, 464, 467, 596
  Dory, 102
  five-lined, 99
  grey, 251–252, 396, 467, 598
  lane, 251–252, 379, 396, 399, 400, 596
  mangrove red, 13
  mutton, 13, 26, 128, 231, 251, 302, 394, 396, 398, 400, 463, 466
  Pacific dog, 143, 464
  pink (*see* silver seabream)
  twinspot, 89, 137, 143, 242, 312, 464–465, 481, 515
  yellow, 38, 202, 398, 400, 460
  yellowtail, 38, 150, 251
Snook, 22, 36, 262
Society for the Conservation of Reef Fish Aggregations (SCRFA), 11, 12, 87, 92, 94, 274, 286, 289, 311, 372, 374, 375, 386, 388, 391, 411, 436, 454, 460, 501, 509, 513, 517, 526, 569
Socioeconomic, 47, 225, 264–269, 376, 382, 410, 538, 574, 576–579
Solomon islands, 92, 111, 232, 247, 335, 343, 345, 346, 348, 352–354, 361, 383, 408, 428, 448, 455, 464, 491–493, 495, 509
Solstice, 142–145, 413
South Africa, 251
South Australia, 478
Spanish mackerel, 389
Sparidae, 22, 36, 242, 250, 252, 472
Sparisoma rubripinne. *See* Redfin parrotfish
Sparisoma viride. *See* Stoplight parrotfish
Spawning
  behaviour, 31, 75, 77, 86, 88, 91, 95–97, 99, 104, 105, 109, 110, 151, 164, 192, 288, 305, 376, 382, 393, 458, 465, 497–501, 504, 525, 546, 571, 581
  biomass, 248, 263, 375, 475, 476
  group, 23, 24, 27, 42, 61, 67, 69–76, 92, 95, 96, 99, 103, 108, 110, 198, 200, 202, 262, 292, 411, 455, 464, 465, 489, 490, 497, 500–502, 504–507, 516, 519, 522, 525, 526, 529, 531–535, 570, 589
  haremic, 154, 590
  pair, 2, 24, 32, 42, 60, 62, 67, 69, 70, 75, 76, 91, 92, 95, 96, 99, 150, 167, 194, 198, 200, 260, 261, 446, 450, 493, 497, 504, 506, 507, 513, 517–519, 522, 525, 526, 528, 530, 531, 533–535, 546, 591
  rush, 2, 31, 39, 67, 70, 71, 73, 74, 165, 202, 262, 499, 523
  season, 10, 24, 26, 28, 36, 39, 42, 46, 48, 49, 136, 139, 140, 142, 199, 209, 211, 212, 215, 218, 226, 230, 236, 238, 243, 244, 248, 251, 264, 266, 293, 352, 361, 377, 378, 382, 395, 408, 409, 413, 425–427, 430, 432, 441, 443, 444, 448, 465, 466, 474, 477–478, 520, 523, 528, 537, 539, 572
  sites, 9, 14, 23, 28, 31, 35, 36, 38, 42, 43, 46, 48, 49, 58, 73, 74, 76, 108, 109, 120, 129, 132, 135–139, 141, 152, 153, 160–162, 168, 171, , 177, 179, 186, 203, 238, 262, 320, 335, 348, 356, 378,

382, 389, 411, 413, 419, 426–428, 435, 441, 443, 444, 449, 457, 489, 490, 538, 541, 570, 572, 575, 581
stupor, 151, 502
tactics, 76
time, 103, 105, 137, 144, 145, 212, 234, 243, 263, 265, 341, 449, 470, 546
Spawning aggregation
definition, 1–18
direct indicator, 570
indirect indicator, 364, 570
recovery, 74, 217, 226, 228, 237, 241, 253, 256, 260, 261, 263, 264, 275, 388, 393, 394, 397–399
site, 3, 13, 22, 24, 25, 27–30, 32–40, 42, 46–49, 121, 127, 152, 161, 163, 174, 177, 236, 238, 252, 322, 359, 374, 390, 395, 407–409, 411, 413, 428, 436, 438, 448, 570
Speckled blue grouper, 72, 346, 600
Sperm
competition, 60, 61, 67, 68, 275, 432, 525, 593
limitation, 69, 70, 200, 260–262, 443, 593
Spotted goatfish, 148
Squaretail coralgrouper
age and growth, 99, 100, 446
aggregations, 34, 36, 92, 99, 146, 148, 208, 211, 241, 252, 259, 348, 409, 411, 446–449
conservation, 148, 232, 235, 253, 411, 446, 449
fisheries, 241, 247, 259, 341, 409, 448
length–weight, 99, 142, 208, 259
male colouration, 31, 409
management, 37, 448, 449
recruitment, 212, 247, 445, 448
reproductive biology, 259, 408, 446–447
Squirrelfishes, 143
Staging area, 26–30, 39, 48, 49, 120, 570, 593
Statement of Concern, 390
St. Croix, 46, 130, 131, 134, 177, 275, 399, 416, 417, 546
Stereolepis gigas. *See* Giant seabass
Sticky water, 15, 60
Stoplight parrotfish, 95, 100, 597
Streamlined spinefoot, 92
Striated surgeonfish, 95, 132, 148
Striped bristletooth, 38, 170, 506, 526–535
Striped mackerel, 150
Striped parrotfish, 67, 72, 95, 502–507
St. Thomas, 25, 29, 131, 134, 177, 399, 400, 416, 417
Stupor, spawning, 151, 502
Surgeonfish
Achilles, 532
blackstreak, 533
blue-banded, 530, 531
blue-lined, 531–532
brown, 29, 42, 61, 68, 69, 95, 132, 148, 151, 170, 209, 263, 319, 506, 526–535
convict, 3, 15, 208, 210, 528, 531
ocean, 95, 517–526
palette, 534, 535
ringtail, 134, 532
spotted, 210, 530–532
Thompson's, 532
white-spotted, 530, 532
Survey, hydroacoustic, 16, 272, 289, 295, 315, 438, 579
Swim bladder, 103, 147, 195–198, 204, 226, 227, 239, 244, 298, 347, 419, 587
Swimming ability, 160, 195, 196, 198, 204, 206, 209, 215
Symphorichthys spilurus. *See* Blue-lined sea bream

T

Tagging
acoustic, 25, 29, 36, 42, 49, 135, 290, 298, 299, 304, 318, 426, 446, 482
conventional, 25, 287, 290, 298, 299, 304, 426
TAG lipids. *See* Triacylglycerol (TAG) lipids
Tan-faced parrotfish, 134
Tang
Achilles, 532, 595
blue, 64, 95, 300, 517–526, 595
brushtail, 530, 534, 595
sailfin, 534
Tarpon, 205, 242, 244, 599
TEK. *See* LEK
Temperature
ranges, 23, 139–142, 146, 152, 154, 178, 218, 413, 427, 446, 529
regimes, 139–142, 152, 218, 319–321, 572
spawning, 10, 23, 119, 135, 139–142, 145, 146, 152, 154, 178, 184, 196, 197, 214, 215, 218, 288, 295, 319–321, 413, 427, 446, 470, 520, 528., 529, 570, 573, 575
Terminal phase, 68, 69, 75, 96, 497, 498, 502–504, 506, 590, 593
Territory, 75, 95, 242, 408, 428, 430, 447, 498, 499, 544
male, 75, 95, 447, 498, 499, 544
Testes, male, 11, 67, 202, 289, 432, 493, 525
Tethering, 298

Thalassoma bifasciatum. See Bluehead wrasse
Thermoclines, 183, 184, 187
Thermographs, 319, 320, 580
Tholichthys, 205
Thompson's surgeonfish, 532, 595
Threadfin bream. *See* Blue-lined sea bream
Threatened species, 244–245, 250, 252–253, 295, 325, 377, 378, 383, 405–406, 435, 444, 578, 593
Tidal
  amplitude, 109, 136, 143, 144, 152, 172, 504, 572, 587
  channel, 134, 136–138, 169, 175, 176, 207, 209–211, 218, 430, 445, 459–461, 497, 503, 504, 509, 513, 572
  currents, 73, 88, 106, 136–138, 144, 147, 160, 171, 172, 175–180, 183, 186, 203, 207, 218, 461, 503, 504, 528, 530, 572
  eddy, 136, 178–181, 187, 461
  excursion distance
  flow, 178
  jets, 88, 136–138, 174–176, 593
Tides, 10, 37, 89, 95, 136–138, 143, 144, 148, 170, 171, 173–182, 185, 186, 203, 214, 295, 296, 317, 319, 320, 332, 344, 346, 460, 481, 484, 491, 497, 504–506, 509, 513, 516, 528, 529, 531, 535, 536, 544
Tiger grouper, 23, 31, 66, 75, 231, 259, 263, 322, 323, 397, 398, 400, 597
Time lapse photography, 305, 307, 308
Titan triggerfish, 91, 545, 546
Totoaba, 238, 244, 245, 254, 267
Totoaba macdonaldi. See Totoaba
Tourism, 226, 241–243, 269, 270, 303, 344, 384, 385, 409, 435, 437, 438, 467, 486, 515, 579
Trachinotus falcatus. See Permit
Tracking
  acoustic, 25, 304, 420, 438
  tag, 25, 290, 291, 304–305, 321
Traditional
  ecological knowledge, 118, 120, 140, 286, 299, 463, 580, 594
  knowledge, 146, 289, 299, 302, 340, 372, 428, 435, 449, 460, 463, 509, 540
  use of aggregation site, 147
Transects, 289, 293, 298, 300, 301, 309–312, 314, 359, 540, 594
Transient aggregation, 7, 8, 18, 38, 42, 89, 97, 119, 121, 131, 135, 154, 166, 167, 171, 176–181, 200, 266, 274–276, 318, 321, 324, 325, 337, 376, 464, 466, 482, 486, 517, 577, 581, 582, 594
  definition, 594
Trevally, bluefin, 42, 71, 512
Triacylglycerol (TAG) lipids, 195, 199
Triggerfish
  finescale, 246, 546, 595
  ocean, 546
  titan, 91, 545.46
  yellowmargin, 31, 33, 91, 92, 542–547, 595
Trophic, 23, 24, 38, 46–48, 88, 96–99, 102, 104, 110, 183–186, 273, 418, 568, 571
Tropical Atlantic, 86, 90, 95, 108, 110, 455
Turks and Caicos, 119, 269, 431, 434–436
Twinspot snapper, 596
  age and growth, 464, 465
  aggregations, 89, 137, 143, 242, 312, 462, 464–465, 481, 596, 599
  fisheries, 464
  food habits, 40, 515
  habitats, 137, 464
  reproductive biology, 89, 137, 461, 464, 465

U

Ulong Channel, Palau, 138–141, 146, 148, 258, 516, 543–545
Ultrasound imaging, 289, 292, 300, 413, 414
Underwater Visual Census (UVC), 15, 16, 271, 289, 300, 314, 350, 352, 353, 355, 357–360, 392, 438, 485, 593
Unicornfish
  bignose, 534
  bluespine, 527, 533–535
  orangespine, 533, 599
  spotted, 534
  whitemargin, 534
United States (USA), 14, 228–230, 237, 238, 242, 243, 246, 259, 261, 267, 270, 302, 325, 379, 417, 421, 422, 429, 435, 436, 439, 442, 445, 487, 488, 517
Upwelling, 179–183, 185, 320
UVC. *See* Underwater Visual Census (UVC)

V

Validation. *See* Aggregation
Vertical mixing, 166, 181

Video recording, 289, 300–301, 306
Vieques Island, 231
Virgin Islands
British, 388, 395, 413, 417
United States, 25, 46, 122, 131, 134, 141, 148, 177, 213, 217, 228, 231, 259, 312, 388, 399, 412, 415, 435, 436, 439, 582
Vortex, 47, 181–185
Vorticity, 177–181, 183, 594

W

Wakes, 168, 181–183, 187
Whale shark, 15–17, 42, 46, 48, 71, 150, 346, 384, 467, 568
White-blotched grouper, 356, 600
White-dotted grouper, 353
Whitemargin unicornfish, 534
White seabass, 238, 272
White-spotted surgeonfish, 530–532, 595
White-streaked grouper, 352, 600
White trevally, 65
Whitsunday Islands, 178
Wrasses, 36, 38, 67, 71, 73, 75, 87, 92–94, 97–100, 103, 104, 108, 170, 194, 198, 204, 209, 213, 263, 486, 491

Y

Yellow and blueback fusilier, 43–44, 65, 147, 201, 202, 595
Yellowfin grouper, 29, 38, 43, 131, 230, 296, 397, 597
Yellow jack, 511
Yellowmargin triggerfish
age and growth, 543
aggregations, 91–92
habitats, 31, 33
reproductive biology, 91, 544
Yellow snapper, 143, 464, 596
Yellowtail snapper, 150, 251, 599
Yellow tang, 534, 535
Yolk sac larvae, 141, 164, 167, 192, 195, 209, 215

Z

Zanclus cornutus. See Moorish idol
Zebrasoma flavescens. See Yellow tang
Zebrasoma scopas. See Brushtail tang
Zebrasoma velliferum. See Sailfin tang
Zooplankton, 37, 146, 171, 181, 182, 185, 187, 194, 200, 204, 206, 214, 317, 505, 527, 533, 547